KINETICS IN MATERIALS SCIENCE AND ENGINEERING

KINETICS IN MATERIALS SCIENCE AND ENGINEERING

Dennis W. Readey
University Emeritus Professor
Colorado School of Mines, Golden, USA

CRC Press is an imprint of the
Taylor & Francis Group, an **informa** business

Cover Image: Courtesy of Jaehyung Lee. From "Vapor Phase Sintering of Hematite in HCl" (PhD Thesis. The Ohio State University, 1984).

CRC Press
Taylor & Francis Group
6000 Broken Sound Parkway NW, Suite 300
Boca Raton, FL 33487-2742

Printed on acid-free paper
Version Date: 20160830

International Standard Book Number-13: 978-1-4822-3566-1 (Pack - Book and Ebook)

Library of Congress Cataloging-in-Publication Data

Names: Readey, D. W., author.
Title: Kinetics in materials science and engineering / Dennis W. Readey.
Description: Boca Raton, FL : CRC Press, Taylor & Francis Group, [2016] |
Includes bibliographical references and index.
Identifiers: LCCN 2016018601| ISBN 9781482235661 (alk. paper) | ISBN
1482235668 (alk. paper)
Subjects: LCSH: Materials science. | Chemical processes. | Dynamics. |
Diffusion.
Classification: LCC TA403.6 .R376 2016 | DDC 620.1/12--dc23
LC record available at https://lccn.loc.gov/2016018601

Visit the Taylor & Francis Web site at
http://www.taylorandfrancis.com

and the CRC Press Web site at
http://www.crcpress.com

To my wife Suzann for her encouragement and patience

Contents

SECTION IV — Diffusion with a Constant Diffusion Coefficient

SECTION V — Fluxes, Forces, and Interdiffusion

Preface

This book is intended to be an undergraduate text for junior- and senior-level students majoring in materials science and engineering (MSE). Most MSE undergraduate programs have a one-semester *kinetics* course in their curricula that has taken the place of a classical physical chemistry course and one or more courses on materials processing. This book is not intended to be a reference work that summarizes and evaluates all of the reactions and phase transformations of importance in MSE, although some unique topics are presented. Given the breadth of the field today—metals, ceramics, electronic materials, polymers, biomaterials, and composites—it is not feasible to cover all of kinetics. Nevertheless, the book does cover considerably more material than can be covered in a one-semester course and includes approaches and topics that are rarely found elsewhere that might be useful to graduate students, researchers, and industrial practitioners. One goal of this book is to incorporate important concepts that all students in MSE should be exposed to at some point in their careers based on the author's personal experience as a student, researcher, practicing engineer, and instructor over a number of years. The primary goal is to develop simple mathematical models, involving process variables such as temperature, pressure, composition, and time, to provide a fundamental model-based understanding of commercially important materials processes critical to the development, fabrication, and application of materials. In doing so, at the expense of additional equations, most of the mathematical steps necessary to develop a model are explicitly presented so the student, or reader, can follow the text without the absolute necessity of an instructor for interpretation and insertion of the missing steps—frequently the case in many textbooks. This, of course, means more pages to cover a given topic.

WHY THIS BOOK

There are several reasons why this book was written. First, there is limited textual material for the junior/senior-level general courses in MSE. There are many ***introduction to materials***, or similarly named, textbooks. Most of these texts do not cover the different types of materials equitably, usually reflecting the author's understandable partiality to a specific material type based on his or her own experience and interests. Unapologetically, some partiality will be found here as well. More importantly, virtually none of introductory texts present kinetics in sufficient depth to be of use in an upper level course. Also, many texts approach important kinetic topics with a ***black box*** or "...it can be shown that..." methodology. For example, most introductory texts have sections on solid state diffusion and they tell you that carbon diffuses into steel, and its concentration, *C*, as a function of distance, *x*, and time, *t*, is given by

$$C = C_s \operatorname{erfc}\left(\frac{x}{\sqrt{4Dt}}\right)$$

where:

C_S is the surface concentration
D is the diffusion coefficient

It is hard to imagine a more opaque black box than "erfc(y)": What is it? Where does it come from? An MSE student should know where it comes from before completing an undergraduate program. This book shows explicitly how to get this result with no more than introductory calculus and serves as an example of one of the several ***kinetic concepts*** an undergraduate MSE student should be exposed to before entering the materials profession as a research scientist or practicing engineer.

Second, MSE programs today incorporate instruction on metals, ceramics, polymers, electronic materials, biomaterials, and composites, all with varying degrees of emphasis. Yet, as discussed in the Chapter 1, most of these programs were, at one point in their history, focused only on metals and called ***metallurgy*** or ***metallurgical engineering***. Today, materials are often separated into ***hard*** materials and ***soft*** materials, the latter including polymers and biomaterials. Biomaterials can be further subdivided into bio-replacement materials—scaffolds and implants—and the study of actual biological materials processes such as bone growth. In one of the author's teaching experiences, with about 100 MSE students taking the kinetics course, over 50% were interested in soft materials. Fifty years ago, ***all*** of the students in this particular program would have been interested only in metals, but not today. This change in interests illustrates the need for exposure to a new and broader range of kinetic processes involving all types of materials used today.

Third, most kinetic processes in materials usually have several possible steps operating in series or in parallel controlled either by material transport (diffusion) or by an actual chemical reaction; for example, the deposition of silicon films from a gas. For hard materials, reactions usually take place at elevated temperatures and are often diffusion controlled. In contrast, reactions in soft materials occur at low temperatures, and a chemical reaction is usually more important. With the growing importance and interest in soft materials—with a significant interest overlap with chemistry, physics, chemical, and bioengineering—inclusion of chemical reaction kinetics in an MSE course is crucial. This text tries to accomplish this as well as brief digression into nuclear chemistry as an example of ideal kinetics. This typically serves as a first exposure to nuclear chemistry to over 80% of the students in spite of its importance. Every MSE major needs some exposure to nuclear reactions given the importance of materials in nuclear power, nuclear waste disposal, nuclear medicine, and other uses of isotopes. Finally, for virtually all of the kinetic models developed, where possible, examples of real industrial processes are presented to demonstrate the applicability of the models.

HOW THIS BOOK DEVELOPED

This book is a textbook on the modeling of kinetic processes for undergraduates majoring in MSE and similar programs. It grew from notes acquired over 30 years of teaching such a course in materials programs at three different academic institutions, each having different student primary interests and faculty research specialties: ceramics, metals, and polymers. The objective is to provide material for a one-semester course (probably two) on kinetics based on the backgrounds expected of juniors and seniors (and some graduate students) in undergraduate programs. This is not a book on ***computational materials science.*** Computational materials science is an important and growing field in materials research but typically involves the use of mathematical or computational techniques that better serve graduate students. This book is also not an extensive review of the original literature. In fact, every effort is made to summarize material that already exists in some form in another text or publication. Similarly, this is ***not*** a physics or chemistry text. In reality, some of the approximations made to models will make physicists, chemists, mathematicians, and probably even some materials scientists, cringe! Nevertheless, where approximations appear, they are intended to make the development of a kinetic model more transparent so that the final—and correct—result is not lost in the details. No single text exists that treats kinetic topics for materials at an introductory, junior–senior level and accomplishes two important goals: (1) goes more deeply into a topic than an introductory materials text does and (2) equitably covers all materials. Although there are many excellent books on chemical kinetics, details of diffusion mechanisms, and solutions to diffusion equations at an advanced level, they are usually focused on a single class of material: metals, ceramics, polymers, semiconductors,* and so on. Today, a student needs exposure to all materials with the *emphasis on areas of commonality*. That is why MSE exists: commonality in the processing, structure, and properties among materials. In this book, an effort is made to include topics and examples for both hard and soft materials because each MSE program has different student and faculty interests. However, for soft materials such as polymers, it is more challenging to easily find commonality with the kinetics of processing of hard materials. The reason for this is probably twofold. First, the useful temperature range for polymers is only a few hundred degrees centigrade, whereas hard materials processing covers a range of over 3000°C. Second, the *microstructure* of hard materials can be varied greatly, largely controlled by kinetic processes at elevated temperatures, the main factor determining properties. This is why steels are heat-treated! In contrast, it is the chemistry, length, and entanglement or cross-linking of polymer molecules that have the largest influence on polymer properties. The greater parts of such topics are more appropriate to a chemistry course.

THE TARGET AUDIENCE

Today, there are about 60–70 accredited undergraduate MSE programs that graduate about 1200 students a year. In addition, there are many chemistry, physics, and engineering programs that have a strong materials orientation or undergraduate option. In the author's experience, many MSE graduate students do not have an MSE undergraduate background and take an undergraduate course, such as kinetics, to become more familiar with MSE topics and terminology. Also, both undergraduate and graduate students from other departments frequently take such a course to gain information about various topics; for example, nuclear engineers need to understand diffusion of neutrons in reactors. The book is not primarily intended as a reference for a practicing engineer or researcher but certainly should be useful to them as a refresher on specific topical areas and introduction to other topics not usually found elsewhere. Please also refer to the sections "For the Instructor" and "For the Student" following this preface.

This book assumes that a student has had a chemical thermodynamics course and an introductory materials course as prerequisites. A mathematical background no higher than a second or third calculus course is all that is required, and a prior course differential equations is not necessary.

HOW THE BOOK IS ORGANIZED

The general organization of this book is to present essentially *classical* chemical reaction kinetics first. Then nucleation and growth phase transitions are presented with control by a chemical reaction rather than by diffusion or transport. Prior to this, a chapter on the surface energy and its effects is offered first because not all students have seen surface energy effects in thermodynamics or chemistry courses. Then diffusion with a constant diffusion coefficient is presented. In the "Table of Contents," the atomistics of diffusion are listed before the solutions to the equations of diffusion. However, this material can be presented the other way around without any loss in continuity. The last section focuses on interdiffusion and more complex geometries. The final chapter is on the kinetics of spinodal decomposition presented with a

* Including *semiconductors* as a type of material is really *comparing apples and oranges* because semiconduction—implying that a material is neither metallic nor insulating—is possible in many different materials other than those used to make integrated circuits and light-emitting diodes. *Semiconduction* really implies a range of values for the conductivity and not a type of material. For example, ionic *semiconductors* are critical for battery and fuel cell applications even though they conduct ions and not electrons. To generalize, there are *really only two classes* of materials, metals and others. Metals have free electrons at absolute zero kelvin—the metallic bond—and all other materials require energy to create free electrons by knocking them out of chemical bonds. If this does not take much energy, the material is a semiconductor: if it takes a lot of energy it is an insulator.

different approach but hopefully a more transparent one for an undergraduate student. So, the organization of the book is, progressing from simpler concepts to more complex ones, as follows:

1. Introduction to Materials Kinetics
2. Reaction Kinetics
3. Phase Transformations
4. Diffusion with a Constant Diffusion Coefficient
5. Fluxes and Forces and Interdiffusion

SUMMARIZING THE MAIN FEATURES OF THIS BOOK

The book is intended to be a junior- to senior-level text for majors in MSE. As such, its main features are as follows:

- Goes beyond the introductory level for MSE students
- Covers topics of interest in all subfields of materials: metals, polymers, and so on
- Addresses kinetics of reactions important for soft materials such as polymers and biomaterials
- Gives real examples of industrial processes and materials used wherever possible
- Discusses both the atomistic and macroscopic aspects of diffusion in solids, liquids, and gases
- Focuses on understanding rather than mathematical rigor, still yielding correct results
- Avoids *black boxes* or equations that *magically* appear
- Explicitly presents intermediate steps in solving equations, allowing for self-study
- Presents all topics with real materials processes in which the results are—or should be—applied in actual industrial practice
- Includes problems for each chapter with the intent of not being *practice* for the student but providing the opportunity to quantify the results of the chapter
- Discusses real processes in a way that shows how the various different serial or parallel steps can be controlled to give different results for the same overall process
- Addresses unique topics such as similarity variable solutions to the partial differential equations of diffusion; simple derivation of the Boltzmann distribution; correct point defect chemistry including the Kröger–Vink notation; the glass transition as a relaxation process; the Kirkendall effect both in metals and compounds; compound formation; a more direct approach to spinodal decomposition

Dennis W. Readey

For the Instructor

This book assumes that students have taken courses on introduction to materials, chemical thermodynamics, and one, two, or three courses in calculus but not differential equations or a higher level engineering mathematics course. Since most students come out of a first thermodynamics course with less than a strong grasp of the topic, thermodynamics and phase equilibria are emphasized wherever possible. Most kinetic models presume an understanding of the thermodynamics of the process. The thermodynamics are explicitly discussed to reinforce the students' understanding of both kinetics and thermodynamics. Many topics involve solving ordinary or partial differential equations. However, a course on differential equations is not a prerequisite because most of the equations can be easily solved, usually by separation of variables, so a second calculus course is sufficient. Nevertheless, at the junior level, many students have either taken, or are taking, differential equations and should be familiar with most of the mathematical manipulations with the possible exception of solutions of partial differential equations. But here again, the relevant equations are solved with no requirements beyond those in a second calculus course. The emphasis is on understanding where equations come from rather than on mathematically rigorous derivations—although simple algebraic manipulation to develop a model is often considered a *derivation* to many students. As a result, a ***plausibility*** argument is sometimes used to obtain an expression or solve an equation; a procedure quite unsatisfactory to a mathematician or physicist, although the final result can be correct.

There are many equations in the text. The reason for this is simply to provide all of the individual steps going from a starting equation to the desired result, so the student (or instructor) does not need several sheets of paper to do the calculations between the beginning and end of a model. This has the advantage that a student can follow the development step by step: nothing is hidden. If the algebra or calculus involves too many steps, they are included in an appendix to not impede the flow of the discussion. Students should have had a course on chemical thermodynamics including phase diagrams and Gibbs energy-composition diagrams. There is a brief review in Chapter 1 of the essentials of thermodynamics that are important in later chapters in this book. Examples of Gibbs energy-composition diagrams and the relation to phase diagrams are given in Chapter 1. A chapter on surfaces is included and a lot of details on the *correct* way to do point defect chemistry in solids are also presented. These topics may have been covered in a prior thermodynamics course but often they are not, yet are crucial for understanding kinetic processes in materials. Also, it is assumed that most students have not had courses on heat transfer or fluid dynamics as most MSE programs do not require these. However, programs whose focus is on extractive or chemical processing might consider these topics essential in a kinetics course, but they are not covered in detail here.

This book is on the *modeling* of kinetic processes important in MSE. In the distant past when MSE programs were mainly metallurgical or ceramic engineering, each program would provide courses very specific to the discipline such as ***heat treatment of steels*** or ***powder processing***. The combined influences of decreases in program credit hours mandated by universities and MSE program inclusion of all materials have either eliminated such specific courses or relegated them to ***areas of specialization*** within a program. There is a great deal of commonality among materials linking structure and properties. However, the processing of each type of material—metals, ceramics, polymers, biomaterials, and semiconductors—has its own unique focus, yet most of these courses disappeared when MSE programs came into existence. It has fallen to courses such as kinetics to be sufficiently broad to fill this lack of processing fundamentals dedicated to individual materials. However, it is challenging to make a course such as kinetics, whose focus is processing, sufficiently inclusive to satisfy the details required for each material. Therefore, principles that apply to all or most materials are emphasized but with the caveat that the temperature ranges—and the importance of temperature—for the processing and performance of metals, ceramics, and semiconductor materials are much larger than for polymers and biomaterials. The main reason for this is that most kinetic processes involved in microstructural changes in hard materials are controlled by diffusion rather than a chemical reaction. In contrast, for polymers and biomaterials, reactions, of necessity, must occur at low temperatures and are controlled by the atomic details of the reaction process. Each of these two areas, diffusion and reaction chemistry, can easily fill more than a semester's worth of kinetics individually, so not all details of each can be easily covered in a single course. Nevertheless, attempts are made to cover kinetic processes that are important for the processing of materials both at high and low temperatures.

The primary goal is to develop models for various kinetic processes so that the rates of reactions can be explicitly written as a function of the process variables; for example, rate = f (temperature, time, pressure, concentration, etc.) where the rate is determined by the rate-controlling step, which is one of many possible steps in the overall process. In most cases, the rate equations developed are ***macroscopic*** in that the relationship between a variable such as temperature and the underlying detailed atomistic mechanism may not be obvious. This is simply because the details of the atomic mechanism are not completely known, that there are just too many possible mechanisms to investigate in detail, or the level of detail—quantum mechanics for example—is beyond the typical background of MSE students at this level. Nevertheless, the goal is to develop fundamentals-based models that can be—and are—used for the control of real materials processing and the understanding of real materials behavior.

The one exception to the more macroscopic approach is the development of atomistic or molecular models of diffusion* in solids, gases, and liquids. For example, it is shown that the diffusion coefficient in solids, *D*, in Fick's first law, $J = -D(dC/dx)$, where J = flux and Units(J) = mol/cm^2-s† and C = concentration with Units(C) = mol/cm^3, depends on temperature as $D = D_0\exp(-Q/RT)$, where D_0 and *Q* come from for diffusion in different materials.

In introductory materials texts, solutions to Fick's second law, $\partial C/\partial t = D(\partial^2 C/\partial x^2)$, such as diffusion into a solid with a fixed surface concentration $C = C_S \mathrm{erfc}\left(x/\sqrt{4Dt}\right)$ are presented, and students are urged to make calculations of concentration as a function of time and distance with this ***black box*** without showing where the black box comes from. For many students, black boxes impede further learning. A major justification for this book is to go beyond introductory texts and to show where expressions such as this come from, that is, how to solve the partial differential equation of Fick's second law. To do this with complete mathematical rigor frequently goes beyond the mathematical experience and maturity of students in junior–senior classes. Here, mathematical rigor and exact expressions are often sacrificed for more transparent approaches that use plausibility arguments, lead to correct solutions, and, hopefully, a better understanding of the fundamental concepts.

Several areas of kinetics important to materials processing are not included. Some of these are: solidification, sintering, grain growth, grain boundary grooving, polymer swelling, interface instability, electron exchange and electrochemical kinetics, and broader coverage of point defect chemistry. There are several reasons for this. One is that some of the topics are too specific to one type of material than for others: solidification and metals; sintering and ceramics; point defect chemistry and electronic materials; and

* Note that *diffusion* is defined as *the spontaneous intermingling or mixing of atoms, ions, molecules, or particles due to random thermal motion.*

† Note that Units(Q) is used in this book to denote *the units of* Q.

swelling and polymers. Another reason is that some topics are best left to graduate courses such as interface instability. Not everything in this book can be covered in a single semester even with the exclusion of these additional topics: choices had to be made. Nevertheless, a few topics such as some nuclear chemistry and ionic conductivity are included simply because MSE students today are probably not exposed to these topics elsewhere but are essential to an undergraduate's education. In reality, major parts of Chapters 2 through 12 can be reasonably completed in a typical semester. The introductory chapter (Chapter 1), Chapter 6 on surfaces, and the discussions of nuclear chemistry and point defect chemistry could be shortened or not covered at all if covered in other courses. Similarly, parts of Chapters 13 through 16 can be included depending on students' backgrounds and the intent of the course. Based on the experience of teaching materials kinetics over many years, there is sufficient material to cover two semesters if all the chapters were to be covered completely. Of course, the sequence of presentation of the various topics certainly does not have to follow the order in the "Table of Contents." For example, *atomistics of diffusion* is listed before solutions to diffusion equations. However, the information can be presented in the opposite order just as well without loss in continuity.

For the Student

This book does not focus on materials *science* exclusively with the assumption that all engineering students will become materials graduate students, so a deeper background is needed. In most MSE programs, only about half the undergraduates go into graduate programs—and many of these are not in MSE—the rest take jobs in industry with the bachelor's degree being the terminal degree, at least for now. As a result, in this book, ***engineering*** or practical applications are emphasized where possible with the philosophy that the development of model-based kinetic processes are those actually used to make or understand real materials and processes. If an engineer knows why variables affect a process in a certain way, then when the process is not working right, the engineer can adjust the variables to make the process perform the way it was designed, eliminating expensive and time-consuming trial and error procedures.

Modeling of kinetic processes involves mathematics, mostly algebra and a little calculus. The intent here is to provide the correct solutions to kinetic problems without relying on students' mathematical exposure beyond the second or third semester of calculus. Although kinetics requires the solution of differential equations, most of these can be solved by integration techniques learned in elementary calculus precluding the need for a formal background in differential equations. Also, ***black boxes***, such as "...it can be shown that...," and other impediments to understanding are largely eliminated. Integrations and intermediate algebraic manipulations are performed step-by-step so that a student can follow the development of a model without the need for working out the intermediate steps with a pencil and paper. Unfortunately, this leads to more lines of equations in the text, but going from one equation to the next requires only a small and easily identified algebraic or calculus step.

There are several things unique about this book. For example, the solution to kinetic processes that go to equilibrium and involve second or higher order forward or back reactions is typically not seen in texts where reaction kinetics are covered. Similarly, the solution for series reactions is usually just given and not modeled. Doing this requires a little more advanced differential equation solving, so it is presented as part of an appendix. The *correct* way to do point defect chemistry is presented, although most texts and published papers do it incorrectly often because the simpler, albeit incorrect, methodology has been passed down from one generation of academics to the next over the past 50 years! Yet the correct version is, in fact, more straightforward once charged species are accounted for and charge neutrality is maintained in the analysis.

A free-volume model of the glass transition is presented that shows that it involves a double exponential with temperature and explains the relative sharpness of the transition. A model for interdiffusion in compounds demonstrates the possibility, although small, for a Kirkendall effect during interdiffusion of compounds that is not available in many other texts. Formation of ternary compounds by diffusion is presented in detail, and the model follows directly from the electrical charge neutrality required of the ion fluxes. Finally, the approach to modeling kinetics of spinodal decomposition is different from the methods generally used and, hopefully, is more direct and plausible to the average undergraduate.

MATHEMATICS

This book is on developing quantitative models of kinetic processes in MSE.* Since this book is intended as a textbook for a course on kinetics for juniors or seniors in MSE programs, it is assumed that the students will have had two or three semesters of calculus but may or may not have had a course on differential equations. The starting point for a model of a kinetic process is the *rate* at which a given process takes place, for example, moles per second, dn/dt. However, what is usually the goal of the model is the ***amount***—the *number of moles,* n—as a function of time and other process variables such as composition, temperature, pressure, and perhaps other parameters. This implies integration of the rate equation or solving the differential equation, dn/dt. The solutions to both linear and relevant partial differential equations are obtained in this book. But fear not! For most of the differential equations of interest, the variables can be separated, so techniques learned in elementary calculus can be used to integrate both sides of the equation to get the desired result. There are a couple of exceptions to this general principle, but these are relegated to appendices in order to not disrupt the flow of model development.

These are not major compromises because even some of the more rigorous developments are still approximations as not all structure and properties of real materials can be easily modeled: this is a main motivation of ***computational materials science.*** For example, the atomistic aspects of diffusion coefficients for gases can be easily and accurately modeled assuming that the atoms are hard spheres. But things get more complicated when the molecules are not spheres, have dipole moments, and so on—typical of real gases. Furthermore, in most cases, particularly for solids, even though the fundamental principles are understood, the uncertainty in the empirically measured values—D_0 and Q in the diffusion coefficient, for example—far exceeds the limits of the model approximations! Therefore, developing an understanding of the fundamentals—both macroscopic and microscopic—and their relationships is the goal of this book rather than detailed theoretical models that explain everything exactly. Nevertheless, every effort is made to focus on phenomena of broad application to all classes of materials.

UNITS, GRAPHS, AND PLOTS

SI units are the preferred units in science today, and they are used in this book where appropriate. However, in many cases, they are not. The use of meters and kilograms rather than centimeters and grams frequently gives numbers that are cumbersome, unfamiliar, and of little use. For example, the densities of most polymers are in the range of 1 g/cm^3. To say that the density of a polyethylene is 10^3 kg/m^3 seems unrealistic unless you are working with cubic meters of material. Certainly in the laboratory and in most technological applications, the quantities dealt with are grams and cm^3 rather than 10^3 kg and m^3. Exceptions exist when dealing with commodity products such as corn, steel ingots, output of cement kilns, or production facilities for raw polyethylene. Then tonnes (a metric ton = 1000 kg) and cubic meters might be appropriate. Therefore, in this book, SI and CGS units (and sometimes English units) are used arbitrarily depending on what the circumstances warrant. Nevertheless, it is important to be able to recognize what units are required for a given parameter in an equation or formula to make its units consistent with all the other terms in the equation. The ability to be able to do ***dimensional analysis*** is crucial. In the 1970s, the United States after years of effort—going back to the founding of the country—made a serious effort to adopt the metric system.† However, it gave up trying in 1982 when the United States Metric Board was abolished by President Reagan after a great deal of popular resistance against the change to a metric system. This led to the rejoicing of many newspapers and commentators on the benefits of *our way* of doing things, the *American Way!* However, with the growth of globalization since the 1970s,

---

* It is worthwhile to note that the development of a mathematical model is not "a derivation." Derivations are things that mathematicians do!

† Marciano, John Bemelmans. 2014. *Whatever Happened to the Metric System?* New York: Bloomsbury.

metric measurements and components have universally appeared so that your toolbox needs sets of both metric and English wrenches. Certainly, there are many industries in the United States that still operate their ovens and furnaces in degrees Fahrenheit, their pressure vessels in pounds per square inch, and use dimensions in mils,* inches, or feet. So a scientist or engineer needs to be able to quickly, and correctly, convert between English and metric units. This ability is emphasized in several of the exercises at the end of each chapter and in the text to highlight this reality. However, today, conversion is easy: all you have to do is ask your cellphone, tablet, or computer to make the conversion for you. He, she, or it will tell you that one bushel = 0.03524 m^3, important for farmers because corn and soybeans are sold by the bushel and crop yields are given in bushels/acre and one acre = 4046.86 m^2. Also, cement, for which it might make sense to report the density in kg/m^3, in the United States is sold by the cubic yard, or more concisely by the *yard*. One cubic yard = 0.7646 m^3. Since about one-half of BS MSE graduates go into industry, they need to be able to quickly convert between different units and no apology is given for not strictly adhering to SI units in this book.

Many of the calculated graphs and plots include the calculated ***data points*** in the graphs. The reason for this is threefold. First, it shows the independent variable values that were used to calculate the graphs. Second, the density of these points along a plot gives some indication of the number of points needed to provide a smooth plot over the data range of interest. Third, and perhaps most important, it gives the student some idea of how a graph should be constructed in terms of proportion, type fonts, line widths, and so on. Before the days of computer-generated graphs, most engineering students were required to take a course in ***engineering drawing*** or ***descriptive geometry*** and be able to generate detailed drawings with multiple views and dimensions from which parts were fabricated in the machine shop and joined on the floor of a factory. In such courses, there were strict rules about line widths, line styles, text fonts—very limited since they were hand-drawn—and their sizes relative to locations on the drawing. So a student became familiar with the generally accepted graphical methods to display technical information. This is no longer the case: courses in engineering graphics are not required and students are frequently not exposed to readily available computer graphing software. In the exercises at the end of most of the chapters in this book, there are requirements for plotting data to give practice in generating ***publication quality*** illustrations that computer graphing software can provide. Hand-drawn curves in the digital age are not acceptable for future engineers and scientists. For most of the graphs in this book, the data points were first calculated with spreadsheet software—frequently having powerful computation capability—and then data transferred to graphics software for plotting.

* 1 mil = 0.001 in.

Acknowledgments

I thank my son, Dr. Michael Readey, for taking time to proofread the final draft of the manuscript for this book while he spent his days as the president of a renewable energy technology company and evenings teaching engineering management at the University of Colorado, Boulder, where he is now a full-time professor. His careful reading has corrected many of my mistakes and any remaining errors are still mine. Also, his background in materials science and engineering gave him the ability to make many useful changes and suggestions to improve the clarity and accuracy of the text. I also thank Luna Han and Judith Simon at Taylor & Francis for their patience during the time it actually took to complete this book: about twice as long as anticipated. Finally, the publisher and I acknowledge and thank the reviewers of the original manuscript for their constructive suggestions during the development process that have helped to improve the final text.

Acknowledgments

Author

Dennis W. Readey is university emeritus professor of metallurgical and materials engineering at the Colorado School of Mines, Golden CO where he served as the H. F. Coors distinguished professor of ceramic engineering and director of the Colorado Center for Advanced Ceramics for 17 years. Prior to that, he served as chairman of the Department of Ceramic Engineering at Ohio State University Columbus, OH. He has been performing research on kinetic processes in materials for almost 50 years and teaching the subject for over 30 years. Before entering academia, he was a program manager in the Division of Physical Research of what is now the Department of Energy, where he was responsible for funding materials research in universities and national laboratories. Earlier, he was also a group leader in the Research Division of the Raytheon Company and in the Materials Division of Argonne National Laboratory. For the last eight years he has been an adjunct professor in MSE at the University of Illinois, Urbana–Champaign.

He had been active in the Accreditation Board for Engineering and Technology (ABET) for a number of years representing TMS (The Mining, Minerals, and Materials Society) and served on several government committees including the Space Sciences Board and the National Materials Advisory Board of the National Academy of Sciences. He is a member of several professional societies and is a fellow of ASM International (formerly the American Society of Metals) and a fellow, distinguished life member, and past-president of the American Ceramic Society.

Dr. Readey's research has involved gaseous and aqueous corrosion of ceramics, the effect of atmospheres on sintering, the properties of porous ceramics, processing and properties of ceramic–metal composites, and the electronic properties of compounds, particularly transparent conducting oxides and microwave and infrared materials. He advised 29 PhD and 42 MS degree theses, which generated about 120 publications and 13 patents. He received a BS degree in metallurgical engineering from the University of Notre Dame, Notre Dame, IN, and a ScD degree in ceramic engineering from MIT, Cambridge, MA.

Section I
Introduction to Kinetics

1

Kinetics and Materials Science and Engineering

1.1 INTRODUCTION

This book is about kinetic processes in materials, and as such, it deals with how fast reactions such as phase transitions occur in materials and is intended as a text for junior-senior level students in undergraduate materials science and engineering (MSE) programs. Included in the scope of materials kinetics are shape changes driven by surface energies and compositional changes driven by concentration gradients. The first question is, "Where does kinetics—and the kinetics topics covered in this book—fit into the broad field of MSE?" Introducing a little modern history of MSE is useful in answering this question, so the first part of this chapter deals with some MSE history.

Before approaching the kinetics of a process, the thermodynamics of the system undergoing change needs to be understood since it is the thermodynamics that drives the direction of the process. It is assumed that readers of this book will have had at least an introductory course in thermodynamics. However, rarely does anyone get a firm grasp on thermodynamic concepts the first time exposed to them. Therefore, the second part of this introductory chapter is a summary of some important thermodynamic principles. Clearly, not everything included in a first course on chemical thermodynamics can be presented in detail here, since there would not be room to discuss kinetics if it were. Therefore, only several of the principles used in later models of kinetic processes are briefly summarized.

1.2 MATERIALS SCIENCE AND ENGINEERING

Frequently, textbooks give the impression that a field of science or engineering suddenly appeared as presented in the text. Of course, this is a completely false impression. Since the field of MSE is relatively new as a separate academic discipline, it is instructive to take some time to review what the field is about, how it got here, and what its main goals are. This is particularly true of MSE because of the breadth of its scope and diversity of its ancestral history.

MSE consists of understanding the relationships between the processing, structure, and properties of materials (Figure 1.1). Many MSE practitioners add *performance* as the fourth corner of a materials tetrahedron, but this makes the relationships hard to draw in two dimensions—*properties* really determine performance anyway, and it usually the goal of the materials engineer to provide a material

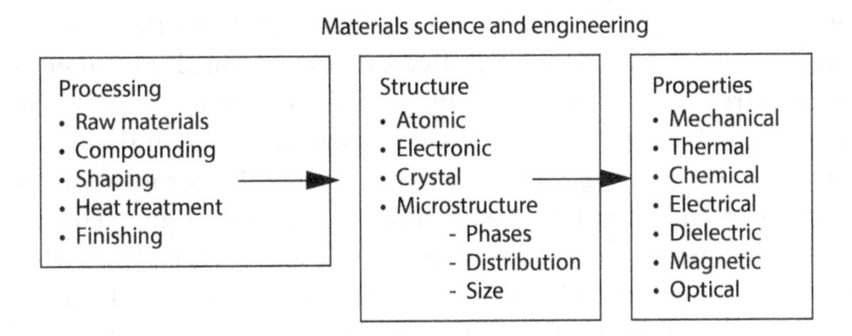

FIGURE 1.1 Illustration of the relationships between the three key areas of materials science and engineering: processing, structure, and properties. Kinetics focuses more on the processing part of these relationships but also includes chemical properties such as corrosion, dissolution, and oxidation.

with a specific set of properties to achieve the required performance of a complex system such as jet turbine or an integrated circuit. More will be said about this later. Some items listed under *structure* in Figure 1.1 need more clarification. The *atomic* part of structure here really means the *composition* of the material in terms of what atoms or ions are present and their relative amounts. The *electronic* structure includes both the type and strength of chemical bonding in a material since they both play important roles in determining properties, such as electrical conductivity. The *crystal structure*, or lack of one, has large effects on most of the properties of interest. Finally, the *microstructure* is one of three concepts central to MSE that justify its existence as a separate discipline. The other two are *crystal structure* and *phase equilibria* (Cahn 2001). Included in the microstructure are the phases present: how many, their compositions, crystal structures—or lack thereof, that is, glasses—their distribution and sizes. For example, small amounts (a few percent) of low melting point second phases segregated to, and covering, the grain boundaries of materials that are mainly crystalline—metals, inorganic compounds, and polymers—can significantly alter the properties of these materials. In contrast, if the second phase were isolated into discontinuous pockets in a material, it may have little or no effect on properties.

The *science* part of MSE has to do with understanding the fundamental principles involved in the relationships between processing, structure, and properties and the *engineering* part is applying these principles to make something useful. For example, Figure 1.2 is focused on a ceramic material—a ceramic being defined as "something useful made from one or more

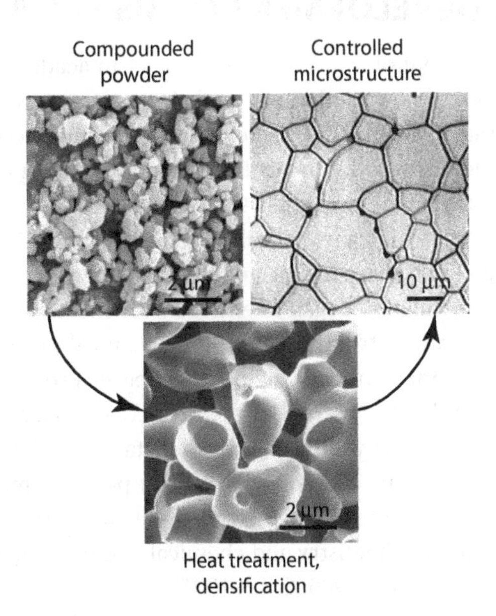

FIGURE 1.2 Relationship between the starting material, heat-treatment (densification), and the final microstructure. The example is yttrium iron garnet, $Y_3Fe_5O_{12}$, a single-phase magnetic material containing several chemical additives to control the final magnetic properties for microwave applications.

inorganic compounds"—simply as an illustration. Figure 1.2 shows the typical processing of a ceramic material beginning with powder particles that are formed into some shape. The part is then *fired* or heat-treated at some high temperature, so that the surface energy associated with the small particles (top left) drives mass transport by diffusion that bonds the particles together (center), produces shrinkage, and eliminates most of the porosity (which starts at about 45 volume percent) to form a dense polycrystalline material (top right). This is a photomicrograph of a magnetic ceramic based on yttrium iron garnet (YIG), $Y_3Fe_5O_{12}$. This is an important material in microwave applications because it is both an electrical insulator, thus transparent to microwave radiation, and magnetic, so the microwaves can interact with the magnetization of the material. By changing the magnetization, the index of refraction to microwaves is changed, which allows the electrical engineer to make interesting devices such as an electronically scanned radar. This particular material is actually a single phase solid solution *oxide alloy* with a composition $Y_{3\text{-}x\text{-}z}Gd_xDy_zFe_{5\text{-}m\text{-}n\text{-}p}Al_mMn_nIn_pO_{12}$. Each of the five additives to YIG gives a different adjustment to the magnetic properties appropriate for the application. Not only does the composition have to be precisely controlled to ensure a single phase solid solution, but so does the microstructure (top right). The average grain size of this material needs to be about 12 μm and the porosity—black spots at grain boundaries—less than 1% by volume. In addition, there can be no other phases present, as these would degrade the desired properties. A proper choice of the starting powders (composition and particle size), firing temperature (~1400°C), firing atmosphere (oxygen-rich), and time must be made and controlled in order to obtain the desired material.

One of the critical steps in the processing of this material is the *calcination* or reaction step that involves heating mixed oxides, Y_2O_3, Gd_2O_3, Dy_2O_3, Fe_2O_3, Al_2O_3, Mn_2O_3, and In_2O_3, causing them to react by solid state diffusion to form the solid solution alloy. This is done at several hundred degrees below the sintering temperature (~1200°C in this case) so that diffusion can occur but the small powder particles do not sinter into a hard dense mass. After calcination, the reacted powder is pulverized to break up any agglomerates and reduce the particle size—0.1–1.0 μm—suitable for subsequent solid-state sintering and densification. In general, *calcination* is a heat treatment process in which a reaction occurs to produce a new solid and is modeled in some detail in Chapter 15.

1.3 HISTORICAL DEVELOPMENT OF MSE AS A DISCIPLINE

Metallurgical engineering grew out of mining engineering as an academic discipline when the need arose to have people trained in the extraction of metals from their ores. In 1894, ceramic engineering also emerged as a separate academic discipline from mining engineering. Here, the motive was to apply the principles of chemistry and physics to the manufacturing of ceramic products, which at that time were limited to *commodity* materials such as paving bricks, roof tiles, refractory bricks, sanitary ware, table ware, and glass. It was not until about the 1930s did ceramic engineering encompass more *enabling*[*] materials such as electronic and optical materials. In addition, at the end of the nineteenth century, physical metallurgy began to emerge with efforts to understand the correlations between the composition and microstructure and the mechanical properties of steel. This led to a rapid increase in the number of metallurgical engineering degree programs to provide metallurgical engineers for the metal-user industries as well as the extractive industry. Although the production of synthetic rubber around 1914 may well have been the start of polymer engineering, it was not until around World War II did the production of synthetic polymer materials become important. Interestingly, a great deal of this work was perceived to be *chemistry* and education in polymers was almost exclusively relegated to chemistry and chemical engineering departments at universities before the past few decades (Morawetz 1985; Cahn 2001).

[*] A *commodity* material such as a building brick is essentially the end product. In contrast, an *enabling* material is one that is a *critical component* of a larger, more complex system, such as the magnetic radar material.

In the early 1950s, there were about 80 *metallurgy* or *metallurgical engineering* degree programs and about 15 *ceramic engineering* programs, and maybe one *polymer engineering program.* In the late 1950s, it was realized by engineers and scientists working in the various materials areas that there was great commonality among their disciplines, particularly the relationships between structure and properties of materials. This ultimately led to the formation of the first MSE graduate program at Northwestern in 1959. By 1970, almost a third of the some 80 metallurgical engineering programs had changed their names—if not their programs—to include *materials* in one form or another. Today, there are about 60 accredited *materials* programs* and only a four metallurgical and two ceramic engineering programs remaining. There are five accredited programs in polymers and none are named simply *polymer engineering.* These polymer programs have names such as *plastics and polymer engineering,* polymer science and engineering, and *polymer and fiber engineering.*

These mergers and changes in department focus did not always occur smoothly. But two events encouraged integrating the different disciplines into single, yet multidisciplinary, MSE programs. However, neither of these was focused on undergraduate programs but primarily graduate programs and research. The first was the establishment of the Materials Research Laboratories (MRLs) by the then Atomic Energy Commission† and later transferred to the National Science Foundation in 1972. The second was the Committee on the Survey of Materials Science and Engineering (COSMAT) report (COSMAT 1974) whose purpose was to determine the nature and scope of materials science and engineering and to encourage interdisciplinary and multidisciplinary research and education. Today, MSE can be considered to be the intersection of four major disciplines: physics, chemistry, chemical engineering, and mechanical engineering, as shown in Figure 1.3. One might argue with this particular selection and certainly other disciplines can be included. The reality is "...the profession has become so broad and merged at the edges with other disciplines is that it is hard to define." (Weertman 2007, 18) Indeed, there are many graduate programs and research activities in some MSE and other academic departments at universities where there are no undergraduate programs. On the other hand, there are both materials undergraduate options and accredited programs in many non-MSE academic departments.

Although there is a strong commonality among materials between structure and properties, there are significant differences between the processing and resulting structure for the different materials: metals and ceramics typically requiring much higher temperatures than polymers, for example. As a result of these mergers and broadening of the scope of materials, the *materials-specific processing*: casting and heat treatment for metals; fine powders, high temperatures and sintering for ceramics; and chemistry and glass transitions for polymers get less coverage than they did in materials-specific degree programs. In addition, with increased availability of electron microscopy and the more recent development of *nanomaterials*, it was clear that the structure–property relationships between man-made materials and those produced in nature, biomaterials, were not too different. As a result, today's field of MSE is a broad one encompassing metals, ceramics, polymers, electronic materials, biomaterials, and composites. Interestingly, some metallurgists recognized similarities between the structures of biological materials and metals and their properties almost a century ago (Goerens 1908).

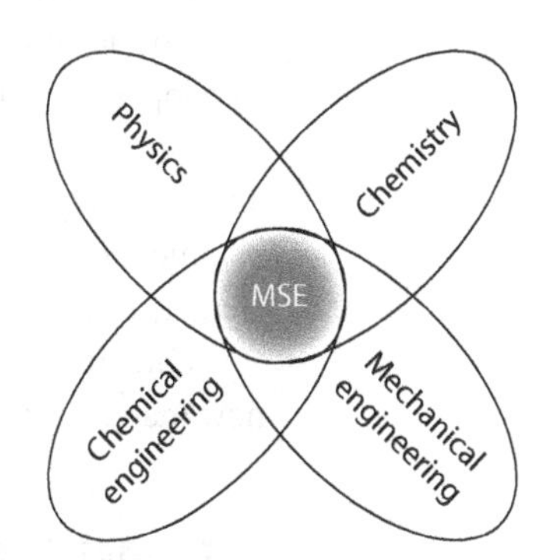

FIGURE 1.3 The intersection between physics, chemistry, chemical engineering, and mechanical engineering to delineate the *fuzzy* boundaries of materials science and engineering.

* Most of these are "materials engineering and science" programs but also includes "metallurgical and materials engineering," "ceramic and materials engineering," and other combination of descriptors.

† Now the U.S. Department of Energy.

1.4 IMPACT OF MSE

Materials engineers and scientists have the most critical position in science and industry since ... *everything has to be made out of something.* It is the availability and properties of existing materials that frequently limits advances in technology. This was one of the motivating factors for the establishment of the MRLs in the late 1950s, slow materials development was perceived to be limiting technology (Cahn 2002). For example, it is very easy for a mechanical engineer or aeronautical engineering to design a turbine engine that would increase the thrust and performance of a modern jet aircraft by several tens of percent. However, this would require raising the operating temperature of the gas turbine well above the typical 1000°C–1100°C range now encountered by the turbine blades to maybe as high as 1500°C. As another example, an article (Chang 2014, D6) discussing the history of the undeveloped X-30 "Aero-Space Plane" initially proposed in 1986 to fly at the edge of space at up to 23 times the speed of sound, states, "For the X-30 to succeed, infant ideas would have had to have been developed into robust, reliable technologies—*materials that could survive intense temperatures,* air breathing engines that could fly faster and higher" (italics added). Unfortunately, even today there are no materials that will reliably perform at such high temperatures. A large government research program tried to develop silicon nitride, Si_3N_4, as a turbine material that could, in principle, operate at that temperature since there are no fundamental material properties preventing it. However, being a ceramic, silicon nitride is brittle, with no plastic deformation before breaking, and design engineers are still learning how to design with materials that are not ductile like metals. Rather than try to design an entirely new engine that might be able to take advantage of the high temperature capability of silicon nitride, the focus was to design a ceramic component as a direct replacement for metal turbine blades. This puts the ceramic part at a big disadvantage because of the differences in processing and properties compared to metals. For these and other reasons, metals are still being used as turbine blades in jet engines. Nevertheless, there is a now a much better understanding of brittle failure and there remains considerable technological and commercial interest in ceramic-matrix composites that have both high-temperature capability and excellent mechanical properties.

Similar comments could be made about nuclear power. Engineers can design a nuclear plant to work at higher powers levels, but the temperature limitations of existing fuel and other materials critical for safe operation do not permit it. Modern microelectronic devices are made from silicon, not because silicon is the best material from a semiconductor properties standpoint—germanium and gallium arsenide are superior—but rather the fact that the native oxide on silicon, amorphous SiO_2, can easily be grown and has superior insulating properties to the native oxides on Ge or GaAs. Yet, as devices such as FETs—field effect transistors—become smaller and the SiO_2 gate oxide becomes so thin—a nanometer—that it no longer has the good insulating properties of bulk SiO_2, other materials, such as HfO_2, hafnium oxide, are being used. As a result, the feasibility of applying other semiconductor materials becomes more attractive.

There is considerable activity with electric motors replacing the internal combustion engines in automobiles reducing the U.S. and world's dependence on the certainly finite supply of fossil fuels with the additional beneficial result of decreasing heat-trapping gases such as CO_2. However, for this to happen, new batteries with high energy storage capacities must be developed. This is one of the most intense areas of materials R&D that exists today: for it is the properties of the anodes, cathodes, and electrolytes—all materials—that determine the energy capacity and performance of electrochemical cells. Here again, it will be the development of new materials that will make widespread electric propulsion a reality.

One article (Moss 2015) suggests that it will be materials engineers and scientists developing new materials that will save the world from various disasters such as global warming!

1.5 MICROSTRUCTURE

A comment about the term *microstructure* is appropriate since it one of the key elements of MSE. The relationship between *microstructure*—structure that can be seen in a microscope—and properties has been a well-established part of MSE for over a 100 years, and one it is one of the three core areas of phase equilibria, crystal structure, and microstructure that distinguish MSE from other fields of science and engineering (Cahn 2002). The historical development of the microstructural investigation of metals has been well documented by Smith (1988). The importance of composition on strength dominated the development of metals, primarily iron and steels, during the late nineteenth century. Since iron and steel were important structural materials, strength was their most important property and remains so for most applications of metals. This ultimately led to the investigation of microstructure by Sorby (Smith 1988) and its relationship to composition and properties. Interestingly, Sorby first studied the microstructure of rocks in transmitted light using thin sections. When he studied metals, he again used thin, mounted samples but examined them in reflected light. Mineralogists quickly adopted Sorby's thin section technique for rocks, while the metallurgists adopted his techniques more slowly (Smith 1988, Preface, v).

It is worth noting that the term *microstructure* originally had nothing to do with the scale of what was being observed in the microscope. Initially, Sorby called what he could see in the optical microscope *microscopical structures* (Smith 1988). By 1900, the term had been shortened to *microstructure*, and was in common use by metallurgists (Goerens 1908; Sauveur 1912), and continued to be used in that way for over a 100 years. It was the correlation between composition (carbon content), heat treatment time and temperature, and microstructure with the strength of steels that was the birth of modern MSE. Today, in materials science, the use of the term *microstructure*, *nanostructure*, and *mesostructure* are used by many to refer to the *size* of the features in a material. In addition, some polymer scientists refer to the configuration of atoms along a polymer chain as the chain's *microstructure* (Sperling 2006). This is very confusing, and hopefully discouraged, since many polymers have both crystalline and glassy phases present in varying amounts and sizes that would be described as the polymer's *microstructure* applying the term's generally accepted meaning. Here, *microstructure* is used in the broader context of *what can be seen with a microscope* regardless of the size of the features or the type of microscope being used to observe them.

1.6 THERMODYNAMICS REVIEW

1.6.1 A Difficult Subject

It is assumed that students taking a course on materials kinetics would have completed a course in chemical thermodynamics and have been exposed to the topics that are summarized here. An elementary knowledge of chemical thermodynamics is a prerequisite for understanding much of the kinetics that will follow. However, the reality is that nobody is really comfortable with his or her understanding of thermodynamics after going through only a one-semester course on the subject. Considerably more exposure to many of the topics and terminology of thermodynamics is required before a level of confidence is attained. In fact, Arnold Sommerfeld (1868–1951), the famous German physicist, supposedly replied in the 1940s to the question "Why didn't he publish a book on thermodynamics?" (Keszei 2012, Preface, v)*

> Thermodynamics is a funny subject. The first time you go through it, you don't understand it all. The second time you go through it, you think you understand it, except for one or two small points. The third time you go through it, you know you don't understand it, but by that time you are so used to it, it doesn't bother you any more (sic).

Sommerfeld was more than your average scientist. He expanded Bohr's theory of the atom to explain spectral lines as well as performed research on X-ray diffraction and crystallography. Although he never won a Nobel Prize himself, he was advisor to the largest number of Noble Prize recipients even to this day. Four of students went on to win Nobel Prizes: Hans Bethe, Werner Heisenberg, Wolfgang Pauli, and Peter Debye. Three of his postgraduate students also won Noble Prizes: Linus

* Searching on the Internet for the first sentence in this quote generates a large number of hits. Curiously, however, the quote appears in only a few texts devoted to physical chemistry or thermodynamics such as the one quoted.

Pauling, Isidor Rabi, and Max von Laue. In addition, many of his students became famous by making significant contributions to physics during their careers (Millar et al. 2002). So, if someone as talented as Sommerfeld had some trouble understanding thermodynamics after his first few exposures to it, it is not surprising that most people have difficulties.

In modeling the kinetics of reactions important in MSE, the thermodynamics of the system must be understood. So, a short refresher on some fundamental thermodynamic concepts is a worthwhile use of time. After all, kinetics typically deals with systems going from a high energy state to a lower one and evaluating the energies of these different states is the goal of thermodynamics. Whereas the goal of kinetics is to determine *how fast* systems go from high to lower energy states usually by either a reaction-controlled or a transport (diffusion)-controlled process. While the thermodynamics of system is crucial for understanding diffusion-controlled processes, it is important in reaction-controlled kinetics as well. Certainly, an extensive review of thermodynamics is neither possible nor attempted here. Selected thermodynamic topics reviewed are those used in many of the kinetic models developed later.

1.6.2 History of Thermodynamics

1.6.2.1 Relation to MSE

The history of how thermodynamics was developed is of particular interest to MSE since both fields had the same motivation: steam engines. Furthermore, the modern field of MSE and its dependence on *phase equilibria* and *phase diagrams* was made possible when chemical thermodynamics was developed toward the end of the nineteenth century, primarily by J. Willard Gibbs, a great American scientist. As noted earlier, the properties of steel only became understandable when phase equilibria were applied to interpretation of the microstructure and its relation to the properties of iron-carbon alloys.

1.6.2.2 Why Thermodynamics Is Difficult

Similar to MSE, the subject of thermodynamics really got its start after the invention of the steam engine in the late 1700s. The field expanded and grew initially by trying to determine the efficiencies of such engines to get the most work out of burning fuel. Thermodynamics was largely developed throughout the 1800s mainly by manipulating differentials of macroscopic quantities such as internal energy, U, temperature, pressure, and volume. In addition new quantities were developed and given strange-sounding names such as *entropy*, S, and *enthalpy*, H (Laidler 1993). The macroscopic equations of thermodynamics were essentially complete before there was general understanding and acceptance of the atomic and molecular structure of matter. As a result, concepts such as the energy content of a substance, U, are rather abstract, and were developed to try to determine how much work could be done by a system given a certain amount of energy input. Today, it is understood that the energy content of a solid, liquid, or gas depends on the vibrations, rotations, and translations of atoms or molecules in a given system. Empirically, the concept of the *entropy* grew out measurements of work done and heat input and the requirement to balance the total energy equations. Today, it is understood that entropy reflects the randomness of a system. When a crystalline solid sublimes to a gas, the entropy of the system increases because the gas atoms are much more randomly arranged compared to the ordered positions of atoms in solids. This was a major topic of controversy. It was not until about 1877 when Boltzmann published his relationship between entropy and probability:

$$S = k_B \ln W$$

where:

W is the number of ways atoms may be distributed over energy states
k_B is Boltzmann's constant.

This made the connection between macroscopic thermodynamic quantities and atomic or molecular structure of a material much clearer. When he proposed this relation, he did not state what the value of the constant was other than it was a constant of proportionality. Boltzmann was sufficiently proud of this result that it is inscribed on his tomb (Figure 1.4) (Lindley 2001). Today, the relationships between macroscopic thermodynamic quantities and atoms and molecules are well understood. This makes terms such as *internal energy* and *entropy*, and the equations relating them, much easier to understand, removing some of the mystery from thermodynamics (Kjellander 2016).

Adding to the complexity of thermodynamic definitions and calculations are the different systems of units used in various countries, mainly the English and the metric systems. In recent years, in efforts to standardize notation, units and, definitions of terms, international bodies such as the International Union of Pure and Applied Chemistry (IUPAC) have suggested standards that require changing some long-used terms and definitions. As a result, there are differences in notation between older and newer references and even differences in contemporary references depending on who has accepted the changes and who has not. In addition, different fields focus on different energies (previously called *free energies*, and still called *free energies* by some): chemists use the Gibbs energy, G, and physicists prefer the Helmholtz energy, A, both of which were *free energies*—the energy available to do work—until the end of the twentieth century.

FIGURE 1.4 Boltzmann's grave in the Central Cemetery, Vienna, Austria, bearing the inscription of Boltzmann's law: S = k lnW. (https://commons.wikimedia.org/wiki/File:Zentralfriedhof_Vienna_-_Boltzmann.JPG)

1.6.2.3 Significant Events in the Development of Thermodynamics

During the development of thermodynamics, there were contributions from French, German, and English scientists. However, sometimes nationalism got in the way of acceptance of the contributions of a colleague from another country (Laidler 1993, 106).

> Today we think of thermodynamics as a rather stolid and established discipline, with little scope for controversy and it comes as a surprise that in the latter part of the nineteenth century there were fundamental and acrimonious disagreements about it, with harsh words written on both sides.

Furthermore, because many scientists were working on similar topics at the same time, it is often difficult to discern who was responsible for a given discovery. For example, Clausius is often given credit for the concept of *entropy*. However, talking about who made contributions to the concept, Lindley (2004, 111) points out that

> Subsequently Rankine, Clausius, and Thomson all made contributions to the statement of a second law, both in its physical conception and in the mathematical demonstration of its universality. Clausius in the end christened the child and most often gets the credit.

As a result, one might conclude that the significance of a discovery or publication in thermodynamics may be more of an observer's personal choice than an absolute reality. Therefore, here are some personal choices:

Lord Rumford (Benjamin Thompson)

Lord Rumford is worthy of being singled out because he was ahead of his time in thermodynamic thinking plus his life story is an extremely fascinating one. Benjamin Thompson was born in Massachusetts in 1753; he was a royalist and British spy during the American Revolution and subsequently went to England when the British lost. There he became an officer in the British army, was knighted, and did some research on gunpowder that led to his election to the Royal Society. Largely through self-promotion and influential contacts, he reappeared in Bavaria where he was made a count and he chose the name *Mumford*, which was the original name of the town he lived in, Concord New Hampshire, when he left the United States. This was a period when even the difference between *heat* and *temperature* were not clearly differentiated and the exact nature of heat was a mystery. After observing the heat generated by boring some cannons in the Munich arsenal in 1797, he did some experiments on the relationship between heat and work. He concluded in 1798 that 5.60 J—in today's units—were equivalent to one calorie of heat (Brown 1999). The accepted value today is 4.184 J/cal (IUPAC 2014). This was only one of his many scientific and engineering pursuits but the one that had the greatest impact on thermodynamics. In 1880, John Tyndall, a renowned scientist of his time, wrote, "When the history of the dynamical theory of heat is completely written, the man, who in opposition to the scientific belief of his time, could experiment, and reason upon experiment, as Rumford did, may count on a foremost place." (Tyndall 1880, 46; Brown 1999, 94) However, it was over 50 years later before others tried to improve on his determination on the equivalence of heat and work.

James Prescott Joule

James Prescott Joule is interesting because he essentially did his research as a private citizen, not in a university or industrial research laboratory, and his measurements were the conclusive results that a skeptical scientific world needed to agree on the equivalence between energy and heat. Furthermore, his name lives on in the *joule* unit of energy. James Prescott Joule was born in England in 1818, son of a wealthy brewing family. This wealth allowed him to begin his research early in life, which he later continued when he ran the beer company. He too did research on the equivalence between work and heat, perhaps to improve the efficiency of the brewing process. In any event, he tried several approaches to heating water by both mechanical and electrical means. In the electrical case, he ran current through a resistor generating I^2R heating—and developed this relationship—that is to this day called *joule heating.* Most significantly, he was able to rotate a paddle wheel driven by a falling weight and found the mechanical equivalent of heat, which he published in 1845. Since he was not an academic, he could not get his results published by the Royal Society but had to publish elsewhere. He was a very careful experimentalist and continued to refine his experiments for greater accuracy. He published additional papers on the mechanical equivalent of one calorie of heat with the final result in 1878 with 772.55 ft-lb of work needed to raise the temperature of a pound of water 1°F at 60°F and sea level (4.144 J/cal a little short of the accepted value today). See Appendix for the conversion calculation). This final value is inscribed on Joule's headstone (Figure 1.5) in the cemetery in Sale England where he spent his entire life (Brown 1999).

FIGURE 1.5 Grave of James Prescott Joule with the inscription for his mechanical equivalent of heat: 772.55 ft-lbs equivalent to 1 calorie. (Photo by Glaukon in the public domain. https://commons.wikimedia.org/wiki/File:James_Prescott_Joule-gravestone.JPG)

Josiah Willard Gibbs

Josiah Willard Gibbs (1839–1903, Figure 1.6) was an American scientist who, in 1863, received a PhD in engineering from Yale University, only the second PhD awarded in the United States up to that time (Millar et al. 2002). He was appointed professor of mathematical physics at Yale in 1871 and spent the rest of his bachelor life there. During the 1870s, he essentially single-handedly developed the field of chemical thermodynamics, including the concept of the Gibbs energy—important because it depends on the easily controlled variables of temperature and pressure—the Gibbs phase rule, and phase diagrams (Laidler 1993). He is considered by some to be one of the greatest American scientists of all time. He remains probably the greates theoretical scientist born in the USA (Millar et al. 2002, 142). "What was the nature of his mind? And the greatness? What was that? What places Gibbs with Newton and Einstein?" (Rukeyser 1942, 7) Unfortunately, he published his ideas and results in rather obscure journals; his writing style was incomprehensible to most other scientists. As a result, the importance of much of his work was not realized until several years after, during which time others had come to similar conclusions. Gibbs and his phase diagrams, of course, enabled metallurgists to understand the relationships between composition, heat treatment, microstructure, and the properties of steel in the early twentieth century. This knowledge of steels permitted the development of boilers that did not explode and bridges that did not collapse bringing full circle the original intent of thermodynamics: increasing the efficiency of steam engines to giving birth to the field of MSE.

FIGURE 1.6 Josiah Willard Gibbs before 1903. (Photograph in the public domain since it was published before January 1, 1923. https://commons.wikimedia.org/wiki/File:Josiah_Willard_Gibbs_-from_MMS-.jpg)

1.7 REVIEW OF THERMODYNAMIC CONCEPTS

1.7.1 Purpose

The goal here is not to justify the various thermodynamic relationships but mainly to review the important conclusions of thermodynamics developed in thermodynamics courses. Most books on physical chemistry cover chemical thermodynamics in detail. In addition, there are several texts devoted to chemical thermodynamics applied to materials (Darken and Gurry 1953; Ragone 1995; DeHoff 2006; Gaskell 2008). The emphasis here is on those relationships that are primarily used in the modeling of kinetic processes that follows.

1.7.2 Four Laws of Thermodynamics

The laws of thermodynamics are essentially empirical laws that have been determined by experiment over the last few hundred years. They are as follows:

1. The first law of thermodynamics states that the energy of the universe is conserved during some kind of process.
2. The second law of thermodynamics states that the *entropy* of the universe increases during a process.
3. The third law of thermodynamics states that there is *an absolute zero of temperature* and the *entropy of all materials is the same* at absolute zero.
4. The zeroth law of thermodynamics, defined by some, is: if three bodies A, B, and C are in equilibrium, then A is equilibrium with B, B is in equilibrium with C, and A is in equilibrium with C.

1.7.3 First Law of Thermodynamics

The *first law* of thermodynamics, which is rather intuitive, is simply as follows: a small increase in the internal energy of a *system*, dU, is given by the sum of the heat input, δq, and the work *done on* the system, δw, the units of all three of them are joules (J). That is,

$$dU = \delta q + \delta w. \tag{1.1}$$

Stating it another way, the first law is simply *the conservation of* energy. Equation 1.1 can also be written as

$$U_2 - U_1 = \Delta U = \Delta q + \Delta w$$

ΔU is the difference in internal energy going from state 1 to state 2; it is referred to as *state function* in that regardless of how the system got from state 1 to state 2, it will have the same set of thermodynamic properties, that is, internal energy. Today, it is known that the internal energy of a system is essentially the vibrational and translational motions of the atoms or molecules that make up the system. But what is a system? A *system* can be defined as a region of space with boundaries and has thermodynamic properties as shown in Figure 1.7.

Now, the internal energy of a system is a *state* function since its value only depends on the values of the *state variables* that affect the system, such as pressure, temperature, and composition. That is, if changes are made to state variables and then they are returned to their original values, the internal energy, a state function, is the same as before the variables were changed as illustrated in Figure 1.8.

The work done on a system can be of several types:

1. Pressure volume work, PdV, where P is the pressure and dV is the volume change.
2. Mechanical work, Fdl where F is the force and dl is the displacement.
3. Surface work, γdA, where γ is the surface energy and dA is the increase in area.
4. Others, including magnetic work, polarization work, and electrical conduction.

It is conventional to write the first law considering only mechanical work, PdV—remember that thermodynamics was developed to understand steam engines—as

$$dU = \delta q - PdV \tag{1.2}$$

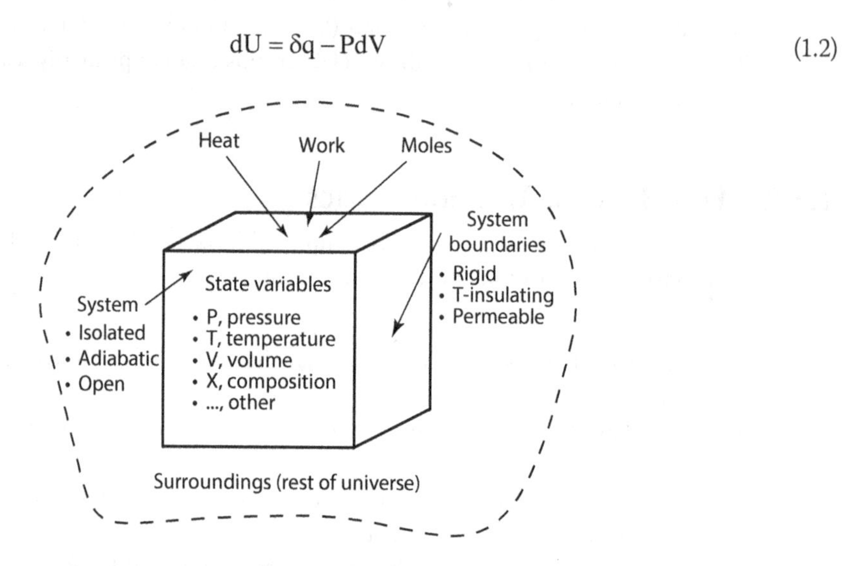

FIGURE 1.7 A thermodynamic system and its surroundings. If the boundaries of the system are rigid, thermally insulating, and impermeable, then the system is *isolated*. If the boundaries are thermally insulating, then the system is *adiabatic*. If both matter and heat can pass through the boundaries then the system is *open*.

where PdV is now pressure-volume work done *by the system* with an expansion dV. Note that PdV is the product of an *intensive variable*, P, and an *extensive variable*, V.* Such pairs of intensive and extensive variables occur frequently in thermodynamics are termed *conjugate variables.* Integrating both sides of Equation 1.2 at *constant pressure* gives†

$$\Delta U = q_P - P\Delta V$$

or the differences in the quantities between two states is

$$\Delta U + P\Delta V = (U_2 + PV_2) - (U_1 + PV_2) = q_P$$

and the *enthalpy*, or heat content, is defined as H = U + PV so $\Delta H = q_P$. Similarly, integrating Equation 1.2 under *constant volume* conditions gives $\Delta U = q_V$.‡ Now, the *heat capacity* at constant volume per mole is

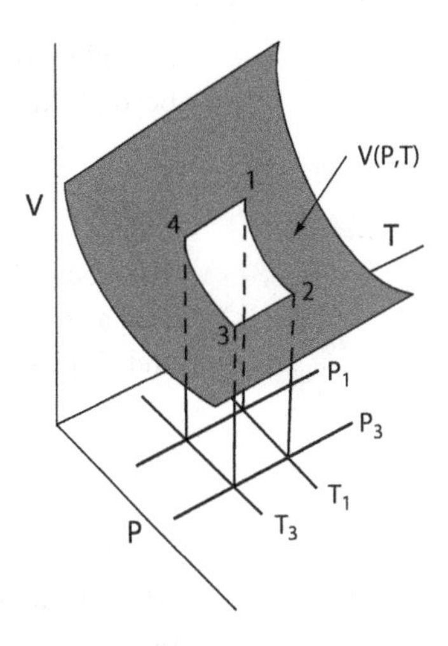

FIGURE 1.8 Schematic representation of a *state function* V(P, T) that is a function of the *state variables* P and T. Starting at any point 1, if the system goes through paths $P_1 \to P_2$, $T_1 \to T_3$, $P_3 \to P_2$, and $T_3 \to T_1$, or any path back to its original P_1 and T_1 the volume will be the same as the starting volume.

$$\bar{C}_V = \frac{q_V}{\Delta T} = \frac{\Delta \bar{U}}{\Delta T}$$
$$\bar{C}_V = \left(\frac{d\bar{U}}{dT}\right)_V \tag{1.3}$$

where the last term is *the molar heat capacity at constant volume*. Similarly, the *molar heat capacity at constant pressure* is given by

$$\bar{C}_P = \frac{q_P}{\Delta T} = \frac{\Delta \bar{H}}{\Delta T}$$
$$\bar{C}_P = \left(\frac{d\bar{H}}{dT}\right)_P. \tag{1.4}$$

Note that for one mole of an ideal gas, its state function is PV = RT, so for an ideal gas

$$\bar{H} = \bar{U} + P\bar{V} = \bar{U} + RT$$

or

$$d\bar{H} = d\bar{U} + RdT$$

so

$$\frac{d\bar{H}}{dT} = \frac{d\bar{U}}{dT} + R$$
$$\bar{C}_P = \bar{C}_V + R \tag{1.5}$$

demonstrating that heat capacities and constant volume and constant pressure are not the same. The ideal gas is a rather extreme case and for most solids there is only a small difference between the constant volume and constant pressure heat capacities. In most tables of data, $\bar{C}_P$ values are given because it is much easier to measure them at constant pressure, one bar, than at constant volume. The relationships between $\bar{C}_P$ and $\bar{C}_V$ can be determined for solid and liquids as well from

* An *extensive* variable is one whose value depends on *how much is present.* On the contrary the value of an intensive variable does not.

† Both U and V are *extensive* variables and can be integrated to get the total internal energy and volume.

‡ If one mole is being considered, then the quantities of interest are *molar* quantities and are indicated with an overhead bar; that is, $\bar{U}$, $\bar{H}$, etc.

a combination of properties by manipulation of various thermodynamic quantities, but this goes beyond this review and can be found in the literature (Darken and Gurry 1953).

1.7.4 Second Law of Thermodynamics

The *second law of thermodynamics* can be most easily stated that for a reversible process, the *entropy*, S, is a *state-function* (only depends in the initial and final states of a system, e.g., P and T like U and H) given by

$$dS = \frac{dq_{rev}}{T} \tag{1.6}$$

where q_{rev} is the amount of heat transferred in a *reversible process*, one that is essentially in equilibrium throughout the process (Gaskell 2008). For an isothermal process, such as melting, at constant pressure,

$$\Delta S_m = \frac{\Delta H_m}{T}$$

where ΔH_m is the enthalpy involved in melting. When the concept of entropy was developed in the mid-1800s it was considered that it was the amount of heat in a system that was not available to do useful work. Today, it is better understood that it has to do with the increase in disorder among the atoms or molecules making up a system. For example, melting takes heat so $\Delta S_m > 0$ as it must be for all spontaneous processes such as melting where the order among the atoms or molecules is much less in the liquid than the very-ordered crystalline solid. So the entropy is a quantitative measure of the disorder of the system or the number of possible states as given by Boltzmann's law,

$$S = k_b \ln W. \tag{1.7}$$

Therefore, the first law can be written

$$dU = TdS - PdV$$

or in its integrated form,

$$U = TS - PV.$$

Stated another way, for any spontaneous process, the entropy of the universe must increase. At equilibrium, a system has its maximum entropy and $dS = 0$ as illustrated in Figure 1.9.

For example, assume that the *system* in Figure 1.7 is one mole of copper at an initial temperature of 300 K, which is placed in a large furnace, the *surroundings*, at a temperature of 500 K and a constant pressure of one bar. Heat will be transferred to the mole of copper from the surroundings until the copper reaches the temperature of the furnace. The amount of heat necessary to heat the copper is given by

$$\Delta Q = \Delta \bar{H} = \int_{300}^{500} \bar{C}_P dT$$

where the overhead bar indicates molar quantities. Assume that $\bar{C}_P$ is constant and equal to 3R, as will be seen later, or $\bar{C}_P = 24.942\,\mathrm{J/mol \cdot K}$ so $\Delta Q = \bar{C}_P \Delta T = 24.942 \times 200 = 4988.4\,\mathrm{J/mol}$. The surroundings is considered to be sufficiently large, so that its temperature does not change with this transfer of heat, so the entropy change of the surroundings is $\Delta S_{surr} = -\Delta Q/T = -4988.4/500 = -9.9768\,\mathrm{J/mol \cdot K}$ and is negative because heat is lost to the system. On the other hand, the entropy change of the copper is

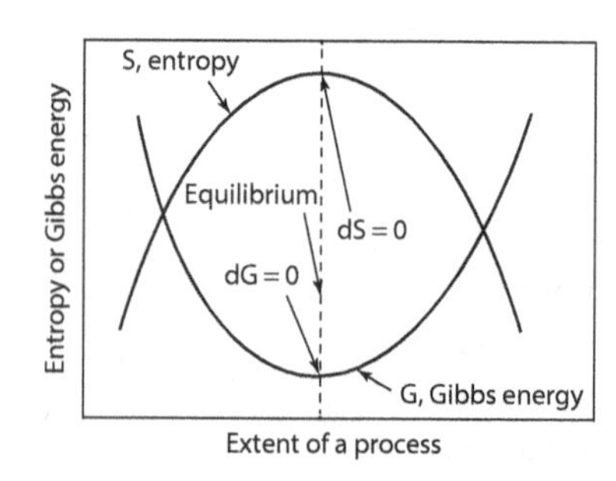

FIGURE 1.9 Schematic showing the relationship between the entropy and Gibbs energy of a system. For a spontaneous process, the entropy always increases while the Gibbs energy decreases. At equilibrium, the entropy is a maximum while the Gibbs energy is a minimum. Therefore, at equilibrium, $dS = 0$ and $dG = 0$.

$$\Delta S_{sys} = \int_{300}^{500} \frac{\bar{C}_P}{T} dT = 24.942 \ln\left(\frac{500}{300}\right) = 12.741\,J/mol \cdot K$$

so the entropy change of the universe is

$$\Delta S_{universe} = \Delta S_{surr} + \Delta S_{sys} = -9.9768 + 12.741 = 2.764\,J/K$$

the *mol* is dropped since one mole of copper is being heated. As predicted, the entropy change for this irreversible process is positive. It is irreversible because heat will not flow back into the surroundings from the copper lowering its temperature. Heat always flows from a high temperature to a low temperature until the temperatures are the same and the system and surroundings, in this case, are in *thermal equilibrium*—their temperatures are equal.

1.7.5 Helmholtz and Gibbs Energies

In order to calculate the amount of work done by systems at constant volume and constant pressure, two additional state functions are defined. The Helmholtz energy, A, given by

$$A = U - TS \tag{1.8}$$

which is the amount of internal energy available to do work and was called the *Helmholtz free energy.* The second is the Gibbs energy* given by

$$G = H - TS = U + PV - TS. \tag{1.9}$$

For chemical thermodynamics, the Gibbs function is most useful. Since the first law for a reversible process is $dU = TdS - PdV$ and taking the total differential of Equation 1.9

$$dG = dU + PdV + VdP - SdT - TdS \tag{1.10}$$

and inserting the first law

$$\begin{aligned} dG &= dU + PdV + VdP - SdT - TdS \\ dG &= (TdS - PdV) + PdV + VdP - SdT - TdS \\ dG &= VdP - SdT \end{aligned} \tag{1.11}$$

where in the last equation, the change in the Gibbs energy depends on changes in pressure and/or temperature, which are easily fixed and controllable experimental variables and the main reason for introducing the Gibbs function in the first place. On the other hand, doing the same with the Helmholtz energy, the change in energy depends on the change in volume, which is a harder variable to control.

At equilibrium, the entropy is a maximum and $dS = 0$. What is the comparable situation for the Gibbs energy at equilibrium? Again,

$$dS_{universe} = dS_{sys} + dS_{surr} \tag{1.12}$$

and

$$dU = TdS - PdV \quad \text{or} \quad dS = \frac{dU}{T} + \frac{PdV}{T}. \tag{1.13}$$

But, $dU_{sys} = -dU_{surr}$ and $dV_{sys} = -dV_{surr}$ so with Equations 1.12 and 1.13, the entropy change is

* For most of the twentieth century, A and G were called *the Helmholtz free energy* and *Gibbs free energy*, respectively. In the 1990s, IUPAC recommended that these be shortened to the *Helmholtz energy or function and the Gibbs energy or function.*

$$dS_{universe} = dS_{sys} - \frac{dU_{sys}}{T} - \frac{PdV_{sys}}{T} \geq 0 \tag{1.14}$$

from the second law. Multiplying Equation 1.14 by (–T) changes the inequality so

$$-TdS_{sys} + dU_{sys} + PdV_{sys} \leq 0.$$

But from Equation 1.11, this is

$$dG_{sys} - V_{sys}dP + S_{sys}dT \leq 0.$$

Therefore, at constant temperature and pressure,

$$dG_{T,P} \leq 0. \tag{1.15}$$

So, the second law requires that the Gibbs energy be a minimum at equilibrium as shown in Figure 1.9; it is just opposite to the variation in entropy and has its *minimum* at equilibrium but, like entropy, $dG = 0$ at equilibrium.

1.7.6 Reactions and Equilibrium Constants

1.7.6.1 Equilibrium

Of great importance in kinetics is "What is the equilibrium or lowest Gibbs energy state of a system?" since kinetics models the route to achieving this lowest energy state. Figure 1.10 shows a ball on a track at various heights above some reference level with different potential energies relative to the reference level.* This also serves as a model of the Gibbs energy as a function of the extent of a reaction. Clearly, in the *metastable* position, the ball (system) will lower its energy by moving to the *metastable equilibrium* position unless it gained enough kinetic energy to go over the barrier of *unstable equilibrium*. If it just has enough kinetic energy, the ball may find itself at the *unstable equilibrium* position where it stops but a slight fluctuation in energy will push it either back to the metastable position or move it forward to the stable equilibrium, the lowest energy state the ball (system) can achieve.

1.7.6.2 Standard States and Equilibrium Constants

What is really of interest is to what extent a reaction will occur. Take the simple reaction of the formation of water vapor by the reaction of hydrogen and oxygen at some constant elevated temperature where all three are gases exist in equilibrium. The reaction is

$$H_2(g) + 1/2\,O_2(g) = H_2O(g). \tag{1.16}$$

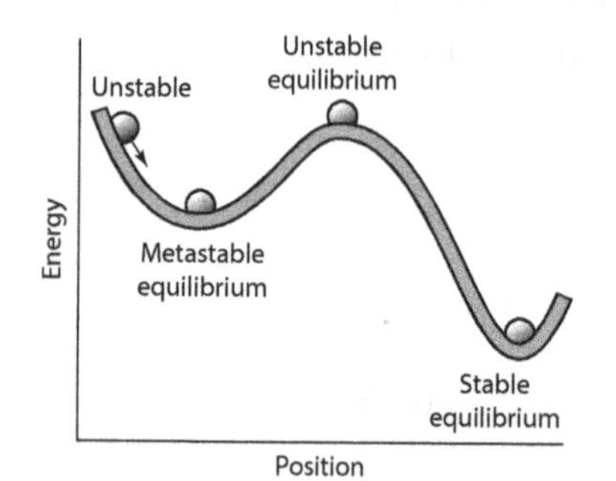

FIGURE 1.10 Schematic showing different kinds of equilibrium for a ball rolling a track with hills and valleys. In this case, the energy is the potential energy and the stable equilibrium position is the lowest point on the track.

For a constant temperature, Equation 1.11 becomes for an ideal gas

$$dG = VdP$$
$$dG = \frac{nRT}{P}dP = nRTd(\ln P) \tag{1.17}$$

where n = number of moles for each of the gases. Integrating this from some standard state

$$G = G^{o} + nRT\ln\left(\frac{p}{P^{o}}\right) \tag{1.18}$$

* Energies are relative. For example, the ball in Figure 1.9 may reach its lowest potential energy on the track but both the ball and track are moving through space at a very high velocity due to the rotations of earth on its axis and in its orbit.

where G^o is the Gibbs energy of the gas in its standard state, P^o: the pure gas at one bar pressure* and whatever temperature is being considered.† The pressure of the gas in a reaction is written with a lowercase P to indicate that it is a partial pressure of a specific gas in a molecular mixture of other gases. Gibbs showed that the change in energies for the reaction in Equation 1.16 could be written

$$\Delta G = G(H_2O) - G(H_2) - 1/2G(O_2) \tag{1.19}$$

$$\Delta G = G^\circ(H_2O) - G^\circ(H_2) - 1/2G^\circ(O_2) + RT\ln\left(\frac{p_{H_2O}/P^\circ}{p_{H_2}/P^\circ\left(p_{O_2}/P^\circ\right)^{1/2}}\right)$$

$$\Delta G = \Delta G^\circ + RT\ln\left(\frac{p_{H_2O}/P^\circ}{p_{H_2}/P^\circ\left(p_{O_2}/P^\circ\right)^{1/2}}\right). \tag{1.20}$$

At equilibrium, $\Delta G = 0$ so

$$\Delta G^\circ = -RT\ln K_e \tag{1.21}$$

where K_e is the *equilibrium constant* and is the term in the brackets in Equation 1.20. Note that the equilibrium constant *has no units*!

Example: Hydrogen-water Gas Mixtures

Hydrogen–water vapor gas mixtures are frequently used to get fixed and low oxygen pressure in various kinds of experiments and processes. For example, consider a gas of almost pure hydrogen, so that $p_{H_2} \cong 1$ bar that has a dew point of about −20°C (easily achievable in practice with a variety of techniques), which corresponds to a $p_{H_2O} \cong 10^{-3}$ bar (Roine 2002). At 500°C, the Gibbs energy for the reaction in Equation 1.19 is $\Delta G^\circ = -205.014$kJ corresponding to an equilibrium constant of $K_e = 7.113\times10^{13}$ (Roine 2002). So, p_{O_2} can be calculated from Equations 1.20 and 1.21; that is,

$$\frac{p_{O_2}}{P^\circ} = \left(\frac{p_{H_2O}}{p_{H_2}K_e}\right)^2 = \left(\frac{10^{-3}}{(1)(7.113\times10^{13})}\right)^2$$

$$p_{O_2} = 1.97\times10^{-34} \text{ bar.}$$

It is common to refer to the pressure of a gas relative to its standard state pressure as its *activity*, a, so that if a gas is in its standard state of one bar, then its activity is a = 1.0. In the above example, the activity of oxygen is $a_{O_2} = 1.97\times10^{-34}$. To handle condensed phases, solids and liquids, their standard states are defined as the pure material at one bar pressure and at whatever temperature is being considered. Therefore, the activities of pure solids and liquids in their standard states is also a = 1.0. So from Equation 1.18, in general, for one mole

$$\bar{G} = \bar{G}^\circ + RT\ \ln a. \tag{1.22}$$

* It used to be a pressure *of one atmosphere* for everyone but the standard state pressure has been redefined by international convention. But there is not much difference between the two: 1 atm = 1.01325 bar.

† Note the difference between the two terms: *standard state* and *standard conditions* of temperature and pressure. Standard conditions are one bar pressure and 25°C, whereas the standard state is the pure material at one bar pressure at whatever temperature is of interest.

Example: Thermodynamic Activity of Water in a Sodium Chloride Solution

At 25°C, the vapor pressure of H_2O in equilibrium with pure water is $p_{H_2O} = 0.031699\,\text{bar}$ (Roine 2002). That is,

$$H_2O(l) \rightleftharpoons H_2O(g); \quad K_e = 0.031699$$

so

$$\frac{p_{H_2O}}{a_{H_2O}} = K_e.$$

Since the activity of pure H_2O is one, the pressure of water vapor over pure H_2O is 0.031699 bar. However, over a saturated solution of NaCl in water, the vapor pressure of water is 0.0236 bar (Haynes 2013). Therefore, the thermodynamic activity of water in the solution is

$$a_{H_2O,soln} = \frac{p_{H_2O,soln}}{K_e} = \frac{0.023855}{0.031699} = 0.7526.$$

1.7.6.3 Gibbs Energy as a Function of Temperature

Equation 1.9 gives the Gibbs energy as $G = H - TS$. As a result, $G^o = H^o - TS^o$ and, likewise, $\Delta G^o = \Delta H^o - T\Delta S^o$. From Equation 1.4, at constant pressure, the enthalpy of a mole of a substance can be obtained by integrating the heat capacity over some temperature region

$$\bar{H}_T^o = \bar{H}_{298.15}^o + \int_{298.15}^{T} \bar{C}_P dT \tag{1.23}$$

where the enthalpy is the heat content per mole—hence the bar over H—if the heat capacity, $\bar{C}_P$, has units of J/mol-K as it usually does. Similarly, the expression for the entropy can be written as

$$\bar{S}_T^o = \bar{S}_{298.15}^o + \int_{298.15}^{T} \frac{\bar{C}_P}{T} dT. \tag{1.24}$$

For a reaction such as $Ti(s) + O_2(g) \rightarrow TiO_2(s)$, the Gibbs energy as a function of temperature is

$$\Delta\bar{G}_{TiO_2}^o = \bar{G}_{TiO_2}^o - \bar{G}_{Ti}^o - \bar{G}_{O_2}^o = \Delta\bar{H}_{TiO_2}^o - T\Delta\bar{S}_{TiO_2}^o. \tag{1.25}$$

Thermodynamic data are determined by experiment and are tabulated and given in various forms (Roine 2002). The data tabulated are values of $\bar{H}_{298.15}^o$ and $\bar{S}_{298.15}^o$ as well as the heat capacities as a function of temperature. The heat capacities, also empirically measured, are usually given in the following form:

$$\bar{C}_P = A + B\times10^{-3}T + C\frac{10^{-5}}{T^2} + D\times10^{-6}T^2 \text{ J/mol}\cdot\text{K} \tag{1.26}$$

and the values of A, B, C, and D are tabulated usually over a finite temperature region. For a given element or compound, these values probably are different over different temperature ranges. Therefore, the enthalpy, entropy, and Gibbs energy can, in principle, be calculated at any temperature of interest and the equilibrium constant for a reaction calculated. However, given the temperature dependence for heat capacities above, the calculations are rather tedious even with a spreadsheet program. Thankfully, software is available that carries out these calculations (Roine 2002). Figure 1.11 shows the enthalpy for rutile TiO_2 by reacting titanium with oxygen. The curvature in these plots indicates that the heat capacities of either the reactants or product or both are not constant over this rather limited temperature range. Any *phase changes* over the temperature range of interest with either the products or reactants will introduce discontinuities to the thermodynamic quantities, G, H, and S at the temperature of the phase transition since each transition adds (subtracts) a finite amount of heat. Figure 1.12 shows the enthalpy of formation of MgO from

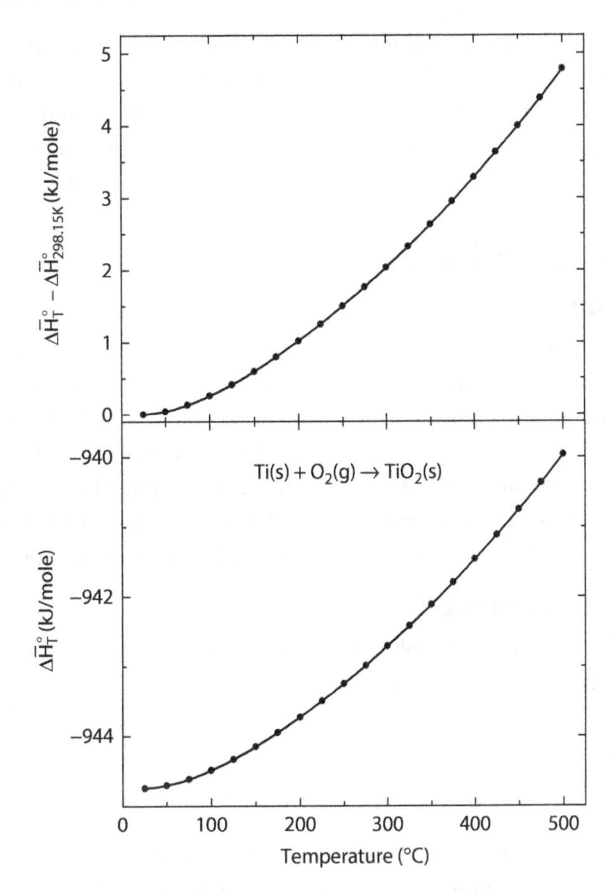

FIGURE 1.11 The bottom part of this figure shows the change in enthalpy for the reaction of titanium with oxygen to form titanium dioxide (rutile) as a function of temperature: $\Delta\bar{H}^\circ_{298} + \int_{298}^{T}\Delta\bar{C}_P dT$ where the $\Delta\bar{C}_P$ is the difference between the molar heat capacities of the products and reactants. And the top curve just shows the change in enthalpy due to the difference in heat capacities. Since the plots are not straight lines, this implies that $\Delta\bar{C}_P$ is not a constant but a function of temperature.

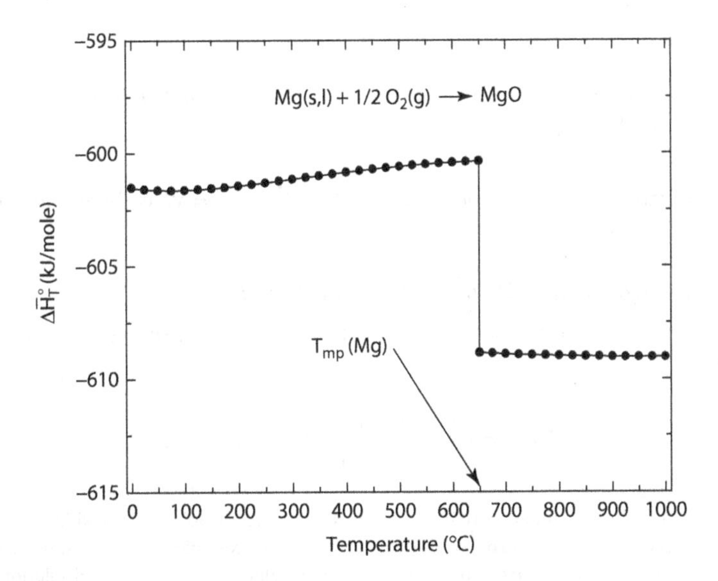

FIGURE 1.12 A plot of the enthalpy versus temperature for the reaction of magnesium with oxygen to form magnesium oxide. In this case, there is an *increase* in enthalpy of the magnesium at its melting point of 650°C due to its enthalpy of fusion, $\Delta\bar{H}_f$. This makes the enthalpy of the reaction more negative and results in a discontinuity of enthalpy versus temperature at this solid–liquid phase transition.

magnesium and oxygen as a function of temperature. There is a discontinuity at the melting point of magnesium since heat is added to the magnesium reactant at its melting point and the enthalpy of reaction for MgO is the difference between the enthalpy of MgO and that of magnesium and oxygen. Consequently, the enthalpy of the MgO product becomes more negative at the point where the enthalpy of one of the reactants, magnesium, becomes more positive.

1.7.7 Multicomponent Systems

1.7.7.1 Introduction

All four of the above examples, namely the hydrogen, oxygen, water vapor system; the water-NaCl solution; the reaction to form TiO_2; and that to form MgO, are all multicomponent systems in that they contain more than one element or chemical species. More often than not, MSE deals with multicomponent systems with two or more components. In what follows, for the sake of simplicity and concreteness, only two-component systems are considered. All results are easily extended to more than two components. With more than one component in a system, there needs to be a convenient way of expressing how each component makes changes when added to a mixture or a solution* of the components.

1.7.7.2 Partial Molar Quantities†

Consider the total volume of a binary solution or mixture of two components, A and B, the partial molar volume of component A is defined by

$$\bar{V}_A = \left(\frac{\partial V}{\partial n_A}\right)_{T,P,n_B} \tag{1.27}$$

where $\bar{V}_A$ is the change in volume of a multicomponent system with the addition on one mole of component A, at constant temperature, pressure, and moles of component B. There is, of course, a similar expression for component B. Therefore, the total volume of a two-component system consisting of n_A moles of A and n_B moles of B is

$$V = n_A\bar{V}_A + n_B\bar{V}_B. \tag{1.28}$$

Dividing by the total number of moles, n, where $n = n_A + n_B$ Equation 1.28 becomes

$$\bar{V} = X_A\bar{V}_A + X_B\bar{V}_B \tag{1.29}$$

where:

$\bar{V}$ is the *molar volume* of the mixture or solution and *the mole fraction*
X_i is simply $X_i = n_i/(n_A + n_B)$.

For a mechanical mixture or an *ideal solution*, $\bar{V}_i = \bar{V}_i^o$ where $\bar{V}_i = \bar{V}_i^o$ is the molar volume of component i in its pure or standard state.

Example: Molar Volumes of an Ideal Mixture of Water and Sand

For water at 25°C, with a molecular weight of 18 g/mol and density of 1 g/cm³, the molar volume is

$$\bar{V}^o_{H_2O} = \frac{M}{\rho} \cong \frac{18\text{ g/mol}}{1\text{ g/cm}^3} \cong 18\text{cm}^3/\text{mol}.$$

* A *mixture* here means that the components are interdispersed on a macroscopic scale in which each of the components can be physically separated (with a tweezers or screen) from one another. Examples would be sand and salt or water and oil. A *solution* consists of the components mixed on the atomic or molecular level such as a solid solution of zinc and copper, brass, or sodium chloride dissolved in water and the components can only be separated at the atomic or molecular level.

† Some of the literature on thermodynamics refers to these as *partial molal* quantities, but this terminology is not accepted by the International Union of Pure and Applied Chemistry (IUPAC 2014), and its use does not help to make thermodynamics less confusing.

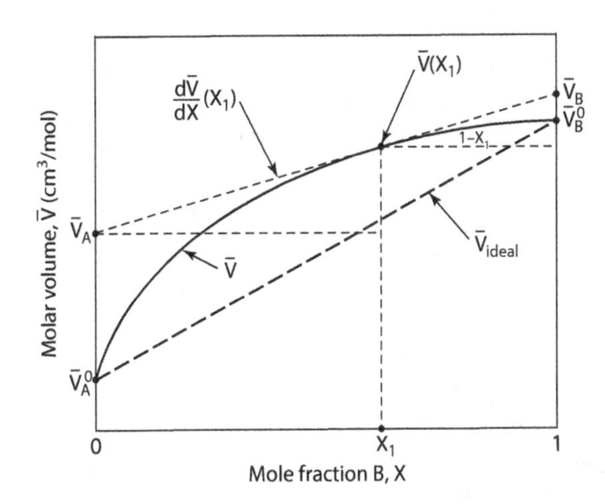

FIGURE 1.13 A schematic plot of molar volume as a function of composition for two situations. The first consists of simple mechanical mixing of the two components A and B resulting in the dashed straight line between their molar volumes in their standard, pure states: $\bar{V}_A^o$ and $\bar{V}_B^o$. The second situation is when they form a solution and the molar volume follows the solid curve. The molar volumes of each constituent are determined from the intersection of a tangent line at a given composition, X_1, and the two vertical axes. As a result, the partial molar volumes of A and B at this composition are given by $\bar{V}_B = \bar{V}(X_1) + (1 - X_1)(d\bar{V}/dX)(X_1)$ and $\bar{V}_A = \bar{V}(X_1) - X_1(d\bar{V}/dX)(X_1)$ as can be seen graphically in the figure.

and for SiO_2 sand grains

$$\bar{V}_{SiO_2}^o = \frac{M}{\rho} \cong \frac{60.085\,g/mol}{2.334\,g/cm^3} \cong 25.8\,cm^3/mol$$

Therefore, a mixture of sand grains and water would follow the $\bar{V}_{ideal}$ straight line in Figure 1.13 and at a composition of $X = 0.5$ the total volume occupied would be*

$$\bar{V} = 0.5 \times 18 + 0.5 \times 25.8 = 21.9\,cm^3/mol.$$

For a solution that does not follow ideal behavior, Figure 1.13 also shows a curve of $\bar{V}$ versus mole fraction. The partial molar values of the two components, A and B, are then given by the intercept of the tangent line $d\bar{V}/dX$ at X_1 at the two vertical axes, $\bar{V}_A$ and $\bar{V}_B$, as shown in Figure 1.13. From Figure 1.13, graphically, at composition X_1,

$$\bar{V}_B = \bar{V}(X_1) + (1 - X_1)\frac{d\bar{V}}{dX}(X_1) \tag{1.30}$$

and

$$\bar{V}_A = \bar{V}(X_1) - X_1\frac{d\bar{V}}{dX}(X_1). \tag{1.31}$$

Example: Molar Volumes of Methanol-Water Solution

With data taken from the literature (Haynes 2013) the molar volume of water-methanol solutions is plotted in Figure 1.14. The partial molar volumes of water and methanol were calculated with

* This calculation ignores the fact that sand grains by themselves would pack like cannon balls and would have about 45% pore volume in between them which the water could fill, so that the apparent volume of the mixture in a calibrated vessel would appear less than this. Nevertheless, this is the *total* volume occupied by the water and sand mixture including the volume in the pore space occupied by water.

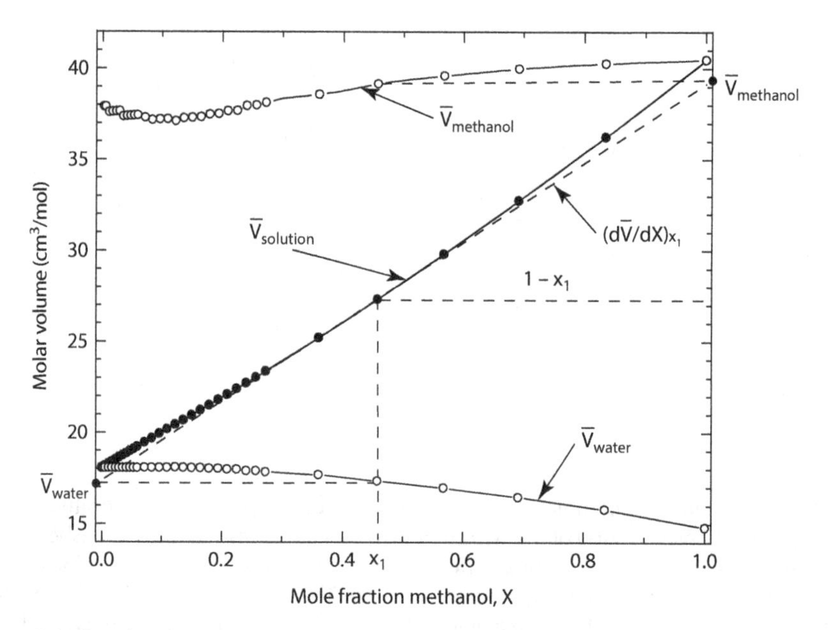

FIGURE 1.14 Calculated molar volumes and partial molar volumes with data taken from the literature (Haynes 2013) for water-methanol solutions. The system is almost ideal although, as the plot shows, the partial molar volumes of water and methanol are not constant. As described in Figure 1.13, the partial molar volumes calculated at $X_1 = X_{methanol} = 0.4576$ are $\bar{V}_{water} = 17.35\,cm^3/mol$ and $\bar{V}_{methanol} = 39.14\,cm^3/mol$ as shown in the figure.

Equations 1.30 and 1.31 on a spreadsheet program and shown in Figure 1.14. Note that the system is almost ideal—a straight line—but clearly the partial molar volumes of both water and methanol are not constant as is necessary for an ideal solution. At the composition $X_1 = 0.4576$ mole fraction methanol, the values of $\bar{V}_{water}$ and $\bar{V}_{methanol}$ calculated with Equations 1.30 and 1.31 are $\bar{V}_{water} = 17.35\,cm^3/mol$ and $\bar{V}_{methanol} = 39.14\,cm^3/mol$ as shown in Figure 1.14 at the intersections of the tangent line at X_1 and the two vertical axes.

1.7.7.3 Chemical Potential and Partial Molar Gibbs Energy

If a system has more than one chemical component, say two,* A and B then the thermodynamic functions need to include the effects of composition since it would be expected that different components would have different internal energies. So the first law can be rewritten as,

$$dU = TdS - pdV + \mu_A dn_A + \mu_B dn_B \tag{1.32}$$

where the chemical potentials, μ_i, are defined as

$$\mu_A = \left(\frac{\partial U}{\partial n_A}\right)_{S,V,\mu_B} \tag{1.33}$$

with a similar expression for component B and the n_i are the number of moles of components A and B. Now, as an exact differential, the Gibbs energy can be written as

$$dG = \left(\frac{\partial G}{\partial P}\right)dP + \left(\frac{\partial G}{\partial T}\right)dT + \left(\frac{\partial G}{\partial n_A}\right)dn_A + \left(\frac{\partial G}{\partial n_B}\right)dn_B \tag{1.34}$$

and the *partial molar Gibbs energy* is defined as,

$$\bar{G}_A = \left(\frac{\partial G}{\partial n_A}\right)_{T,P,n_B} \tag{1.35}$$

* Only two-component systems are considered but, again, certainly more components are possible.

and similarly for component B. Therefore, from Equation 1.11 and these definitions, Equation 1.34 becomes

$$dG = VdP - SdT + \bar{G}_A dn_A + \bar{G}_B dn_B. \tag{1.36}$$

Substituting Equation 1.32 into Equation 1.10 now gives,

$$dG = VdP - SdT + \mu_A dn_A + \mu_B dn_B. \tag{1.37}$$

so the *partial molar Gibbs energies and the chemical potentials must be equal*; that is,

$$\bar{G}_A = \left(\frac{\partial G}{\partial n_A}\right)_{T,P,n_B} = \mu_A \tag{1.38}$$

with a similar expression for the chemical potential of component B. *A system is in chemical equilibrium when the chemical potential of each component is uniform throughout the system.*

1.7.7.4 Gibbs–Duhem Relation

At constant temperature and pressure Equation 1.37 becomes

$$dG = \mu_A dn_A + \mu_B dn_B \tag{1.39}$$

and integrating over the extensive variables n_A and n_B gives

$$G = \mu_A n_A + \mu_B n_B. \tag{1.40}$$

Taking the complete differential of Equation 1.40 and subtracting Equation 1.39, the result is

$$0 = n_A d\mu_A + n_B d\mu_B \tag{1.41}$$

and dividing by $n = n_A + n_B$

$$0 = X_A d\mu_A + X_B d\mu_B. \tag{1.42}$$

This is the Gibbs–Duhem equation and is used frequently in many of the models developed later.

1.7.7.5 Solutions

Equation 1.22 is

$$\bar{G} = \bar{G}^o + RT \ln a \tag{1.43}$$

since the units of $\bar{G}$ are clearly J/mol. In general, if two components A and B form a solution, then the Gibbs energy of mixing can be written as

$$\Delta G_{mix} = (n_A \bar{G}_A + n_B \bar{G}_B) - (n_A \bar{G}_A^o + n_B \bar{G}_B^o) = n_A RT \ln a_A + n_B RT \ln a_B \tag{1.44}$$

where the terms within the first set of parentheses are the Gibbs energies in the solution state and the second set are the Gibbs energies of the components in their unmixed state. An *ideal solution* is one that follows *Raoult's law*, $a_i = X_i$, that is, the activity equals the mole fraction. So the Gibbs energy of mixing of an ideal solution of two components A and B is given by

$$\Delta G_{mix,id} = n_A RT \ln X_A + n_B RT \ln X_B \tag{1.45}$$

and dividing by $n = n_A + n_B$ gives

$$\Delta \bar{G}_{mix,id} = RT(X_A \ln X_A + X_B \ln X_B). \tag{1.46}$$

From Equation 1.11, at constant pressure

$$\left(\frac{dG}{dT}\right)_P = -S$$

Therefore,

$$\Delta\bar{S}_{mix,id} = -R\left(X_A \ln X_A + X_B \ln X_B\right) \tag{1.47}$$

which is exactly the same result obtained with the statistical approach with $S = k_B \ln W$ in Appendix A.3, Equation A.6.

Note, however, if two compounds such as Al_2O_3 and Cr_2O_3 form an ideal liquid or solid solution, then the cations are distributed over $2N_A$ lattice sites and the ideal entropy of mixing is just twice that given in Equation 1.47, namely,

$$\bar{S}_{mix,id} = -2R\left(X_{Al_2O_3} \ln X_{Al_2O_3} + X_{Cr_2O_3} \ln X_{Cr_2O_3}\right)$$

as shown in Appendix A.3, Equation A.8.

For a nonideal solution, of course, the mole fractions must be replaced by activities and there may be a $\Delta\bar{H}_{mix}$ as well. In general, if solutions are not ideal, then the activities may be represented as $a_i = \gamma_i X_i$ where γ_i is the *activity coefficient* for component i, and it can vary with the composition.

Example: Activity Coefficient of NaCl in a NaCl-H_2O Solution

In Section 1.7.6.2 it was found that the activity of water in a saturated sodium chloride solution at 25°C was $a_{H_2O} = 0.7526$. At 25°C, a saturated aqueous solution of sodium chloride is 6.248 *molal*; that is, there are 6.248 moles of NaCl per 1000 g of solution (Roine 2002). The molecular weight of NaCl is 58.443 g/mol. The number of grams of NaCl in the 1000 g of saturated solution is $m_{NaCl} = 6.248\,\text{mol} \times 58.443\,\text{g/mol} = 365.15\,\text{g}$ so the number of moles of water in the saturated solutions is $n_{H_2O} = (1000 - 365.15)\text{g}/18.015\,\text{g/mol} = 35.240\,\text{mol}$. Therefore, the mole fraction of H_2O in the saturated solution is $X_{H_2O} = 35.240/(35.240 + 6.248) = 0.8494$. As a result, the activity coefficient for water in a saturated solution of NaCl at 25°C is

$$\gamma_{H_2O} = \frac{0.7526}{0.8494} = 0.8860.$$

Now the mole fraction of NaCl is $X_{NaCl} = 6.248/(6.248 + 35.240) = 0.1506$ which checks since $X_{H_2O} + X_{NaCl} = 1.0000$. But the activity of NaCl must be $a_{NaCl} = 1.0$ since it is a saturated solution; that is, it is equilibrium with pure NaCl that has an activity of 1.0. Therefore, the activity coefficient of sodium chloride in the saturated solution is,

$$\gamma_{NaCl} = \frac{1.0}{0.1506} = 6.640.$$

1.7.8 Phase Rule and Phase Diagrams

1.7.8.1 Phase Rule

Gibbs derived the phase rule, where the number of phases present plus the number of degrees. of freedom is equal to the number of components plus 2, or more simply: $P_{hases} + F_{reedom} = C_{omponents} + 2$.* This is the mathematical result of

$$\text{unknowns}(F) = \text{variables}(C + 2) - \text{equations}(P)$$

* A useful mnemonic for remembering the phase rule is *Police Force* = *Cops* + 2.

the last term comes from the fact that the chemical potentials must be equal in all of the phases in equilibrium, so there are P equations with the chemical potentials. The number of variables is the number of chemical components, C, plus temperature and pressure, for the extra 2. It should be noted that a *phase* can be defined as *a physically and/or chemically distinct, mechanically separable region of matter.* The key phrase here is *mechanically separable.* For example, a mixture of salt and sand particles consists of two chemically distinct regions, NaCl and SiO_2. They are also physically distinct in that the salt particles may be small cubes where the sand grains are probably somewhat larger and less uniform. They certainly are mechanically separable since, if the mixture were sieved, the smaller salt particles might pass through the screen while the larger sand grains would not. If worse came to worse, the sand grains could be separated from the salt particles with a magnifying glass and a small pair of tweezers. On the other hand, if the salt were dissolved in water, the salt ions and the water molecules are mixed at the molecular level and a magnifying glass and a pair of tweezers will not be of much use. Another example might be ice in equilibrium with water at one bar and 273.15 K, the two phases are chemically the same but physically distinct: one is a solid and one is a liquid, and they certainly are mechanically separable, just knock over the glass!

1.7.8.2 Phase Diagrams

Phase diagrams are extremely useful because they visually display a large amount of thermodynamic information. For a one-component, *unary,* system, the axes of the two dimensional diagram are pressure and temperature. For a two-component, *binary,* system the pressure is held fixed, usually one bar, and the temperature and composition are the axes of the diagram. For a three-component, *ternary,* system the pressure is also fixed and a composition triangle forms the base, with the percentages of the three components along the axes, and the temperature is plotted vertically to give a three-dimensional diagram. Useful thermodynamic data are obtained from planar, isothermal sections parallel to the base of the diagram. Additional details of ternary diagrams are not covered here since they are not used in this book. For systems containing more than three components, the diagrams become more complex and considerably less visually useful in displaying thermodynamic data so they are ignored here.

1.7.8.3 Single Component System

A schematic of the phase diagram for the H_2O system is given in Figure 1.15. Consider a point someplace in the middle of the liquid phase region. In this case, C = 1 (H_2O), P = 1 (liquid), therefore

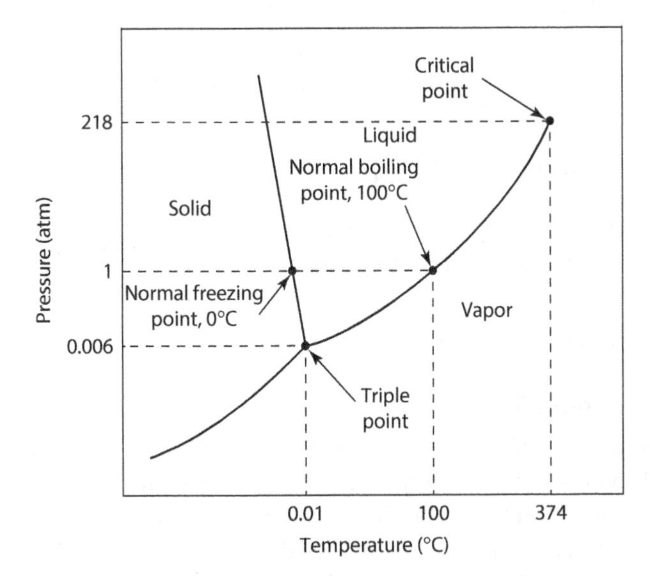

FIGURE 1.15 P-T phase diagram for pure H_2O that shows the triple point, the melting and boiling points at 1 atm pressure, and the critical temperature and pressure where the gas and liquid are no longer distinguishable. The negative slope of the solid–liquid equilibrium is due to the larger volume of ice at any given temperature than that of water. This is discussed in detail in the text.

the degrees of freedom, F, must be F = 3 − 1 = 2. That is, there are two degrees of freedom in that the temperature and pressure can be varied independently and a point stays within the single phase liquid field. The same is true in both the vapor or the solid phase regions as well. However, along the equilibrium lines between the phases, for example, solid–liquid, then P = 2 and F = 1. Neither the pressure nor temperature can be varied independently and stay on the equilibrium curve between the two phases: if the temperature changes, so must the pressure as given by the curve between the phases. Finally, at the triple point, all three phases are in equilibrium, so F = 0, the three phases can exist in equilibrium at only a single value of temperature and pressure. Note that there is also a *critical point* where the gas and the liquid phases merge to become a single phase. This is not too surprising since, as will be seen, the only real difference between a liquid and a gas is the packing density of the atoms or molecules, the number per unit volume.

Note that for the ice–water transition, the temperature of the transition goes to lower temperatures as the pressure increases. This is not surprising since ice is one of few the solids that has a greater volume in the solid state than in the liquid state due to the directional hydrogen bonding in solid H_2O making the solid structure more open than in the liquid. Silicon with its directed covalent bonding behaves the same way. This behavior can be made quantitative by applying Equation 1.11 to the equilibrium between ice and water:

$$d\bar{G}_L = d\bar{G}_S$$

$$\bar{V}_L dP - \bar{S}_L dT = \bar{V}_S dP - \bar{S}_S dT$$

$$\left(\bar{V}_L - \bar{V}_S\right) dP = \left(\bar{S}_L - \bar{S}_S\right) dT.$$

Rearranging the last equation gives

$$\frac{dP}{dT} = \frac{\left(\bar{S}_L - \bar{S}_S\right)}{\left(\bar{V}_L - \bar{V}_S\right)} = \frac{\Delta\bar{H}_m}{T_m\left(\bar{V}_L - \bar{V}_S\right)} \tag{1.48}$$

this is a form of *Clausius–Clapeyron* equation. Taking values from the literature (Roine 2002),

$$\bar{V}_L = \frac{M}{\rho_L} = \left(\frac{18.015}{0.997}\right) \times 10^{-6}\,\text{m}^3/\text{cm}^3 = 18.06 \times 10^{-6}\,\text{m}^3/\text{mol}$$

$$\bar{V}_S = \frac{M}{\rho_S} = \left(\frac{18.015}{0.917}\right) \times 10^{-6}\,\text{m}^3/\text{cm}^3 = 19.65 \times 10^{-6}\,\text{m}^3/\text{mol}$$

and $\Delta\bar{H}_m = 6.07\,\text{kJ/mol}$ and at $T_m = 273.15$ K,

$$\frac{dP}{dT} = \frac{\Delta\bar{H}_m}{T_m\left(\bar{V}_L - \bar{V}_S\right)} = \frac{6079 \times 10^6}{273.15(18.08 - 19.65)} = -139.8\,\text{bar/K}$$

so the slope dP/dT is negative as shown in Figure 1.15.

1.7.8.4 Two-Component System

Consider the two-component phase diagram for A and B in Figure 1.16, a simple eutectic system with some terminal solid solutions at both A and B. Usually, such phase diagrams refer to a constant pressure of one bar. Therefore, with the pressure variable constant, the phase rule becomes $F = C - P + 1$. For point 1 in a single phase field, here the liquid phase, F = 3 − 1 = 2 degrees of freedom. As was the case in a single phase field for the one component system, both *temperature* and *composition* can be varied independently and a point remains in the single phase field. At point 2 in the liquid + alpha solid solution two-phase field, there is only 1 degree of freedom. The compositions of the two phases in equilibrium are given by the ends of the *tie-line* intersecting the *liquidus*

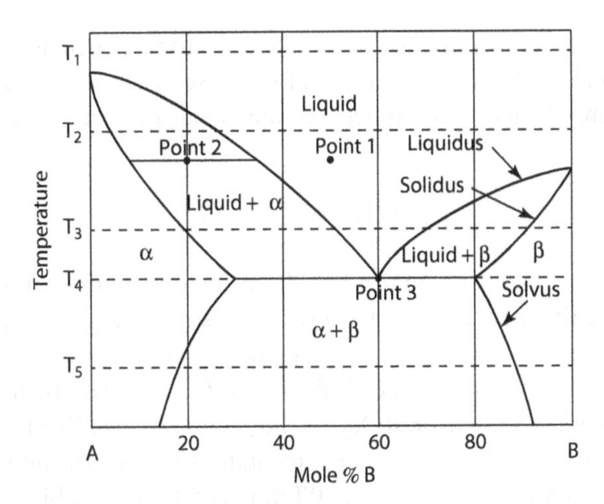

FIGURE 1.16 Binary phase diagram of two components A and B that exhibits a eutectic and point 3 and terminal solid solutions alpha, α, and beta, β. The Gibbs energy-composition plots in Figure 1.18a–e are constructed at the temperatures labeled 1–5.

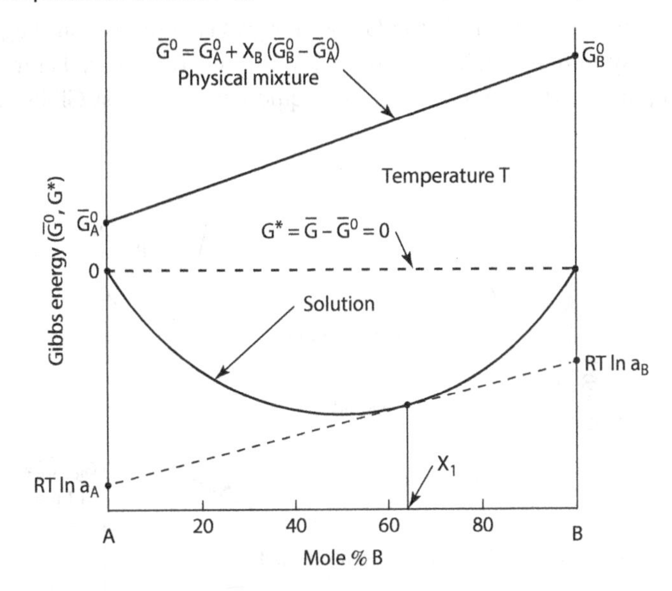

FIGURE 1.17 Schematic diagram showing how a new function G^* can be calculated that removes the standard state part of the equation for mixing two components A and B. Plotting $G^* = \bar{G} - \bar{G}^{\circ}$ versus composition is often convenient since the tangent intercepts at the two vertical axis give $\mu_A - \mu_A^{\circ}$ and $\mu_B - \mu_B^{\circ}$.

and *solidus* lines* and are fixed at a given temperature (Figure 1.16). If the temperature is raised or lowered, the tie-line moves and the compositions of the two phases in equilibrium changes. At point 3, the eutectic point, $F = 3 - 3 = 0$, there are no degrees of freedom. This is a point fixed in temperature and composition and is called an *invariant* point in the phase diagram. Note that going horizontally across a binary phase diagram at a constant temperature, say T_3, the phase regions follow the pattern: one-phase, two-phase, one-phase, two-phase, and so on.

1.7.8.5 Phase Diagrams and Gibbs Energy-Composition Diagrams

One of the most useful things about phase diagrams is that there is a one-to-one correspondence between them and the molar Gibbs energies of the phases that in the diagram. Consider the Gibbs energy versus composition diagram in Figure 1.17. The upper straight line represents the Gibbs energy

* The *liquidus* is the line separating the liquid phase region from the liquid + solid region. The *solidus* is the line separating the two-phase liquid plus solid region and the solid region. The *solvus* is the line separating the two solid phase region from the single-phase solid region. The names for these lines were given in the late 1800s when a *Latin sounding* name was assumed to lend more scientific credibility to something.

of two components A and B that only form a mechanical mixture such as sand and water, salt and pepper, or a polymer and a solid filler material. In these cases, there is no reaction between the two phases, and there is simple linear relation between any molar quantity of A and B, $\bar{G}$ in the figure, namely,

$$\bar{G} = \bar{G}_A^o + X_B\left(\bar{G}_B^o - \bar{G}_A^o\right) \tag{1.49}$$

where the molar Gibbs energies of A and B remain in their standard states. It is sometimes useful, however, to remove this part involving the standard states when working with Gibbs energy-composition diagrams, so that it is easier to evaluate the role of solutions. Therefore, if Equation 1.49 is relabeled $\bar{G}^o$ and a new Gibbs function G^* is defined as $G^* = \bar{G} - \bar{G}^o$. Then when the two components do not form a solution, $G^* = 0$ for all X_B, the horizontal dashed line in Figure 1.17. If A and B do form a solution, as shown in the bottom curve of Figure 1.17, then the intersection of the tangent line at a point X_1 with the two vertical axes gives $RT \ln a_A$ and $RT \ln a_B$, which can be useful, where a_A and a_B are the thermodynamic activities of A and B, respectively.

Example: Gibbs Energy-Composition Diagrams for the Simple Eutectic System of Figure 1.16.

From the phase diagram in Figure 1.16, the Gibbs energy, G^*, composition diagrams can be constructed. At T_1, the system is liquid for any composition. Therefore, Figure 1.18a shows the G^*-composition diagram for this temperature. The liquid has the lowest Gibbs energy of the three

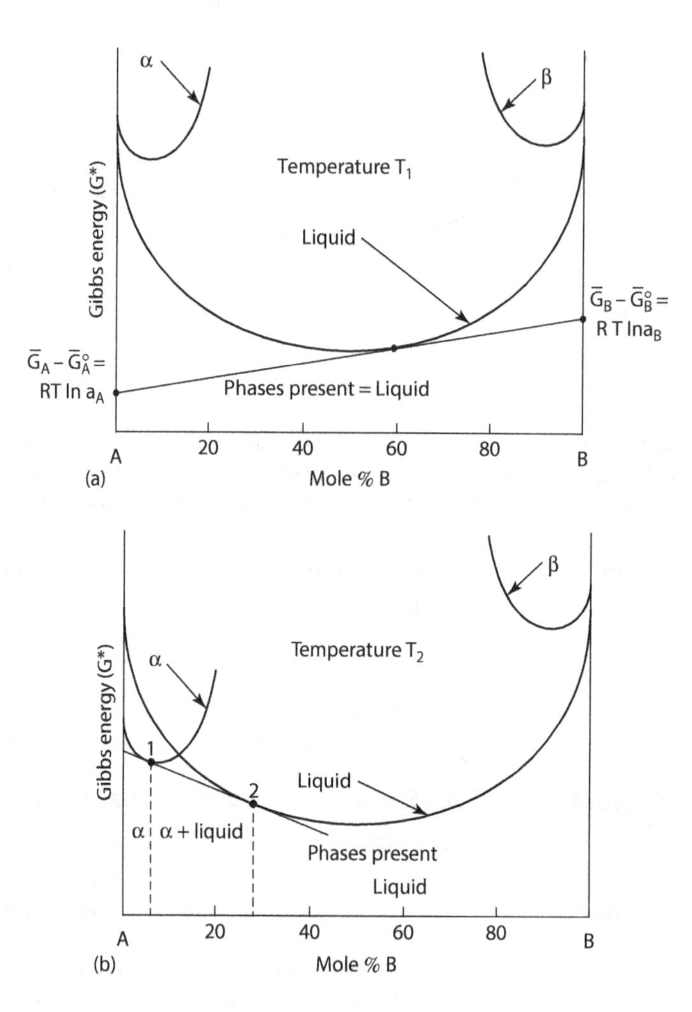

FIGURE 1.18 (a) Gibbs energy-composition diagram constructed from the phase diagram in Figure 1.16 at T_1. (b) Gibbs energy-composition diagram constructed from the phase diagram in Figure 1.16 at T_2. *(Continued)*

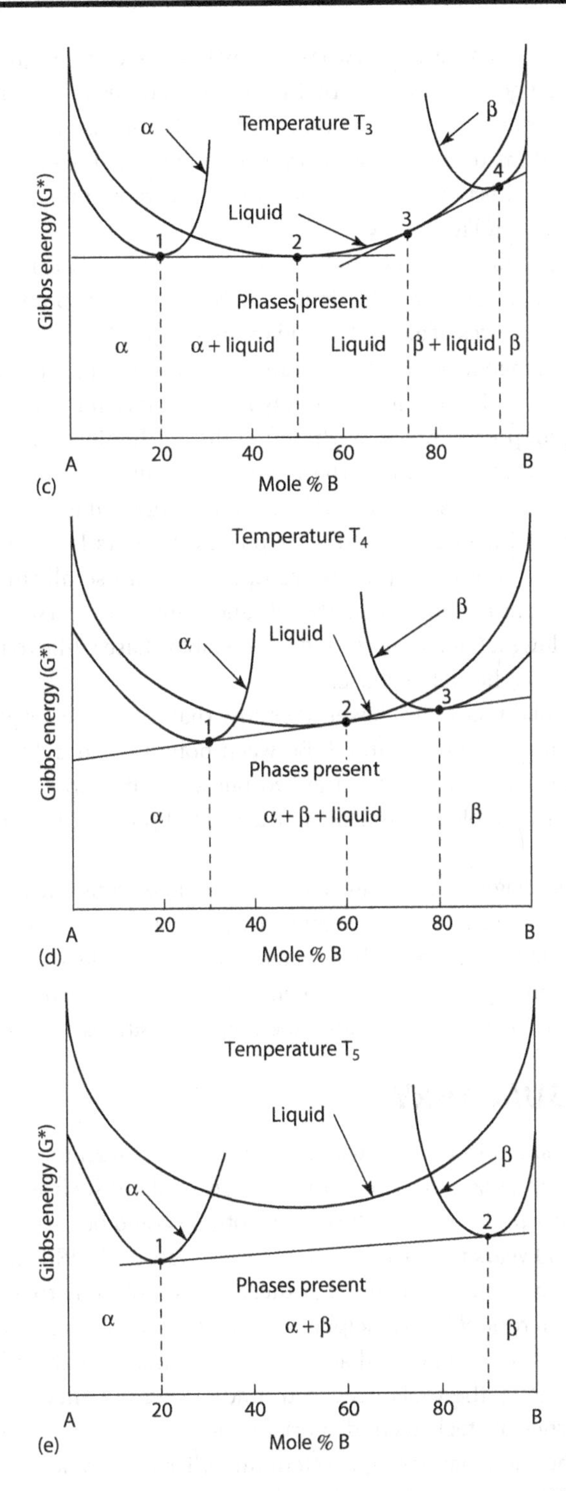

FIGURE 1.18 (Continued) (c) Gibbs energy-composition diagram constructed from the phase diagram in Figure 1.16 at T_3. (d) Gibbs energy-composition diagram constructed from the phase diagram in Figure 1.16 at T_4. (e) Gibbs energy-composition diagram constructed from the phase diagram in Figure 1.16 at T_5.

phases (liquid, solid alpha, and solid beta) so it is the stable phase at all compositions with the tangent line at some composition giving the G^* for A and B at the intersection of the two vertical axes.

At T_2, Figure 1.18b, the α phase has a lower energy than the liquid over a finite region. A common tangent line to the two phases, α and liquid, at points 1 and 2, respectively, gives the G^* values for A and B again at the intersection of the tangent line with the two vertical axes. Of course, if these two phases are in equilibrium, then the Gibbs energies of both A and B must be the same in both phases; hence, the common tangent. However, between the contacts, points 1 and 2 of the common tangent with the two phases, a mixture of two phases with compositions given by the tangent points has a lower energy, the tangent line, than either of the two phases in equilibrium. Therefore, these two tangent points determine the composition range of a two-phase region at this temperature in Figure 1.16. For compositions to the left of the tangent point 1 on the α phase, the α phase has the lowest Gibbs energy and so this tangent point determines the maximum composition of single phase alpha on the phase diagram in Figure 1.16.

At T_3, Figure 1.18c, two tangent lines can be drawn: one intersecting the alpha and liquid phases at points 1 and 2, and one intersecting the liquid and beta phases at points 3 and 4. As at T_2, G^* is lower for the two-phase mixtures of alpha + liquid (points 1 and 2) and liquid + beta (points 3 and 4) between these common tangent points on the tangent lines. Again, the extensions of the tangent lines to the vertical axes give the activities (actually RT ln a) of A and B for the compositions of the three phases at the tangent points. Again, to the left of the α + liquid, two-phase region (point 1) the single phase α has the lowest energy and is the stable phase. Similarly, for compositions to the right of point 4, tangent to the beta phase, beta has the lowest energy and is the stable phase.

At T_4, a single tangent line intersects all three phases at points 1, 2, and 3. This is the invariant eutectic temperature where the three phases are in equilibrium so all three must have the same G_A^* and G_B^* where the tangent line intersects the left and right vertical axes, respectively. To the left of point 1, alpha has the lowest Gibbs energy and is the stable phase while to the right of point 3, beta has the lowest energy and is the stable phase.

Finally, at T_5, both solid phases are lower in energy than the liquid as given by the tangent line between them intersecting at points 1 and 2. Between points 1 and 2, the tangent has the lowest energy so these two points give the limits of the two-phase α + β region. To the left of point 1, α has the lowest energy and is the stable phase while to the right of point 2, β has the lowest Gibbs energy and is the stable phase.

Conversely, the phase diagram shown in Figure 1.16 could be constructed from the Gibbs energy-composition diagrams in Figure 1.18. This example demonstrates the very close relationship between Gibbs energy-composition diagrams and phase diagrams. It is possible to create either from the other. Understanding this relationship is invaluable in being able to use phase diagrams to initiate development of kinetic models for the rates of reactions and phase transitions.

1.8 CHAPTER SUMMARY

The purpose of this introductory chapter was twofold. First, a student is presented a topic and too often, along with it, the impression that out of the cloud, disciplines such as MSE or thermodynamics suddenly appeared in their present state. Of course, nothing could be further from the truth. It took thermodynamics over 100 years to reach a rather mature state and MSE almost as long. Therefore, a brief history of each is presented to show that these fields took time to get to their present state. Furthermore, the common root of both disciplines, steam engines, is emphasized. This culminated in the chemical thermodynamics and phase diagrams of Gibbs that gave the metallurgist at the beginning of the twentieth century the tools needed to understand the effect of composition and heat treatment on the properties of steel. Second, a brief review of important topics in thermodynamics is presented since the thermodynamics of a system underlies the development of kinetic models of reactions and phase transformations, the subject of this book.

APPENDIX

A.1 JOULE'S MECHANICAL EQUIVALENT OF HEAT

As early as 1824, the calorie was defined as the amount of heat needed to raise the temperature of 1 kg of water 1°C. This is now designated as the kilogram-calorie or calorie. During the 1800s, there was confusion between this and the definition of the gram-calorie or the amount of heat needed to raise 1 g of water 1°C. The difference between the two is now recognized and during Joule's measurements, he was probably not concerned with either since they were terms used in France, while in his England, the unit of heat was the British thermal unit, the amount of heat necessary to raise 1 lb of water 1°F! Today, the energy unit is a joule, so the goal is to convert Joule's measurement of the energy equivalent of heat into joules and compare that to a calorie. His best measurement of the equivalence of work and heat was 722.55 ft-lb/BTU. There are 2.2046 lb/kg. Therefore, carrying out the conversion step by step:

$$1\,\text{ft} = 1\,\text{ft} \times 12(\text{in./ft}) \times 2.54(\text{cm/in.}) \times \frac{1}{100(\text{cm/m})} = 0.3048\,\text{m}$$

$$1\,\text{lb} = 1\,\text{lb} \times \frac{1}{2.2046\,\text{lb/kg}} = 0.4525\,\text{kg}.$$

Therefore,

$$1\,\text{ft} \cdot \text{lb} = 0.3048\,\text{m} \times 0.4525\,\text{kg} \times 9.8\,\text{m/s}^2 = 1.35164\,\text{J}$$

where 9.8 m/s^2 is the acceleration of gravity necessary to change mass into a force. So 772.55 ft-lb are

$$772.55\,\text{ft} \cdot \text{lb} \times 1.35164\,\text{J/ft} \cdot \text{lb} = 1.04421\,\text{J} \times 10^3$$

is the amount of heat to raise the temperature of 1 lb of water 1°F. Therefore, the number of joules required to heat 1 g of water 1°C is

$$1.04421 \times 10^3 (\text{J/lb} \cdot {}^\circ\text{F}) \times 2.2046(\text{lb/kg}) \times \frac{1}{1000(\text{g/kg})} \times \frac{9}{5}({}^\circ\text{F}/{}^\circ\text{C}) = 4.144(\text{J/g} \cdot {}^\circ\text{C}) = 1\ \text{calorie}$$

Since the amount of heat necessary to raise 1 g of water depends on the starting temperature, the IUPAC has determined, based on more recent measurements, that the value of the thermodynamic calorie is

$$1\ \text{calorie} = 4.184\text{J}.$$

So James Prescott Joule's measurement in his laboratory, almost 150 years ago, was off by only about 1% from the currently accepted value.

A.2 Stirling's Approximation

A.2.1 Plausibility Argument

Stirling's approximation is

$$\ln(N!) \cong N \ln N - N \tag{A.1}$$

which is made plausible from the following very non rigorous approach:

$$\ln(N!) = \ln(1 \times 2 \times 3 \times 4 \times \cdots N) = \ln 1 + \ln 2 + \cdots \ln N = \sum_1^N \ln n \tag{A.2}$$

and approximating the sum with an integral (here N and n are just integers and *dummy* variables)

$$\sum_1^N \ln n \cong \int_1^N \ln n\, dn = |n \ln n - n|_1^N \tag{A.3}$$
$$= N \ln N - N + 1 \cong N \ln N - N$$

$$\ln(N!) \cong N \ln N - N \tag{A.4}$$

Substituting the integral for the sum certainly seems valid if N is large. The integral is evaluated by integration by parts where the value of the integral at the lower limit is just 1 which can be ignored compared to any reasonably large value of N.

A.2.2 How Good Is Stirling's Approximation?

For example, even for $N = 100$, $N! = 9.3 \times 10^{157}$ so $\ln N! = 363.7$ while $N \ln N - N = 100 \times \ln 100 - 100 = 360.5$ or a difference of about 0.9%. Therefore, for the large numbers involved in solids, $N \cong 10^{22}\,\text{cc}^{-1}$ Stirling's approximation is certainly good enough!

A.3 Statistical Approach to the Entropy of Mixing

A.3.1 Molar Entropy of Mixing of Two Crystalline Elements

The entropy of mixing of an ideal solution can be modeled from Boltzmann's equation for the entropy

$$S = k_B \ln W$$

where W, in this case, is the number of ways a total number of N atoms consisting of a atoms, N_a, and N_b b atoms can be distributed among N lattice sites, where $N = N_a + N_b$. Therefore,

$$W = \frac{N!}{N_a!N_b!} \tag{A.5}$$

is the number of ways that the a and b atoms can be arranged over the N total sites. Taking the natural logarithm of both sides of Equation A.5

$$\ln W = \ln N! - \ln N_a! - \ln N_b!.$$

And with the application of Stirling's approximation this becomes,

$$\ln W = N \ln N - N - N_a \ln N_a + N_a - N_b \ln N_b + N_b$$

$$\ln W = -N_a \ln\left(\frac{N_a}{N}\right) - N_b \ln\left(\frac{N_b}{N}\right)$$

and if $N = N_A$, Avogadro's number, one mole of atoms, then

$$\ln W = -N_A\left(\frac{N_a}{N_A}\right)\ln\left(\frac{N_a}{N_A}\right) - N_A\left(\frac{N_b}{N_A}\right)\ln\left(\frac{N_b}{N}\right)$$

so the entropy becomes,

$$\begin{aligned} \bar{S}_{mix,id} &= -k_B N_A\left(X_a \ln X_a + X_a \ln X_a\right) \\ \bar{S}_{mix,id} &= -R\left(X_a \ln X_a + X_a \ln X_a\right) \end{aligned} \tag{A.6}$$

where X_a and X_b are the mole fractions of a and b, respectively. This is the same as Equation 1.47 in the text obtained through a strictly thermodynamic path. For a 50–50 solution, $\bar{S}_{mix,id} = -R \ln 0.5 = -5.762\,\text{J/mol} \cdot \text{K}$.

A.3.2 Molar Entropy of Mixing of Two Compounds

NiO and MgO form almost an ideal solid solution. Therefore, the ideal molar entropy of mixing of these two compounds is also given by Equation A.6 since the two cations are distributed over a total of N_A lattice sites. However, what is not always appreciated is that the molar entropy of mixing *is different* if the two compounds are Al_2O_3 and Cr_2O_3 that also form an almost ideal solid solution. In this case, the total number of sites that the aluminum and chromium ions are distributed over is not N_A but $2N_A$ since there are two cations for each mole of the two constituent compounds. Therefore, now $2N_A = N_{Al} + N_{Cr}$ and

$$W = \frac{(2N_A)!}{N_{Al}!N_{Cr}!} \tag{A.7}$$

and taking the natural logarithm of both sides of the equation

$$\ln W = \ln(2N_A!) - \ln(N_{Al}!) - \ln(N_{Cr}!)$$

and applying Stirling's approximation

$$\ln W \cong 2N_A \ln(2N_A) - 2N_A - N_{Al}\ln N_{Al} + N_{Al} - N_{Cr}\ln N_{Cr} + N_{Cr}$$

$$\ln W \cong (N_{Al} + N_{Cr})\ln(2N_A) - N_{Al}\ln N_{Al} - N_{Cr}\ln N_{Cr}$$

$$\ln W \cong -N_{Al}\ln\left(\frac{N_{Al}}{2N_A}\right) - N_{Cr}\ln\left(\frac{N_{Al}}{2N_A}\right)$$

$$\ln W \cong -N_{Al}\left(\frac{2N_A}{2N_A}\right)\ln X_{Al_2O_3} - N_{Cr}\left(\frac{2N_A}{2N_A}\right)\ln X_{Cr_2O_3}$$

$$\ln W \cong -2N_A\left(X_{Al_2O_3}\ln X_{Al_2O_3} + X_{Cr_2O_3}\ln X_{Cr_2O_3}\right).$$

Therefore, the molar entropy of mixing is

$$\bar{S}_{mix,id} = -2R\left(X_{Al_2O_3}\ln X_{Al_2O_3} + X_{Cr_2O_3}\ln X_{Cr_2O_3}\right) \tag{A.8}$$

just *twice* that for the molar entropy of mixing for a solution of elements. So, for a 50–50 solid or liquid solution, $\bar{S}_{mix,id} = -2R\ln 0.5 = -11.526\,\text{J/mol}\cdot\text{K}$. Obviously, if the solution consisted of Al_2O_3 and Cr_2S_3—but not very likely—then there are 5 N_A *sites* over which the cations and anions can be distributed and the molar entropy would be

$$\bar{S}_{mix,id} = -5R\left(X_{Al_2O_3}\ln X_{Al_2O_3} + X_{Cr_2S_3}\ln X_{Cr_2S_3}\right).$$

EXERCISES

1.1 The value of the gas constant, R, is R = 8.314 J/mol-K. However, it is often more convenient to have it in different units. Convert R to the following units:

  a. cal/mol-K
  b. liter-atm/mol-K
  c. cm^3-atm/mol-K
  d. m^3-MPa/mol-K
  e. in^3-psi/mol-°F
  f. cm^3-torr/mol-°F

1.2 On the other hand, when dealing with energies of atoms or molecules rather than moles, it is frequently more convenient to use Boltzmann's constant, $k_B = R/N_A$ where N_A is Avogadro's number.

  a. Calculate Boltzmann's constant in J/K
  b. Calculate Boltzmann's constant erg/°F

1.3 Physicists like to use energies of atoms and molecules in terms of electron volts (eV) which is the energy necessary to move an electron across a potential of 1 V.
   a. Calculate the value of an eV in J.
   b. Calculate the value of 1.7 eV/atom in J/mole.
   c. Calculate a value of Boltzmann's constant in eV rather than Joules.

1.4 Dichlorosilane, SiH_2Cl_2, can be used to make high purity silicon for integrated circuit processing via the following reaction:

$$SiH_2Cl_2(g) = Si(s) + 2\,HCl(g)$$

The Gibbs energies at 1400 K for dichlorosilane and HCl are, respectively, $\Delta G°(SiH_2Cl_2) = -183{,}410$ J/mol and $\Delta G°(HCl) = -102{,}822$ J/mol.
   a. Calculate the value of the equilibrium constant at 1400 K.
   b. Calculate the equilibrium partial pressures of dichlorosilane and HCl over silicon at 1400 K if the starting pressures are $p(SiH_2Cl_2) = 0.01$ atm and $p(HCl) = 0$.

1.5 Gas mixtures of methane, CH_4 and H_2 are frequently used in surface hardening of steels to increase the carbon content at the surface. The Gibbs energy for the reaction at 700°C is given by (Roine 2002):

$$C(s) + 2H_2(g) \rightleftharpoons CH_4(g); \quad \Delta G^\circ_{700°C} = 16.312\,\text{kJ/mol}$$

   a. It is desired that the surface of the steel in this atmosphere contain 1.0 w/o C. If the atomic weight of carbon is 12.01 g/mol and that of iron 55.85 g/mol, calculate the mole fraction of carbon in 1 w/o steel.
   b. If the mole fraction is the same as the activity of carbon, calculate the necessary pressure (bar) of methane in the gas to achieve this surface concentration if the pressure of hydrogen is fixed at one bar.

1.6 The molar heat capacity of solid copper is given by (Kubaschewski et al. 1993)

$$\bar{C}_P = A + B \times 10^{-3}T + C\frac{10^{-5}}{T^2} + D \times 10^{-6}T^2 \text{ J/mol}\cdot\text{K}$$

where A = 30.29, B = −10.71, C = −3.22, and D = 9.47 and Units (T) = K.
   a. Calculate and plot the molar heat capacity of solid copper from 0°C to 1000°C at 100° intervals.
   b. Calculate and plot $\Delta\bar{H}^\circ_T - \Delta\bar{H}^\circ_{273.15}$ of solid copper from 0°C to 1000°C at 100° intervals.
   c. Do the same for the entropy.

1.7 The melting point of copper is about $T_m = 1083°C$, its density at 20°C is $\rho = 8.96$ g/cm^3, its atomic weight is M = 63.55 g/mol, its heat of fusion is 13.02 kJ/mol, and its linear thermal expansion coefficient is $\alpha = 17.0 \times 10^{-6}$ K^{-1} (Brandes and Brook 1992).
   a. Calculate the density of solid copper at its melting point. Assume that the volume expansion coefficient $\cong 3\alpha$.
   b. The density of liquid copper at the melting point is $\rho = 8.00$ g/cm^3. Calculate the slope of the P-T line between solid and liquid copper at 1 atm.

1.8 The melting point of silicon is about $T_m = 1410°C$, its density at 20°C is $\rho = 2.34$ g/cm^3, its atomic weight is M = 28.09 g/mol, its heat of fusion is 50.66 kJ/mol, and its linear thermal expansion coefficient is $\alpha = 7.6 \times 10^{-6}$ K^{-1} (Brandes and Brook 1992).
   a. Calculate the density of solid silicon at its melting point. Assume that the volume expansion coefficient $\cong 3\alpha$.
   b. The density of liquid silicon at the melting point is $\rho = 2.51$ g/cm^3. Calculate the slope of the P-T line between solid and liquid silicon at 1 atm.

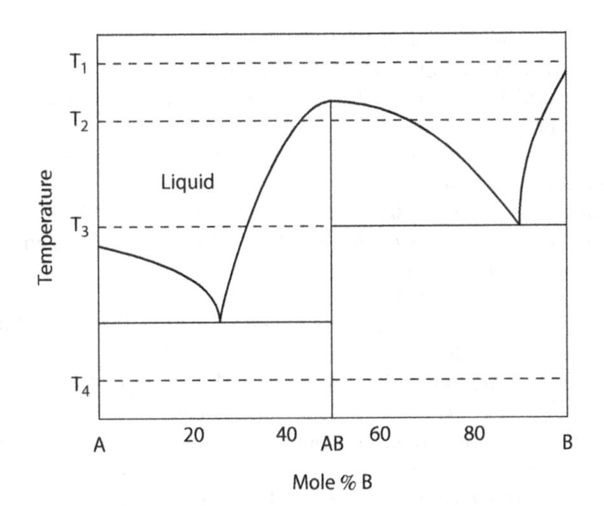

FIGURE E.1 Phase diagram that shows an intermetallic compound AB and no terminal solid solutions at components A and B.

1.9 From Boltzmann's relation $S = k_B \ln W$ show that the ideal molar entropy of mixing of aluminum sulfide (Al_2S_3) and aluminum selenide (Al_2Se_3) is given by

$$\bar{S}_{mix,id} = -3R\left(X_{Al_2S_3} \ln X_{Al_2S_3} + X_{Al_2Se_3} \ln X_{Al_2Se_3}\right)$$

1.10 The phase diagram in Figure E.1 is similar to that for many III-V compounds that have useful electronic properties allowing them to be used for light-emitting diodes (LEDs) and other applications.

a. Label all of the phase fields in the diagram.
b. Sketch the Gibbs energy-composition plots for temperatures T_1 to T_4 each on a separate plot.
c. If the Gibbs energy of formation of the compound AB is −120 kJ/mol at T_4, calculate the thermodynamic activity of B in AB at the 40 m/o B overall composition.
d. Do the same for A in AB at the 60 m/o B overall composition.

REFERENCES

Brown, G. I. 1999. *Count Rumford.* Stroud, Gloucestershire: Sutton Publishing.

Cahn, R. W. 2001. *The Coming of Materials Science.* London: Pergamon.

Chang, K. 2014. Time Travel. *New York Times,* October 24, 2014.

Committee on the Survey of Materials Science and Engineering (COSMAT). 1974. *Materials and Man's Needs.* Washington, DC: National Academy of Sciences.

Darken, L. S. and R. W. Gurry. 1953. *Physical Chemistry of Metals.* New York: McGraw-Hill Book.

DeHoff, R. 2006. *Thermodynamics in Materials Science,* 2nd ed. Boca Raton, FL: Taylor & Francis Group.

Gaskell, D. R. 2008. *Introduction to the Thermodynamics of Materials,* 5th ed. New York: Taylor & Francis Group.

Goerens, P. 1908. *Introduction to Metallography,* translated by F. Ibbotson. London: Longmans, Green.

Haynes, W. M., Editor-in-Chief. 2013. *CRC Handbook of Chemistry and Physics,* 94th ed. Boca Raton, FL: CRC Press.

International Union of Pure and Applied Chemistry (IUPAC). 2014. *Compendium of Chemical Terminology.* Gold Book. goldbook.iupac.org.

Keszei, E. 2012. *Chemical Thermodynamics.* Berlin, Germany: Springer-Verlag.

Kjellander, R. 2016. *Thermodynamics Kept Simple.* Boca Raton, FL: CRC Press.

Kubaschewski, O., C. B. Alcock, and P. J. Spencer. 1993. *Materials Thermochemistry,* 6th. ed. Oxford: Pergamon Press.

Laidler, K. J. 1993. *The World of Physical Chemistry.* Oxford: Oxford University Press.

Lindley, D. 2001. *Boltzmann's Atom.* New York: The Free Press.
Lindley, D. 2004. *Degrees Kelvin.* Washington, DC: John Henry Press.
Millar, D., I. Millar, J. Millar, and M. Millar. 2002. *The Cambridge Dictionary of Scientists,* 2nd ed. Cambridge: Cambridge University Press.
Morawetz, H. 1985. *Polymers, The Origins and Growth of a Science.* New York: Wiley.
Moss, S. 2015. Materials Scientists to the Rescue. *MRS Bull.* 40(3): 296.
Ragone, D. V. 1995. *Thermodynamics of Materials,* Vols. I and II. New York: Wiley.
Roine, A. 2002. *Outokumpu HSC Chemistry for Windows,* Ver. 5.11, thermodynamic software program, Outokumpu Research Oy, Pori, Finland.
Rukeyser, M. 1942. *Willard Gibbs.* Woodbridge, CT: Ox Bow Press.
Sauveur, A. 1912. *The Metallography of Iron and Steet.* Cambridge, MA: Sauveur and Boylston.
Smith, C. S. S. 1988. *A History of Metallography,* Cambridge, MA: MIT Press.
Sperling, L. H. 2006. *Introduction to Physical Polymer Science,* 4th ed. New York: Wiley.
Tyndall, J. 1880. *Heat: A Mode of Motion.* London: Longmans, Green.
Weertman, J. 2007. The Evolution of the Materials Science Profession and Professional over the Past 50 Years. *JOM.* 59(2): 18–19.
Wheeler, L. P. 1951. *Josiah Willard Gibbs.* Woodbridge, CT: Ox Bow Press.

Section II
Reaction Kinetics

2

Introduction to Kinetic Processes in Materials

2.1 INTRODUCTION: MATERIAL TRANSPORT AND REACTION RATES

A major consideration in presenting kinetics in materials science and engineering (MSE) is what topics should be covered and to what depth? Certainly, mass transport—diffusion—is an important topic for all materials. Indeed, the second half of this book is devoted to diffusion: *the spontaneous intermingling or mixing of atoms, ions, molecules, or particles due to random thermal motion.*

The macroscopic diffusion equations, how they are solved, and their relation to kinetic processes are important and general areas for all materials and are central to developing model-based process control. However, many texts only cover kinetic processes involving phase changes in the solid state and ignore transport in liquids and gases—fluids—or between liquids and gases, and solids and gases. The solutions to the macroscopic diffusion equations are not always applied to material processes involving solids with liquids or gases. Furthermore, often only the atomistic mechanisms of diffusion in solids are discussed, and those in liquids and gases are not. MSE is much broader than this, and, in many cases, the interactions between solids, gases, and liquids are of critical importance in materials processing and the development of structure and properties. Therefore, this chapter introduces the classical concepts of macroscopic reaction kinetics by defining reaction rates, their temperature dependence, the order of a reaction, and other terminology such as fraction reacted, half-life, and relaxation time. In this chapter, only zero- and first-order reactions are covered with a significant discussion about nuclear reactions and decay because they are ideal first-order reactions and introduce some nuclear chemistry and physics important to materials. The following two examples illustrate the importance in understanding reaction kinetics in MSE.

2.1.1 EXAMPLE: DEPOSITION OF SILICON FROM THE GAS

A singularly important example is the chemical vapor deposition (CVD) of silicon (Si) thin films from a Si-containing gas for semiconductor and photovoltaic device applications. *CVD is the formation of a solid on some suitable solid substrate by the reaction of gases only.* There are several Si-containing gases that are used to deposit Si. One process is

$$SiHCl_3(g) + H_2(g) \rightleftharpoons Si(s) + 3\,HCl(g)$$

illustrated in Figure 2.1. In this case, the reactant gases—trichlorosilane ($SiHCl_3$) and hydrogen—must first be transported to the surface of the solid substrate before the reaction can take place. Usually, this happens by a gas phase diffusion process. Then the reactants must actually react at the surface to deposit the Si. Such surface reactions can involve many individual reaction steps (as will be seen later in this chapter) before the reaction is complete. Finally, the product gas, HCl, must be transported away from the surface, again by diffusion. This process illustrates the broader range of kinetic processes that today's materials scientists and engineers need to be familiar with. Older texts on materials frequently discuss only diffusion in the solid state because only kinetic processes in solids were considered important. The situation has changed, and the very important technological

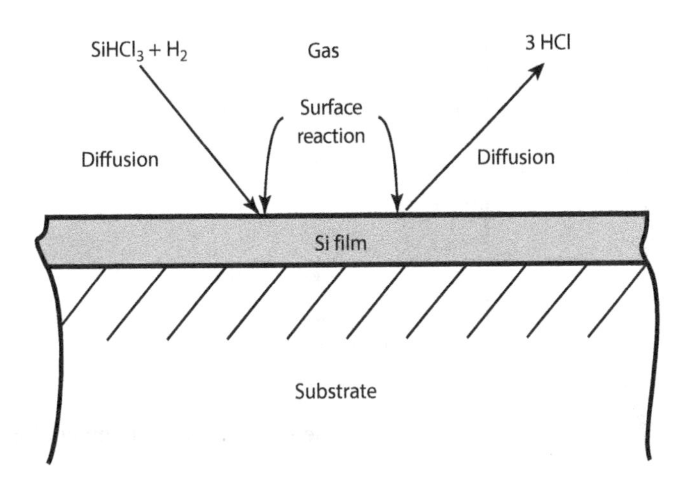

FIGURE 2.1 Schematic showing the deposition of a silicon film on a substrate by chemical vapor deposition from trichlorosilane ($SiHCl_3$) in a hydrogen carrier gas. The three series steps are as follows: (1) transport of the reactant gases to the surface, (2) the actual reaction at the surface, and (3) transport or diffusion of the product gases away from the surface. The slowest of the three steps controls the overall kinetics of the deposition.

process in MSE of Si deposition illustrates that diffusion in the gas phase and reaction kinetics need to be part of MSE education today.

The process variables under a materials engineer's control that determine the rate of Si deposition are the various gas pressures, gas flow rates, and temperatures. Reactant diffusion to the surface, followed by a surface reaction, finally followed by product diffusion away from the surface are three kinetic steps in series. As will be demonstrated later, the slowest step in the sequence determines the overall rate of a series sequential step process. Usually, at low temperatures, the surface reaction rate is the slowest step and limits the rate of deposition. In contrast, at high temperatures, diffusion is slower and limits the rate of deposition. Therefore, in order to develop a rational model of this process, it is essential to understand *both* the kinetics of diffusion in the gas and the kinetics of surface and other reactions. The first part of this book focuses on what could be called "classical kinetics" that might be taught in a physical chemistry course. But most MSE students do not take physical chemistry as a separate course today, so there is a need to expose the student to the terminology and concepts of classical reaction kinetics. This chapter is the first of several chapters that attempts to do this.

2.1.2 Example: Dissolution Rates of NaCl and Al_2O_3, Contrast Diffusion and Reaction Control

The kinetics of dissolution of solids in aqueous or other solutions, an area certainly important in the processing and/or performance of all materials—metals, ceramics, polymers, and biomaterials—illustrates the effect of the relative speed of these two series processes: reaction and transport. For example, sodium chloride (NaCl) is thermodynamically quite soluble in water and dissolves by the reaction:

$$NaCl(s) \rightleftharpoons Na^{+}(solution) + Cl^{-}(solution); \quad \Delta G^{\circ}(0^{\circ}C) = -7.84\ kJ/mole$$

which gives for the equilibrium constant (Roine 2002):

$$\frac{[Na^{+}][Cl^{-}]}{a_{NaCl}} = K_e(0^{\circ}C) = e^{-\frac{\Delta G^{\circ}}{RT}} = 31.53$$

(where $[Na^+]$ and $[Cl^-]$ are concentrations in mol/kg-soln, that is, *molal* concentrations). If the solid is pure NaCl, its thermodynamic activity is $a_{NaCl} = 1$, so the equilibrium concentrations of Na^+ and Cl^- are $[Na^{+}]_e = [Cl^{-}]_e = \sqrt{31.53} = 5.62$ mol/kg-soln at 0°C. The NaCl-H_2O phase diagram with data taken from a variety of sources is sketched in Figure 2.2, showing the solubility of NaCl in H_2O as a function of temperature, and the depression of the freezing point of water down to

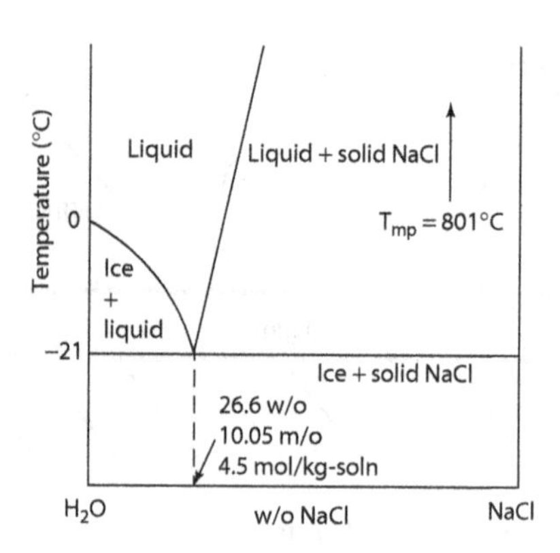

FIGURE 2.2 Sodium chloride–water phase diagram based on data from a number of sources.

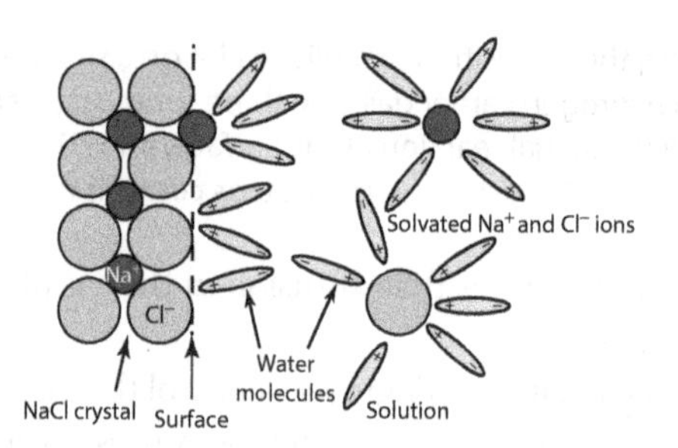

FIGURE 2.3 The dissolution of a sodium chloride crystal in water with the solvation energy of the water molecules with the Na^+ and Cl^- ions providing the necessary free energy for dissolution. The rate of dissolution of NaCl is generally controlled by liquid diffusion of the solvated ions away from the surface.

the eutectic temperature of −21.1°C at 26.6 w/o (w/o = weight percent), or 10.05 m/o (m/o = mole percent), or 4.5 mol/kg-soln. Figure 2.3 schematically shows the sequential reaction and diffusion steps. In this case, the Na^+ and Cl^- ions give up their nearest neighbor anion and cations in the solid surface for the polar H_2O molecules in solution. The enthalpy for the reaction at 0°C is actually positive, 8.132 kJ/mol, because the solvation energy of the ions is less than the energy of the ions in the crystal. But it is the high increase in entropy, 58.462 J/mol-K, going from the ordered solid to the disordered liquid,

$$\Delta G = \Delta H - T\Delta S = 8.132\times10^3 - 273.15\times58.462 = -7.837\,\text{kJ/mol}$$

that produces the high solubility (Roine 2002). This reaction step is followed by the diffusion of the solvated ions away from the surface. For the dissolution of NaCl in water, the rate of dissolution is controlled by diffusion in the liquid, which is easily demonstrated by stirring the solution—essentially decreasing the diffusion distance—which increases the rate of dissolution. Experience shows that salt (NaCl) dissolves quite rapidly in water, at least at room temperature and above.

In contrast, the dissolution or corrosion of Al_2O_3 in acidic solutions occurs by the following reaction:

$$Al_2O_3(s) + 6H^+(\text{solution}) \rightarrow 2Al^{+3}(\text{solution}) + 3H_2O\,(\text{solution}); \quad \Delta G°(30°C) = -101.5\,\text{kJ/mol}$$

with an equilibrium constant of (Roine 2002)

$$\frac{\left[Al^{+3}\right]^2 a_{H_2O}^3}{\left[H^+\right]^6 a_{Al_2O_3}} = K_e(30°C) = 3.09\times10^{17}.$$

The thermodynamic activities of water and alumina are 1.0 (not exactly true for the water but close enough for estimation purposes). If the solution is acidic, so that $[H^+] \cong 1$ mol/kg-soln (pH $\cong$ 0), this would make the equilibrium concentration of $[Al^{+3}]$ (or Al_2O_3) on the order of 10^8 molal, which is clearly ridiculous because the molar concentration of pure water is only

$$[H_2O] = \frac{1000}{M} = \frac{1000}{18} = 55.6\,\text{mol/kg}$$

and that of pure Al_2O_3,

$$[Al_2O_3] = \frac{1000}{102} = 9.9\,\text{mol/kg}.$$

Nevertheless, it is obvious that, thermodynamically, Al_2O_3 is quite soluble in acidic solutions. However, the dissolution of Al_2O_3 in acids occurs so exceedingly slowly that, for all practical purposes, Al_2O_3 does not dissolve in acids! Why not? Diffusion of the ions in the solution to and from the surface of the solid should occur at about the same rate as the diffusion of the ions that controls the rate of dissolution of NaCl because diffusion coefficients in most liquids are about the same ($D_L \approx 10^{-5}$ cm^2/s)—as will be shown later. However, for Al_2O_3, the rate of dissolution is controlled by the rate of the surface chemical reaction between the H^+ ions in the solution and the aluminum and oxygen ions at the surface of the alumina. This reaction rate is vanishingly small in spite of the fact that the thermodynamics for the dissolution are quite favorable. Here, stirring does not help because the reaction at the alumina surface controls the dissolution rate under all conditions and decreasing the distance that the ions diffuse by stirring does not affect the dissolution rate.

2.1.3 Conclusions from Examples

These examples of two identical processes for two solids—dissolution in an aqueous solution—illustrate the importance of the combined effects of the relative rates of the surface chemical reaction and diffusion in the liquid. They demonstrate the necessity of studying both chemical reaction kinetics and diffusion processes to understand more broadly the "phase transformation kinetics" of solids. These examples also illustrate two other important points: (1) *the thermodynamics of the process must be known before the kinetics can be studied* and (2) *for reactions in series, sequential steps—in these examples, a surface reaction followed by liquid diffusion—it is the slower of the two steps that controls the overall rate of the dissolution reaction.* Therefore, it is essential to investigate both chemical reaction kinetics—particularly at solid surfaces and interfaces—and the macroscopic and atomistic aspects of diffusion in solids, liquids, and gases in order to develop a firm grasp of kinetic processes in materials.

2.2 HOMOGENEOUS AND HETEROGENEOUS REACTIONS

There are some differences between "classical" chemical kinetics and kinetics of material processes. For example, *chemical kinetics* can be defined as (Laidler and Meiser 1999) "that branch of physical chemistry that deals with the rates of reaction and with the factors on which the rate depends: time, temperature, pressure, composition, etc." With materials, the kinetics are sometimes referred to as *kinetics of phase transformations* (Porter et al. 2009) because the concern is not only with the rate of a reaction but with the rates of change in a crystal or glass structure, the number of phases, the phase assemblage (microstructure), and the size and/or shape of the material. In all cases, the driving force for the reaction is a reduction or minimization of the Gibbs energy of the system. For solids, the Gibbs energy can take the form of the actual energy for a reaction, an energy difference because of compositional differences, or differences due to stress or surface curvature—surface energy. The goal is to understand how the controllable variables affect the kinetics of a materials process to develop a fundamentals-based model that predicts how the process will respond to changes in the variables. For example, the rates of many solid-state reactions depend exponentially on temperature and linearly on time. With a quantitative model of the system, it is possible to predict which would be economically better to take a commercial process to completion: increase the reaction time, increase the pressure, or increase the reaction temperature.

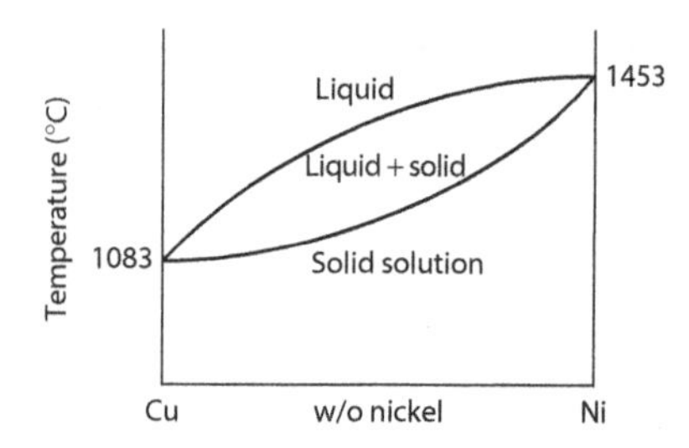

FIGURE 2.4 The copper-nickel phase diagram showing essentially ideal behavior with complete solid solubility. (After Brandes and Brook 1992.)

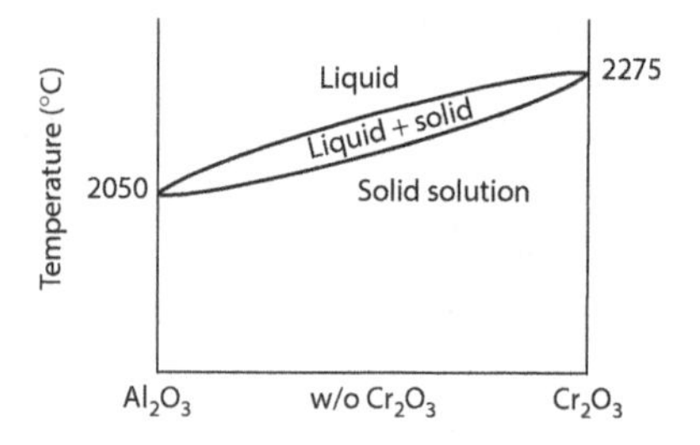

FIGURE 2.5 The Al_2O_3–Cr_2O_3 phase diagram showing essentially ideal behavior with complete solid solubility. (After Levin et al. 1964.)

Frequently, classical chemical kinetics is concerned primarily with *homogeneous* reactions defined *as reactions in which the reactants and products are all in the same phase*, that is, no phase change or transformation. For example,

- $H_2(g)+I_2(g) \rightleftharpoons 2\,HI(g)$, reactants and products in the same gas phase
- Fe^{+2} (solution) + Cl_2(solution) $\rightleftharpoons$ Fe^{+3}(solution) + 2 Cl^-(solution) in the same liquid phase
- $x\,Ni(s) + (1-x)\,Cu(s) \rightarrow Ni_xCu_{1-x}(s)$, because Cu and Ni form a continuous series of solid solutions, same solid phase (Figure 2.4) and similarly
- $x/2\,Cr_2O_3(s) + (1-x/2)\,Al_2O_3(s) \rightarrow Cr_xAl_{2-x}O_3(s)$

because Cr_2O_3 and Al_2O_3 also form a continuous solid solution as well (Figure 2.5).* Whether it is acceptable to call these last two homogeneous "reactions" is just a question of how far to stretch the concept of a "reaction." In these latter two cases, there is merely a compositional change, but it is driven by the entropy of mixing of the two end-member compositions—pure Ni and Cu and pure Al_2O_3 and Cr_2O_3—and, at the temperatures where these reactions occur, they are certainly controlled by solid-state diffusion. Homogeneous reactions in fluids—gases and liquids—are the kinds of reactions that chemists and chemical engineers primarily deal with.

In contrast, in materials kinetics, *heterogeneous* reactions are more common. A *heterogeneous reaction is one in which* more than one phase *is involved in the reaction.* The previous examples of CVD of Si and dissolution of NaCl and Al_2O_3 are examples of heterogeneous reactions. Another familiar example is the oxidation of a metal (Ni) or a nonmetal (silicon carbide (SiC), polyethylene). In the oxidation of a metal such as nickel (or silicon, iron, etc.),

$$Ni(s) + \frac{1}{2}O_2(g) \rightleftharpoons NiO(s)$$

three phases are involved: solid Ni, gaseous oxygen, and solid NiO. Figure 2.6 schematically shows this oxidation and the series and parallel kinetic steps that must take place for the oxidation to occur. Oxygen must be (1) transported through the gas phase to the NiO–gas interface where it (2) can react with (3) Ni ions diffusing through the NiO layer to the NiO–gas interface. Another possibility, oxygen molecules are (1) transported to the NiO–gas interface where they (2) separate into oxygen atoms, (3) pick up two electrons to become O^{-2} ions that then (4) diffuse through the NiO layer to (5) react with the Ni at the NiO–Ni interface. In the first option there are three steps in series and in the second option there are five steps in series. For series steps, the slowest reaction step controls the overall rate. In reality, the slowest step in oxidation is usually the diffusion of either the cation (Ni^{+2}) or the anion (O^{-2}) through the oxide layer that controls the overall rate or reaction, oxidation in this case. However, the diffusion of nickel out and the diffusion of oxygen in are parallel steps, and the faster of the two controls the rate.† In reality, in the oxidation of nickel—and many metals—it is the diffusion of the cation that is faster and controls the rate of oxidation: at least at elevated temperatures. Oxidation kinetics will be examined more closely later.

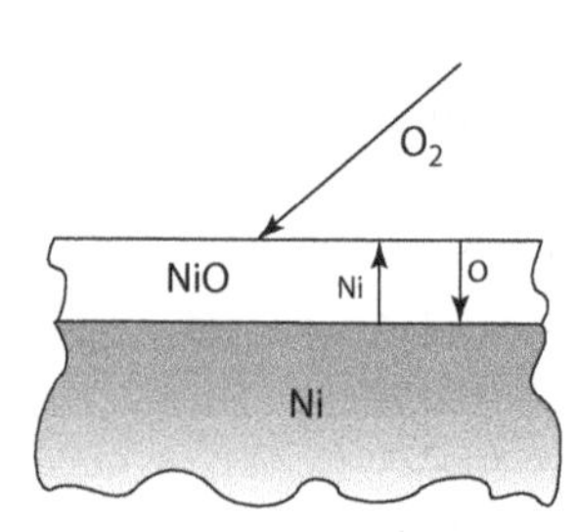

FIGURE 2.6 Schematic representation of the oxidation of nickel with the formation of nickel oxide on the surface. This represents a heterogeneous reaction because three phases are involved: gas, NiO, and Ni. In addition, the figure shows that the NiO may be formed by Ni^{2+} ions diffusing out to the NiO–O_2 interface or O^{2-} ions diffusing in toward the Ni–NiO interface.

* A few percent of chromium oxide in aluminum oxide is "ruby" if the aluminum oxide is a single crystal.

† Another parallel step is the diffusion of electrons, which is usually quite fast and does not produce mass transfer for the reaction but is necessary for charge flux balance as will be seen later.

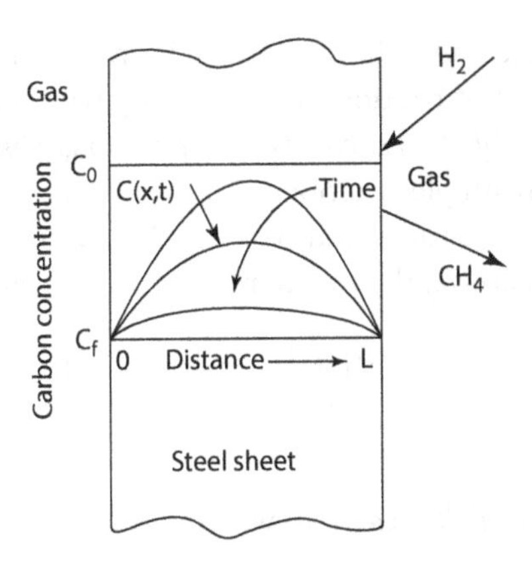

FIGURE 2.7 Decarburization of a steel sheet of thickness "L" in a hydrogen atmosphere by diffusion of carbon to the surface of the steel sheet to form CH_4 in the gas phase. By controlling the $[H_2]/[CH_4]$ in the gas, the carbon content (thermodynamic activity) can be set to a specific value. The C(x,t) are shown with C_0 (the concentration at zero time) and C_f (the final concentration as time goes to infinity).

Another example of a heterogeneous reaction is the degassing or decarburizing a steel sheet—which has the identical kinetic model as removal of a solvent from a polymer sheet or drying a cast ceramic tape shown in Figure 2.7. In this case, the reaction is

$$C(Fe) + 2H_2(g) \rightleftharpoons CH_4(g)$$

and there are two phases involved, the steel sheet and the gas phase. The reverse of this reaction is *carburization—increasing the carbon content*—of a steel sheet. Both processes lead to the same model only with different *initial* carbon contents and gas compositions.

2.3 HOMOGENEOUS REACTION RATES

2.3.1 Why Study?

Homogeneous *fluid* reactions are studied mainly by chemists and chemical engineers because many chemical processes of interest to them, such as combustion, occur in a single phase (Laidler and Meiser 1999). These reactions are explored here in order to provide a concrete platform for some common kinetic terminology that can be applied to heterogeneous reactions as well. Consider the following homogeneous gas reaction—one that has been well studied because the kinetics are sufficiently slow so that the reaction can be monitored with modest equipment over easily accessible time periods—hours rather than nanoseconds (Laidler and Meiser 1999):

$$H_2(g) + I_2(g) \rightleftharpoons 2HI(g).$$

The *reaction rate* is usually defined by the change in concentration or pressure of either one of the reactants or one of the products:

$$\frac{d[H_2]}{dt} = \frac{d[I_2]}{dt} = -\frac{1}{2}\frac{d[HI]}{dt} \tag{2.1}$$

where again "[]" denotes "concentration" (mol/cm³, mol/L, mol/m³, etc.). Because these rates are related by the stoichiometry of the reaction, any one of them can be used to measure the rate of reaction.

2.3.2 Reaction Order

For the gas phase formation of hydrogen iodide (HI), it would not be surprising that the rate of reaction depends on the frequency of collisions between hydrogen and iodine molecules in the gas. As a result, the rate intuitively depends on the concentration, the number of molecules per unit volume, of each of these reactants. In this particular reaction, this is a good assumption and fits the reality. However, a word of caution: It is generally true that *the dependence of the reaction rate on the concentrations of the reactants is not obvious simply by inspecting the reaction equation.* As will be seen, even simple reactions can involve multiple steps that are not reflected in the stoichiometric reaction equation, and it is one of these steps that will control the rate. Referring again to the reaction above, the rate of reaction can be given by

$$\text{Rate} = \frac{d[H_2]}{dt} = -k[H_2]^{\alpha}[I_2]^{\beta} \tag{2.2}$$

where:

$\alpha + \beta$ is the (overall) *order of the reaction,* where α and β are just numbers and are, but not always, integers
α is the order with respect to $[H_2]$
β is the order with respect to $[I_2]$.

Usually, the main interest is the *concentration* of either the products or reactants as a function of time, temperature, etc. and not their *rate* of appearance or disappearance. Therefore, differential rate equations, such as Equation 2.2, are usually integrated. So the order of the reaction leads to different expressions for the concentrations as a function of the experimental variables, time and concentrations. Zero-, first-, and second-order reactions are examined more closely simply because many elementary reactions follow one of these types of reactions. The *reaction rate constant,* k, typically has the form of an Arrhenius equation:

$$k = k_0 e^{-\frac{Q}{RT}}$$

where:

Q = the activation energy in kJ/mol
k_0 = the pre-exponential (the units of k_0 depend on the order of the reaction)
R = gas constant = 8.314 J/mol-K
T = absolute temperature in kelvin (Svante Arrhenius).

This exponential temperature dependence of reaction rate constants ensures that the rate of a reaction is a strong function of temperature. The basis for this exponential dependence on temperature and the justification for the activation energy are explored in detail in Chapter 4.

To generalize, simplify, and also make the notation more transparent, assume that there are two gases A and B that react as

$$A(g) \rightleftharpoons B(g)$$

with a rate of reaction given by

$$\text{Rate} = \frac{d[A]}{dt} = -k[A]^{\alpha} \tag{2.3}$$

where α is the order of the reaction (with respect to the reactant A). If $\alpha = 0$, the reaction is *zero order*; if $\alpha = 1$, the reaction is *first order*; and if $\alpha = 2$, the reaction is *second order*, etc. As mentioned above, what is usually of interest, is the integrated form of the rate expression to give the concentration of A, [A], as a function of time.

2.3.3 Units and Symbols

Of course, either gas pressures or concentrations can be used as they are proportional to each other, because of PV = nRT for an ideal gas (all gases in this book are considered to be ideal). So, for some gas A at a pressure P(A),

$$[A] = \frac{n}{V} = \frac{P(A)}{RT},$$

where the units of concentration depend on the units of pressure and those of R. In this book, the format *Units*(x) is used to signify the units of whatever symbol is inside the parentheses, ().* So, for the Units(R):

$$\text{Units}(R) = \frac{J}{mol \cdot K} = \frac{N \cdot m}{mol \cdot K} = \frac{N}{m^2} \frac{m^3}{mol \cdot K} = \frac{Pa \cdot m^3}{mol \cdot K}$$

so that if the pressure is in pascals, then the Units([A]) = mol/m^3. However, often, Units (P) = atmospheres (or bars) and sometimes, for convenience, Units([A]) in mol/cm^3 or mol/L are more useful. So,

$$\text{Units(R)} = 8.314 \frac{Pa\ m^3}{mol\ K} = \frac{8.314\ Pa\ m^3}{1.01325 \times 10^5 Pa/atm} \times 10^6 \frac{cm^3}{m^3} \times \frac{1}{mol\ K} = 82.05 \frac{atm\ cm^3}{mole\ K}$$

or

$$\text{Units(R)} = 8.314 \frac{Pa\ m^3}{mol\ K} = \frac{8.314\ Pa\ m^3}{1.01325 \times 10^5 Pa/atm} \times 10^3 \frac{L}{m^3} \times \frac{1}{mol\ K} = 0.08205 \frac{atm\ L}{mole\ K}$$

are the other two commonly used units for R. Because 1 bar = 10^5 Pa and 1 atm = 760 torr = 760 mm of Hg, a number of other possible values for Units(R) could be calculated. Note that

$$1\ atm = h(Hg) \times \rho(Hg) \times g$$

$$1\ atm = 76.0\ cm \times 13.6\ g/cm^3 \times 980\ cm/s^2 \times \frac{(10^4\ cm^2/m^2)(1\ m/10^2\ cm)}{1000\ g/kg}$$

$$1\ atm = 1.013 \times 10^5 \frac{N}{m^2} = 1.013\ bar$$

Also, it should be pointed out that

$$\frac{R}{N_A} = \frac{8.314 J/mol \cdot K}{6.022 \times 10^{23}\ atoms/mol} = 1.38 \times 10^{-23}\ J/atom \cdot K = k_B$$

or Boltzmann's constant (frequently given by just k, but here k_B is used).

For example, putting numbers into some of the above expressions, for example, expressing [A] in mol/cm^3, [A] = n/V = P/RT, P = 1 atm, and R = 82.04 cm^3-atm/mol-K, then at 30°C, $[A] = 4.02 \times 10^{-5}$ mol/cm^3. Multiplying this by Avogadro's number, $N_A = 6.022 \times 10^{23}$ atoms or molecules per mole, gives $\eta = 2.4 \times 10^{19}$ molecules/cm^3 in a gas at room temperature and 1 atm pressure.†

It is interesting to contrast the number of atoms/cm^3 in a gas to the number of atoms—or ions or molecules—in a solid or liquid. For example, the density, ρ, of Al_2O_3 is ρ = 4.00 g/cm^3 and its molecular weight is 100.96 g/mol (Haynes 2013), so its molar volume is

$$\bar{V}(Al_2O_3) = \frac{M}{\rho} = \frac{101.96}{4.00} = 25.49\ cm^3/mol.$$

The number of Al_2O_3 "molecules" or "formula units" per unit volume is given by

$$\eta(Al_2O_3) = \frac{\rho}{M} N_A = \frac{4.00}{101.96} \times 6.022 \times 10^{23} = 2.36 \times 10^{22}\ cm^{-3}$$

* "Units(x)" is used to designate the "units of x" while "[x]" designates the "concentration of x."

† The Greek letter eta, η, is used for *number concentrations* in this book, or *number density*, simply because n and N are used to represent so many other quantities and η at least looks like an n. The units of η are, Units(η) = atoms/cm^3 or atoms/m^3. Of course, the number of molecules per unit volume is proportional to the pressure for an ideal gas.

so the number of oxygen ions per cm^3, $\eta(O) = 3\ \eta(Al_2O_3) = 7.09 \times 10^{22}/cm^3$ and the number of aluminum ions per cm^3, $\eta(Al) = 2\ \eta(Al_2O_3) = 4.72 \times 10^{22}/cm^3$. Note that for virtually all condensed phases—solids and liquids—the number of atoms, ions, or molecules per cm^3 is *always something times 10^{22}*. So the number density in condensed phases is about three orders of magnitude higher than the number density in gases at room temperature and 1 atm pressure. The number density in condensed phases does not vary much with temperature and reasonable pressures. In contrast, of course, the number density in gases varies significantly with pressure and temperature.

2.4 ZERO-ORDER REACTION

Going back to Equation 2.3, for a zero-order reaction, $\alpha = 0$, so

$$\frac{d[A]}{dt} = -k \tag{2.4}$$

so Units(k) = mol/cm^3 s, if the concentration of A is in moles per cm^3. This is readily integrated to give

$$[A] = -kt + C$$

where C is just an integration constant. When $t = 0$, $[A] = [A]_0$, the starting concentration of A, which, after substitution, gives the simple linear relation

$$[A] = [A]_0 - kt \tag{2.5}$$

and the reaction goes to completion ($[A] = 0$) when $t = [A]_0/k$ as shown in Figure 2.8.

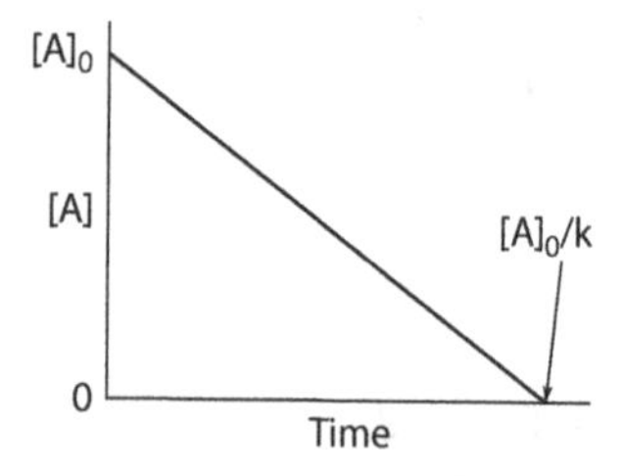

FIGURE 2.8 The concentration of reactant A, [A], as a function of time for a zero-order reaction with a reaction rate constant k and an initial concentration at zero time of $[A]_0$.

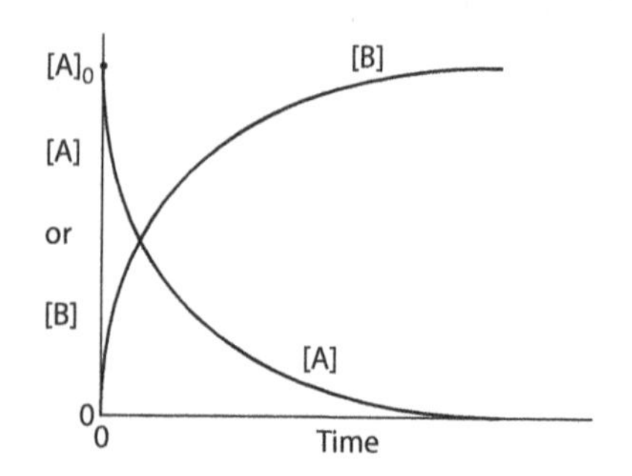

FIGURE 2.9 The concentrations of reactant A, [A], and product B, [B], for the first-order reaction $A \rightarrow B$ as a function of time with initial concentrations of $[A]_0$ and zero for A and B, respectively.

2.5 FIRST-ORDER REACTION

2.5.1 Integrated Equation

First-order reactions are particularly important because they occur frequently in MSE. For the $A \rightleftharpoons B$ reaction in Section 2.3.2, for a first-order reaction, the differential equation that needs to be solved is

$$\frac{d[A]}{dt} = -k[A] = -\frac{d[B]}{dt}. \tag{2.6}$$

Note that the units for k are Units(k) = s^{-1} and are different from those for the zero-order reaction. Integrating,

$$\ln[A] = -kt + C \quad \text{or} \quad [A] = Ce^{-kt}$$

where C is a constant. At $t = 0$, $[A] = [A]_0 = C$, which leads to

$$[A] = [A]_0 e^{-kt}. \tag{2.7}$$

The concentrations of both A and B are plotted as a function of time in Figure 2.9. An easy way to obtain k from experimental data is, of course, to plot the concentration of A on a semilog plot as shown in

Figure 2.10. Note that the plot $\log_{10}[A]$ versus time is a straight line with a slope of $-k/2.303$.* (Note that if $x = 10^{\log_{10} x}$, then $\ln x = \log_{10} x \times \ln 10$. Because e = 2.718, $10 = 2.718^{2.303}$, so ln 10 = 2.303 and $\ln x = 2.303 \log_{10} x$ and $\log_{10} x = \ln x/2.303$.)

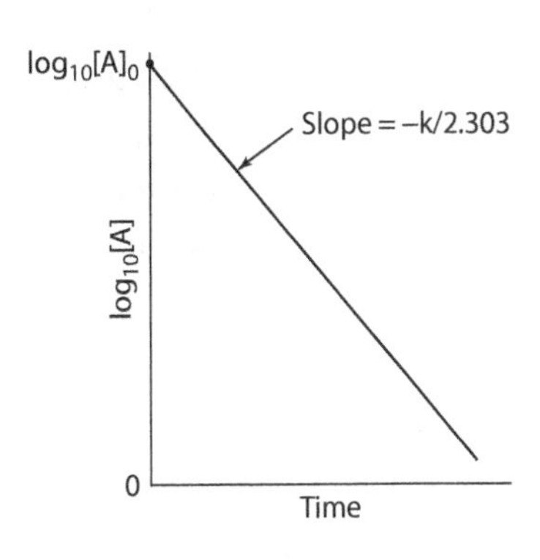

FIGURE 2.10 Concentration of the reactant A, [A] on a semilog plot versus time for a first-order reaction. From the slope, the value of the reaction constant k can be determined. $[A]_0$ is again the initial concentration of A.

2.5.2 Fraction Reacted

A few commonly used terms need defining. The *fraction reacted* can be calculated for any order reaction. In the specific case of a first-order reaction, call the *fraction unreacted*, f_U, then f_U = (amount of reactant remaining)/(initial amount of reactant) or

$$f_U = \frac{[A]}{[A]_0} = e^{-kt} \tag{2.8}$$

and the *fraction reacted*, f_R, is

$$f_R = 1 - f_U = 1 - \frac{[A]}{[A]_0} = 1 - e^{-kt}$$

or (2.9)

$$f_R = \frac{[A]_0 - [A]}{[A]_0} = 1 - e^{-kt}.$$

2.5.3 The Half-Life of a Reaction

Another expression relating to the extent of a reaction is the *half-life or* $t_{1/2}$, which is the amount of time for one-half of the original reactant disappear or react, or *the time necessary for the reaction to go half way to completion.* The half-life is very important in nuclear decay reactions because they are all first-order reactions and characterized by their half-lives, which gives an indication of how intensively radioactive they are. For a first-order reaction, the time for $[A] = [A]_0/2$ is the half-life, or

$$\frac{[A]}{[A]_0} = 0.5 = e^{-kt_{1/2}} \tag{2.10}$$

where $t_{1/2}$ is the half-life of the reaction. Therefore $-kt_{1/2} = \ln 0.5 = -0.693$, so $t_{1/2} = 0.693/k$. For example, if $k = 10^{-3}\,s^{-1}$, then $t_{1/2} = 693$ s.

2.5.4 Relaxation Time

One final term requiring definition is the *relaxation time*, τ, or the *time for the reaction to proceed to 1/e = 1/2.718 = 0.37 of its final value*—zero for Equation 2.7. That is $[A]/[A]_0 = e^{-1}$, which implies that $k\tau = 1$ or $\tau = 1/k$. For example, if $k = 10^{-3}$ s, then $\tau = 10^3$ s and

$$[A] = [A]_0 = e^{-t/\tau}. \tag{2.11}$$

* *Note: Never, never, ever make a semilog plot with natural logarithm values on the vertical axis if the results are going to be presented to someone else such as in a publication, report, or paper.* The reason for this is that everyone is familiar with quantities to the base 10, but few can translate quantities to the base e into base 10 values. For example, if one of the points is on the vertical axis at −6 on a $\log_{10}$ plot, then the value at this point is 10^{-6}, or one part per million. Everyone has a reasonable idea of what 10^{-6} means. However, if this were a natural logarithm plot, then the −6 would be $e^{-6} = 2.718^{-6}$ = ????, whatever this is! However, the penalty that you have to pay for plotting an exponential function as $\log_{10}$ on a semilog plot is that the slope is not −k as it would be on natural logarithm plot but $-k/(\ln 10) = -k/2.303$.

2.5.5 Example of First-Order Reaction: $COCl_2$ Decomposition

Carbonyl chloride, $COCl_2$, phosgene, was introduced during the First World War as a poison gas but today is used in large quantities in the production of certain polymers. $COCl_2$ gas undergoes decomposition—at elevated temperatures—by the following reaction:

$$COCl_2(g) \rightleftharpoons CO(g) + Cl_2(g).$$

The equilibrium constant, K_e, is given by

$$K_e = \frac{p(CO)p(Cl_2)}{p(COCl_2)}$$

and $K_e = 217$ at 1200 K (Roine 2002). At this temperature, thermodynamics indicates that essentially all of the $COCl_2$ is decomposed at equilibrium. How far the reaction goes to completion can be estimated by assuming that $p(CO) = p(Cl_2) \cong 1$ bar, so $p(COCl_2) \cong 1/217 = 4.6 \times 10^{-3}$ bar, or about 99.5% of the $COCl_2$ is decomposed.

At this point, the kinetics become interesting.* This is a first-order reaction with a first-order reaction constant (NIST Kinetics Database):

$$k = k_0\, e^{-\frac{Q}{RT}} = 1.45 \times 10^{14}\, e^{-\frac{345\,\text{kJ/mol}}{RT}}\ s^{-1}. \tag{2.12}$$

At 1200 K, Equation 2.12 gives $k = 1.45 \times 10^{14} \exp(-(345{,}000/(8.314 \times 1200)))\ s^{-1} = 0.139\ s^{-1}$, which gives a half-life of $t_{1/2} = 0.693/0.139 = 4.99$ s, and a relaxation time of $\tau = 1/k = 7.19$ s, a fairly rapid reaction.

In contrast, $COCl_2$ is pretty stable at low temperatures and at 300 K, (NIST-JANAF Thermochemical Tables; NIST Kinetics Database)

$$\frac{p(Cl_2)p(CO)}{p(COCl_2)} = K_e = 1.12 \times 10^{-12}. \tag{2.13}$$

For example, if at the start of the reaction, pure $COCl_2$ is enclosed in a fixed volume at 300 K at one bar pressure and is allowed to go to equilibrium, the pressure of $COCl_2$ does not change much because of the small equilibrium constant, Equation 2.13, and the amount of Cl_2 and CO can be estimated from $p_e(CO) = p_e(Cl_2) = (1.1210^{-12})^{1/2} \cong 1 \times 10^{-6}$ bar, or only about one part per million (ppm) of the $COCl_2$ decomposes. This is a sufficiently small amount that it would be doubtful if there were much interest in the kinetics of decomposition. Making it even less interesting, the first-order reaction constant is calculated to be $k = 1.45 \times 10^{14} \exp(-345{,}000/(8.314 \times 1200))\ s^{-1} = 1.23 \times 10^{-46}\ s^{-1}$! Not only does it not decompose very much, but it would take a long time for even this small amount of decomposition (1 year $\cong 3 \times 10^7$ s, and the age of the universe is only 13.8 billion years or 4.14×10^{17} s).†

2.6 RADIOACTIVE DECAY AND RELATED NUCLEAR REACTIONS

2.6.1 Ideal First-Order Reaction

Radioactive decay is the prime example of a first-order reaction because the number of disintegrations per unit time, dm*/dt, (Units(dm*/dt) = becquerel (Bq), 1 Bq = one disintegration per second), is directly proportional to the amount of radioactive material present, m*, that is,

* The NIST Kinetics Database provides a great deal of information on gas reaction kinetics, whereas the Thermochemical Tables are a primary source of thermodynamic data for gases, liquids, and solids. These thermodynamic data are also included in many available software packages (HSC Chemistry Software) that utilize the NIST data and other data sources to calculate a variety of thermodynamic properties for a large number of compounds and thermodynamic data for many reactions.

† However, it should be pointed out that the database suggests that the parameters in the reaction rate constant, k_0 and Q, are only valid between 1400 and 2000 K, both of which are above the temperatures used in this example. So the extrapolation of the given data down to the temperatures of interest may not give very accurate results. Nevertheless, these are the best data available, so there is little choice but to use them: a situation frequently faced by a materials scientist or engineer. Also note the very large value of Q (345 kJ) that produces the very strong temperature dependence of the reaction rate constant.

$dm^*/dt = -km^*$. Furthermore, radioactive decay and other nuclear reactions are extremely important in MSE considering the critical role that radioactive materials play in nuclear power generation, nuclear waste disposal, and the use of radioactive isotopes to study reactions and transport in materials. Therefore, a short discussion of some important nuclear reactions and nuclear chemistry is germane to the overall theme of materials kinetics as well as examples of ideal first-order reactions.

2.6.2 Uranium Isotopes

The first example is the decay of ${}^{238}_{92}U$ by the emission of an alpha particle, a helium nucleus, ${}^{4}_{2}He$, along with 4.25 MeV (million electron volts of energy)* to ${}^{234}_{92}Th$. ${}^{238}U$ is the most common isotope of uranium found in uranium ore deposits and represents 99.3% of natural uranium with ${}^{235}U$, the fissionable isotope—the one that can be split by neutrons while generating large amounts of energy—present at about 0.7%. The half-life of 4.51 billion years is shown for the decay to ${}^{234}Th$ over the arrow. The ${}^{234}Th$ is also radioactive and undergoes several disintegrations, eventually decaying to one of the stable lead isotopes such as ${}^{208}_{82}Pb$.

$$ {}^{238}_{92}U \xrightarrow{4.51\times10^{9}\text{ year}} {}^{234}_{90}Th + {}^{4}_{2}He + 4.25\text{ MeV} $$

$$ {}^{234}_{90}Th \rightarrow \text{eventually to Pb.} $$

Uranium 235 also undergoes alpha particle decay with a half-life of about 700 million years:

$$ {}^{235}_{92}U \xrightarrow{7.0\times10^{8}\text{ year}} {}^{231}_{90}Th + {}^{4}_{2}He + 4.68\text{ MeV} $$

$$ {}^{231}_{90}Th \rightarrow \text{eventually to Pb.} $$

Uranium is usually "enriched" to about 5% ${}^{235}_{92}U$ so that a nuclear chain reaction obtained by fission can be sustained in a nuclear reactor—"reactor grade uranium." For a nuclear weapon, the enrichment needs to be much higher, greater than 85%—"weapons grade uranium." This enrichment in the past was accomplished with UF_6 gas in very large—several kilometers on a side—gaseous diffusion plants, in which the ${}^{235}UF_6$ and ${}^{238}UF_6$ were separated by the small difference in their gaseous diffusion rates that depend on $1/M^{1/2}$, where M is the molecular weight—as is discussed in Chapters 8 and 9. Such plants were hard to hide from satellite observation because they were so large—diffusion distances had to be long to separate gases whose diffusion coefficients varied by only about 0.4%. However, today the enrichment can be done with high-speed centrifuges in a much smaller and harder to detect facility. Hence, the concern when a country announces an enrichment program: What is the level of enrichment it intends, reactor grade, or weapons grade? On the other hand, an unannounced program is even more insidious, because it may go undetected. The ${}^{238}U$ left over is called "depleted uranium" and because of its high density—19.05 g/cm^3—it is used for armor piercing projectiles. It can be placed in a reactor to produce fissionable ${}^{239}Pu$ that can also be used in a reactor or a weapon. In a reactor, ${}^{238}U$ undergoes the following reaction and decay sequence to produce ${}^{239}Pu$, a man-made element:

$$ {}^{238}_{92}U + {}^{1}_{0}n \rightarrow {}^{239}_{92}U \xrightarrow{23.4\text{ minute}} {}^{239}_{93}Np + \beta^- + 1.28\text{ MeV} $$

$$ {}^{239}_{93}Np \xrightarrow{2.35\text{ day}} {}^{239}_{94}Pu + \beta^- + 0.723\text{ MeV.} $$

A β^- particle is a high-energy electron resulting from the decay of a neutron into a proton, which carries most of the energy of the reaction. ${}^{239}Pu$ also undergoes radioactive decay with emission of an alpha particle similar to ${}^{238}U$,

$$ {}^{239}_{94}Pu \xrightarrow{24{,}000\text{ year}} {}^{235}_{92}U + {}^{4}_{2}He + 5.2\text{ MeV} $$

$$ {}^{235}_{92}U \rightarrow \text{eventually to Pb.} $$

* 1 eV = 1 electron volt = energy to move the charge on one electron across a potential of 1 volt = 1.602×10^{-19} C × 1 V = 1.602 10^{-19} J since 1 V = 1 J/C and for 1 mole = 1eV × N_A = $1.602 \times 10^{-19} \times 6.022 \times 10^{23}$ = 96,472 J/mol ≅ 100 kJ/mol on the order of the energy of an atomic bond. Physicists like to deal in eV per atomic reaction event, although chemists more commonly use kJ/mol when talking about moles of reactants and products. Materials people use whatever suits the situation best.

Even with an apparently long half-life of 24,000 years, the rate of energy generated in the disintegration of ^{239}Pu, makes a piece of pure plutonium metal actually warm to the touch.

2.6.3 Radioisotope Power Sources

On the other hand, this heating can be useful and another radioactive isotope of plutonium, ^{238}Pu, is used as a radioisotope heat source because the half-life is short enough to provide sufficient power to reach temperatures in excess of 1000°C from the reaction:

$$^{238}_{94}\mathrm{Pu} \xrightarrow{86\ \mathrm{year}} {}^{4}_{2}\mathrm{He} + {}^{234}_{92}\mathrm{U} + 5.59\ \mathrm{MeV}$$

Such heat sources are made from PuO_2 spheres or cylinders encased in a welded iridium or platinum–rhodium shell and used for satellite—particularly military—and deep space probes where solar panels do not provide enough power, Figure 2.11 (radioisotope power systems). Several of the modular units shown in Figure 2.11 are used, in order to supply the necessary power for a satellite. Each of the iridium-encased PuO_2 modules contain about 34 Ci* of $^{238}_{94}Pu$, that generates about 1 W of heat. This heat is used to generate electricity with thermoelectric elements (thermocouples) and the entire subassembly is called a radioisotope thermoelectric generator (RTG) or radioactive power source. The United States has launched many satellites with these power sources, with powers ranging from a few hundred watts, to a few kilowatts with some containing tens of pounds of plutonium. The *New Horizons* mission to Pluto in 2015 used a radioisotope power system. Supposedly, the PuO_2 will not get dispersed into the environment even if the satellite were to be aborted because of the iridium encapsulation. There have been a few accidents, but purportedly no radioactive ^{238}Pu was released to the environment (Radioisotope Power Systems). The concern with plutonium is that, because the isotopes have significant decay rates and emit energetic alpha particles, only small concentrated amounts—micrograms—can do considerable tissue damage, particularly in the lungs, that potentially lead to cancer.

2.6.4 Nuclear Reactors and Fission Products

The other concern with plutonium is that it can be separated chemically from the uranium and fission products in nuclear reactor fuel—a much simpler process than the physical processes necessary for uranium enrichment—providing the operating personnel can be protected from the extremely high level of radiation from partially used nuclear fuel. Also, the process can be carried out in small facilities that can be easily hidden, if the intent is to use the plutonium for something other than reactor fuel. It only takes a few 100 cm^3 of ^{239}Pu to make a nuclear weapon! As a result, the United States decided in the 1970s that it would not pursue nuclear fuel reprocessing to recover fissionable

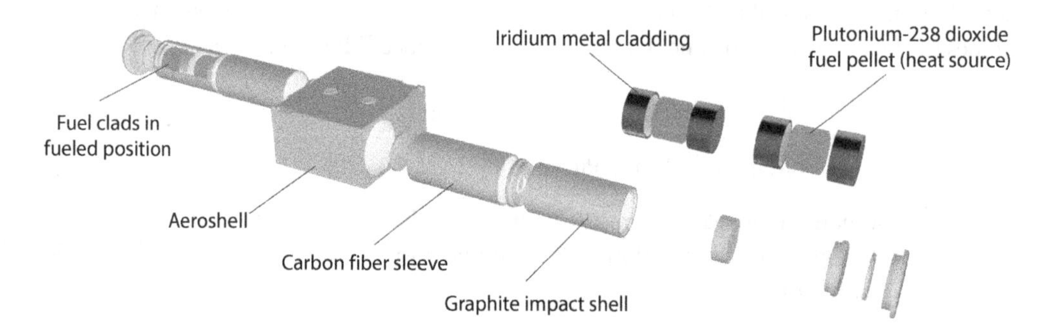

FIGURE 2.11 General purpose heat source module showing how the iridium-clad $^{238}_{94}PuO_2$ pellet heat sources are inserted into various carbon parts to form the module. NASA photograph (Radioisotope Power Systems). The original color photograph was color inverted, converted to grayscale, and contrast-enhanced with Photoshop®.

* Note, *1 Ci = 1 Curie = 3.7 × 10^{10} disintegrations/s = 3.7 × 10^{10}Bq*, which is roughly the number of disintegrations per second from a gram of radium, the element that was discovered by Marie and Pierre Curie and the unit was named after them.

plutonium, unused ^{235}U and ^{238}U from nuclear power reactors for fear of proliferation of nuclear weapons, because of the availability of ^{239}Pu. However, from an energy generation perspective, with the geological abundance of ^{238}U, that can be converted into ^{239}Pu, there is virtually an unlimited supply of energy from the nuclear power potential in the uranium ore.

The potential for energy from nuclear reactions comes from the fission process of the uranium and plutonium isotopes that can easily fission. Both ^{235}U and ^{239}Pu are fissionable materials in that when bombarded with neutrons their nuclei split into two isotopes of similar, but not equal, atomic weights, plus more neutrons, plus a large amount of energy, which is the basis for nuclear power generation:

$$^{235}_{92}U + ^{1}_{0}n \rightarrow ^{90}_{38}Sr + ^{143}_{54}Xe + 3\,^{1}_{0}n + 205\text{ MeV}$$

with a similar reaction for ^{239}Pu. In both cases, several radioactive fission products are produced, such as ^{90}Sr, ^{131}I, and ^{127}Cs:

$$^{90}_{38}Sr \xrightarrow{28.1\text{ year}} ^{90}_{39}Y + \beta^- + 0.546\text{ MeV}$$

$$^{90}_{39}Y \xrightarrow{64\text{ hour}} ^{90}_{40}Zr + \beta^- + 2.95\text{ MeV}$$

^{90}Zr is a stable isotope, meaning that it does not undergo further radioactive decay;

$$^{131}_{53}I \xrightarrow{8.07\text{ day}} ^{131}_{54}Xe + \beta^- + 0.97\text{ MeV}$$

and ^{131}Xe is a stable isotope;

$$^{127}_{55}Cs \xrightarrow{62\text{ hour}} ^{127}_{54}Xe + \beta^+ + 2.1\text{ MeV}$$

$$^{127}_{54}Xe \xrightarrow{36.4\text{ day}} ^{127}_{53}I + \gamma + 0.44\text{ MeV}$$

Here, the ^{127}I is stable and the last reaction occurs by an "electron capture" process in which an orbital electron enters the nucleus dropping the number of protons from 54 to 53. The β^+ is a high-energy positron—antielectron. These radioactive *fission products* are unwelcome additions to the human body because iodine concentrates in the thyroid and strontium in bones, and their high-energy emissions lead to broken chemical bonds, cell damage, and, potentially, cancer.

2.6.5 Radiocarbon Dating

A useful example of radioactive decay and a first-order reaction is the dating of objects from the decay of carbon-14 (Taylor 2000; "Radiocarbon Dating"). Willard Libby did the first work on radiocarbon dating in 1949 when he was a professor at the University of Chicago, research for which he received the Nobel Prize in Physics in 1960 (Willard Libby). The natural abundances of carbon isotopes are

$$^{12}_{6}C \quad 98.89\%$$

$$^{13}_{6}C \quad 1.11\%$$

$$^{14}_{6}C \quad \sim 1\times10^{-12}.$$

The concentration of $^{14}_{6}C$ is estimated from literature data (Taylor 2000). Here ^{12}C and ^{13}C are stable, whereas ^{14}C decays as

$$^{14}_{6}C \xrightarrow{5730\text{ year}} ^{14}_{7}N + \beta^- + 0.156\text{ MeV} \quad (2.14)$$

and so its concentration can be measured by counting the beta particle radiation. Given a sample of carbon, by counting the rate of disintegration, the concentration of ^{14}C, [^{14}C], can be determined. Of course, with a half-life of 5730 years, virtually all of the ^{14}C that existed when the Earth was formed would have long ago decayed to the stable ^{14}N, given the 4.5 billion year age of the Earth. However,

^{14}C is constantly being regenerated in the upper atmosphere by cosmic rays that generate neutrons by nuclear reactions that react with nitrogen in the atmosphere,

$${}^{14}_{7}N + {}^{1}_{0}n \rightarrow {}^{14}_{6}C + {}^{1}_{1}H$$

and forms ${}^{14}_{6}CO_2$, and the ^{14}C in the atmosphere reaches a steady-state value that has been more or less constant over time (about 1 part per trillion of stable carbon, 1 part in 10^{12}).* Corrections have been made by other techniques such as tree-ring dating—which gives the $\left[{}^{14}_{6}C\right]$ as a function of time because each tree ring counts as 1 year. Atomic bomb testing during the 1950s and 1960s almost doubled the concentration of carbon 14, but it now has dropped to close to its earlier—more constant—level. A living organism that is constantly in contact with the atmosphere and, using its biological processes to emit or absorb CO_2, will maintain in its body this constant concentration of ^{14}C in equilibrium with the atmosphere. However, when the organism dies, it no longer exchanges CO_2 with the atmosphere, and the ^{14}C in its remains will start to decay. As a result, by measuring the ^{14}C in something that previously had been alive, the time since its death can be determined from

$$\frac{\left[{}^{14}_{6}C\right]}{\left[{}^{14}_{6}C\right]_0} = e^{-kt} \tag{2.15}$$

where $\left[{}^{14}_{6}C\right]_0$ is the constant ^{14}C concentration in the atmosphere and $k = -\ln(0.5/5730) = 1.21 \times 10^{-4}$ years, as noted earlier. So if a mammoth tusk is found that has a carbon-14 ratio of 0.25, then from above, $t = -\ln(0.25/1.21 \times 10^{-4}) = 11{,}460$ years. So the mammoth lived and died around 9500 BC.

Today, mass spectrometry is also used to find the concentration of ^{14}C, so a smaller amount of material can be used. For example, in 1988, radiocarbon dating was used to attempt to determine the age of the Shroud of Turin, the venerated and reputed burial cloth of Jesus, which many believe shows his face and body outline, shown in Figure 2.12. The analysis gave results that suggested that the Shroud was only about 690 years old, about the time when it first appeared on the scene in the early 1300s. It was dated from the flax used to make the cloth, and the same age was found independently by three different laboratories. The radiocarbon data suggest that the shroud was, in reality, a painting made in the early 1300s. However, these results have been disputed for a number of reasons including the possibility of repairs to the shroud over time and that the sampling was done on cloth added more recently as a repair. Another suggestion is that bacteria were present in the cloth in more recent times, and their presence led to a higher ^{14}C content and a younger apparent age than the actual age. Several other suggestions have been made, as well, that could explain a higher initial ^{14}C than is normally present in the atmosphere leading to an apparent younger age for the cloth (Wilson 1998; Turney 2006; Shroud of Turin). Clearly, rather than giving definitive results in this case, the radiocarbon dating of the Shroud probably increased the mystery and speculation surrounding its origin. For example, the image on the shroud is essentially a *negative image*! It is hard to explain how a fourteenth century painter could have made a painting that was the exact negative of the actual image as shown in Figure 2.13. That the image was a negative was not discovered until the late 1800s after the discovery of photography.

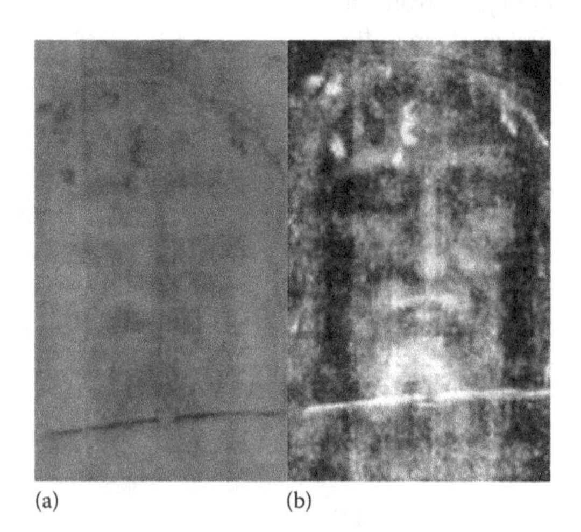

(a) (b)

FIGURE 2.12 Photographs of the Shroud of Turin. (a) positive. (b) negative. Photo from Wikipedia Commons, public domain (Shroud of Turin_2). Original color image converted to grayscale and contrast-enhanced with Adobe Photoshop.®

Another example of recent radiocarbon dating was showing that the mummified human found in the snow and ice in the Alps in 1991—Ötzi, after the region where the body was found—died in 3300 BC. When the body was found, the discoverers thought it was someone who had died recently in a skiing or climbing accident. The artifacts found with the body, such as a 99.7% copper axe, have led some to speculate that he might have been involved with copper smelting and have given anthropologists a great deal of information about the inhabitants of Europe in a period for which no history of the region exists (Ötzi the Iceman).

* This is an example of a steady-state concentration of an intermediate reaction product of two series reactions that will be discussed in Chapter 3.

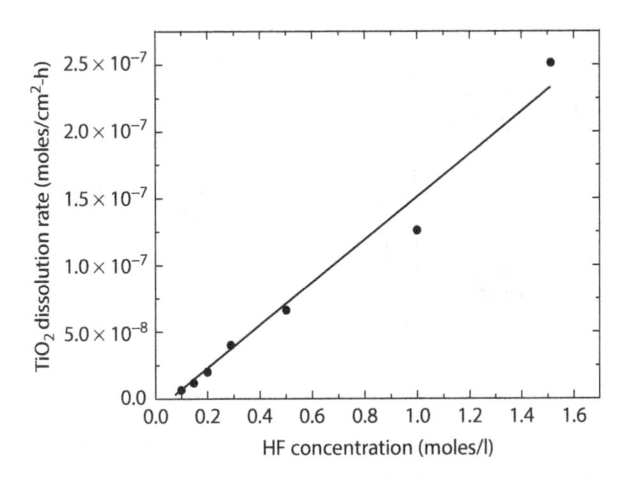

FIGURE 2.13 Linearized rate of TiO_2 dissolution at 95°C in a [HCl] = 0.5 mole/L solution as a function of the [HF]. (Data from Bright and Readey 1987.)

A mammoth tusk found on an island of the Southern California coast is undergoing radiocarbon dating to determine its age and whether or not it is, indeed, a mammoth tusk (Chang 2009). Finally, radiocarbon data were used to find the source of the "brown cloud" of pollution that covers much of South Asia. The question was as follows: *Is the brown cloud due to burning fossil fuels or to burning forests and grasslands?* If fossil fuels were the source, then the ^{14}C content in the soot contributing to the *brown cloud* would be low. If the ^{14}C content were near the normal atmospheric concentration, then the source would be the burning of forests and grasslands. The conclusion was the latter and so those concerned with improving the environment in the area now know which problem to address (Fountain 2009). Another example is the dating of some pages of the Quran, found in a British university library in 2015, to 568–645 about the time when Muhammad received the revelations for the Quran, between 610 and 632, the year he died (Bilefsky 2015). In spite of some controversy over these conclusions and their implications, many feel that this discovery demonstrates that the Quran has existed in its present form since the time of Muhammad.

In summary, radiocarbon dating is a good example of first-order kinetics and shows how these kinetics can be applied to determine the age of carbon-containing materials to address a number of archaeological, anthropological, materials, and societal questions.

2.7 IMPORTANCE OF FIRST-ORDER REACTIONS IN MATERIALS

First-order reactions are important in materials because many gas–solid and liquid–solid reactions are found to be first order. This is likely because the surface or interfacial reactions involve many series steps and it is only a single—perhaps final—first-order step involving one reactant species that determines the rate of reaction. This conclusion is drawn from many empirical results on surface reactions with solids. An example is the dissolution of TiO_2, titanium dioxide—titania, in HF solutions, which is very similar to the dissolution of Al_2O_3 in HF solutions discussed in Section 2.12 (Bright and Readey 1987). The overall reaction is

$$TiO_2(s) + 6HF(solution) \rightleftharpoons TiF_6^{2-}(solution) + 2H^+(solution) + 2H_2O(solution)$$

and from the stoichiometry of the reaction, it would be impossible to predict the order of the reaction. Certainly, 6 HF molecules are not going to simultaneously converge at a single Ti^{4+} surface ion to produce the TiF_6^{2-} ion. The probability of such a multimolecular event is just too low. However, as shown in Figure 2.13, the order of the reaction with respect to the [HF] is very close to 1.0 because the dissolution rate is essentially proportional to the HF concentration, [HF]. These approximate first-order kinetics suggest that the rate-controlling step may be the "squeezing" in of the last HF molecule at a titanium site on the surface to reduce the surface bond between the

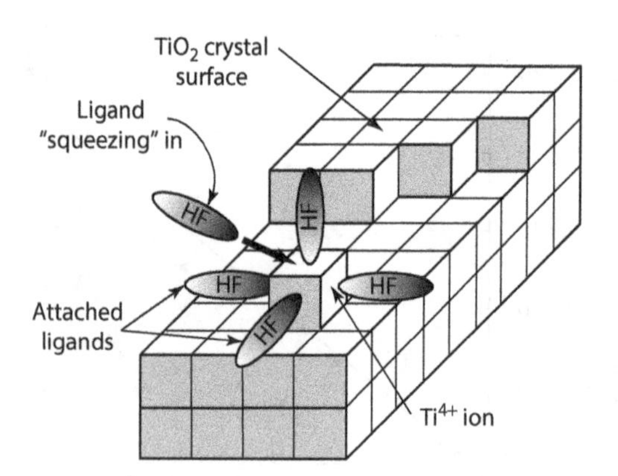

FIGURE 2.14 Postulated possible mechanism for the first-order rate dependence on [HF] for the dissolution of TiO_2 in 0.5 molar HCl solutions at 95°C. The figure shows an isolated Ti^{4+} ion at the surface of the dissolving TiO_2 surrounded by several attached HF ligands (the exact nature of and number of the ligands is not known, possibly F^-) and one final HF trying to "squeeze" into this coordination sphere that would provide the necessary bonding for the Ti^{4+} ligands (TiF_6^{2-}?) to leave the surface and enter the solution. (After Bright and Readey 1987.)

titanium ion—with its other surrounding ligands—to allow it to leave the surface and enter the solution. This is illustrated in Figure 2.14.

2.8 CHAPTER SUMMARY

This chapter introduces several concepts about reaction kinetics. Is the reaction controlled by transport—diffusion—or an actual chemical reaction? Reactions can occur within single phases, homogeneous reactions, or more common in MSE, heterogeneous reactions involving more than one phase: the deposition of Si by CVD and the oxidation of a metal are examples. When discussing actual chemical reaction kinetics, the order of the reaction is important because knowing it can lead to understanding the actual atomistics of the chemical reactions taking place. The dissolution of TiO_2 in HF is presented as a possible example of this. Zero-order reactions are the ones that depend neither on the concentrations of the reactants nor those of the products and are physically not very likely. First-order reactions are important because they occur frequently, particularly for heterogeneous reactions. In this case, the reaction rate is proportional to one of the reactants with the reaction rate constant, k, the proportionality constant. The reaction rate constant follows an Arrhenius equation (Svante Arrhenius), namely, it is exponentially temperature dependent—strongly temperature dependent—and the rate is largely defined by the value of the activation energy, Q. The concepts of fraction reacted, half-life, and relaxation time are introduced for first-order equations. Radioactive decay is used as an example of first-order reactions in radioisotope power sources and radiocarbon dating along with some nuclear chemistry associated with fissionable materials for nuclear reactors. The next chapter goes further into reaction kinetics discussing second-order reactions, multistep reactions including serial and parallel reactions, and reactions that go to equilibrium. In reality, most reactions—even simple gas phase homogeneous reactions such as the formation of water vapor from hydrogen and oxygen—consist of multiple series and parallel reactions that preclude analytic solution and dictate the need for numerical computation of concentrations and reaction rates.

EXERCISES

2.1 The reaction rate constant for a zero-order reaction for the decomposition of another liquid dissolved in water is given by $k = k_0 \exp(-Q/RT)$ for the reaction

$$A \rightarrow B.$$

a. Write the equation for the reaction rate for the decomposition of A.
b. Give the units of k_0 if the units of the reactant concentration [A] are millimol/cm³.

c. If $Q = 30$ kJ/mol and $k_0 = 10^{-2}$, calculate the reaction rate in millimol/liter at 25°F.
d. Calculate how long it takes for the reaction to go to completion if $[A]_0$, the reactant concentration at time zero, is one mole per liter.
e. Neglecting any temperature effects on the density of water, which can be assumed to be 1.0 g/cm^3, calculate the number of moles per cm^3 of water in pure water.
f. Assuming that the molar volumes of water and the reactant are the same, calculate the m/o, w/o, and v/o of A in water at 25°F at the start of the reaction if the molecular weight of the reactant is 60 g/mol.

2.2 a. The reaction rate constant $k\,(k = k_0\exp(-Q/RT))$ for a first-order reaction is $5.00 \times 10^{-3}\,s^{-1}$ at 500°C and is 10.00 s^{-1} and 1000°C. Calculate the activation energy for this reaction.
b. Calculate the preexponential term, k_0, for the reaction rate constant and give its units.
c. Calculate k at 750°C.
d. Calculate the relaxation time for this reaction.
e. Calculate the half-life of this reaction.
f. Make a plot of $\log_{10}$(rate) versus $10^3/T$ (K^{-1}) from 100°C to 1200°C.
g. Calculate how long it takes (hours) at 750°C for the unreacted fraction to be 0.001.
h. If the reaction is $A \rightarrow 2B$, plot the concentrations of A and B versus time from time zero up to the time that was calculated in (g).

2.3 A well-known generalization (for back of the envelope calculations) for biochemical reactions near room temperature (300 K) is that their rates double for every 10 K rise in temperature.
a. Determine the activation energy (J/mol) for such reactions implied by this rule.
b. Calculate the value of the activation energy in eV/molecule.

2.4 a. The half-life for a first-order reaction is 1000 s. Calculate the reaction rate constant (s^{-1}) for this reaction.
b. Calculate the relaxation time (s) for this reaction.

2.5 a. The half-life for a first-order reaction is 1200 s. Calculate the reaction rate constant (s^{-1}) for this reaction.
b. Calculate the relaxation time (s) for this reaction.
c. Plot the fraction unreacted versus time for 0–10,000 s. Indicate the half-life and the relaxation time on your plot.

2.6 a. Plutonium-238, ${}^{238}_{94}Pu$, spontaneously decays by alpha particle, ${}^{4}_{2}He$, emission with a half-life of 86 years. The nuclear disintegration reaction is

$$ {}^{238}_{94}Pu \rightarrow {}^{234}_{92}U + {}^{4}_{2}He + 5.234\,\text{Mev}. $$

This disintegration rate is sufficiently fast to produce enough heat to be used as a power source in satellites. Calculate the first-order reaction rate constant for the nuclear disintegration of plutonium-238.
b. Calculate the number of becquerels at time = 0 with one mole of pluonium-238.
c. Calculate the number of curies at time zero in one mole of pure ${}^{238}_{94}Pu$.
d. Calculate the rate of energy generated in the plutonium-238 (watts) for one mole of plutonium-238 at time zero.
e. Calculate how long (years) it will take for the rate of energy generation to drop to 10% of its initial value.
f. If the ${}^{238}_{94}Pu$ were in the form of PuO_2 rather than pure plutonium, calculate the number of curies in 5 cm^3 of PuO_2 at time zero if the density of PuO_2 is 11.5 g/cm^3.
g. Calculate the number of watts that this 5 cm^3 of ${}^{238}_{94}Pu$-containing PuO_2 generates.
h. If this 5 cm^3 of PuO_2 were perfectly insulated, calculate how long it takes (hours) for the 5 cm^3 of PuO_2 to reach its melting point of 2390°C if its molar heat capacity is 75 J/mol K.

2.7 A reaction obeys the following stoichiometric equation:

$$A + B = Z.$$

The rates of formation of Z at various concentrations of A and B are given in the following table.

[A] (mol/L)	[B] (mol/L)	Rate (mol/L-s)
2.00×10^{-2}	4.20×10^{-2}	3.53×10^{-9}
6.50×10^{-2}	4.80×10^{-2}	4.87×10^{-8}
1.10×10^{-1}	1.22×10^{-1}	9.00×10^{-7}

Determine the values of α, β, and k in the rate equation: rate = $k\,[A]^{\alpha}[B]^{\beta}$. Give the units for k as well. You may use programs such as Excel and Mathematica, or by hand, to get the solution. Just clearly indicate what operations were performed, that is, copy that part of the Excel spreadsheet or the Mathematica page that shows the calculations.

2.8 In 2012, what were thought to be the bones of King Richard III of England were found under a parking lot in England and later confirmed by DNA evidence in 2014 that the bones were indeed his. The bones were dated by radiocarbon dating and confirmed that the date of death of the skeleton was 1485, the date of death of the king. Calculate the ratio of carbon-14 remaining in the bones.

2.9 If the number of disintegrations per minute per gram of carbon produced by pyrolyzing a piece of recently cut wood is 12.5 dis/min g, calculate the fraction of ^{14}C in carbon of age zero if the half-life of ^{14}C is 5730 years. (The wood is assumed to have an age of zero.)

REFERENCES

Bilefsky, D. 2015. Pieces of the Quran, Perhaps as Old as the Faith. *New York Times*, July 23.

Brandes, E. A. and G. B. Brook, eds. 1992. *Smithells Metals Reference Book,* 7th ed. Oxford, UK: Butterworth-Heinemann.

Bright, E. and D. W. Readey. 1987. Dissolution Kinetics of TiO_2 in HF-HCl Solutions. *J. Am. Ceram. Soc.* 70(12): 900–907.

Chang, A. 2009. Possible Mammoth Tusk Found on So Cal Island. *San Diego Tribune*, January 14.

Fountain, H. 2009. Study Pinpoints The Main Source of Asia's Brown Cloud. *New York Times,* January 27.

Haynes, W. M. Editor-in-Chief. 2013. *CRC Handbook of Chemistry and Physics,* 94th ed. Boca Raton, FL: CRC Press.

HSC Chemistry Software. http://www.hsc-chemistry.net/.

Laidler, K. J. and J. H. Meiser. 1999. *Physical Chemistry,* 3rd ed. Boston, MA: Houghton Mifflin.

Levin, E. M., C. R. Robbins, and H. F. McMurdie. 1964. *Phase Diagrams for Ceramists.* Columbus, OH: The American Ceramic Society.

NIST-JANAF Thermochemical Tables. http://kinetics.nist.gov/janaf/.

NIST Kinetics Database. http://kinetics.nist.gov/kinetics.

Ötzi the Iceman. *Wikipedia.* http://en.wikipedia.org/wiki/%C3%96tzi_the_Iceman.

Porter, D. A., K. E. Easterling, and M. Y. Sherif. 2009. *Phase Transformations in Metals and Alloys,* 3rd ed. Boca Raton, FL: Taylor & Francis Group.

Radiocarbon Dating. *Wikipedia.* http://en.wikipedia.org/wiki/Radiocarbon_dating.

Radioisotope Power Systems. http://solarsystem.nasa.gov/rps/types.cfm.

Roine, A. 2002. *Outokumpu HSC Chemistry for Windows,* Ver. 5.11, Thermodynamic software program, Outokumpu Research Oy, Pori, Finland.

Shroud of Turin. *Wikipedia.* http://en.wikipedia.org/wiki/Shroud_of_Turin.

Shroud of Turin_2. Butko, Andrew. 2005. USPD photograph of PD piece of art. https://commons.wikimedia.org/wiki/File:Shroud_of_Turin_001.jpg.

Svante Arrhenius. *Wikipedia.* http://en.wikipedia.org/wiki/Svante_Arrhenius.

Taylor, R. E. 2000. Fifty Years of Radiocarbon Dating. *Am. Sci.* 88(1): 60–67.

Turney, C. 2006. *Bones, Rocks, and Stars.* London: MacMillan.

Willard Libby. *Wikipedia.* http://en.wikipedia.org/wiki/Willard_Libby.

Wilson, I. 1998. *The Blood and the Shroud.* New York: Free Press.

3

Second-Order and Multistep Reactions

3.1 INTRODUCTION

This chapter introduces second-order reactions as examples of higher order reactions. Different initial concentrations of reactants in second-order reactions leads to quite complex results. In contrast, starting with equal initial concentrations greatly simplifies the results and the steps leading to them. This suggests that when trying to deduce the mechanism of the reaction from the kinetics, it is much easier if the initial concentrations are equal. This is also true when back reactions are considered. Reactions with back reactions go to equilibrium—the usual case for most reactions—and not to completion. Analytical solutions for first-order forward and back reactions are quite easily obtained. However, things become much more mathematically tedious when the forward or back reaction, or both, is a second-order reaction. Nevertheless, modeling a second-order equation that goes to equilibrium is useful largely because such solutions are rarely found elsewhere. The chapter then examines parallel reactions demonstrating that the faster or fastest step in parallel reactions determines the overall reaction rate and product concentrations. For series reactions, it is shown that the situation is the opposite in that the slower or slowest reaction step controls the overall reaction rate. These results are important and apply to diffusion controlled processes as well as reaction control. In addition, the steady-state approximation for two reactions in series has several important applications. Finally, most reactions depend on a combination of several series and parallel steps that are best solved by numerical techniques rather than attempting to obtain analytical expressions for the changes in concentrations with time and temperature.

3.2 SECOND-ORDER REACTIONS

3.2.1 One Reactant

Again, assume the simple reaction of $A \xrightarrow{k} B$ but in this case, the reaction is second order with respect to [A]; that is,

$$\frac{d[A]}{dt} = -\frac{d[B]}{dt} = -k[A]^2 \tag{3.1}$$

this integrates to

$$-\frac{1}{[A]} = -kt + C$$

and if $[A] = [A]_0$ when $t = 0$, then

$$\frac{1}{[A]} = \frac{1}{[A]_0} + kt = \frac{1 + [A]_0 kt}{[A]_0}$$

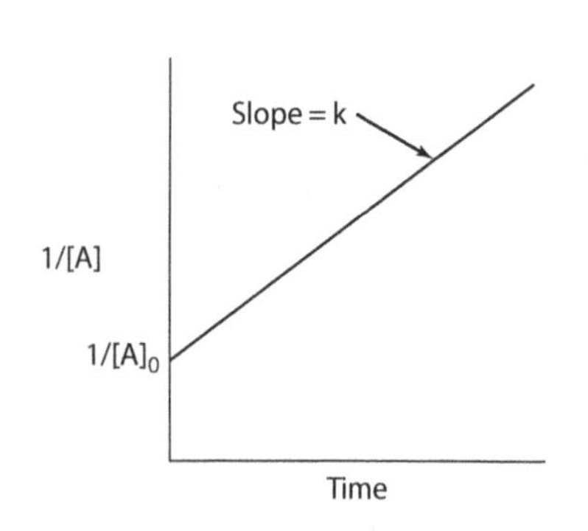

FIGURE 3.1 1/[A] versus time for a second-order reaction.

which has the form $y = mx + b$, and is linear when plotted as 1/[A] versus temperature as is shown in Figure 3.1. Rearranging to obtain [A] explicitly,

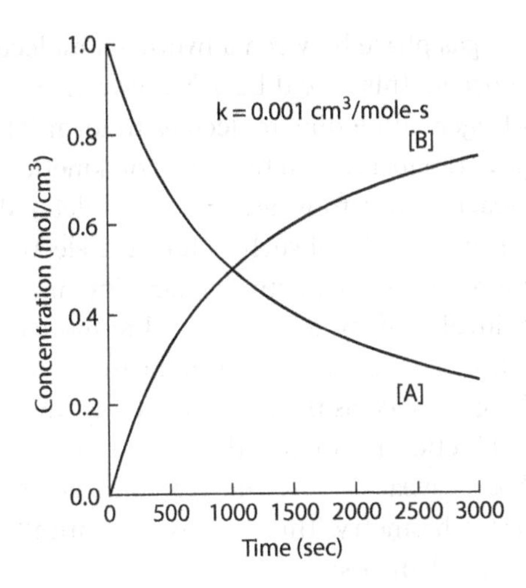

FIGURE 3.2 Concentrations versus time for a second-order reaction of A → B that will eventually go to completion for $k = 10^{-3}\ cm^3/mols$.

$$[A] = \frac{[A]_0}{1 + [A]_0 kt} \tag{3.2}$$

as shown in Figure 3.2. If $[B]_0 = 0$, then [B] is the amount of A reacted, so

$$[B] = [A]_0 - [A] = [A]_0 - \frac{[A]_0}{1 + [A]_0 kt}$$

or,

$$[B] = \frac{[A]_0^2 kt}{1 + [A]_0 kt} \tag{3.3}$$

also shown in Figure 3.2 with $k = 10^{-3}\ cm^3/mol\ s$.*

3.2.2 Two Reactants

Assume that there are two reactants, A and B reacting to give a product, C:

$$A + B \xrightarrow{k} C \tag{3.4}$$

where C can be one or more products. Also assume that the reaction is first order with respect to both A and B resulting in overall second-order kinetics:

$$\frac{d[A]}{dt} = -k[A][B] \tag{3.5}$$

and is the case for the reaction of hydrogen and iodine to produce hydrogen iodide mentioned in Chapter 2†: $H_2(g) + I_2(g) \rightleftharpoons 2HI(g)$ and

$$\frac{d[H_2]}{dt} = -k[H_2][I_2].$$

For this reaction, there is a relation between the stoichiometry of the reaction and the kinetics because one mole of hydrogen gas and one mole of iodine gas react to give two moles of hydrogen iodide gas, and the kinetics are first order with respect to each of the reactants. This is not

* Note that the units of k in this case are Units(k) = $cm^3/mol\ s$ or $m^3/mol\ s$ or $L/mol\ s$.

† However, some references (Laidler 1987) suggest that the kinetics for this reaction are more complex than simple second order.

unexpected if a collision in the gas phase between a hydrogen molecule and an iodine molecule is necessary for the reaction to occur. This would be a *bimolecular* reaction and, most importantly, the collision between the hydrogen and iodine molecules to form HI would represent the *mechanism* of the reaction, which is a principal reason that reaction kinetics are studied: to determine the mechanism or model of the reaction rate to whatever level of detail desired, as a function of temperature and concentration or in more detail such as how the electron orbitals interact. However, a simple relation between the reaction stoichiometry, kinetics, and mechanism are not the usual case and most reactions may involve many steps that lead to more complex reaction kinetics. For example, the very similar reaction of hydrogen and bromine to produce hydrogen bromide does not follow simple second-order kinetics such as the formation of hydrogen iodide (Moore and Pearson 1981). Nevertheless, reaction kinetics are frequently studied to determine the molecular mechanism of the reaction but with the caveat (Moore and Pearson 1981, 3) "that the mechanism cannot be predicted from the overall stoichiometry. This point is emphatically reiterated in the case of gas-phase formation of water from its elements:

$$2H_2(g) + O_2(g) \rightarrow 2H_2O(g).$$

The reaction certainly does not involve simultaneous, trimolecular collisions of $2H_2$ with O_2. As many as 40 elementary steps have been suggested for the mechanism, and about 15 steps are needed to account for the slow reaction under simplified conditions". The formation of HBr, HCl, and H_2O, all of which exhibit more complex reaction kinetics will be examined later.

Back to the reaction in Equation 3.4,

$$\frac{d[A]}{dt} = \frac{d[B]}{dt} = -\frac{d[C]}{dt} \tag{3.6}$$

where the concentrations of A and B can be expressed as

$$\begin{aligned} [A] &= [A]_0 - ([A]_0 - [A]) \\ [B] &= [B]_0 - ([B]_0 - [B]) \\ &= [B]_0 - ([A]_0 - [A]) \\ [B] &= ([B]_0 - [A]_0) + [A] \end{aligned} \tag{3.7}$$

because the amount of B produced must be the same as the amount of A reacted at any given time and the initial concentrations of A and B at time = 0 are $[A]_0$ and $[B]_0$, respectively. If the initial concentrations are the same, $[B]_0 = [A]_0$, then Equation 3.5 becomes Equation 3.1 and $[A] = [B]$ are given by Equation 3.2. If they are not equal, things become a little bit more complex but the equations can easily be integrated to give (Appendix A.1.1, where it is assumed that $[B]_0 > [A]_0$ simply for convenience while doing the algebra)

$$[A] = \frac{([B]_0 - [A]_0)[A]_0 e^{-([B]_0 - [A]_0)kt}}{[B]_0 - [A]_0 e^{-([B]_0 - [A]_0)kt}} \tag{3.8}$$

and

$$[B] = ([B]_0 - [A]_0)\left\{\frac{[B]_0}{[B]_0 - [A]_0 e^{-([B]_0 - [A]_0)kt}}\right\} \tag{3.9}$$

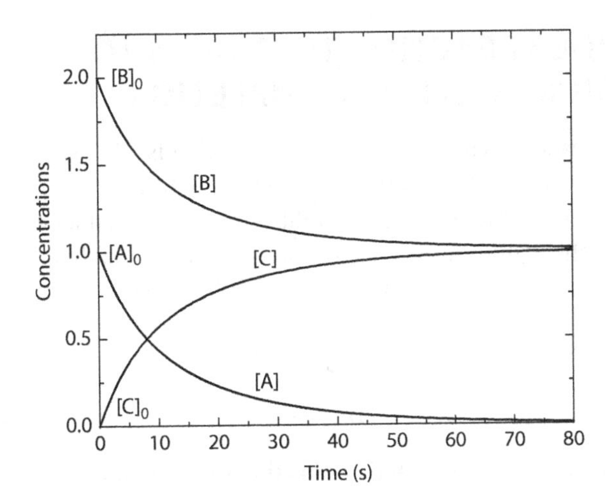

FIGURE 3.3 Second-order reaction with two reactants A and B going to completion—all of A is reacted—to C. Note that there are no units on concentration in this and subsequent figures in this chapter because they depend either on the units of k or on the units of concentration chosen which need not be specified.

and, of course, from Equation 3.6

$$[C] = [A]_0 - [A]. \tag{3.10}$$

These three concentrations are plotted in Figure 3.3 for initial values of $[B]_0 = 2.0$, $[A]_0 = 1.0$, and $[C]_0 = 0$ with $k = 0.05$. (Note: the Units(k) = Units([A]/s) so in Figure 3.3 [and subsequent figures] the "concentration" axis is whatever units of concentration chosen and that will determine the units of k. For example, in Equation 3.8 Units([A]) = mol/cm^3, then Units(k) = cm^3/mol s. If the Units([A]) = mol/m^3, then Units(k) = m^3/mol s, etc.)

There are two extreme possibilities for this result: $[B]_0 \cong [A]_0$, that is, starting concentrations of the two reactants are essentially the same, and Equations 3.8 and 3.9 give the expected result, Equation 3.2, as shown in Appendix A.1.2:

$$[A] \cong \frac{[A]_0}{1 + [A]_0 kt}.$$

3.2.3 Pseudo First-Order Reaction

The other possibility is that $[B]_0 \gg [A]_0$, that is, the starting concentration of B is much larger than that of A, and Equation 3.8 gives

$$[A] \cong [A]_0 e^{-[B]_0 kt} \tag{3.11}$$

which is referred to as a *pseudo* first-order reaction (Houston 2001). For example, if the HI reaction were being carried out with a great excess of hydrogen compared to the concentration of iodine, then the hydrogen concentration essentially does not change and leads to apparent first-order kinetics but with a modified reaction rate constant, $k[B]_0$.

When experimentally determining the reaction kinetics, the reaction rate constant, and the reaction mechanism, it clearly makes sense to start with the same initial concentrations of A and B to make the reaction kinetics simpler and easier to analyze. However, there may be constraints in an industrial process, in which A and B are—by necessity—not equal. In that case, the more complex kinetics of Equations 3.8 through 3.10 must be used, and the main reason that they are included here is to raise awareness of this possibility.

3.3 FIRST-ORDER REACTION THAT GOES TO EQUILIBRIUM AND NOT COMPLETION

More often than not, a reaction will not go to completion, that is, $[A] \rightarrow 0$, but will reach some equilibrium concentration.* A first-order reaction in which $A \rightarrow B$ that does not go to completion will eventually reach some concentration, $[A]_e$, in equilibrium with a product concentration, $[B]_e$, where the subscript e denotes the equilibrium concentrations. This implies that there is a back reaction of $B \rightarrow A$, that could also be first order (but does not have to be), so

$$\begin{aligned} A &\xrightarrow{k_1} B \\ B &\xrightarrow{k_2} A \end{aligned} \tag{3.12}$$

and the rate of reaction is the forward reaction plus the back reaction to give $d[A]/dt$,

$$\frac{d[A]}{dt} = -k_1[A] + k_2[B]. \tag{3.13}$$

Again, for the sake of simplicity, assume that at $t = 0$, $[B]_0 = 0$ and $[A] = [A]_0$ so that $[B] = [A]_0 - [A]$. The rate of reaction Equation 3.13 becomes

$$\begin{aligned} \frac{d[A]}{dt} &= -k_1[A] + k_2[A]_0 - k_2[A] \\ &= -(k_1 + k_2)[A] + k_2[A]_0 \\ &= -(k_1 + k_2)\left([A] - \frac{k_2}{k_1 + k_2}[A]_0\right). \end{aligned} \tag{3.14}$$

At equilibrium, the reaction rate must go to zero, so

$$\frac{d[A]}{dt} = 0 = -k_1[A]_e + k_2[B]_e. \tag{3.15}$$

Rearranging Equation 3.15 gives

$$\frac{[B]_e}{[A]_e} = \frac{[A]_0 - [A]_e}{[A]_e} = \frac{k_1}{k_2} = K_e \tag{3.16}$$

because $[B] = [A]_0 - [A]$, the amount of B is the amount of A reacted, and where K_e is the equilibrium constant for the reaction, which—of course—can be calculated from the thermodynamics for the reaction. Solving Equation 3.16 for $[A]_0$,

$$k_2[A]_0 - k_2[A]_e = k_1[A]_e$$

$$k_2[A]_0 = (k_1 + k_2)[A]_e$$

or

$$[A]_0 = \frac{k_1 + k_2}{k_2}[A]_e$$

* Nuclear decay being an exception in that the reactions *do* go to completion; that is, the concentration of the reactant goes to zero.

and substituting this for $[A]_0$ in Equation 3.14 gives

$$\frac{d[A]}{dt} = -k_1[A] + (k_1 + k_2)[A]_e - k_2[A]$$
$$\frac{d[A]}{dt} = -(k_1 + k_2)([A] - [A]_e). \tag{3.17}$$

Let $x = [A] - [A]_e$, so that Equation 3.17 takes the form $dx/x = -(k_1 + k_2)dt$ and integrating

$$\int_{x_0}^{x} \frac{1}{x} dx = \ln x - \ln x_0 = \ln\left(\frac{x}{x_0}\right) - (k_1 + k_2)t$$

and noting that $[A] = [A]_0$ when $t = 0$ gives

$$\frac{[A] - [A]_e}{[A]_0 - [A]_e} = e^{-(k_1 + k_2)t} \tag{3.18}$$

which is the rather nonsurprising result—after a lot of fun algebra—that the *reaction stops when equilibrium is reached* and reverts to the equation for a first-order reaction that goes to completion when $[A]_e = 0$, Equation 2.5, $[A] = [A]_0 e^{-kt}$. Of course $k_1 = k_2K_e$ from Equation 3.16 and $k_1 + k_2$ can be written as just k and the time dependence is again exponential, as shown in Figure 3.2. It is perhaps surprising that k should be a constant, that is, independent of concentration.

What this means is that the original equation for the reaction rate *must include* the equilibrium concentration in the differential equation when the reaction goes to equilibrium (which it almost always does with the notable exception of nuclear decay) because the reaction must stop at equilibrium. As a result, the original first-order rate expression could have been written more generally as

$$\frac{d[A]}{dt} = -k([A] - [A]_e) \tag{3.19}$$

and give the same result, Equation 3.18.

In general, the concentration of B, [B], is the initial concentration of B, $[B]_0$, plus the amount of A reacted to form B, $[A]_0 - [A]$, so

$$[B] = [B]_0 + ([A]_0 - [A]) = [B]_0 - ([A] - [A]_0)$$

and makes the equilibrium concentration of B,

$$[B]_e = [B]_0 + ([A]_0 - [A]_e) \quad \text{or} \quad ([B]_0 - [B]_e) = -([A]_0 - [A]_e)$$

and

$$[B] - [B]_e = [B]_0 + ([A]_0 - [A]) - [B]_0 - ([A]_0 - [A]_e)$$
$$[B] - [B]_e = -([A] - [A]_e)$$

so the equivalent expression for [B] can be obtained simply by substituting these relations between [B] and [A] in Equation 3.18 to give

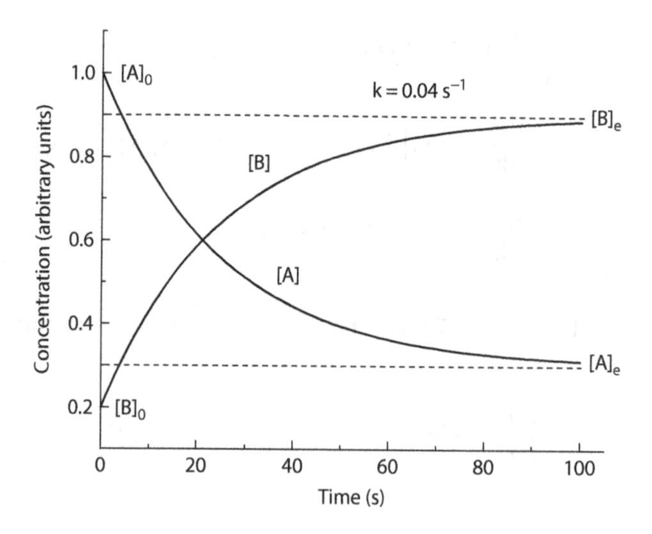

FIGURE 3.4 First-order reaction that goes to *equilibrium*—one that has a first-order back reaction.

$$\frac{[B]-[B]_e}{[B]_0-[B]_e}=e^{-kt}. \tag{3.20}$$

Furthermore, Equation 3.13 can be written more generally as

$$\begin{aligned}\frac{d[A]}{dt}&=-k_1([A]-[A]_e)+k_2([B]-[B]_e)\\&=-k_1([A]-[A]_e)-k_2([A]-[A]_e)\\&=-(k_1+k_2)([A]-[A]_e)\end{aligned} \tag{3.21}$$

and Equation 3.19 is obtained as before. Also, Equation 3.21 shows immediately that $k = k_1 + k_2$. Figure 3.4 shows the concentrations as a function of time for a first-order reaction that goes to equilibrium.

Equations 3.19 and 3.21 show that the rate of reaction, d[A]/dt is equal to the product of a *kinetic factor*, k, and a *thermodynamic factor*, $([A] - [A]_e)$. The latter is a thermodynamic factor because it depends on the equilibrium value of [A], namely $[A]_e$, that can be manipulated by the overall composition of the fluid phase while keeping k fixed. The reaction rate constant is, of course, exponentially temperature dependent. This product of a kinetic factor and a thermodynamic factor whose value determines the overall rate of a reaction or phase transformation will appear repeatedly in discussions of other kinetic processes.

3.4 SECOND ORDER ⇔ SECOND ORDER THAT GOES TO EQUILIBRIUM

3.4.1 General

Whether the reaction is a second order forward and second order back, or first order forward and second order back, or second order forward and first order back, the results are all similar and the procedures used to get the solutions are essentially the same for all three cases. The reason for this is that the rates of reaction, dx/dt (x here is concentration to minimize symbol complexity), are all proportional to a second-order polynomial in x. From the results in Section 3.3.2, it is tempting to write for a second-order reaction,

$$\frac{d[A]}{dt} = -k\left([A]-[A]_e\right)^2. \tag{3.22}$$

Unfortunately, this is *incorrect* and solutions to second-order equations that go to equilibrium unfortunately are more complex.

The detailed procedure carried out in Appendix A.2 works for second order ⇔ second order, second order ⇔ first order, and first order ⇔ second order because all involve the solution of quadratic equations. In Appendix A.2.4, the general solution for second order ⇔ second order is found. The reaction is assumed to be the simple reaction:

$$A \rightleftharpoons B$$

or

$$\frac{d[A]}{dt} = -k_1[A]^2 + k_2[B]^2. \tag{3.23}$$

To minimize the notation, let $[A] = x$, and $[A]_0 = x_0$ and $[B]_0 = 0$ so that Equation 3.23 becomes

$$\frac{dx}{dt} = -k_1x^2 + k_2\left(x_0 - x\right)^2. \tag{3.24}$$

Now this is a relatively simple differential equation to solve because the variables are separable, all the "xs" on the left, and "ts" on the right, so that standard integration gives a solution without recourse to special differential equation solution techniques.

3.4.2 Equilibrium

At equilibrium, $x = x_e$ so Equation 3.24 becomes

$$-k_1x_e^2 + k_2\left(x_0 - x_e\right)^2 = 0$$

or

$$k_1x_e^2 = k_2\left(x_0 - x_e\right)^2$$

$$\pm\sqrt{\frac{k_1}{k_2}} = \frac{x_0 - x_e}{x_e}.$$

Note the ± sign in front of the term on the left that is necessary because the square root is being taken. This leads to two mathematically possible equilibrium values as will be seen. At this point, the initial concentration of A, x_0, could be obtained and substituted into Equation 3.24 and eliminated as was done for the first-order reaction in Section 3.3. However, this does not help much to solve this problem. In any event, the equilibrium concentration can be obtained from

$$x_e \pm \sqrt{\frac{k_1}{k_2}}x_e = x_0$$

$$x_e\left(\frac{\sqrt{k_2} \pm \sqrt{k_1}}{\sqrt{k_2}}\right) = x_0$$

which leads to

$$x_e = x_0\left(\frac{\sqrt{k_2}}{\sqrt{k_2} \pm \sqrt{k_1}}\right) \tag{3.25}$$

and the algebra is already getting messy and the differential equation, Equation 3.24, has yet to be solved. The general solution, involving much algebra, is left to the Appendix.

3.4.3 Specific Example

To make things simpler, and hopefully more transparent, start with a specific case of $k_1 = 2$, $k_2 = 1$, and $x_0 = 1$. So Equation 3.25 becomes

$$x_e = x_0\left(\frac{\sqrt{k_2}}{\sqrt{k_2} \pm \sqrt{k_1}}\right) = 1.0\left(\frac{1.0}{1.0 \pm \sqrt{2}}\right)$$

which gives $x_e = 0.41421$ and -2.41421. Clearly, x_e must be greater than zero so the first $x_e = 0.41421$ must be the equilibrium value. (Note that if $k_2 > k_1$, then $x_e > x_0$, which is physically impossible if $[B]_0 = 0$ as assumed.) The differential equation, Equation 3.24 with these numerical values, becomes,

$$\frac{dx}{dt} = -2x^2 + (1-x)^2$$

$$\frac{dx}{dt} = -2x^2 + 1 - 2x + x^2 \tag{3.26}$$

$$\frac{dx}{dt} = -x^2 - 2x + 1.$$

This is easy to solve by finding the roots to the quadratic equation on the right and one of these roots will be the equilibrium concentration of A from which the equilibrium concentration of B can be calculated. So

$$x = -\frac{b \pm \sqrt{b^2 - 4ac}}{2a}$$

or

$$x = \frac{2 \pm \sqrt{4+4}}{-2}$$

$$x = -2.41421 \text{ and } 0.41421.$$

Not surprisingly, the roots of the quadratic equation are the equilibrium concentrations. As a result, Equation 3.26 can be written in terms of its factors:

$$\frac{dx}{dt} = (2.41421 + x)(0.41421 - x). \tag{3.27}$$

Just as a check, multiplying these two factors gives

$$2.41421 \times 0.41421 - 2.41421x + 0.41421x - x^2$$

$$0.99999 - 2x - x^2$$

which is pretty close to $1 - 2x - x^2$! The integration of Equation 3.27 is now straight forward. First, separate variables:

$$\frac{dx}{(2.41421 + x)(0.41421 - x)} = dt. \tag{3.28}$$

Now separate the denominator into two terms that are easily integrated, the method of *partial fractions,*

$$\frac{1}{(2.41421+x)(0.41421-x)}=\frac{A}{(2.41421+x)}+\frac{B}{(0.41421-x)}$$

multiplying both sides by the denominator of the left-hand side

$$A(0.41421-x)+B(2.41421+x)=1$$

and equate the left- and right-hand factors of x and the constant gives $-Ax+Bx=0$ or $A=B$ because there is no x-term on the right-hand side, and $0.41421A+2.41421A=2.82842A=1$ or $A=1/2.82842=B$. Equation 3.28 can now be written as

$$\frac{dx}{(2.41421+x)}+\frac{dx}{(0.41421-x)}=2.82842\,dt$$

and integrated to

$$\ln(2.41421+x)-\ln(0.41421-x)=2.82842t+C$$

where C is just a constant of integration. Taking the exponent of both sides of the equation gives

$$\frac{(2.41421+x)}{(0.41421-x)}=C'e^{2.82842t}$$

where C' is just another constant. When $t=0$, $x=x_0$, so

$$C'=\frac{(2.41421+x_0)}{(0.41421-x_0)}=\frac{(2.41421+1)}{(0.41421-1)}=-5.8284$$

and the solution becomes

$$\frac{(2.41421+x)}{(0.41421-x)}=-5.8284\times e^{2.82842t}=D \tag{3.29}$$

where at any given time t the right-hand side of this equation is a constant D. It is now necessary to manipulate Equation 3.29 to get "x" explicitly. So, Equation 3.29 can be written as

$$(2.41421+x)=D\times(0.41421-x)$$

$$2.41421+x=0.41421\times D-Dx$$

$$x(1+D)=0.41421\times D-2.41421$$

$$x=\frac{0.41421\times D-2.41421}{1+D}$$

$$x=\frac{0.41421-2.41421/D}{1/D+1}$$

and replacing Equation 3.29 for D gives the sought for solution

$$x=\frac{0.41421+0.41421\times e^{-2.82842t}}{1-0.17157\times e^{-2.82842t}} \tag{3.30}$$

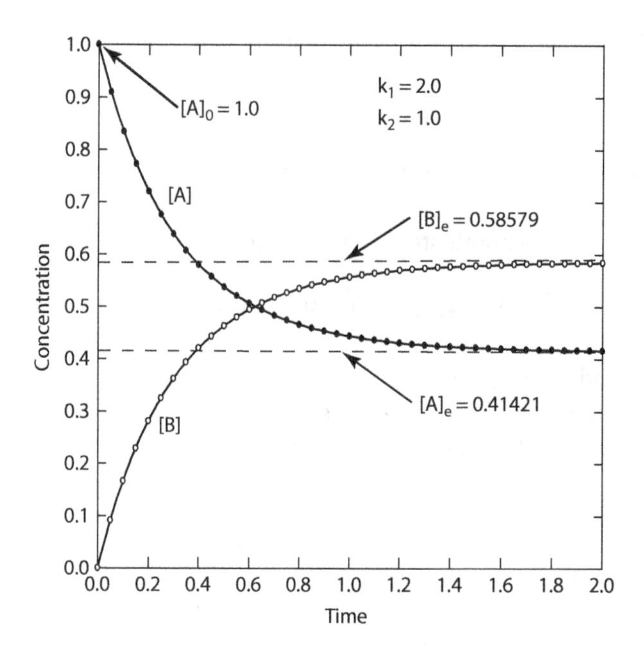

FIGURE 3.5 Second-order reaction with a second-order back reaction $A \rightleftharpoons B$ with the specific values for the constants in the equation: $[A]_0 = 1.0$, $[B]_0 = 0$, $k_1 = 2.0$, and $k_2 = 1.0$. The calculated points and the equilibrium concentrations are shown.

when $t = 0$, $x = 0.99999$ which is pretty close to 1.0 and when $t = \infty$, $x = x_e = 0.41421$. The concentration of B, [B], is just $[B] = x_0 - x$ and $[B]_e = 0.58579$. Both [A] and [B] are plotted in Figure 3.5.

3.4.4 The General Solution

Following exactly the same procedure as in Section 3.4.3 for the specific case, the general solution to Equation 3.24 is solved in Appendix A.2.4 and is

$$x = \frac{x_0\sqrt{k_2}/\left(\sqrt{k_2}+\sqrt{k_1}\right)\left\{1+e^{-\left(2x_0\sqrt{k_2k_1}\right)t}\right\}}{1-\left(\left(\sqrt{k_1}-\sqrt{k_2}\right)/\left(\sqrt{k_1}+\sqrt{k_2}\right)\right)e^{-\left(2x_0\sqrt{k_2k_1}\right)t}}. \tag{3.31}$$

3.4.5 Numerical Solution

Given the amount of algebra and elementary calculus involved in obtaining the general solution to a relatively simple differential equation of importance to kinetics, what about higher order reactions and noninteger orders? In general, it is generally neither reasonable nor possible to solve such problems analytically—at least not in the time it might take—so numerical techniques are used. As simple example of one of these numerical procedures is the so-called *Euler's method* (Celia and Gray 1992; Blanchard et al. 2012) that involves treating the derivative as a ratio of differences: $dx/dt = \Delta x/\Delta t$. This is an example of a *finite difference approximation* that can be used to solve both ordinary and partial differential equations under certain conditions (Celia and Gray 1992; Kee et al. 2003). It is a particularly useful technique because the solutions can frequently be obtained with simple spreadsheet calculations.

For example, consider the specific differential equation given in Equation 3.26

$$\frac{dx}{dt} = -x^2 - 2x + 1 \tag{3.32}$$

which required a fair amount of algebra and calculus to get the final result, Equation 3.30 with the initial condition that $x = 1.0$ at $t = 0$. Putting Equation 3.32 into *difference* form,

$$\frac{\Delta x}{\Delta t} = \frac{(x_{t+\Delta t} - x_t)}{\Delta t} = -x_t^2 - 2x_t + 1$$

and rearranging gives

$$x_{t+\Delta t} = x_t + \Delta t\left(-x_t^2 - 2x_t + 1\right). \tag{3.33}$$

That is, for a uniform time-step, Δt, the value of $x_{t+\Delta t}$ at the next time-step, is just the value at the previous time-step, x_t, plus Δt times the value of the differential equation at the previous time-step. For the plots of [A] and [B] in Figure 3.5 from Equation 3.30, a relatively large time-step of $\Delta t = 0.05$ s was used to calculate these curves. This is somewhat larger than desirable for the evaluation by Euler's method but it will be used to illustrate the method. So, because $x_0 = 1.0$, x_1 at Δt is calculated by

$$x_1 = x_0 + \Delta t\left(-x_0^2 - 2x_0 + 1\right)$$
$$x_1 = 1.0 + 0.05(-1 - 2 + 1)$$
$$x_1 = 0.9$$

and x_2 at $t = 2\Delta t$,

$$x_2 = x_1 + \Delta t\left(-x_1^2 - 2x_1 + 1\right)$$
$$x_2 = 0.9 + 0.05(-(0.9)^2 - 2(0.9) + 1)$$
$$x_2 = 0.9 - 0.0805$$
$$x_2 = 0.8195$$

and so on. The values of x(t) at these two times calculated with the exact solution, Equation 3.30, are $x_1 = x(0.05) = 0.9092$ and $x_2 = x(0.1) = 0.8342$, less than 2% difference between the numerical and exact solution. This difference could be made considerably smaller with a smaller Δt. Figure 3.6 shows how the finite difference approximation can be easily and quickly calculated with a spreadsheet. The first column is just n increasing going down the column as far as desired; the second column is time, $n\Delta t$, where Δt is fixed (but does not have to be); and column three, x is calculated with Equation 3.33. A powerful feature of a spreadsheet solution is only the first row for n = 0 needs to be inserted and then simply copied vertically down to a desired end point. Also, Δt can be introduced as a separate constant and varied to get as close as possible to the analytical solution. Figure 3.7 gives plots of [A]

$\frac{dx}{dt} = -x^2 - 2x + 1$ Differential equation

$x_{n+1} = x_n + \Delta t(-x_n^2 - 2x_n + 1)$ Differential equation

n	Time	x	
0	0	x_0	
1	$1\Delta t$	x_1	$x_1 = x_0 + \Delta t(-x_0^2 - 2x_0 + 1)$
2	$2\Delta t$	x_2	
⋮	⋮	⋮	
n	$n\Delta t$	x_n	
n+1	$(n+1)\Delta t$	x_{n+1}	$x_{n+1} = x_n + \Delta t(-x_n^2 - 2x_n + 1)$
⋮	⋮	⋮	

Spreadsheet columns

FIGURE 3.6 Spreadsheet layout for the numerical calculation of the concentrations as a function of time, for the specific reaction equation plotted in Figure 3.5, with Euler's finite difference method of approximation. The values of x_n from the difference equation are plotted in the x column.

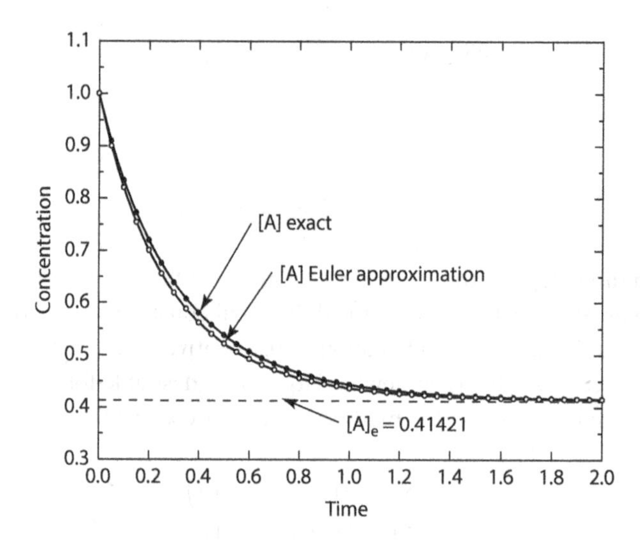

FIGURE 3.7 The exact concentration [A] from Figure 3.5 plotted along with Euler's method approximation of [A] with the time-step $\Delta t = 0.05$. The calculated points are shown for both curves. Although a smaller time-step would have made the approximation closer to the analytical solution, even with this relatively large time-step, the curves do not differ by more than 2% at any given time.

for both the exact and a spreadsheet numerical solutions to Equation 3.32 with $\Delta t = 0.5$. The two solutions are pretty close and get closer at longer times because, in this case, the derivative is getting smaller. Clearly, the numerical solution is the easier and faster way to solve this relatively simple problem and numerical solutions become mandatory as the kinetics become more complex, as indeed they do, even for simple reactions.

3.5 PARALLEL AND SERIES REACTIONS

3.5.1 Introduction

Even more complex situations can be modeled such as parallel or series reactions, any order, with or without back reactions, with and without initial product concentrations, and so on, depending on the level of enthusiasm for doing algebra and solving ordinary differential equations! However, a brief look at simple first-order parallel and series reactions is useful for reasons other than just obtaining a solution.

3.5.2 Parallel Reactions

Consider first a parallel reaction such that A can react to form either B or C. As mentioned in Section 3.1, this leads to the important result that the *faster* of two parallel reactions will determine how quickly the reaction goes to completion and how much B and C are formed. For this parallel reaction, assume no back reactions and that the initial concentrations of B and C are both zero ($[B]_0 = [C]_0 = 0$) to avoid hiding the principles in the forest of the notation:

$$A \xrightarrow{k_1} B \text{ and } A \xrightarrow{k_2} C \quad \text{so} \quad \frac{d[B]}{dt} = k_1[A] \text{ and } \frac{d[C]}{dt} = k_2[A] \text{ so} \tag{3.34}$$

$$\frac{d[A]}{dt} = -\frac{d[B]}{dt} - \frac{d[C]}{dt} = -(k_1 + k_2)[A]$$

and therefore

$$[A] = [A]_0 e^{-(k_1+k_2)t}. \tag{3.35}$$

Substitution of Equation 3.35 in the rate expression for B gives

$$\frac{d[B]}{dt} = k_1[A] = k_1[A]_0 e^{-(k_1+k_2)t}$$

and integrating $\int_0^{[B]} d[B] = k_1[A]_0 \int_0^t e^{-(k_1+k_2)t} dt$ to give

$$[B] = \frac{k_1}{k_1+k_2}[A]_0 \left\{1 - e^{-(k_1+k_2)t}\right\}. \tag{3.36}$$

Similarly, $\frac{d[C]}{dt} = k_2[A] = k_2[A]_0 e^{-(k_1+k_2)t}$ and is integrated to give

$$[C] = \frac{k_2}{k_1+k_2}[A]_0 \left\{1 - e^{-(k_1+k_2)t}\right\} \tag{3.37}$$

which clearly shows that the faster reaction—the larger k—controls the rate of reaction and determines which product is preferentially created, as shown in Figure 3.8. This is a very general result: for parallel reactions the *faster* (fastest if more than two products) controls the overall rate of the reaction. For the k-values used to plot Figure 3.8, [C] is always about 20% of [B].

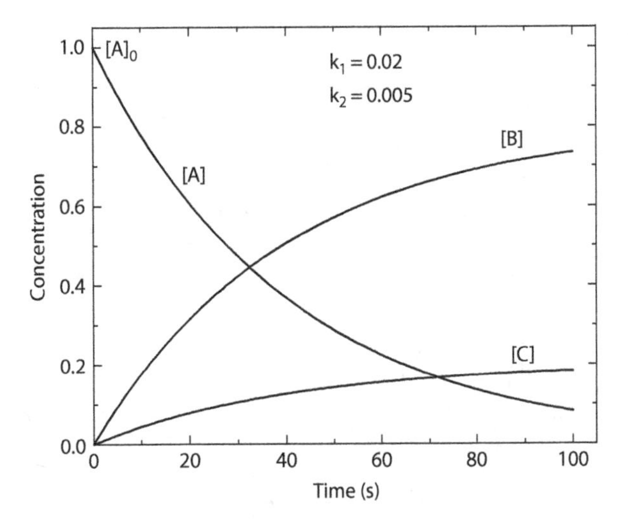

FIGURE 3.8 Concentrations as a function of time for parallel reactions that go to completion: A → B and A → C. For parallel reactions, the *faster* (fastest) of the rates, k_i, determines the rate of the overall reaction and the relative amount of the major product, B.

3.5.3 Series Reactions

3.5.3.1 Exact Solution

Consider a series reaction $A \xrightarrow{k_1} B \xrightarrow{k_2} C$ in which the *slower* of the two reactions—the smaller of k_1 and k_2—determines the overall rate, or [C](t). The rates of reactions are given by

$$\frac{d[A]}{dt} = -k_1[A]$$
$$\frac{d[B]}{dt} = k_1[A] - k_2[B] \tag{3.38}$$
$$\frac{d[C]}{dt} = k_2[B].$$

The solutions to these three simultaneous ordinary differential equations are given in many physical chemistry texts without the mathematical details that require some techniques for solving differential equations (but given in Appendix A.3):

$$[A] = [A]_0 e^{-k_1 t} \tag{3.39}$$

$$[B] = \frac{[A]_0 k_1}{k_2 - k_1}\left(e^{-k_1 t} - e^{-k_2 t}\right) \tag{3.40}$$

$$[C] = [A]_0\left\{1 + \frac{1}{k_2 - k_1}\left(k_1 e^{-k_2 t} - k_2 e^{-k_1 t}\right)\right\}. \tag{3.41}$$

Figure 3.9 shows a case where $k_1 \gg k_2$ so A quickly forms B, which reacts more slowly to form C. Figure 3.10 shows the opposite case where $k_2 \gg k_1$ so that hardly any B is present at any time because it quickly reacts to form C. In Appendix A.3.5, it is shown that it is the smaller of the two "ks"—the slower reaction—controls the overall reaction rate. Note that [C](t) does not depend on which of the two rates is the faster, which is obvious from Equation 3.41 because the reversal of k_1 and k_2 gives the same rate equation for [C]. For example, with $k_1 = 0.05$ and $k_2 = 0.3$ as in Figure 3.10, Equation 3.41 becomes

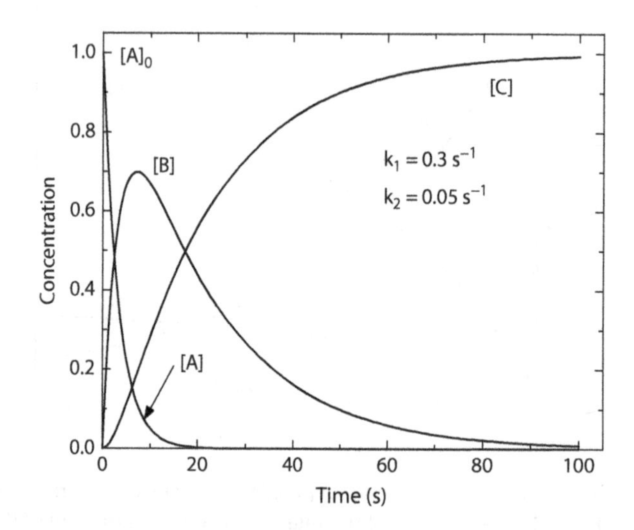

FIGURE 3.9 Series reaction in which the reaction of A to B, k_1, is considerably faster than the reaction of B to C, k_2. Because the rate of reaction of A to B is fast, A quickly reacts to form B and it is the slow reaction of B going to C that controls the reaction. As a result, the slower controls the rate of the overall reaction to C.

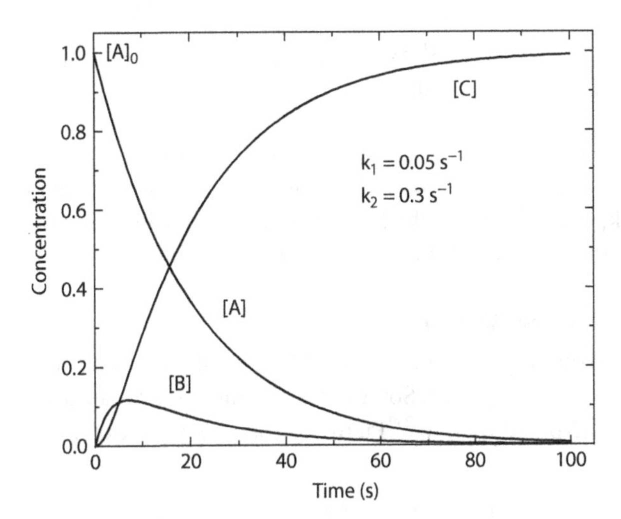

FIGURE 3.10 Same series reaction as in Figure 3.9 only the reaction rates are reversed. Now B converts quickly to C and it is the slower rate of A reacting to B that controls the overall rate of reaction: the formation of C. Note also that the concentration of C, [C], as a function of time in both this figure and Figure 3.9 are the same because the slower reaction controls the overall rate, the concentration of C as a function of time.

$$[C] = [A]_0 \left\{ 1 + \frac{1}{0.3 - 0.05}\left(0.05e^{-0.3t} - 0.3e^{-0.05t}\right) \right\} = [A]_0 \left(1 + 4\left(0.05e^{-0.3t} - 0.3e^{-0.05t}\right)\right).$$

Reversing k_1 and k_2 so that $k_1 = 0.3$ and $k_2 = 0.05$, the equation becomes

$$[C] = [A]_0 \left\{ 1 + \frac{1}{0.05 - 0.3}\left(0.3e^{-0.05t} - 0.05e^{-0.3t}\right) \right\} = [A]_0 \left(1 + 4\left(0.05e^{-0.3t} - 0.3e^{-0.05t}\right)\right)$$

which—not too surprisingly—is the same as the previous equation and [C](t) is the same in both Figures 3.9 and 3.10. However, [B](t) depends strongly on which is the faster reaction.

In the limit of $k_2 \gg k_1$, Equations 3.31 and 3.32 become respectively:

$$[B] \cong \frac{[A]_0 k_1}{k_2} e^{-k_1 t} = \frac{k_1}{k_2}[A] \tag{3.42}$$

$$[C] \cong [A]_0 \left(1 - e^{-k_1 t}\right). \tag{3.43}$$

These results clearly show that the rate of reaction that determines the concentrations of A, B, and C depend on the smaller reaction rate constant, k_1: that is, *for a series reaction, the slower (slowest) reaction controls the rate of the overall reaction.*

3.5.3.2 Steady-State Assumption

An approach to get an approximate solution to these series reaction equations, Equation 3.38, is to make the *steady-state assumption* for the concentration of B. It is extremely important to note that *steady-state* and *equilibrium* have two very different meanings in reactions even though the terms are often used—incorrectly—interchangeably. The term *equilibrium* means that the *reaction has stopped.* In contrast, *steady-state* implies that *concentrations are no longer changing with time* but the reaction is still taking place and may be a long way from equilibrium.

The steady-state assumption is particularly useful for analyzing complex reactions having multiple series and parallel steps (Laidler 1987). That is,

$$\frac{d[B]}{dt} = k_1[A] - k_2[B] = 0 \tag{3.44}$$

this leads immediately to Equations 3.42 and 3.43 and is, of course, tantamount to the assumption that $k_2 \gg k_1$, for which, as Equation 3.42 shows, [B] does not vary rapidly with time because it depends on $e^{-k_1 t}$ and k_1 is small. Of course, Equations 3.42 and 3.44 are mathematically inconsistent, but the result is a reasonable approximation if $k_2 \gg k_1$.

3.5.3.3 Example: Nuclear Decay

In a nuclear reactor, the roughly $^{235}_{92}U$ enriched to 4% or 5% of the uranium present undergoes fission generating energy and several neutrons. Some of these neutrons are captured by the 96% $^{238}_{92}U$ to give $^{239}_{92}U$, which decays to $^{239}_{93}Np$ and then to $^{239}_{94}Pu$ by the following reactions:

$$\begin{aligned} ^{239}_{92}U &\xrightarrow{23.5\,\text{min}} {}^{239}_{93}Np + \beta^- + 1.28\,\text{MeV} \\ ^{239}_{93}Np &\xrightarrow{2.35\,\text{d}} {}^{239}_{94}Pu + \beta^- + 0.23\,\text{MeV} \\ ^{239}_{94}Pu &\xrightarrow{24{,}000\,\text{yr}} {}^{235}_{92}U + {}^{4}_{2}He + 5.23\,\text{MeV}. \end{aligned} \tag{3.45}$$

If a reactor is operating at a constant power level for a long period of time, typically several years, then the amount of $^{239}_{92}U$ will stay at some small time-independent value as will the amount of $^{239}_{93}Np$ because of their relatively short half-lives. On the other hand, the amount of $^{239}_{94}Pu$ in the reactor core will continually increase. However, it too will reach a steady-state concentration because $^{239}_{94}Pu$ is also fissionable and can be used as fuel in nuclear power reactors, as it is in France. Nevertheless, it would be expected that the relative steady-state concentrations of $^{239}_{92}U$ and $^{239}_{93}Np$, which from Equations 3.44 and 3.45, would be $\left[^{239}_{93}Np\right]/\left[^{239}_{92}U\right] = 2.35\,\text{d}/23.5\,\text{min} = 720$.

Of course, the concentration in the atmosphere of radioactive $^{14}_{6}C$ used for radiocarbon dating is a perfect example of a steady-state concentration between its formation by cosmic rays and its own nuclear decay.

3.6 COMPLEXITY OF REAL REACTIONS

3.6.1 Hydrogen Iodide

In all of the previous discussions, the kinetic expressions are for homogeneous reactions and are therefore most applicable to gases and liquids but they can be applied to heterogeneous reactions and solids as well. The objective has been to introduce some of the terminology and types of reactions that can take place and the effects of time and concentration on the reactions. The focus has been on homogeneous gas-phase reactions because they have been the most studied and they are—in principle—the simplest type of reaction. The questions are "How complex are real gas reactions?" and "Where can data be found on gas-phase reaction kinetics?" The answer to the second question is that the National Institute of Standards and Technology (NIST) of the U.S. Department of Commerce has an online database NIST, Kinetics Database (NIST 2013) of gas reaction kinetic parameters. These data show that the hydrogen iodide reaction is indeed a simple second-order reaction:

$$\frac{d[H_2]}{dt} = -k[H_2][I_2] \tag{3.46}$$

with the *reaction rate constant* given by $k = A\exp(-E_A/RT)$, which is the typical exponential temperature dependence of the reaction rate constant discussed in Chapter 4. In this case, the *preexponential term* $k_0 = A = 3.22 \times 10^{-10}\,\text{cm}^3/\textit{mol}\ \text{s}$ which means that the concentrations would be expressed in mol/cm³. $E_A = Q$, the *activation energy* in this case is $E_A = 171$ kJ/mol.* However, the reality even for most gas-phase reactions is that they are not simple single-step reactions but consist of many serial and parallel steps.

3.6.2 Chain Reactions

Similar to the reaction for the production of HI, the formation of HCl can be written as

$$H_2(g) + Cl_2(g) \rightleftharpoons 2HCl(g).$$

However, for this reaction the kinetics are more complex (Kee et al. 2003). In fact, the kinetics of this reaction represent a *chain reaction*. In a chain reaction, there is an *initiation step*, a *chain-propagating step*, and a *termination step*. For this reaction

Initiation: $Cl_2 \rightleftharpoons Cl^\bullet + Cl^\bullet$
Propagation: $Cl^\bullet + H_2 \rightleftharpoons HCl + H^\bullet$ and $Cl_2 + H^\bullet \rightleftharpoons HCl + Cl^\bullet$
Termination: $Cl^\bullet + Cl^\bullet \rightleftharpoons Cl_2$ and $Cl^\bullet + H^\bullet \rightleftharpoons HCl$.

Clearly, the propagation step leads to a continuation of the reaction by producing hydrogen and chlorine *radicals* (the species with the "dot") to react with Cl_2 and H_2, respectively, to continue the reaction to HCl.† In principle, the differential equations for the concentrations of all of these species can be developed and, because the reaction rate constants for the reactions exist in the NIST Kinetics database, the equations for the concentrations of the various species could be solved as a function of time. However, rather than solving the equations analytically, it is easier to solve these simultaneous differential equations numerically. In fact, commercial software programs exist to perform complex gas-phase kinetic calculations. Such a program was developed at the Sandia Livermore Laboratory and is now available commercially (Kee et al. 2003; CHEMKIN). A similar series of reactions are involved in the formation of HBr by the reaction (Chang 2000)

$$H_2(g) + Br_2(g) \rightleftharpoons 2HBr(g).$$

3.6.3 Chain Polymerization

3.6.3.1 Mechanism

There are many different mechanisms for polymerization of monomers. In fact, a central feature of polymer science is the types of polymerization processes that can take place, how they differ for different polymers, the effects of temperature and concentration, and—most important—the role of various catalysts. The intention here is not to try to summarize all of this information that is best left to polymer chemistry. However, one mechanism that fits very nicely into the current discussion, and gives an example of one of the processes, *radical chain polymerization* that includes steps of initiation, propagation, and termination to form polymer chains. The process is covered in many polymer textbooks (Billmeyer 1984; Saunders 1988; Young and Lovell 1991; Elias 1997).

---

* "A" and "E_A" are the symbols used in the NIST Database for the "k_0" and "Q" terms used here.

† The "dot" superscript represents a reactive *radical*, an atom or molecule with an unsatisfied electron bond—called *free radicals* in older literature. The literature is somewhat inconsistent in the designation of a radical. Sometimes the dot is placed directly on top of the symbol (Kee et al. 2003), sometimes to the side (Silbey and Alberty 2001), and sometimes it is not used at all (Laidler 1987; Houston 2001), but most frequently it is raised and to the right (or left of the atom with the unsatisfied bond) (Cowie 1991) as used here.

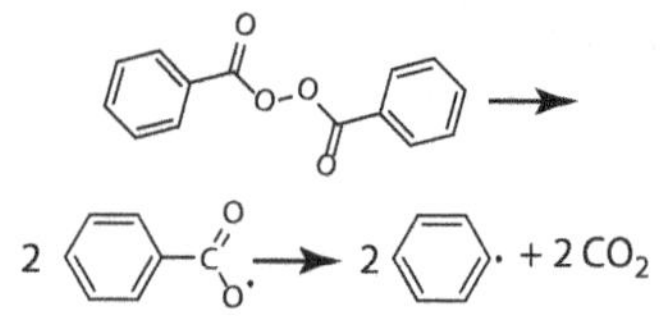

FIGURE 3.11 Benzoyl peroxide, a typical chain initiator for radical polymerization undergoing thermolysis—decomposition by heat—with the breaking of the oxygen-oxygen bond and release of CO_2 to give two radicals, that can act as chain initiators, $I^\bullet$.

Figure 3.11 shows a typical initiator material, benzoyl peroxide, that on heating, splits into two parts, gives off CO_2, and provides two radicals, $I^\bullet$, that can act as chain initiation for polymerization. This radical reacts with a monomer molecule

$$\text{Initiation step: } I^\bullet + CH_2 = CRH \Rightarrow I - CH_2 - C^\bullet RH$$

$$\text{Propagation step: } I - CH_2 - C^\bullet RH + CH_2 = CRH \Rightarrow$$
$$I - CH_2 - CRH - CH_2 - C^\bullet RH\text{, etc.}$$

$$\text{Termination step: } \leftarrow - CH_2 - C^\bullet RH + C^\bullet RH - CH_2 - \rightarrow$$
$$\leftarrow - CH_2 - CRH - CRH - CH_2 - \rightarrow$$

In these equation, R = H, Cl, CH_3, and so on and the single arrows indicate the extension of the polymer chain in both the positive and negative directions in the termination step.

3.6.3.2 Rate of Polymerization

Initiation can be represented by two steps:

$$I \xrightarrow{\text{slow}} nI^\bullet$$
$$I^\bullet + M \xrightarrow{\text{fast}} IM_i^\bullet \tag{3.47}$$

where M is the monomer molecule. Let R_I be the rate of initiator formation $R_I = d[I^\bullet]/dt$ and k_p the rate of chain propagation: $M_i + M \xrightarrow{k_p} M_{i+1}$ so the rate of consumption of the monomer is

$$-\frac{d[M]}{dt} = k_p[M_1^\bullet][M] + k_p[M_2^\bullet][M] + \cdots [M_i^\bullet][M]$$

$$-\frac{d[M]}{dt} = k_p[M]\sum_{i=1}^{\infty}[M_i^\bullet] = k_p[M][M^\bullet] \tag{3.48}$$

where $[M^\bullet]$ is the total concentration of all the radical species. The rate of termination is given by $M_i^\bullet + M_j^\bullet \xrightarrow{k_t} M_{i+j}$ where k_t is the rate of termination so

$$-\frac{[M^\bullet]}{dt} = 2k_t[M^\bullet]^2. \tag{3.49}$$

The 2 comes from the fact that two radical species are destroyed in a termination reaction. Steady-state conditions are assumed; that is, $d[I^\bullet]/dt = -(d[M^\bullet]/dt)$ so $R_I = 2k_t[M^\bullet]^2$ or $[M^\bullet] = (R_I/2k_t)^{1/2}$. Substituting this into Equation 3.48 and calling $-d[M]/dt = R_p$, the "rate of polymerization," the result is

$$R_p = \left(\frac{k_p}{2^{1/2}k_t^{1/2}}\right)R_I^{1/2}[M] \tag{3.50}$$

which indicates that the rate of monomer consumption is proportional to the monomer concentration with the proportionality consisting of the various rate constants. This is a first-order equation, at least in the specified steady state.

3.6.4 Chain Branching Reactions

Many combustion reactions include chain branching reactions and can lead to explosive mixtures. For the very familiar simple reaction of the combustion of hydrogen with oxygen to form water vapor

$$H_2(g) + 1/2O_2(g) \rightarrow H_2O(g)$$

the reaction has a total of 19 identified separate steps including the following as examples (Li et al. 2004):

$$\text{Initiation: } H_2 + O_2 \rightleftharpoons H^{\bullet} + HO_2^{\bullet}$$

$$\text{Propagation: } H_2 + O_2 \rightleftharpoons H^{\bullet} + HO_2^{\bullet}$$

$$^{\bullet}OH + H_2 \rightleftharpoons H_2O + H^{\bullet}$$

$$\text{Branching: } H^{\bullet} + O_2 \rightleftharpoons O^{\bullet} + {}^{\bullet}OH$$

$$\text{Termination: } HO_2^{\bullet} + {}^{\bullet}OH \rightleftharpoons H_2O + O_2$$

the solution of which, of course, requires some sort of numerical solution (Kee et al. 2003) to obtain the rate of the reaction and/or the concentrations of the various possible gas species as a function of time and temperature.

3.7 SUMMARY

In this chapter, second-order, primarily homogeneous reactions, as well as multistep reactions are considered. The focus remains on homogeneous reactions purely because the reactions are conceptually simpler and lead more readily to transparent solutions. Furthermore, in the materials world of today, what goes on in the gas or liquid phase is critical to many important commercial materials processes. The objectives of this chapter are several-fold. One is to demonstrate how rapidly reaction solutions can become complex strongly suggesting numerical rather than analytical solutions. A second objective is to demonstrate how some of these more complex reactions can be solved analytically with elementary calculus techniques. Another objective is to demonstrate the rate-controlling step in series and parallel reactions.

The reactions investigated in this chapter include: simple second-order reactions; second-order reactions with different initial concentrations of reactants; reactions that go to equilibrium—those with back reactions; serial and parallel reactions; chain reactions; and branching reactions. It is demonstrated that the initial reaction rate for a first-order reaction that goes to equilibrium could be written as

$$\frac{d[A]}{dt} = -k\left([A] - [A]_e\right) \tag{3.51}$$

where $[A]_e$ is the equilibrium concentration. Equation 3.51 shows the very general result that the reaction rate is the product of a kinetic factor, k, and a thermodynamic factor, $([A]-[A]_e)$, and this is generally seen in the kinetics of other reactions. For second- and higher order reactions, even unpretentious looking reaction rates quickly become difficult to integrate analytically. As a result, numerical techniques are easy to implement and the simple case of Euler's method applied to a second order ⇔ second order equation demonstrates the utility of a numerical solution to all but the simplest of reactions that go to equilibrium. Simple series and parallel reactions are then examined that show that the slowest step controls the overall rate of a series reaction while the fastest step controls the rate and product concentrations of parallel reactions. Finally, the potential complexity of ostensibly simple reactions, such as the reaction of hydrogen and chlorine to form HCl, and hydrogen and oxygen to form water vapor, are examined. These examples demonstrate the need for numerical solution to the several parallel and series kinetic steps probable in real reactions to understand the effects of composition and ambient conditions on reaction rates and possible rate-controlling reaction steps. Up to this point, little has been said about the temperature dependence of the reaction rate constant. This is the subject of the next chapter.

APPENDIX

A.1 Second-Order Reaction with Different Initial Concentrations

A.1.1 General Result

In this case, assume that there are two reactants, A and B, reacting to give a product, C:

$$A + B \xrightarrow{k} C \tag{A.1}$$

where C in this case can be one or more products. In addition, assume that the reaction is first order with respect to both A and B resulting in overall second-order kinetics:

$$\frac{d[A]}{dt} = -k[A][B] \tag{A.2}$$

and that the reaction goes to completion (either [A] and/or [B] goes to zero at $t = \infty$) the concentrations of A and B can be expressed as*:

$$\begin{aligned}
[A] &= [A]_0 - ([A]_0 - [A]) \\
[B] &= [B]_0 - ([B]_0 - [B]) \\
&= [B]_0 - ([A]_0 - [A]) \\
\therefore [B] &= ([B]_0 - [A]_0) + [A]
\end{aligned} \tag{A.3}$$

because the amount of B reacted must be the same as the amount of A reacted at any given time and the initial concentrations of A and B at time = 0 are $[A]_0$ and $[B]_0$, respectively. So Equation A.2 becomes

$$\frac{d[A]}{dt} = -k[A]\left[([B]_0 - [A]_0) + [A]\right]$$

or

$$\frac{d[A]}{[A]\left[([B]_0 - [A]_0) + [A]\right]} = -kdt \tag{A.4}$$

this can be integrated by the standard method of partial fractions (Edwards and Penney 1986). This means that constants D and E need to be found so that

$$\frac{1}{[A]\left[([B]_0 - [A]_0) + [A]\right]} = \frac{D}{[A]} + \frac{E}{([B]_0 - [A]_0) + [A]}.$$

* To reduce the amount of algebra, assume that $([B]_0 - [A]_0)$.

Multiplying both sides of the equation by the denominator on the left-hand side of this equation gives $1 = D([B]_0 - [A]_0) + D[A] + E[A]$, and equating like powers of [A] on both sides of the equation: $D[A] + E[A] = 0$ so $E = -D$; because there is no term in x on the left and $1 = D([B]_0 - [A]_0)$ so $D = 1/([B]_0 - [A]_0)$, Equation A.4 now becomes

$$\frac{d[A]}{([B]_0 - [A]_0)[A]} - \frac{d[A]}{([B]_0 - [A]_0)\left[([B]_0 - [A]_0) - [A]\right]} = -kdt$$

multiplying both sides by $([B]_0 - [A]_0)$ and integrating from t = 0 to t gives

$$\ln\left(\frac{[A]}{[A]_0}\right) - \ln\left(\frac{([B]_0 - [A]_0) + [A]}{([B]_0 - [A]_0) + [A]_0}\right) = -k([B]_0 - [A]_0)t. \tag{A.5}$$

This last equation essentially forces $[B_0]$ to be greater than $[A_0]$ because when $[A] \rightarrow 0$, $([B_0] - [A_0])$ would be negative and the $\log_e$ of a negative number is not defined. Taking the exponent of both sides of the equation and moving a few things around, Equation A.5 becomes

$$\frac{[A]}{([B]_0 - [A]_0) + [A]} = \frac{[A]_0}{[B]_0} e^{-([B]_0 - [A]_0)kt} = f(t)$$

so

$$[A] = ([B]_0 - [A]_0) f(t) + [A] f(t)$$

or

$$[A] - [A] f(t) = [A](1 - f(t)) = ([B]_0 - [A]_0) f(t)$$

and rearranging gives the final rather ugly result:

$$[A] = \frac{([B]_0 - [A]_0)[A]_0 e^{-([B]_0 - [A]_0)kt}}{[B]_0 - [A]_0 e^{-([B]_0 - [A]_0)kt}}. \tag{A.6}$$

Solving for [B], because

$$[B] = ([B]_0 - [A]_0) + [A]$$

$$[B] = ([B]_0 - [A]_0) + \frac{([B]_0 - [A]_0)[A]_0 e^{-([B]_0 - [A]_0)kt}}{[B]_0 - [A]_0 e^{-([B]_0 - [A]_0)kt}}$$

factoring out $([B]_0 - [A]_0)$ and getting a common denominator, this equation becomes

$$[B] = ([B]_0 - [A]_0)\left\{\frac{[B]_0}{[B]_0 - [A]_0 e^{-([B]_0 - [A]_0)kt}}\right\}. \tag{A.7}$$

When $t = 0$, $[B] = [B]_0$ and when $t = \infty$, $[B] = ([B]_0 - [A]_0)$ as it should.

A.1.2 Initial Concentrations Equal

From Equation A.6, when $[B]_0 \approx [A]_0$, or the initial concentrations are about equal, then [A] is indeterminate, that is, [A] = 0/0. Therefore, to get a value for [A], the approximation is made that $e^x = 1 + x + (x^2/2!) + \cdots \cong 1 + x$, that is, only the first two terms in the infinite series expression for e^x are used when $x \cong 0$. Making this approximation, the right-hand side of Equation A.6 becomes

$$\frac{([B]_0 - [A]_0)[A]_0(1 - ([B]_0 - [A]_0)kt)}{[B]_0 - [A]_0(1 - ([B]_0 - [A]_0)kt)}$$

$$= \frac{([B]_0 - [A]_0)[A]_0(1 - ([B]_0 - [A]_0)kt)}{([B]_0 - [A]_0) + [A]_0(([B]_0 - [A]_0)kt)}$$

and canceling the $([B]_0 - [A]_0)$ term gives

$$[A] = \frac{[A]_0(1 - ([B]_0 - [A]_0)kt)}{1 + [A]_0 kt}$$

which—after all of this algebra—because $[B]_0 - [A]_0 \cong 0$ gives the result

$$[A] \cong \frac{[A]_0}{1 + [A]_0 kt} \tag{A.8}$$

and the result is the same as when [A] = [B], Equation 3.2, as it should.

A.2 Second Order ⇔ Second Order

A.2.1 Beginning

The general procedure carried out here works for second order ⇔ second order, second order ⇔ first order, and first order ⇔ second order because all involve the solution of quadratic equations. The reaction is assumed to be the simple reaction

$$A \rightleftharpoons B$$

or

$$\frac{d[A]}{dt} = -k_1[A]^2 + k_2[B]^2. \tag{A.9}$$

Again, to minimize the notation, let $[A] = x$, and $[A]_0 = x_0$ and $[B]_0 = 0$ so that Equation A.9 becomes

$$\frac{dx}{dt} = -k_1 x^2 + k_2(x_0 - x)^2. \tag{A.10}$$

Now, this is a relatively simple differential equation to solve because the variables are separable—all the "xs" on the left and "ts" on the right—so that standard integrations give a solution without recourse to special differential equation solution techniques. As shown in this chapter, the equilibrium concentrations, x_e (only one of which gives a solution), are given by

$$x_e = x_0\left(\frac{\sqrt{k_2}}{\sqrt{k_2} \pm \sqrt{k_1}}\right). \tag{A.11}$$

At this point, several options are available to proceed with the solution: (1) just solve Equation A.10 as is; (2) solve Equation A.11 for x_0 and substitute it into Equation A.10 and solve; or (3) solve Equation A.11 for k_2 and substitute it into Equation A.10 and solve (Laidler 1987). Here the first approach is used to minimize the symbol complexity. The procedure used is exactly that for the specific case in Section 3.4.3 where specific numbers were used to simplify the system even further. The difference here is that the solution in terms of the parameters x_0, k_1, and k_2 (and/or x_e) is sought so that it can be applied to any such reaction.

A.2.2 Finding Roots (Factors)

As in Section 3.4.3, the two roots, x_1 and x_2, of the right-hand side of Equation A.10 are sought, namely,

$$x_i = \frac{-b \pm \sqrt{b^2 - 4ac}}{2a}.$$

Expanding Equation A.10,

$$\frac{dx}{dt} = -k_1x^2 + k_2\left(x_0^2 - 2xx_0 + x^2\right)$$

$$\frac{dx}{dt} = -\left(k_1 - k_2\right)x^2 - 2k_2x_0x + k_2x_0^2$$

$$\frac{1}{k_2}\frac{dx}{dt} = -\left(\frac{k_1}{k_2} - 1\right)x^2 - 2x_0x + x_0^2$$

$$\frac{1}{k_2}\frac{dx}{dt} = ax^2 + bx + c.$$

So the roots of the right-hand side of the quadratic equation are

$$x_i = \frac{2x_0 \pm \sqrt{4x_0^2 + 4((k_1/k_2) - 1)x_0^2}}{-2((k_1 - k_2)/k_2)}$$

$$x_i = -\frac{x_0\left(\left(\sqrt{k_2} \pm \sqrt{k_1}\right)/\sqrt{k_2}\right)}{\left((k_1 - k_2)/\left(\sqrt{k_2}\sqrt{k_2}\right)\right)}$$

$$x_i = \frac{x_0\sqrt{k_2}\left(\sqrt{k_2} \pm \sqrt{k_1}\right)}{\left(\sqrt{k_2} + \sqrt{k_1}\right)\left(\sqrt{k_2} - \sqrt{k_1}\right)}.$$

So the two roots are, as before, Equation A.11,

$$\begin{aligned} x_1 &= x_0\sqrt{k_2}/\left(\sqrt{k_2} + \sqrt{k_1}\right) \\ x_2 &= x_0\sqrt{k_2}/\left(\sqrt{k_2} - \sqrt{k_1}\right) \end{aligned} \tag{A.12}$$

and $x_1 > 0$ while $x_2 < 0$ if $k_1 > k_2$. If the latter is not the case and $[B]_0 = 0$, there is no reaction, and if $[B]_0 > 0$, then the reverse reaction would occur, $B \rightarrow A$. So k_1 must always be greater than k_2 if the initial concentration of B is assumed to be zero. As a result, x_1 is the equilibrium point of interest physically because the concentrations must always be positive.

A.2.3 Checking the Factoring

So two factors for the right-hand side of the differential equation, Equation A.10 are $(-x_2+x)$ and (x_1-x). Is that all that is necessary? That is, if these factors are multiplied do they reproduce the right-hand side of Equation A.10? Checking this product,

$$\begin{aligned}
\text{Product} &= (-x_2+x)(x_1-x) \\
&= -x_2x_1 + x_2x + x_1x - x^2 \\
&= -x_2x_1 + (x_2+x_1)x - x^2 \\
&= -\left(\frac{x_0\sqrt{k_2}}{\left(\sqrt{k_2}-\sqrt{k_1}\right)}\right)\left(\frac{x_0\sqrt{k_2}}{\left(\sqrt{k_2}+\sqrt{k_1}\right)}\right) + \left(\frac{x_0\sqrt{k_2}}{\left(\sqrt{k_2}-\sqrt{k_1}\right)} + \frac{x_0\sqrt{k_2}}{\left(\sqrt{k_2}+\sqrt{k_1}\right)}\right)x - x^2 \\
&= \frac{x_0^2k_2}{k_1-k_2} + x_0\sqrt{k_2}\left(\frac{\sqrt{k_2}+\sqrt{k_1}+\sqrt{k_2}-\sqrt{k_1}}{k_2-k_1}\right)x - x^2 \\
\text{Product} &= \frac{x_0^2k_2}{k_1-k_2} - \frac{x_0k_2}{k_1-k_2}x - x^2.
\end{aligned}$$

To obtain the right-hand side of Equation A.10, this last equation must be multiplied by (k_1-k_2) and divided by k_2. So the complete factoring of the right-hand side leads to the differential equation:

$$\frac{1}{k_2}\frac{dy}{dx} = \frac{1}{k_2}(k_1-k_2)(-x_2+x)(x_1-x)$$

$1/k_2$ cancels and the differential equation to be solved is now,

$$\frac{dy}{dx} = (k_1-k_2)(-x_2+x)(x_1-x). \tag{A.13}$$

A.2.4 Solution of the Differential Equation

Again, Equation A.13 is the equation to be integrated and follows the procedure used in this chapter. Separating variables

$$\frac{dx}{(-x_2+x)(x_1-x)} = (k_1-k_2)dt$$

and letting, where A and B are now integration constants here,

$$\frac{1}{(-x_2+x)(x_1-x)} = \frac{A}{(-x_2+x)} + \frac{B}{(x_1-x)}$$

and solving for A and B, $Ax_1 - Ax - Bx_2 + Bx = 1$ so $A = B$ and $Ax_1 - Bx_2 = 1$ or $A = B = 1/(x_1-x_2)$. So, the equation that needs to be integrated is now

$$\frac{dx}{(-x_2+x)} + \frac{dx}{(x_1-x)} = (x_1-x_2)(k_1-k_2)dt$$

and integrating gives

$$\ln(x-x_2)-\ln(x_1-x)=(x_1-x_2)(k_1-k_2)t+C$$

$$\frac{(x-x_2)}{(x_1-x)}=C'e^{(x_1-x_2)(k_1-k_2)t}$$

where C and C′ are both constants of integration. And when $t = 0$, $x = x_0$ so

$$C'=\frac{(x_0-x_2)}{(x_1-x_0)}$$

and the solution becomes

$$\frac{(x-x_2)}{(x_1-x)}=\frac{(x_0-x_2)}{(x_1-x_0)}e^{(x_1-x_2)(k_1-k_2)t}=\frac{1}{D} \quad \text{(A.14)}$$

where D is a constant for a given time t. Equation A.14 needs to be solved for x so inverting Equation A.14 and multiplying by $(x - x_2)$ gives $x_1 - x = -x_2D + xD$ and solving for x

$$x=\frac{x_1+x_2D}{1+D}. \quad \text{(A.15)}$$

Substituting in the values of x_1, x_2, and D, the solution is *finally* obtained. But wait! The answer really is desired in terms of k_1, k_2, and x_0, the real parameters of the problem. So it is worthwhile to see what the final equation looks like in terms of these parameters. Attacking the exponent first,

$$k=-(x_1-x_2)(k_1-k_2)$$

$$=-\left\{\frac{x_0\sqrt{k_2}}{\left(\sqrt{k_2}+\sqrt{k_1}\right)}-\frac{x_0\sqrt{k_2}}{\left(\sqrt{k_2}-\sqrt{k_1}\right)}\right\}(k_1-k_2)$$

$$k=x_0\sqrt{k_2}\left\{\frac{\sqrt{k_2}-\sqrt{k_1}-\sqrt{k_2}-\sqrt{k_1}}{k_1-k_2}\right\}(k_1-k_2)$$

which reduces to the resultant reaction rate constant k, which has the appropriate units of $cm^3/mol\ s$:

$$k=-2x_0\sqrt{k_2k_1}. \quad \text{(A.16)}$$

Now for the factor in front of the exponent in D so

$$\frac{(x_1-x_0)}{(x_0-x_2)}=\frac{\left(\left[x_0\sqrt{k_2}/\left(\sqrt{k_2}+\sqrt{k_1}\right)\right]-x_0\right)}{\left(x_0-\left[x_0\sqrt{k_2}/\left(\sqrt{k_2}-\sqrt{k_1}\right)\right]\right)}$$

$$=\frac{\left(\left(\sqrt{k_2}-\sqrt{k_2}-\sqrt{k_1}\right)/\left(\sqrt{k_2}+\sqrt{k_1}\right)\right)}{\left(\left(\sqrt{k_2}-\sqrt{k_1}-\sqrt{k_2}\right)/\left(\sqrt{k_2}-\sqrt{k_1}\right)\right)}$$

with the result

$$\frac{(x_1-x_0)}{(x_0-x_2)}=-\left(\frac{\sqrt{k_1}-\sqrt{k_2}}{\sqrt{k_1}+\sqrt{k_2}}\right). \quad \text{(A.17)}$$

The other factor that needs to be calculated is

$$\frac{(x_1-x_0)}{(x_0-x_2)}\times x_2=-\left(\frac{\sqrt{k_1}-\sqrt{k_2}}{\sqrt{k_1}+\sqrt{k_2}}\right)\times\left(\frac{x_0\sqrt{k_2}}{\left(\sqrt{k_2}-\sqrt{k_1}\right)}\right)$$

which is

$$\frac{(x_1 - x_0)}{(x_0 - x_2)} \times x_2 = \left(\frac{x_0 \sqrt{k_2}}{\sqrt{k_1} + \sqrt{k_2}} \right). \tag{A.18}$$

Putting x_1 and Equations A.16 through A.18 into A.15 gives the final result—the solution:

$$x = \frac{x_0 \sqrt{k_2} / \left(\sqrt{k_2} + \sqrt{k_1} \right) \left\{ 1 + e^{-\left(2x_0\sqrt{k_2 k_1}\right)t} \right\}}{1 - \left((\sqrt{k_1} - \sqrt{k_2}) / (\sqrt{k_1} + \sqrt{k_2}) \right) e^{-\left(2x_0\sqrt{k_2 k_1}\right)t}} \tag{A.19}$$

which is a disagreeably messy solution, to say the least, for an apparent simple differential equation.

A.2.5 Checking the Solution

When $t \to \infty$, only the first term in the numerator remains and it is x_1 or the equilibrium composition, Equation A.12 which checks. When $t = 0$,

$$x = \frac{x_0 \sqrt{k_2} / \left(\sqrt{k_2} + \sqrt{k_1} \right) \{1 + 1\}}{1 - \left((\sqrt{k_1} - \sqrt{k_2}) / (\sqrt{k_1} + \sqrt{k_2}) \right)}$$

$$= \frac{2x_0 \sqrt{k_2} / \left(\sqrt{k_2} + \sqrt{k_1} \right)}{(\sqrt{k_1} + \sqrt{k_2} - \sqrt{k_1} + \sqrt{k_2}) / (\sqrt{k_1} + \sqrt{k_2})}$$

$$x = x_0$$

which also checks. The result can also be checked with the result of Equation 3.30 in which $k_1 = 2$, $k_1 = 1$, and $x_0 = 1$. In this case, the exponent is

$$k = -2x_0\sqrt{k_2 k_1}$$

$$k = -2(1)\sqrt{(1)(2)}$$

$$k = -2.82842$$

which checks with Equation 3.30. The first factor in the numerator of Equation A.19 is simply x_1 given by

$$x_1 = \frac{x_0 \sqrt{k_2}}{\left(\sqrt{k_2} + \sqrt{k_1} \right)}$$

$$x_1 = \frac{1\sqrt{1}}{\sqrt{1} + \sqrt{2}}$$

$$x_1 = 0.41421$$

which also checks. So it is concluded that Equation A.19 is the general solution to Equation A.10. One last check could be made and that is when $k_2 = 0$, no back reaction, does Equation A.19 go to Equation A.8? This is left as an exercise.

A.3 Solution for Two Reactions in Series

A.3.1 Introduction

The problem is to find the solution for two first-order reactions in series; namely:

$$A \xrightarrow{k_1} B \xrightarrow{k_2} C \tag{A.20}$$

so the rate equations become

$$\frac{d[A]}{dt} = -k_1[A]$$

$$\frac{d[B]}{dt} = k_1[A] - k_2[B] \qquad (A.21)$$

$$\frac{d[C]}{dt} = k_2[B]$$

with the conservation of matter: $[A]_0 = [A] + [B] + [C]$ at any time t with $[A]_0$ being the initial concentration of A and the initial concentrations of B and C at $t = 0$ are $[B]_0 = [C]_0 = 0$ to make the calculations more manageable. The solution to this series of equations is given in many physical chemistry texts but they do not give the details of the integration process and simply indicate that "... the following is obtained by integration ..." (Laidler and Meiser 1995; Silbey and Alberty 2001). The result is certainly not obvious but can easily be obtained although it requires some familiarity with ordinary differential equations. For completeness, and to satisfy the curiosity of those who are more conversant with ordinary differential equations, the details of the calculations are given.

A.3.2 Concentration of A

The solution to the first of the above equations for [A] is simply the solution for any first-order equation,

$$[A] = [A]_0 e^{-k_1 t}. \qquad (A.22)$$

A.3.3 Concentration of B

The differential equation for [B]—the second equation above—now becomes

$$\frac{d[B]}{dt} = k_1[A]_0 e^{-k_1 t} - k_2[B] \qquad (A.23)$$

which can be rewritten as

$$\frac{d[B]}{dt} + k_2[B] = k_1[A]_0 e^{-k_1 t} \qquad (A.24)$$

which is a *nonhomogeneous* differential equation—the right-hand side of the equation is nonzero. The solution to such an equation is to first solve the *homogeneous* equation (make the right-hand side = 0)

$$\frac{d[B]}{dt} + k_2[B] = 0 \qquad (A.25)$$

and then add to the homogeneous solution a particular solution that satisfies the right-hand side of the equation, Equation A.24 in this case. The complete solution is the sum of the homogeneous and particular solutions (Kreyszig 1988). To simplify the notation, let [B] = y, so the homogeneous equation (subscript y) becomes

$$\frac{dy_h}{dt} + k_2 y_h = 0 \qquad (A.26)$$

where the subscript h refers to the homogeneous equation. This equation has been encountered several times already along with its solution. However, because more advanced differential equations are being solved here, the formal method for solution of such equations will be followed as a demonstration. The formal way to solve such an equation is by the *method of characteristics* because it is a linear differential equation with constant coefficients; that is, let $y_h = C_1 e^{\lambda t}$, where C_1 is just a constant of integration. Substituting this in the differential equation for y_h, Equation A.26,

$$C_1\lambda e^{\lambda t} + C_1 k_2 e^{\lambda t} = 0 \tag{A.27}$$

which gives $\lambda = -k_2$ and the solution, $y_h = C_1 e^{-k_2 t}$ as expected. And the equation for d[B]/dt, Equation A.24, suggests a particular solution, y_p, of the form: $y_p = C_2 e^{-k_1 t}$, where C_2 is just another integration constant. Substituting y_p for [B] in the differential equation gives

$$y_p' + k_2 y_p = k_1[A]_0 e^{-k_1 t}$$

$$-k_1 C_2 e^{-k_1 t} + k_2 C_2 e^{-k_1 t} = k_1[A]_0 e^{-k_1 t}.$$

Obviously, $e^{-k_1 t}$ cancels in all the terms leaving $C_2(k_2 - k_1) = k_1[A]_0$ that gives for y_p:

$$y_p = \frac{k_1[A]_0}{k_2 - k_1} e^{-k_1 t}. \tag{A.28}$$

The complete solution is simply $y = y_h + y_p$, which is

$$[B] = C_1 e^{-k_2 t} + \frac{k_1[A]_0 e^{-k_1 t}}{k_2 - k_1} \tag{A.29}$$

At $t = 0$, $B = 0$ so that $C_1 = -k_1[A]_0/(k_2 - k_1)$ and the solution for [B] is

$$[B] = \frac{[A]_0 k_1}{k_2 - k_1}\left(e^{-k_1 t} - e^{-k_2 t}\right). \tag{A.30}$$

A.3.4 Concentration of C

Because $[C] = [A]_0 - [A] - [B]$, inserting Equations A.22 and A.30 for [A] and [B],

$$[C] = [A]_0 - [A]_0 e^{-k_1 t} - \frac{[A]_0 k_1}{k_2 - k_1}\left(e^{-k_1 t} - e^{-k_2 t}\right) \tag{A.31}$$

and rearranging

$$[C] = [A]_0\left\{1 - \frac{k_2 e^{-k_1 t} + k_1 e^{-k_1 t} - k_1 e^{-k_1 t} + k_1 e^{-k_2 t}}{k_2 - k_1}\right\}$$

gives for [C]

$$[C] = [A]_0\left\{1 + \frac{1}{k_2 - k_1}\left(k_1 e^{-k_2 t} - k_2 e^{-k_1 t}\right)\right\} \tag{A.32}$$

Equations A.22, A.30, and A.32 are the solutions to the differential equations Equation A.21.

A.3.5 Slower Reaction Controls [C]

If $k_2 \gg k_1$ in Equation A.30, [B], and Equation A.32, [C], so that the reaction from B to C is much faster than the reaction from A to B, then the following approximation results from Equation A.30:

$$[B] \cong \frac{k_1}{k_2}[A] \tag{A.33}$$

which implies that, at any given time, the concentration of B will be much less than that of A, and [C] becomes from Equation A.32

$$[C] \cong [A]_0\left(1 - e^{-k_1 t}\right) = [A]_0 - [A]. \tag{A.34}$$

This implies that C comes almost directly from A and the rate depends on the smaller k, in this case, k_1. The concentration of B is always small because it reacts very fast to form C so it is the *slower* step that controls the rate of reaction to form C from A; that is, the reaction of A going to B, which is the slower step.

Conversely, if $k_1 \gg k_2$, then from Equations A.30 and A.32

$$[B] \simeq [A]_0 e^{-k_2 t} \tag{A.35}$$

$$[C] \simeq [A]_0 \left(1 - e^{-k_2 t}\right) \simeq [A]_0 - [B] \tag{A.36}$$

which imply that A goes to B very quickly, and it is now the reaction of B to C, or k_2—the slower series step—that controls the rate of formation of C. So as stated earlier, with series steps in a reaction process, it is the *slower* (or slowest step if there are more than two) that controls the rate of the overall reaction: in this case, A going to C.

EXERCISES

3.1 a. The molecular weight of HI is 127.9 g/mol. Calculate the concentration of Hl in mol/cm^3 at 800°C if its partial pressure is p(HI) = 1 atm.

b. Calculate the number of HI molecules per cm^3 at this temperature and pressure.

c. If the Gibbs energy for the reaction: $H_2(g) + I_2(g) = 2HI(g)$ at 800°C is $\Delta G^\circ = -26{,}400$ J/mol, calculate equilibrium partial pressures (atm) of HI, H_2, and I_2 at 800°C assuming that a closed container initially contained only H_2 and I_2 gases each at a partial pressure of 1 atm at 800°C.

3.2 a. From the NIST database, the reaction rate constant for the reaction $H_2(g) + I_2(g) = HI(g)$ is given by $k_0 = 3.22 \times 10^{-10}$ (cm^3/molecule s) and Q = 171 (kJ/mol) and it is a second-order reaction. Calculate the reaction rate constant (cm^3/molecule s) at 800°C.

b. Calculate the $[H_2]_0$ and $[I_2]_0$ in mol/cm^3 at the starting pressure and temperature.

c. For the reaction $H_2(g) + I_2(g) = 2HI(g)$ at 800°C, H_2 and I_2 reach an equilibrium pressure calculated in Exercise 3.1 above. Calculate the $[H_2]_e$, $[I_2]_e$, and $[HI]_e$ in mol/cm^3 at this temperature.

d. Calculate and plot $[H_2] = [I_2]$ and [HI] (on the same graph) at 800°C versus time (on a linear plot) until the $[HI] = 0.90\ [HI]_e$ where "[]" is in mol/cm^3.

3.3 A hypothetical second-order reaction that goes to completion with two reactants A and B going to C, that is, $A + B \rightarrow C$

$$\frac{d[C]}{dt} = k[A][B]$$

with $k = 10^{-3}\ L/mol \cdot s$, $[A]_0 = 0.1\ mol/L$, $[B]_0 = 0.5\ mol/L$, $[C]_0 = 0.1\ mol/L$. Calculate and plot [A], [B], and [C] as a function of time from t = 0 to [A] = 0.001 mol/L all on the same plot.

3.4 For a simple reaction that goes to equilibrium $A \rightleftharpoons B$ by forward and backward first-order reactions,

$$A \xrightarrow{k_1} B$$

$$B \xrightarrow{k_2} A$$

with $k_1 = 5 \times 10^{-3} s^{-1}$ and $k_2 = 1 \times 10^{-3} s^{-1}$, if $[A]_0 = 1.0\ mol/L$ and $[B]_0 = 0$:

a. Calculate the equilibrium constant for this reaction.

b. Calculate the equilibrium concentrations of A and B, $[A]_e$ and $[B]_e$.

c. On the same graph, plot [A] and [B] from t = 0 until $[A] = 0.01[A]_e$.

3.5 Show that the general solution for a reaction $A \rightleftharpoons B$ that goes to equilibrium with both forward and back second-order reactions where $x = [A]$ and $[B]_0 = 0$,

$$x = \frac{x_0\sqrt{k_2}/\left(\sqrt{k_2}+\sqrt{k_1}\right)\left\{1+e^{-\left(2x_0\sqrt{k_2k_1}\right)t}\right\}}{1-\left((\sqrt{k_1}-\sqrt{k_2})/(\sqrt{k_1}+\sqrt{k_2})\right)e^{-\left(2x_0\sqrt{k_2k_1}\right)t}}$$

goes to the simple solution for a second-order reaction with no back reaction—goes to completion, that is,

$$x \cong \frac{x_0}{1+x_0k_1t}$$

when $k_2 \rightarrow 0$. Hint: When $k_2 \rightarrow 0$ let $e^{-y} \rightarrow 1$ in the numerator and let $e^{-y} \rightarrow 1-y$ in the denominator.

3.6 A simple reaction $A \rightleftharpoons B$ that goes to equilibrium with both forward and back second-order reactions where $x = [A]$, $x_0 = [A]_0$, and $[B]_0 = 0$ gives the differential equation:

$$\frac{d[A]}{dt} = -k_1[A]^2 + k_2[B]^2$$

$$\frac{dx}{dt} = -k_1x^2 + k_2(x_0 - x)^2$$

a. Calculate the equilibrium concentrations of A and B in terms of $[A]_0$ when the forward and back reactions have the same rate constant; that is, $k_1 = k_2 = k$.
b. Calculate the exact solution for x when $k_1 = k_2 = k$.
c. Use Euler's method to approximate the solution if $k = 3 \times 10^{-2}$ mol/L-s, $[A]_0 = 1$ mol/L, for n = 1 to n = 100 with a time-step of $\Delta t = 1.0$.
d. Plot the results of parts b and c on the same graph with the parameters in part c.

3.7 A simple reaction $A \rightleftharpoons B$ that goes to equilibrium with a forward second-order reaction and a first-order back reaction, where $x = [A]$, $x_0 = [A]_0$, and $[B]_0 = 0$, gives the differential equation:

$$\frac{d[A]}{dt} = -k_1[A]^2 + k_2[B]$$

$$\frac{dx}{dt} = -k_1x^2 + k_2(x_0 - x)$$

If $k_1 = 2$, $k_2 = 1$, and $x_0 = 1$

a. Calculate the equilibrium concentrations of A and B.
b. Derive an expression for x, the concentration of A, as a function of time for this specific set of conditions.
c. Make a plot of [A] and [B] versus time on the same graph from t = 0 until $[A] = 0.99\,[A]_e$.

3.8 Derive the general solution for x from the differential equation in number 7 above; that is,

$$\frac{dx}{dt} = -k_1x^2 + k_2(x_0 - x)$$

for a second-order forward reaction and a first-order back reaction.

3.9 For parallel first-order reactions of $A \xrightarrow{k_1} B$ and $A \xrightarrow{k_2} C$, where $k_2 = 0.1 \times k_1$, and $k_1 = 0.01\,s^{-1}$ plot the concentrations of the three species on the same graph as a function of time up until $[A] = 0.01[A]_0$ with $[A]_0 = 1$ at $t = 0$ and the reactions go to completion; that is, no back reactions.

3.10 For a series reaction $A \xrightarrow{k_1} B \xrightarrow{k_2} C$, the equations for the concentrations of A, B, and C are

$$[A] = [A]_0 e^{-k_1 t}$$

$$[B] = \frac{[A]_0 k_1}{k_2 - k_1}\left(e^{-k_1 t} - e^{-k_2 t}\right)$$

$$[C] = [A]_0 \left\{1 + \frac{1}{k_2 - k_1}\left(k_1 e^{-k_2 t} - k_2 e^{-k_1 t}\right)\right\}$$

a. Make two separate plots of [A], [B], and [C] versus time up to $t = 10^4$ s assuming that $[A]_0 = 1$ and the initial concentrations of B and C are zero. In the first plot, $k_1 = 5 \times 10^{-4}\,s^{-1}$ and $k_2 = 5 \times 10^{-3}\,s^{-1}$ and in the second plot, the values of the constants are just reversed, $k_1 = 5 \times 10^{-3}\,s^{-1}$ and $k_2 = 5 \times 10^{-4}\,s^{-1}$.

b. Derive approximate expressions for these three equations if $k_2 \gg k_1$.

c. Calculate the expressions for the three concentrations as a function of time if $k_1 = k_2$.

REFERENCES

Billmeyer, F. W. Jr. 1984. *Textbook of Polymer Science,* 2nd ed. New York: Wiley Interscience.

Blanchard, P., R. L. Devaney, and G. R. Hall. 2012. *Differential Equations,* 4th ed. Boston, MA: Brooks/Cole (Cengage Learning).

Celia, M. A. and W. G. Gray. 1992. *Numerical Methods for Differential Equations.* Englewood Cliffs, NJ: Prentice Hall.

Chang, R. 2000. *Physical Chemistry for the Chemical and Biological Sciences.* Sausalito, CA: University Science Books.

CHEMKIN. http://www.ca.sandia.gov/chemkin/. Program available from: Reaction Design, San Diego, CA.

Cowie, J. M. G. 1991. *Polymers: Chemistry & Physics of Modern Materials,* 2nd ed. New York: Chapman and Hall.

Edwards, C. H. and D. E. Penney. 1986. *Calculus and Analytical Geometry,* Englewood Cliffs, NJ: Prentice Hall.

Elias, H.-G. 1997. *An Introduction to Polymer Science.* New York: VCH Publishers.

Houston, P. L. 2001. *Chemical Kinetics and Reaction Dynamics.* Mineola, NY: Dover.

Kee, R. J., M. E. Coltrin, and P. Glarsborg. 2003. *Chemically Reacting Flow.* Hoboken, NJ: Wiley.

Kreyszig, E. 1988. *Advanced Engineering Mathematics,* 6th ed. New York: Wiley.

Laidler, K. J. 1987. *Chemical Kinetics,* 3rd ed. New York: Harper and Row.

Laidler, K. J. and J. H. Meiser. 1995. *Physical Chemistry,* 2nd ed. Boston, MA: Houghton Mifflin.

Li, J., Z. Zhao, A. Kazakov, and F. L. Dryer. 2004. An Updated Comprehensive Kinetic Model of Hydrogen Combustion. *Int. J. Chem. Kinetics.* 36: 566–575.

Moore, J. W. and R. G. Pearson. 1981. *Kinetics and Mechanism.* New York: Wiley.

NIST. 2013. *NIST Chemical Kinetics Database.* Standard Reference Database 17, Version 7.0 (Web Version). Release 1.6.7. Data Version 2013.03. http://kinetics.nist.gov/kinetics/index.jsp.

Saunders, K. J. 1988. *Organic Polymer Chemsitry,* 2nd ed. New York: Chapman and Hall.

Silbey, R. J. and R. A. Alberty. 2001. *Physical Chemistry,* 3rd ed. New York: Wiley.

Young, R. J. and P. A. Lovell. 1991. *Introduction to Polymers,* 2nd ed. London: Chapman and Hall.

4

Temperature Dependence of the Reaction Rate Constant

4.1 INTRODUCTION

The main purpose of this chapter is to demonstrate why the Arrhenius equation depends exponentially on temperature. This is not a first principles approach, but rather it is based on the barometric formula, the effect of altitude on atmospheric pressure, that assumes the equation of state of an ideal gas: PV = nRT. This result leads naturally to the Boltzmann distribution for the energy of atoms and molecules as a function of temperature. Because reaction rates depend exponentially on temperature, if a reaction is exothermic, releasing heat, then its rate can continually increase resulting in very high temperatures: combustion. Examples are given showing both the bad and useful effects of this exponential temperature dependence. Finally, a thermodynamic argument leading to an exponential temperature is also presented that leads to the concepts of an *activated state*, an *activation energy*, and how catalysts affect reaction rates.

4.2 THE ARRHENIUS EQUATION

4.2.1 ORIGINS

As was indicated earlier, reaction rate constants are exponential functions of temperature,* $k = k_0\exp(-Q/RT)$ or $k = A\exp(-E_A/RT)$, the so-called *Arrhenius equation*, named after Svante Arrhenius (Figure 4.1). Arrhenius was a Swedish chemist who received the 1903 Noble Prize in chemistry for his work on electrolytes (Worek 2008). However, he also did research on chemical kinetics. Van't Hoff originally proposed the exponential temperature dependence of reaction rates in the late 1800s based on his equation proposed in 1884 for the temperature dependence of equilibrium constants,

$$\frac{d\ln K}{dT} = \frac{\Delta H^\circ}{RT^2} \quad \text{or} \quad \frac{d\ln K}{d\,1/T} = -\frac{\Delta H^\circ}{R} \tag{4.1}$$

where:
K is thermodynamic equilibrium constant
ΔH° is the standard enthalpy for the reaction (J/mol)
R is the gas constant (J/mol-K).

FIGURE 4.1 Svante Arrhenius (1859–1927). (Photograph in the public domain because the copyright has expired, http://commons.wikimedia.org/wiki/File:Arrhenius2.jpg.)

As was seen in Chapter 3, when forward and back reactions with rate constants k_1 and k_2 go to equilibrium, the equilibrium constant is just the ratio of these two rate constants: $K = k_1/k_2$. Van't Hoff realized that, if K is exponentially dependent on 1/T, the reaction rate constants must also have the same temperature dependence and therefore must be of the form

$$k = Ae^{\frac{-E_A}{RT}}$$

* More precisely, the rate constant is an exponential function of 1/T.

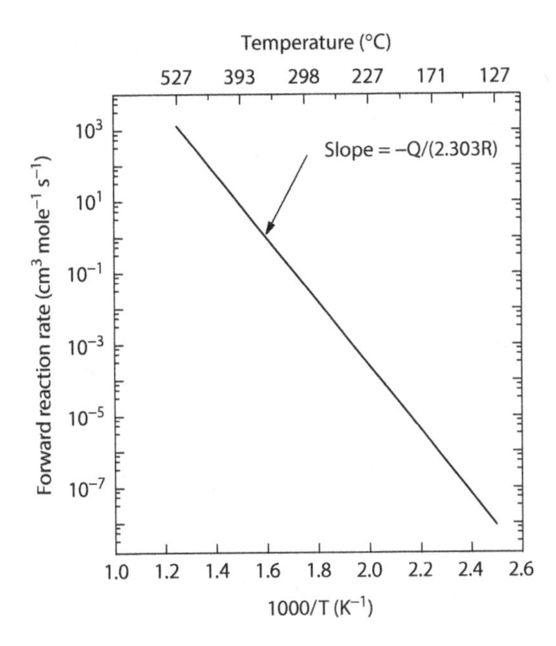

FIGURE 4.2 The reaction rate constant as a function of T for the reaction $H_2(g) + I_2(g) \rightleftharpoons 2HI(g)$. (Data from the NIST Kinetics Database, http://kinetics.nist.gov/kinetics/index.jsp.)

where:

A is the constant, the *preexponential* term, the units of which depend on the order of the reaction

E_A is the *activation energy* (J/mol) (Laidler 1984).

This is what might be termed a *plausibility argument*, in that what is known about the temperature dependence of the equilibrium constant, and the relationship between it and the reaction rate constants, makes the suggestion of the Arrhenius equation for the rate constants entirely reasonable and consistent with the known facts. Arrhenius's subsequent contribution was that he felt that the temperature dependence of colliding atoms or molecules in gases and liquids was too small to give the observed values of activation energies. So he proposed an equilibrium between reactive molecules and the others based on the van't Hoff equation (Laidler 1984). Of course, later experiments have shown that reaction rate constants do follow an Arrhenius equation and are indeed exponentially dependent on 1/T (there is often a mild temperature dependence of the preexponential term as well). Figure 4.2 shows the exponential temperature dependence for the second-order reaction rate constant (note the units for k) for the forward reaction of $H_2(g) + I_2(g) \rightleftharpoons 2HI(g)$. In this case (NIST 2013),

$$k_1 = 1.94 \times 10^{14}\left(cm^3 mol^{-1} s^{-1}\right) e^{-\frac{171 kJ/mol}{RT}}$$

Note that the reaction rate increases by about *11 orders of magnitude* over a span of only about 400°C. By measuring the rate of a reaction as a function of temperature, the two important parameters of the rate equation, Q and k_0, can be determined.

4.2.1.1 Example: Calculating Q and k_0 from Kinetic Reaction Rate Constants and Temperature

Given the following rate constants for some reaction at two different temperatures:

$$k_1 = 5.7 \times 10^{-5}\ s^{-1} \text{ at } 25°C$$

$$k_2 = 1.64 \times 10^{-4}\ s^{-1} \text{ at } 40°C$$

what are the values for Q and k_0? By taking the ratio k_1/k_2, k_0 is eliminated and Q can be obtained

$$\frac{k_1}{k_2} = \frac{e^{-\frac{Q}{RT_1}}}{e^{-\frac{Q}{RT_2}}} = e^{-\frac{Q}{RT_1}} e^{\frac{Q}{RT_2}} = e^{-\frac{Q}{R}\left(\frac{1}{T_1}-\frac{1}{T_2}\right)} \quad \text{so} \quad \ln\frac{k_1}{k_2} = -\frac{Q}{R}\left(\frac{1}{T_1}-\frac{1}{T_2}\right)$$

$$Q = -\frac{R\ln(k_1/k_2)}{((1/T_1)-(1/T_2))} = \frac{-8.314\ln[(5.7\times10^{-5})/(1.64\times10^{-4})]}{((1/298)-(1/313))} = 54{,}636\,\text{J/mol-K}$$

k_0 can now be obtained from either of the two given rates because $k = k_0\exp(-Q/RT)$, so $k_0 = k\exp(Q/RT)$

$$k_0 = 5.7\times10^{-5}\,e^{\frac{54{,}636}{8.314\times298}} = 2.15\times10^5\,\text{s}^{-1}$$

or

$$k_0 = 1.64\times10^{-4}\,e^{\frac{54{,}636}{8.314\times313}} = 2.15\times10^5\,\text{s}^{-1}.$$

4.3 THE HINDENBURG DISASTER

4.3.1 Combustion of Hydrogen

Because reaction rates are typically exponentially temperature dependent, once a reaction that has a large exothermic heat of reaction—essentially the thermodynamic enthalpy ΔH°—initiates, if this heat is not dissipated, it will continuously increase the temperature of the reactants and products, leading to an ever increasing reaction rate and temperature that can result in an explosion. This was the situation that befell the Hindenburg in May of 1937 when its hydrogen buoyancy gas ignited (Figure 4.3). This picture of the Hindenburg burning immediately comes to mind and generates concern when hydrogen is being considered as a motor vehicle fuel, possibly in a fuel cell. However, it should be noted that gasoline has almost three times the energy per unit volume than liquid hydrogen, so it is potentially more explosive. On the other hand, gasoline does not evaporate rapidly, whereas hydrogen–air mixtures are explosive between 5 and 95 v/o (volume percent) hydrogen.

4.3.2 Hydrogen and the Hindenburg

Modern dirigibles or airships and balloons are filled with inert helium rather than the explosive hydrogen that destroyed not only the Hindenburg but the future of passenger travel by lighter than

FIGURE 4.3 Hindenburg hydrogen-filled zeppelin accident, Lakehurst, New Jersey, May 6, 1937. (U.S. Navy photograph in the public domain, http://commons.wikimedia.org/wiki/File:Hindenburg burning.jpg.)

air airship or dirigible.* Why was not the Hindenburg filled with helium? Helium was not discovered until 1868 when studies on the spectrum of the sun found it in the atmosphere of the sun as the product of the nuclear fusion reactions such as

$$^{3}_{1}H + ^{2}_{1}H \Rightarrow ^{4}_{2}He + ^{1}_{0}n + 17.6\,MeV. \tag{4.2}$$

The new element was named "helium" after the Greek god of the sun, Helios (Emsley 2001). In reality, there is a lot of helium around because most of the visible matter in the universe consists of about 75% hydrogen and 25% helium created at the time of the *Big Bang*, 13.75 billion years ago when the universe was born. However, on the earth, helium is found only at about 8 ppb (parts per billion) in the earth's crust where it occurs because of the alpha particle decay of some elements. Recall the decay of uranium 238 from Chapter 2:

$$\begin{aligned} &^{238}_{92}U \xrightarrow{4.51\times10^{9}\,year} {}^{234}_{90}Th + {}^{4}_{2}He + 4.25\,MeV \\ &^{234}_{90}Th \rightarrow \text{eventually to Pb.} \end{aligned} \tag{4.3}$$

Helium had been detected in small amounts in uranium minerals in the late 1800s. In 1903, helium was found in significant quantities in natural gas wells in the United States and its potential for use in lighter than air airships was immediately recognized.

But there is considerably more to the Hindenburg story (Provan 2011). Hydrogen was used to inflate gas balloons as early as the 1850s. Count von Zeppelin proposed a hydrogen-filled airship in 1895 and built one by 1900. These hydrogen-filled Zeppelins flew successfully for the next 40 years.† However, with the recent availability of much safer helium as a buoyancy gas, the Hindenburg was designed to use it. At that time (the 1930s), the United States was the world's largest producer of natural gas—and helium as a by-product. Because of helium's potential for military airship applications, its low abundance and high cost, and the virtual monopoly on its supply, the United States banned the export of helium in 1927. Germany thought that the United States would lift its ban and allow it to purchase helium for the Hindenburg class of commercial airships. However, the United States refused to lift the ban because Germany had used airships for propaganda purposes as well as the concern of many in the United States and other parts of the world—probably with good reason—that these airships were being used for mapping by the German military for use in the not too distant future. Without access to helium, the Hindenburg was redesigned and reconstructed to use hydrogen. After all, Germany had been using its airships for over 40 years without a major mishap and, besides, it was cheaper than helium and had more lifting ability so that more passengers could be carried making the venture more profitable (Provan 2011).

4.3.3 A Materials Failure?

What caused the Hindenburg to crash and burn has been studied repeatedly for over 70 years (Provan 2011). Was it a gas leak? Was it a lightning strike? There were several thunderstorms in the Lakehurst New Jersey crash area on May 6, 1937, and the landing was delayed until the weather improved. No witness claims to have seen a lightning strike. Was it sabotage, perpetrated by a disgruntled employee or someone opposed to the National Socialist German Workers' Party who then governed Germany? Was it a spark produced by a buildup of electrostatic charge on the covering of the airship? The external covering was a coating with a cellulose nitrate paint embedded with aluminum flakes to make it conductive and prevent charge buildup. Subsequent investigations have suggested that a spark ignited the very flammable paint, and its burning caused the rapid spread of

* The initiation of trans-oceanic airplane travel was really a more important factor in the demise of airship travel.

† Interestingly, it was the discovery of age-hardenable aluminum alloys, duralumin, around the turn of the century, with a high strength and light weight that allowed the construction of large passenger-carrying dirigibles such as the Hindenburg. This is another example of the *enabling* of a technology by materials.

flames rather than burning of hydrogen. Was there an insufficient or a nonuniform distribution of the aluminum pigment to provide the necessary electrical conductivity to bleed off any charge accumulation? If so, then the loss of the Hindenburg could be attributed to a twofold materials failure: the conductive paint did not perform as expected and the polymer binder in the paint was too flammable to be used as the final encapsulation for such a large amount of highly combustible hydrogen.*

4.4 CONTROLLED COMBUSTION REACTIONS

4.4.1 Hydrogen–Oxygen Combustion

The fact that hydrogen and oxygen react and generate high temperatures is used for combustion processes in gas torches and fuel for propulsion. Figure 4.4 shows the liftoff of the Columbia space shuttle launch on STS-1† in 1981 on the first shuttle mission. In 1992, on STS 50, the Columbia carried the U.S. Materials Laboratory experiments and two materials researchers, Dr. Bonnie Dunbar who was an astronaut and Dr. Eugene Trinh a mission specialist. The main engines of the space shuttle are fueled by liquid oxygen–liquid hydrogen fuel stored in the large main tank directly under the shuttle.‡

Hydrogen–oxygen flames are also used for cutting torches because the *adiabatic flame temperature*—all of the enthalpy in the reaction heats the products and is not lost to the surroundings—is about 4500°C (Butts 1943). Figure 4.5 is a temperature–enthalpy plot for the formation of water vapor from the reaction of hydrogen and oxygen:

$$H_2(g)+\frac{1}{2}O_2(g)\rightarrow H_2O(g). \tag{4.4}$$

The enthalpy change for each of the reactants and products in this figure is essentially $H_T = H_{398} + C_P(T-398)$ if molar heat capacities are constant and no phase changes occur, as assumed in this figure. That is why the starting temperature was taken as 100°C to avoid the heat of vaporization of H_2O for this graph. In reality, the heat capacities are not constant as discussed in Chapter 1, but close enough to being constant to demonstrate the principle involved. At all temperatures, the enthalpy for the reaction is about −250 kJ/mol of H_2O, an exothermic reaction. If no heat is lost to

FIGURE 4.4 The space shuttle Columbia taking off on STS-1 on April 12, 1981 on its initial flight. Its final flight, STS-107, ended in the destruction of the vehicle on re-entry. (NASA photograph in the public domain. Original color image converted to grayscale image, http://www.NASA.gov/centers/kennedy/shuttleoperations/orbiters/columbia_info.html.)

* The zeppelins built both before and after the Hindenburg used a less flammable polymer binder.

† STS stands for "Space Transportation System" which the shuttle program was originally called. Each shuttle mission was given an STS designation more or less in sequence of launch date.

‡ Tragically, on its final mission, STS-107, the Columbia was destroyed and the crew lost on re-entering the earth's atmosphere.

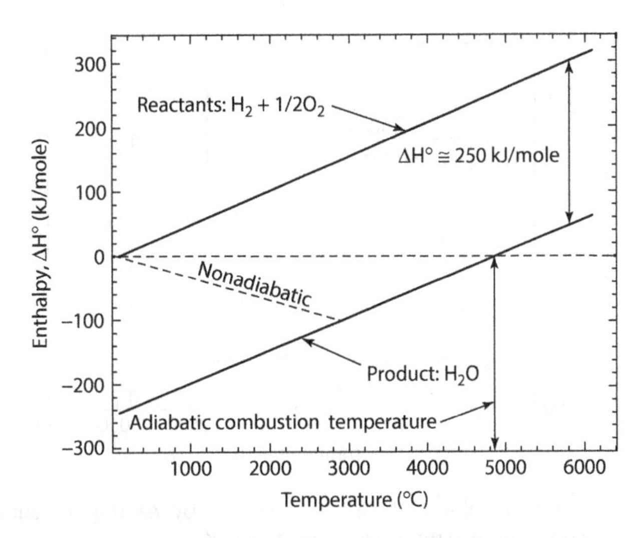

FIGURE 4.5 Enthalpy versus temperature plot for the reaction: $H_2(g) + 1/2O_2(g) \rightarrow H_2O(g)$. This figure shows that if the reaction takes place at 100°C, then the adiabatic temperature is around 4800°C.

the surroundings during the reaction,—*adiabatic conditions*—then the *adiabatic combustion temperature* is reached—the intersection of the horizontal dashed line in the figure with the product enthalpy. The higher the value of the reaction rate constant, the more likely combustion will be rapid and adiabatic. With the data used for this graph, if the combustion occurs at 100°C, an adiabatic temperature of about 4800°C is reached. As a result, the pressure generated by the high temperature gas is the driving force for the liquid fuel main engines on the space shuttle. The downward sloping line labeled "nonadiabatic" is what happens when some heat is lost to the surroundings by conduction, convection, and radiation, much more likely if the reaction rate is slow. In the example shown, the temperature reached would be about 3000°C, still hot enough to melt most metals.

4.4.2 Combustion Synthesis

The two rockets on either side of the shuttle that are producing the majority of the brightness and all of the smoke in Figure 4.4 are solid fuel boosters that make use of the reaction of powdered aluminum with a strong oxidizer, namely, ammonium perchlorate,

$$4Al(s) + 2NH_4ClO_4(s) \rightarrow 2Al_2O_3(s) + 2H_2O(g) + N_2(g) + 2HCl(g) + H_2(g) \tag{4.5}$$

generating a lot of expanding gas at high temperatures and pressures for propulsion because of the very high enthalpy and rapid rate of the reaction. The smoke primarily consists of (probably nanosize) particles of aluminum oxide. The formation of Al_2O_3 by this high temperature combustion process is an example of a materials synthesis process called *combustion synthesis* or *self-propagating high temperature synthesis (SHS)* that has been used to produce a wide variety of materials. However, producing them in the exhaust of the shuttle is probably not a cost-effective way of making materials! Interest in the process began in the old Soviet Union when solid fuel rocket scientists and engineers decided to apply their expertise to synthesizing materials after the end of the Cold War in the 1980s when rocket design became a less popular career choice.

The principle of combustion synthesis is fairly straightforward. Figure 4.6 shows the enthalpy as a function of temperature for the reactants and products of a reaction to produce a high temperature and a high melting point material; in this case, $Ti(s) + 2B(s) \rightarrow TiB_2(s)$. In this figure, the enthalpy of melting is shown for both the products and reactants. If two solids are mixed, such as Ti and B powders, formed into a cylinder, and ignited at some point, then the enthalpy for the reaction will cause the temperature to become sufficiently high to continue the reaction. In this example, an adiabatic temperature above the melting points of both reactants

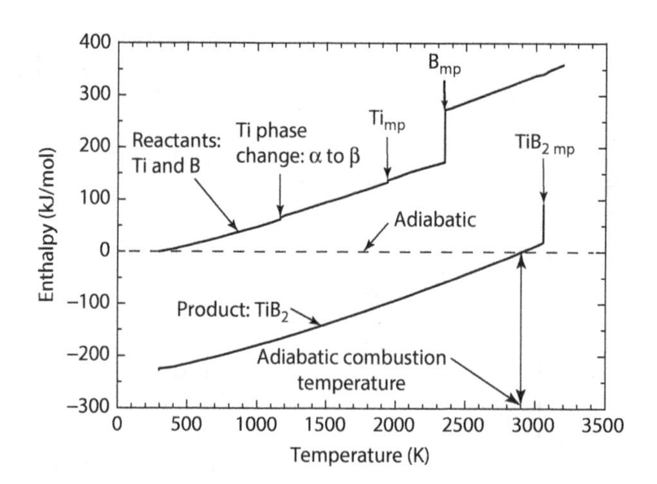

FIGURE 4.6 Temperature–enthalpy plot for the formation of TiB_2 by reacting titanium and boron mixtures. The adiabatic combustion temperature in this case is about 2900 K.

is produced. However, the actual ignition temperature may be much higher than room temperature, producing an even higher adiabatic reaction temperature. This reaction zone is essentially a "flame" and the flame front will propagate from the point of ignition throughout the compacted mixed reactant powders. In many cases, the powders are formed into a cylinder and ignited at the top, and the flame front propagates down the length of the cylinder by heat transfer to the unreacted material completing the reaction throughout as shown in Figure 4.7. This is why the process is called SHS because the enthalpy from the reaction produces sufficiently high temperatures to produce a "self-propagating" flame front in the material as the reaction proceeds through the mixed reactants. The heat necessary to get the reaction started—increase the rate of reaction—can either be applied within a furnace, with a torch, or by some other heat source such as a spark. The combustion synthesis process has some advantages particularly if very high purity materials are needed. For example, SiC furnace parts that are used in silicon integrated circuit processing must be very pure. Starting with silicon and carbon of very high purity—electronic grade—SiC can be produced compared to the considerably less pure silicon carbide produced by the conventional Acheson process of reacting SiO_2 and carbon in a carbon-arc furnace to make abrasive and refractory materials where high purity is not essential.

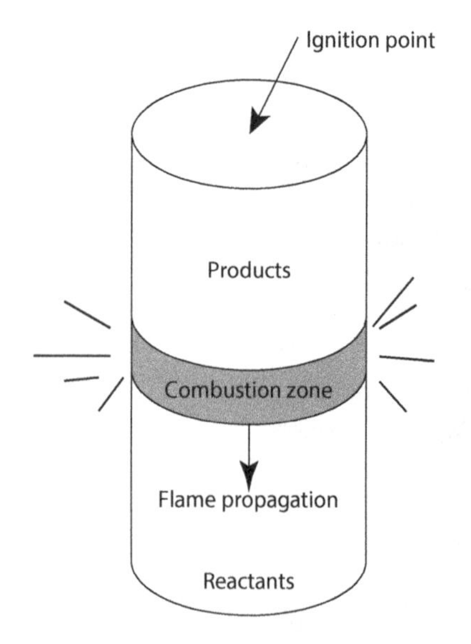

FIGURE 4.7 Schematic drawing showing the process of combustion synthesis or SHS—self-propagating high temperature synthesis.

Combustion reactions to make useful materials are not a new process. The *thermite* (themit) process was invented in the late 1800s and was/is used to weld steel railroad rails by the reaction of mixed powders of iron oxide and aluminum (Cary 1979):

$$Fe_2O_3(s) + 2\ Al(s) \rightarrow Al_2O_3(s) + Fe(l) \tag{4.6}$$

which liberates almost a million joules of heat per mole and an adiabatic flame temperature of 4273°C (Roine 2002), more than enough to form molten iron above its melting point of 1538°C, which then flows into a mold forming the weld.

4.5 BAROMETRIC FORMULA

4.5.1 Development

A transparent approach to gain insights into the Arrhenius equation and why there is an exponential temperature dependence of reaction rates is to start with the *barometric formula* (Silbey and Alberty 2001). The barometric formula gives the pressure at higher altitudes relative to the pressure at sea level—or whatever reference point is chosen and can be developed relatively simply.

One of few things in the world that needs to be memorized is $F = ma$. So taking a slice of a column of atmosphere of cross-sectional area A extending from sea level upward, depicted in Figure 4.8,

$$dF = -g\,dm$$

where g is the acceleration of gravity and m is the mass of the volume Adx and dF is negative because the force is downward. Now,

$$dm = \rho\, dV = \rho A\, dx$$

where:

ρ is the density of air
dV is the small volume of air slice.

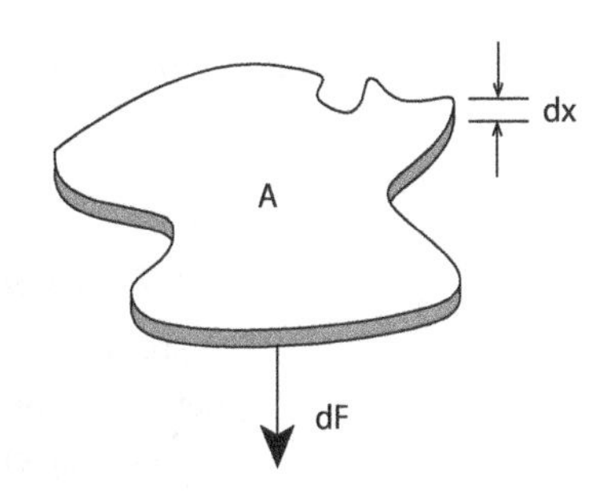

FIGURE 4.8 Schematic representation of cut through a column of air of area A and thickness dx extending from the surface of the earth out to infinity. The small slice of the column exerts a vertical downward force of dF.

Therefore the pressure produced by this slice is

$$dP = \frac{dF}{A} = -\frac{g\rho A\,dx}{A} = -g\rho\,dx \tag{4.7}$$

For an ideal gas, $PV = nRT$ (one of the other things that is useful to memorize), so that

$$\rho = M_A \frac{n}{V} = \frac{M_A P}{RT}$$

where M_A is the atomic or molecular weight of the gas species. A Combination of these equations gives

$$\frac{dP}{P} = -\frac{M_A g}{RT} dx.$$

Integrating from sea level, $P = P_0$ to some height above the earth, h, where the pressure is P

$$\int_{P_0}^{P} \frac{dP}{P} = -\frac{M_A g}{RT} \int_0^h dx$$

leads to the barometric formula that gives the pressure as a function of altitude above sea level:

$$P = P_0 e^{-\frac{M_A g h}{RT}} = P_0 e^{-\frac{m_A g h}{k_B T}} \tag{4.8}$$

where m_A is the atomic or molecular mass of species A ($m_A = M_A/N_A$). Of course, Boltzmann's constant, k_B, is just equal to $k_B = R/N_A = 8.314$ J/mol/6.022×10^{23} atoms/mol $= 1.381 \times 10^{-23}$ J/atom. Actually, Equation 4.8 gives the pressure difference between two elevations separated by a distance, h.

4.5.1.1 Example: Atmospheric Pressure in Denver

Denver is 5280 ft above sea level*, what is the pressure in Denver relative to sea level? If sea level pressure is 1 atm, then the pressure in Denver is

$$P = 1 \times e^{-\frac{(28.84\times10^{-3}\,\text{kg/mol})(9.8\,\text{m/s}^2)(5280\times12\times2.54/100\,\text{m})}{(8.314\,\text{J/mol-K})(298\,\text{K})}}$$
$$= 0.833 \text{ atm}$$

and since 1 atm = 1013.2 millibar, $P = 1013.2 \times 0.833 = 844.9$ millibar.

* Denver is called the "mile-high city" and not the "1.6 km-high city." Hence, the use of English units.

The average molecular weight of air used for this calculation is given by

$$\bar{M} = 0.79\,M(N_2) + 0.21\,M(O_2) = (0.79 \times 28) + (0.21 \times 32) = 28.84\ \text{g/mol}$$

4.5.1.2 Example: Atmospheric Pressure at the International Space Station

The space station is about 220 miles high as it orbits the earth—high enough to miss most atmospheric friction but within the protection of the earth's magnetic field to minimize radiation from solar mass ejection of energetic particles and cosmic rays. The pressure at this altitude is

$$P = 1 \times e^{-\frac{(28.84\times10^{-3}\,\text{kg/mol})(9.8\,\text{m/s}^2)(220\,\text{mi}\times1609\,\text{m/mi})}{(8.314\,\text{J/mol-K})(298\,\text{K})}}$$

$$= 3.79 \times 10^{-18}\ \text{atm}$$

The actual pressure is 10^{-11} atm because there are temperature variations in the atmosphere as a function of altitude and the temperature is not a constant 298 K as assumed here. In any event, the International Space Station (ISS) operates in a pretty good vacuum.

4.5.1.3 Example: The Thickness of Earth's Atmosphere

Assume that the thickness of the atmosphere is that distance from sea level to where the atmospheric pressure drops to 0.01 atm or 1% of the sea level pressure. So

$$h = -\frac{RT}{\bar{M}g}\ln 0.01 = -\frac{8.314\times300\times\ln 0.01}{28.84\times10^{-3}\times9.8} = 4.06\times10^4\,\text{m} \sim \frac{4.06\times10^4\,\text{m}}{1609\text{m/mi}} \simeq 25\,\text{mi}$$

Now the radius of the earth is about 4000 mi, so that the thickness of the atmosphere is about 0.6% of the earth's radius. You might remember Al Gore's picture in *An Inconvenient Truth* (Gore 2006) showing the thickness of the atmosphere compared to the radius of the earth, and its thinness was striking. This is why! And it *is* thin! This was part of his argument about the effect of CO_2 emissions on global warming. In reality, the "ocean of air" above us is really not so deep. Interestingly, Arrhenius, over 100 years ago, was worrying about the effect of atmospheric CO_2 produced by industrial production on the earth's temperature and predicted man-made global warming not greatly different from today's more rigorous calculations (Arrhenius 1896).

4.6 BOLTZMANN DISTRIBUTION

What is the relation between the Arrhenius equation and the barometric formula? Notice that M_Agh or m_Agh is the potential energy of the atoms at the higher elevation compared to the energy of the atoms at the lower elevation. As such, this is a *Boltzmann distribution* (Silbey and Alberty 2001). Again, since PV = nRT, the barometric formula could be rewritten as

$$n = n_0 e^{-\frac{E}{k_B T}} = n_0 e^{-\frac{Q}{RT}} \tag{4.9}$$

that is, the number of moles in some energy state E equals the number in the ground state n_0, where E = 0, times an exponential factor that depends on energy difference between the two states relative to thermal energy, the *Boltzmann factor.* Both sides can be multiplied by Avogadro's number to give the number of atoms or molecules in the two states, N and N_0. So this expression could now be written as

$$N(E) = N_0 e^{-\frac{E}{RT}}. \tag{4.10}$$

If the total number of atoms or molecules available is N_T, then

$$N_T = \int_0^\infty N(E)dE = \int_0^\infty N_0 e^{-\frac{E}{k_B T}} dE = N_0 k_B T \tag{4.11}$$

which can be easily integrated by just letting $x = E/k_BT$. So the fraction of atoms or molecules having energy E, f(E), is just

$$f(E) = \frac{N(E)}{N_T} = \frac{1}{k_BT} e^{-\frac{E}{k_BT}} \tag{4.12}$$

which is the Boltzmann distribution. This result could also be obtained by taking the fraction of molecules, f(E), having energies between E and E + dE written as $f(E) = (1/N)(dN/dE)$ or

$$f(E) = \frac{1}{N}\frac{dN}{dE} = A'e^{-\frac{E}{RT}}$$

and can be integrated to get the value of A′:

$$\int_0^\infty \frac{dN}{N} = 1 = A'\int_0^\infty e^{-\frac{E}{RT}} dE = A'RT\int_0^\infty e^{-x}dx =$$

$$-A'RT\left[e^{-x}\right]_0^\infty = -A'RT\left[e^{-\infty} - e^{-0}\right] = A'RT.$$

So $A' = 1/RT$ and

$$f(E) = \frac{dN(E)}{N} = \frac{1}{RT} e^{-\frac{E}{RT}} dE \tag{4.13}$$

as before. Now, if molecules need a certain minimum energy to react, call it Q, then only those molecules or atoms with $E \geq Q$ will react, or the fraction of the molecules reacting will be

$$f(E \geq Q) = \frac{1}{k_BT}\int_Q^\infty e^{-\frac{E}{k_BT}} dE = -e^{-\frac{E}{k_BT}}\Bigg]_Q^\infty$$

or

$$f(E \geq Q) = e^{-\frac{Q}{k_BT}}. \tag{4.14}$$

Hence, the exponential dependence of the rate constants with temperature. Equation 4.14 is shown in Figure 4.9.

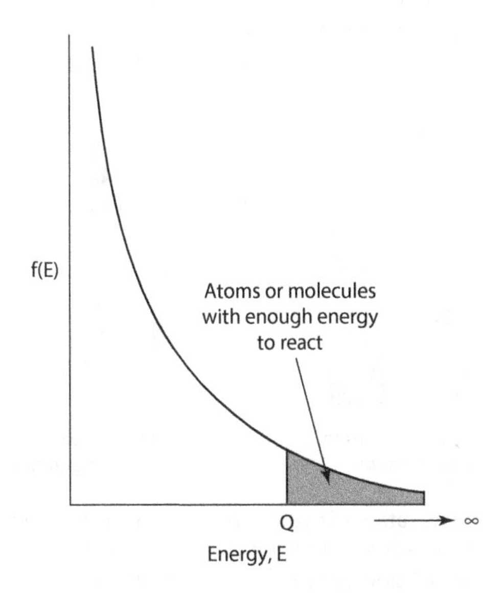

FIGURE 4.9 Boltzmann distribution of energy showing that molecules with an energy greater than some minimum for reaction, Q, will react: shaded region.

4.7 THE ACTIVATED STATE

4.7.1 Activation Energy

An alternate way to justify the Arrhenius equation is to use a thermodynamic approach and assume that only certain atoms or molecules have energies in excess of the activation energy, E_A, and that they are in some *activated state*. For example, for the reaction of H_2 and I_2 to form HI the hydrogen and iodine molecules will collide with each other in the gas phase, so there will be Z such collisions per second (Figure 4.10). However, only some of these molecules will have enough energy to react, namely, those in an *activated state*, H_2^*, for example. Assuming a first-order reaction in hydrogen for simplicity, the reaction rate could be written as

$$\frac{d[H_2]}{dt} = -Z[H_2^*] \tag{4.15}$$

where $[H_2^*]$ is the concentration of hydrogen molecules that have enough energy to react or the concentration of hydrogen molecules in the "activated state." There is an equilibrium between the molecules in the activated state and those in a normal state, determined by an *activation Gibbs energy*, ΔG^*,

$$\Delta G^* = \Delta H^* - T\Delta S^*$$

and an equilibrium constant can be written for the reaction $H_2 \rightleftharpoons H_2^*$

$$\frac{[H_2^*]}{[H_2]} = K^* = e^{-\frac{\Delta G^*}{RT}} = e^{\frac{\Delta S^*}{R}} e^{-\frac{\Delta H^*}{RT}} \tag{4.16}$$

so that the rate of reaction becomes

$$\frac{d[H_2]}{dt} = -Z[H_2]e^{\frac{\Delta S^*}{R}} e^{-\frac{\Delta H^*}{RT}} = -\left(Ze^{\frac{\Delta S^*}{R}}\right)\left(e^{-\frac{\Delta H^*}{RT}}\right)[H_2]$$

$$= -\left(k_0 e^{-\frac{Q}{RT}}\right)[H_2] = -k[H_2] \tag{4.17}$$

which means that the activation energy, Q, is *the activation enthalpy*, ΔH^*, and $k_0 = Z\exp(\Delta S^*/R)$ in $k = k_0 \exp(-Q/RT)$. The *activation entropy*, ΔS^*, probably includes some orientation effects suggested in Figure 4.10, namely, the reacting molecules have to be oriented in a particular spatial relation, or close to it, in order to react. Two possible orientations are shown at the bottom of the figure.

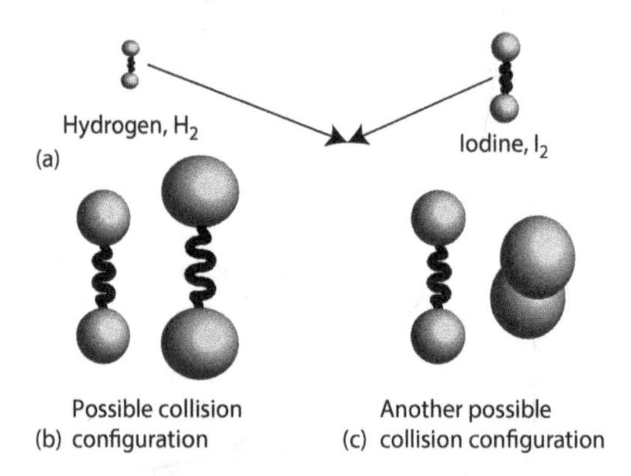

FIGURE 4.10 (a) Schematic representation of hydrogen and iodine molecules colliding in the gas phase to react to form 2 HI molecules. (b) Possible orientation configuration of the molecules when they collide. (c) Another possible orientation on collision. Note that the bonds between the atoms are represented by springs giving rise to the vibrational energy of the diatomic molecules. However, the distances between molecules and the lengths of the spring are greatly exaggerated.

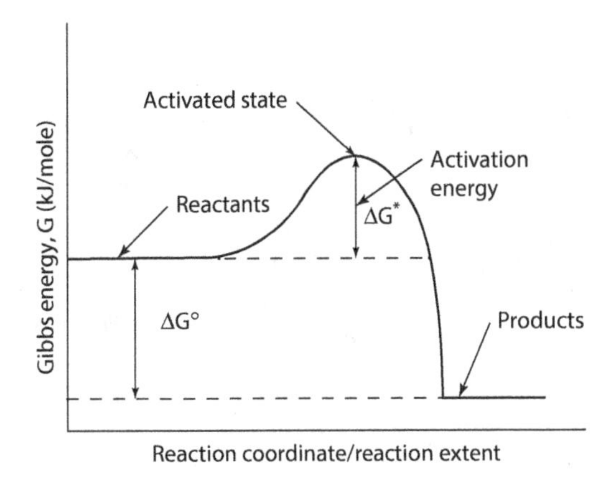

FIGURE 4.11 Free energies as a function of the extent of the reaction from reactants to products. In order to react, the reactants must reach some activated state or activation free energy, ΔG^*.

However, it should be noted that all of these diatomic molecules are rapidly rotating and vibrating that makes the details of their interaction rather complex.

Another way of representing this activated state is on a plot of the Gibbs energy as a function of the *extent of the reaction* or the *reaction coordinate* shown in Figure 4.11. $\Delta G°$ is the energy for the reaction, but for it to occur, molecules must be excited into the *activated state* having an energy ΔG^* above the energy of the reactants. Therefore, there is an *activation energy* or *activation barrier*, ΔG^*, that must be overcome for the reaction to proceed.

4.7.2 Role of a Catalyst

It is usually stated that the role of a catalyst—*something that speeds up(usually) the rate of a reaction but does not enter the reaction*—is to lower the activation energy for reaction. For example, platinum is used to speed up the reaction between oxygen in hydrogen in a proton exchange membrane (PEM) fuel cell.* Platinum is also used as a catalyst to speed up the combination of oxygen ions diffusing through ionically conducting zirconia, ZrO_2, to form oxygen gas in automobile oxygen sensors and, of course, it is used in large quantities as the oxidation catalyst in vehicle catalytic converters to oxidize CO and any unburned hydrocarbons in exhaust gases.

However, catalysts usually play a more complex role than merely reducing the activation energy for a reaction. More generally, a catalyst provides an entirely different reaction pathway for a reaction to occur, most likely containing several different steps as shown schematically in Figure 4.12 (Haim 1989). As a result, to suggest that the primary role of a catalyst is to simply lower the activation energy for reaction is too simplistic. Although the intent is not to spend a lot of time on the atomistic or molecular details of reactions at surfaces or interfaces, one illustration is given for the simple case of what is thought to occur at the surface of platinum when it catalyzes the simple reaction of (Dickerson and Geis 1976; Friend 1993)

$$H_2(g) + D_2(g) \rightleftharpoons 2HD(g)$$

where D stands for deuterium or ^2H. Figure 4.13 attempts to show the five steps envisioned to occur at the surface of the platinum. *Note that because these are series steps, the slowest will control the overall rate.* The steps are the following: (1) H_2 and D_2 adsorb onto the surface of the platinum. (2) The diatomic bonds in the H_2 and D_2 molecules are broken by the electronic attraction—and larger separation distance—of the electron orbitals of the platinum surface atoms. Eventually, most of the Pt sites on the surface are covered with single H or D atoms. When another molecule is absorbed to one of the few remaining empty sites, as in step 3, it can no longer be separated, but it is attracted to a neighboring H or D atom. The molecule can then exchange one of its atoms to the adsorbed

* A PEM fuel cell is called either a polymer electrolyte membrane or proton exchange membrane fuel cell.

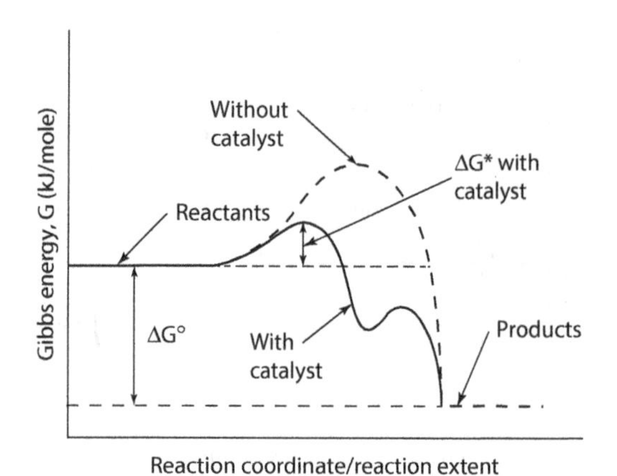

FIGURE 4.12 The effect of a catalyst on the reaction rate, activation energy, and activated state. The reality is that a catalyst frequently leads to a different pathway for the reaction that may show several steps, two in series are shown here, and not just lower the activation energy for the reaction.

atom forming another molecule—on the average, an HD molecule. The HD molecule is shown on the platinum surface in step 4. Finally, the HD molecule desorbs from the surface as shown in step 5 releasing this site for further reaction. In reality, the H and D atoms are probably diffusing around on the surface, so that the locations of the open sites are constantly changing.

Clearly, the electronic surface properties of the platinum are extremely important. There should be a sufficiently strong attraction between the Pt and the gas molecules to break the bond and separate the molecule into atoms, but not strong enough to form a permanent bond with either the diatomic molecule or a single atom at the surface. Understanding the details of the electronic structures of surfaces and their roles in catalysis is an active area of surface physics and chemistry research. It is very attractive to be able to tailor the electronic properties of an alloy or compound consisting of more readily available and cheaper materials than platinum or other noble metals. There have been instances in the past in which new catalyst compounds were thought to have been found only to have subsequent research show that the materials were contaminated with small platinum particles that were actually doing the catalysis.

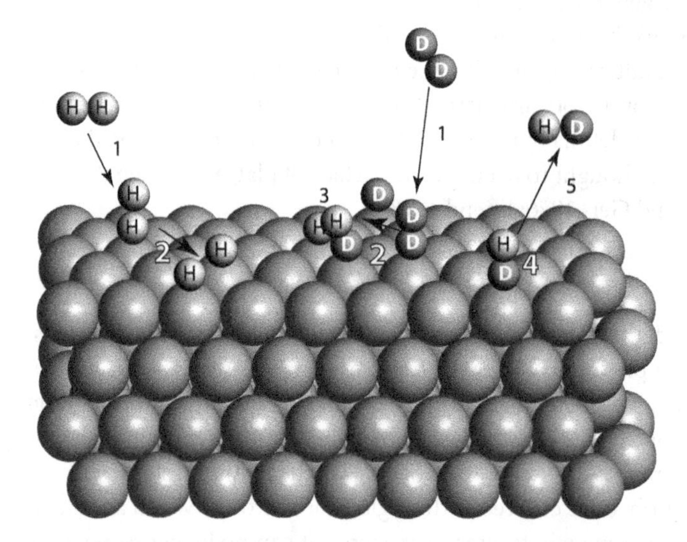

FIGURE 4.13 Schematic representation of the reaction $H_2(g) + D_2(g) \rightleftharpoons 2HD(g)$ on the surface of a noble metal such platinum: adsorption of H_2 and D_2 on the surface (1), breaking of the molecular bond and separation of the atoms on the surface (2), the molecule interacts with a surface atom forming an HD molecule (3), HD molecule attached to the surface (4), and desorption of HD from the surface (5).

Other good examples where catalysts provide more than just a decrease in activation energies are biological processes in which enzymes serve as catalysts. Take for example the oxidation of sucrose, sugar, $C_{12}H_{22}O_{11}$. Sugar will burn in air being oxidized to

$$C_{12}H_{22}O_{11}(s)+12O_2(g)\rightarrow 12CO_2(g)+11H_2O(g) \tag{4.18}$$

and the large enthalpy for the reaction will sustain combustion just like any SHS reaction. In fact, a mixture of sugar and a good oxidizer such as potassium nitrate, KNO_3, makes a nice solid fuel model rocket propellant. However, in the human body, enzymes take the sugar molecules apart slowly piece by piece generating the energy necessary for life. Clearly, the overall reaction pathways for combustion and biological oxidation of sugar are very different, and the enzymes do much more than just lower the activation energy of the reaction.

Another example is the use of Ziegler–Natta catalysts (Claverie and Schaper 2013) on the polymerization of olefins, unsaturated hydrocarbon molecules with one or more carbon double bonds such as ethylene, $H_2C{=}CH_2$. In this case, these catalysts—both homogeneous and heterogeneous—are designed to control, usually minimize, the branching of the polymer chain. As was seen in Chapter 3, *radical polymerization* is frequently used to make nominally linear polymers such as polyethylene, $[CH_2CH_2]_n$. However, during polymerization, the radical may end up someplace along the chain, rather than at its end, forming a side branch and a branched polymer molecule. In contrast, Ziegler–Natta catalysts contain transition metal atoms, and the interaction of the monomers with the d-orbitals of these atoms adds additional monomers to the polymer molecule only at one site, essentially at the beginning of the molecule. As a result, these catalysts minimize branching. The lack of branching leads to better packing of the polymer molecules and a larger degree of crystallinity in polymers such as *high density polyethylene*, HDPE. These catalysts can also be designed to control the stereochemistry of the polymer chain, the *tacticity*, by determining how the various side groups arrange along the length of the polymer chain. Ziegler–Natta catalysts clearly demonstrate that the role of a catalyst is not merely to lower the activation energy for reaction but actually change the overall configuration of the polymer molecules.

4.8 SUMMARY

The goal of this chapter has been to justify the exponential temperature dependence of reaction rate constants and how exothermic reaction rates can increase rapidly because of this strong temperature dependence. Strongly temperature exothermic rates can be taken advantage of in propulsion, welding torches, and the production of materials by combustion synthesis as examples. The first approach developed the barometric formula leading to the Boltzmann distribution for atoms and molecules as a function of temperature and the meaning of the activation energy. The thermodynamic approach to the activated state is presented, and the activation energy and the activated state are introduced. Finally, the role of catalysts in reaction kinetics is briefly discussed emphasizing that the role of catalysts is not merely to lower the activation energy for a reaction, but, more generally, achieve this by changing the paths or steps that take place in the catalyzed reaction compared to those in the absence of the catalyst. This chapter and the preceding two chapters have focused on homogeneous reactions to introduce the concepts of reaction kinetics. In the next chapter, heterogeneous reaction kinetics are introduced, specifically gas–solid reactions that play huge roles in many processes in materials science and engineering.

EXERCISES

4.1 a. If $k = 2.16 \times 10^{-6}\ s^{-1}$ at 25°C and $k = 5.35 \times 10^{-2}\ s^{-1}$ at 100°C for a gas reaction, calculate the values of the preexponential factor and the activation energy.

b. Make a plot of $\log_{10}$(rate) versus 1000/T over this temperature range.

4.2 Make a sketch showing the enthalpy versus temperature for both the products and reactants for a combustion reaction such as $C_2H_2(g) + 3/2\ O_2(g) \rightleftharpoons H_2O(g) + 2\ CO_2(g)$.
Show a temperature at which combustion could initiate and the adiabatic combustion temperature.

4.3 a. The average heat capacities for various gases are $C_P(H_2O)$ = 50 J/mol-K, $C_P(N_2)$ = 33 J/mol-K, $C_P(CO_2)$ = 55 J/mol-K. If the ΔH° at 30°C for the reaction $H_2(g) + 1/2\ O_2(g) \rightarrow H_2O(g)$ is −241.8 kJ/mol per mole of H_2, calculate the adiabatic temperature for this reaction if ignition is at 30°C in pure oxygen.
b. Do the same if the reaction occurs in air.

4.4 a. If the ΔH° at 30°C for the reaction $CH_4(g) + 2O_2(g) \rightarrow CO_2 + 2H_2O(g)$ is −802.5 kJ/mol per mole of CH_4, calculate the adiabatic temperature for this reaction if ignition is at 30°C in pure oxygen.
b. Do the same if the reaction occurs in air.

4.5 a. If the ΔH° at 30°C for the reaction $C_2H_2(g) + 5/2O_2(g) \rightarrow 2CO_2 + H_2O(g)$ is −1256 kJ/mol per mole of C_2H_2, calculate the adiabatic temperature for this reaction if ignition is at 30°C in pure oxygen.
b. Do the same if the reaction occurs in air.

4.6 Calculate the air pressure (atmospheres) at the top of Mount Everest (29,029 ft). Do the same in millibar.

4.7 Calculate the number of atoms in an energy state 150 kJ/mol above the ground state having 6.022×10^{23} atoms at both 300K and 1300K.

4.8 Calculate the number of molecules that can react if the activation energy is 1 eV above the ground state containing 6.022×10^{23} molecules at both 300K and 1300K.

4.9 Calculate how many times a catalyst increases the reaction rate at both 300K and 1300K if the activation energy without a catalyst is 120 kJ/mole, while with a catalyst it is just one-half this value. The preexponential term is the same in both cases.

REFERENCES

Arrhenius, S. 1896. On the Influence of Carbonic Acid in the Air upon the Temperature of the Ground. *Phil. Mag. J. Sci.* 41(4): 237–276.

Butts, A. 1943. *Metallurgical Problems,* 2nd ed. New York: McGraw-Hill.

Cary, H. B. 1979. *Modern Welding Technology.* Englewood Cliffs, NJ: Prentice Hall.

Claverie, J. P. and F. Schaper. 2013. Ziegler-Natta catalysis: 50 Years After the Nobel Prize. *MRS Bull.* 38(3): 213–218. (Note: the focus of the issue is Ziegler-Natta Catalysts).

Dickerson, R. E. and I. Geis. 1976. *Chemistry, Matter and the Universe: Integrated Approach to General Chemistry.* New York: Benjamin-Cummings.

Emsley, J. 2001. *Nature's Building Blocks.* New York: Oxford.

Friend, C. M. 1993. Catalysis on Surfaces. *Scientific American.* April: 74–79.

Gore, A. 2006. *An Inconvenient Truth.* New York: Roble Books.

Haim, A. 1989. Catalysis: New reaction Pathways, Not Just a Lowering of the Activation Energy. *J. Chem. Ed.* 66(11): 935–937.

Laidler, K. J. 1984. The Development of the Arrhenius Equation. *J. Chem. Ed.* 61(6): 494–498.

NIST. 2013. *NIST Chemical Kinetics Database.* Standard Reference Database 17, Version 7.0 (Web Version). Release 1.6.7. Data Version 2013.03. http://kinetics.nist.gov/kinetics/index.jsp.

Provan, J. 2011. *The Hindenburg.* http://www.Amazon.com.

Roine, A. 2002. *Outokumpu HSC Chemistry for Windows,* Ver. 5.11, thermodynamic software program, Outokumpu Research Oy, Pori, Finland.

Silbey, R. J. and R. A. Alberty. 2001. *Physical Chemistry,* 3rd ed. New York: Wiley.

Worek, M., ed. 2008. *Nobel: A Century of Prize Winners.* Buffalo, NY: Firefly Books.

5

Heterogeneous Reactions

Gas–Solid Reactions

5.1 INTRODUCTION

In the previous chapters, the focus was on homogeneous reactions, with the exception for catalysts, and mainly on reactions in gases and liquids. In this chapter, the focus is gas–solid reaction kinetics, although many of the same principles apply to gas–liquid reactions. This is an area of great commercial importance in both the processing and performance of *hard* materials: metal, ceramics, composites, and electronic materials. It is also important in the processing and behavior of soft materials in a wide variety of environments. However, the focus is on hard materials because the various steps in the overall process are better understood with these materials. Moreover, these systems are engineered to give different characteristics depending on the requirements for the material or the environment it is being exposed to. In this chapter and several chapters to follow, the kinetic behavior of gas atoms and molecules are used extensively beginning with some kinetic theory of gases that apply directly to simple gas–solid or gas–liquid reactions such as condensation and sublimation that do not require any chemical reaction for the process to occur. The information obtained from these simple processes is useful when applied to chemical reactions such as the deposition of silicon from gas-phase reactants. The deposition of silicon from the gas phase is an excellent example of a gas–solid reaction of extreme importance in the fabrication of silicon integrated circuits, which, of course, have transformed the world we live in. A better example of materials enabling a system could not be found. The general concepts of heterogeneous reactions are illustrated by deposition from gaseous constituents. The major concepts to be encountered in this chapter are some aspects of the kinetic theory of gases, the factors important in gas–solid reactions, how gas–solid reactions are important in both deposition and corrosion processes, the wide variety of real processes to which these principles apply, and the interaction between kinetic and thermodynamic factors on solid growth kinetics.

5.2 GAS KINETICS

5.2.1 Temperature and Speeds of Gas Molecules

It is useful to begin by looking at the temperature dependence of the energies of atoms, ions, and molecules. Consider a box of with sides of equal length, L, with a monatomic gas atom having mass m and an x-component of velocity, v_x, as shown in Figure 5.1. Assume that the walls of the box are impenetrable so the atom will hit a wall and rebound completely elastically in the negative x-direction. When it does so, the momentum change is $\Delta(mv) = 2\,mv_x$—one for coming and one for going in the opposite direction with the same x-velocity. Therefore the force exerted by this single wall collision is simply

$$F = ma = \frac{d(mv)}{dt} \tag{5.1}$$

or $\Delta F = \Delta(mv)/\Delta t$ where Δt = time between collisions of this atom with this wall and is just the time it takes for the atom to travel to the back wall and return; that is, $\Delta t = 2L/v_x$. Therefore, the force per atom on the wall is

$$\Delta F = \frac{\Delta(mv_x)}{\Delta t} = \frac{2mv_x}{2L/v_x} = \frac{mv_x^2}{L}. \tag{5.2}$$

If N is the total number of atoms in the box, then they all strike the wall in the time Δt so the total force, F, is just $N\Delta F$ and the pressure, P, is given by $P = F/L^2$, so

$$P = \frac{F}{L^2} = \frac{N\Delta F}{L^2} = \frac{Nmv_x^2}{L^3} = \frac{Nmv_x^2}{V} \tag{5.3}$$

where $V = L^3$ = volume of the box. Rearranging this, for an ideal gas

$$PV = Nmv_x^2 = nRT = \frac{N}{N_A}RT = Nk_BT \tag{5.4}$$

and therefore $mv_x^2 = k_BT$ where N_A is the Avogadro's number and k_B is the Boltzmann's constant. The kinetic energy in the x-direction, KE_x, for the atom is defined by

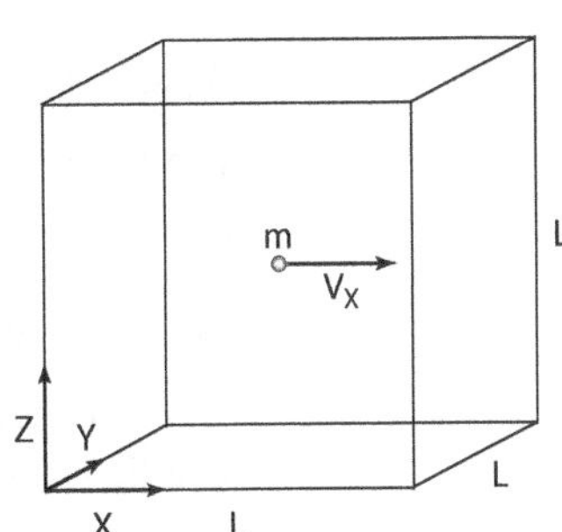

FIGURE 5.1 Model of a gas molecule moving in a cubic box of side L in the x-direction with a velocity v_x used to calculate the translational kinetic energy of molecules.

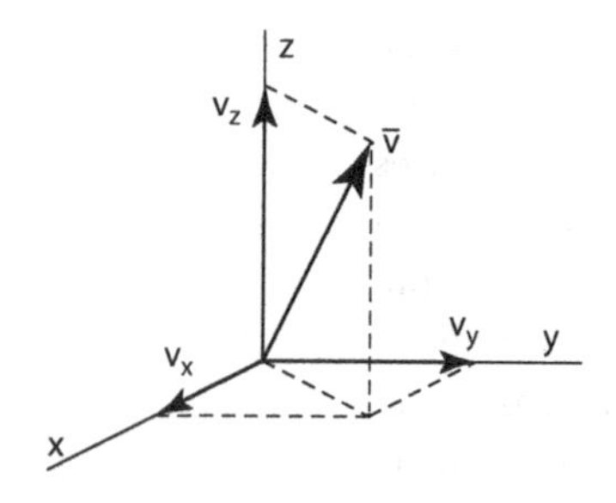

FIGURE 5.2 Schematic showing the x-, y-, and z-components of the velocity of a molecule that has a vector velocity $\bar{v}$.

$$KE_x = \frac{1}{2} m v_x^2 = \frac{1}{2} k_B T. \tag{5.5}$$

The velocity vector has three-components, one in each of the x, y, and z directions as shown in Figure 5.2. Therefore,

$$v^2 = v_x^2 + v_y^2 + v_z^2$$

and the total kinetic energy, KE, is given by the sum of the kinetic energies in each direction

$$KE = \frac{1}{2} m v_x^2 + \frac{1}{2} m v_y^2 + \frac{1}{2} m v_z^2 = \frac{1}{2} m v^2$$
$$KE = \frac{1}{2} k_B T + \frac{1}{2} k_B T + \frac{1}{2} k_B T = \frac{3}{2} k_B T. \tag{5.6}$$

The second of these last two equations is a statement of the principle of *equipartition of energy* (Silbey and Alberty 2001) which essentially says that the microscopic modes—mechanisms—for absorbing energy will all have the same energy, 1/2 $k_B T$, when things are in *thermal equilibrium.* For a monatomic gas, the only energy-absorbing modes are the translational energies in the three directions leading to KE = 3/2 $k_B T$. This assumes that the temperature is not high enough to excite electrons into higher energy states, which does not happen at temperatures of normal interest.* From above, the *RMS—root mean square*—speed of gas molecules is defined as

$$v_{rms} = \sqrt{\frac{3k_B T}{m}} = \sqrt{\frac{3k_B T}{m} \frac{N_A}{N_A}} = \sqrt{\frac{3RT}{M}} \tag{5.7}$$

where M is the molecular or atomic weight.

Take He gas as an example, M = 4 g/mol, at 300 K,

$$v_{rms} = \sqrt{\frac{3(8.314\ \text{J/mol}\cdot\text{K})(300\ \text{K})}{4.00 \times 10^{-3}\text{kg/mol}}} = 1368\ \text{m/s}$$

$$= \frac{1368\ \text{m/s}}{1609\ \text{m/mi}} \times 3600\ \text{s/hour} = 3060\ \text{mph}$$

$$v_{rms} = \frac{3060\ \text{mph}}{761.2\ \text{mph/Mach}} = 4.02\ \text{Mach}$$

where Mach 1 = 761.2 mph, the speed of sound at sea level. Even at room temperature, the speed of gas atoms is indeed very fast.

5.2.2 Maxwell–Boltzmann Speed Distribution

In most cases, the interest is in the *average* speed of the atoms and not in their RMS speed. This can be calculated from the *Maxwell–Boltzmann* speed distribution. In Chapter 4, it was shown

* For example, at T = 1000 K: $k_B T = 1.38 \times 10^{-23}$ J/K × 1000 K = 1.38×10^{-20} J = 1.28×10^{-20} J/1.602×10^{-19} J/eV = 0.08 eV which is considerably less than the several eV necessary for electron excitations. Note that at room temperature, T = 298 K, $k_B T \cong 0.025$ eV ≅ 1/40 eV.

that the distribution of energies for N_T total atoms or molecules, or their fraction in a given energy state is

$$\frac{1}{N_T}\frac{dN}{dE} = \frac{1}{k_BT}e^{-\frac{E}{k_BT}}. \tag{5.8}$$

This is an extension of the barometric formula that gives the number of molecules as a function of height above sea level, that is, their potential energy. The equivalence between the potential and kinetic energies of gas molecules is not surprising. For example, suppose an atom had a kinetic energy $KE_x = 1/2mv_z^2$ at sea level in the vertical direction, which is taken to be z. If it did not undergo any collisions with other gas molecules, then it would rise to a height where its initial kinetic energy would just reach its potential energy, its kinetic energy would go to zero, and then it would be accelerated back to earth ending with its initial kinetic energy. Therefore, the number of atoms at any given height, or potential energy, says something about the Maxwell distribution of the kinetic energy of atoms at sea level moving with a vertical velocity, v_z; that is,

$$\frac{1}{N_z}\frac{dN_z}{dv_z} = A'e^{-\frac{mv_z^2}{2k_BT}}$$

which considers molecules moving in the z-direction only, a Gaussian distribution (Figure 5.3) where A′ is just some constant (Moore 1955).* To evaluate the constant, A′, for molecules moving in the x-direction, this expression needs to be integrated; that is,

$$A'\int_{-\infty}^{\infty} e^{-\frac{mv_x^2}{2k_BT}}\,dv_x = \frac{\int_{-\infty}^{\infty} dN_x}{N_x} - 1. \tag{5.9}$$

Let $mv_x^2/2k_{BT} = y^2$, or $v_x = \sqrt{(2k_BT/m)}\;y$, and the integral becomes

$$A'\sqrt{\frac{2k_BT}{m}}\int_{-\infty}^{\infty} e^{-y^2}dy = 1.$$

Therefore, the key is evaluating the integral $\int_{-\infty}^{\infty} e^{-y^2}dy$. How this is done is shown in Appendix A.2, and everyone needs to see this integration at least once in his or her career. In any event, from Appendix A.2

$$\int_{-\infty}^{\infty} e^{-y^2}dy = \sqrt{\pi}. \tag{5.10}$$

So

$$A' = \sqrt{\frac{m}{2\pi k_BT}}$$

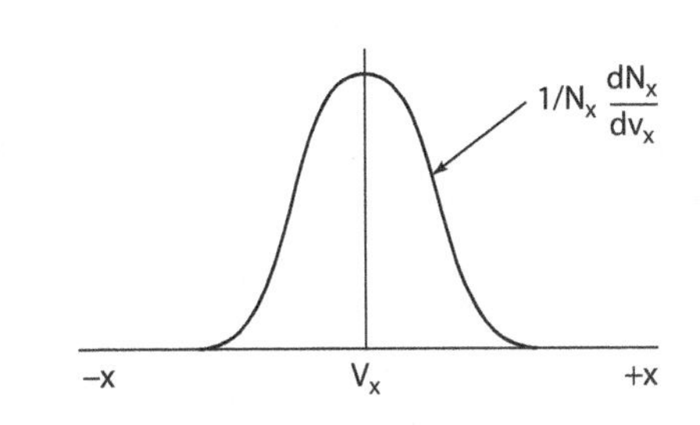

FIGURE 5.3 One-dimensional Gaussian distribution of gas molecule speeds.

* Of course, in reality, no single molecule or atom travels very far (only about 0.1 μm at 1 bar and 300 K) before colliding with another gas atom or molecule, as will be seen when diffusion in gases is discussed later. Nevertheless, there is simply an energy exchange between atoms or molecules and there must have been some of them that had this energy at sea level and transfer their energy through multiple collisions to other atoms that made it to the higher potential energies.

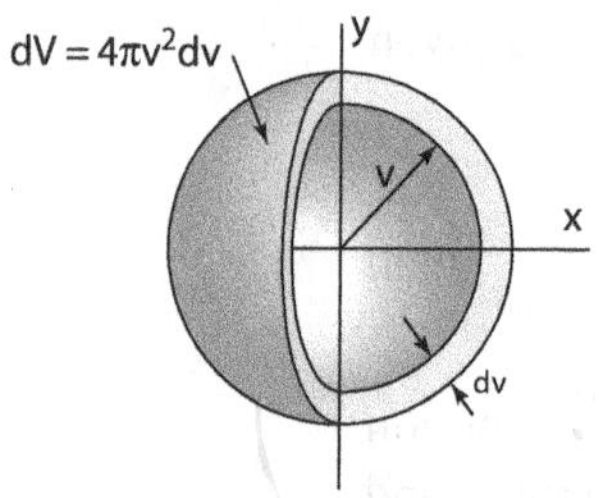

FIGURE 5.4 The volume element in spherical coordinates depending only on the radius, v, the molecular speed.

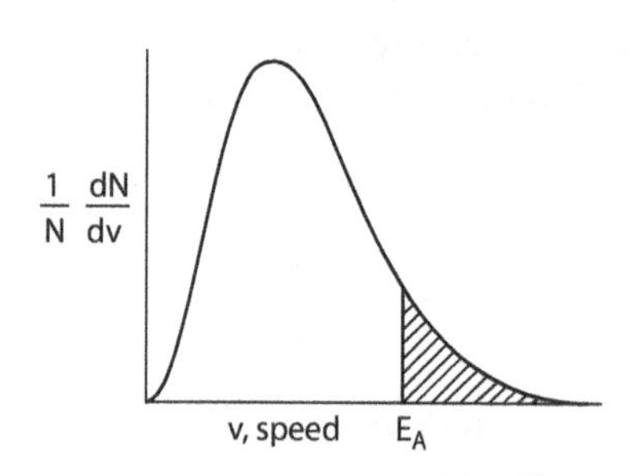

FIGURE 5.5 Maxwell–Boltzmann distribution for the molecular speed showing those molecules or atoms having energies above E_A in the exponential tail of the distribution.

and

$$\frac{dN_x}{N_x} = \sqrt{\frac{m}{2\pi k_B T}}\, e^{-\frac{mv_x^2}{2k_B T}}\, dv_x. \tag{5.11}$$

This is only in one dimension and similar expressions exist for the other two directions (y and z) and what is of interest is the speed, v, which is $v = \sqrt{v_x^2 + v_y^2 + v_z^2}$, as shown in Figure 5.2. The fraction of molecules in three dimensions is just the product of the fractions in each direction, namely

$$\begin{aligned} \frac{dN}{N} &= \left(\frac{dN_x}{N_x}\right)\left(\frac{dN_y}{N_y}\right)\left(\frac{dN_z}{N_z}\right) \\ &= \left(\frac{m}{2\pi k_B T}\right)^{3/2} e^{-\frac{m(v_x^2+v_y^2+v_z^2)}{2k_B T}}\, dv_x dv_y dv_z. \end{aligned} \tag{5.12}$$

Make the substitution for the speed, v, above, and change the volume element from $dV = dv_x dv_y dv_z$ to a spherical shell of radius v and thickness dv, so the volume element becomes $dV = 4\pi v^2 dv = dv_x dv_y dv_z$, Figure 5.4, with the change in coordinate system shown in Appendix A.3. This gives the *Maxwell–Boltzmann distribution* for the speeds of gas atoms and molecules in thermal equilibrium:

$$\frac{dN}{N} = \left(\frac{m}{2\pi k_B T}\right)^{3/2} 4\pi v^2 e^{-\frac{mv^2}{2k_B T}}\, dv \tag{5.13}$$

shown in Figure 5.5. The high speed tail end of this distribution, the shaded region in Figure 5.5, is essentially an exponential function. Note that Equation 5.13 could also be written in terms of the kinetic energy of the atom or molecule, $\varepsilon = (1/2)mv^2$ as

$$\frac{dN}{N} = \frac{2\pi}{(\pi k_B T)^{3/2}}\, \varepsilon^{1/2} e^{-\varepsilon} d\varepsilon.$$

5.2.3 Mean Molecular Speed

By definition, the *mean speed*, $\bar{v}$, is given by

$$\bar{v} = \int_0^\infty v \frac{dN}{N} = \left(\frac{m}{2\pi k_B T}\right)^{3/2} \int_0^\infty v 4\pi v^2 e^{-\frac{mv^2}{2k_B T}}\, dv \tag{5.14}$$

and substitution of $z = mv^2/2k_B T$ (or $v = \sqrt{(2k_B T/m) z^{1/2}}$) in the integral gives

$$\begin{aligned} \int_0^\infty v 4\pi v^2 e^{-\frac{mv^2}{2k_B T}} dv &= 4\pi \int_0^\infty \left(\frac{2k_B T}{m}\right)^{3/2} z^{3/2} \frac{1}{2}\left(\frac{2k_B T}{m}\right)^{1/2} z^{-(1/2)} e^{-z} dz \\ &= \frac{4\pi}{2}\left(\frac{2k_B T}{m}\right)^2 \int_0^\infty z e^{-z} dz \end{aligned}$$

$\int_0^\infty z e^{-z^2} dz$ can be easily integrated by parts or found in most calculus books (Edwards and Penney 2002)

$$\int_0^\infty v4\pi v^2 e^{-\frac{mv^2}{2k_BT}} dv = \frac{4\pi}{2}\left(\frac{2k_BT}{m}\right)^2 \left|-e^{-z}(z+1)\right|_0^\infty$$

$$= \frac{4\pi}{2}\left(\frac{2k_BT}{m}\right)^2 (-0+1)$$

so

$$\bar{v} = \left(\frac{m}{2\pi k_BT}\right)^{3/2} \frac{4\pi}{2}\left(\frac{2k_BT}{m}\right)^2$$

and gives the desired result for the mean speed,

$$\bar{v} = \sqrt{\frac{8}{\pi}\frac{k_BT}{m}} = \sqrt{\frac{8}{\pi}\frac{RT}{M}} \quad (5.15)$$

which is about 7% smaller than the RMS speed, Equation 5.7.

5.2.4 Flux of Gas Atoms Impinging on a Surface

The number of gas atoms that strike a solid or liquid surface per unit area, per unit time is a quantity of interest. Figure 5.6 shows how this can be approximated. Consider a column of gas of height $h = v_{RMS}$ that contains a number of gas atoms per unit volume, η, determined by the pressure. At any point in time, one-sixth of the gas atoms are moving in the $-z$ direction and in 1 second all of these atoms in the column will strike the surface, so the approximate number of atoms striking the surface per unit area per unit time, I, the rate of *impingement is* (Hudson 1998)

$$I \cong \frac{1}{6}\eta v_{RMS}. \quad (5.16)$$

A better approximation would use the mean velocity, $\bar{v}$, but this still is only an approximation. Equation 5.16 assumes that the atoms are only traveling in the three mutually perpendicular directions. The reality is the gas atoms can strike the area A from any direction. When this is taken into consideration, along with the mean velocity, the exact relation—as shown in Appendix A.4—is

$$I = \frac{1}{4}\eta\bar{v}. \quad (5.17)$$

Substituting Equation 5.15 for $\bar{v}$ and $\eta = N_AP/RT$ gives

$$I = \frac{1}{4}\left(\frac{N_AP}{RT}\right)\sqrt{\frac{8}{\pi}\frac{RT}{M}}$$

$$I = \frac{N_AP}{\sqrt{2\pi RTM}}$$

or as it normally found in the literature,

$$I = \frac{P}{\sqrt{2\pi mk_BT}} \quad (5.18)$$

where Units(I) = atoms/m²-s, Units(m) = kg/atom, and Units(P) = Pa. Not surprisingly, the impingement rate depends on the concentration of gas atoms, which is proportional to the pressure.

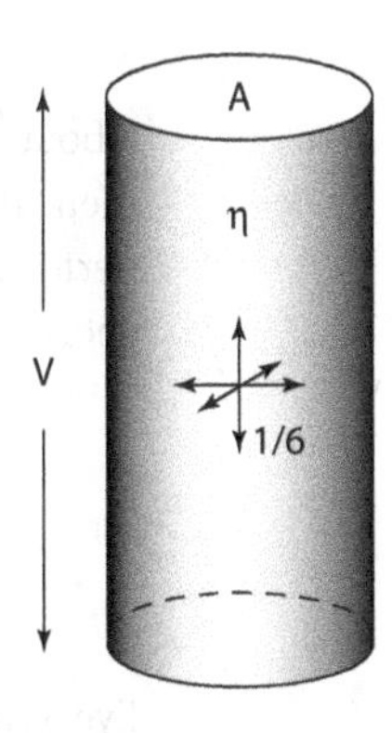

FIGURE 5.6 A model to calculate the approximate number of gas molecules hitting a surface of area A per unit time with a mean speed v. Only 1/6 of the atoms per unit volume, η, are moving in the z-direction at any given time and only those strike the surface.

5.2.5 Need for High Vacuum in Surface Studies

Take air with a mean molecular weight of $\bar{M} = 28.84$ g/mol (Chapter 4) or 28.84×10^{-3} kg/mol ($\bar{M} = 28.84$ at a temperature of 25°C or 298 K at a pressure of 1 atm or 1.0132×10^5 Pa). Then,

$$I = \frac{P}{\sqrt{2\pi m k_B T}} = \frac{1.0132 \times 10^5}{\sqrt{2(3.1416)\left(\frac{28.84 \times 10^{-3}}{6.022 \times 10^{23}}\right)(1.38 \times 10^{-23})(298)}} \tag{5.19}$$

$$I = 2.88 \times 10^{27} \text{ atoms/m}^2 \cdot \text{s} = 2.88 \times 10^{23} \text{ atoms/cm}^2 \cdot \text{s}$$

Suppose that an experimentalist wanted to study the surface properties of pure nickel, Ni, without any adsorbed oxygen impurities. If the atmosphere over the nickel were air, it might be expected that all of the oxygen molecules hitting the surface will react to form at least a monolayer—one atom thick—of nickel oxide, NiO, since it is stable in air at room temperature. The number of oxygen atoms impinging on the surface would simply be

$$I_O = 2I_{O_2} = 2 \times 0.21 \times 2.88 \times 10^{23} \text{ atoms/cm}^2 \cdot \text{s} = 1.21 \times 10^{23} \text{ oxygen atoms/cm}^2 \cdot \text{s}$$

since air is 21% oxygen. For nickel, M = 58.693 g/mol and the density $\rho = 8.90$ g/cm^3. So the number of Ni atoms per cm^3 is

$$\eta_{Ni} = \frac{\rho}{M} N_A = \frac{(8.90)}{(58.693)}(6.022 \times 10^{23})$$

$$\eta_{Ni} = 9.13 \times 10^{22} \text{ atoms/cm}^3$$

typical of most solids and liquids—something times 10^{22}/cm^3. The approximate number of nickel atoms per unit area is given by

$$\text{Ni/cm}^2 \cong \eta_{Ni}^{2/3} = (9.13 \times 10^{22})^{2/3}$$

$$\text{Ni/cm}^2 \cong 2.03 \times 10^{15} \text{cm}^{-2}.$$

So the time it would take for the entire surface to be covered with a monolayer of NiO would simply be

$$\frac{\text{Ni/cm}^2}{I_O} = \frac{2.03 \times 10^{15}}{1.21 \times 10^{23}} = 1.68 \times 10^{-8} \text{s!}$$

About 10 ns, this means that the experimentalist must be pretty quick in carrying out the experiment if it is to be done on a surface essentially free of oxygen atoms! A more reasonable pace to perform such an experiment might be about 1 hour during which time, say, only 1% of the surface gets covered with oxygen. In that case, I_O must be substantially reduced and can be estimated by

$$I_O = \frac{\text{Ni/cm}^2}{\text{time}} = \frac{(0.01)2.03 \times 10^{15}}{3600}$$

$$I_O = 5.64 \times 10^9 \text{ O/cm}^2 \cdot \text{s}.$$

Everything else being equal, this means that the pressure of air in the system would have to be reduced to about $5.64 \times 10^9 / 1.21 \times 10^{23} \cong 4.66 \times 10^{-14}$ atm. As seen in Chapter 4, the experiment might be best done on the outside of the International Space Station where the vacuum has about this value. However, modern vacuum systems can reach vacuums on the order of 10^{-11} torr* or in the

* The quality of a vacuum is usually given in the units of torr = mm of mercury. One atmosphere = 760 torr.

order of 10^{-14} atm. Therefore, experiments on *clean* surfaces must be done in *ultra-high vacuum* on the order of 10^{-11} torr.

5.2.6 Condensation and Evaporation or Sublimation

5.2.6.1 The Model

Equation 5.18 also gives the rate of condensation or evaporation of atoms from the surface of a solid or liquid at equilibrium if the equilibrium pressure is p_e,

$$I = \frac{p_e}{\sqrt{2\pi m k_B T}}. \tag{5.20}$$

If the solid or liquid is not in equilibrium with the gas that has a partial pressure p of the solid or liquid molecules or atoms in the gas phase, the Equation 5.20 needs to be modified to (Pound 1972),

$$J' = \frac{\alpha(p - p_e)}{\sqrt{2\pi m k_B T}} \tag{5.21}$$

where J′ is the atom flux density to—or away from—the solid or liquid surface depending on the relative values of p and p_e: it $p > p_e$, condensation will occur; if $p < p_e$ then evaporation or sublimation happens. Alpha, α, is called the *accommodation coefficient* or *sticking coefficient* (Adamson 1982) and is a positive quantity less than or equal to 1.0. If $\alpha \neq 1.0$ then the rate of condensation or evaporation is slower than if every atom or molecule that hit the surface stuck or remained there. Also, alpha less than one, implies that some kind of surface reaction must be occurring as well. An alpha of zero means that the atoms do not stick and just bounce off the surface. This situation will be explored in more detail later in this chapter. Also, the value of alpha may be different on evaporation than it is on condensation if the events are occurring far from equilibrium. However, for many substances, metals, inorganic compounds, and organics, $\alpha \cong 1.0$ when there is no chemical reaction but simply condensation or evaporation (sublimation) when near equilibrium (Pound 1972).

5.2.6.2 Example: Sublimation of 1,4-Dichlorobenzene

Today, mothballs consist of 1,4-dichlorobenzene (Figure 5.7) that has a melting point of 53.5°C, a molecular weight of 147.00 g/mol, and a vapor pressure of $p_e = 3.27 \times 10^{-3}$ bar at 30°C (NIST Webbook). If there is no pressure of the compound in the gas phase, then dichlorobenzene will sublime. Assuming $\alpha = 1.0$, then J′ becomes,

$$J' = \frac{p_e}{\sqrt{2\pi m k_B T}} = \frac{(3.27\times10^{-3}\text{ bar})(10^5\text{Pa/bar})}{\sqrt{2(3.14.6)((147.00\times10^{-3}\text{ kg/mol})/(6.022\times10^{23}\text{ atoms/mol}))(1.38\times10^{-23})(303)}}$$

$$J' = 4.08\times10^{24}\text{ molecules/m}^2\cdot\text{s} = 4.08\times10^{20}\text{ molecules/cm}^2\cdot\text{s}$$

and

$$J = \frac{J'}{N_A} = \frac{4.08\times10^{20}}{6.022\times10^{23}} = 6.78\times10^{-4}\text{ mol/cm}^2\cdot\text{s}.$$

Therefore, the dichlorobenzene is subliming as expected. Of practical interest is the rate of sublimation in terms of the thickness sublimed in a given amount of time. The density of dichlorobenzene $\rho = 1.25$ g/cm³ (Haynes 2013), so

$$J(\text{mol/cm}^2\cdot\text{s}) = \left(\frac{\rho(\text{g/cm}^3)}{M(\text{g/mol})}\right)\left(\frac{1}{A(\text{cm}^2)}\right)\left(\frac{dV(\text{cm}^3)}{dt(\text{s})}\right). \tag{5.22}$$

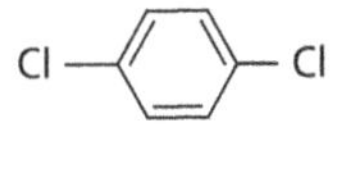

Cl — ⬡ — Cl

FIGURE 5.7 The chemical structure of 1,4-dichlorobenzene.

For a sphere or radius a, $V = (4/3)\pi a^3$, $dV/dt = 4\pi a^2(da/dt)$, and $A = 4\pi a^2$ so

$$J = \frac{\rho}{M}\frac{da}{dt}.$$

Therefore, the rate of evaporation, da/dt, is

$$\frac{da}{dt} = \frac{M}{\rho}J = \frac{147.02}{1.25}\left(6.78 \times 10^{-4}\right)$$

$$\frac{da}{dt} = 7.97 \times 10^{-2}\ \text{cm/s}.$$

This seems very fast compared to how long mothballs last, suggesting $\alpha \neq 1.0$, so there is some kind of surface reaction taking place and/or diffusion through the gas phase is rate controlling, the latter being more likely case.

5.2.7 Langmuir Adsorption Isotherm

In Chapter 4, the role of catalysts in reactions was discussed: specifically, the adsorption of hydrogen and deuterium onto platinum and the catalytic role of the platinum surface in the exchange of hydrogen and deuterium atoms. A reasonable question to ask is, "What might determine the surface concentration of the adsorbed gases?" A simple way to approach this is through the equilibrium between the adsorbed gas and the atmosphere such as

$$\text{empty surface site, S} + \text{gas atom} \rightleftharpoons \text{occupied surface site, S}'; \Delta G^\circ \tag{5.23}$$

where ΔG° is the Gibbs energy of the adsorption reaction of the gas atoms (molecules) on the solid surface. An equilibrium constant for the reaction in Equation 5.23 can be written as

$$\frac{[S']}{[S]P} = K_L = e^{-\frac{\Delta G^\circ}{RT}} \tag{5.24}$$

where $[S'] = \theta$, the *fraction of occupied* surface sites by the adsorbed gas, $[S] = 1 - \theta$, the fraction of *unoccupied* surface sites, and P = pressure. So Equation 5.24 becomes

$$\frac{\theta}{1-\theta} = K_L P$$

$$\theta = K_L P - \theta K_L P$$

$$\theta\left(1 + K_L P\right) = K_L P$$

with the final result, the Langmuir isotherm (Adamson 1982; Hudson 1998):

$$\theta = \frac{K_L P}{1 + K_L P}. \tag{5.25}$$

If P = 1 bar and $K_L = 1$ at 298 K, then $\theta = 0.5$ or one-half of the sites are occupied and $\Delta G^\circ \cong 0$. Now, $\Delta G^\circ = \Delta H^\circ - T\Delta S^\circ$ and, of course,

$$K_L = e^{\frac{\Delta S^\circ}{R}} e^{-\frac{\Delta H^\circ}{RT}}$$

where ΔS° is the entropy difference between the vibrations and translations of the atoms or molecules on the surface and the gas phase and ΔH° is the enthalpy of adsorption. If the adsorbed atoms stick at one spot and do not translate on the surface, an estimate for the entropy change for water molecules adsorbing at 25°C and one bar is the entropy of sublimation of ice, $\Delta S^{\circ}_{298} \cong -141.6$ J/mol·K (Roine 2002). In order to have $K_L = 1.0$, ΔH° would have to be about −42,200 Jmol. On the other hand, if there is 99% coverage ($\theta = 0.99$), then $\Delta G^{\circ} \cong -11{,}385$ J/mol or $\Delta H^{\circ} \cong -53{,}568$ J/mol.* If $\theta \cong 0.01$, then $\Delta H^{\circ} = -11{,}435$ J/mol. All else being the same, as the pressure decreases, the fraction adsorbed decreases. Since the value of the bonding enthalpy, ΔH°, determines what fraction of the surface is covered by an adsorbing gas, the boundary between *physical adsorption* and *chemical adsorption* or *chemisorption* becomes rather arbitrary and is around $\Delta H^{\circ} = -10$ kJ/mol: less negative for physical adsorption (Hudson 1998).

5.3 GAS–SOLID CORROSION REACTIONS

Passive corrosion is the formation of a protective (usually oxide, but not always) layer that slows down the rate of further corrosion since this requires diffusion through the solid state, which is a slow process. For example, Figure 5.8 shows the passive oxidation of silicon to SiO_2 that occurs at most oxygen pressures:

$$Si(s) + O_2(g) \rightarrow SiO_2(s). \tag{5.26}$$

Further oxidation occurs by either oxygen diffusing from the gas through the SiO_2 to the Si–SiO_2 interface or silicon diffusing from the Si–SiO_2 interface to react with oxygen at the SiO_2–gas interface. As a result, as the oxide layer gets thicker, the rate of oxidation becomes slower because of the greater distance the atoms (ions) must diffuse (to be discussed in greater detail in Chapters 10 and 14): the reason that it is called *passive oxidation* is that the rate decreases with time.

In contrast is *active corrosion*. In this case, there is a gas–solid reaction with no formation of a protective solid layer, only gaseous products are formed with silicon at low oxygen pressures to form SiO, silicon monoxide gas,† as shown in Figure 5.9 and Equation 5.27

$$Si(s) + \frac{1}{2}O_2(g) \rightarrow SiO(g). \tag{5.27}$$

This is called *active corrosion* because the rate of corrosion is constant—there is no protective layer developed to keep the solid from oxidizing or corroding further. Another example of active corrosion includes the oxidation of silicon carbide at low oxygen pressures

$$SiC(s) + O_2(g) \rightleftharpoons SiO(g) + CO(g). \tag{5.28}$$

Silicon carbide heating elements are frequently used in electric furnaces that operate up to about 1600°C. In an oxidizing atmosphere, a protective or passive oxide layer of SiO_2 is formed on the SiC that slows oxidation. However, in reducing or very low oxygen pressure atmospheres, active gas corrosion will occur greatly reducing the life of the SiC heating elements. Therefore, such furnaces should never be used with reducing or low oxygen chemical potential atmospheres, to avoid frequent and expensive replacement of the heating elements.

Another example is the active oxidation of chromium metal in an oxidizing atmosphere with the production of CrO_3 gas:

$$2Cr(s) + 3O_2(g) \rightleftharpoons 2CrO_3(g). \tag{5.29}$$

An example that everyone is familiar with is the reaction of a hot tungsten filament in an incandescent bulb with air if the bulb is cracked or broken. The tungsten reacts to form WO_3 gas that condenses on the cooler inside of the remaining bulb forming a white powder:

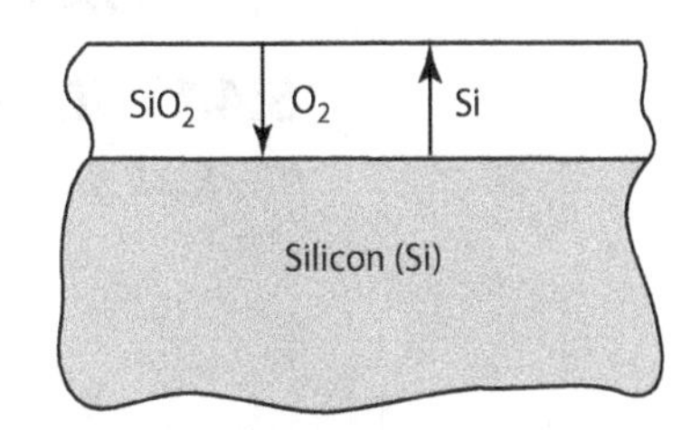

FIGURE 5.8 Model for the *passive* oxidation of silicon with the formation of a protective SiO_2 layer through which either silicon or oxygen atoms must diffuse to continue the oxidation.

* See the discussion about catalysts and bonding in Chapter 4.

† SiO is a compound that exists only as a gas. On condensation it decomposes to solid SiO_2 and Si.

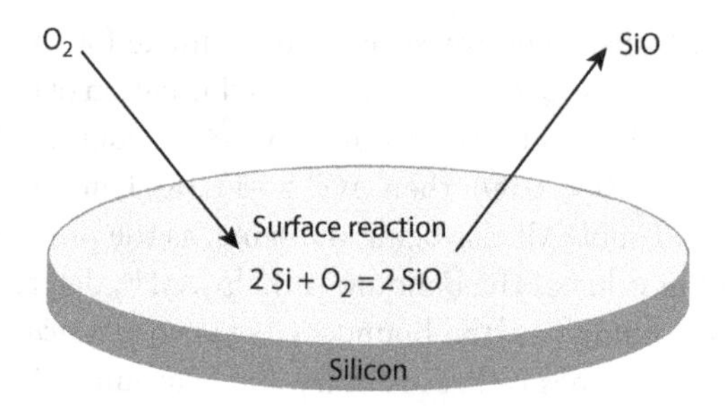

FIGURE 5.9 *Active* oxidation—gaseous corrosion—of silicon at low oxygen pressures with the formation of SiO gas and no SiO_2 protective layer.

$$2W(s) + 3O_2(g) \rightarrow 2WO_3(g) \rightarrow 2WO_3(s). \tag{5.30}$$

Finally, the chlorides of most metals are volatile and metals used in furnaces to incinerate dangerous chlorine-containing compounds can corrode: examples include the chemical weapon phosgene, $COCl_2$, and PCBs—polychlorinated biphenyls ($C_{12}H_{10-x}Cl_x$, where x = 1–10) once used as coolants in transformers. PCBs were banned in the 1970s because of their carcinogenic risk, but they still are found in older transformers and other electrical equipment. In addition, incinerators that burn PVC—polyvinylchloride—bags and other chlorinated polymers are subject to similar corrosion as are furnaces operating near sea water where NaCl in the atmosphere can be incorporated into the combustion gases. In each of these cases, the hot gases contain gaseous HCl or some other corrosive chlorine-containing gas. Most metals, nickel, for example, will undergo active gas corrosion to form volatile chlorides:

$$Ni(s) + 2HCl(g) \rightleftharpoons NiCl_2(g) + H_2(g). \tag{5.31}$$

Even the oxide refractory insulation in such furnaces is subject to similar active gas corrosion

$$MgO(s) + 2HCl(g) \rightleftharpoons MgCl_2(g) + H_2O(g). \tag{5.32}$$

In many cases, these active gas corrosion reactions are controlled by the surface reaction. As discussed Section 2.1.1 and shown again in Figure 5.9, in order for the active oxidation of silicon to occur, oxygen must: (1) be transported by diffusion through the gas to the solid surface (2) where it reacts with the silicon surface atoms and (3) the SiO product gas must diffuse away from the surface. Again, these are series steps and the slowest one controls the rate. In this chapter, the rates of surface reactions are considered to be slowest, and rate-controlling, and examined in detail. While active corrosion usually limits the lives of many components, the process can also be used to commercial advantage, as discussed in Section 5.4.

5.4 GAS–SOLID REACTIONS IN MATERIALS PROCESSING

5.4.1 Titanium Metal and TiO_2 Pigments

The *Kroll process* for making titanium consists of two important gas–solid reaction steps:

$$TiO_2(s) + 4HCl(g) \rightleftharpoons TiCl_4(g) + 2H_2O(g) \tag{5.33}$$

$$TiCl_4(g) + 2Mg(s) \rightarrow MgCl_2(s) + Ti(s). \tag{5.34}$$

TiO_2 powders for paint pigments are made in extremely large quantities throughout the world by the reaction of $TiCl_4(g)$—commonly referred to as *tickle*—produced by the reaction of TiO_2 similar to the Kroll process with the subsequent re-oxidation of the $TiCl_4$ to TiO_2 (Equation 5.33 in reverse) in a flame whose temperature is generated by the large enthalpy of oxidation of $TiCl_4$ to TiO_2. The Cl_2 produced in the oxidation is recycled to produce more $TiCl_4$

$$TiCl_4(g) + O_2(g) \rightleftharpoons TiO_2(g) + 2Cl_2(g) \tag{5.35}$$

by essentially the reverse of this reaction, an active gas corrosion process.

5.4.2 Siemens Process

Another example of these reversible gas–solid processes is the *Siemens process* for making *electronic grade silicon* (EGS), which is close to 9-nines pure Si—only a few parts per billion (ppb) impurities—from *metallurgical grade silicon* (MGS)—about 98% pure

$$Si_{MGS}(s) + 3HCl(g) \xrightleftharpoons{700K} SiHCl_3(g) + H_2(g). \tag{5.36}$$

$SiHCl_3$, *trichlorosilane*, is collected as a liquid that has a boiling point around 33°C allowing it to easily be distilled several times to obtain a very high purity final liquid. Multiple distillation is an important chemical engineering unit process. Because of the tremendous value added to pure trichlorosilane for electronic materials production, the cost of distillation is a minor part of the total cost of the final EGS. At higher temperatures, the reverse reaction takes place depositing high purity silicon

$$SiHCl_3(g, \text{high purity}) \xrightleftharpoons{1300K} Si_{EGS}(s) + 3HCl(g) \tag{5.37}$$

either directly onto electronic devices or circuits or as polycrystalline material that is subsequently melted and grown into single crystals from the molten EGS. These crystals are then cut into wafers and processed into chips for electronic devices.

5.4.3 Optical Fibers

Another extremely important example of a gas–solid reaction is the production of optical fibers of high purity silica for fiber-optic applications, Figure 5.10. All of us are familiar with the green color of

FIGURE 5.10 Silicon dioxide fibers for fiber optics applications. Original color photo changed to grayscale. Original photo attributed to BigRiz at Optical fiber, http://en.wikipedia.org/wiki/Fiber-optic.

window glass when viewed edge on. The greenish hue is caused by the presence of mainly iron ions dissolved in the glass that come from impurities in the starting SiO_2 sand. These impurities absorb light, greatly reducing light transmission over long distances. As such, commercial, low cost glass would not be suitable as optical fibers now used in communications and many other devices, particularly in the medical field, because of the light loss over long path lengths. In order to achieve transparency over long distances—kilometers—very high purity SiO_2 glass is required. To achieve this, semiconductor grade SiO_2 (at least six-9s pure) is formed as a powder in the gas phase and deposited either inside or outside of a fused SiO_2 tube by reacting $SiCl_4$ with oxygen. SF_6 gas is added to volatilize any residual impurities—for example, iron—as fluorides. Other oxides—GeO_2 and P_2O_5—can be added in different radial concentrations to change the index of refraction across the radius of the fiber to ensure that light does not escape at the fiber surface. This porous SiO_2 powder *preform* is then sintered to high density and drawn into fibers. This gas–solid deposition process requires the use of very high purity reagents—electronic grade—to produce the high purity SiO_2 glass free of impurities that would limit its optical transmission.

5.4.4 Halogen Bulbs

One final example used in certain high-intensity, tungsten-filament, halogen lamps where a halogen gas reacts to form a volatile halide and then undergoes the reverse reaction to deposit solid tungsten. For example,

$$W(s) + I_2(g) \rightleftharpoons WI_2(g) \tag{5.38}$$

where the reaction goes to the right at low—2500°C temperatures—and to the left at high—2800°C—temperatures, which allows a bulb to operate at a higher filament temperature and higher light output. This process is discussed in more detail later.

5.5 CHEMICAL VAPOR DEPOSITION OF SILICON

5.5.1 Introduction

An example of a gas–solid reaction that is surface reaction controlled is the deposition of silicon on a substrate from trichlorosilane, $SiHCl_3$. This process is used to demonstrate the steps in developing a gas–solid surface reaction model since it is an important commercial process used for silicon integrated circuits and is well understood. The silane-related compounds that have been used in this process include the following (Haynes 2013).

Silane	SiH_4	Boiling point (°C) = −185
Monochlorosilane	SiH_3Cl	Boiling point (°C) = −30.4
Dichlorosilane	SiH_2Cl_2	Boiling point (°C) = 8.3
Trichlorosilane	$SiHCl_3$	Boiling point (°C) = 33
Tetrachlorosilane (silicon tetrachloride)	$SiCl_4$	Boiling point (°C) = 57.6

5.5.2 Overview of Various Deposition Processes

Before discussing the details of silicon deposition, a brief summary of the different types of deposition processes and how they are used to produce a range of materials is presented.

5.5.2.1 Chemical Vapor Deposition

Chemical vapor deposition (CVD) can be defined as *the production of a solid from gas phase reactants.* There are many variants in the CVD process used for the deposition of solids that include

APCVD: atmospheric pressure CVD
LPCVD: low-pressure CVD
PECVD: plasma-enhanced CVD
MOCVD: metal organic CVD

In the latter case, compounds such as gallium arsenide, GaAs, are deposited for light-emitting diodes (LEDs) and vertical cavity surface-emitting lasers (VCSELs) by processes such as

$$Ga(CH_3)_3(g) + AsH_3(g) \rightarrow GaAs(s) + 3CH_4(g) \tag{5.39}$$

where the reactants are trimethyl gallium, $Ga(CH)_3$, and arsine, AsH_3, and are considered to be *metal-organic* compounds: hence, MOCVD. All of these are variants of essentially the same process except for PECVD, where the plasma generates ionic species and radicals that can speed up the reaction and give a very different reaction pathway. The reason for mentioning these variations on the CVD process is that they are frequently referred to in the materials literature only by their acronyms, so materials scientists and engineers need to be familiar with them.

Another variant of the CVD process is *chemical vapor infiltration*—CVI—which is used, for example, to deposit graphite inside of a bundle of carbon fibers to form a carbon–carbon composite such as that used on the leading edge of the space shuttle, whose failure lead to the tragic destruction of the shuttle Columbia on re-entry. The process for making carbon–carbon composites is discussed in more detail later.

In this example of silicon deposition, it is assumed that the surface reaction is controlling the deposition rate. In a real process, LPCVD is typically used to ensure this, since the rates of gas diffusion increase as the pressure decreases, making the surface reaction the slowest step in the sequence of the three series steps illustrated in Figure 5.11. This process is used to produce epitaxial layers of single crystal silicon on substrates of single crystal silicon—to change the additive or *doping* concentration—or on substrates of single crystal aluminum oxide, Al_2O_3—sapphire—to produce *silicon on sapphire* (SOS or SOI—silicon on insulator). The use of sapphire as a substrate is to electrically insulate or isolate the various electronic devices made on the silicon film since sapphire is a much better electrical insulator than silicon.

5.5.2.2 Epitaxy

Epitaxy can be defined as the formation of a crystalline layer on a crystalline substrate in which there are *definite crystallographic orientations* between the crystal axes of the substrate and those of the layer. *Homoepitaxy* occurs when the layer is the same material as the substrate, that is, Si on Si. *Heteroepitaxy* occurs when they are different, that is, Si on Al_2O_3.

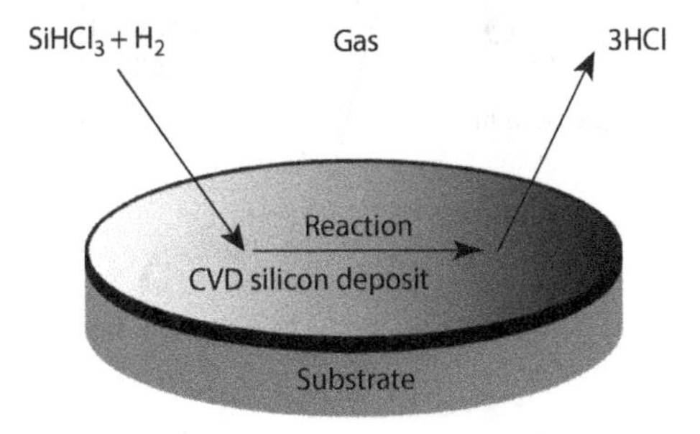

FIGURE 5.11 Schematic description of the chemical vapor deposition, CVD, of silicon from trichlorosilane, $SiHCl_3$, and hydrogen that shows the three series steps necessary for the deposition: (1) transport of the reactants to the surface; (2) the surface reaction; and (3) transport of the products, HCl, away from the surface.

5.5.2.3 Sputtering

Chemical vapor deposition is frequently used to deposit *thin films*—layers that are a micrometer thick or less. As such, CVD is very different than the other common process to deposit thin films, namely, *sputtering*, a form of *physical vapor deposition* (PVD). Sputtering and other PVD processes are briefly discussed here to contrast them with CVD. This is not to negate the importance of PVD processes because they are extremely important not only for forming electronic circuits but also are used for many other applications, including the partially reflective coatings on windows. There are other PVD processes for depositing thin films besides sputtering. These involve a source material at high temperature so that its vapor pressure is sufficiently high to produce a rather large flux of atoms to the substrate essentially as described by Equation 5.21 in an vacuum chamber similar to that shown in Figure 5.12. For example, aluminum and gold depositions are easily performed by heating the metals on a tungsten filament to temperatures high enough so the metals melt, wet the tungsten, and evaporate at a rate predicted by Equation 5.21. If the vacuum is sufficiently high, the atoms go directly from the filament and deposit on the target without any collisions with residual gas atoms in the vacuum deposition chamber.

In contrast, *sputtering* can be defined as a process in which *atoms, ions, or molecules are physically removed from a target material (composed of the same elements or compounds desired in the deposited layer) by bombardment with high energy ions or atoms and the target atoms deposit on a substrate*. As such, the process is carried out in a vacuum with a high voltage between the target and substrate to give the bombarding atoms enough energy to remove atoms from the target. The *sputtered* atoms travel from the target and deposit on a substrate forming a thin film, as shown schematically in Figure 5.12. A vacuum is necessary so that the bombarding and sputtered atoms can travel far enough without colliding with other gas atoms before they reach the target or substrate. As shown in the sketch, argon ions are frequently used as the bombarding ions, and they are accelerated by the high voltage to gain enough energy to remove atoms—break the chemical bonds—from the target. This PVD process is usually slower than CVD and is typically limited to thin films. In contrast, CVD can be used to build up very thick layers such as the polycrystalline EGS silicon in the Siemens process that can reach tens of centimeters in thickness. It should also be noted that there are several variations of sputtering just as there are for CVD outlined above. For example, *RF sputtering* uses a radio frequency alternating field of about 14 MHz to prevent charge build-up on the target material and is particularly useful for sputtering insulators. Another variation is *magnetron sputtering* in which permanent magnets

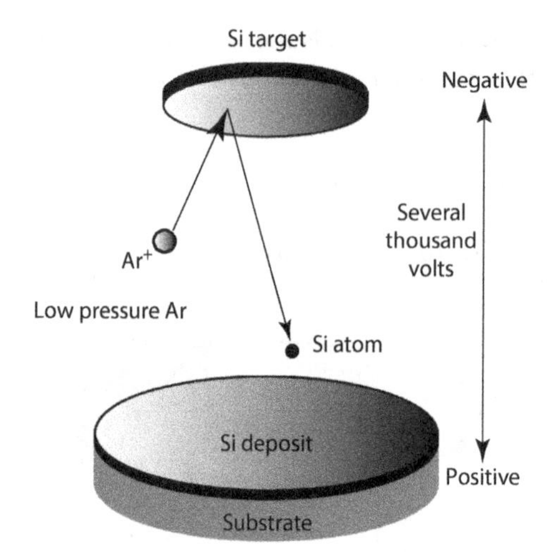

FIGURE 5.12 Schematic showing the formation of a thin-film of silicon by the process of DC sputtering. At low pressures and high voltages, argon atoms are ionized and accelerated at high energy to the silicon target where they physically eject silicon atoms from the surface. These atoms travel to, and condense on, the substrate material forming a silicon layer.

are used to produce a magnetic field that focuses the sputtering ions onto the target to achieve higher sputtering rates.

5.5.3 THE CVD OF SILICON

Returning now to the details of silicon deposition by CVD, the modeling starts with the reaction of trichlorosilane with the hydrogen carrier gas, Equation 5.37.

5.5.3.1 Series Steps

Again, for this reaction, there are three transport processes in series:

- Diffusional transport of the $SiHCl_3$ to the Si surface
- The actual reaction at the surface
- Diffusional transport of the HCl and H_2 product gases away from the surface

These are series or sequential steps, and as was shown before, and again later, for this particular process, the *slowest* of series steps controls the overall rate of reaction. The transport of the gaseous reactants to and the products away from the surface are usually gaseous diffusion processes. Later, Chapters 8–12 are devoted to diffusion processes and CVD of silicon by diffusion control is covered in Chapter 10.

5.5.3.2 Fluxes or Flux Densities

Since a surface reaction is being considered as the controlling step, it makes sense to consider the rate of reaction in terms of the rate per unit area of the surface. It is convenient to do this in terms of fluxes or flux densities, J, where the units of J are Units(J) = mol/cm^2/s. For the reaction in question, the stoichiometry implies that the relation between the various fluxes is $J(SiHCl_3)$ to the surface = $J(H_2)$ to the surface = $-J(Si)$ at the surface = $-J/3(HCl)$ away from the surface. The 1/3 comes from the fact the there are three moles of HCl produced for every mole of $SiHCl_3$ reacting and means that the flux of HCl away will be three times greater than the $SiHCl_3$ flux to the surface.

5.5.3.3 Possible Rate-Controlling Steps on the Surface

The actual surface reaction can consist of many possible molecular or atomistic series steps—similar to some of those presented for the $H_2 + D_2$ reaction on Pt presented in Chapter 4—which could include (among many other possibilities):

- Adsorption or *sticking* of the $SiHCl_3$ molecules on the surface
- Adsorption of H_2 molecules on the surface
- Splitting of the $SiHCl_3$ and H_2 into H, Si, and Cl *atoms* on the surface
- Migration (diffusion) of Si to the reaction site
- Reaction of Si at the *reaction site*
- Migration (diffusion) of H and Cl *atoms* to their reaction sites
- Reaction of H and Cl to form HCl somewhere on the surface
- Diffusion of HCl away from their reaction sites
- Desorption or removal of the HCl molecules from the surface

As shown in Figure 5.13—a schematic illustration in which the atoms are depicted as cubes—the surface of a solid consists of partially complete atomic planes of atoms. The *most likely site for surface reactions to take place is the* kink *site* since atoms at this position have more *unsatisfied* bonds or fewer nearest neighbors than surface atoms on steps (ledges) or terraces. In principle, by changing the crystallographic orientation of the surface, the number of kink sites can be increased or decreased, thereby changing the rate of reaction since any surface will consist of low-surface energy planes. It is commonly observed that the rate of surface reactions on solids varies with crystallographic orientation and the variation of the number of kink sites with orientation can sometimes

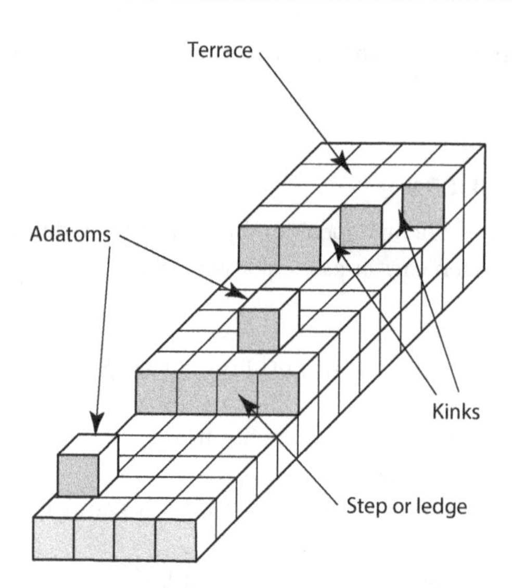

FIGURE 5.13 The classical picture of a surface where each cube in the drawing represents an atom. The important surface feature are named: terraces, steps or ledges, adatoms, and kinks, where the latter are the most likely spots for reactions to occur or atoms to stick.

explain this effect. The adatom shown is basically unstable and will diffuse to a kink site, or the edge of a step or *ledge*, or perhaps leave the surface. Again, all of these atomistic steps are mainly in series since they must follow one another. For gas–solid reactions, the details of most such reactions are not known well enough to identify the slowest and singular *rate-controlling* step among these—and other—imaginable steps. Therefore, unless there is good experimental evidence to suggest otherwise, the assumption is made that the *surface reaction rate is a simple first order reaction.*

5.5.4 Deposition of Silicon from $SiHCl_3$

5.5.4.1 Assumptions for the Model

Any one of the silane compounds could be, and have been, used for the deposition of silicon by CVD. However, trichlorosilane is a liquid at room temperature and very near its boiling point, so it is easy to control its vapor pressure by slightly cooling the liquid and thereby controlling the pressure of $SiHCl_3$ in the hydrogen carrier gas. In addition, its reaction rate covers a convenient range of temperatures for silicon deposition.

To develop a model, several assumptions are made: (1) a surface reaction is the rate-controlling step—which implies that diffusion to and from the surface of the reactant and product gases is infinitely fast—that is, there are *no* concentration gradients in the gas phase; (2) the reaction goes to equilibrium; (3) a flowing reactor is used so that the pressure of $SiHCl_3$ is constant in time; (4) the geometry is a flat surface (it could as easily be curved and the rate of deposition will be the same for a reaction-controlled process); (5) the HCl pressure in the gas is controlled to define the equilibrium pressure of $SiHCl_3$; and (6) the surface reaction is first order. For gas–solid reactions, a first-order reaction is frequently observed and, if you had to guess, guessing a first-order reaction is probably correct.* This is the model for the reaction with these assumptions. How valid these assumptions are can only be determined by comparing model and experiment. If they do not compare favorably, then the model needs to be changed. However, it is just that, a model, and so there is freedom to assume whatever parameters are deemed necessary and important to model the real process. Most of science and engineering consists of trying to find better models of reality!

5.5.4.2 Thermodynamics of the Reaction

Since the reaction is assumed to be first order, then the rate, expressed in terms of a flux, is

* A first-order reaction is predicted by, and is consistent with, Equation 5.21.

$$J(SiHCl_3) = -k\left([SiHCl_3]_0 - [SiHCl_3]_e\right)$$

$$= -\frac{k}{RT}\left(p_0(SiHCl_3) - p_e(SiHCl_3)\right) \tag{5.40}$$

which depends on the equilibrium pressure of the $SiHCl_3$ at the surface. So this needs to be calculated. Assume a temperature of 1100°C and a $p_0(SiHCl_3) = 0.1$ bar, $p(H_2) = 1$ bar as the *carrier gas*, and $p(HCl) = 0.03$ bar.* Note the similarity between Equations 5.40 and 5.21. These are very close to the values used in the actual CVD of silicon from $SiHCl_3$. The overall reaction is

$$SiHCl_3(g) + H_2(g) \rightleftharpoons Si(s) + 3HCl(g). \tag{5.41}$$

The Gibbs energy for this reaction at 1100°C is $\Delta G° = 43{,}280$ J/mol (Roine 2002). Note that even though the free energy for the reaction is positive, it will still occur to some extent with these conditions. The equilibrium constant, K_e, is given by

$$K_e = e^{-\frac{\Delta G°}{RT}} = e^{-\frac{43280}{8.314 \times 1373}} = 2.26 \times 10^{-2}$$

Therefore,

$$\frac{a_{si} p_{HCl}^3}{p_{SiHCl_3} p_{H_2}} = K_e$$

or

$$p_e(SiHCl_3) = \frac{a_{si} p_{HCl}^3}{p_{H_2} K_e} = \frac{(1.0)(0.03)^3}{(1.0)(2.26 \times 10^{-2})} = 1.19 \times 10^{-3} \text{ bar}.$$

Notice that even though the Gibbs energy for the reaction is positive, about 99% of the $SiHCl_3$ is used up in the deposition $(p_0 - p_e = 0.1 - 0.001 = 0.099)$. The activity of Si is 1.0 since it is pure silicon and the total pressure in the system is slightly above one bar, which can be neglected.

5.5.4.3 Implications of Infinitely Fast Gas Diffusion

Since some of the $SiHCl_3$ reacts, some HCl will be generated at the gas–solid surface. However, the assumption is that diffusion is infinitely fast compared to the rate of surface reaction. As a result, the hydrogen, HCl, and the $SiHCl_3$ pressures must remain *constant* throughout the gas phase right up to the silicon surface—there can be *no pressure gradient* in the gas since the diffusion coefficient is assumed to be infinite (Figure 5.14). If just the opposite were true, that is, the surface reaction is infinitely faster than diffusion, then all of the surface pressures reach equilibrium and must be calculated from

$$\frac{a_{Si}\left(p_{HCl}^0 + 3x\right)^3}{\left(p_{SiHCl_3}^0 - x\right)\left(p_{H_2}^0 - x\right)} = K_e \tag{5.42}$$

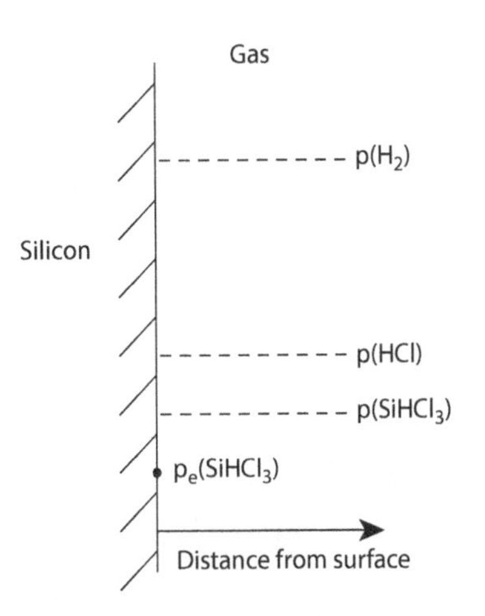

FIGURE 5.14 Reactant and product gas pressures as a function of distance from the surface for the CVD of silicon from $SiHCl_3$ and H_2 with a surface reaction controlling the rate of deposition. Although some of the reactants are depleted and the HCl is increased in the reaction, their pressures or concentrations remain constant with distance since it is assumed that diffusion is infinitely fast and therefore there can be no pressure or concentration gradients in the gas phase.

* In reality, the pressure can probably be given in either *atm* or *bar* since all older thermodynamic data was based on a standard state of one atmosphere and much of the data has not been changed to a standard state of one bar. In reality, the thermodynamic data for most reactions are not known sufficiently accurately for the difference between one bar and one atmosphere to have a significant effect anyway. The NIST JANAF Tables, 3rd ed. (Chase et al. 1985) show the conversion factors used between the old and new definitions of the standard pressure state.

where x is the pressure of $SiHCl_3$ consumed (since this is an ideal gas, the pressure and the number of moles are proportional). This equation can easily solved numerically by iteration on a spreadsheet with the results:

$$p_e(SiHCl_3) = 7.13 \times 10^{-2} \text{ bar}$$

$$p_e(H_2) = 0.9713 \text{ bar}$$

$$p_e(HCl) = 0.1161 \text{ bar.}$$

With diffusion control, only about 30% of the $SiHCl_3$ reactant is consumed. This is fine in theory to calculate the two extremes, but for a real process, neither the diffusion flux nor the surface reaction will be infinitely faster than the other, and this leaves the pressures at the interface somewhat uncertain. The combined effects or reaction and diffusion on the concentration-distance curves and surface concentrations are explored later in a model that takes into consideration both diffusion and reaction.

Since HCl is a product of the reaction, adding HCl to the initial gas stream provides some control over the equilibrium pressures and the rate of deposition. For example, it is quite possible for the reaction to take place in the gas phase with the formation of small particles of silicon. In fact, silicon, silicon carbide, SiC, and silicon nitride, Si_3N_4, as well as other powders can be made in this way (Ring 1996). Putting HCl into the gas stream provides some control to prevent gas phase formation of silicon particles. Generating powders in the gas is desirable if the interest is in producing nanoparticles of high purity silicon, but it is not if the intent is to deposit a single crystal epitaxial layer of silicon.

5.5.4.4 Deposition Rate Model

The flux of trichlorosilane to the surface determines the deposition rate, and this can be expressed as

$$J(SiHCl_3) = -k\left([SiHCl_3]_0 - [SiHCl_3]_e\right)$$

$$= -\frac{k}{RT}\left(p_0(SiHCl_3) - p_e(SiHCl_3)\right)$$

where the [] represents concentration and the lower equation comes from $[SiHCl_3] = n/V = p/RT$. Whether concentrations or pressures are used makes no difference. It just depends on what the value and units of k are. In this case, since the units of J are Units(J) = mol/cm²-s and the units of concentration are mol/cm³, so the units of k must be Units(k) = cm/s. And for this reaction,

$$k = k_0 e^{-\frac{Q}{RT}}$$

the values of $k_0 = 36.7$ cm/s and $Q = 67{,}200$ J/mol can be calculated from data in the literature on the deposition of silicon (Campbell 1996, Wolf and Tauber 2000).

It is necessary to relate the flux of $SiHCl_3$ at the surface to the growing thickness of the silicon layer since this is really what is of interest, "How fast does the silicon layer grow?" This is simply a mass balance: the molar flux density of Si to the surface J that is producing the deposited silicon layer is

$$J(Si) = \frac{1}{A}\frac{dn}{dt} \tag{5.43}$$

where:

A is the area as shown in Figure 5.15

n is the moles of silicon.

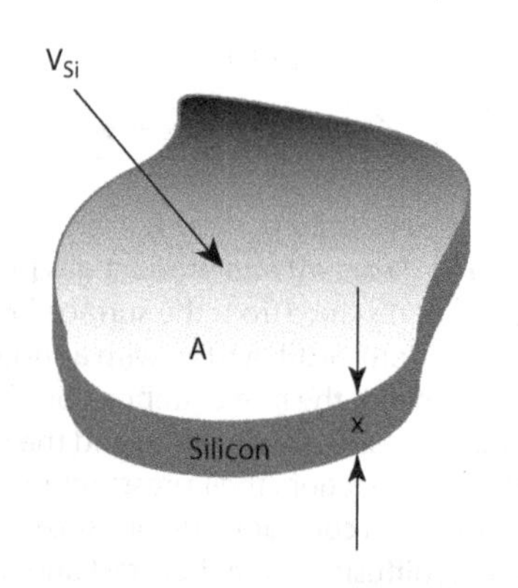

FIGURE 5.15 Geometry of CVD silicon deposit used to calculate the rate of deposition, dx/dt.

Now,

$$n = \frac{\rho}{M} V_{Si}$$

where:

ρ is the density of silicon = 2.33 g/cm^3
M is the molecular weight of silicon = 28.09 g/mol
V_{Si} is the volume of the silicon layer Ax, where x = thickness.

Note that $M/\rho = \bar{V}$ = the molar volume of Si in cm^3/mol. Making the mass balance—per unit area per unit time, as in Equation 5.22—and since Si is deposited as $SiHCl_3$ is consumed,

$$J(Si) = -J(SiHCl_3)$$

$$\frac{1}{A}\frac{\rho}{M}\frac{dV}{dt} = \frac{1}{\cancel{A}}\frac{\rho}{M}\cancel{A}\frac{dx}{dt} = \frac{k}{RT}\left(p_0(SiHCl_3) - p_e(SiHCl_3)\right)$$

or

$$\frac{dx}{dt} = \frac{Mk}{\rho RT}\left(p_0(SiHCl_3) - p_e(SiHCl_3)\right) \tag{5.44}$$

which is the *rate of deposition*, frequently reported, and can be integrated to give the thickness x of the deposited silicon layer as a function of time:

$$x = \frac{Mk}{\rho RT}\left(p_0(SiHCl_3) - p_e(SiHCl_3)\right)t. \tag{5.45}$$

The flux density of the silicon is opposite to the flux density of the trichlorosilane since it is increasing in thickness while the $SiHCl_3$ is disappearing. This sign difference is also correct in a directional sense since the thickness is increasing in the positive vertical direction while the $SiHCl_3$ is disappearing in the negative direction.

5.5.4.5 Deposition Rate Calculation

To begin, collect all the parameters for the process in the above equations:

T = 1373 K	Chosen deposition temperature
ρ = 2.33 g/cm^3	Density of silicon
M = 28.09 g/mol	Molecular(atomic) weight of silicon
$p_0(SiHCl_3)$ = 0.1 bar	Initial trichlorosilane pressure
$p_e(SiHCl_3) = 1.19 \times 10^{-3}$ bar	Equilibrium trichlorosilane pressure
$p(H_2)$ = 1 bar	Hydrogen pressure, assumed constant
$k = k_0 e^{-(Q/RT)}$	Reaction rate constant
k_0 = 36.7 cm/s	Pre-exponential
Q = 67,200 J/mol	Activation energy
R = 8.314 J/mol-K	To calculate k
R = 83.14 cm^3-bar/mol-k	In the denominator of Equation 5.45

Calculating a value for k at 1373 K: $k = k_0 \exp(-Q/RT) = 36.7 \exp[-67{,}200/(8.314 \times 1373)] = 0.102$ cm/s so the rate of deposition dx/dt is

$$\frac{dx}{dt} = \frac{28.09}{2.33}\frac{0.102}{83.14 \times 1373}\left(0.1 - 1.19 \times 10^{-3}\right) = 1.06 \times 10^{-6}\ \text{cm/s}. \tag{5.46}$$

Since the rate of deposition is constant, the layer thickness is just the product of the rate and time. Therefore, the time to grow a 10 μm layer would be

$$\frac{x}{dx/dt} = t$$

$$\frac{10\times 10^{-4}\ \text{cm}}{1.06\times 10^{-6}\ \text{cm/s}} \times \frac{1}{60\ \text{s/min}} = 15.7\,\text{min}. \tag{5.47}$$

These rates and times are in the range of those that are actually used in industrial practice.

5.6 ACTIVE GAS CORROSION OF Si IN HCl AND H_2

Notice in the above equation for the deposition rate, Equation 5.44,

$$\frac{dx}{dt} = \frac{Mk}{\rho RT}\left(p_0\left(SiHCl_3\right) - p_e\left(SiHCl_3\right)\right)$$

that if the $p_0(SiHCl_3)$ were actually less than $p_e(SiHCl_3)$, then the rate of deposition would be negative. That is, the silicon surface would be undergoing *active gas corrosion* or *etching*. As was shown earlier in Chapter 3, for a first-order reaction with a first order back reaction leading to equilibrium, the overall reaction rate constant is the same for both directions of the reaction: $k = k_1$(forward) + k_2(back) So the same value of k holds for corrosion and if the assumption is made that $p_0(SiHCl_3) = 1\times 10^{-3}$ atm—less than the equilibrium pressure, then

$$\frac{dx}{dt} = \frac{28.09}{2.33}\frac{0.102}{83.14\times 1373}\left(0.001 - 1.19\times 10^{-3}\right) = -2.05\times 10^{-9}\ \text{cm/s}. \tag{5.48}$$

This demonstrates that exactly the same type of surface reaction kinetics apply to both active gas corrosion and CVD reactions. Which occurs depends on the concentrations of the reacting gas species and how fast either occurs depends on the value of the reaction rate constant for deposition or corrosion.

5.7 HALOGEN BULBS

There are certain types of halogen lamps that use a combination of both active gas corrosion and CVD to allow operation at higher temperature and, therefore, a higher light output or brightness (halogen bulbs). These lamps contain a normal tungsten filament (Figure 5.16) that typically consists of a double

FIGURE 5.16 Tungsten filament from a burned-out bulb. The diameter of the tungsten wire is about 45 μm and consists of a double helix.

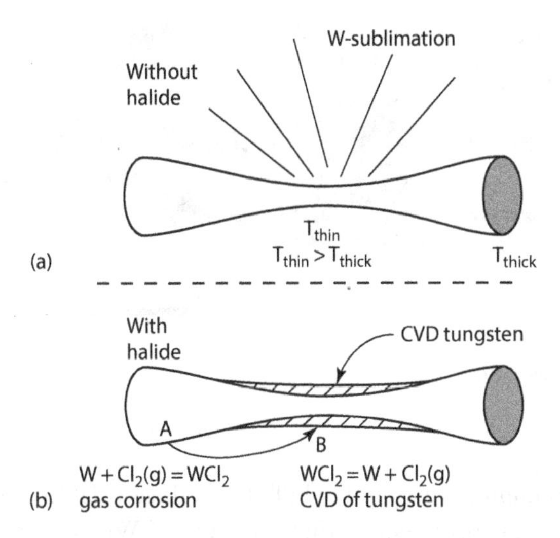

FIGURE 5.17 (a) how a tungsten filament can fail in an incandescent bulb due to evaporation of tungsten from the thinner and hotter part of the filament. (b) the active gas corrosion reaction of tungsten with a halide gas at lower temperatures and the reverse reaction, CVD of tungsten, at the higher temperatures or thinner part of the filament. This prevents thinning of the filament allowing it to be used at higher temperatures giving off more light.

helix of tungsten wire about 45 μm in diameter and a small amount of halogen gas or substance that produces a gas such as I_2, Br_2, or Cl_2, so that the following reversible reaction can take place:

$$W(s) + Cl_2(g) \underset{\text{high T}}{\overset{\text{low T}}{\rightleftarrows}} WCl_2(g).$$

Regular incandescent bulbs eventually burn out because a hot spot forms at a thinner region in the filament as shown in Figure 5.17. At the high operating temperatures of these filaments—in the neighborhood of 2800°C—the bulbs burn out by sublimation of the hot tungsten onto the glass bulb causing darkening or dark areas on the inside of the bulb. As a filament region thins, it gets hotter because its electrical resistance goes up and sublimes tungsten at a faster rate, getting thinner, which in turn increases its electrical resistance and gets hotter until the thin region opens and the bulb fails (Figure 5.18). However, the tungsten filament in a halogen lamp (Figure 5.19) will

FIGURE 5.18 Burned-out incandescent bulb showing blackening of the inside of the bulb due to evaporation and deposition of tungsten on the inside of the glass envelope leading to eventual bulb failure.

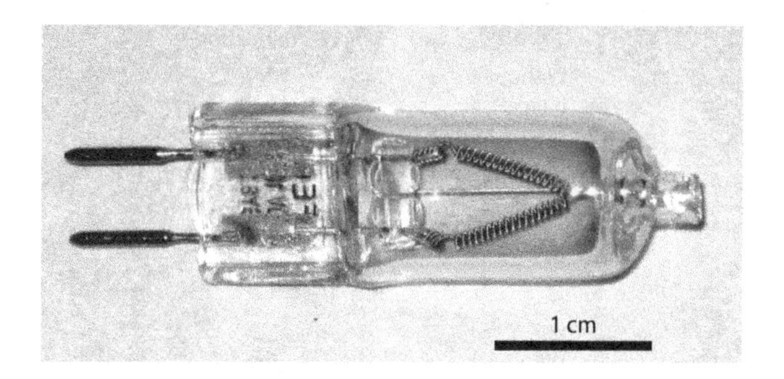

FIGURE 5.19 A small halogen bulb.

interact with halogen-containing gases in the bulb to produce volatile tungsten halides as shown in Figure 5.20. This figure gives the equilibrium pressure of WCl_2(g)—and those of Cl(g), WCl, and WCl_3, the species with the highest vapor pressures—over tungsten in chlorine. The important point to note is that the WCl_2(g) pressure is *higher* at lower temperatures than it is at high temperatures. This means that W can be transported from lower temperature regions via WCl_2—active gas corrosion—and deposited at higher temperature regions—CVD—as shown in Figure 5.17, keeping the hot regions from getting thinner and preventing bulb burnout. As a result, such halogen bulbs are not only brighter but also have a longer life—taking advantage of the sublimation and deposition phenomena, which itself is determined by the thermodynamics and kinetics between the halogen gas and tungsten.

These halogen bulbs are typically small, so that the glass bulb gets quite hot. As a result, the bulbs are made from amorphous SiO_2 or *fused quartz*—SiO_2 glass—that has a much higher glass transition temperature, about 1200°C—and softening temperature—than conventional soda-lime-silica glass bulbs with glass transition temperatures around 500°C. However, because of the much higher operating temperature of the silica envelope, care must be taken not to contaminate the surface, particularly with the sodium from fingerprints. The sodium enhances *devitrification*—crystallization—of the silica, which appears as white *powdery* regions on the exterior of a bulb reducing light output and forming a potential source for cracking of the silica.

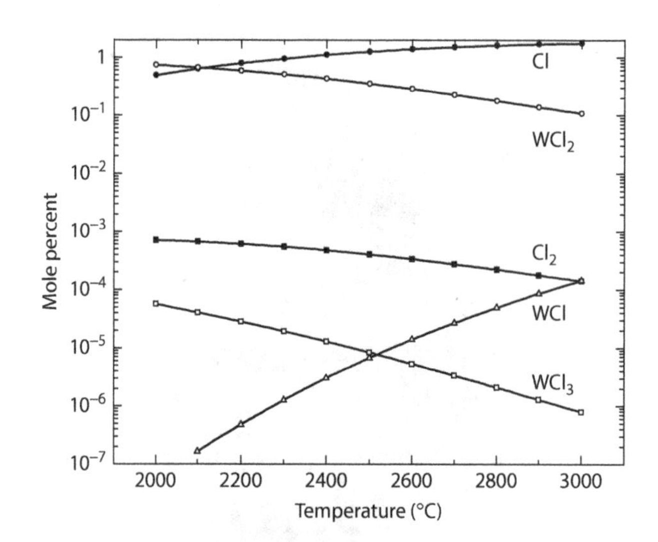

FIGURE 5.20 Calculated equilibrium gas pressures of the major tungsten-containing gases over tungsten in an inert atmosphere containing 1m/o Cl_2 (Roine 2002). Note that major species, WCl_2, has a higher pressure at lower temperatures. The operating temperature of such a bulb is in the neighborhood of 2500°C–2800°C.

5.8 CARBON–CARBON COMPOSITES

5.8.1 INTRODUCTION

The destruction of the space shuttle *Columbia* during its reentry into the atmosphere in 2003 (Figure 5.21) is believed to have been caused by fracture of the carbon–carbon composite on the leading edge of the wing by a piece of polymer foam falling from the fuel tank on lift-off (Columbia Accident Investigation Board 2003). Since carbon–carbon composites are complex materials that often involve the use of the chemical vapor infiltration, CVI, process to bind the carbon fibers together, it is appropriate to say a little about carbon–carbon composite processing.

FIGURE 5.21 Columbia prior to its final launch of January 16, 2003. Note the black carbon–carbon composite leading edges of the wings. NASA Photograph, Kennedy Space Center (Columbia Accident Investigation Board 2003). Original color photograph transformed to grayscale.

5.8.2 CARBON–CARBON COMPOSITES ON THE SHUTTLE

Carbon–carbon composites are used on the leading edges of the wings, fin, and nose of the shuttle—the black areas seen in Figure 5.21—because they are light weight and can withstand temperatures over 2000°C with no degradation in their mechanical properties. These composites are used in these locations because of the potential for aerodynamic heating on re-entry reaching temperatures on the order of 1600°C. Unfortunately, carbon–carbon composites are brittle. In previous shuttle flights, there was evidence that some of the carbon–carbon composite parts had undergone damage—chunks of the material removed by brittle failure from impacts with some type of foreign objects. The investigation into the accident determined that it was likely that insulating foam hitting the leading edge of the shuttle wing on lift-off ultimately lead to its destruction on reentry. As part of the accident investigation, pieces of foam were fired from an air gun at the likely failed composite parts with the result that large holes—over 1 ft in size—could be produced in these parts as Figure 5.22 demonstrates (Columbia Accident Investigation Board 2003).

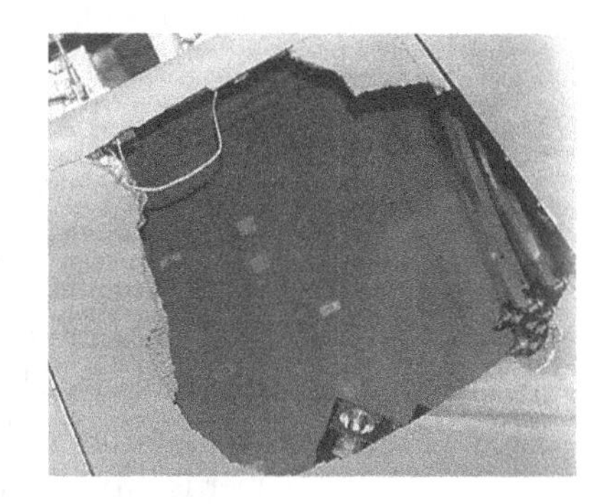

FIGURE 5.22 Large hole produced in one of the shuttle wing leading edge carbon–carbon composites by a piece of insulating foam fired from an air gun to mimic the piece of foam that hit the orbiter upon lift-off. NASA Photograph (Columbia Accident Investigation Board 2003). Original color photograph transformed to grayscale.

5.8.3 CVD OF GRAPHITE: PYROLYTIC GRAPHITE

Pyrolytic graphite can be made into free-standing parts or coatings several centimeters thick for a number of high-temperature applications. These components are produced by the CVD process with a reaction such as the decomposition of methane (Pierson 1999),

$$CH_4(g) \xrightarrow{\text{temperature}} C(s) + 2H_2(g). \tag{5.49}$$

The actual reaction kinetics are presumably more complicated than suggested by Equation 5.49 with several intermediate parallel and series steps. Nevertheless, so called *pyrolytic graphite*, polycrystalline graphite, can be deposited by this overall reaction at temperatures of 1100°C and above over a range of total pressures from 10^{-3} to 1.0 bar. The result is a *columnar grain*—long along the growth direction—structure as depicted in Figure 5.23. Presumably grains or crystallites having the best orientation for growth grow the fastest and produce the columnar structure. For pyrolytic graphite, the sheets of graphene (a sheet of hexagonally-arranged carbon atoms one atom thick in graphite) are lined up parallel to the substrate surface and so is the a-axis of the graphite structure while the c-axis is parallel to the growth direction or parallel to the long axes of the columnar grains. This leads to a strong anisotropy of many properties such as thermal conductivity due to the covalent bonding along the a-axis and van der Waals bonding in the c-direction. For example,

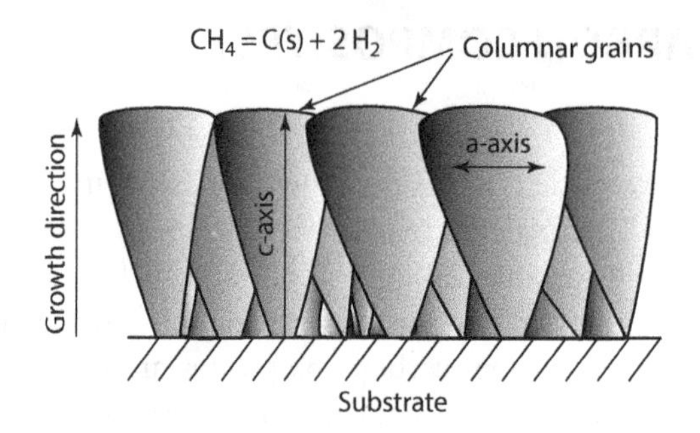

FIGURE 5.23 Schematic microstructure of a pyrolytic graphite deposit depicting the columnar grain structure from the decomposition of methane. The c- and a-axes of the graphite crystals are shown. The columnar grains extend through the thickness of the deposited layer that may be as thick as several centimeters. Because of this strong preferred orientation, properties such as thermal conductivity along the a-axis may be as much as 100 times larger than that along the c-axis.

the thermal conductivity along the a-axis of the structure is on the order of 100 W/m-k while in the c-direction it is on the order of 1 W/m-k. This anisotropy in properties can be utilized to optimize the performance of a part such as a rocket nozzle. This is the kind of carbon or graphite deposition that is used in CVI to infiltrate and bond together the carbon fibers in a carbon–carbon composite.

5.8.4 Fabrication of Carbon Fibers

By far, the majority of carbon fibers are used to reinforce polymers to form carbon-fiber-reinforced-polymers, CFRPs. These composites are now used not only in military aircraft but also in civilian aircraft structural parts, as automotive parts, and as high-technology bicycle frames as well. They are produced by many manufacturers by largely proprietary processes. Nevertheless, there are several principles that are fairly common in making fibers. They are all made from carbon-rich polymer resins, an example being polyacrylonitrile (PAN), as shown in Figure 5.24. The polymer is mixed with a solvent to form a viscous material that is drawn into fibers. This drawing process helps to align the polymer molecules and, what will become, the graphite rings parallel

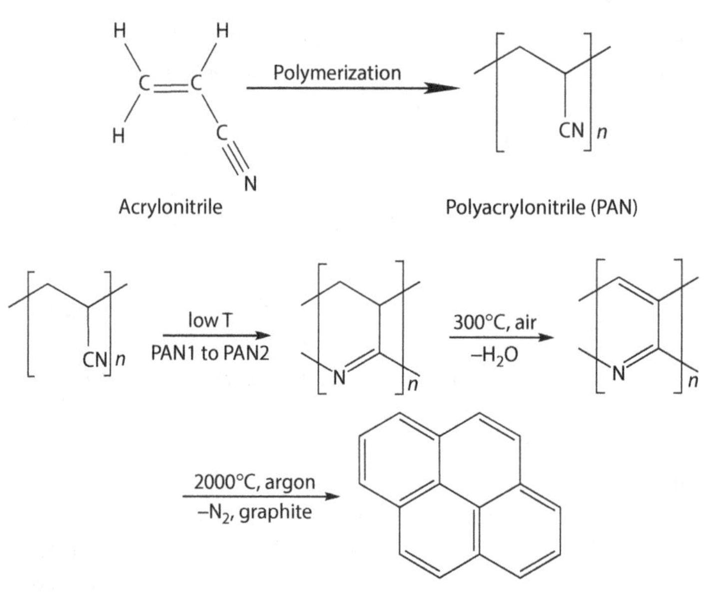

Acrylonitrile —Polymerization→ Polyacrylonitrile (PAN)

low T, PAN1 to PAN2 → 300°C, air, $-H_2O$ → 2000°C, argon, $-N_2$, graphite

FIGURE 5.24 The formation of polyacrylonitrile (PAN) and its multistep thermal transformation to graphene sheets in carbon/graphite fibers.

to the long axis of the fiber. The solvent is then removed by extraction in another solvent or by evaporation. The polymer fiber is given a low temperature treatment in oxygen or air to remove hydrogen and finally a high temperature graphitization reaction in excess of 2000°C to remove the nitrogen from the structure and form graphene sheets whose a-axes lie parallel to the fiber axis. During heat treatment, tension is applied to the fiber to enhance the alignment of the graphite sheets along the fiber axis increasing the modulus of elasticity of the fiber (Buckley and Edie 1992). The finished fibers are about 8 μm in diameter, have elastic moduli of about 300 GPa (43×10^6 psi) and fracture strengths on the order of 3.1 GPa (443,000 psi) (Matthews and Rawlings 1994). Compare these values to those of mild carbon steel with an elastic modulus about 200 GPa (26×10^6 psi) and a tensile strength around 430 MPa (61,000 psi) (Ashby and Jones 1980). Carbon fibers are sold commercially in bundles or yarns or *tows* consisting of several thousand individual carbon fibers.

5.8.5 Composite Fabrication

Carbon–carbon composites are made from woven carbon fibers yarns—as bundles or fabric—that are formed into the shape of the part to be made—the *prepreg* (Savage 1993). One process consists of infiltrating the porous *fabric* with a phenolic resin (or any number of carbon-rich resins)—a thermosetting resin produced by the reaction of formaldehyde, CH_2O, and phenol, HOC_6H_5—cured and pyrolyzed (heated in a nonoxidizing atmosphere to drive all volatile gases leaving only carbon behind) to graphite at some elevated temperature in excess of 2000°C. The graphite forms a rigid bond between the fibers producing a solid piece.

A variant is to use a CVD process to deposit the graphite in the interstices between the carbon fibers, such as in Equation 5.49 (Savage 1993):

$$CH_4(g) \xrightarrow{\text{temperature}} C(s) + 2\,H_2(g); \quad \Delta G^\circ(1500°C) = -103\,\text{kJ/mol}.$$

This process is *chemical vapor infiltration* (CVI) and is a variation of the CVD process in which the conditions are controlled, so that *surface reaction* is the slower step and is rate controlling. This ensures that the reactant gas concentrations are essentially uniform throughout the fibrous part—since diffusion is relatively fast—and the deposition rate will be the same throughout the part. This also implies that low pressures are preferred since diffusion in gases varies inversely with the pressure as will be seen later. Figure 5.25 shows a schematic microstructure of a typical carbon–carbon composite infiltrated by CVI. Usually, several infiltrations by CVI are necessary as the surface reaction is still sufficiently fast to close the surface porosity and stop the reaction. As a result, the part must be cooled and the surface ground to remove the dense graphite layer exposing the porosity in the structure to allow additional CVI.

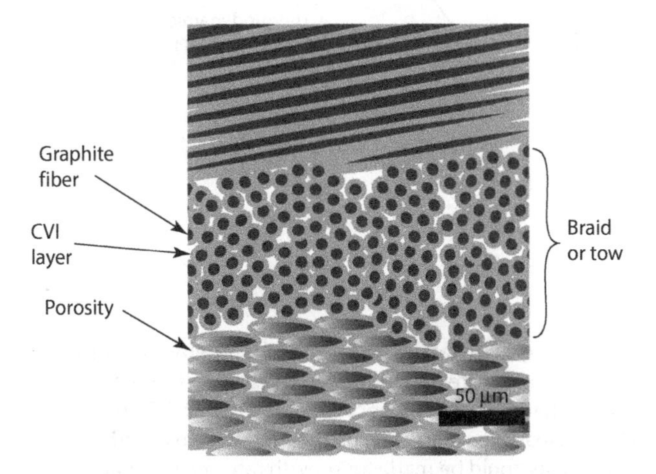

FIGURE 5.25 Schematic microstructure of a carbon–carbon composite showing three braids or tows of carbon fibers woven together in different directions showing: carbon fibers, CVI carbon layer, and residual porosity.

For the carbon–carbon composites used on the space shuttle, the outer surfaces of the parts are given a CVD coating of SiC by a similar reaction to close any remaining porosity and slow down the rate of oxidation of the carbon–carbon

$$SiHCl_3(g) + CH_4(g) \rightleftarrows SiC(s) + H_2(g) + 3\ HCl(g); \quad DG^\circ(1200°C) = -100\ kJ/mol.$$

Here, *diffusion control* of the reaction is important, so that all of the reaction takes place at the surface and sealing it. Unfortunately, SiC has a larger thermal expansion coefficient than that of graphite, and the SiC layer cracks on cooling. The cracks are filled with TEOS—tetraethyl orthosilicate or tetraethoxysilane—$Si(C_2H_5O)_4$, which reacts with water at room temperature to form SiO_2 (glass) that seals the cracks at high temperatures.* When completed, a typical carbon–carbon composite still contains about 15 volume percent porosity (Buckley and Edie 1992).

5.9 GENERAL OBSERVATIONS ABOUT REACTION-CONTROLLED GROWTH

5.9.1 Kinetic and Thermodynamic Factors

Looking again at the equation for the growth of silicon from $SiHCl_3$ (Equation 5.44)

$$\frac{dx}{dt} = \frac{Mk}{\rho RT}\left(p_0(SiHCl_3) - p_e(SiHCl_3)\right)$$

it should be noted that—other than the constants—the rate of growth, dx/dt, is essentially a product of a *kinetic factor*, k, and a *thermodynamic factor*, $[p_0(SiHCl_3) - p_e(SiHCl_3)]$. The kinetic factor decreases exponentially with decreasing temperature while the thermodynamic factor increases linearly with the difference between the ambient pressure and the equilibrium pressure—which in many cases decreases with temperature—as sketched in Figure 5.26. Therefore, the product of these two factors frequently leads to a *maximum* in the growth rate at some temperature below the equilibrium temperature as evident in Figure 5.26.

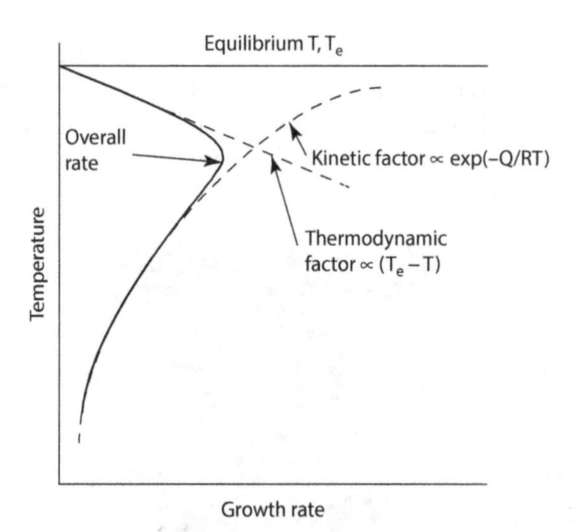

FIGURE 5.26 Schematic of the growth rate as a function of temperature for a phase transition below the equilibrium temperature. The growth rate depends on two factors, a thermodynamic factor, proportional to $(T - T_e)$ and a kinetic factor that typically is proportional to exp(−Q/RT) where Q is the activation energy. The product of the two factors leads to a maximum in the growth rate at some temperature below the equilibrium temperature. Similar plots could be made for growth rate versus composition or vapor pressure, etc.

* TEOS is used in the semiconductor industry to produce SiO_2 glass insulating layers by CVD on silicon microchips.

5.9.2 Reaction Rate Constant and Accommodation Coefficient

If the rate of evaporation, sublimation, or condensation of a substance does not follow Equation 5.18

$$I = \frac{p}{\sqrt{2\pi m k_B T}}$$

then an *accommodation coefficient,* α, is inserted to give Equation 5.21

$$J' = \frac{\alpha(p - p_e)}{\sqrt{2\pi m k_B T}}$$

where J′ is the atomic or molecular and Units(J′) = atoms/m^2-s. Dividing Equation 5.21 by N_A, gives the molar flux, J, with Units(J) = mol/m^2-s

$$J = \frac{J'}{N_A} = \frac{\alpha(p - p_e)}{\sqrt{2\pi MRT}}. \tag{5.50}$$

If $\alpha \neq 1.0$, then there must be some kind of surface reaction taking place,

$$J = k\left([A] - [A]_e\right) \tag{5.51}$$

where:

A is some gaseous species

A_e is the equilibrium concentration for condensation, sublimation, or evaporation.

As seen in Chapter 2 and as Equation 5.50 suggests, the surface reactions in Equation 5.51 are usually first order. In this equation, it is convenient to express concentrations in mol/cm^3 so that Units(k) = cm/s. The question is, "What is the relationship between α and k?" Expressing Equation 5.51 in terms of pressures, it becomes

$$J = \frac{k}{RT}(p - p_e). \tag{5.52}$$

Equating Equations 5.50 and 5.52 gives the relationship between the accommodation coefficient, α, and the reaction rate constant, k,

$$J = \frac{\alpha(p - p_e)}{\sqrt{2\pi MRT}} = \frac{k}{RT}(p - p_e). \tag{5.53}$$

If the units of pressure are Pa, the left-hand side of this equation has units of mol/m^2-s. If the units of k are Units(k) = cm/s, then the units of R must be Units(R) = (Pa-m^3)/(mol-K) so the right-hand side must be divided by 100 cm/m, that is,

$$\text{Units}(J) = \frac{(\text{cm/s})\text{Pa}}{8.314[\text{Pa}\cdot\text{m}^3/(\text{mol-K})]\times \text{K}}\,\frac{1}{100\ \text{cm/m}} = \frac{1}{100R}\,\frac{\text{mol}}{\text{m}^2\cdot\text{s}}.$$

So the relationship between α and k becomes

$$\alpha = 10^{-2}\sqrt{\frac{2\pi M}{RT}}\,k. \tag{5.54}$$

This assumes that the pressure units in Equations 5.50 and 5.52 are the same and Units(p) = Pa. So if the units of p in Equation 5.52 are something other than pascals, then a different value of the gas constant in Equation 5.52 appropriate to the pressure units must be used.

5.9.2.1 Example: Condensation of Water Vapor to Ice

The growth of ice crystals in an atmosphere containing water vapor is important in environmental science and meteorology to understand cloud formation kinetics at high altitudes.

Figure 5.27 shows some calculated data of growth of ice crystals from the gas phase as a function of temperature where the ambient H_2O pressure is at equilibrium at −5°C, assuming surface-reaction limited growth (Equation 5.52). In this case,

$$k = k_0 e^{-\frac{Q}{RT}}$$

with $k_0 = 0.6$ cm/s and $Q = 10$ kJ/mol arbitrarily chosen—and not necessarily the real values—in order to give a growth curve that clearly shows the product of the thermodynamic and kinetic factors illustrated in Figure 5.27. These data clearly indicate that a maximum in the rate occurs at about −25°C and, that above −5°C, the crystal actually sublimes since its vapor pressure is greater than that of the ambient H_2O pressure. How realistic are these values for k_0 and Q? There have been many determinations of the accommodation coefficient for condensation of water vapor on ice and ice sublimation and the approved value for these temperatures is $\alpha \cong 0.7$, although it is somewhat temperature dependent and wide variations in its value have been reported by different investigations (IUPAC Data Sheet 2009). In this case, taking the data at the maximum growth temperature, −25°C, where the calculated value of k is $k = 4.70 \times 10^{-3}$ cm/s, the calculation of the accommodation coefficient gives,

$$\alpha = 10^{-2}\sqrt{\frac{2\pi M}{RT}}k$$

$$\alpha = 10^{-2}\sqrt{\frac{2\pi(18 \times 10^{-3})}{8.314 \times 248}}(4.70 \times 10^{-3})$$

$$\alpha = 3.48 \times 10^{-7}$$

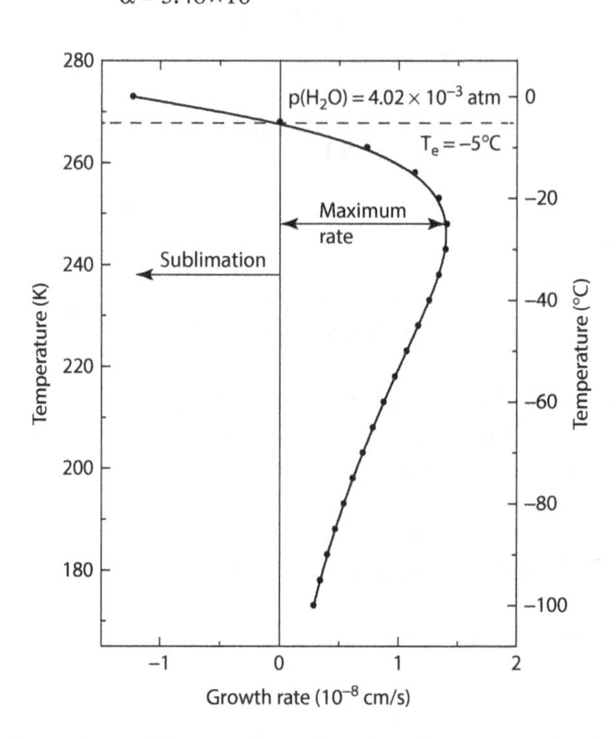

FIGURE 5.27 A plot of growth rate of ice crystals as a function of temperature in an atmosphere containing water vapor at $p(H_2O) = 4.02 \times 10^{-3}$ atm. In this plot, the activation energy for condensation was assumed to be $Q = 10$ kJ/mole, which is much too large to give the generally accepted value for the condensation coefficient of $\alpha = 0.7$ (International Union of Pure and Applied Chemistry 2009) but was used anyway to emphasize the concept of the growth rate being the product of both the thermodynamic and kinetic factors.

which is about a factor of 10^6 smaller than the literature value, suggesting that k really is much larger, $k \cong 2 \times 10^6$ cm/s. It should be noted that small values of α in the order of $\alpha \cong 6 \times 10^{-3}$ have been observed (Magee et al. 2006) and would imply $k \cong 1$ cm/s However, the goal here is not to accurately calculate the accommodation coefficient or the growth rate for ice but to demonstrate the general feature of growth involving the product of a kinetic factor and a thermodynamic factor. In order to do this, a larger activation energy is assumed. Such growth behavior as a function of temperature is typical of solids or liquids from the vapor or solids from liquids or solids from other solids. Growth of a second or new phase will be examined further when the topic of *nucleation and growth* is discussed.

5.10 SUMMARY

In order to illustrate heterogeneous reactions, gas–solid reactions are chosen because of their importance in materials science and engineering. To understand the parameter dependence of these reactions, some gas kinetics and the interaction between gas atoms/molecules and a solid surface are described. The Langmuir isotherm is developed for the adsorption of gases onto solid (or liquid) surfaces. The importance of gas solid reactions in many industrial materials processes and applications is discussed. One area of particular importance in industrial practice is the process of CVD. The CVD process is contrasted with PVD and the important thin film deposition process of sputtering is described. The technologically important process of CVD of silicon is shown in detail with results that are close to those used in actual practice. The importance and analysis of active gas corrosion is also discussed. The example of halogen bulbs in which CVD and active gas corrosion occur simultaneously at different temperatures to enhance the lifetime and operating temperature of such lamps is presented. Some discussion about carbon–carbon composites and their fabrication is offered as another example of a gas–solid deposition process, chemical vapor infiltration, CVI. Finally, growth of a solid or a liquid from a gas depends on the product of a thermodynamic and a kinetic factor that typically lead to some maximum in the growth rate at some temperature (or other variable) below equilibrium conditions. These concepts reappear when discussing the process of phase transformations in Chapter 7.

APPENDIX

A.1 Transforming from Cartesian to Polar Coordinates

It is often useful to transform from Cartesian or x and y coordinates to polar coordinates, r and θ. For polar coordinates, this can be done by inspection, as indicated in Figure A.1. However, a general technique to transform from one set of coordinates to another is valuable. This is particularly true for situations where the transformation is not so obvious, such as spherical or other three-dimensional coordinates. To transfer from an area element, dA, in Cartesian coordinates to polar coordinates, the formal mathematical method is

$$dA = dxdy = \left|\frac{\partial(x,y)}{\partial(r,\theta)}\right| drd\theta \qquad \text{(A.1)}$$

where

$$\frac{\partial(x,y)}{\partial(r,\theta)} = \begin{vmatrix} \frac{\partial x}{\partial r} & \frac{\partial x}{\partial \theta} \\ \frac{\partial y}{\partial r} & \frac{\partial y}{\partial \theta} \end{vmatrix} = J \qquad \text{(A.2)}$$

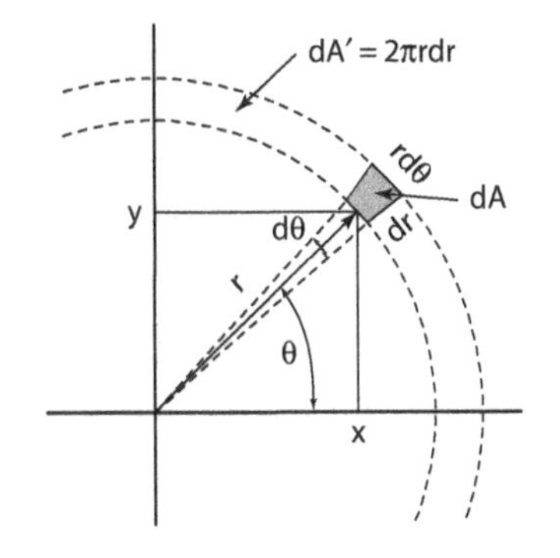

FIGURE A.1 Polar coordinates (r, θ) showing two differential area elements dA = rdrdθ and the theta-integrated area of dA′ = 2πrdr.

where:

J is the *Jacobian* matrix
|J| is the absolute value of the Jacobian (Kreyszig 2011).

For polar coordinates, $x = r\cos\theta$, $y = r\sin\theta$, and Equation A.2 becomes,

$$\begin{vmatrix} \frac{\partial x}{\partial r} & \frac{\partial x}{\partial \theta} \\ \frac{\partial y}{\partial r} & \frac{\partial y}{\partial \theta} \end{vmatrix} = \begin{vmatrix} \cos\theta & -r\sin\theta \\ \sin\theta & r\cos\theta \end{vmatrix} = r(\cos^2\theta + \sin^2\theta) = r$$

so, in polar coordinates, dA becomes

$$dA = rdrd\theta. \qquad (A.3)$$

Since θ is independent of r, the differential area element dA can be changed to dA′ by integrating θ over the range of interest, 0 to 2π,

$$dA' = \int_0^{2\pi} rdrd\theta = rdr\int_0^{2\pi} d\theta = 2\pi rdr$$

which is now a circular ring of thickness dr and circumference $2\pi r$ as shown in Figure A.1

A.2 Integration of $\int_{-\infty}^{\infty} e^{-x^2}dx$

Now $\int_{-\infty}^{+\infty} e^{-x^2}dx$ is a definite integral and is just some number, call it I. Then I^2 can be written as (this is the crucial step in the process)

$$I^2 = \left(\int_{-\infty}^{+\infty} e^{-x^2}dx\right)\left(\int_{-\infty}^{+\infty} e^{-y^2}dx\right) = \int_{-\infty}^{\infty}\int_{-\infty}^{\infty} e^{-(x^2+y^2)}dxdy \qquad (A.4)$$

where the last step is valid since x and y are independent *dummy* variables and the integration is over all space, $-\infty \le x \le \infty$ and $-\infty \le y \le \infty$. It is much more convenient to change to polar coordinates so that $x^2 + y^2 = r^2$ and $dA = dxdy = rdrd\theta$ as shown in Figure A.1, and integrate over all space from $0 \le r \le \infty$ and $0 \le \theta \le 2\pi$.

$$I^2 = \int_{-\infty}^{\infty}\int_{-\infty}^{\infty} e^{-(x^2+y^2)}dxdy$$

$$I^2 = \int_0^{\infty}\int_0^{2\pi} re^{-r^2}d\theta dr = 2\pi\int_0^{\infty} re^{-r^2}dr$$

$$I^2 = 2\pi\left(-\frac{1}{2}e^{-r^2}\right)_0^{\infty}$$

$$I^2 = 2\pi\left(-\frac{1}{2}(0-1)\right) = \pi.$$

Therefore, taking the square root,

$$I = \int_{-\infty}^{+\infty} e^{-x^2}dx = \sqrt{\pi}. \qquad (A.5)$$

In several situations, $\int_0^{\infty} e^{-x^2}dx$, is used. Now, e^{-x^2} is a symmetrical or even function about the origin—$f(-x) = f(x)$—as shown in Figure 5.3. And for an even function (Kreyszig 2011), $\int_{-\infty}^{\infty} e^{-x^2}dx = 2\int_0^{\infty} e^{-x^2}dx$.

This is easily shown to be the case since,

$$\int_{-\infty}^{\infty} e^{-x^2}dx = \int_{-\infty}^{0} e^{-x^2}dx + \int_{0}^{\infty} e^{-x^2}dx. \quad \text{(A.6)}$$

In the first integral, make the substitution, y = −x so

$$\int_{-\infty}^{0} e^{-x^2}dx = -\int_{\infty}^{0} e^{-y^2}dy = \int_{0}^{\infty} e^{-y^2}dy = \int_{0}^{\infty} e^{-x^2}dx.$$

since exchanging the limits of integration changes the sign (Edwards and Penney 2002) and x and y are dummy variables. Therefore, Equation A.6 becomes

$$\int_{-\infty}^{\infty} e^{-x^2}dx = \int_{0}^{\infty} e^{-x^2}dx + \int_{0}^{\infty} e^{-x^2}dx = 2\int_{0}^{\infty} e^{-x^2}dx$$

so

$$\int_{0}^{\infty} e^{-x^2}dx = \frac{\sqrt{\pi}}{2}. \quad \text{(A.7)}$$

A.3 Changing Coordinates for Spherical Coordinates

Figure A.2 shows the volume element in spherical coordinates,

$$dV = r^2 \sin\theta\, dr\, d\theta\, d\varphi \quad \text{(A.8)}$$

as can be seen graphically from the figure. However, again it can be obtained from the Jacobian for the spherical coordinates of Figure A.2

$$x = r\sin\theta\cos\varphi$$
$$y = r\sin\theta\sin\varphi \quad \text{(A.9)}$$
$$z = r\cos\theta.$$

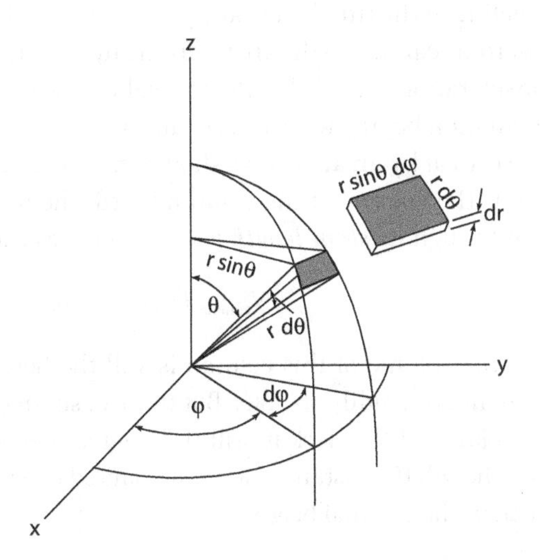

FIGURE A.2 Polar coordinates (r, θ, φ) and differential area, $dA = r^2 \sin\theta d\theta d\varphi$, and differential volume, $dV = r^2 \sin\theta d\theta d\varphi dr$, elements.

In this case, the Jacobian is given by

$$J = \begin{vmatrix} \frac{\partial x}{\partial r} & \frac{\partial x}{\partial \theta} & \frac{\partial x}{\partial \varphi} \\ \frac{\partial y}{\partial r} & \frac{\partial y}{\partial \theta} & \frac{\partial y}{\partial \varphi} \\ \frac{\partial z}{\partial r} & \frac{\partial z}{\partial \theta} & \frac{\partial z}{\partial \varphi} \end{vmatrix} = r^2 \sin\theta \tag{A.10}$$

where the details of evaluating this Jacobian to get this result are left to the Exercises at the end of the chapter. So this means that the volume element, dV, transforms as

$$dV = dx\,dy\,dz = r^2 \sin\theta dr\,d\theta\,d\varphi \tag{A.11}$$

and since $0 \le \theta \le \pi$ and $0 \le \varphi \le 2\pi$ and r, θ, and φ are independent of each other, a new volume element, dV, can be determined by integrating dV with respect to θ and φ

$$dV = r^2 dr \int_0^{\pi} \sin\theta d\theta \int_0^{2\pi} d\varphi$$

$$dV = r^2 dr\left(-\cos\theta\Big|_0^{\pi}\right)(2\pi)$$

$$dV = r^2 dr\left(-[-1-1]\right)(2\pi)$$

or

$$dV = 4\pi r^2 dr. \tag{A.12}$$

A.4 Exact Result for Gas Atoms Colliding with a Surface

Figure 5.6 shows an approximate way to calculate the approximate number of gas atoms or molecules hitting a surface, per unit area, per second:

$$Z = \frac{1}{6}\eta\bar{v}.$$

This is only an approximation since this assumes that the atoms are only traveling in the three mutually perpendicular directions. The reality is the gas atoms can strike the area A from any direction. When this is taken into consideration, a slightly different value of Z is obtained. A more accurate calculation begins with a gas column again of length, v, but inclined from the vertical by an angle θ as shown in Figure A.3 (McQuarrie and Simon 1997). The volume of the column is still the height—in this case, v cos θ, even though its *slant length* is now v—times the area

$$V_{column} = A\,v\cos\theta. \tag{A.13}$$

That the volume of this column is still the height, v cos θ times the area, might not be readily obvious. But think of sliding a deck of cards at an angle, the volume of the deck is still the original height times the area of a card even though the distance along the slanted edge might be considerably longer than the original height.

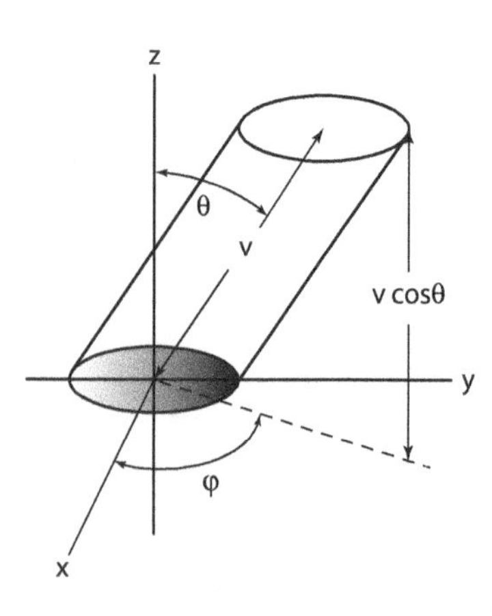

FIGURE A.3 Column of gas atoms of speed v striking a surface at some angle θ.

Now, the fraction of atoms moving along the axis of this tilted cylinder must be taken into account. It is assumed that atoms can be moving in any directions so the probability of them crossing the surface of a sphere is equally probable in any angular direction. So that the fraction of atoms in any direction at angles θ and φ is the small area of the sphere

$$dA = (r\sin\theta d\theta)(rd\varphi) \tag{A.14}$$

as shown in Figure A.2. The total area of the sphere is $4\pi r^2$ and therefore the fraction of the atoms moving in the direction of θ and φ is

$$d\Omega = \frac{r^2 \sin\theta d\theta d\varphi}{4\pi r^2} = \frac{\sin\theta d\theta d\varphi}{4\pi}. \tag{A.15}$$

Therefore, combination of these gives dZ = (number of atoms/vol) (volume of cylinder) (fraction of atoms traveling down the cylinder axis) or

$$dZ = (\eta)(v \cos\theta)\left(\frac{\sin\theta d\theta d\varphi}{4\pi}\right) \tag{A.16}$$

and integrating with respect to θ and φ

$$Z = \frac{1}{4\pi}\eta v \int_0^{2\pi} d\phi \int_0^{\pi/2} \sin\theta\cos\theta d\theta$$

$$= \frac{1}{4\pi}\eta v(2\pi)\left(\frac{1}{2}\sin^2\theta\Big|_0^{\pi/2}\right)$$

with the final result

$$Z = \frac{1}{4}\eta v. \tag{A.17}$$

A.5 Mean Vertical Distance Traveled

Later, there will be a requirement to calculate the mean *vertical distance* an atom travels, $\bar{z}$. The vertical distance an atom at an angle θ travels is simply v cos θ and is independent of φ. So if Equation A.16 is integrated with respect to φ only, then dZ(θ) is obtained which is the fraction of atoms coming from an angle θ. So

$$dZ(\theta) = \frac{1}{2} nv\sin\theta\cos\theta d\theta$$

and the mean vertical distance traveled is

$$\bar{z} = \frac{\int_0^{\pi/2} (v\cos\theta)dZ(\theta)}{Z} \tag{A.18}$$

which becomes on making the substitutions and integrating,

$$\bar{z} = \frac{\int_0^{\pi/2} (v\cos\theta)dZ(\theta)}{Z}$$

$$\bar{z} = \frac{(1/2)nv^2}{(1/4)nv}\int_0^{\pi/2} \cos^2\theta\sin\theta d\theta$$

$$\bar{z} = 2v\left(-\frac{1}{3}\cos^3\theta\Big|_0^{\pi/2}\right) = -\frac{2}{3}v(0-1)$$

to give the mean distance as

$$\bar{z} = \frac{2}{3}v. \tag{A.19}$$

EXERCISES

5.1 Calculate the root-mean-square speed, the mean speed, and the most probable speed (the maximum in the Maxwell–Boltzmann distribution) for hydrogen molecules at 1000°C.

5.2 a. Calculate the number of hydrogen molecules/cm^3 at 1000°C and 1 atm pressure.
  b. Calculate the approximate hydrogen impingement rate (molecules/cm^2-s) from $I = (1/6)\eta v_{rms}$ under the same conditions as in a.
  c. Calculate the exact impingement rate (molecules/cm^2-s) under the same conditions from $I = p/\sqrt{2\pi m k_B T}$.

5.3 The number of atoms or molecules in a gas at a gas pressure p that attach themselves to a surface (or are evaporated or sublimed from), or any surface, is given by

$$J' = Z\alpha$$

where Units(J′) = atoms/cm^2-s and α = *sticking coefficient*, which goes from zero to one depending on the surface reaction necessary for the atoms to stick. If there is no surface reaction, then $\alpha = 1.0$ and J′ is

$$J' = Z = \frac{p}{\sqrt{2\pi m k_B T}}$$

  a. If J is the molar flux, mol/cm^2-s show that: $J = 10^{-4} p/\sqrt{2\pi MRT}$ where Units(p) = Pa and Units(M) = kg/mol.
  b. From a. derive an expression for the rate of sublimation, dr/dt, (cm/s) of a graphite wire lamp filament (a long cylinder) at elevated temperatures where r is the wire radius.
  c. Calculate how long it would take (years) at 1800°C to completely sublime a graphite lamp filament 3 mils in diameter if the Gibbs energy for the reaction: C(s) = C(g) is $\Delta G^\circ = 716{,}292 - 157\,T(K)$ J/mol (Roine 2002) and the density of the graphite is $\rho = 2.5$ g/cm^3 and the molecular weight is M = 12.011 g/mol.

5.4 Given the following thermodynamic data at 1227°C (Roine 2002), calculate the water vapor partial pressure in a flowing hydrogen gas of 1 atm below which active gas corrosion of silicon occurs; that is, SiO(g) forms and SiO_2(s) does not.

Active corrosion: Si(s) + 1/2 O_2(g) ⇒ SiO(g)
Passive corrosion: Si(s) + O_2(g) ⇒ SiO_2(s)

Compound	Gibbs Energy, ΔG° (J/mol)
$SiO(g)$	−227,501
$SiO_2(s)$	−643,026
$H_2O(g)$	−164,419

5.5 a. Thin films of titanium nitride, TiN, (the gold coating seen on drill bits and other tools are either TiN or ZrN for wear resistance) can be deposited by CVD at 1000 K by the following reaction:

$$2TiCl_4(g) + N_2(g) + 4H_2(g) = 2TiN(s) + 8HCl(g)$$

With the following thermodynamic data (Roine 2002), calculate the equilibrium constant at 1000 K.

Compound	ΔG° (J/mol)
$TiCl_4(g)$	−642,872
$TiN(s)$	−242,969
$HCl(g)$	−100,805

b. Calculate the equilibrium partial pressure of $TiCl_4$ if the initial $TiCl_4$ pressure is 0.01 atm and the pressure of N_2 and H_2 are held constant at 0.5 atm each.

c. The reaction for the CVD of TiN (mol/cm²-s) is given by

$$J_{TiCl_4} = -k\left([TiCl_4]_0 - [TiCl_4]_e\right)$$

where the subscript 0 indicates the concentration (mol/cm³) in the gas and the subscript e indicates the equilibrium concentration. Derive a general expression for the rate of deposition (cm/s) for TiN in terms of the $TiCl_4$ pressures in atmospheres.

d. Calculate the reaction rate constant at 1000 K if $k = k_0 \exp(-Q/RT)$ with $k_0 = 5.49 \times 10^4$ cm/s and Q = +120 kJ/mol.

e. If the density and molecular weight of TiN are 5.22 g/cm³ and 61.89 g/mol, respectively, calculate how long (hr) it takes to deposit a layer of TiN 1 μm thick at 1000 K.

5.6 The following thermodynamic data at 1027°C are taken from an old database (Note units!).

Compound	ΔG° (cal/mol)
$FeCl_2(g)$	−46,596
$HCl(g)$	−24,534

a. Iron is undergoing active gas corrosion in an inert atmosphere containing 0.1 atm HCl by the following reaction:

$$Fe(s) + 2\,HCl(g) = FeCl_2(g) + H_2(g)$$

Calculate the equilibrium constant at this temperature.

b. Calculate the exact equilibrium pressures (atm) at 1027°C of all of the gas species. Assume that all of the hydrogen is generated by the reaction.

c. If the first-order reaction rate constant for the active gas corrosion of iron by hydrogen chloride gas is given by $k = k_0\exp(-Q/RT)$ with $k_0 = 0.1$ cm/s and Q = 20 kJ/mol, calculate the value of the reaction rate constant at 1027°C.

d. Assuming the following first-order reaction for HCl with iron and a flowing gas atmosphere, derive an expression for the thickness of the iron reacted as a function of time in terms of the HCl pressures.

$$J_{HCl} = -k\left([HCl]_0 - [HCl]_e\right)$$

In the above, Units(J) = mols/cm²-s, 0 refers to the HCl concentration (mol/cm³) in the gas stream and e refers to the equilibrium concentration.

e. If the density and atomic weight of iron are 7.86 g/cm³ and 55.85 g/g-atom respectively, calculate the rate of active gas corrosion (cm/s) at 1027°C under these conditions.

f. Calculate the depth of corrosion (mils) of the iron in 1 day under these conditions.

5.7 A tungsten filament is reacting in a halogen lamp with the Cl_2 in the lamp. There are several gaseous chlorides of tungsten produced, WCl_2, WCl_4, and WCl_6. For this problem, it will be assumed that only WCl_2 is produced. The following data are available:

Melting point of W: T_{mp} = 3410°C
Density of W: $\rho(W)$ = 19.35 g/cm³
Molecular weight: M(W) = 183.85 g/mol
Filament length: 40 in.
Filament diameter: 3 mil (note: 1 mil = 0.001 in.)
$\Delta G°$ (WCl_2, 2500 K): −142,787 J/mol
$\Delta G°$ (WCl_2, 3000 K): −167,328 J/mol
Bulb volume: 1 cm³
Initial Cl_2 pressure: 10^{-2} atm
and the reaction to consider is

$$W(s) + Cl_2(g) = WCl_2(g)$$

a. Calculate the exact pressures of Cl_2 and WCl_2 at equilibrium at both 2500 and 3000 K assuming that the Cl_2 does not decompose.

b. If the first-order reaction rate constant for active gas corrosion of tungsten by chlorine is given by $k = k_0\exp(-Q/RT)$ with k_0 = 0.1 cm/s and Q = 20 kJ/mol, where $J = -k\{[Cl_2] - [Cl_2]_e\}$, calculate the values of the reaction rate constant at both 2500 and 3000 K.

c. Calculate how long it takes for bulb to reach 99% of equilibrium if the temperature of the tungsten filament is rapidly raised from 2500 to 3000 K. Assume that the gas in the bulb is at a uniform temperature and that the cold SiO_2 glass bulb wall has no effect. (This is not a good assumption. However, this is a model and any assumptions are acceptable. Whether they fit reality or not, is something that has to be examined.) Use the $p(Cl_2)$ at 2500 K and neglect any change in total pressure with T.

d. Calculate the *initial* rate of deposition of W in units of in/s if the filament in part c. is suddenly heated to 3000 K. Assume that the reaction rate constant for the deposition of tungsten by chemical vapor deposition from WCl_2 is the same as for the active gas corrosion of tungsten by chlorine. Neglect any change in total pressure on heating.

5.8 Show that the Jacobian for spherical coordinates, Equation A.10, is indeed equal to $r^2 \sin\theta$.

REFERENCES

Adamson, A. W. 1982. *Physical Chemistry of Surfaces*, 4th ed. New York: John Wiley & Sons.

Ashby, M. F. and D. R. H. Jones. 1980. *Engineering Materials: An Introduction to Their Properties and Applications.* Oxford: Pergamon Press.

Buckley, J. D. and D. D. Edie, eds. 1992. *Carbon-Carbon Materials and Composites.* NASA Reference Publication 1254. Washington, DC: National Aeronautics and Space Administration.

Campbell, S. A. 1996. *The Science and Engineering of Microelectronic Fabrication.* Oxford: Oxford University Press.
Chase, M. W., Jr. et. al. 1985. JANAF Thermochemical Tables, Third Edition. *J. of Physical and Chemical Reference Data.* 14. Supplement No. 1.
Columbia Accident Investigation Board. 2003. *Columbia Accident Investigation Board,* Report Volume 1. Washington, DC: National Aeronautics and Space Administration.
Edwards, C. H. and D. E. Penney. 2002. *Calculus,* 6th ed. Englewood Cliffs, NJ: Prentice Hall.
Halogen Lamp. *Wikipedia Encyclopedia.* http://en.wikipedia.org/wiki/Halogen_lamp.
Haynes, W. M., editor-in-chief. 2013. *Handbook of Chemistry and Physics,* 94th ed. Boca Raton, FL: CRC Press.
Hudson, J. B. 1998. *Surface Science.* New York: John Wiley & Sons.
International Union of Pure and Applied Chemistry. 2009. *IUPAC Task Group on Atmospheric Chemical Kinetic Data Evaluation-Data Sheet V.A1.6.HI6.* http://iupac.pole-ether.fr/htdocs/datasheets/pdf/H2O+ice_V.A1.6.pdf.
Kreyszig, E. 2011. *Advanced Engineering Mathematics,* 10th ed. New York: John Wiley & Sons.
Magee, N., A. M. Moyle, and D. Lamb. 2006. Experimental determination of the deposition coefficient of small cirrus-like ice crystals near −50°C. *Geophys. Res. Lett.* 33: L178113.
Matthews, F. L. and R. D. Rawlings. 1994. *Composite Materials: Engineering and Science.* London: Chapman and Hall.
McQuarrie, D. A. and J. D. Simon. 1997. *Physical Chemistry.* Sausalito, CA: University Science Books.
Moore, W. J. 1955. *Physical Chemistry,* 2nd ed. Englewood Cliffs, NJ: Prentice Hall.
NIST Chemistry Webbook. http://webbook.nist.gov.
Optical Fiber. *Wikipedia Encyclopedia.* http://en.wikipedia.org/wiki/Fiber-optic.
Pierson, H. O. 1999. *Handbook of Chemical Vapor Deposition,* 2nd ed. Norwich, NY: Noyes Publications.
Pound, G. M. 1972. Selected Values of Evaporation and Condensation Coefficients for Simple Substances. *J. Phys. Chem. Ref. Data.* 1(1): 135–146.
Ring, T. A. 1996. *Fundamentals of Ceramic Powder Processing and Synthesis.* San Diego, CA: Academic Press.
Roine, A. 2002. *Outokumpu HSC Chemistry for Windows,* Ver. 5.11, thermodynamic software program, Outokumpu Research Oy, Pori, Finland.
Savage, G. 1993. *Carbon-Carbon Composites.* London: Chapman and Hall.
Silbey, R. L. and R. A. Alberty. 2001. *Physical Chemistry,* 3rd ed. New York: John Wiley & Sons.
Wolf, S. and R. N. Tauber. 2000. *Silicon Processing for the VLSI Era,* Volume 1-Process Technology, 2nd ed. Sunset Beach, CA: Lattice Press.

Section III

Phase Transformations

6

Thermodynamics of Surfaces and Interfaces and Some Consequences

6.1 INTRODUCTION

The thermodynamics of surfaces and interfaces may seem to be out of place in a book on kinetics because this is a topic that would normally be covered in a book or course on thermodynamics. Yes (DeHoff 2006) and no (Gaskell 2008)! These are two very popular books used in thermodynamics courses in MSE: one covers surface properties and the other does not. Thermodynamics is a broad and difficult subject when encountered for the first time and surface phenomena may or may not be covered in a first thermodynamics course. Nevertheless, in the analysis of most kinetic processes, it is assumed that the thermodynamics of the system are known and usually the starting point for understanding the kinetics. Indeed, the thermodynamics of surfaces plays an important role as the driving force in many materials kinetic processes such as sintering, grain growth, nucleation, particle coarsening, and so on. The role of surface energies is critical in understanding these and other processes. Furthermore, as the size of a particle decreases, its surface-to-volume ratio increases and the surface energy has an even greater effect. This is particular true for processes that involve particles on the nanoscale. The thermodynamics and behavior of surfaces covers several phenomena and there are numerous books on surfaces and surface effects that cover virtually all of the topics discussed here, as well as many more (Adamson 1982; Adamson and Gast 1997; Miller and Neogi 1985; Hiemenz 1986; Hudson 1998; Evans and Wennerström 1999, to list but a few). The emphasis in this chapter is on those surface- and interface-related topics of solids and liquids that affect energies, vapor pressures, solubilities, and other properties that impact microstructural development and other kinetic processes important in materials science and engineering.

6.2 SURFACE ENERGY, TENSION, AND STRESS

6.2.1 Surface Energy and Surface Tension

Free surfaces of condensed phases—solids and liquids—have energies associated with them over and above the internal energy of the bulk material due to unsatisfied bonds at surfaces. Figure 6.1 depicts the situation for broken covalent bonds at the surface of silicon. In reality, this is oversimplified because the atoms or molecules at the surface adjust their positions—and bonding—to generate a lower energy structure—a process known as surface *reconstruction*. Investigation of such surface reconstructions is an active area of surface science research. Nevertheless, there still is a residual surface energy, γ. From Chapter 1, the differential of the Gibbs energy, dG, including a surface work term, γdA, for a two-component system is

$$dG = SdT - VdP + \mu_1 dn_1 + \mu_2 dn_2 + \gamma dA \tag{6.1}$$

so γ can be represented thermodynamically (macroscopically) as

$$\gamma = \left(\frac{\partial G}{\partial A}\right)_{T,p,n_i} \tag{6.2}$$

where G = Gibb's energy and γ = surface energy (sometimes called the "Gibbs surface energy"). The units of surface energy are Units(γ) = J/m^2 or J/cm^2. Note that because 1 J = 1 Nm, the units of surface energy are also Units(γ) = N/m, or a force per unit length. As a result, γ is more commonly referred to as the *surface tension* and appears to be the preferred terminology over *surface energy* (IUPAC 2014) and is now used in most books describing surface phenomena. However, the terminology causes confusion because there is also a *surface stress* in solids that is indeed different as discussed below. This multiplicity of terms is best summarized by Adamson in his book on surfaces (Adamson 1982, 6; with similar wording in Adamson 1997, 5):

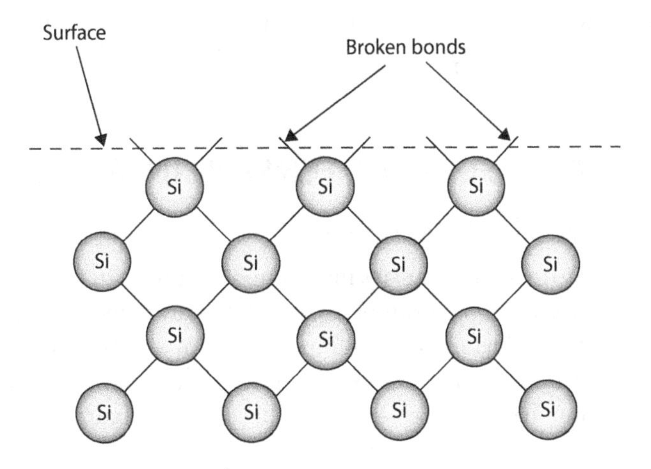

FIGURE 6.1 Broken or unsatisfied bonds at the surface of a solid or liquid are the source of surface energy, illustrated here with the surface of crystalline silicon.

...so that the decision as whether *surface tension* or *surface free energy* is the more fundamental concept becomes somewhat a matter of individual taste. The two terms generally are used interchangeably in this book.

The term surface tension is the earlier of the two; it goes back to early ideas that the surface of a liquid had some kind on contractile "skin." More subtly, it can convey the erroneous impression that extending a liquid surface somehow stretches the molecules in it. In contrast, the term surface free energy implies only that work is required to form more surface, that is, to bring molecules from the interior of the phase into the surface regions. For this reason, and also because it ties more readily into conventional chemical thermodynamic language, this writer considers the surface free energy concept to be preferable if a choice must be made.

Adamson's point is a good one, because γ appears in Equation 6.1 as a thermodynamic work term, *surface energy* seems to be the less misleading term.* Here, it is assumed that surface energy and surface tension refer to the same excess energy at surfaces and both terms will be used depending on what is being discussed. For example, treating surface energy as a surface tension—as a force per unit length—permits the simple development of several useful relations between the energies of different surfaces and their microstructural effects more easily than with surface energies and surface areas.

6.2.2 Surface Stress and Surface Tension

Adamson is not the only one concerned about the use of the term surface tension. Several other authors feel strongly that the term *surface tension* is confusing because there exists another surface term, *surface stress*, which is defined differently and can be the most important surface term is solids, particularly at low temperature (Cammarata 1994). Note that *surface tension* and *surface stress* are not the same for solids (Cammarata 1997). An actual surface stress, f, exists and can be smaller or larger than γ and occurs *when a strain is applied to the bonds between the atoms at the surface* of a solid: $f = \gamma + (\partial\gamma/\partial\varepsilon)$, where ε is the strain. The rationale for this expression is developed in Appendix A.1. This equation applies to isotropic or highly symmetrical crystalline materials. More generally, surface stress and strain are second-rank tensors. Of course, in liquids, no strain in atomic bonds can exist at the surface because bulk atoms simply move into the surface to relieve the strain, so there really is no *surface stress* and, therefore, *surface tension* and *surface energy* are the same. This is also true for solids if the temperature is high enough so that atoms from the bulk can move into the surface to relieve the surface strain if sufficient time is allowed for diffusion to occur. Because much of the kinetics where surface tension plays a role in solids is at high temperatures, *the latter situation is assumed in all further discussions on the effects of surface tension*: that is, *surface stress* will not be considered. Nevertheless, it should be noted that for solids at low temperatures, many of the relations developed between γ and a given property should more appropriately be expressed in terms of f. Unfortunately, it is considerably more difficult to measure f and few experimental values are available, and theoretically calculated values must frequently be used (Cammarata 1994).

6.3 SURFACE TENSION IN MATERIALS

6.3.1 Importance

Surface tension plays many important roles in materials science and engineering. One example is very familiar; namely, the formation of certain crystallographic bounding planes on crystals grown from the melt or from solution, such as the amethyst crystals—quartz with iron and water impurities that give the purple color—in Figure 6.2. These crystals slowly

FIGURE 6.2 Centimeter-size quartz (amethyst) crystals grown from solution showing low surface energy faces.

* As mentioned in Chapter 1, the IUPAC has also thrown out the "free" in free energy and now G is simply the "Gibbs energy" or the "Gibbs function" (IUPAC 2014).

grown from solution have crystal faces or facets that are usually those that have the lowest surface energies. In many metal-joining processes involving a liquid phase, there is the requirement for the liquid to wet the solid. These processes include the following: *soldering*—joining metals with a filler metal alloy that has a melting temperature less than about 450°C (lead-tin alloys are the most common solders); *brazing*—joining metals with a molten filler metal with a melting temperature above about 450°C (copper alloys are common); and *welding*—actual melting of the metals to be joined occurs at the joint with or without a filler metal.

Surface tension is the driving force for sintering or densification of metal powders—powder metallurgy—as well as ceramic materials and polymers made from powders. Surface tension is also the driving force for coarsening or the growth of particles dispersed in gases, liquids, or solids—the process known as *Ostwald ripening* named after Wilhelm Ostwald, 1909 Nobel Prize in Chemistry (Laidler 1993; Ratke and Voorhees 2002). In addition, the surface or interfacial tension associated with grain boundaries in polycrystalline materials is responsible for the growth of these grains by diffusion at high temperatures. The relative interfacial energies between solids and liquids can also determine the distribution and shapes of these phases in the microstructures of multiphase materials and how the phases are distributed strongly influences the properties. Finally, surface tension plays a critical role in rates of phase transformations.

6.3.2 Values of Surface Tension

Values of surface tension are not always easy to find in the literature and exist in several disparate sources representing numerous scientific disciplines. Table 6.1 gives some values of surface tensions for a variety of solids and liquids gathered from several references. Note that for many solids at their melting points, the surface tensions of the liquids and the solids are similar, which is reasonable because the number of broken or unsatisfied chemical bonds at the surface is not too different between liquid and solid near the melting point. Also, note that surface tensions vary by about an order of magnitude from somewhere around 0.1 J/m^2 for polymers to about 1 J/m^2 for high-melting metals and compounds. For *back-of-the-envelope calculations*, if one had to assume a value for surface tension for a metallic, ionic, or covalently bonded solid or liquid, a value of 500 mJ/m^2 would not be a bad guess: within a factor of 2 in most cases. For materials such as polymers whose surfaces represent unsatisfied van der Waals bonds, values on the order of 50 mJ/m^2 are reasonable.

6.3.3 Surface Energies and Broken Bonds

Surface energies essentially arise from unsatisfied bonds at the surface of solids and liquids. The details of the surface structure of solids can be complex because the atoms tend to move from their normal lattice sites and *reconstruct* the surface to minimize the surface tension or surface stress if given the opportunity for atomic movement. However, as shown in Table 6.1, the surface tensions of both solid and liquid copper and SiO_2 glass and liquid SiO_2 are not too different near the melting point, indicating that the detailed atomic structure at the surface does not have an overwhelming effect on the value of the surface tension, at least for these two materials.

6.3.3.1 Example: Copper

The surface energy or surface tension of copper can be approximated by the number of broken bonds at the surface. The density of solid copper is $\rho = 8.92$ g/cm^3 and its atomic weight is M = 63.55 g/mol, so the number of atoms per cm^3 in solid copper, $\eta(\mathrm{Cu})$, is

$$\eta(\mathrm{Cu}) = \frac{\rho}{M} N_A = \frac{8.92}{63.55} \times 6.022 \times 10^{23} = 8.45 \times 10^{22}\ \mathrm{cm}^{-3}.$$

TABLE 6.1
Values of Surface Tensions

Material	T (°C)	Melting T (°C)	γ (mJ/m²)	Source
H_2O	25	0	72.8	1, 2, 5
Ethanol (C_2H_5OH)	25	−117.3	22.3	1, 5
Acetone (CH_3CH_3CO)	25	−95.4	23.7	1, 5
n-hexane	25	−95	18.4	1, 5
Polyethylene	20	139	35.7	4
Poly(ethylene oxide)	20	66	42.9	4
Polytetrafluoroethylene	20	330	23.9	4
Hg(l)	25	−38.87	498	3
Pb(l)	350	327	468	3
Cu(s)	1080	1083	1430	2
Cu(l)	1083	"	1285	3
Cu(l)	1120	"	1270	2
Ag(s)	750	961	1140	2
Ag(l)	961	"	903	3
Fe(l)	1536	1535	1872	3
Pt(l)	1769	1769	1800	3
Si(l)	1410	1410	865	3
NaCl(100)(s)	25	801	300	2
NaCl(l)	801	"	114	2
NaCl(l)	1000	"	98	3
LiF(100)(s)	25	845	340	2
LiF(l)	1050	"	229	3
CaF_2(111)(s)	25	1423	450	2
SiO_2(s,glass)	1100	1723	295	6
SiO_2(l)	1800	1723	307	7
Al_2O_3(s)	1850	2050	905	2
Al_2O_3(l)	2080	"	700	2
MgO(s)	25	2852	1000	2
TiC(s)	1100	3140	1190	2
V_2O_5(l)	1000	690	86	3
B_2O_3(l)	900	450	80	2
FeO(l)	1420	1369	585	2

Sources: 1. Chang (2000), liquids; 2. Kingery et al. (1976), nonmetals; 3. Brandes and Brook (1992), molten metals and molten salts; 4. Sperling (2006), polymers; 5. Haynes (2013), extensive lists of organic liquids and molten elements; 6. Parikh (1958); 7. Kingery (1959).

The number of Cu atoms per square centimeter is roughly $\eta^{2/3} = 1.93 \times 10^{15}$ cm^{-2} or 1.93×10^{19} m^{-2}. The number of broken bonds times about half of the single-bond energy gives a rough estimate of the surface tension. About half of the bond energy is used because fracture in a piece of copper creates two surfaces, in each of which the atoms have about half of the bond energy between copper atoms. For copper, if the bond strength is roughly 1 eV, then $\gamma \cong 1.93 \times 10^{19}$ eV/m^2 $\times$ 1.602×10^{-19} J/eV/2 $\cong$ 1.55 J/m^2, which is probably fortuitously close to the measured values in Table 6.1; certainly close enough for a "back of the envelope" calculation.

6.3.3.2 Example: Polymer

On the other hand, for polymers, there are really few broken chemical bonds at surfaces, just van der Waals forces, which have energies on the order of 0.1 eV/bond. Take, for example, polyethylene, $(C_2H_4)_n$. In this case the density is $\rho \cong 0.9$ g/cm^3 and the molecular weight for a mer unit is 28 g/unit. Therefore, $\eta \cong (\rho/M)N_A \cong (0.9/28)\times 6.022\times 10^{23} = 1.93\times 10^{22}$ cm^{-3}. So, the number of mer units at the surface is $\eta^{2/3} \cong (1.93\times10^{22})^{2/3} \cong 1.81\times10^{15}\ \text{cm}^{-2} = 1.81\times10^{19}\ \text{m}^{-2}$. Now if there are roughly four atoms (more than one anyway) in each mer on the surface, the total surface tension is $\gamma \cong (4\times1.81\times10^{19})(0.1)(1.602\times10^{-19})/2 \cong 58\ \text{mJ}/\text{m}^2$. Here again, the rough, calculated value is again reasonably close to measured values! These two calculations are good demonstrations that the surface tension of solids and liquids—in which any surface strain can be relaxed by atoms moving to the surface—is due to unsatisfied chemical or physical bonds at the surface.

6.3.4 Anisotropy of Surface Tension and the Shape of Crystals

6.3.4.1 Introduction

Surface tensions in crystalline solids are not isotropic and usually vary with different crystal faces or planes. However, in many of the discussions that follow, to simplify things, the assumption is that the surface tension for both liquids and solids is isotropic. This assumption does not invalidate any of the basic results obtained. However, in real systems, surface tension anisotropy may introduce unwanted complexities in both measuring and interpreting surface effects. In the following few paragraphs, some simple—and rather old—ideas concerning surface tension anisotropy are presented mainly to demonstrate how anisotropy can affect the shapes of small crystals. What follows is entirely a thermodynamic approach that determines the shapes of crystals, particularly small ones in which the distance between the faces is sufficiently small to ensure that material transport processes are fast enough to move material around the crystal surfaces to ensure that all the faces are in chemical equilibrium. For larger crystals, the faces or *facets* that are observed may or may not be related to thermodynamics but could also be determined by the different growth rates—from gases, liquids, or solids—of different crystal faces.

6.3.4.2 Simple Cubic Crystal

In Figure 5.13 of Chapter 5, the classical picture of the structure of the surface of a crystalline solid was schematically represented by the stacking of small cubes representing atoms that leads to steps, terraces, and kinks on the surface of a solid. Figure 6.3 again schematically shows the

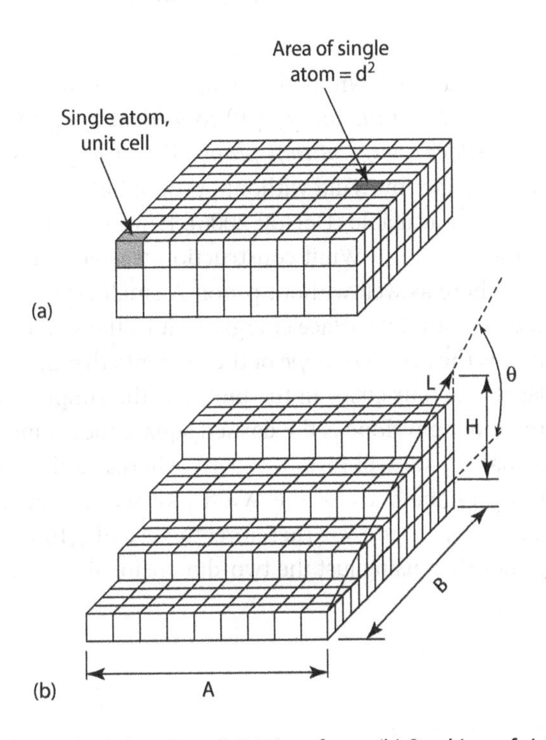

FIGURE 6.3 (a) Simple cubic crystal showing all {100} surfaces. (b) Stacking of simple cubic planes to give a different surface at an angle θ with the horizontal used to develop a Wulff construction of surface energy versus angle.

simple cubic packing of atoms with cubes representing the atoms of the solid. Only in this case, this representation is used to calculate the anisotropy of the surface tension. The surfaces of the simple cubic* crystal at the top of the figure have a number of broken bonds that determine their surface energies. Now if a new surface is constructed with these simple cubic crystal sheets stacked so that the terrace length is four unit cells and the ledge height is one unit cell, then this stacking represents a new surface at an angle θ relative to the (001) flat plane of the crystal, as shown in Figure 6.3. The angle θ is determined by $\tan\theta = (\text{ledge atoms/terrace atoms}) = 1/4 = 14.04°$. From the triangle, $\tan\theta = (H/B)$ or $H = B\tan\theta$ and $B = L\cos\theta$ or $L = B/\cos\theta$. The area of the plane defined by L and A is $\text{Area}_{LA} = LA = BA/\cos\theta$. If Ω is the number of broken bonds on this area, then

$$\Omega = \frac{BA + HA}{d^2} = \frac{BA + BA\tan\theta}{d^2}$$

$$\Omega = \frac{BA}{d^2}(1 + \tan\theta)$$

and if the energy per broken bond is ε, then the surface energy of the new surface at an angle θ is

$$\gamma = \varepsilon\frac{\Omega}{\text{Area}_{LA}} = \varepsilon\frac{BA/d^2(1+\tan\theta)}{BA/\cos\theta}$$

$$\gamma = \frac{\varepsilon}{d^2}(\cos\theta + \sin\theta)$$

If $\gamma(100) = \varepsilon/d^2$, then

$$\gamma(\theta) = \gamma(100)(\cos\theta + \sin\theta).$$

This is plotted in Figure 6.4 and is known as the *Wulff construction* or *Wulff plot* (Wulff 1901). *The length of a line from the center of the plot of γ(θ) to a plane tangent to γ(θ) is proportional to the surface energy of that plane.* This was proposed in the original—very long—paper by Wulff without proof. It has since been proven and the proof is readily available (DeHoff 2006). However, the proof is rather tedious and if the use of the Wulff construction was good enough for Wulff, it will be used here as well without proof. A small crystal assumes a shape giving it the lowest total surface energy. That is, the shape of the crystal in the Wulff plot is the *inner envelope* of the tangents that intersect and are perpendicular to the four cusps in the plot. For the simple cubic crystal in Figure 6.4, the inner envelope is the dashed square indicating that a small two-dimensional crystal will form a square, whereas a three-dimensional crystal will form a cube because the Wulff plot will be the same in each coordinate plane: xy, xz, and yz. Therefore, the use of γ(100) is justified in the drawing rather than using just the two-dimensional version, γ(01).

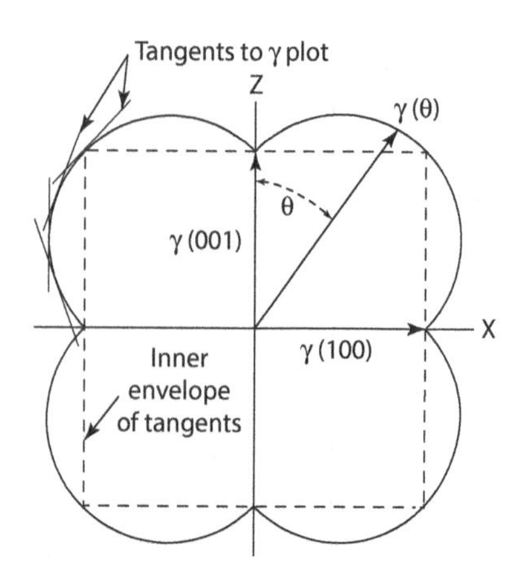

FIGURE 6.4 Wulff construction of surface energy versus angle for stacking of simple cubic planes to give surfaces with different surface energies as a function of angle (see Figure 6.3). The surface energy is plotted as the distance from the center of the plot; tangents are drawn to at each point; the inner envelop of tangents (dotted square) gives the crystal shape.

* It should be noted that a simple cubic crystal does not require only one atom per cell. Perovskite crystals such as barium titanate, $BaTiO_3$, are simple cubic but have only one formula unit ($BaTiO_3$) per unit cell.

6.3.4.3 A More Complex Crystal

In Figure 6.5, Wulff plots are constructed with two values of the surface energy, γ(01) and γ(11). Two-dimensional crystals are considered first because they are easier to visualize and certainly easier to draw without any loss in generality. Figure 6.5a shows a 1 cm^2 area (in the plane of the page) small square crystal with γ(01) = 500 mJ/m^2 (along the edges) and a total surface energy of 0.2 mJ.* Figure 6.5b shows a similar construction with γ(11) = 400 mJ/m^2. In this case, a 1 cm^2 crystal has a total surface energy of 0.16 mJ. Putting Figure 6.5a and b together into Figure 6.5c gives a schematic plot of γ(θ), the Wulff construction, which shows that the cusps in γ(θ) on the {01} and {11} families of surfaces have the lowest values of γ and form the *inner envelope of the tangents* that generates the shaded area, the shape of the crystal. A 1 cm^2 version of this crystal with the length of its faces drawn to scale is shown in Figure 6.5d and gives a total surface energy from the edges of 0.1579 mJ demonstrating that this crystal with the two different crystallographic faces does indeed give the lowest surface energy. Figure 6.5e shows a sketch of what the three-dimensional crystal would look like and, interestingly, Figure 6.5f shows some small crystals of NiO that have been equilibrated in an HCl atmosphere via vapor transport of $NiCl_2(g)$: $NiO(s) + 2HCl(g) \rightleftharpoons NiCl_2(g) + H_2O(g)$

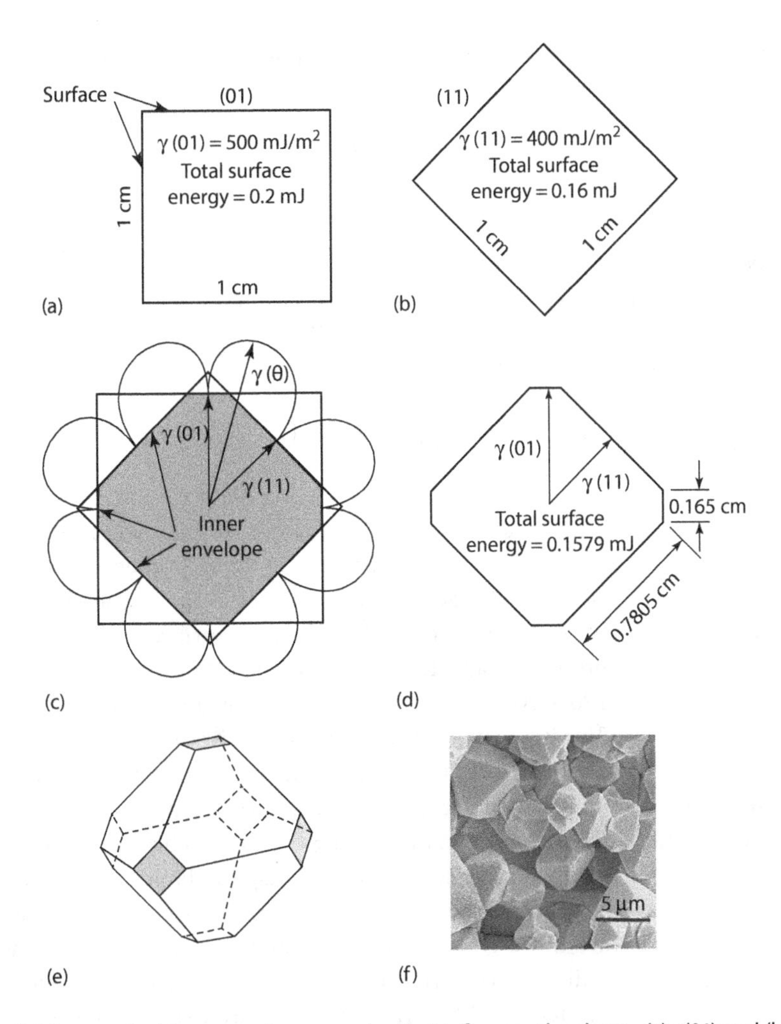

FIGURE 6.5 Cubic crystal with two surface energies that influence the shape: (a) γ(01) and (b) γ(11) in two dimensions; (c) the Wulff construction; (d) the two-dimensional crystal shape; (e) the three-dimensional crystal shape; and (f) NiO crystals showing similar shapes. (Photomicrograph courtesy of Steve Landin, CoorsTek, unpublished research, Colorado School of Mines, 1997.)

* Note that the surface energy of the crystal face in the plane of the drawing is not being considered in this two-dimensional construction. This does not change the principles involved in the construction. The 0.2 mJ comes from summing γ(01) around the four edges. Total $= 4 \times 500 \times 10^{-3}\ J/m^2 \times m^2/10^4 cm^2 = 0.2 mJ/cm^2$ or 0.2 mJ for the 1 cm^2 crystal.

(with permission of Steven Landin, CoorsTek, unpublished research). These crystals clearly show a surface shape similar to the theoretical one in Figure 6.5e. However, combinations of other planes such as the {111} family could give a similar shape. Without knowing the crystallographic orientation of a crystal relative to its faces, or the angles between the faces, whether they are actually {100} and {110} planes cannot be confirmed. Nevertheless, these crystals do show that the surface energy of NiO is anisotropic at the temperatures where these crystals were equilibrated, about 1100°C.

6.4 SURFACE TENSION AND SURFACE CURVATURE

In many cases, the combination of surface curvature and surface tension provides the driving force—Gibbs energy difference—for microstructural changes. For example, Figure 6.6 shows a particle or drop that is sectioned through its center. If the surface tension is a force per unit length, then it produces a total downward force, F_{-z}, around the perimeter of the cut equal to

$$F_{-z} = 2\pi r\gamma \tag{6.3}$$

The drop is not coming apart, so there must be an equal opposing force in the F_{+z} direction

$$F_{+z} = A\Delta P = \left(\pi r^2\right)\Delta P \tag{6.4}$$

where:

A is the cross-sectional area of the drop
ΔP is an internal pressure minus the ambient pressure external to the drop.

Equating these two, the *important result* is obtained:

$$\Delta P = \frac{2\gamma}{r}. \tag{6.5}$$

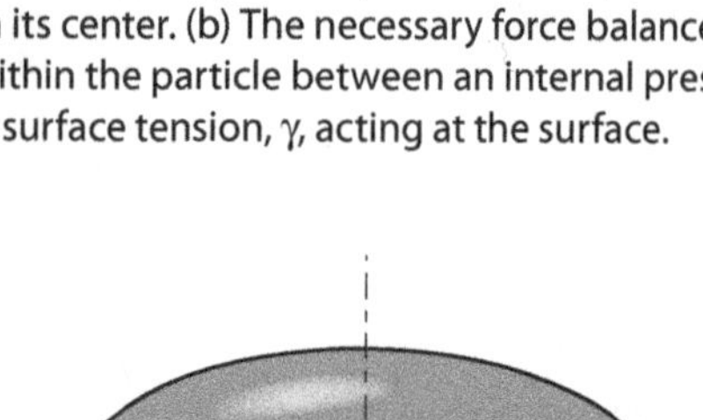

FIGURE 6.6 (a) A spherical particle or drop sliced through its center. (b) The necessary force balance that exists within the particle between an internal pressure, Δp, and the surface tension, γ, acting at the surface.

This equation shows that there is an internal pressure—compression in this case—in the particle or drop produced by the surface tension and the radius. For example, if a particle of Al_2O_3 had a radius of 0.1 μm = 10^{-7} m at 1850°C with a surface tension of 900 mJ/m², the pressure inside the particle would be $\Delta P \cong 2 \times 900 \times 10^{-3}\,J/m^2/10^{-7}\,m \cong 180 \times 10^5\,N/m^2 = 180 \times 10^5\,Pa \cong 180\,atm = 14.7\,psi \times 180 \cong 2650$ psi! This is a significant stress that can drive atom motion and shape change and is the driving force for many kinetic processes in materials. The important point is that by using surface tension as a force per unit length the influences of that surface tension can be expressed in terms of pressures and stresses that lead to shape and microstructural changes.

Not all surfaces of interest are perfect spheres, however. There are more general shapes as shown in Figure 6.7 for which several radii of curvature can be defined. At the inner surface of the toroid in Figure 6.7 two radii of curvature exist: r_1, which is the radius of the circular cross section, and r_2, which is the radius from the center to the inner surface of the toroid. Similarly, at the outer surface of the toroid, the two radii are r_1 and r_3. The inverse of a radius of curvature to a surface is called the *curvature*, $\kappa = 1/r$. Now, at every point on a surface, two mutually perpendicular radii of curvature, r_1 and r_2, or curvatures, κ_1 and κ_2, can be constructed so that one of the curvatures is the smallest and the other is the largest that can

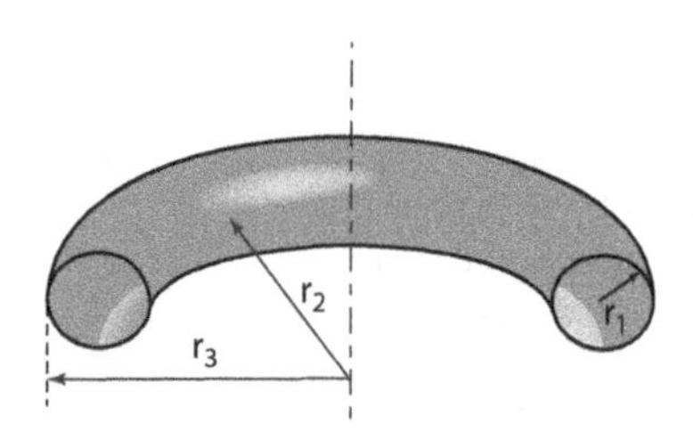

FIGURE 6.7 A more general solid shape that has more than one radius of curvature at each point: three (r_1, r_2, and r_3) are shown.

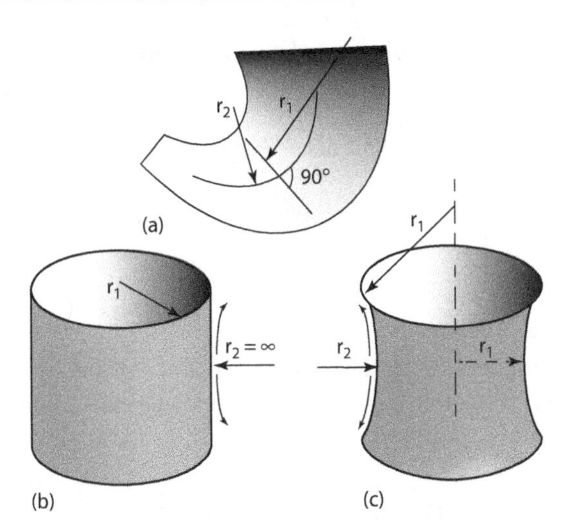

FIGURE 6.8 (a) General curved surface showing principal radii of curvature, r_1 and r_2 in mutually perpendicular planes that are both perpendicular to the surface where they intersect. (b) Radii of curvature for a right circular cylinder. (c) Radii of curvature for a concave circular cylinder, right: r_2 is the outer radius of curvature and is constant while r_1 is the inner radius of curvature and varies along the height of the cylinder since it must remain perpendicular to r_1 at a given point.

be constructed at that point (Figure 6.8a). These are called the *principal curvatures* and a more general form of Equation 6.5 for ΔP becomes

$$\Delta P = \gamma\left(\frac{1}{r_1}+\frac{1}{r_2}\right) = \gamma(\kappa_1+\kappa_2) = \gamma 2H \tag{6.6}$$

where H is called the *mean curvature* and $H = (1/2)(\kappa_1 + \kappa_2)$ (Casey 1996; Green 2005; DeHoff 2006). Of course, for a sphere $r_1 = r_2 = r$. Two other examples are given in Figure 6.8. In the case of the right circular cylinder in Figure 6.8b, $r_2 = \infty$ because the side of the cylinder is straight, so $\Delta P = \gamma/r_1$. This would be the situation inside of a round wire. Where the surface of the cylinder is concave (Figure 6.7c), the radius of curvature r_2 is actually *negative* because the end of the radius arrow lies *outside* of the solid, so at the middle of the cylinder where one of radii as shown is r_1

$$\Delta P = \gamma\left(\frac{1}{r_1}-\frac{1}{r_2}\right). \tag{6.7}$$

Also note that r_1—because it has to be drawn perpendicular to the surface—will vary along the height of this convex cylinder constant, as shown by the two different r_1 radii in contrast to the right circular cylinder, where r_1 is constant. Such a geometry is important in modeling shape changes because it suggests that if the magnitude of r_2 is less than that of r_1, then there will be a *negative pressure*—a tension—at the surface trying to pull the surface out from the center of the cylinder.

6.5 CURVATURE AND VAPOR PRESSURE

6.5.1 The Model

Because the surface tension or energy generates an internal pressure in a particle or a small drop, its Gibbs energy will be different from that of a completely flat surface; that is, one where $r = \infty$.

The chemical potential, μ, or the partial molar Gibbs energy, $\bar{G}$, is given by

$$d\bar{G} = d\mu = \bar{V}dP - \bar{S}dT \tag{6.8}$$

where $\bar{V}$ and $\bar{S}$ are the partial molar volume and entropy, respectively. At a constant temperature, $dT = 0$ and

$$d\bar{G} = \bar{V}dP$$

and can be integrated

$$\int_{\bar{G}^o}^{\bar{G}_r} d\bar{G} = \bar{V}\int_{P_\infty}^{P_r} dP. \tag{6.9}$$

Here $\bar{G}^o$ is the Gibbs energy of a flat surface where the ambient pressure is P_∞ and the radius of curvature is infinity and $\bar{G}_r$ is the free energy for a particle of radius r with an internal pressure P_r. This gives

$$\bar{G}_r - \bar{G}^o = \bar{V}(P_r - P_\infty) = \bar{V}\frac{2\gamma}{r}. \tag{6.10}$$

That is, the free energy of a particle or drop or radius r is greater than that of a flat surface.

If the equilibrium vapor pressure of the material in the particle or drop—say water—over a flat surface is p^o (note: lower case p is being used here simply to distinguish *vapor pressure* from the applied pressure P), then, by definition of the *thermodynamic activity a*,

$$\bar{G}_r - \bar{G}^o = RT\ln a = RT\ln\left(\frac{p}{p^o}\right) \tag{6.11}$$

where p is now the vapor pressure under an applied pressure P. Combining these last two equations gives the effect of surface tension, γ, and particle size, r, on the vapor pressure compared with the equilibrium vapor pressure over a flat surface:

$$\bar{V}\frac{2\gamma}{r} = RT\ln\left(\frac{p}{p^o}\right)$$

or

$$p = p^o e^{\frac{2\gamma\bar{V}}{rRT}} \tag{6.12}$$

this means that the vapor pressure over a particle of radius r is higher than that over a flat surface (note: sometimes $2\gamma\bar{V}/RT = d_p$ is referred to as the *capillary length* [Green 2005]). Furthermore, the smaller the particle size, r, or the larger the surface tension, γ, the greater the increase in vapor pressure.

6.5.1.1 Example: Water Drop at 25°C

As an example, consider a drop of water that has a radius of say r = 0.1 μm at 25°C. The equilibrium vapor pressure of water at 25°C is (Haynes 2013) p^o = 23.8 mm Hg. Because 1 atm = 760 mm Hg, the equilibrium vapor pressure—100% humidity—is 23.8/760 = 3.13×10^{-2} atm × 1.013×10^5 Pa/atm = 3.17×10^3 Pa. The molar volume of water is $\bar{V} = M(H_2O)/\rho(H_2O)$ = 18 (g/mol)/(1 g/cm³) = 18 cm³/mol = 18×10^{-6} m³/mol and from the Table 6.1, $\gamma(H_2O) = 72.8 \times 10^{-3}$ J/m². Let the exponent in Equation 6.12 be x, so

$$x = \frac{2\gamma\bar{V}}{rRT} = \frac{2\left(72.8\times10^{-3}\,\mathrm{J/m^2}\right)\left(18\times10^{-6}\,\mathrm{m^3/mol}\right)}{\left(1\times10^{-7}\,\mathrm{m}\right)\left(8.314\,\mathrm{J/mol\cdot K}\right)\left(298\,\mathrm{K}\right)}$$

$$x = 1.05\times10^{-2}$$

This is pretty small. Because $e^x = 1 + (x/1!) + (x^2/2!) + \cdots$, for x small, $e^x \cong 1 + x$ so that $p \cong 1.01\, p^\circ$ or that the vapor pressure is only increased by 1%! This may seem small, but it is large enough to lead to significant changes of water vapor particle size distributions—a fog—with time.*

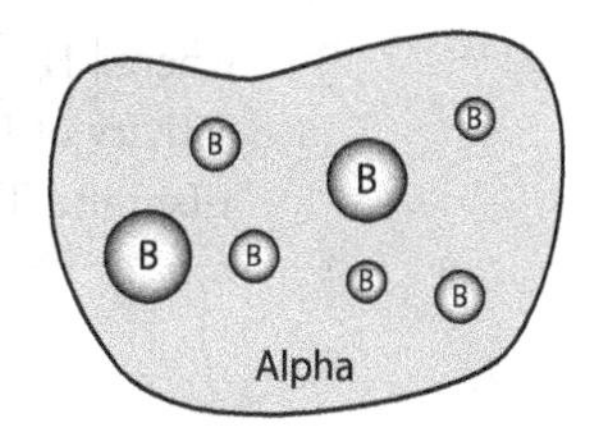

FIGURE 6.9 Schematic showing isolated particles of B in a continuous matrix of solution alpha.

6.6 CURVATURE AND SOLUBILITY

Figure 6.9 shows particles of B distributed in a continuous solution, α. Alpha could either be a liquid or a solid solution. The question is: "How does curvature and surface tension affect the solubility of B in alpha?" Assume that alpha is a solid solution and the particles of B are solid as well. They do not have to be: the same argument would hold if they were liquids. The phase diagram sketched at the bottom of Figure 6.10 shows that the equilibrium solubility of B in alpha at some temperature T_e is $[B_\infty]$, where the subscript ∞ is used to indicate that this is the solubility for B over a flat surface—an infinite radius of curvature. The equilibrium between the solution and the solid B can be written as

$$B \rightarrow B_{\text{solution},\infty} \tag{6.13}$$

with an equilibrium constant

$$K_e = \frac{[B_\infty]}{a_B} \tag{6.14}$$

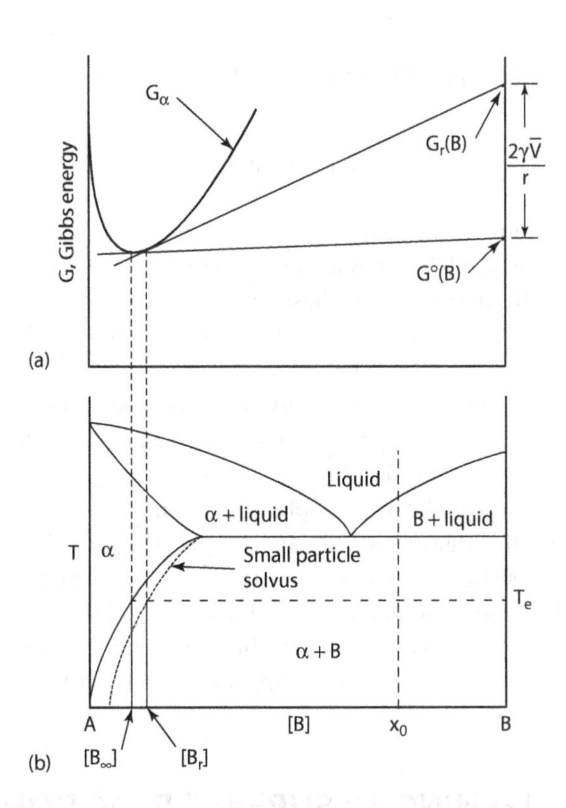

FIGURE 6.10 (a) Gibbs energy-composition diagram showing the difference in solubility as a function of the particle size. (b) Schematic phase diagram of components A and B showing the solid solubility of B in alpha, X_B, at some temperature T_e and the effect of particle size on the solubility—the solvus curve, and its relationship to the energy-composition diagram.

* Note that $e^x = e^{1.05\times10^{-2}} = 1.011$ showing that the $e^x \cong 1 + x$ approximation is more than satisfactory for this x.

where $[B_\infty]$ = concentration of B in α over a flat surface of B and where a_B = the thermodynamic activity of B. If it is pure B, as in Figure 6.10, then $a_B = 1.0$ by definition of the standard state of P = 1 bar and T. This means that

$$\Delta G^o = -RT_e \ln K_e = -RT_e \ln[B_\infty]. \tag{6.15}$$

One way to handle this for curved surfaces is as follows: for a curved surface, in this case spheres, the Gibbs energy of the spherical particle relative to the standard state, G°(B), is given by

$$G_B(\text{pure}) = G_B^o(\text{pure}) + \frac{2\gamma\bar{V}}{r} + RT \ln a_B(\text{pure})$$

$$G_B(\text{sol}) = G_B^o(\text{sol}) + RT \ln[B_r]$$

so

$$\Delta G = \Delta G^o - \frac{2\gamma\bar{V}_B}{r} + RT_e \ln\left(\frac{[B_r]}{a_B}\right)$$

$$= -RT_e \ln K_e - \frac{2\gamma\bar{V}_B}{r} + RT_e \ln[B_r]$$

$$\Delta G = -RT_e \ln[B_\infty] - \frac{2\gamma\bar{V}_B}{r} + RT_e \ln[B_r]$$

where $[B_r]$ is the solubility for a particle of radius r. At equilibrium, $\Delta G = 0$, so

$$[B_r] = [B_\infty] e^{\frac{2\gamma\bar{V}_B}{rRT^e}} \tag{6.16}$$

Here, $\bar{V}_B$ is the molar volume of B and $\gamma = \gamma_{\alpha B}$, or the interfacial energy between B and alpha. Equation 6.16 shows that the solubility of B increases as the particle radius decreases. $[B_r]$ and $[B_\infty]$ are both plotted on the bottom axis on the phase diagram. Because of the higher solubility of the small particles, the solvus curve is actually shifted to higher solubilities of B. This is just one example of how surface tension can have an effect on a phase diagram.

The top part of Figure 6.10 is the Gibbs energy versus composition plot that shows the energy of the alpha phase as a function of composition and the energies of the large particles, G°(B) and the small particles G_r(B). The separation of these points is greatly exaggerated to emphasize the relation between the G versus [B] plot and the phase diagram. The tangents to the G_α curve from the two values of G(B) for the large and small particles give the solubilities on the phase diagram as shown by the vertical dashed lines between the two drawings. The top drawing in Figure 6.10 shows the effect of particle size on solubility and solvus curve on the phase diagram. The bottom drawing graphicallys shows why this occurs with the Gibbs energy-composition diagram. Shifts in other phase boundary curves in phase diagrams also occur with small particles for similar reasons.

6.7 OSTWALD RIPENING BY SURFACE REACTION CONTROL

Because the solubility of a particle depends on its radius, if there is a distribution of particle sizes of B as shown in Figure 6.9, this is an unstable configuration and the smaller particles will dissolve and the larger particles will grow. As a result, the average particle size will increase with time: particle *coarsening* or *Ostwald ripening.* For particles of B to grow, several series steps are required: the smaller ones must dissolve, B must diffuse to the larger particles and then B atoms must attach themselves to the larger particles. The three kinetic steps in series are (1) surface reaction-controlled dissolution of smaller particles; (2) solid-state or liquid diffusion of B; and (3) surface reaction-controlled growth of larger B particles. Diffusion control of the process will be examined later. For now, assume that growth of the larger particles by reaction at their surfaces controls the rate of particle growth, step 3. As shown earlier, the rate of growth—or shrinkage—of the particles can be obtained from the molar flux to—or from—the surface of the particles. With the assumption of spherical particles, where r = particle radius and a first-order reaction with reaction rate constant k (m/s):

$$J = k\left([B]_{\bar{r}} - [B]_r\right) \tag{6.17}$$

where:

$\bar{r}$ is the mean particle size of the distribution

$\left[B_{\bar{r}}\right]$ is the concentration of B in the solution (mol/cm³) determined by the mean particle size

$\left[B_r\right]$ is the equilibrium concentration of B (mol/cm³) at the surface of a particle or radius r.

Equation 6.17 implies that particles with radii less than $\bar{r}$ will shrink and those with radii greater than $\bar{r}$ will grow. This will lead to an increase in the average particle radius, $\bar{r}$, with time. All powders, particles, and precipitates show a range of particle size that can be described by some sort of particle size distribution. If the all of the particles are changing size, the distribution and the mean particle size are also changing. That makes modeling particle size changes with time a very difficult problem. This complex problem was solved over half a century ago, and the complicated part was developing the equation for the steady-state particle size distribution (Wagner 1961). Several other solutions with different particle size distributions based on volume constraints have been developed since then (Ratke and Voorhees 2002).

The messy mathematical details of developing the appropriate particle size distribution will not be attempted here. Nonetheless, insight into the solution is realized by examining the behavior of a simple two-size distribution: one in which there are some very large particles all of the same radius and also some very small particles, again, all with the same smaller radius. An example of this might be fog particles over a lake where there is only one large particle, the lake with a radius of curvature of infinity. This is not an unreasonable distribution. If a composition, x_0, in Figure 6.10 were cooled from the liquid phase, solid B would form when the temperature reached the liquid + B phase boundary. Because these B particles are growing from the liquid, they could be "large" because diffusion in the liquid is rapid. When the eutectic temperature is reached, all of the remaining liquid freezes to a eutectic mixture of alpha + B. If the rate of cooling were fast enough, these eutectic B particles could be much smaller than those formed earlier. As a result, a two-particle size distribution is produced. If this composition were reheated to some temperature below the eutectic temperature, then the small particles would dissolve and B would deposit on the large particles, increasing their size. Because the large B particles are much larger than the smaller particles, essentially then $\left[B_{\bar{r}}\right] \cong \left[B_\infty\right]$, so Equation 6.17 becomes

$$J = k\left([B]_\infty - [B]_r\right). \tag{6.18}$$

Working on the left-hand side of the flux equation for the change in radius with time:

$$J = \frac{\rho}{M}\frac{1}{A}\frac{dV}{dt} = \frac{\rho}{M}\frac{1}{4\pi r^2}\frac{d\left(4/3\pi r^3\right)}{dt} = \frac{\rho}{M}\frac{dr}{dt} \tag{6.19}$$

so, combining Equations 6.18 and 6.19

$$\frac{dr}{dt} = \frac{Mk}{\rho}\left([B]_\infty - [B]_r\right) \tag{6.20}$$

where:

M is the molecular weight of B (g/mol)
ρ is the density of B (g/cm³)
k is the first-order reaction rate constant (cm/s)
$[B]_r$ is the concentration of B at the surface of a particle of radius r (mol/cm³)
$[B]_\infty$ is the concentration of B over a flat surface.

Now, $M/\rho = \bar{V}_B$ so that the rate of particle growth becomes

$$\frac{dr}{dt} = k\bar{V}_B\left([B]_\infty - [B]_r\right)$$

$$\frac{dr}{dt} = \bar{V}_B k\left([B]_\infty - [B]_\infty e^{\frac{2\gamma\bar{V}_B}{rRT}}\right) \cong \bar{V}_B k[B]_\infty\left(1 - 1 - \frac{2\gamma\bar{V}_B}{rRT}\right) \tag{6.21}$$

$$= -\bar{V}_B k[B]_\infty \frac{2\gamma\bar{V}_B}{rRT}$$

and leads to

$$r\frac{dr}{dt} = -\frac{2\bar{V}_B^2 k[B_\infty]\gamma}{RT} \tag{6.22}$$

and is integrated to give

$$r^2 - r_0^2 = -\frac{4\bar{V}_B^2 k[B]_\infty \gamma}{RT}t = -At \tag{6.23}$$

where r_0 is the initial particle size at time = 0. This result basically says that the smaller particles will continue to get smaller and finally disappear. Wagner's (1961) result for his particle size distribution is

$$\bar{r}^2 - \bar{r}_0^2 = \left(\frac{8}{9}\right)^2 \frac{\bar{V}_B^2 k[B]_\infty \gamma}{RT}t = A't \tag{6.24}$$

where the mean particle size *increases* with time. The result for a two-size distribution, Equation 6.23, and that for a complex particle size distribution, Equation 6.24, have the same terms with two exceptions: in the complete solution, r^2 is $\bar{r}^2$, the mean particle radius squared, and instead of the 4 in Equation 6.23, the more accurate solution in Equation 6.24 has this factor as $(8/9)^2$. The exact solution takes into account a particular particle size distribution produced by the particle coarsening and that the rate of growth of a particle of a given radius depends on the difference between its solubility and that of the *average* particle size in the distribution, which changes with time. Determination of the particle size distribution produced by the coarsening makes model development much more complex. But notice that all the other parameters and predicted time dependence are exactly the same in Equations 6.23 and 6.24! The main point is that the particle size grows or shrinks roughly as the square root of time and is directly proportional to the thermodynamic solubility (over a flat surface), $[B]_\infty$, the interfacial energy, γ, and the reaction rate constant, k.

In wet chemistry, Ostwald ripening or coarsening is used to "age" precipitates by waiting a while after precipitation before filtering to a get a larger particle size, making it easier to filter out the solid particles. This was one of the original observations examined by Ostwald. In addition, particle coarsening can occur via transport through the vapor phase because small particles have a higher vapor pressure—or reaction product pressure—than larger particles. For example, studies on the effect of atmospheres

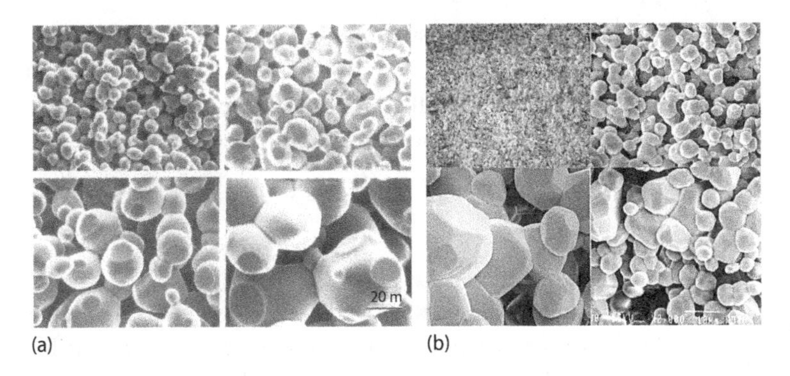

FIGURE 6.11 (a) Particle coarsening or Ostwald ripening of Fe_2O_3 powders in 10% HCl gas at 1200°C: top left, 10 min; top right, 30 min; bottom left, 100 min; bottom right, 300 min. The starting particle size was about 0.1 μm and exceeds 50 μm after 300 min. The flat surfaces are grain boundary fractures. (J. Lee photograph, Lee 1984) (b) Particle coarsening or Ostwald ripening of Al_2O_3 in pure HCl gas at 1500°C. Time increasing clockwise from 0 to 1084 min. (Photomicrograph by Marc Ritland, unpublished research, Colorado School of Mines, 1998.)

on sintering have shown that Ostwald ripening occurs when the product gas pressures over solids are enhanced in a reactive atmosphere. Figure 6.11a shows this for iron oxide in HCl (Lee 1984)

$$Fe_2O_3(s) + 6\ HCl(g) = 2\ FeCl_3(g) + 3\ H_2O(g)$$

whereas Figure 6.11b shows the same for Al_2O_3 (Marc Ritland, unpublished research)

$$Al_2O_3(s) + 6\ HCl(g) = 2\ AlCl_3(g) + 3\ H_2O(g).$$

It is the equilibrium metal chloride pressures—$FeCl_3$ and $AlCl_3$—for these two materials that are enhanced by the curvature: a higher pressure over small particles compared with that over large particles. Note, particularly for the largest particle size, there are crystal faces or *facets* on the surfaces of the aluminum oxide grains, but the iron oxide grains are uniformly rounded except at the fracture surfaces, where particles were bonded together—grain boundaries. This shows that the surface tension of Al_2O_3 is still quite *anisotropic* at 1500°C, whereas that of Fe_2O_3 is *isotropic* at 1200°C. The anisotropy in surface tension would be expected to decrease with temperature simply because $G = H - TS$ holds for surface energies as well. The main thing of interest about this coarsening is that no shrinkage or densification takes place while heating in the reactive HCl atmosphere. If the same starting compacted powders—with about 45% initial porosity—were held at the same temperatures and times in air, the resulting sintered powders would achieve densities on the order of 95% of the theoretical crystalline density—or only about 5% residual porosity by volume. The main reason that densification is significantly decreased in a reactive atmosphere is that the particles grow rapidly and reduce the surface tension, the driving force for sintering, which, of course, is the same driving force for particle growth. This process of sintering solids in a reactive atmosphere can be used to make porous materials of controlled porosity (amount: 0% to about 50% by volume) and pore size (proportional to the grain size) for filter applications.

6.8 FREEZING POINT DEPRESSION OF SMALL PARTICLES

6.8.1 Application of the Clausius–Clapeyron Relation

Figure 6.12 is the pressure–temperature phase diagram for H_2O and shows that the melting point of ice decreases as the pressure increases. This is because ice has a lower density than water and over the range from 0°C to about −20°C, ice has about 9% greater volume than water as shown in Chapter 1. A variation of the Clausius–Clapeyron equation (Silbey and Alberty 2001) can be developed to

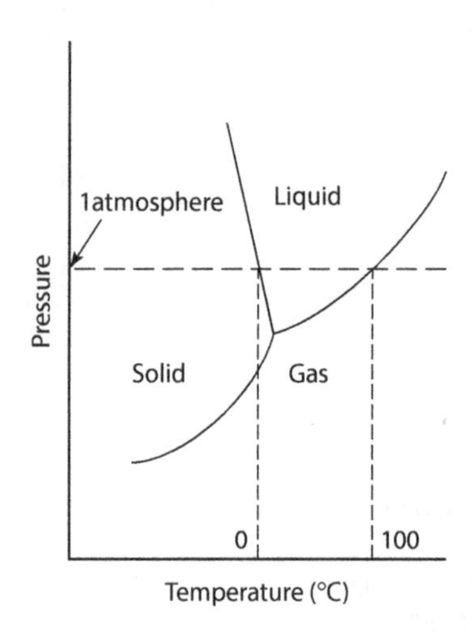

FIGURE 6.12 Schematic pressure–temperature diagram for H_2O.

show the effect of pressure on the melting point, and more specifically, the effects of surface tension on the melting point of small particles. For liquid and solid water, the differential partial molar Gibbs energies—differential chemical potentials—of each can be written as

$$\begin{aligned} d\bar{G}_L &= \bar{V}_L dP - \bar{S}_L dT \\ d\bar{G}_S &= \bar{V}_S dP - \bar{S}_S dT \end{aligned} \tag{6.25}$$

where L refers to liquid water and S to ice—solid H_2O—and the *barred* quantities are the partial molar values. At equilibrium along the line in Figure 6.12 between solid and liquid, the partial molar free energies are equal, so

$$\bar{V}_L dP - \bar{S}_L dT = \bar{V}_S dP - \bar{S}_S dT \tag{6.26}$$

and rearranging

$$\begin{aligned} \left(\bar{V}_L - \bar{V}_S\right) dP &= \left(\bar{S}_L - \bar{S}_S\right) dT \\ &= \frac{\Delta H_{fusion}}{T} dT \end{aligned}$$

where ΔH_{fusion} is obviously the enthalpy or heat of fusion, and integrating

$$T = T_0 e^{\frac{(\bar{V}_L - \bar{V}_S)\Delta P}{\Delta H_{fusion}}} \tag{6.27}$$

which for particles of radius r with a surface tension γ becomes

$$T = T_0 e^{\frac{(\bar{V}_L - \bar{V}_S)}{\Delta H_{fusion}} \frac{2\gamma}{r}}. \tag{6.28}$$

6.8.1.1 Example: One-nm Water Drops

Take a particle size of $r = 1$ nm $= 1 \times 10^{-9}$ m, $\gamma(H_2O) = 72.8 \times 10^{-3}$ J/m^2, and, from Chapter 1, $V_L = 18.08 cm^3/mol$ and $V_S = 19.65 cm^3/mol$, so $\left(\bar{V}_L - \bar{V}_S\right) \cong (18.08 - 19.65) = -1.57 cm^3/mol \times 10^{-6} m^3/cm^3$

$$\Delta H_{fusion} = (333.5 \text{ J/g})(18 \text{ g/mol}) = 6000 \text{ J/mol}.$$

Because $T_0 = 273$ K,

$$T = 273 \times \exp\left(\frac{-1.57 \times 10^{-6}}{6000} \frac{2 \times 72.8 \times 10^{-3}}{1 \times 10^{-9}}\right) = 263 \text{K}.$$

In other words, the *equilibrium* freezing point of 1 nm radius drops is about −10°C.

6.9 SPECIFIC SURFACE AREA

Because particle size is being discussed, this is an appropriate time to introduce the concept of the *specific surface area* = S. The units of S are m^2/g or cm^2/g and S is frequently used to give a measure of the average particle size of a powder. S is defined as the area per unit mass. So taking a sphere of diameter d,

$$S = \frac{A}{m} = \frac{A}{\rho v} = \frac{\pi d^2}{\rho(\pi d^3/6)} = \frac{6}{\rho d} \tag{6.29}$$

where:

m is the mass (g)
A is the area (m^2)
v is the volume (m^3)
ρ is the density (g/m^3)
d is the particle diameter (m).

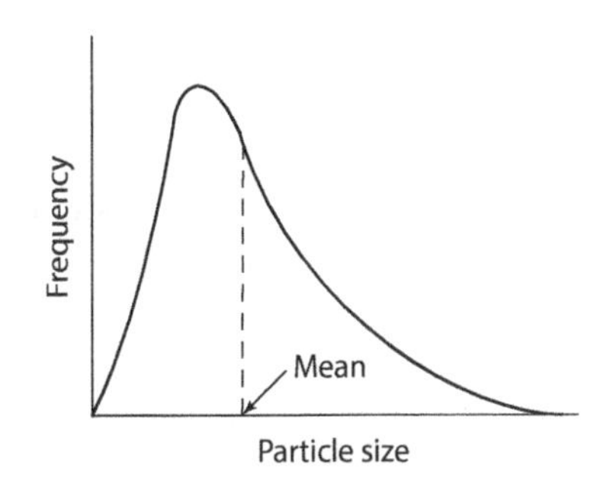

FIGURE 6.13 Typical particle size distribution, usually a log-normal distribution, showing the mean particle size.

For example, take Al_2O_3, with a density of $\rho = 4.0\,g/cm^3$ (Haynes 2013) with $S = 200\,m^2/g$; then $d = 6/\rho S = 6/\left(4.0 \times 10^6\,g/m^3 \times 200\,m^2/g\right) = 7.5 \times 10^{-9}\,m$ or 7.5 nm, roughly 50 atoms along an edge. The specific surface area, S, is only a rough estimate of the particle size because most powders will have a particle size distribution similar to that shown in Figure 6.13: one that is skewed to the smaller particle sizes typically and fits a *lognormal distribution*, that is, if plotted versus $\log_{10}d$, the distribution would look like a normal distribution symmetrical about $\mu = \overline{\log_{10} d}$, where μ is the lognormal mean of the lognormal size distribution.

Sometimes the specific surface area per mole, $\overline{S}$ (keeping the convention of barred quantities for molar quantities), is used rather than per gram. In this case

$$\overline{S} = \frac{A}{m}M = \frac{A}{\rho v}M = \frac{\pi d^2}{\rho(\pi d^3/6)}M = \frac{6}{d}\overline{V} \tag{6.30}$$

where:

M is the molecular weight
$\overline{V}$ is the molar volume (m^3/mol).

For the alumina example above, with a molecular weight of M = 102 g/mol, $\overline{S} = 2.0 \times 10^4\,m^2/mol$.

6.10 WETTING

As seen above, treating surface tensions as if they were forces acting along the line of intersecting surfaces is a useful way of establishing relationships between surface tensions or surface energies and the shapes of condensed phases. The same can be said for the wetting of a liquid (or solid) on a solid (or liquid) surface. In Figure 6.14, a liquid drop in the shape of spherical cap is resting on a solid surface and the system is at equilibrium. Where the liquid surface meets the solid surface, a force balance can be established treating the relative surface—interfacial to be precise—energies as if they were forces acting along the curve of intersection of the surfaces. From Figure 6.14, this becomes

$$\gamma_{SL} + \gamma_{LV}\cos\theta = \gamma_{SV}$$

or

$$\cos\theta = \frac{\gamma_{SV} - \gamma_{SL}}{\gamma_{LV}} \tag{6.31}$$

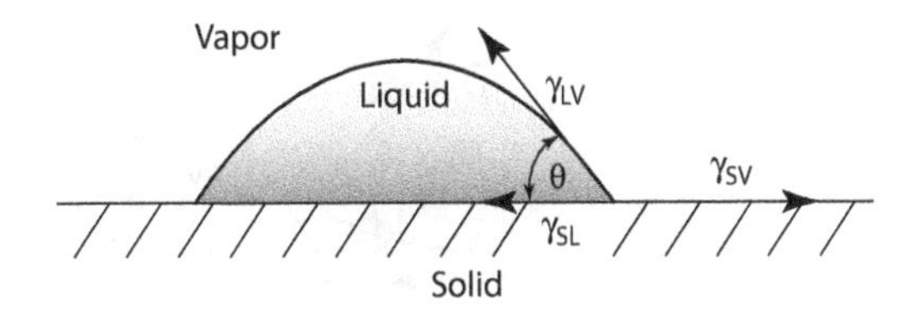

FIGURE 6.14 A liquid drop on a solid surface with vapor, solid, and vapor in "equilibrium."

which says that the angle at the surface depends on the relative values of the three interfacial energies. This is *Young's equation*. If $\theta < 90°$, then the liquid *wets* the solid. If $\theta > 90°$, the liquid is *nonwetting*. And if $\theta = 0$, the liquid *spreads* on the surface or is *spreading*. So, the *spreading coefficient*, S (an unfortunate redundant symbol), can be defined as

$$S = \gamma_{SV} - (\gamma_{LV} + \gamma_{LS}) \tag{6.32}$$

and for $S > 0$, the liquid spreads or covers the surface of the solid. For $S < 0$, droplets form on the surface. This is obvious from the relative interfacial energies determining S. If the sum of the interfacial energies between the liquid and solid and liquid and vapor are less than that of the solid–vapor interface, then the total interfacial energy is lowered by spreading. As an example of partial wetting, take $\gamma_{SV} = 1$ J/m^2, $\gamma_{LV} = 0.3$ J/m^2, and $\gamma_{SL} = 0.8$ J/m^2, so

$$\cos\theta = \frac{1.0 - 0.8}{0.3} = 0.67$$

$$\theta = 48°$$

which would look very similar to the drop in Figure 6.14. Also, $S = 1 - (0.8 + 0.3) = -0.1$, which is less than zero, so droplets should form as they do.

There is only one problem with this analysis. Clearly, in Figure 6.14, the vertical forces are not balanced! This has been pointed out by others (Hiemenz 1986; Hudson 1998) as one problem with Equation 6.31, among others, but it is used, nevertheless, to measure interfacial energies. On the other hand, if all three phases—gas, solid, and liquid—were in complete equilibrium, a situation much like that shown in Figure 6.15 might result. Here, some of the solid has dissolved in the liquid and redeposited to form the angle φ at the three-phase interface allowing equilibrium between the vertical surface forces as well as the horizontal. There are now two equations that must be solved, a modified equation 6.31,

$$\gamma_{SL}\cos\varphi + \gamma_{LV}\cos\theta = \gamma_{SV}\text{: horizontal} \tag{6.33}$$

and

$$\gamma_{LV}\sin\theta = \gamma_{SL}\sin\varphi\text{: vertical.} \tag{6.34}$$

Given the three interfacial energies, the two equations can, in principle, be solved for the two angles: However, this might not be so straightforward and may require numerical iteration. Similarly, if both angles can be measured, then two of the interfacial energies can be calculated, and so on.

The geometry in Figure 6.15 would be expected for two immiscible liquids such as a drop of oil floating on water and both the horizontal and vertical components would balance. Also, this equilibrium geometry could be envisioned for a water drop on the surface of a piece of sodium chloride, NaCl. The solid would dissolve in the drop until the liquid reached saturation and the equilibrium shape shown in Figure 6.15 would develop. However, complete equilibrium would never be reached for a water drop on SiO_2 glass or on PTFE, polytetrafluoroethylene, Teflon® because the rate of dissolution would be too slow or thermodynamically not possible. Nevertheless, on the basis of

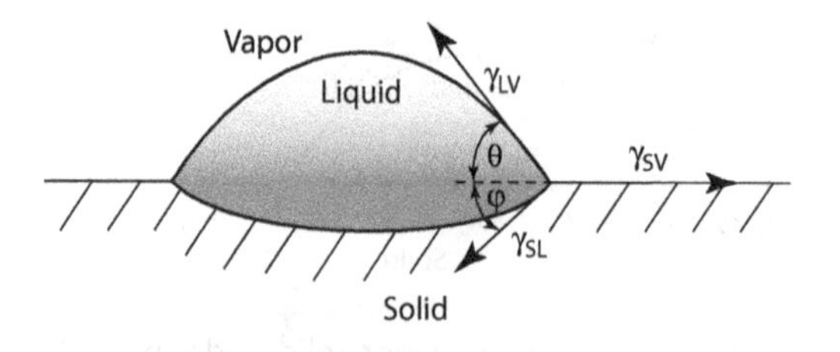

FIGURE 6.15 Liquid drop on solid surface with true equilibrium between the solid, liquid, and gas.

thermodynamic arguments, others (Adamson 1982; Miller and Neogi 1985; DeHoff 2006) conclude that Equation 6.31 is valid even if complete equilibrium is not reached, that is, the solid is essentially inert. In Appendix A.2, applying a composite of two approaches in the literature, a thermodynamic argument, it is shown that Young's equation does hold for an inert solid surface. Because the Young equation has been used for over 200 years, it must give reasonably correct answers!

6.11 INTERFACIAL ENERGIES AND MICROSTRUCTURE

Phase diagrams indicate whether two or more phases are in equilibrium but say nothing about how those phases are distributed in the microstructure of a solid (or a solid plus liquid or two liquids—maybe in microgravity of space to prevent convection). Figure 6.16 schematically depicts a second phase, β, dispersed in a solid phase, α (β could simply be a pore in a ceramic), with β shown at the grain boundary of α. Again, application of a force balance between the interfacial energies leads to

$$\gamma_{\alpha\alpha} = 2\gamma_{\alpha\beta}\cos\frac{\varphi}{2} \tag{6.35}$$

or

$$\cos\frac{\varphi}{2} = \frac{\gamma_{\alpha\alpha}}{2\gamma_{\alpha\beta}}$$

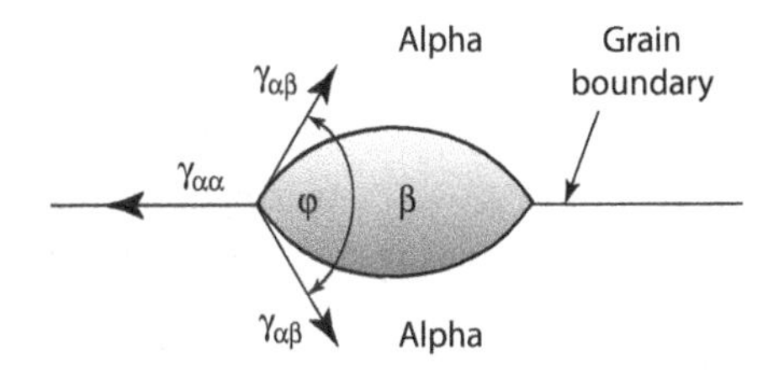

FIGURE 6.16 Interfacial energy force balance on a second phase particle of beta, β, at a grain boundary in alpha, α.

where $\gamma_{\alpha\alpha}$ is the grain boundary energy and φ is the *dihedral angle*—that is, it is bisected by the line of the grain boundary. If $\gamma_{\alpha\alpha}/2\gamma_{\alpha\beta} = 0$, the $\varphi/2 = 90°$ or $\varphi = 180°$ and beta forms spherical particles and there is no tendency for β to prefer grain boundaries. On the other hand, if $\gamma_{\alpha\alpha}/2\gamma_{\alpha\beta} = 1.0$, then β completely wets the grain boundaries of α, as shown in Figure 6.17.

How the two phases are dispersed has a significant effect on properties. For example, assume that beta is copper and alpha is Al_2O_3. The electrical conductivity of copper is about $\sigma \cong 10^6$ S/cm and that of Al_2O_3 about $\sigma \cong 10^{-12}$ S/cm, or about 18 orders of magnitude different! For the microstructure of Figure 6.17, continuous copper, the low-frequency* conductivity would be about $\sigma \cong 10^3$ S/cm with only about 0.1 v/o Cu. For the microstructure of Figure 6.16, in which the copper is isolated in lenses at grain boundaries, the copper would have hardly any effect on the low-frequency conductivity and $\sigma \cong 10^{-12}$ S/cm, about that of the alumina. Clearly, the relative interfacial energies of phases have a significant effect on the microstructure of a material and its consequent properties.

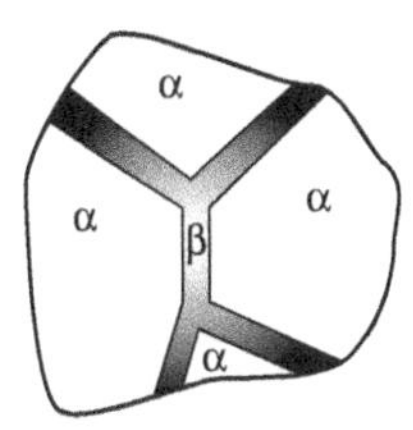

FIGURE 6.17 Two phase microstructure of beta, β, completely wetting the grain boundaries of alpha, α, and forming a continuous phase.

An industrial example where only a small amount of a grain boundary wetting phase has a significant impact on properties is the *hot-shortness* of steel. Hot-shortness refers to rolling of steel at elevated temperatures with cracking along the steel grain boundaries. The reason for this is sulfur impurities in the steel form molten FeS_2 (melting point = 1172°C) that wets the grain boundaries with a continuous liquid film causing the grains to separate. To avoid this, manganese, Mn, is added to the steel, which preferentially forms MnS rather than FeS_2 and has a melting point of 1530°C, above the rolling temperature of the steel so no liquid phase is formed during rolling and cracking does not occur.

A corollary to the situation for the two-phase system in Figures 6.16 and 6.17 is a single-phase polycrystalline system where at three-grain junctions, as shown in Figure 6.18, the grain boundaries must form 120° angles.

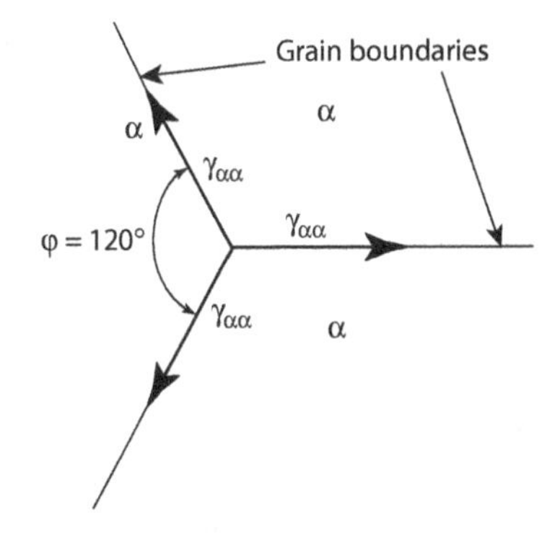

FIGURE 6.18 Three-grain junction in a single-phase crystalline material. Interfacial energy balance requires 120° angles at the junctions.

* High frequencies would give different results.

6.12 INTERFACIAL ENERGIES AND SURFACE SHAPE

Chemical etchants are frequently used to delineate the details of the microstructures of metals. These etchants can either etch different crystallites or grains at different rates in different crystallographic directions or simply form an oxide layer of differing thickness in different crystallographic directions depending on the relative rate of the etchant–surface reaction kinetics. In either case, each grain or crystallite reflects differently in a microscope (optical or scanning electron microscope, SEM), so the microstructure is visible. For compounds such as ceramics, very strong acids at relatively high temperatures are needed to etch—dissolve—the crystallites. Also, of course, for oxides, oxidizing etchants will not work. Therefore, chemical etchants are not as useful for ceramic materials as for metals. However, the technique of *thermal etching* is far superior. In this case, the surface of the material is polished just as in the preparation of a surface for chemical etching. The sample is then heated at some elevated temperature, usually a few hundred degrees or more below the sintering or densification temperature where the atoms or ions can diffuse and cause shape changes and allow the surface tension forces to equilibrate as shown in Figure 6.19. In this case, at equilibrium

$$\gamma_{SS} = 2\gamma_{SV}\cos\frac{\varphi}{2} \quad \text{or} \quad \cos\frac{\varphi}{2} = \frac{\gamma_{SS}}{2\gamma_{SV}} \tag{6.36}$$

where γ_{SS} = grain boundary energy. Figure 6.20 shows an actual grain boundary groove formed on a thermally etched surface of MgO at 1200°C measured with an *atomic force microscope* (*AFM*). The AFM measures the vertical height on a surface by scanning a sharp point along the surface, which

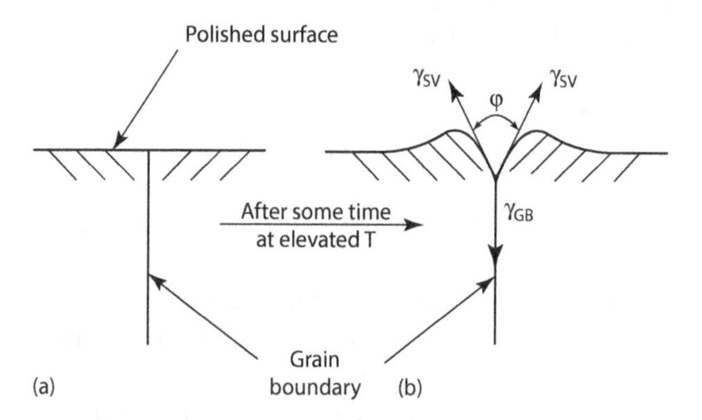

FIGURE 6.19 The formation of a grain boundary groove at the surface of a polished polycrystalline solid. (a) Polished surface near a grain boundary. (b) Grain boundary-surface energy equilibration at elevated temperatures.

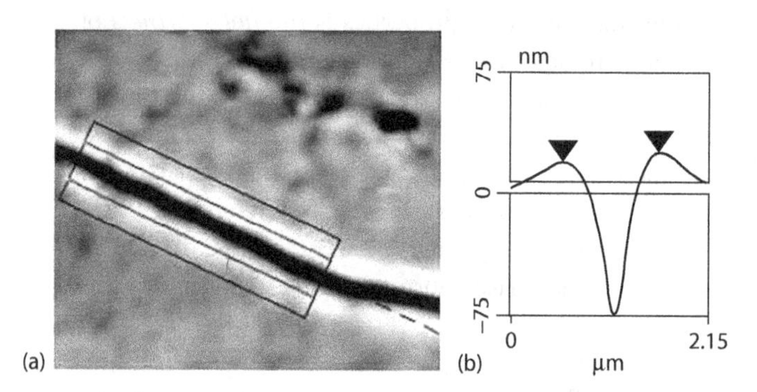

FIGURE 6.20 Grain boundary groove on polycrystalline MgO formed at 1200°C after 2 h in air imaged in an atomic force microscope (AFM). (a) Image produced by scanning and (b) shape across the grain boundary, average of the scans in the rectangle in the image on the left. Note that the groove shape is distorted by a 15:1 vertical to horizontal magnification. (Y. Yeo photograph; Yeo 2007.)

measures the attractive force between the point and the surface. Note that the vertical magnification in Figure 6.20 is about 15 times that of the horizontal magnification. In reality, these grooves are rather shallow and their angles large. Assume that $\gamma_{SS} \cong 0.5\ \gamma_{SV}$. Then, cos (φ/2) = 0.25 and φ/2 = 75.6° or φ = 151.2°. A rather shallow groove. The "hump" at the side of the groove is caused by material moving away from the grain boundary by diffusion forming this characteristic groove shape. Figure 6.21 is again the thermally etched microstructure of the yttrium-iron-garnet magnetic oxide shown earlier, which also shows the "humps" or built-up regions in the vicinity of all of the grain boundaries. The "black" spots are pores, mainly at the grain boundaries; however, there is a single pore within a grain at the top center of the photograph. Note that the pore within the grain is essentially spherical, suggesting that the surface tension is isotropic at the temperatures at which this material was densified, about 1400°C. On the other hand, the pores at the grain boundaries are distorted into "football" shapes (upper right) by the equilibration of the surface and grain boundary energies (Figure 6.16).

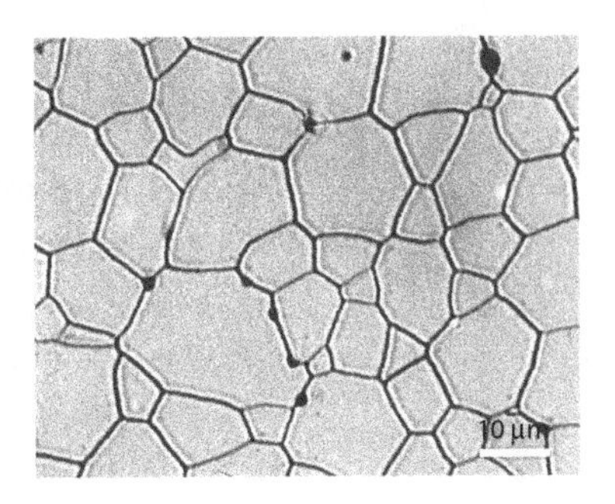

FIGURE 6.21 Thermally etched polycrystalline garnet, $Y_3Fe_5O_{12}$. Note the hills on the sides of the grain boundaries and the shapes of the pores both within the grain (top) and at grain boundaries. (D. W. Readey photograph.)

6.13 CAPILLARY RISE

6.13.1 The Model

Figure 6.22 sketches the rise of a wetting liquid in a capillary tube of radius r. The force balance in this case is the upward pull of the surface tension of the liquid wetting the tube and the weight of the liquid in the capillary pulled up to a height h

$$\begin{gathered} F_{+z} = F_{-z} \\ 2\pi r \gamma_{LV} \cos\theta = \pi r^2 \rho g h \end{gathered} \tag{6.37}$$

where ρ = liquid density (g/cm³) and g = acceleration of gravity = 9.8 (m/s²). Rearranging this equation leads to

$$\gamma_{LV} = \frac{r\rho g h}{2\cos\theta} \tag{6.38}$$

and can be used to measure the surface tension of liquids and it is particularly useful for measuring the surface tensions of high-temperature liquids. The major difficulty is measuring the angle theta, θ.

The above approach to the simple result in Equation 6.38 is unfortunately invalid, but it gives the correct answer anyway! The proper way to get to Equation 6.38 is to consider the pressure difference over the curved liquid surface and its balance with the amount of liquid raised by the pressure as shown in Figure 6.23. Because of the angle theta, θ, made by the intersection of the liquid with the surface of the capillary, the radius of curvature of the surface of the liquid, R, is actually negative (its center is outside of the curved liquid surface), so there is a negative pressure ΔP = −2γ/R pulling the liquid up balanced by the weight of the liquid pulling down. From the geometry of Figure 6.23, r = R cosθ, so,

$$F_{+z} = (\text{area}) \times \Delta P = \pi r^2 \Delta P = \pi r^2 \frac{2\gamma}{R} = \pi r^2 \frac{2\gamma\cos\theta}{r} = 2\pi r \gamma \cos\theta \tag{6.39}$$

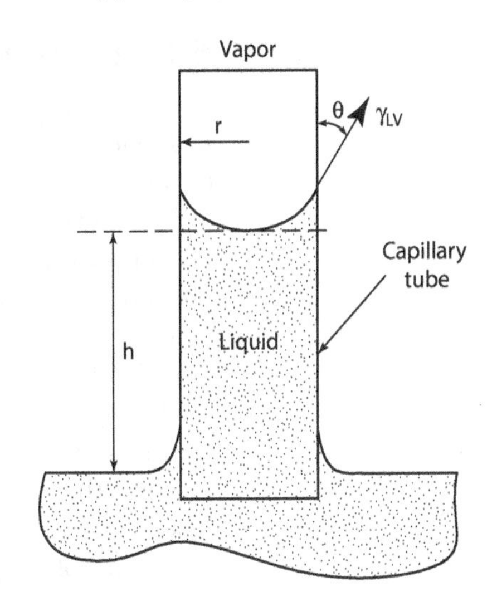

FIGURE 6.22 Rise of a liquid in a small diameter capillary due to wetting.

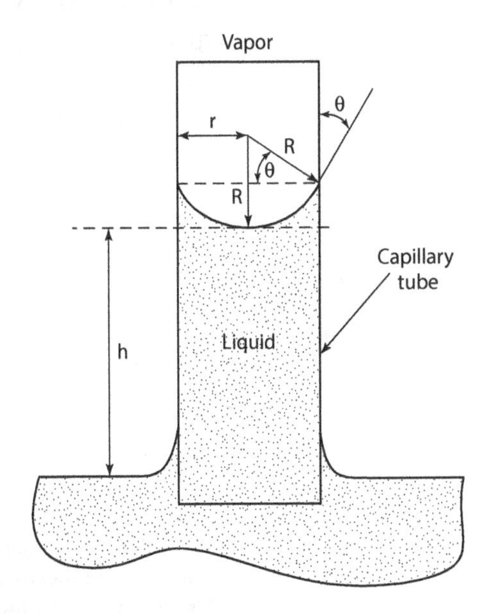

FIGURE 6.23 Same as Figure 6.22 but with other geometrical features identified to use in the correct method for calculating capillary rise.

which is exactly the same F_{+z} obtained from Figure 6.22. Equation 6.39 leads immediately to Equations 6.37 and 6.38.

6.13.1.1 Example: Liquid FeO at 1369°C

Molten FeO completely wets solid Al_2O_3 (melting point Al_2O_3 = 2050°C) at 1369°C, so $\theta = 0$ and rises to a height of 4.18 cm in an alumina tube of r = 0.5 mm = 5×10^{-4} m. The density of molten FeO is $\rho = 5.7$ g/cm³ × (10^6 cm³/m³)/1000 g/kg = 5.7×10^3 kg/m³. Therefore,

$$\gamma_{FeO} = \frac{r\rho g h}{2} = \frac{\left(5\times10^{-4}\right)\left(5.7\times10^{3}\right)\left(9.8\right)\left(4.18\times10^{-2}\right)}{2}$$

$$\gamma_{FeO} = 0.584\ \mathrm{J/m^2}.$$

6.13.2 Mercury Porosimetry

A very useful measurement technique based on the phenomenon of capillary rise is that of *mercury porosimetry,* which is used to measure pore size distributions in porous materials. In this case, mercury has a surface tension of $\gamma_{LV} = 0.476$ J/m² and essentially does not wet nonmetals; so $\theta = 180°$C and mercury must be forced into pores under positive pressure. From the value of the pressure, the pore size can be determined, and the volume of mercury forced into the porous body at a given pressure is the volume of pores of the size corresponding to this pressure. Suppose a porous rigid polymer has a pore diameter of 0.1 μm or r = 5×10^{-8} m. Therefore, $\Delta P = 2\gamma/r = 2 \times 0.476/5 \times 10^{-8} = 1.9 \times 10^7$ Pa = 19 MPa ≅ 2700 psi. If 0.1 cm³ of Hg were forced into the sample at this pressure in a total sample of 1 cm³, then the sample would have 10 v/o porosity of pore size 0.1 μm. By gradually changing the pressure and observing the volume of mercury forced into the sample as a function of pressure, the volume of porosity as a function of pore size can be obtained. Of course, this assumes that all of the porosity is open to the outer surfaces, so that the mercury can get to all of the porosity. If some of the porosity is closed porosity, mercury porosimetry will not detect it and additional techniques must be used to determine the amount of closed porosity.

6.13.3 Other Measurement Techniques

There are many other techniques used to measure surface tension of both liquids and solids. Some examples are shown in Figure 6.24. The *sessile* drop technique relies on a liquid drop that does not wet the surface. From the density and the shape of the drop—d and h—the surface tension of the liquid

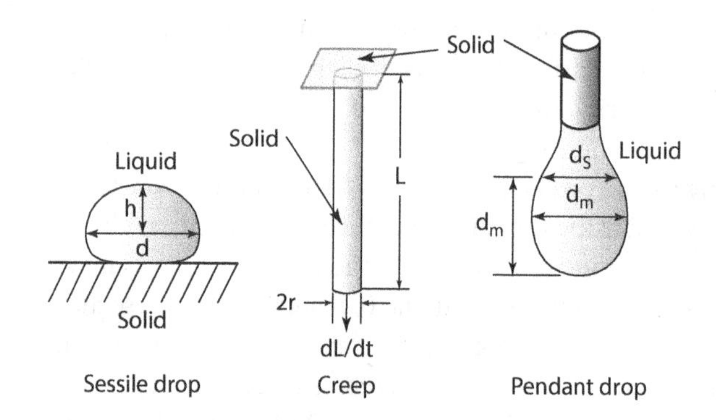

FIGURE 6.24 Other techniques used to measure surface energies that are particularly useful at high temperatures are shown. In the sessile and pendant drop techniques for liquids, the shapes are determined by the density and surface tension of the liquid and the latter is determined by comparing the shape to empirically determined shape parameters in the literature. For solids, the surface tension is balanced by the weight of the wire or fiber plus added weight when the creep rate is zero. From the weight of the fiber and added weight at zero creep, the surface tension can be calculated.

can be calculated from empirical shapes determined with liquids of known surface tension (Miller and Neogi 1985). This technique is particularly useful for making measurements at high temperatures (Humenik and Kingery 1954). A similar technique, the *pendant drop*, measures the dimensions of a suspended liquid drop as shown in Figure 6.24 and compares its shape to pendant drops of liquids of known surface tension (Adamson 1997). This technique is also particularly useful at high temperatures up to over 2000°C (Kingery 1959). Finally, a technique that is particularly useful for both inorganic glasses and amorphous polymers is *zero creep*, also shown in Figure 6.24. In this technique, thin strands or wires of different lengths of the material to be tested are suspended with or without weights and the creep rate is measured. The stress causing creep is the difference between the surface tension forces trying to distort the wire into a sphere making it deform in the upward direction and the downward weight of the wire itself (and added weights). When there is zero creep, these stresses just balance and the surface tension of the wire can be calculated. Again, this is a very useful technique for making high-temperature measurements on inorganic glasses and other solids (Udin et al. 1949; Parikh 1958).

6.14 SURFACE SEGREGATION AND SURFACE TENSION

6.14.1 Gibbs Adsorption Isotherm

The *Gibbs adsorption isotherm* in a two-component system in which the solvent is A and the solute is B is given by (Appendix A.3):

$$d\gamma_A = -\Gamma_{B,A} d\mu_B = -\Gamma_{B,A} d\bar{G}_B$$
$$= -\Gamma_{B,A} RT d\ln a_B$$
$$= -\Gamma_{B,A} RT d\ln X_B$$

or

$$d\gamma_A = -\Gamma_{B,A} RT d\ln C_B \tag{6.40}$$

where γ_A = the surface tension of A (J/m^2), $\mu_B = \bar{G}_B$ = the chemical potential or partial molar Gibbs energy of B (J/mol), a_B = activity of B and $a_B = f_B X_B$, f_B = activity coefficient of B,* X_B = mole fraction of B, and C_B = the concentration of B = $X_B / \bar{V}_{soln}$, where $\bar{V}_{soln}$ = the molar volume of the solution of this composition. Note: because $\ln a_B = \ln f_B + \ln X_B$, $d\ln a_B \cong d\ln X_B \cong d\ln C_B$ if the composition range is small so that f_B and $\bar{V}_{soln}$ are essentially constant, so the equation in the box is more or less valid. Now, $\Gamma_{B,A}$ is the *excess surface concentration of* B at the surface of the solution and Units($\Gamma_{B,A}$) = mol/m^2. $\Gamma_{B,A}$ is the difference between the actual surface concentration of B minus the concentration that would be expected if the concentration of B were the same in the bulk and in the surface. As shown in Appendix A.2, the surface excess concentration of B in A, $\Gamma_{B,A}$, can also be expressed as

$$\Gamma_{B,A} = \left(\Gamma_B - \frac{X_B}{X_A} \Gamma_A \right) \tag{6.41}$$

where:

- Γ_B is the actual surface concentration of B (mol/m^2)
- Γ_A is the surface concentration of A in pure A (mol/m^2)
- X_A and X_B are the mole fractions of A and B, respectively, in the bulk solution.

Now $(X_B/X_A)\Gamma_A$ is the surface concentration of B if it had the same surface concentration as the concentration in the bulk. So, if $(X_B/X_A)\Gamma_A$ has the same value as Γ_B, then there is no surface excess of B—no segregation—and no effect of B on the surface tension. Therefore, Equation 6.40 states that the surface tension of the solution is lowered if the solute B segregates to the surface so that

* Normally, the activity coefficient is given the symbol γ, but because this symbol is being used for the surface tension (energy), f will be used instead because it is only used briefly here.

$\Gamma_{B,A} > 0$. Atoms, ions, or molecules that *segregate to the surface lower the surface tension* and are called *surfactants*. Conversely, things that have a lower concentration in the surface than predicted from the bulk concentration, desegregate and raise the surface tension of A (Evans and Wennerström 1999).

6.14.1.1 Example: Water with Sodium Dodecylsulfate, $C_{12}H_{25}OSO_3Na$, in Solution

About 8 × 10⁻³ mol/L of Sodium dodecylsulfate (SDS) in solution will lower the surface tension of water considerably, about a factor of 2, to about 36 × 10⁻³ J/m² (Evans and Wennerström 1999). First, calculate the surface concentration of water in pure water,

$$\frac{1}{\bar{V}} = \frac{\rho}{M} = \frac{1\,\text{g/cm}^3}{18\,\text{g/mol}} = 5.56 \times 10^{-2}\,\text{mol/cm}^3 \ldots$$

$$\times\, 10^6\,\text{cm}^3/\text{m}^3 = 55600\,\text{mol/m}^3 \times 6.022\text{e}^{23} = 3.348 \times 10^{28}\ \text{molecules/m}^3$$

which gives

$$\Gamma_{H_2O} = \frac{\left(3.348 \times 10^{28}\right)^{2/3}}{6.022 \times 10^{23}} = 1.725 \times 10^{-5}\ \text{mol}\ H_2O/\text{m}^2$$

for the number of moles at the surface of pure water. Another method is to calculate a monolayer thickness, d, at the surface of H_2O. This is

$$d = \frac{1}{(3.348 \times 10^{28})^{1/3}} = 3.103 \times 10^{-10}\,\text{m}.$$

So,

$$\Gamma_{H_2O} = \frac{d}{\bar{V}} = 3.103 \times 10^{-10}\,\text{m}\left(55{,}600\ \text{mol/m}^3\right)$$

$$= 1.725 \times 10^{-5}\,\text{mol/m}^2$$

which is the same. For 8 × 10⁻³ mol/L of SDS in the bulk (8 × 10⁻³ mol/L × 10³ L/m³ = 8 mol/m³), the same calculation would give a surface concentration of

$$\Gamma^0_{SDS} = 3.103 \times 10^{-10}\,\text{m} \times 8\ \text{mol/m}^3$$

$$= 2.482 \times 10^{-9}\ \text{mol SDS/m}^2$$

if there were no surface segregation. This can also be obtained from

$$\frac{X_B}{X_A}\Gamma_A = \frac{8}{55600}\left(1.725 \times 10^{-5}\right)$$

$$= 2.482 \times 10^{-9}\ \text{mol/m}^2$$

because $X_B/X_A = n_B/n_A$ as both the number of moles are divided by $n_A + n_B$, the total number of moles, to get the mole fractions. Calculating Γ_{SDS,H_2O} from the lowering of the surface tension of water:

$$\Gamma_{SDS,H_2O} = -\frac{\Delta\gamma_{H_2O}}{RT\Delta \ln C_{SDS}}$$

$$\Gamma_{SDS,H_2O} = -\frac{36 \times 10^{-3}}{(8.314)(300)\ln\left(8 \times 10^{-3}\right)}$$

$$\Gamma_{SDS,H_2O} = 3.0 \times 10^{-6}\ \text{mol/m}^2$$

This value of the SDS surface concentration would be three orders of magnitude higher than that if the SDS were uniformly dispersed throughout the system. The excess surface concentration of SDS, $\Gamma_{SDS,H_2O} = \Gamma_{SDS} - \Gamma^0_{SDS} \cong \Gamma_{SDS}$, which is the surface concentration of SDS. Dividing this result by "d" gives the molar concentration at the surface

$$[SDS] = \frac{\Gamma_{SDS}\left(mol/m^2\right)}{d(m)} \times \frac{1}{10^3 l^3/m^3}$$

$$[SDS] = \frac{3.0\times10^{-6}\left(10^{-3}\right)}{3.103\times10^{-10}}$$

$$[SDS] = 9.7\ mol/l$$

about a factor of about 1000 higher than in the bulk! Clearly, the SDS segregates to the surface.

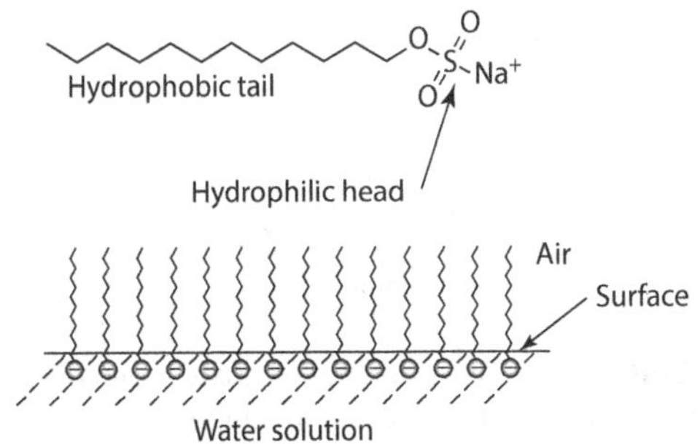

FIGURE 6.25 Sodium dodecylsulfate (SDS) molecule and how the SDS molecules segregate to the surface of an aqueous solution so that the hydrophobic part of the molecule points out of the solution surface.

6.14.2 Other Examples

6.14.2.1 Surface Segregation in Other Materials

In the above example, SDS has a nonpolar, *hydrophobic*—water-hating—tail, $C_{12}H_{25}$, which would like to be at the surface and out of the water, the reason for its surface segregation, and the anionic *hydrophilic*—water-loving—head, $-OSO_3^-$, which would prefer to be solvated in the polar water (Figure 6.25). In contrast, sodium chloride, NaCl, in aqueous solution raises the water surface tension because the sodium and chlorine ions prefer the *polar* bulk of the water rather than to be near the air interface and *desegregate* (Evans and Wennerström 1999).

Similar effects occur in other liquids. Oxygen and sulfur segregate to the surface of molten iron and lower its surface tension (Figure 6.26), whereas carbon does not segregate and has virtually no effect on γ_{LV} of iron (Kingery et al. 1976). Oxygen segregates to the surface of molten copper and silver and lowers their surface tensions as well. Finally, metals that are strong oxide formers, such as titanium, chromium, and zirconium, when added to molten metals such as Cu and Ni lower the interfacial energy, γ_{SL}, between these molten metals and solid oxides such as Al_2O_3 and increase wetting of the oxide. In principle, the strong tendency for Ti, Cr, and Zr to form oxides causes them to segregate to the molten metal–solid oxide interface, thereby lowering the interfacial energy because they segregate to the interface. This concept is used to formulate metal–ceramic brazing alloys (Schwartz 1990).

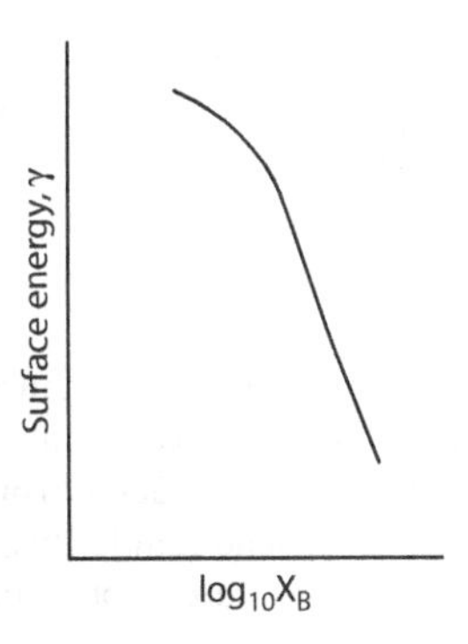

FIGURE 6.26 The Gibbs adsorption isotherm shows the effect of segregation to the surface on the surface energy of A, γ_{SV}, as a function of the mole fraction of an additive B, X_B. Typical of the behavior of oxygen and sulfur in liquid iron and oxygen in liquid copper as examples. (Kingery et al. 1976.)

There are other examples. The segregation to the interface between molten copper and solid aluminum oxide by oxygen lowers the interfacial energy between the molten copper and the alumina, which allows the copper to wet the alumina (O'Brien and Chaklader 1974), This makes possible the fabrication of a *near net shape* Cu–alumina composite that has the mechanical properties and machinability of a steel but with a lower density and a much higher thermal conductivity. The former because of the alumina and the latter due to the presence of a continuous phase of copper: its microstructure is shown in Figure 6.27.

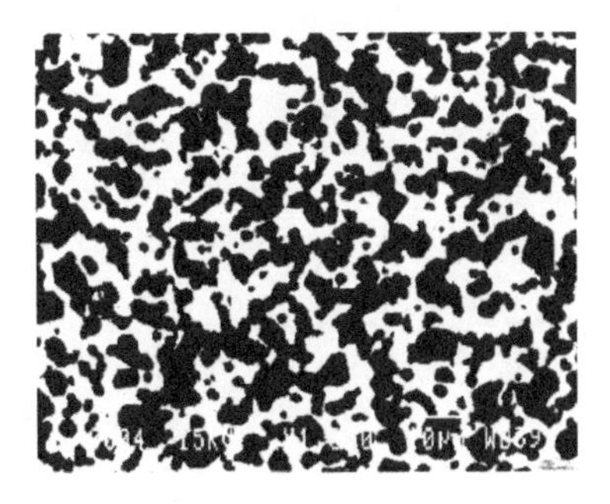

FIGURE 6.27 Backscatter SEM micrograph of a Cu-Al2O3 composite formed by the infiltration of molten copper into porous aluminum oxide. Cu is the light phase in this case and the Al_2O_3 the dark one. (S. Larpkiattaworn photograph; Larpkiattaworn, 1995.)

6.14.2.2 Froth Flotation

Another important example where surface segregation plays an important role in materials processing is *froth flotation*, which is used to separate ores, particularly sulfides, from other minerals. The ore is ground into

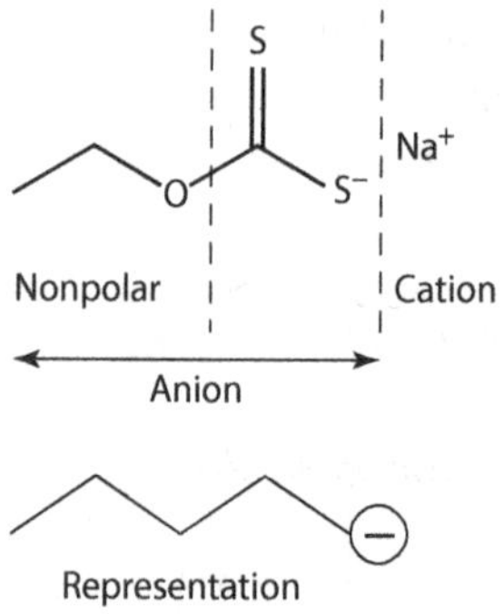

Nonpolar
Cation
Anion
Representation

FIGURE 6.28 Sodium ethyl xanthate molecule used for froth flotation of sulfide minerals.

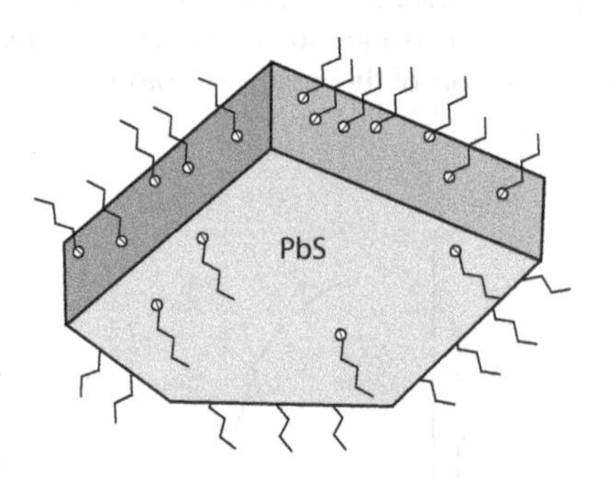

FIGURE 6.29 Schematic representation of how xanthate ions attach themselves to sulfide ore particles allowing separation. Sodium ethyl xanthate specifically attaches to lead sulfide, PbS, particles allowing them to be separated from other sulfides.

fine powder and dispersed in a water suspension containing sodium xanthate (Figure 6.28), which again is a surfactant that has an ionic head and a hydrophobic organic tail. The ionic sulfur head attaches itself to the sulfide powder particles and the organic tail points away from the surface as shown in Figure 6.29. Air is bubbled though the suspension, producing a "froth" of bubbles, and the hydrophobic tail of the xanthate ion segregates to the bubble surface and both the bubble and mineral rise to the surface of the suspension, separating the sulfides from the other minerals. The froth bubbles are collected and the sulfide mineral is washed from the xanthate solution, which can then be used again.

6.14.2.3 Coupling Agents

Another example of the tailoring of surfaces in materials science and engineering is the use of *coupling agents* (Plueddemann 1991) to bond silicate glass and other fibers (or other shapes) to polymers. In this case, silicon-containing compounds with organic tails—R—are used as shown in Figure 6.30. The CH_3O– groups on the silane molecule, methacryloxypropyltrimethoxysilane here, react with the OH^- ions on the surface of silicate glass to form Si–O–Si surface bonds with the formation of CH_3OH, which goes into the solution. The OH is on the surface because of the reaction of the ambient atmospheric water vapor with the silicate surface: –Si–O–Si– + $H_2O \rightarrow$ 2 –Si–OH. The organic tail of the silane now makes the inorganic surface of the silicate appear as a nonpolar polymer surface, allowing it to readily attach itself to another polymer or other nonpolar material. Such coupling agents are used in polymer–fiber glass composites and on the surface of silicon wafers to help bond the organic *photoresist* material to the native silicon dioxide on silicon. The photoresist is used to pattern the semiconductor device structures into the surface of silicon.

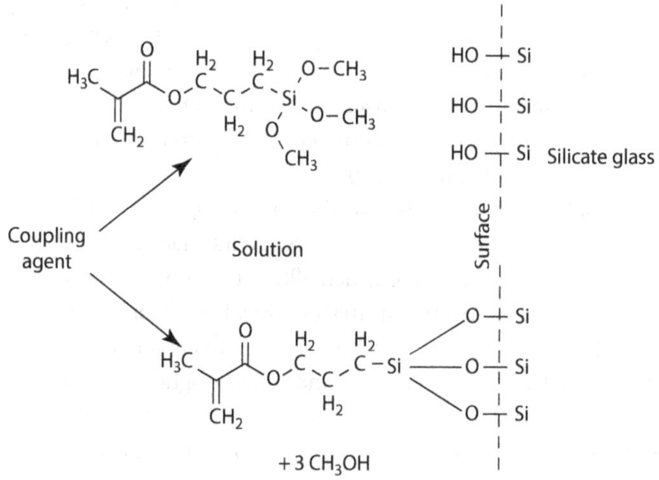

Coupling agent
Solution
Surface
Silicate glass
+ 3 CH_3OH

FIGURE 6.30 Methacryloxypropyltrimethoxysilane is just one of many silane coupling agents (Plueddemann 1991). It is shown here to demonstrate how the methoxy groups react with the OH groups on the surface of the silicate forming chemical bonds and leaving the "organic" part of the molecule facing the solution, which makes it easier to bond the silicate glass to nonpolar materials such as polymers.

6.15 WETTING OF MICROPOROUS SURFACES

In contrast to making surfaces more wettable, with the availability of nanoscale materials, there is an interest in using these small particles to create nonwettable surfaces (Yang et al. 2013). In the wetting equation, Equation 6.31

$$\cos\theta = \frac{\gamma_{SV} - \gamma_{SL}}{\gamma_{LV}}$$

if the area of the surface the drop is covering (Figure 6.14) can be made larger than the geometric area of the bottom of the drop by roughening the surface producing a *roughness factor*, χ, where χ is simply (Wenzel 1936)

$$\chi = \text{roughness factor} = \frac{\text{actual surface area}}{\text{geometric surface area}}$$

then Equation 6.31 can be modified to

$$\cos\theta_D = \frac{\chi(\gamma_{SV} - \gamma_{SL})}{\gamma_{LV}} \quad (6.42)$$

where θ_D is the contact angle on the rough surface to distinguish it from that on a smooth surface, θ. For example, assume that the liquid does not wet the solid and $\gamma_{SV} = 332\,\text{mJ/m}^2$, $\gamma_{SL} = 400\,\text{mJ/m}^2$, and $\gamma_{LV} = 200\,\text{mJ/m}^2$, then

$$\cos\theta = \frac{332 - 400}{200} = -0.34$$

$\theta = 110°$. If the roughness factor were $\chi = \sqrt{2} = 1.414$, then $\cos\theta = -0.481$ and $\theta = 119°$, as shown in Figure 6.31. If the roughness factor were $\chi = 2.08$, then $\cos\theta = -1$ and $\theta = 180°$ or no wetting at all! Wenzel found indeed that the contact angle was greater on a rough surface compared with a smooth surface of the same material.

Cassie and Baxter (1944) carried the analysis further to include porosity as well as roughness with the intent of minimizing wettability with a designed surface structure, which included both a higher surface area and porosity as shown in Figure 6.32. They were interested in waterproof

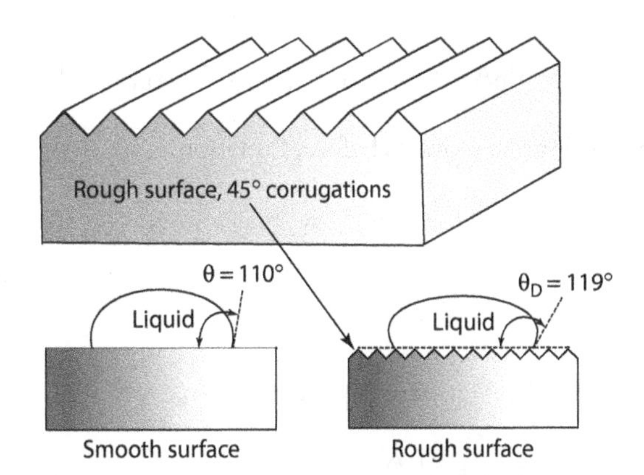

FIGURE 6.31 The difference in wetting angle for a given liquid–solid system between a smooth and rough surface. The surface roughness increases the liquid–solid interfacial area and decreases the wetting angle as described in the text.

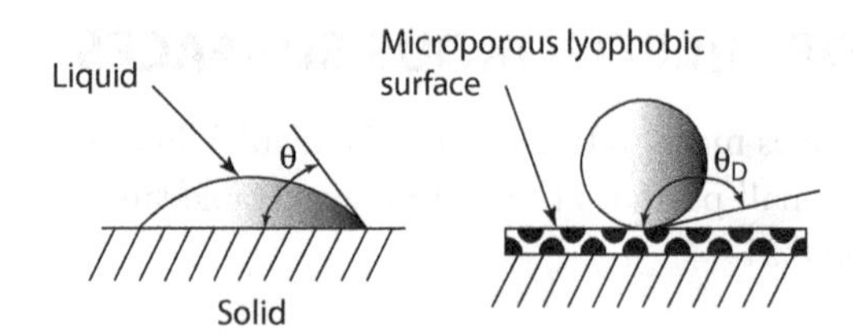

FIGURE 6.32 The difference in wetting angle for a given liquid–solid system between a flat surface and one with roughness and porosity. With surface layers of nanoparticles with lyophobic coatings, so-called *super lyophobic* surfaces, can be made for a number of applications.

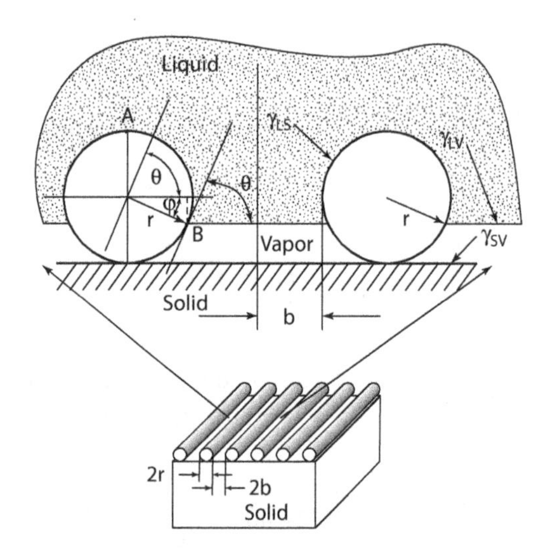

FIGURE 6.33 Parallel rods of Cassie and Baxter (1944) used to model the apparent wetting angle, θ_D, for a real wetting angle of θ for a rough, microporous surface as shown at the top. The model is given in detail in the text.

clothing, and Figure 6.33 shows the envisioned structure of parallel cylinders that simulates cloth fibers. Figure 6.33 also shows the details of the structure made of parallel cylinders of unit length perpendicular to the page with a radius r and a separation distance of 2b (the porosity). Let A_{LS} be the area of the liquid–solid interface when a drop of liquid is placed on this composite surface. Likewise, let A_{LV} be the area of the liquid–vapor interface *in the pore area*. The energy in forming a total area A_D is given by

$$E_D A_D = A_{LS}(\gamma_{LS} - \gamma_{SV}) + A_{LV}\gamma_{LV} \tag{6.43}$$

where E_D is the energy per unit area. Now, as before, Equation 6.31,

$$\cos\theta = \frac{\gamma_{SV} - \gamma_{SL}}{\gamma_{LV}}$$

so

$$E_D A_D = A_{LS}(-\gamma_{LV}\cos\theta) + A_{LV}\gamma_{LV}$$

$$E_D = \gamma_{LV}\left(\frac{A_{LV}}{A_D} - \frac{A_{LS}}{A_D}\cos\theta\right)$$

$$E_D = \gamma_{LV}(f_{LV} - f_{LS}\cos\theta)$$

where:

f_{LV} is the fractional area of the liquid–vapor interface
f_{LS} is the fractional area of the liquid–solid interface.

Now Equation 6.31 can be written as

$$\cos\theta = \frac{-E}{\gamma_{LV}}$$

because $(\gamma_{LS} - \gamma_{LV})$ is the energy, E, to form the unit area of the liquid–solid interface. Therefore, by analogy, for a drop sitting on a porous surface (Figure 6.32), a contact angle for a drop on the porous surface can be written as

$$\cos\theta_D = -\frac{E_D}{\gamma_{LV}} = f_{LS}\cos\theta - f_{LV}. \tag{6.44}$$

If the surface is just rough and not porous, $f_{LV} = 0$, and the result is Wentzel's equation for a rough surface (Equation 6.42).

For the model in Figure 6.33, because the structure is one unit long perpendicular to the paper surface, a unit of area is $A_D = 1\times(r+b)$, so

$$f_{LS} = \frac{1}{(r+b)}\left(\pi r - \pi r\frac{\theta}{180°}\right)$$

$$f_{LS} = \left(\frac{\pi r}{r+b}\right)\left(1-\frac{\theta}{180}\right). \tag{6.45}$$

Also,

$$f_{LV} = \frac{1}{(r+b)}(b + r - r\cos\varphi).$$

From the figure,

$$\cos\varphi = \cos(90-\theta) = \cos(90)\cos\theta + \sin(90)\sin\theta$$

$$\cos\varphi = 0\times\cos\theta + 1\times\sin\theta$$

$$\cos\varphi = \sin\theta$$

therefore,

$$f_{LV} = \frac{1}{(r+b)}(b + r - r\sin\theta)$$

$$f_{LV} = 1 - \frac{r}{r+b}\sin\theta.$$

Figure 6.34 shows how the wetting angle θ_D increases as a function of the ratio of $(r+b)/r$ with the wetting angle for a smooth interface, θ, as a parameter. Clearly, as the separation between the

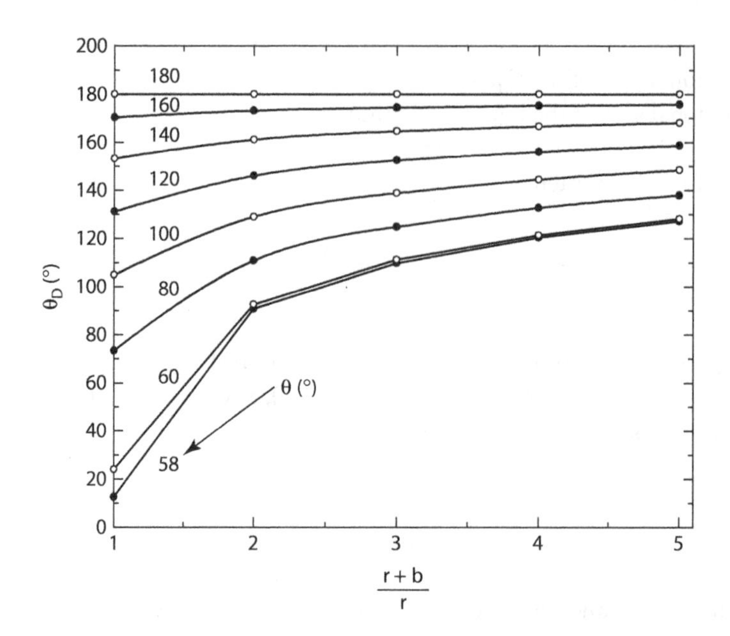

FIGURE 6.34 The apparent wetting angle versus essentially the porosity for the parallel rod model of Figure 6.33 with the real wetting angle as a parameter. For wetting angles, the apparent wetting angle—the lyophobicity—increases as the amount of porosity increases.

cylinders increases, the amount of porosity and the wetting angle increases. Therefore, it is the structure of the surface that is producing the lack of wettability. This is exactly the reason water "rolls off a duck's back" (Cassie and Baxter 1944) and why water beads up on lotus leaves. It is also the driving force for the interest in making hydrophobic surfaces with nanoparticles because the structure can be designed and fabricated into many different mixtures of particles and porosities. Interestingly, more careful modeling and observation of the wettability of such surfaces leads to almost identical results as these two older studies described above (Wang et al. 2015).

6.16 CHAPTER SUMMARY

Although this chapter is mainly concerned with the thermodynamics of surfaces, surface tension, and its consequences, many of the concepts presented are important in kinetics, particularly because of the increasing interest is small or "nanoparticles" and their intrinsic instability because of their large surface areas. Surface tensions arise due to broken or *unsatisfied* atomic bonds at surfaces. As a result, ionic, metallic, and covalently bonded materials have high surface tensions or surface energies, whereas materials with weaker intermolecular bonds, such as polymers, have lower surface tensions. For solids, in addition to the surface energy, there exists a surface stress as well, which may not be relaxed at low temperatures and needs to be considered when surface energy effects are important. Surface tensions can be related to surface curvature and lead to changes in vapor pressures, solubilities, reactivities, and other thermodynamic properties of condensed phases. This leads to particle coarsening in particulate systems because of the higher total energies of the surfaces of small particles relative to larger ones, the kinetic phenomenon known as *Ostwald ripening*. In addition, the relationships between the various interfacial energies determine wetting behavior and microstructures of multiphase materials. This phenomenon is useful in measuring surface energies and can be used to measure porosity in materials by mercury porosimetry because of the nonwetting behavior of mercury on many surfaces. Additives or impurities that segregate to the surface of a condensed phase, solid or liquid, lower the surface tension, as described by the Gibbs adsorption isotherm, and have several important commercial applications. Finally, by structuring the surface on a micro- or nanoscale, the wettability of the surface can be decreased, which is important in designing waterproof or hydrophobic coatings on a variety of surfaces.

APPENDIX

A.1 Surface Stress

The demonstration of the existence of a surface stress and its relation to the surface energy or the surface tension is presented in several places in the literature (Cammarata 1997; Hudson 1998). Figure A.1 illustrates the various quantities and steps involved in the derivation. As shown in the figure, consider a cube of dimension a. Then, if the cube is extended in the x-direction by an amount da, the work done, W_0 (this should be a dW or ΔW, but differential and difference symbols are dropped for clarity), is

$$W_0 = Fda. \tag{A.1}$$

Now if the cube is cut in two at its midsection, the total forces acting on both the top and bottom are the bulk applied force F/2 plus the *surface stress*, which acts along the length of the cut a and perpendicular to the direction of a, is a second-rank tensor (like all stresses) and typically given the symbol f_{xx}. Also, like the surface energy and surface tension, it too has units of N/m! This new work term, W′, is given by

$$W' = 2\frac{F}{2}dA + 2f_{xx}ada = W_0 + 2f_{xx}ada. \tag{A.2}$$

Rearranging this equation and defining the strain in the x-direction, $\varepsilon_{xx} = da/a$, the surface stress is given by

$$f_{xx} = \frac{W' - W_0}{2ada} = \frac{1}{a^2}\frac{W' - W_0}{2da/a} = \frac{1}{a^2}\frac{W' - W_0}{2\varepsilon_{xx}}. \tag{A.3}$$

The task is to find expressions for W′ and W_0 in terms of the surface tension on the two faces of the cut and relate them to the surface stress.

First consider the top sequence in the figure, designated (a), and leading to result W_I. Here, the cube is first extended by the applied force and then cut. The surface energy term for each face on the cut is the product of the new surface tension of the extended cut, $\gamma + \Delta\gamma$, and the extended surface, $A + \Delta A$ (note that Δs and differentials are being used more or less indiscriminately, but it makes no difference). So, the total work for sequence (a) is given by

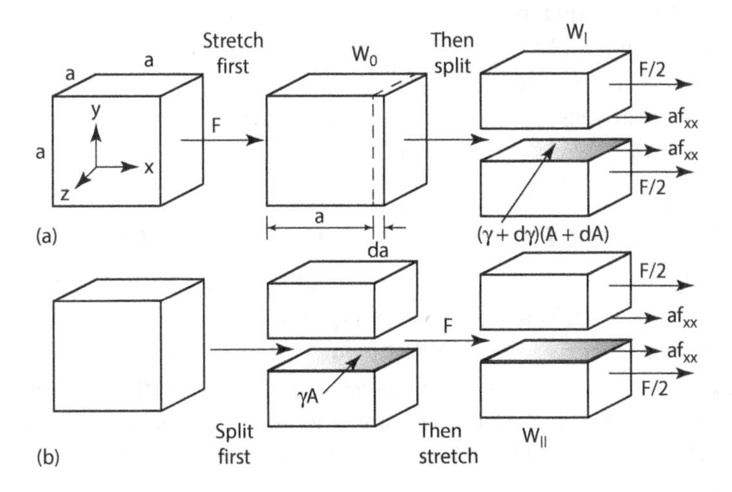

FIGURE A.1 Model (Hudson 1998; Cammarata 1997) used to show that the surface stress, f_{xx}, is given by $f_{xx} = \gamma + (d\gamma/d\varepsilon_{xx})$. (a) The cube is stretched first then split. (b) The cube is split first then stretched. The details of the model are described in Appendix A.1.

$$W_I = W_0 + 2(\gamma + \Delta\gamma)(A + \Delta A)$$

$$W_I = W_0 + 2(\gamma + \Delta\gamma)(a^2 + a\Delta a)$$

$$W_I = W_0 + 2(\gamma + \Delta\gamma)(a^2 + a(a\varepsilon_{xx}))$$

$$W_I = W_0 + 2a^2(\gamma + \Delta\gamma)(1 + \varepsilon_{xx})$$

this leads to

$$W_I \cong W_0 + 2a^2\gamma + 2a^2\Delta\gamma + 2a^2\gamma\varepsilon_{xx} \quad (A.4)$$

ignoring the small $\Delta\gamma\varepsilon_{xx}$ second-order product. In the second case, (b), the cube is cut first, giving two surface energy terms, $a^2\gamma$, and then extended giving the work, W′. The resulting total work is W_{II}

$$W_{II} = 2a^2\gamma + W'. \quad (A.5)$$

If both processes are carried out reversibly, then the total work must be equal for the two processes:

$$W_I = W_{II}. \quad (A.6)$$

Therefore, combining Equations A.4 and A.5 gives

$$W_0 + 2a^2\gamma + 2a^2\Delta\gamma + 2a^2\gamma\varepsilon_{xx} = 2a^2\gamma + W'$$

and can be rearranged to give

$$\frac{W' - W_0}{2a^2\varepsilon_{xx}} = \gamma + \frac{\Delta\gamma}{\varepsilon_{xx}} = \gamma + \frac{d\gamma}{d\varepsilon_{xx}}. \quad (A.7)$$

Substitution of Equation A.7 into A.3 gives the desired result:

$$f_{xx} = \gamma + \frac{d\gamma}{d\varepsilon_{xx}} \quad (A.8)$$

that is, the surface stress is equal to the surface energy or surface tension, γ, plus the change in γ with the surface strain. In liquids and solids at high temperatures, where atoms or molecules from the bulk can easily move into surface positions, the surface stress is the same as the surface tension or surface energy. Only in solids at low enough temperatures, where diffusion cannot relieve the surface strain, is the surface stress an important factor.

A.2 A Thermodynamic Argument for Young's Equation and Wetting

Because the vertical component of the force balance in Equation 6.34 is not satisfied, that is, $\varphi = 0$, can the horizontal component, Young's equation, Equation 6.31, be justified in other ways? A thermodynamic approach is most reasonable. What follows is a modified composite of two approaches in the literature used to show that Young's equation holds (Miller and Neogi 1985; DeHoff 2006). For constant temperature, volume, and composition, for a liquid drop, L, on a *flat* solid surface, S, both in contact with the vapor, V (Figure 6.14), the differential Helmholtz energy, dF, can be written as (Miller and Neogi 1985):

$$dF = \gamma_{SV}dA_{SV} + \gamma_{SL}dA_{SL} + \gamma_{LV}dA_{LV}. \quad (A.9)$$

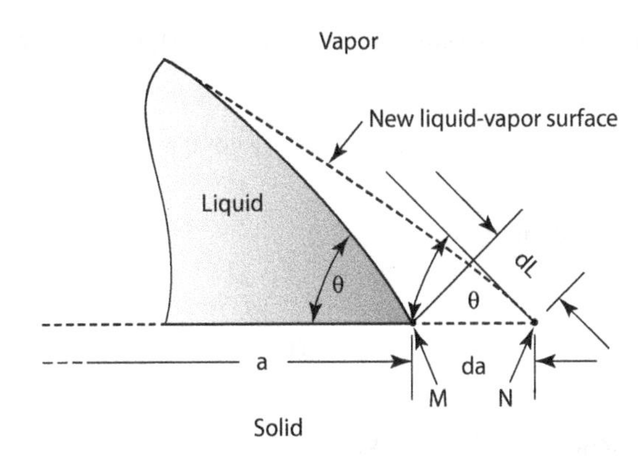

FIGURE A.2 Geometry used to obtain Young's equation, Equation 6.31, by a thermodynamic argument. The triple interface at M is moved a distance da to point N. The change in area of each of the interfaces is determined and related to dA_{SL} and then the Helmholtz energy is minimized reproducing Young's equation.

Figure A.2 shows the relation between the differential areas of the interface and the liquid drop. Adopting a modified procedure that considers an arbitrary displacement of the three-phase interface from point M to point N (DeHoff 2006)* in two dimensions. In this case, only changes in the x-direction are important because the vertical components can not be in equilibrium if the liquid drop is sitting on a flat, inert solid surface. Consider da to be a vector $\overrightarrow{da}$ from point M to point N. Then, the dot product of $\overrightarrow{da}$ with direction of the solid–vapor interfacial energy will give the x-component of the change in area of the solid–vapor interface when multiplied by some length along the three-phase interface, which, for simplicity, will be called $\Delta\rho$. So, $dA_{SV} = -\Delta\rho da$, for example, because

$$-\overrightarrow{da}\cdot\vec{i} = -da\cos(0) = -da$$

where $\vec{i}$ is a unit vector in the x-direction and the dot product gives the horizontal or x-component (and only component) for A_{SV}. The minus sign is necessary because the area is decreased between the solid and the vapor. Similarly, for the solid–liquid interface,

$$-\overrightarrow{da}\cdot(\vec{i}) = -da\cos(180) = da$$

not terribly surprising: $dA_{SL} = \Delta\rho da = -dA_{SV}$. For the *x-component* of the change in liquid–vapor area,

$$\overrightarrow{-da}\cdot\overrightarrow{dL} = -da\cos(180-\theta)$$

$$\overrightarrow{-da}\cdot\overrightarrow{dL} = -da(\cos 180\times\cos\theta + \sin 180\times\sin\theta)$$

if $\overrightarrow{dL}$ is a unit vector tangent to the liquid–vapor interface at points M and N and the double-angle formula can be found in many places in the literature. So,

$$dA_{LV} = \Delta\rho da\cos\theta = dA_{SL}\cos\theta. \tag{A.10}$$

Substituting the values for the areas in Equation A.9

$$dF = -\gamma_{SV}dA_{SL} + \gamma_{SL}dA_{SL} + \gamma_{LV}dA_{SL}\cos\theta$$

* DeHoff's (2006) thermodynamic approach of calculating the change in entropy of the shift of the triple line between the phases to give changes in area of the phases reaches exactly the same conclusion as the force balance, Equation 6.33—with slight differences in sign and angle conventions.

At equilibrium, the Helmholtz energy must be a minimum with respect to dA_{SL}, so

$$\frac{\partial F}{\partial A_{SL}} = \gamma_{SL} - \gamma_{SV} + \gamma_{LV}\cos\theta = 0$$

or Young's equation is obtained, namely,

$$\cos\theta = \frac{\gamma_{SV} - \gamma_{SL}}{\gamma_{LV}}.$$

A.3 The Gibbs Adsorption Isotherm

Consider a liquid in equilibrium with its vapor and a "surface layer" in between both, in equilibrium with both, and where the properties—such as concentration—change continuously between the liquid and the vapor. If the liquid is a two-component system in which there is some solute, component 2, dissolved in the liquid of component 1. Then, the first law of thermodynamics for the liquid and vapor phases are

$$\begin{aligned} dU_{vap} &= TdS_{vap} - PdV_{vap} + \mu_1 dn_{1,vap} + \mu_2 dn_{2,vap} \\ dU_{liq} &= TdS_{liq} - PdV_{liq} + \mu_1 dn_{1,liq} + \mu_2 dn_{2,liq}. \end{aligned} \tag{A.11}$$

A similar equation for the interface region is obtained by substituting the surface work term, γdA, for the PdV term because the surface really has no volume, and this term is positive because it is work done on the surface "phase":

$$dU_\sigma = TdS_\sigma + \gamma dA + \mu_1 dn_{1,\sigma} + \mu_2 dn_{2,\sigma} \tag{A.12}$$

where σ refers to the surface. Integrating either of Equations A.11 over the extensive variables, S, V, and n_i, gives

$$U_{liq} = TS_{liq} - PV_{liq} + \mu_1 n_{1,liq} + \mu_2 n_{2,liq}$$

and taking the total differential

$$dU_{liq} = TdS_{liq} + S_{liq}dT - PdV_{liq} - V_{liq}dP + \mu_1 dn_{1,liq} + n_{1,liq}du_1 + \mu_2 dn_{2,liq} + n_{2,liq}d\mu_2 \tag{A.13}$$

and subtracting Equation A.11 from A.13 gives (see Chapter 1)

$$0 = S_{liq}dT - V_{liq}dP + n_{1,liq}du_1 + n_{2,liq}d\mu_2. \tag{A.14}$$

At constant temperature and pressure, Equation A.14 becomes the Gibbs–Duhem equation,

$$0 = X_1 du_1 + X_2 d\mu_2 \tag{A.15}$$

because $X_{1or2} = n_{1or2}/(n_1 + n_2)$. With the same procedure applied to the surface, Equation A.12, the following is obtained for the surface region:

$$0 = S_\sigma dT + Ad\gamma + n_{1,\sigma}du_1 + n_{2,\sigma}d\mu_2. \tag{A.16}$$

At constant temperature and dividing by the area, A,

$$0 = d\gamma + \Gamma_1 d\mu_1 + \Gamma_2 d\mu_2$$

where the $\Gamma = n/A$ = the surface concentrations with Units(Γ) = mol/m^2. Substitution of Equation A.15, the Gibbs–Duhem equation, for du_1, gives

$$d\gamma = -\Gamma_{2,1} d\mu_2 = -\left(\Gamma_2 - \frac{X_2}{X_1}\Gamma_1\right) d\mu_2 \tag{A.17}$$

where $\Gamma_{2,1}$ is the "excess surface concentration of component 2 in component 1." That is, if the concentration in the surface is different from what would be expected from the bulk composition, the surface energy of the solution will either increase or decrease. Of course, if $\Gamma_2/\Gamma_1 = x_2/x_1$, that is, the surface concentrations are in the same ratio or are the same as in the bulk, then there is no surface excess and $d\gamma = 0$. Now

$$\mu_2 = \mu_2^0 + RT \ln a_2$$

by definition. So, $d\mu_2 = RTd \ln a_2 \cong RTd \ln X_2 \cong RTd \ln C_2$ particularly if the concentration, C_2, is not very large. Therefore, the *Gibbs adsorption isotherm becomes*

$$d\gamma = -RT\,\Gamma_{2,1} d \ln C_2 = -RT\left(\Gamma_2 - \frac{X_2}{X_1}\Gamma_1\right) d \ln C_2 \tag{A.18}$$

which states that if the solute, component number 2, segregates to the surface, $\Gamma_{2,1} > 0$, then the surface energy, γ, decreases. Conversely, if component two desegregates, $\Gamma_{2,1} < 0$, then the surface energy increases.

EXERCISES

6.1 a. At 1200°C, for liquid copper, γ_{LV} = 1285 mJ/m^2, for solid Al_2O_3 γ_{SV} = 900 mJ/m^2, and for the interfacial free energy between liquid copper and solid Al_2O_3 at this temperature, γ_{SL} = 1200 mJ/m$_2$. Calculate the wetting angle, θ (degrees), for a liquid copper drop on an alumina plate.

b. Calculate the spreading coefficient for molten copper on solid alumina.

c. Make a hand sketch of this drop on the alumina plate.

6.2 a. Addition of oxygen in solution in the copper increases the wetting of the molten copper on alumina. If 1 w/o copper changes interfacial energy to γ_{SL} = 900 mJ/m^2, calculate the wetting angle (degrees) for this oxygen–copper alloy on alumina.

b. Calculate the spreading coefficient for this oxygen–copper alloy on the solid alumina.

c. Make a hand sketch of this drop on the alumina plate.

6.3 a. Finally, enough oxygen is added to the molten copper to make it react with the solid alumina, so that γ_{SL} = −600 J/m^2. Calculate the spreading coefficient for this oxygen–copper alloy.

b. Make a hand sketch of this drop on the alumina plate.

6.4 a. Small particles (0.1 μm in diameter at t = 0 seconds) of hematite, Fe_2O_3, are undergoing coarsening by gas transport through the vapor phase and a surface reaction in an HCl-containing atmosphere through the following reaction:

$$Fe_2O_3(s) + 6\,HCl(g) \rightleftharpoons 2\,FeCl_3(g) + 3\,H_2O(g)$$

If the Gibbs energy, $\Delta G°$, for this reaction at 1200°C is $\Delta G°$ = 131,376 J/mole, calculate the equilibrium constant at this temperature.

b. The coarsening is taking place at 1200°C in argon gas containing 10 % HCl with a total pressure of 1 atm. All of the water in the gas is produced by the reaction. Calculate the equilibrium $FeCl_3$ pressure at this temperature in this gas.
c. Calculate the $[FeCl_3]$ in the gas phase in moles/m^3 under these conditions.
d. If the density of Fe_2O_3 is 5.24 g/cm^3 and its molecular weight is 159.69 g/mole, calculate the molar volume of Fe_2O_3 in m^3/mole.
e. For the surface reaction rate constant k, $k_0 = 7.20 \times 10^4$ m/s and $Q = 120{,}000$ J/mole. Calculate the value of the surface reaction rate constant (m/s) at 1200°C.
f. If particle coarsening is taking place by surface reaction control, calculate the average particle size (diameter in μm) of the Fe_2O_3 after a total reaction time of 3 h if the surface energy of Fe_2O_3 is 900 mJ/m^2. Neglect the starting particle size if possible.
g. If the initial particle size had been 0.1 μm in diameter, calculate the number of the initial particles that have been incorporated into a single coarsened particle after 3 h.

6.5 The following data exist for Cu in the literature (Brandes and Brook 1992): melting point, $T_{mp} = 1083$°C; heat of fusion, $\Delta H_m = 13.02$ kJ/mole; density at 20°C, $\rho = 8.96$ g/cm^3; density of the liquid at the melting point, $\rho_l = 8.00$ g/cm^3; linear thermal expansion coefficient, $\alpha_L = 17.0 \times 10^{-6}$ K^{-1}; molecular weight, $M = 63.55$ g/mol; and the liquid surface energy at the melting point, $\gamma = 1285$ mJ/m^2.
a. Calculate the molar volume, $\bar{V}_0$(m^3/mol), of copper at 20°C.
b. Assuming that the change in volume of the solid with heating to the melting point is given by $\Delta\bar{V} = \bar{V}_0 3\alpha_L \Delta T$, calculate the molar volume of the solid at the melting point of copper (m^3/mol).
c. Calculate the molar volume of the liquid copper at the melting point (m^3/mol).
d. Calculate the melting temperature of 10 nm diameter copper particles.

6.6 The following data exist for silicon (Si) in the literature (Brandes and Brook 1992): melting point, $T_{mp} = 1410$°C; heat of fusion, $\Delta H_m = 50.66$ kJ/mol; density at 20°C, $\rho = 2.34$ g/cm^3; density of the liquid at the melting point, $\rho_l = 2.51$ g/cm^3; linear thermal expansion coefficient, $\alpha_L = 7.60 \times 10^{-6}$ K^{-1}; molecular weight, $M = 28.09$ g/mol; and the liquid surface energy at the melting point, $\gamma = 865$ mJ/m^2.
a. Calculate the molar volume, $\bar{V}_0$ (m^3/mol), of silicon at 20°C.
b. Assuming that the change in volume of the solid with heating to the melting point is given by $\Delta\bar{V} = \bar{V}_0 3\alpha_L \Delta T$, calculate the molar volume of the solid at the melting point of silicon (m^3/mol).
c. Calculate the molar volume of the liquid silicon at the melting point (m^3/mol).
d. Calculate the melting temperature (°C) of 10 nm diameter silicon particles.

6.7 A liquid silicate with surface tension of 500 mJ/m^2 makes contact with a polycrystalline oxide with an angle of $\theta = 40$ ° on the surface of the oxide. If mixed with the oxide, it forms discrete liquid globules at three grain intersections with the average dihedral angle $\varphi = 80$°. If the interfacial tension of the oxide grain boundary is 900 mJ/m^2,
a. Sketch these two microstructures and show the interfacial energies acting as forces.
b. Compute the solid–liquid surface energy.
c. Compute the solid–vapor surface energy.
d. Is this a reasonable value for the solid surface energy?

6.8 Figure E.1 schematically depicts a typical snowflake about 1 mm in diameter. The hexagonal shape reflects the hexagonal symmetry of the ice crystal, and the dendritic (branched, tree-like) structure reflects the details and anisotropy of the *kinetics* of the ice crystal growth process from the water vapor in the atmosphere after the snowflake had been nucleated in the center (probably by a dust particle) and not the anisotropy of the surface energy. The radius of curvature at the end of one of the hexagonal tips is about 10 μm. The vapor pressure of ice over a flat surface at 23°F is 3.01 mm of Hg (760 mm of Hg = 1 atm) and the density of ice at this temperature is 0.917 g/cc (Haynes 2013).

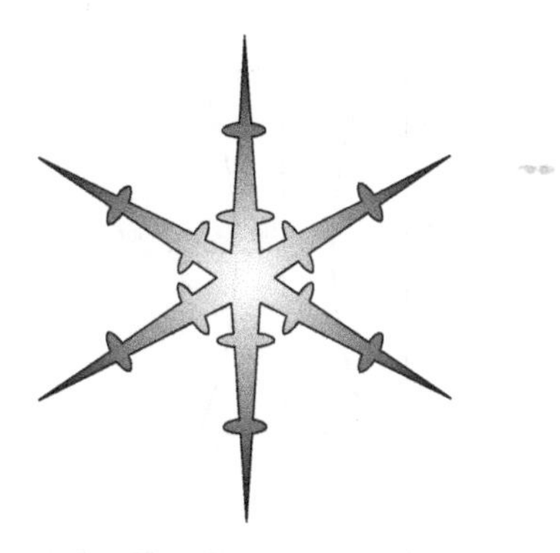

FIGURE E.1 Schematic of a snowflake that shows the sharp points at the ends of each of its six arms.

a. Calculate the molar volume of ice at 23°F.
b. Convert 23°F into K.
c. Calculate the percent increase of the vapor pressure of water at the end of one of the hexagonal tips at 23°F compared to a flat surface.
d. What do the results of the calculation in (c) imply about the long-term stability of the shape of the snowflake assuming that the surface energy of ice is isotropic?

6.9 Ion implantation of helium in a particle accelerator deposits He interstitial atoms in a certain metal. If annealed, the helium atoms diffuse and form helium bubbles more or less uniformly throughout the metal thickness that cause the metal to swell and reduce its density to 0.95 of its original value (i.e., 5 volume percent porosity). When the porous metal is dissolved in acid, 2.26 cm^3 at STP (standard temperature and pressure) of He is given off for each cubic centimeter of porous metal dissolved. Helium is quite insoluble in most metals, so assume that all the He given off was in bubbles and none was left in solid solution in the metal. Microscopic examination shows that the He bubbles are of a uniform size with a diameter of 0.5 micrometer.
a. Calculate the He pressure in the bubbles.
b. Calculate the surface energy of the metal.
c. Is this a reasonable value? Why?
d. How important is the assumption that all of the He evolved was in the bubbles?

6.10 The surface energy of poly(methyl methacrylate), PMMA, is $\gamma_{PMMA} = 41.1$ mJ/m^2. (Sperling 2006). Water forms a contact angle, θ, of 71° on PMMA.
a. Calculate the interfacial energy (mJ/m^2) between water and PMMA.
b. 0.1 w/o sodium stearate, $C_{18}H_{35}NaO_2$, a common surfactant, is added to the water and segregates to both the water–air and the water–PMMA interfaces and lowers both the interfacial energies by one-half when the concentration of sodium stearate goes from 0.01 to 0.1 w/o. Calculate the new contact angle in degrees.
c. Calculate the bulk concentration of sodium stearate in mol/L.
d. With the Gibbs adsorption isotherm, calculate the excess surface concentration of sodium stearate in mol/L.
e. From (d), calculate the actual surface concentration of sodium stearate in mol/L.

6.11 Experimental data on interfacial properties of the Ni(liquid, m.p. = 1453 °C)-Al_2O_3(solid, m.p. = 2050°C) system at 1850° are the following (Kingery 1954): $\gamma_{LV}(Ni) = 1480$ mJ/m^2; grain boundary groove dihedral angle for Al_2O_3 and air, $\varphi = 151.8°$ (Figure 6.13); grain boundary groove dihedral angle for a grain boundary groove formed under liquid Ni, $\varphi_{Ni} = 166.4°$; and the wetting angle of molten Ni on solid Al_2O_3, $\theta = 130°$. From these data, calculate the following at 1850 °C:
a. The alumina-Ni interfacial energy, γ_{SL}
b. The alumina grain boundary energy, γ_{GB}
c. The solid alumina surface energy, γ_{SV}

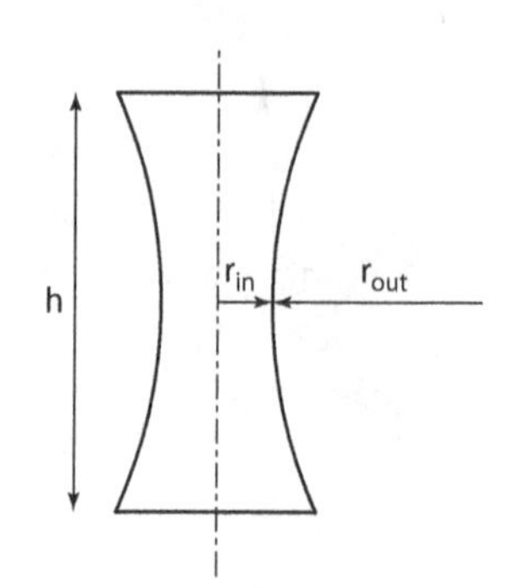

FIGURE E.2 A concave right circular cylinder with a *constant* external radius of curvature, r_{out}, and an inner radius of curvature, r_{in}, that *varies* along the height, h, of the cylinder.

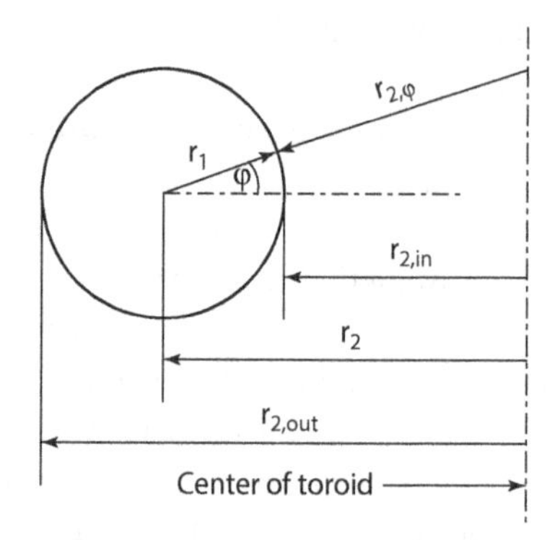

FIGURE E.3 Cross section through a toroid with a *constant* inner radius of curvature, r_1, and an outer radius of curvature, $r_{2\varphi}$, that varies with the angle from $r_{2,in}$ for $\varphi = 0$ to $r_{2,out}$ for $\varphi = 180°$.

6.12 Figure E.2 shows a concave circular cylinder with the following dimensions: h = 8 cm, $r_{out} = 6$ cm is the radius of curvature of the vertical surface of the cylinder, and the diameter at the top = 5 cm.

a. Calculate the inner radius of curvature at 0.5 cm intervals from the middle of the cylinder to the top (see Figure 6.8).

b. Calculate the mean curvature at each of these points.

6.13 Figure E.3 sketches part of a toroid with an inner radius r_1, the radius from the center of revolution of the toroid to the center of the circular part of the toroid, r_2, the radius from the center of revolution to the inner surface of the toroid, $r_{2,in}$, and the radius from the center of revolution of the toroid to the outer surface of the toroid, $r_{2,out}$. Also is shown $r_{2,\varphi}$ = the second radius of curvature that must be normal to the surface.

a. If $r_1 = 1$cm and $r_2 = 10$ cm, calculate $r_{2,\varphi}$ as a function of the angle φ for $0 \le \varphi \le \pi$ at 10° steps.

b. From the results in (a), plot the mean curvature as a function of the angle φ for $0 \le \varphi \le \pi$ at 10° steps.

c. From the results in (b), how would the shape of the toroid change if the temperature were high enough for atoms to move in response to the surface energy forces?

REFERENCES

Adamson, A. W. 1982. *Physical Chemistry of surfaces*, 4th ed. New York: Wiley.

Adamson, A. and A. Gast. 1982. *Physical Chemistry of surfaces*, 7th ed. New York: Wiley-Interscience.

Brandes, E. A. and G. B. Brook, eds. 1992. *Smithells Metals Reference Book*, 7th ed. Oxford: Butterworth-Heinemann.

Cammarata, R. C. 1994. Surface and interface stress in thin films. *Progress in Surface Science.* 46(1): 1–38.

Cammarata, R. C. 1997. Surface and interface stress effects on interfacial and nanostructured materials. *Materials Science and Engineering* A237: 180–184.

Casey, J. 1996. *Exploring Curvature.* Wiesbaden, Germany: Vieweg and Sohn.

Cassie, A. B. D. and S. Baxter. 1994. Wettability of porous surfaces. *Transactions of the Faraday Society.* 40: 546–550.

Chang, R. 2000. *Physical Chemistry for the Chemical and Biological Sciences.* Sausalito, CA: University Science Books.

DeHoff, R. 2006. *Thermodynamics in Materials Science,* 2nd ed. Boca Raton, FL: CRC Press, Taylor & Francis.

Evans, D. F. and H. Wennerström. 1999. *The Colloidal Domain.* New York: Wiley-VCH.

Gaskell, D. R. 2008. *Introduction to the Themrodynamics of Materials,* 5th ed. New York: Taylor & Francis.

Green, P. F. 2005. *Kinetics, Transport, and Structure in Hard and Soft Material.* Boca Raton, FL: Taylor & Francis.

Haynes, W. M., editor-in-chief. 2013. *CRC Handbook of Chemistry and Physics,* 94th ed. Boca Raton, FL: CRC Press.

Hiemenz, P. C. 1986. *Principles of Colloid and Surface Chemistry,* 2nd ed. New York: Marcell Dekker.

Hudson, J. B. 1998. *Surface Science.* New York: Wiley.

Humenik, M., Jr. and W. D. Kingery. 1954. Metal-ceramic interactions: III. Surface tension and wettability of metal-ceramic systems. *Journal of the American Ceramic Society* 37(1): 18–23.

International Union of Pure and Applied Chemistry (IUPAC). 2014. *Compendium of Chemical Terminology, Gold Book,* Version 2.3.2. 2014-02-24. IUPAC.

Kingery, W. D. 1954. Metal-ceramic interactions: IV. Absolute measurement of metal-ceramic interfacial energy and the interfacial adsorption of silicon from iron-silicon alloys. *Journal of the American Ceramic Society* 87(2): 42–45.

Kingery, W. D. 1959. Surface tension of some liquid oxides and their temperature coefficients. *Journal of the American Ceramic Society* 45(1): 6–10.

Kingery, W. D., H. K. Bowen, and D. R. Uhlmann. 1976. *Introduction to Ceramics,* 2nd ed. New York: Wiley.

Laidler, K. J. 1993. *The World of Physical Chemistry.* New York: Oxford University Press.

Larpkiattaworn, S. 1995. *Oxidation Behavior of Al_2O_3-Cu Composites.* MS Thesis. Colorado School of Mines. Golden, Colorado.

Lee, J. 1984. *Vapor Phase Sintering of Hematite in HCl.* PhD Thesis. The Ohio State University, Columbus, OH.

Miller, C. A. and P. Neogi. 1985. *Interfacial Phenomena.* New York: Marcell Dekker.

O'Brien, T. E. and A. C. D. Chaklader. 1974. Effect of oxygen on the reaction between copper and sapphire. *Journal of the American Ceramic Society* 57(8): 329–332.

Parikh, N. M. 1958. Effect of atmosphere on surface tension of glass. *Journal of the American Ceramic Society* 41(1): 18–22.

Plueddemann, E. 1991. *Silane Coupling Agents,* 2nd ed. New York: Plenum Press.

Ratke, L. and P. W. Voorhees. 2002. *Growth and Coarsening.* Berlin, Germany: Springer.

Schwartz, M. M. 1990. *Ceramic Joining.* Materials Park, OH: ASM International.

Shuttleworth, R. 1950. The surface tension of solids. *Proceedings of the Physical Society* A63: 444–457.

Silbey, R. J. and R. A. Alberty. 2001. *Physical Chemistry,* 3rd ed. New York: Wiley.

Sperling, L. H. 2006. *Introduction to Physical Polymer Science,* 4th ed. Hoboken, NJ: Wiley.

Udin, H., A. J. Shaler, and J. Wulff. 1949. The surface tension of solid copper. *Metal Transactions* 1(2): 186–190.

Wagner, C. 1961. Theorie der alterung von niederschlagen durch umlosen. *Zeitschrift fur Elektrochemie* 65(7–8): 581–591.

Wang, Y., et. al. 2015. Evaluation of macroscale wetting equations on a microrough surface. *Langmuir* 31: 2342–2350.

Wenzel, R. N. 1936. Resistance of solid surfaces to wetting by water. *Industrial and Engineering Chemistry* 28(8): 988–994.

Wulff, G. 1901. Zur frage der geschwindigkeit des wachsthums und der auflösung der krystallflächen. *Zeitschrift für Krystallographie und Mineralogie* 34(5/6): 449–530.

Yang, S., X. Jin, K. Liu, and L. Jiang. 2013. Nanoparticles assembly-induced special wettability for bio-inspired materials. *Particuology* 11: 361–370.

Yeo, Y. 2007. *Gas Corrosion of MgO Studied by Grain Boundary Grooving and Pore Smoothing.* PhD. Thesis. Colorado School of Mines, Golden, Colorado.

7

Phase Transitions

7.1 INTRODUCTION

One of the goals of this chapter is to describe and model the kinetics of different types of phase transformations including changes in state. The primary focus will be the types of phase changes or *phase transitions* that occur in solids. The largest number of phase transformations in all materials occur by a *nucleation and growth* process that is easy to understand and can be simply modeled in detail to predict the effects of times and temperatures on the extent of the phase transformation. Both *homogeneous* and *heterogeneous* nucleation are covered in this chapter. Heterogeneous nucleation is more import in real materials but homogeneous nucleation is easier to model and readily leads to the heterogeneous model. *Spinodal decomposition* is described qualitatively here relative to phase equilibria with the detailed kinetic model development relegated to Chapter 16. The other major type of phase transformation is the so-called *diffusionless* or *displacive* transitions that occur primarily in crystalline metals and ceramics. These are also referred to as *martensitic* transformations and have been studied extensively in metals primarily because of the hardening effect martensite has in steels. Here, the phenomenon will be discussed more qualitatively and described for several materials systems of technical importance. Finally, the *glass transition* is discussed, a topic of major importance in polymers, many inorganic elements and compounds, and some metals. A simple free-volume model for the glass transition is presented that makes a strong plausibility argument for many features of the glass transition. The chapter includes a description of the *Challenger* space shuttle accident whose proximate cause was the glass transition or brittle transition in the solid booster rocket polymer seals. Finally, particle size effects on phase transitions are discussed.

7.2 THERMODYNAMICS OF PHASE TRANSITIONS

Frequently in kinetics of materials, the interest is in the rate of a transformation, or a phase change, of phase alpha, α, to a phase beta, β. These transformations or transitions include changes of state—solid, liquid, or gas—as well. Examples of phase transformations or phase transitions in single and multicomponent systems include

- Polymorphic (allotropic) transformation: structure 1 $\Rightarrow$ structure 2; for example, white tin $\Rightarrow$ gray tin
- Solidification; liquid $\Rightarrow$ crystal; for example, freezing of copper
- Crystallization; glass $\Rightarrow$ crystal; for example, crystallization of amorphous Si or polyethylene; devitrification (crystallization) of a silicate glass
- Condensation; gas $\Rightarrow$ liquid; for example, rain
- Condensation; gas $\Rightarrow$ solid; for example, snow
- Recrystallization; strained solid $\Rightarrow$ strain-free solid; for example, rolled Cu

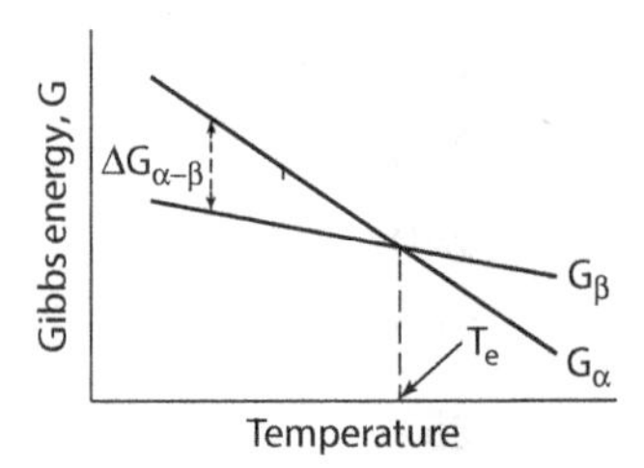

FIGURE 7.1 Gibbs energy for two phases undergoing a phase transformation from α to β. T_e is the equilibrium temperature.

- Gelation: liquid ⇒ gel; for example, polymer plus liquid; cell walls
- Precipitation: liquid or solid ⇒ solid; for example, Fe_3C in Fe-C alloys

In general, for any of these phase transitions to occur, the Gibbs energy of the product phase(s) must be lower than that of the reactant phase(s) as shown schematically in Figure 7.1. Figure 7.2 shows the Gibbs energy versus temperature for zinc liquid and zinc vapor and Figure 7.3 shows the Gibbs energy versus temperature for liquid, solid, and gaseous H_2O calculated from thermodynamic data (Roine 2002) as two examples for real materials. In both cases, the phase with the lowest Gibbs energy at a given temperature is the equilibrium state. Note that in both of these examples, $\Delta G = \Delta G°$ since all of the phases are in their standard states—activity of the solid and liquid = 1.0 and pressure of the gas = 1 bar. If the Gibbs energy $\Delta G^o_{\alpha \to \beta}$ is negative, then the reaction is thermodynamically favorable. However, it may or may not take place depending on the details of the kinetics of the *phase transition* or *phase transformation*—as it is frequently referred to. If the reaction does take place, how fast it happens depends on the details of the kinetics of the phase transition. In this chapter, the rates of phase transformations controlled by a surface reaction are considered. Similar results hold for diffusion-controlled reactions modeled in Chapters 8 to 16. Diffusion control is the more important of the two processes for the high temperature transformations involving *hard materials*—metals, ceramics, and electronic materials—compared to polymers and *some* biomaterials, which are considered *soft materials.* The low temperature limitation of these materials frequently leads to a reaction controlling the rate of phase transition.

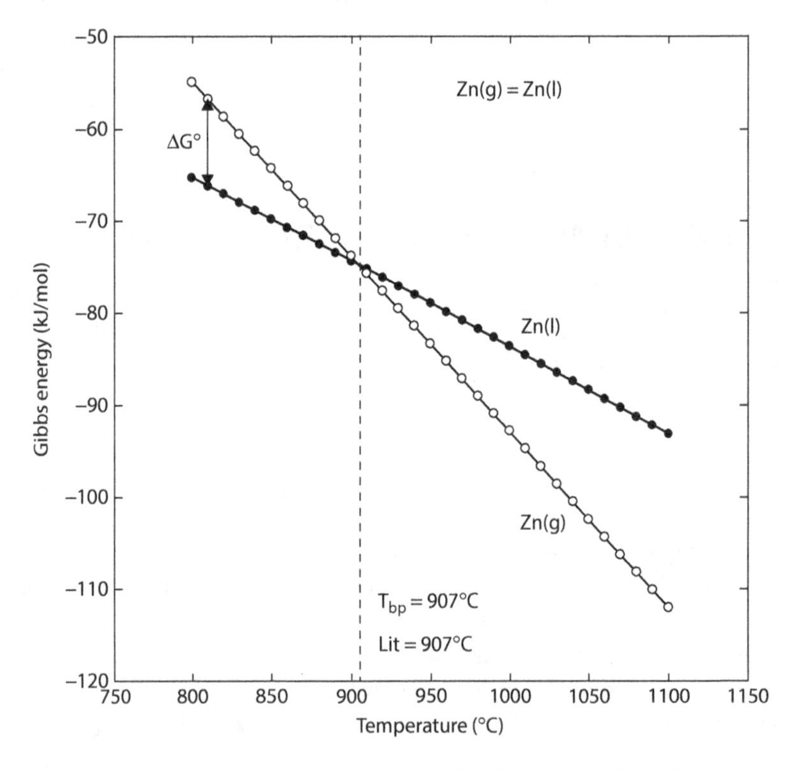

FIGURE 7.2 Gibbs energy versus temperature for zinc liquid and zinc vapor. (Data from Roine 2002.) T_{bp} is the boiling point at the intersection of the two lines and it is the value of the boiling point given in the literature, for example, Emsley (1998).

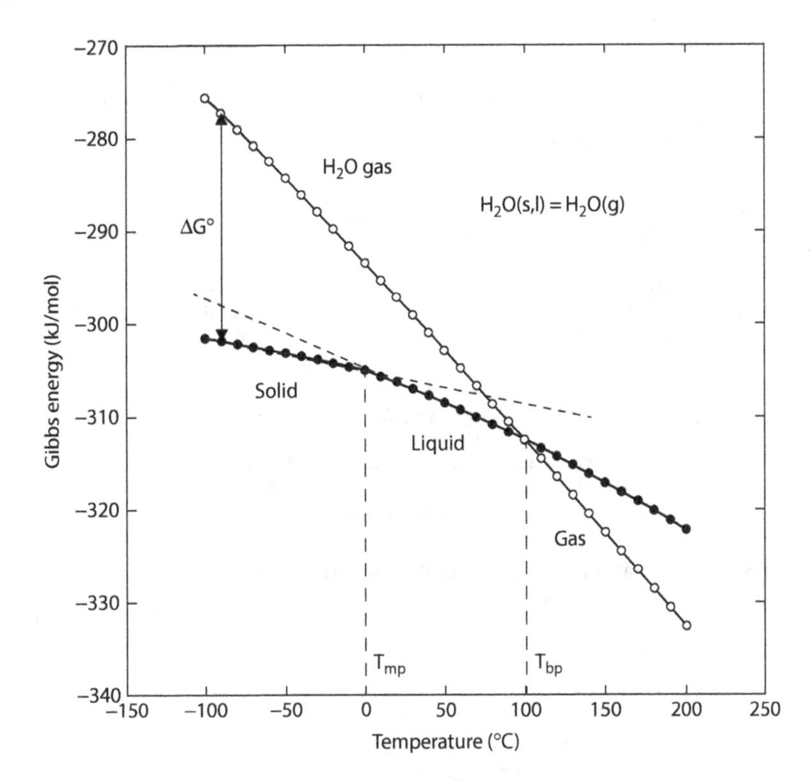

FIGURE 7.3 Gibbs energy versus temperature for liquid, solid, and gaseous H_2O. The dashed lines are the extrapolation of the data for the liquid and solid states beyond their region of stability. Note that the phase with the lowest Gibbs energy at any temperature is the stable phase. (Data from Roine 2002)

7.3 RATES OF PHASE TRANSFORMATIONS

The primary interest here is to develop an understanding of the rates of various phase transformations. However, it is impossible to cover all types of phase transitions completely. Only the *nucleation and growth* process is examined in detail since it can be controlled by both surface reactions, that have been examined earlier, and diffusion, and thus applies to a wide spectrum of materials and transitions. Furthermore, it is the most common type of transformation and has important technological implications.

To put things in perspective, it is instructive to examine the iron–carbon system—steels—since the kinetics of phase transformations in this system have been studied exhaustively, not only to develop a better understanding of phase transitions or transformations but also, and perhaps more importantly, to produce heat treatment procedures that are used to optimize the properties of a given steel: very important to the industrial materials engineer. As emphasized earlier, materials science and engineering essentially began when the relationships between the chemistry, processing, microstructure, and properties of steel became the focus of the materials engineer. Boiler explosions and bridge collapses were a major impetus for achieving a better understanding of the mechanical properties of steel. By the early 1900s, a reasonably good understanding of these relationships had been developed. Furthermore, this system exhibits several phase transformations that are important commercially. The iron–carbon system up to Fe_3C, cementite, is given in Figure 7.4 (Brandes and Brook 1992). Figure 7.5 gives the *time–temperature–transformation* (*TTT*) *diagram* for this steel (American Society for Metals 1973). This TTT diagram is for 1080 steel—which is almost the eutectoid composition in Figure 7.4—where the alloy is heated into the single-phase austenite region and then rapidly cooled into the two-phase ferrite + cementite field and held at a fixed temperature while the austenite Þ ferrite + cementite phase transformation occurs. This is an *isothermal* transformation diagram in that the data are taken by

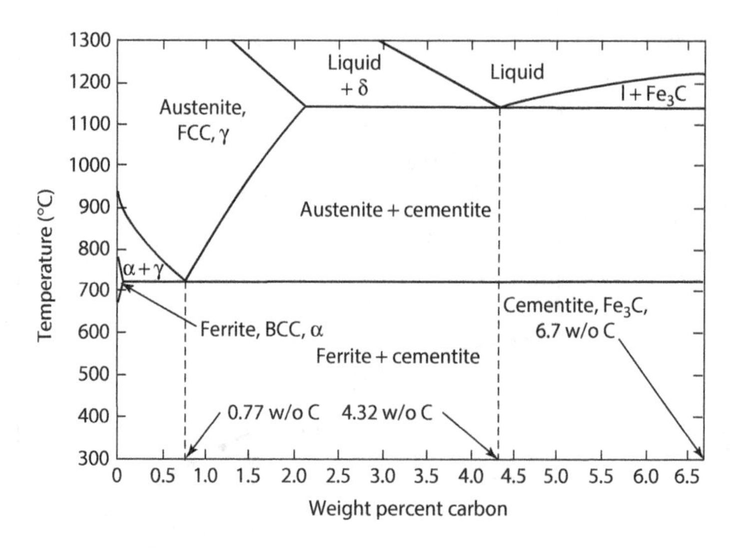

FIGURE 7.4 Part of the iron–carbon diagram after Brandes and Brook 1992.

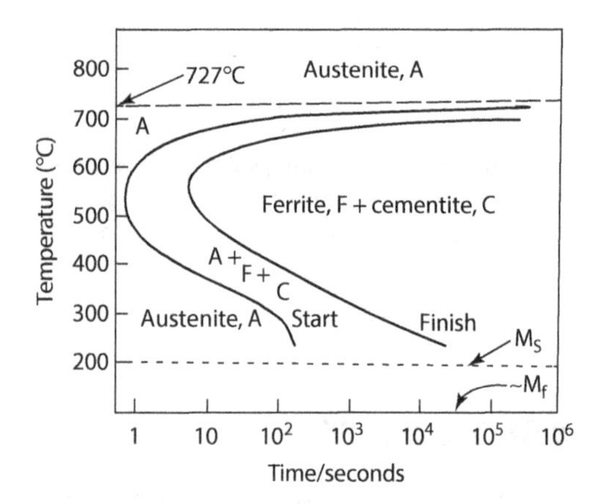

FIGURE 7.5 Part of the isothermal transformation (TTT) of 1080 steel after American Society for Metals 1973. M_S is the temperature at which martensite begins to form and M_f is the temperature at which martensite formation stops.

cooling rapidly from above the As temperature 727°C (austenite, γ-iron-C alloy, face-centered cubic [FCC] lattice) to some temperature—say 650°C—and then holding at this temperature and observing when the presence of the ferrite (α-iron, body-centered cubic [BCC]) and cementite (Fe_3C) first start to appear in the microstructure. This is the leftmost C-shaped curve in the diagram, which shows that this will occur in about 7 seconds at this temperature. The phase transformation continues and finishes at the intersection with the final curve in about 1 minute. The utility of the TTT diagram for steel is that it defines the time and temperature necessary to attain a certain microstructure, which in turn predicts the mechanical properties of the resulting heat-treated steel. Such diagrams are of great practical importance to the practicing industrial materials engineer. This transformation is a classic example of a nucleation and growth transformation in that at some point below the equilibrium temperature, T_E = 727°C, at some locations in the austenite, the new phases will start to form. These locations are the *nuclei*—or the beginning sites for the new phase(s)—that then grow to completely transform the solid as shown schematically in Figure 7.6, for the more general case of alpha phase transforming completely to beta phase.

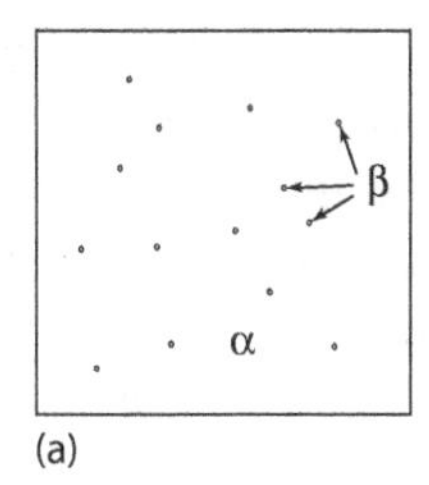

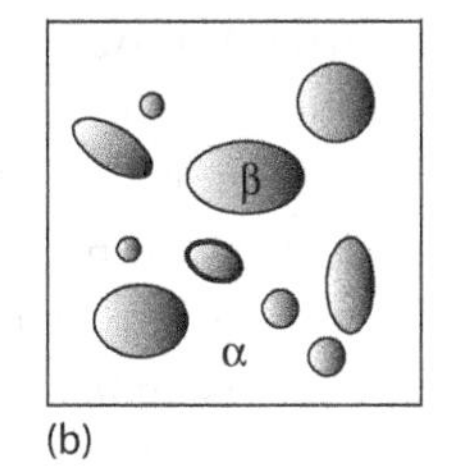

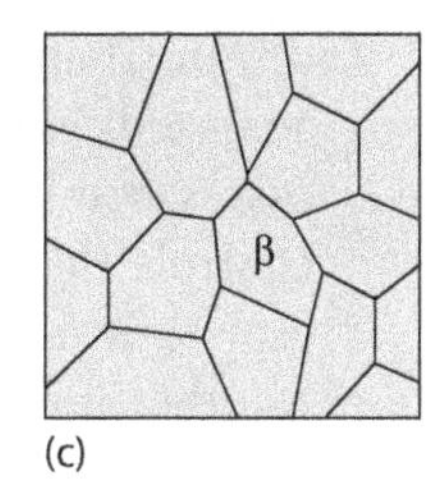

FIGURE 7.6 Schematic showing the stages in a nucleation and growth phase transition for the transformation of α to β for times (a) $t \cong 0$, (b) $0 < t < \infty$, and (c) $t = \infty$.

Referring again to Figure 7.1, above an equilibrium temperature, T_E, alpha is the stable phase and below T_E, beta is the stable phase. So below T_E in a transformation from α to β, there is a negative Gibbs energy change; that is, $\Delta G_{\alpha\text{-}\beta} = G(\beta) - G(\alpha)$. Although the Gibbs energy of β is lower than that of α, the phase transformation does not occur instantly. Most phase transformations require a density or composition change and the generation of some phase interface surface energy all of which require either a diffusion process or a surface reaction or both (and sometimes heat transfer) that take time. In addition, the formation of a second phase, beta, requires that the new phase starts somewhere in the old phase, alpha. These particles of the new phase, perhaps very small containing only tens or hundreds of atoms, are the *nuclei* of the new phase. Even though the new phase is thermodynamically more stable, these nuclei have a surface or interfacial energy that needs to be overcome before they are stable and can grow. Once the stable nuclei of the second phase are formed, they grow by either a diffusion control or surface reaction control (Chapter 5), eventually transforming all of the original alpha to beta. As noted earlier, nucleation and growth is a major mechanism of phase transformations of interest in materials science and engineering. Figure 7.7 shows the nucleation and growth of silicon crystals in an amorphous thin film of silicon. Now, nucleation can be either *homogeneous* or *heterogeneous. Homogeneous nucleation is the process where the small nuclei form spontaneously throughout the bulk of the material undergoing the phase change.* In reality, homogeneous nucleation is hard to achieve in practice but it is easier to model. In contrast, heterogeneous nucleation is more common and it occurs when the new phase forms on particles of dust or another phase that allows the nucleus to lower its interfacial energy thereby allowing the nucleus to form more easily. Therefore, *heterogeneous nucleation is the process where the small, stable particles of the new phase are formed at the interface with another phase (could be the free surface) that lowers the interfacial energy of the nuclei.* An important example of heterogeneous nucleation is the process of *cloud seeding* with silver iodide, AgI, smoke particles that act as heterogeneous nuclei for ice formation, which will ultimately lead to rain (hopefully) during drought. Another classic and visual result of heterogeneous nucleation is the intentional production of *crystalline glazes* on ceramic art pieces where the heterogeneous nuclei are dust particles or small pieces of the kiln—furnace—insulation that fall onto the glaze surface serving as heterogeneous nucleation sites for crystals (Figure 7.8). The large round areas are the crystals of willemite, Zn_2SiO_4, that have grown out of the glaze. Notice the deeper coloration of the crystals compared to the glass glaze. This is due to the optical absorption of electrons in the d-shells of transition metal ions, such as cobalt, that have different optical absorption characteristics depending on their local electronic environments—number and proximity of anions—that are different in the crystal and glass.

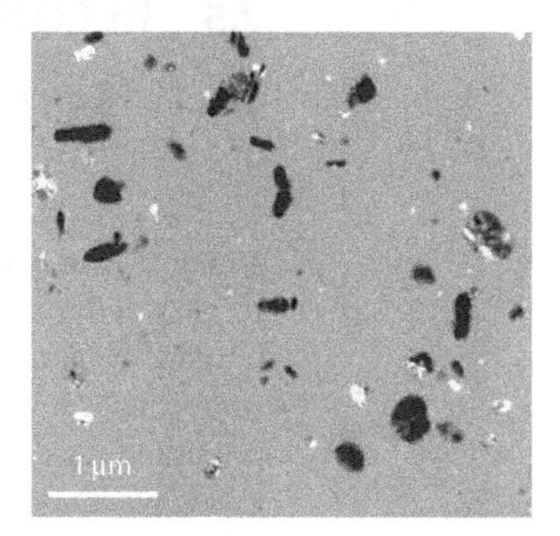

FIGURE 7.7 Crystallization of an amorphous thin film of silicon by a nucleation and growth phase transition. The gray area is the glass and the white and black regions are crystals or grains of silicon. (Roy 2001)

FIGURE 7.8 Crystalline glaze vase from Taiwan on which large acicular crystals of zinc silicate, willemite, have grown from the glass by heterogeneous nucleation and growth.

In Chapter 5, growth processes controlled by surface reactions are analyzed. Now the process of nucleation is modeled and the two processes combined to demonstrate how nucleation and growth process controls the rate of phase transformation in materials. Nucleation and growth is not the only path by which phase transformations can occur. Another important one is *spinodal decomposition* that will be modeled in Chapter 16.

7.4 TYPES OF PHASE TRANSFORMATIONS

There are several different terms that are used, sometimes loosely, to describe different phases that might be undergoing a phase transition or transformation. These include the following examples:

- *States of matter*: Solid, liquid, and gas.
- *Allotropes*: Different forms (usually crystal structures) of an element: for example, tetragonal (white) and diamond cubic (gray) tin; BCC and FCC iron; O_2 and O_3 gas; amorphous and crystalline silicon.
- *Polymorphs*: Different crystal structures of the same compound or element: for example, white and gray tin (also allotropes); cubic, monoclinic, and tetragonal ZrO_2.

In general, there are two kinds of phase transitions in crystalline solids that involve only changes in crystal structure and not composition: (1) displacive or diffusionless transitions and (2) reconstructive or diffusion-controlled transitions. These can occur in both single and multicomponent systems. For both, the phase transition takes place by a *nucleation and growth* process. However, the growth or speed of the new phase interface in displacive or diffusionless transitions is roughly the speed of sound. As a result, the controlling kinetics are neither a surface reaction nor diffusion but perhaps the motion of dislocations. Furthermore, this rapid rate of transformation is difficult to study experimentally and results in a less than satisfactory understanding of the kinetic details of diffusionless transitions. Several important displacive transformations are presented later without any attempt to model their kinetics.

7.5 RECONSTRUCTIVE TRANSFORMATIONS

7.5.1 SILICATES

In reconstructive transformations, atoms must diffuse to different positions in order for crystal structures to change. Figure 7.9 shows a schematic and purely hypothetical example for a silicate structure that goes from a *ring* arrangement of the SiO_4^{4-} tetrahedra into a chain structure by a reconstructive transformation as well as a displacive transformation through a simple rotation of tetrahedra. Clearly, bonds must be broken and ions must diffuse to make the reconstructive structural change but not the displacive one. As a result, reconstructive transformations take time to occur and rapid cooling below the equilibrium temperature may prevent the transformation from taking place at all. The kinetics of these transformations is controlled by a nucleation and growth process. Figure 7.10 shows the various reconstructive and displacive phase transformations that occur for pure silica, SiO_2 (Kingery et al. 1976).* In firing traditional ceramic porcelains that contain

* There are references in the literature to several other displacive transformations in both tridymite and cristobalite but without unanimous agreement (Eitel 1954; Sosman 1965; Wells 1984).

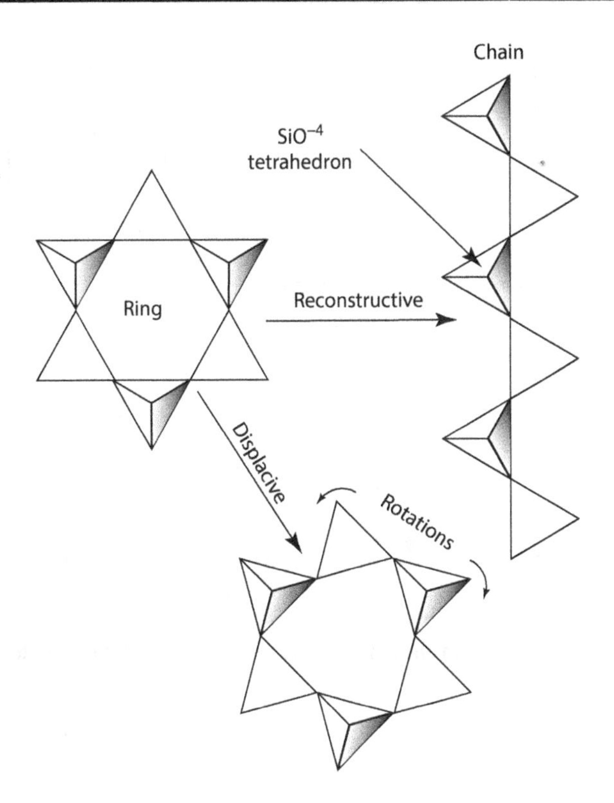

FIGURE 7.9 Schematic showing how displacive and reconstructive phase transformation take place in silicates where the SiO_4^{-4} tetrahedral share corners.

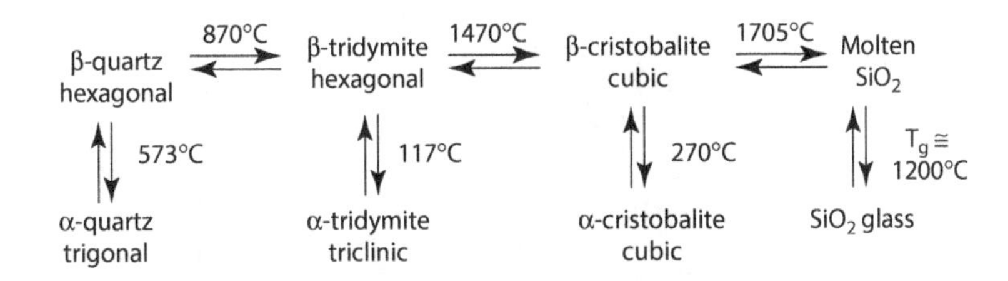

FIGURE 7.10 Phase transformations in SiO_2. The horizontal transformations are reconstructive while the vertical lines indicate displacive transformations. The glass transition temperature is included as well. (Eitel 1954, Wells 1984, Sosman 1965)

silica as a filler material, depending on the firing temperature and time—a reconstructive phase transition is involved—since quartz, cristobalite, or tridymite (or some of each) is present in the final fired piece. All of these phases undergo displacive phase transitions on cooling with volume changes that can lead to internal cracking that either lowers the strength or actually causes cracking of the fired ceramic piece (or ware).

7.5.2 Metals

Titanium goes from a hexagonal crystal structure to BCC at about 900°C. And one of the classic examples, is that of white tin—tetragonal—going to gray tin—diamond cubic below room temperature with about a 21% increase in volume (Brandes and Brook 1992). Figure 7.11 shows the calculated Gibbs energies for white and gray tin predicting a phase transformation temperature of about 25°C although the actual observed temperature is considerably lower, about 13.2°C

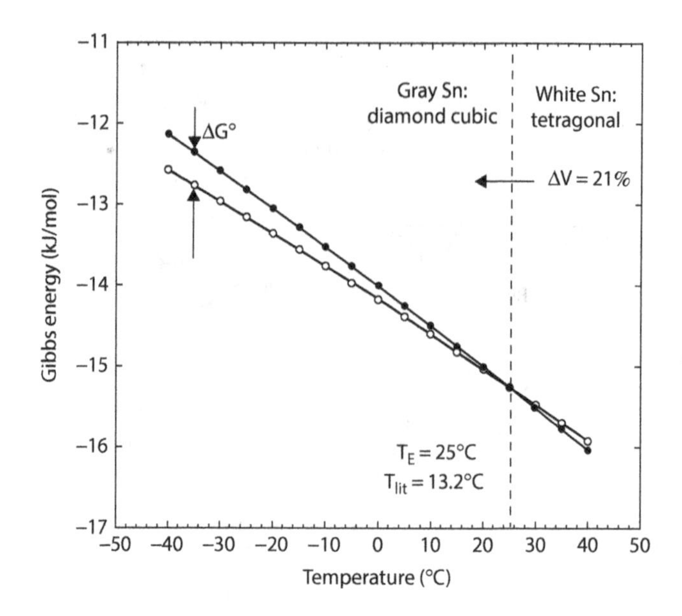

FIGURE 7.11 Free energies of gray and white tin versus temperature. (Data from Roine 2002)

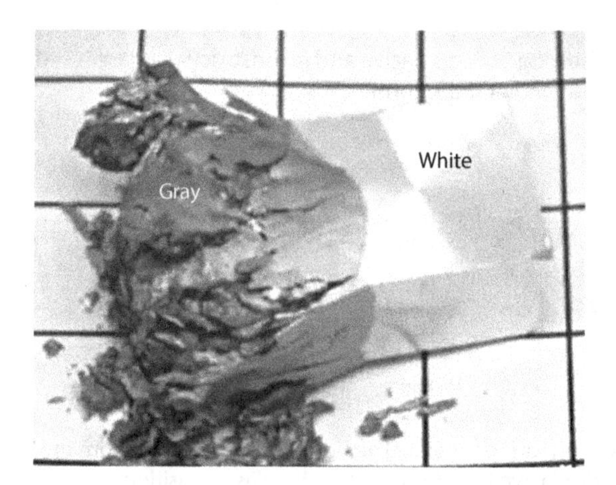

FIGURE 7.12 White tin undergoing the very destructive phase transformation to gray tin—*tin pest*—at −40°C. (Frame from YouTube video: http://www.youtube.com/watch?v=sXB83Heh3_c uploaded by http://www.periodictable.ru.)

(Roine 2002). The difference is probably caused by uncertainties in the thermodynamic data since it is very difficult to get bulk samples of gray tin to measure properties such as heat capacity necessary for accurate values of the Gibbs energy. Figure 7.12 shows how disruptive this transition can be at −40°C. It took about 20 hours for this roughly 1.5 in. long bar of white tin to completely transform—the markers are 1 cm. This loss in mechanical integrity is known as *tin pest* or *tin plague* and was observed to occur in tin buttons and on organ pipes in churches after a very cold winter in Russia in 1850 (Emsley 2001).

Tin pest has been cited—perhaps apocryphally—as one of the determining factors in the defeat of Napoleon in the Russian winter of 1812. In June of 1812, Napoleon marched into Russia with an army of about 690,000. In September of that year he entered Moscow, which, shortly thereafter was burned to the ground by the Russians in what is known as a *scorched earth policy*, denying the enemy access to captured buildings, materiel, and food. Not having shelter, the French army left

Moscow in October and started to return to France. Deprived of supplies, by December 14, 1812, the remaining 22,000 of Napoleon's troops were forced out of Russian territory (French Invasion of Russia). Moscow can become bitterly cold in the winter and while records indicate that the winter of 1812 was not that cold during most of the retreat (the temperature did not go much below about 10°C), there were periods where the temperature dropped below 0°C. This led to one of the great stories in materials science that one of the factors that impacted Napoleon's defeat in Russia was that his troops' uniform buttons were made of white tin which underwent the white–gray transition leading to degradation and powdering of the tin. As a result, the French troops were supposedly too busy holding up their pants to reload their muskets and, as a result, lost the war! This serves as another example of the critical role that materials have played—and continue to play—in world history!

This is a nice story, but not very likely to be true. It is suggested that any tin buttons made at that time were likely to have been impure and less likely to transform than pure tin (tin pest). Furthermore, the transformation takes place more rapidly the lower the temperature below the transformation temperature and during most of French army's time in Russia, the temperature was not that low for long periods of time. Perhaps even more convincing is the evidence that the buttons on the French troops were actually made of bone and those of the officers were made of brass! (Emsley 2001). In contrast, there is strong evidence that Robert Scott's expedition to the South Pole in 1910–1912 was a disaster partly because the cans of kerosene that were left behind for the return trip were found to be empty. These cans were soldered with tin and it has been suggested that tin pest caused the soldered joints to leak leaving no fuel to be used to provide heat for the expedition members who all subsequently froze (Emsley 2001; Tin Pest).

7.6 TRANSITIONS WITH PRODUCT PHASES OF DIFFERENT COMPOSITIONS

7.6.1 Nucleation and Growth Transitions

Some phase transitions discussed above occur in multicomponent solids where there is no change of composition over large distances, only a change in structure. However, many phase transformations of interest in materials science and engineering are of the type where a single phase of a given composition decomposes into two or more other phases of quite different compositions. The decomposition of austenite to ferrite plus cementite in steels is a primary example of such a system. As a result, diffusion must be one of the series steps in the transformation in order to achieve the large changes in concentrations among the phases involved. Figure 7.13 schematically depicts a two component system with a complete solid solution, α, that decomposes to γ and β at the eutectoid temperature. If an alloy of composition x_0 is rapidly cooled from the solid solution region at temperature T_1 to a temperature T_2, precipitation of β can occur having a composition x_β while α will equilibrate at composition x_α. This will occur by a nucleation and growth process that involves diffusion and possibly a surface reaction. For hard materials, at high temperatures, diffusion is likely the rate-controlling step. However, for soft materials or at low temperatures, surface reaction may very well control the overall rate of decomposition. Nucleation and growth is described in detail shortly.

7.6.2 Spinodal Decomposition

In contrast to nucleation and growth, certain multicomponent phase transitions can occur by a process called *spinodal decomposition* in which there is no sharp interface developed by the new phases, although the process is controlled entirely by diffusion. Figure 7.14 schematically shows a system that can undergo spinodal decomposition. If an alloy of composition x_0 is cooled from the γ solid solution range down to T_e and held there to allow the reaction to take place, x_0 will decompose into the equilibrium compositions x_1 and x_2. This process takes place entirely by diffusion

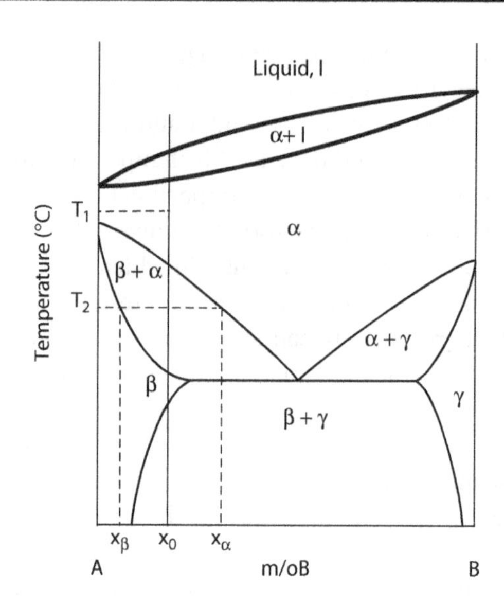

FIGURE 7.13 A schematic binary eutectoid system in which a complete solid solution, α, of composition X_0 at T_1 precipitates β of composition X_β at T_2 with accompanying change in α to X_α.

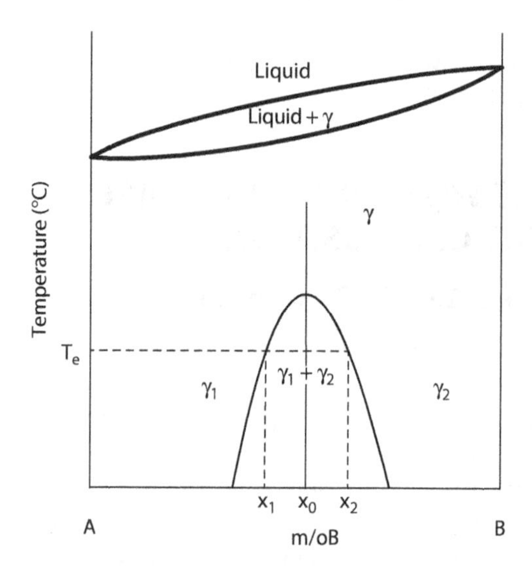

FIGURE 7.14 Schematic phase diagram in which a solid solution alloy of composition X_0 undergoes phase separation into new compositions X_1 and X_2 at temperature T_e.

with the concentration of B atoms increasing in γ_2 and the concentration of A atoms increasing in γ_2 but without the formation of an initial sharp interface between the two new phases. Since spinodal decomposition occurs entirely by diffusion in a thermodynamically nonideal system, a more complete analysis will be left to Chapter 16 after some discussion about diffusion in thermodynamically nonideal systems.

7.7 NUCLEATION AND GROWTH

7.7.1 Introduction

Since the nucleation and growth process of phase transformations can be controlled by a surface reaction, it is useful to model the process since all of the tools necessary for the development of the model have been discussed in Chapters 2, 4 and 6. It should be noted, however, that diffusion can also control the rate of a nucleation and growth process and so the process can be analyzed in terms of diffusion control as well. Diffusion control and reaction control essentially lead to similar results with minor changes in the meaning and values of some of the parameters of the process. Therefore, the nucleation and growth process is examined here assuming that surface reactions are the rate controlling steps.

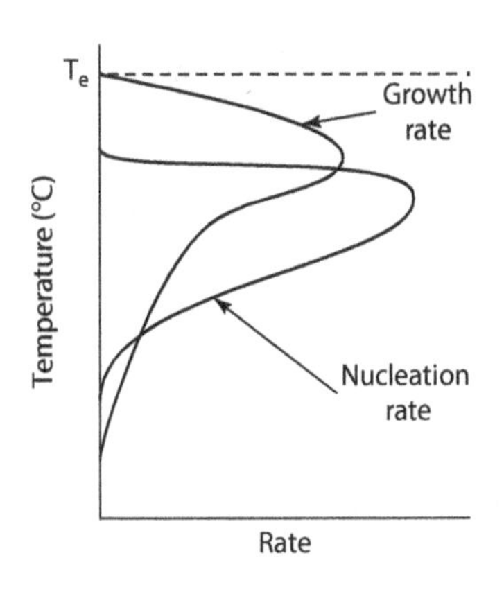

FIGURE 7.15 Schematic drawing of the rates of growth and nucleation for a nucleation and growth phase transformation as a function of temperature below the equilibrium temperature T_e where the new phase is stable.

7.7.2 Qualitative Description of Nucleation and Growth

Before going into the mathematical details of the nucleation and growth process, it is useful to examine qualitatively features of the transformation process that leads to TTT kinetics and diagrams, such as Figure 7.5, that have been developed extensively for steels. Figure 7.15 shows the rate of growth and the rate of nucleation as a function of temperature below the equilibrium temperature, T_e, where the new phase(s) is the thermodynamically stable phase. Recalling the results from Chapter 5, the rate of growth is a product of a thermodynamic term that gets larger more or less linearly as the temperature decreases below T_e. This term is multiplied by the kinetic factor that exponentially decreases with temperature because of the activation energy for the reaction. The product of the two gives the shape of the curve shown in Figure 7.15 with a maximum in the growth rate at some temperature below T_e. As will be seen, similar considerations lead to a nucleation rate that also goes through a maximum at some temperature below T_e—also the product of a thermodynamic term and a kinetic term. At T_e, the nucleation rate must be zero since the thermodynamic driving force for the transformation must be sufficiently large to overcome the interfacial energy created with the formation of the new phase as well as any strain energy generated between the new phase and the matrix phase. The overall rate of the transformation is the product of the growth rate and the nucleation rate and leads to the curve shown in Figure 7.16, which also shows a maximum at some temperature below the equilibrium temperature, T_e. By choosing three temperatures as shown, the overall percent of the transformation as a function of time can be calculated and these are shown in Figure 7.17. These *S-shaped* or *sigmoidal* curves of *percent transformed as a function of time* are characteristic of a nucleation and growth transformation. By choosing certain percent transformed at the various temperatures, the time for this percentage can be read from these curves and a TTT curve can be constructed as shown in Figure 7.18.

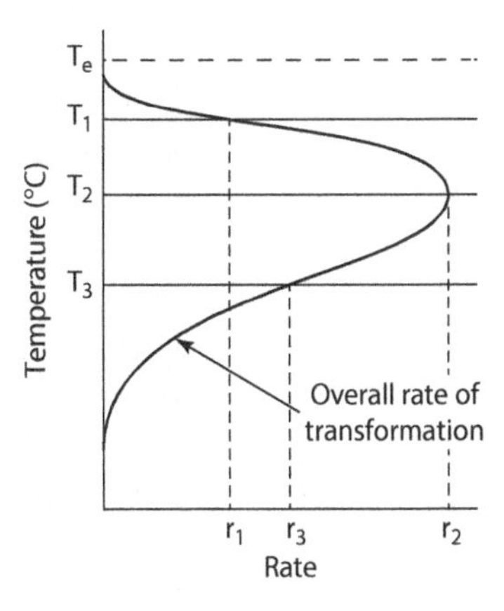

FIGURE 7.16 Schematic drawing of the overall rate of phase transformation as a function of temperature (product of the nucleation and growth rates of Figure 7.15) with three temperatures T_1, T_2, and T_3 chosen to construct a time–temperature–transformation (TTT) diagram.

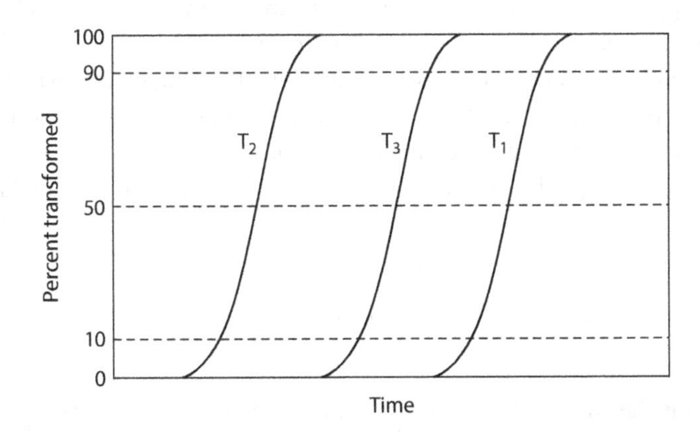

FIGURE 7.17 Percent transformed for the three temperatures of Figure 7.16: one above the maximum rate, T_1; one at the maximum rate, T_2; and one below the maximum rate, T_3.

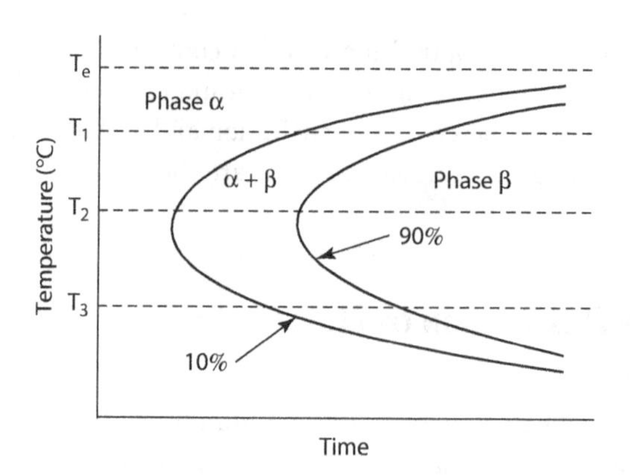

FIGURE 7.18 TTT diagram drawn from the percent transformed versus time curves of Figure 7.17.

7.7.3 Homogeneous and Heterogeneous Nucleation

Nucleation and growth processes require the formation of a stable nucleus of the new phase that then grows to complete the phase transition. *A nucleus is a stable small particle of the new phase consisting of only a small number of atoms or molecules—hundreds or so—whose Gibbs energy decreases when it grows. Homogeneous nucleation occurs when nuclei form throughout the bulk of the material* because the interfacial and strain energies of the new phase are overcome by the Gibbs energy of the transformation. *Heterogeneous nucleation occurs when the nuclei form at some surface or interface where the interfacial and strain energies of the nucleus are lower for a given size nucleus compared to what it would be if the nucleus were forming in the bulk of the material transforming.* Hopefully, this distinction between homogeneous and heterogeneous nucleation will become clearer as these two processes are explored more fully below. In reality, homogeneous nucleation is very hard to achieve because of the usual presence of interfaces or impurity particles that can serve as sites for heterogeneous nucleation. However, homogeneous nucleation is somewhat easier to understand and it is modeled first. Going from homogeneous to heterogeneous nucleation is relatively straight forward and relies on the relationships between the relative interfacial energies discussed in Chapter 6.

7.8 HOMOGENEOUS NUCLEATION

7.8.1 Critical Nucleus Size

For a nucleus of new phase to form, not only must the volume Gibbs energy, ΔG_V, be favorable for the formation of a new phase—ΔG_V must be negative—but must be sufficiently large to overcome the new interfacial energy of the nucleus (and perhaps strain energy in a solid). Below the equilibrium temperature, T_e (or above an equilibrium vapor pressure), where the energy to form the new phase is negative, atomic or molecular fluctuations within the transforming phase form *embryos* of the new phase where an *embryo is a small cluster of atoms or molecules* that form randomly and spontaneously throughout the bulk of the transforming phase. For a spherical embryo—another shape could be chosen but does not change the principles of the argument—of radius r, the Gibbs energy of the embryo is ΔG_r is

$$\Delta G_r = \frac{4}{3}\pi r^3 \Delta G_V + 4\pi r^2 \gamma \tag{7.1}$$

where the first term is the volume energy for the transformation for a spherical particle and is negative below some equilibrium temperature, T_e, since ΔG_V—the Gibbs energy per unit volume for the transformation to the new phase—is negative ($\Delta G_V = \Delta G/\bar{V}$ where ΔG is the molar energy for the

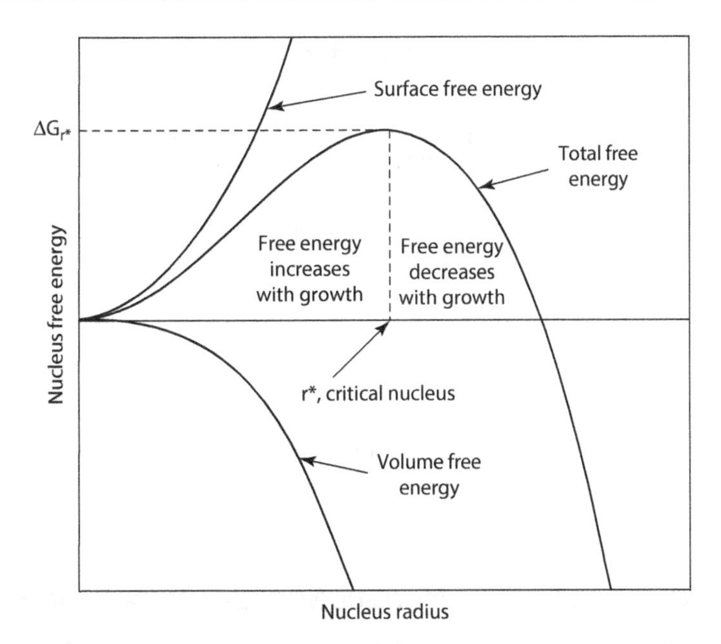

FIGURE 7.19 The energy of a nucleus, ΔG_r, as a function of the nucleus radius, r, showing: the surface term increasing with r^2, the volume term decreasing with r^3, and the net energy with a maximum at the critical nucleus size, r*.

transition and $\overline{V}$ is the molar volume of the nucleating phase). In solids, ΔG_v may contain a strain energy term determined by any differences in molar volumes and the elastic properties of the parent or matrix phase and the nucleus. In what follows, the strain energy will assumed to be zero since its value depends strongly on the specific system and the goal here is to keep the final results as general as possible. The second term in Equation 7.1 is just the surface energy associated with a spherical nucleus of radius r.

As shown in Figure 7.19, the volume energy term is negative and varies as r^3 while the surface energy term is positive and varies with r^2, where γ is the interfacial energy between the new and old phases. The resulting sum leads to a maximum in ΔG_r at the *critical nucleus size,* r*. This critical nucleus size is very important for the following reasons. If the radius of the embryo is less than the critical nucleus size, r*, then any increase in the size of the embryo leads to a net *positive Gibbs energy change,* and the embryo is unstable, cannot exist, and will disappear. However, if the embryo has a radius greater than r*, then any increase in size leads to a net *negative change in Gibbs energy* so this embryo becomes a stable nucleus for the formation of the new phase. The critical nucleus size, r*, is obviously obtained by differentiating the above equation and setting it equal to zero:

$$\frac{d\Delta G_r}{dr} = 0 = 4\pi r^{*2}\Delta G_V + 8\pi r^{*}\gamma \tag{7.2}$$

and leads to a value for the critical nucleus size of

$$r^{*} = -\frac{2\gamma}{\Delta G_V} \tag{7.3}$$

(which looks similar to the effect of surface energy of the Gibbs energy of a spherical particle that was seen in Chapter 6) and, of course, r* is positive since ΔG_V is negative. This makes intuitive sense in that high surface energies lead to a much larger critical nucleus size and, if the surface energy term is small, then the critical nucleus size is small.

It follows that, the Gibbs energy for the critical nucleus, ΔG_{r^*}, is obtained by substituting Equation 7.3 into Equation 7.1:

$$\Delta G_{r^*} = 4/3\pi r^{*3}\Delta G_V + 4\pi r^{*2}\gamma$$

$$= \frac{4}{3}\pi\left(\frac{-2\gamma}{\Delta G_V}\right)^3 \Delta G_V + 4\pi\left(\frac{-2\gamma}{\Delta G_V}\right)^2 \gamma$$

$$= -\frac{4}{3}\pi\frac{8\gamma^3}{\Delta G_V^3}\Delta G_V + 4\pi\frac{4\gamma^2}{\Delta G_V^2}\gamma$$

$$\Delta G_{r^*} = -\frac{32}{3}\frac{\pi\gamma^3}{\Delta G_V^2} + 16\frac{\pi\gamma^3}{\Delta G_V^2}$$

with the result

$$\Delta G_{r^*} = \frac{16}{3}\pi\frac{\gamma^3}{\Delta G_V^2}. \tag{7.4}$$

7.8.2 Example of the Critical Nucleus Size for Water Condensation

Consider the case of water vapor condensing as a liquid at 300 K. The reaction is

$$H_2O(g) = H_2O(l);\ \Delta G^o$$

Now, $\Delta G = \Delta G^o + RT\ \ln K$, and in this case, $K = a(H_2O)_L/p(H_2O) = 1/p(H_2O)$ where $a(H_2O)_L$ is the activity of the H_2O and of course is 1.0 for the pure liquid. At equilibrium, $\Delta G = 0$, and

$$p_e(H_2O) = p_e = e^{\frac{\Delta G^o}{RT}}$$

$$\text{or} \quad \Delta G^o = RT\ln p_e \ \text{so} \quad \Delta G = RT\ln p_e - RT\ln p$$

and

$$\Delta G = -RT\ln\left(\frac{p}{p_e}\right) \tag{7.5}$$

and dividing by the molar volume, $\bar{V}_{H_2O}$ gives ΔG_V. Putting in some numbers for water at 300 K (Haynes 2013)

$$r^* = -\frac{2\gamma}{\Delta G_V} = -\frac{2\gamma}{\frac{\Delta G}{\bar{V}}} = \frac{2\gamma\bar{V}}{RT\ln(p/p_e)}$$

$$= \frac{2(72\times10^{-3}\text{J/m}^2)(18\times10^{-6}\text{m}^3/\text{mol})}{(8.314\text{J/mol}\cdot\text{K})(300\text{K})\ln(p/p_e)}$$

leads to

$$r^* = \frac{1.04\times10^{-9}\text{m}}{\ln(p/p_e)} \tag{7.6}$$

where p/p_e = *supersaturation*, often given by the symbol S. If $p = p_e$ then the supersaturation = 1.0% or 100% humidity. Note that if the supersaturation is 1.0, then the critical nucleus is infinitely large! Table 7.1 gives the critical nucleus size and the number of H_2O molecules comprising it as a function of the supersaturation. Supersaturation for homogeneous nucleation can reach values on the order of 4.0 (Silbey and Alberty 2001) and would suggest that a nucleus at this supersaturation would contain 59 H_2O molecules and would be about 1.5 nm in diameter.

TABLE 7.1
Supersaturation of Water Vapor and Critical Water Nucleus Size

p/pe	r*(m)	Molecules
1.0001	1.040×10^{-5}	1.58×10^{14}
1.001	1.041×10^{-6}	1.58×10^{11}
1.01	1.045×10^{-7}	1.60×10^{8}
1.1	1.091×10^{-8}	1.82×10^{5}
1.2	5.704×10^{-9}	2.60×10^{4}
1.3	3.964×10^{-9}	8.73×10^{3}
1.4	3.091×10^{-9}	4.14×10^{3}
1.5	2.565×10^{-9}	2.36×10^{3}
1.6	2.213×10^{-9}	1.52×10^{3}
1.7	1.960×10^{-9}	1.06×10^{3}
1.8	1.769×10^{-9}	776
1.9	1.620×10^{-9}	596
2	1.500×10^{-9}	473
3	9.466×10^{-10}	119
4	7.502×10^{-10}	59.2
5	6.462×10^{-10}	37.8
6	5.804×10^{-10}	27.4
7	5.345×10^{-10}	21.4
8	5.001×10^{-10}	17.5
9	4.733×10^{-10}	14.9
10	4.517×10^{-10}	12.9

7.8.3 Concentration of Critical Nuclei

A chemical equation can be written for the formation of a critical nucleus of n atoms of species A:

$$nA \rightleftharpoons A_n : \Delta G_{r^*} \tag{7.7}$$

where A_n is the critical nucleus. If the assumption is made that the concentrations of atoms and critical nuclei are small so that the activities of each are the same as their mole fractions (as will be seen, this is a good assumption), an equilibrium constant can be written as

$$\frac{X_{A_n}}{X_A^n} = e^{-\frac{\Delta G_{r^*}}{k_B T}}$$

where:

X_{A_n} is the mole fraction of nuclei of size n
X_A is the mole fraction of atoms (or molecules) A.

Note that Boltzmann's constant, k_B, is used in the exponent rather than R since ΔG_{r^*} is the actual energy for just a few atoms and nuclei and not moles of them. For the sake of simplicity, consider a gas and let η_A = number of gas atoms/m^3—about 10^{25} at STP—standard temperature and pressure—(one bar and 298 K) then η_{r^*} = number of critical nuclei per m^3. Assuming $\eta_{r^*} \ll \eta_A$ (as will be seen to be true) so that:

$$X_{A_n} = \frac{\eta_{r^*}}{\eta_{r^*} + \eta_A} \simeq \frac{\eta_{r^*}}{\eta_A}$$

$$X_A = \frac{\eta_A}{\eta_{r^*} + \eta_A} \simeq \frac{\eta_A}{\eta_A} = 1.$$

Therefore,

$$\frac{\eta_{r^*}}{\eta_A} = K_e = e^{-\frac{\Delta G_{r^*}}{k_B T}}$$

or the concentration of critical nuclei, η_{r^*}, is

$$\eta_{r^*} = \eta_A e^{-\frac{\Delta G_{r^*}}{k_B T}}. \tag{7.8}$$

7.8.4 Rate of Nucleation

7.8.4.1 Limitations

Modeling of the nucleation rate is a little more difficult and, although there have been many attempts at doing this accurately over the past 100 years, the results are still somewhat unsatisfactory in that the models frequently differ from experiment by large amounts. Furthermore, the values of experimental parameters such as the accommodation coefficient, α, discussed in Chapter 5, may vary considerably in the literature. Nevertheless, the main methodology of these approaches is illustrated to give nucleation rates that are reasonable if not precise. The fact that the model does not predict exact numerical results is not critical. It is the modeled *behavior* of the nucleation rate, and the concepts involved, as a function of the experimental parameters, such as the supersaturation or supercooling leading to the formation of a new phase, that is instructive and essentially correct.

7.8.4.2 Nucleation of a Liquid or Solid from a Gas

A very important conceptual point is that a *critical nucleus will grow and become a stable nucleation site with the addition of only a* single atom or molecule *since the addition of the single atom or molecule to the critical nucleus will lead to a lower Gibbs energy* (Figure 7.19). Therefore, the nucleation rate depends on how fast gas atoms can attach themselves to critical nuclei to make them grow and generate the phase transformation. It should be noted that several different symbols are used for nucleation rate in the literature including I, J, and N, among others. For the sake of being different and not using the same letter or symbol for a number of entities, here the nucleation rate is given by Θ, capital theta. For the case of a gas forming liquid or solid nuclei, this process can be approximated by (Reed-Hill and Abbaschian 1992)

$$\Theta = J' A_{r^*} \eta_{r^*} \tag{7.9}$$

where:

Θ is the nucleation rate with Units(Θ) nuclei/m^3-s
J' is the atomic or molecular flux.

Units(J') = atoms/m^2-s and $J' = N_A J$ where J is the molar flux density, Units(J) = mol/m^2-s that was modeled earlier for first order gas–solid (or liquid) surface controlled reactions; A_{r^*} = the area of the critical nucleus Units(A_{r^*}) = m^2; and η_{r^*} = the number of critical nuclei per unit volume, Units(η_{r^*}) = m^{-3}, Equation 7.8. As was as shown in Chapter 5 for condensation from a gas

$$J' = \frac{\alpha(p - p_e)}{\sqrt{2\pi m k_B T}} \tag{7.10}$$

where:

α is the accommodation or *sticking* coefficient
p is the pressure of the condensing gas
p_e is the equilibrium pressure of the condensing gas.

If $\alpha < 1.0$, this implies some sort of surface reaction must occur for the atoms or molecules to stick to the surface as discussed in Chapter 5.

7.8.4.3 Nucleation of a Liquid from a Supersaturated Gas

Putting everything together, the rate of nucleation is given by

$$\Theta = \left(\frac{(p - p_e)}{\sqrt{2\pi m k_B T}}\right)\alpha\left(4\pi r^{*2}\right)\eta e^{-\frac{\Delta G_{r^*}}{k_B T}} \tag{7.11}$$

Or

$$\Theta = \Theta_0 e^{-\frac{\Delta G_{r^*}}{k_B T}} \tag{7.12}$$

where

$$\Theta_0 = \left(\frac{(p - p_e)}{\sqrt{2\pi m k_B T}}\right)\alpha\left(4\pi r^{*2}\right)\eta$$

and, from above,

$$r^* = -\frac{2\gamma}{\Delta G_V} \quad \text{and} \quad \Delta G_{r^*} = \frac{16}{3}\pi\frac{\gamma^3}{\Delta G_V^2}.$$

7.8.4.4 Example: Nucleation of Water Drops from Gas at Equilibrium at 300 K by Supersaturation

The reaction to form water droplets from a supersaturated atmosphere is

$$H_2O(g) = H_2O(l).$$

As mentioned in Section 7.8.2, homogeneous nucleation is observed when the supersaturation, S, $S = p/p_e$ is about 4. At 300 K, $p_e(300\text{ K}) = 3.53 \times 10^{-2}$ bar (Roine 2002). From Chapter 6, $\gamma(H_2O) = 72\text{ mJ/m}^2$; and the molar volume, $\bar{V} = 18\text{cm}^3/\text{mol}\left(1/10^6\text{ m}^3/\text{cm}^3\right) = 18 \times 10^{-6}\text{ m}^3/\text{mol}$. The value for the accommodation coefficient for water and water vapor is somewhat uncertain and varies by about an order of magnitude or more (Pound 1972; Davidovits et al. 2004). The most likely value is some place between 0.4 and 1.0, so for the sake of calculation, a value of $\alpha \cong 0.7$, is used here. Some additional assumptions are made that simplify the calculations and do not make a major difference in the final result. Making the calculation for ΔG_V to calculate Θ_0, a constant supersaturation of $S = 4.0$ is used so that:

$$\Delta G_V = -\frac{RT\ln(p/p_e)}{\bar{V}}$$

$$\Delta G_V = -\frac{(8.314)(300)\ln 4}{18 \times 10^{-6}} \tag{7.13}$$

$$\Delta G_V = -1.92 \times 10^8\text{ J/m}^3$$

and for the critical nucleus radius,

$$r^* = -\frac{2\gamma}{\Delta G_V}$$

$$r^* = \frac{2\left(72 \times 10^{-3}\right)}{1.92 \times 10^8} \tag{7.14}$$

$$r^* = 7.5 \times 10^{-10}\text{m}$$

which, from Table 7.1, is about 60 water molecules. It should be noted that the increase in the equilibrium pressure of water over drops this small, Chapter 6, has been neglected as well. But here again, this will not make a major difference the result. The area of the critical nucleus is then

$$A_{r^*} = 4\pi r^{*2} = 4(3.1416)(7.5\times10^{-10})^2$$
$$A_{r^*} = 7.07\times10^{-18}\,m^2. \tag{7.15}$$

The number of H_2O molecules in the gas phase, η is given by

$$\eta = \frac{N_A p_e S}{RT}$$
$$\eta = \frac{(6.022\times10^{23}\,atoms/mol)(3.53\times10^{-2}\,bar)(10^6\,cm^3/m^3)(4)}{(83.14\,cm^3\cdot bar/mol\cdot K)(300K)} \tag{7.16}$$
$$\eta = 3.45\times10^{24}\,H_2O\ molecules/m^3$$

and the Gibbs energy for the critical nucleus,

$$\Delta G_{r^*} = \frac{16}{3}\pi\frac{\gamma^3}{\Delta G_V^2}$$
$$\Delta G_{r^*} = \frac{16}{3}(3.1416)\frac{(72\times10^{-3}\,J/m^2)^3}{(1.92\times10^8\,J/m^3)^2} \tag{7.17}$$
$$\Delta G_{r^*} = 1.69\times10^{-19}\,J$$

and, finally, the value of Θ_0,

$$\Theta_0 = \left(\frac{(p-p_e)}{\sqrt{2\pi m k_B T}}\right)\alpha A_{r^*}\eta$$
$$\Theta_0 = \left(\frac{(4-1)(3.53\times10^{-3}\,bar)(10^5\,Pa/bar)}{\sqrt{2(3.1416)\left(\frac{18\times10^{-3}\,kg/mol}{6.022\times10^{23}\,atoms/mol}\right)(1.38\times10^{-23}\,J/atom\cdot K)(300K)}}\right) \tag{7.18}$$
$$\times(0.7)(7.07\times10^{-18}\,m^2)(3.45\times10^{24}\,atoms/m^3)$$
$$\Theta_0 = 1.137\times10^{34}\,nuclei/m^3\cdot s$$

leading to a nucleation rate of

$$\Theta = \Theta_0 e^{-\frac{\Delta G_{r^*}}{k_B T}}$$
$$\Theta = (1.137\times10^{34})e^{\frac{-1.6\times10^{-19}}{(1.38\times10^{-23})(300)}} \tag{7.19}$$
$$\Theta = 2.12\times10^{16}\,nuclei/m^3\cdot s.$$

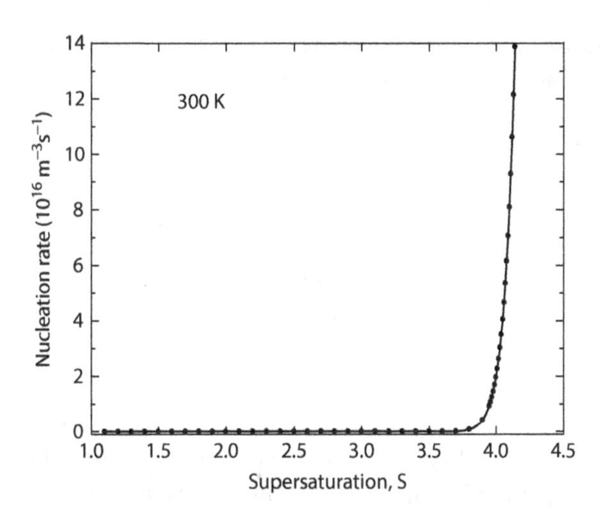

FIGURE 7.20 Nucleation rate of water drops at 300 K as a function of the supersaturation.

The nucleation rate as a function of supersaturation is shown in Figure 7.20 where ΔG_V was calculated as a function of supersaturation, S, from,

$$\Delta G_V = -\frac{RT\ln(p/p_e)}{\bar{V}} = -\frac{RT\ln S}{\bar{V}}. \tag{7.20}$$

Notice in Figure 7.20 how rapidly the rate of nucleation rises with supersaturation. Also note that the number of nuclei per unit volume is much smaller (≈8 orders of magnitude) than that of the gas atoms, $\eta = 3.45 \times 10^{24}$ m^{-3}. Finally, even such low concentrations of nuclei, ≈ 10^{16} m^{-3} implies that they are roughly only 5 μm distant from each other!

7.8.4.5 Example: Nucleation of Water Drops from Gas at Equilibrium at 300 K by Supercooling

The Gibbs energy for the reaction $H_2O(g) = H_2O(l)$ can be written:

$$\Delta G = \Delta H_{vap} - T\Delta S_{vap} \tag{7.21}$$

where subscript *vap* refers to the enthalpy and entropy of vaporization. Now, if the supercooling is not too large (so the enthalpy doesn't change much over the temperature range) then this equation can be written:

$$\Delta G = \Delta H_{vap} - T\frac{\Delta H_{vap}}{T_E}$$

$$= \frac{\Delta H_{vap}}{T_E}(T_E - T)$$

where $T_E - T$ is the degree of *supercooling* and leads to the volume Gibbs energy for the transformation:

$$\Delta G_v = \frac{\Delta H_{vap}(T_E - T)}{T_E\bar{V}} \tag{7.22}$$

where going from the gas to the liquid ΔH_{vap} is negative and, of course $T < T_E$. For water that is at equilibrium at 300 K (i.e., the water vapor pressure at 300 K is the equilibrium pressure for water and water vapor, S = 1 at 300 K), this leads to a supercooling of about 25°C or down to about 275. Again, to simplify the equation, the factors in Θ_0, and Θ_0 itself, are calculated assuming this degree of undercooling. The factors in Θ_0 are calculated as before for a fixed supersaturation of S = 4, with the only exception was that ΔG_V calculated from the supercooling from Equation 7.22 with $\Delta H_{vap} =$ 45 kJ/mol (Haynes 2013) and $\alpha = 0.7$ again at $\Delta T = 25$ K,

$$\Delta G_v = \frac{\Delta H_{vap}(T_E - T)}{T_E \bar{V}}$$

$$\Delta G_v = \frac{(45,000)(25)}{300(18\times10^{-6})} \tag{7.23}$$

$$\Delta G_v = 2.08\times10^8\ J/m^3.$$

Calculation of r* as in Equation 7.14 gives $r^* = 6.92\times10^{-10}$m and so from Equation 7.15 $A_{r^*} = 5.96 \times 10^{18}\ m^2$. Likewise, from Equation 7.16 $\eta = 8.63\times10^{23}$ molecules m^{-3} which is a factor of four smaller than in the case of supersaturation that was estimated to be about four. However, for supercooling to about 275 K, the equilibrium water vapor at this temperature is $p_e(H_2O) \cong 7\times10^{-3}$ atm which is about 0.2 of the equilibrium pressure at 300 K. So for all practical purposes, $p(300) - p(275) \cong 0.8p(300) \cong p(300) = 3.53\times10^{-2}$ atm in Θ_0. These values give from Equation 7.18 $\Theta_0 = 5.99\times10^{32}$. nuclei/m³·s. Note that Θ_0 does not include its variation with temperature. The reality is that this can be ignored given all the other approximations in the calculations. For example, the fractional difference in Θ_0 between 300 and 275 K is roughly, $(275/300)^{3/2} \cong 0.9$ or only about a 10% difference that is clearly negligible given the other uncertainties in the terms of Θ_0, particularly the value of the accommodation coefficient, α. And finally, from Equation 7.17 $\Delta G_{r^*} = 1.45\times10^{-19}$ J which gives a nucleation rate at 275 K from Equation 7.19 of $\Theta = 1.52\times10^{16}$ nuclei/m³·s. The nucleation rate versus temperature is plotted in Figure 7.21 with the temperature plotted vertically to correspond to the earlier plots during the qualitative discussion of nucleation and growth processes. It should be noted that since the accommodation coefficient, α, is assumed constant, the nucleation rates in Figures 7.26 and 7.27 constantly increase. However, if the accommodation coefficient were strongly temperature dependent because of the nature of the surface reaction, that is, $\alpha = \alpha_0 \exp(-Q/RT)$, then the nucleation rate would go through a maximum as the temperature decreases as shown in Figure 7.15.

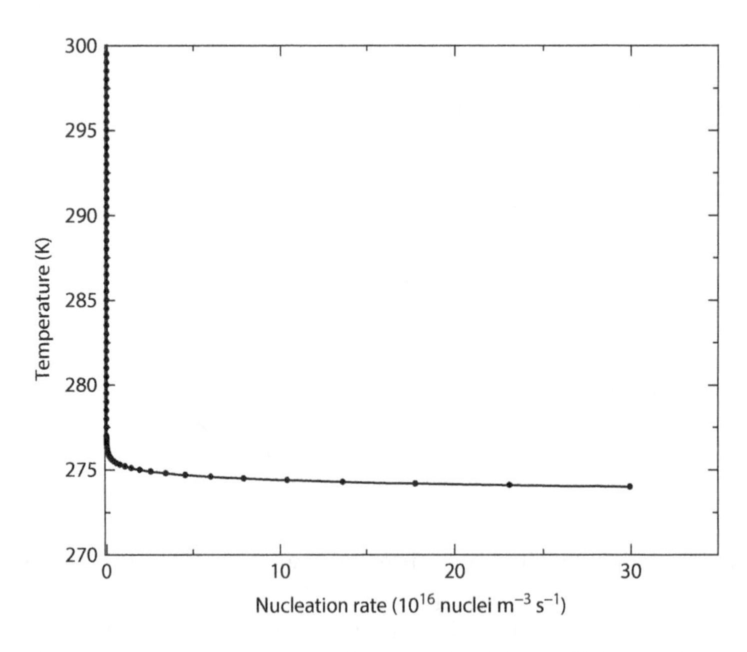

FIGURE 7.21 Nucleation rate versus temperature for water that is in equilibrium with its vapor at 300 K.

7.8.5 Homogeneous Nucleation of Solids from Liquids and Solids from Solids

For the nucleation of solids from liquids or solids from solids everything is essentially the same as described above for gas condensation to liquids (or solids) with the major difference in the terms in, and the more difficult interpretation of, Θ_0. Here, Θ_0 is determined by something close to a diffusion process since the atoms neighboring the critical nucleus only have to move one interatomic distance. Nevertheless, Θ_0 in these cases could be interpreted similar to the above gas for gases, Equation 7.11; for example,

$$\Theta = \left(\frac{1}{4}\bar{v}\eta\right)\alpha\left(4\pi r^{*2}\right)\eta e^{-\frac{\Delta G_r^*}{k_B T}} \tag{7.24}$$

where η is now the number of atoms per unit volume in the solid or liquid and all the other terms are defined as above with the exception of the mean speed, $\bar{v}$. The atom speed, $\bar{v}$ can be interpreted as the speed of an atom moving roughly one interatomic distance from the surrounding liquid or solid to the nucleating solid. This is kind of like a diffusion process as will be seen in Chapter 8 in that the speed could be written as

$$\bar{v} = \frac{\text{distance}}{\text{time}} \cong a\left(fe^{-\frac{Q}{RT}}\right)$$

where:

- a is the jump distance, roughly equal to an interatomic distance
- f is the frequency or how often a neighboring atom tries to jump and is essentially equal to the atomic vibration frequency, $f \cong 10^{12}\,s^{-1}$
- Q is some activation energy the atom must have to make the jump from the liquid/solid to the solid nucleus.

So Equation 7.24 could be written,

$$\Theta = \left(\frac{1}{4}a\eta\right)\left(fe^{-\frac{Q}{RT}}\right)\alpha\left(4\pi r^{*2}\right)\eta e^{-\frac{\Delta G_r^*}{k_B T}}$$

and $a\eta \cong \eta^{2/3}$ roughly the number of atoms per unit area surrounding the critical nucleus. As a result, for solid nucleation,

$$\Theta_0 \cong \left(\frac{1}{4}\eta^{2/3}\right)\left(fe^{-\frac{Q}{RT}}\right)\alpha\left(4\pi r^{*2}\right)\eta. \tag{7.25}$$

If the nucleation is controlled by a surface reaction, α could also be exponentially temperature dependent as well; if diffusion-controlled, $\alpha \cong 1.0$. In either case, Θ_0 will be exponentially temperature dependent and the nucleation rate will go through a maximum as shown in Figure 7.15 for the nucleation of solids.

7.8.6 Overall Rate of Phase Transformation

Before discussing *heterogeneous* nucleation it is useful to model the relationships between nucleation and growth rates and the resulting overall rate of phase transformation. As described later, the results are the same whether the nucleation is homogeneous or heterogeneous the only difference being the rate of nucleation in the final expression. Figure 7.22 schematically depicts an initial volume V of phase alpha, α, undergoing a phase transformation to phase beta, β. The resulting model holds regardless whether the transformation is gas to liquid or solid, liquid to solid, or solid to solid as long as the transition is a nucleation and growth process.

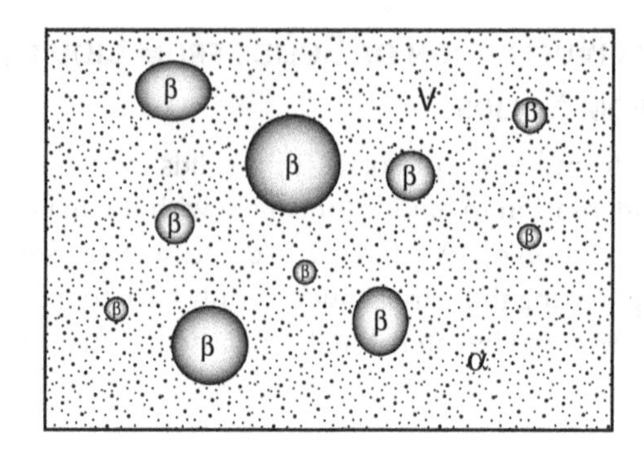

FIGURE 7.22 Schematic showing a volume V of phase α transforming to phase β by a nucleation and growth process.

Referring to Figure 7.22, let n_β = the number of β particles in V so that

$$dn_\beta = \Theta V_\alpha d\tau$$

where:

Θ is the nucleation rate
dτ is a small change in time.

Now the volume of a single β particle, v_β, is simply

$$v_\beta = \frac{4}{3}\pi a^3 = \frac{4}{3}\pi\left(G(t-\tau)\right)^3$$

where:

a is the particle radius
G is the growth rate = da/dt = constant (surface reaction)
t-τ is the time the particle has had to grow since it was nucleated at time = τ.

Therefore, the total change in the volume of the new phase β is given by

$$dV_\beta = v_\beta dn_\beta = \frac{4}{3}\pi V_\alpha \Theta G^3 (t-\tau)^3 d\tau.$$

But $V_\alpha = V - V_\beta$ so that

$$\frac{dV_\beta}{V_\alpha} = \frac{dV_\beta}{V - V_\beta} = \frac{d(V_\beta/V)}{1-(V_\beta/V)} = \frac{df_t}{1-f_t}$$

where f_t is the volume fraction transformed. Combining these last two equations gives the differential equation that needs to be solved:

$$\frac{df_t}{1-f_t} = \frac{4}{3}\pi\Theta G^3 (t-\tau)^3 d\tau.$$

This is integrated to give—assuming Θ and G are constant:

$$-\ln(1-f_t) = \frac{4}{3}\pi\Theta G^3 \int_0^t (t-\tau)^3 d\tau.$$

Let $z = t - \tau$ so $z^3 = (t - \tau)^3$, $dz = -d\tau$, and when $\tau = 0$, $z = t$, and when $\tau = t$, $z = 0$. Therefore, the equation becomes

$$\ln(1-f_t) = -\frac{4}{3}\pi\Theta G^3 \int_t^0 z^3(-dz)$$

$$= -\frac{4}{3}\pi\Theta G^3 \int_0^t z^3 dz$$

$$\ln(1-f_t) = -\frac{1}{3}\pi\Theta G^3 t^4$$

and leads to

$$f_t = 1 - e^{-\frac{1}{3}\pi\Theta G^3 t^4}. \tag{7.26}$$

which gives the fraction reacted as a function of time with the nucleation and growth rates as parameters. This is known as the Johnson–Mehl–Avrami–Kolmogorov (JMAK) equation having been derived in different fields of science and engineering: Robert Mehl was a metallurgist and one-time head of the then Metallurgy Department at what is now Carnegie Mellon University. This leads to the *S-shaped* or sigmoidal curves discussed in Section 7.7.2 and shown in Figure 7.23. From these curves, as was shown earlier, TTT curves can be obtained for the phase transformation. Note that for small values of the exponent since $e^{-y} \approx 1 - y$, so

$$f_t \cong 1 - \left(1 - \frac{1}{3}\pi\Theta G^3 t^4\right) \approx \frac{1}{3}\pi\Theta G^3 t^4 \tag{7.27}$$

or for short times or small fractions transformed, $f_t \propto t^4$.

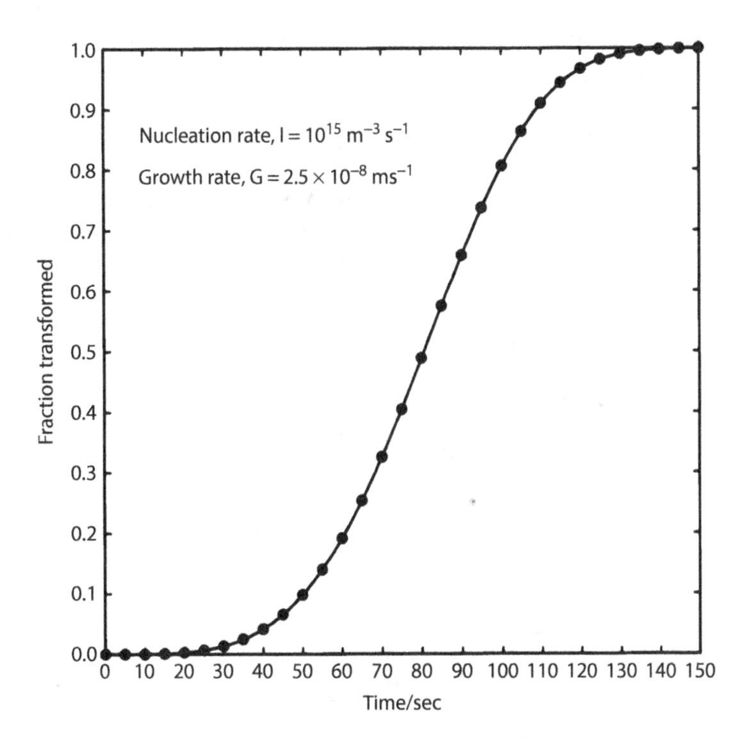

FIGURE 7.23 Fraction transformed as a function of time calculated for the growth and nucleation rates indicated.

7.9 HETEROGENEOUS NUCLEATION

7.9.1 Examples

Heterogeneous nucleation occurs much more readily and much more often than homogeneous nucleation. Obvious examples are the formation of dew and frost long before any fog or ice crystals form in the air. From the ice crystal or frost patterns that form on the surface of an automobile, for example, it is clear that the ice crystals are forming on dirt particles, wipe marks, and other artifacts on the surface. Another common example is the formation of CO_2 bubbles on the surface of a glass of beer, champagne, or soda. The bubbles emanate from specific sites on the glass that are either small surface defects—cracks or flaws—in the glass or—hopefully not—dirt particles. Heterogeneous nucleation is used in *cloud seeding* in which silver iodide, AgI, smoke particles are used to nucleate raindrops during drought periods. AgI has the hexagonal wurtzite, ZnS, structure with each anion and cation surrounded by four nearest neighbors satisfying each atom's four covalent bonds in ABAB packing. The wurtzite structure is analogous to the hexagonal diamond structure—that exists under special conditions—in which both Ag and I are replaced by a C atom covalently bonded to four other C atoms. This is similar to the crystal structure of *hexagonal ice* (ice I), the stable crystalline form at one bar pressure and zero centigrade. The idea is to have nuclei of ice form heterogeneously on the AgI particles which later grow to serve as nuclei for raindrops. The similar crystal structures of hexagonal ice and hexagonal AgI minimize the interfacial energy between the ice nucleus and AgI particle, enhancing the nucleation by a heteroepitaxy process.

7.9.2 Modeling Heterogeneous Nucleation

The analysis of heterogeneous nucleation is a straightforward extension of homogeneous nucleation but with a reduction in the overall interfacial energies since the nucleus forms on/at another surface, as shown schematically in Figure 7.24. This figure shows a partial wetting of the new phase on the heterogeneous nucleus surface. For the sake of concreteness, assume that the new phase is liquid and the heterogeneous nucleus is solid but this does not restrict the model to only this case: it could also be liquid on liquid, solid on solid, liquid or solid on a free surface, and so on. Now the Gibbs energy of the nucleating phase, ΔG_{het}, is

$$\Delta G_{het} = V_L \Delta G_V + A_{LV}\gamma_{LV} + A_{LS}(\gamma_{LS} - \gamma_{SV}) \tag{7.28}$$

where V_L = the volume of spherical liquid cap, and the A_i are the areas of interest and the interfacial energies are those discussed in Chapter 6. Also, from Chapter 6, equating the x-components of the surface energies:

$$\gamma_{SV} = \gamma_{LS} + \gamma_{LV}\cos\theta.$$

From the geometry in Figure 7.30,

$$A_{LS} = \pi r^2 \sin^2\theta$$

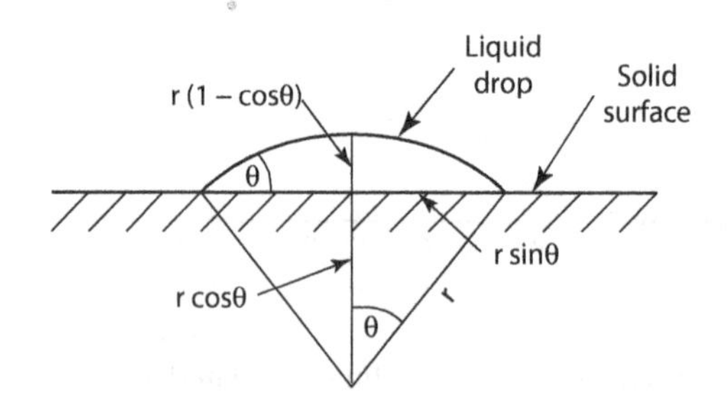

FIGURE 7.24 Spherical cap of the heterogeneously nucleating phase forming on the free surface of a solid heterogeneous nucleation site that shows partial wetting of the surface by the new phase.

and

$$A_{LV} = A_{cap} = 2\pi r^2 (1 - \cos\theta)$$

$$V_L = V_{cap} = \frac{\pi}{3} r^3 (2 - 3\cos\theta + \cos^3\theta)$$

which are taken from the literature (Lipschutz et al. 2012). (But could easily be obtained by integration of the surface area and volume of the spherical cap.) Putting in these areas and volumes into the equation for ΔG_{het}, Equation 7.28

$$\Delta G_{het} = \frac{\pi}{3} r^3 (2 - 3\cos\theta + \cos^3\theta)\Delta G_V + 2\pi r^2 (1 - \cos\theta)\gamma_{LV} + \pi r^2 \sin^2\theta(\gamma_{LS} - \gamma_{SV}). \tag{7.29}$$

Notice that if $\theta = 180°$, $\cos\theta = -1$, and $\sin\theta = 0$, there is no wetting, the spherical cap now becomes a spherical drop resting on the surface and the equation for ΔG_{het} becomes

$$\Delta G_{het} = \frac{4}{3}\pi r^3 \Delta G_V + 4\pi r^2 \gamma_{LV}$$

which is the equation for a spherical drop used for homogeneous nucleation. This is, of course, what it should be since with $\theta = 180°$ there is no wetting and reduction of surface energy by the presence of the solid particle and only homogeneous nucleation will take place. Substitution of the relationships between the interfacial energies, gives

$$\Delta G_{het} = \frac{\pi}{3} r^3 (2 - 3\cos\theta + \cos^3\theta)\Delta G_V + 2\pi r^2 (1 - \cos\theta)\gamma_{LV} + \pi r^2 \sin^2\theta(-\gamma_{LV}\cos\theta).$$

The rest is just algebra. Combination of the last two terms gives

$$2\pi r^2 (1 - \cos\theta)\gamma_{LV} + \pi r^2 \sin^2\theta(-\gamma_{LV}\cos\theta)$$

$$= \pi r^2 (2 - 2\cos\theta - \cos\theta\sin^2\theta)\gamma_{LV}$$

$$= \pi r^2 (2 - 2\cos\theta - \cos\theta(1 - \cos^2\theta))\gamma_{LV}$$

$$= \pi r^2 \gamma_{LV} (2 - 3\cos\theta + \cos^3\theta)$$

and finally

$$= 4\pi r^2 \gamma_{LV} \frac{(2 - 3\cos\theta + \cos^3\theta)}{4} \tag{7.30}$$

while the first term can be written as

$$\frac{4\pi}{3} r^3 \Delta G_V \frac{(2 - 3\cos\theta + \cos^3\theta)}{4}. \tag{7.31}$$

Combining Equations 7.30 and 7.31 gives the *final*, and desired result:

$$\Delta G_{het} = \left\{\frac{4\pi}{3} r^3 \Delta G_V + 4\pi r^2 \gamma_{LV}\right\}\left\{\frac{(2 - 3\cos\theta + \cos^3\theta)}{4}\right\}. \tag{7.32}$$

The first term is simply the Gibbs energy for the homogeneous nucleus and the second term depends on the extent of wetting, θ. Therefore, as Equation 7.32 shows, the energy for the heterogeneous nucleus is the same as for the homogeneous nucleus times a function of wetting angle:

$$\Delta G_{het} = \Delta G_{hom} f(\theta)$$

where

$$f(\theta) = \frac{\left(2 - 3\cos\theta + \cos^3\theta\right)}{4}.$$

As a result, the previous discussion about the critical radius and the implications for nucleation rate and the overall rates of transformation follow the exact same procedure as for homogeneous nucleation. The only difference is now the dependence of the wetting angle, f(θ), which is just a number for a given system that varies between zero and one. Figure 7.25 shows how f(θ) varies with theta and Figure 7.26 shows how the resulting nucleation rate varies with supersaturation or supercooling. Of course, if $\theta \to 0$, $f(\theta) \to 0$ and there is no barrier to nucleation and it proceeds without any supersaturation or supercooling. On the other hand, if $\theta \to 180°$, $f(\theta) \to 1$ and nucleation proceeds by homogeneous nucleation. Similar relations can be developed for other situations of heterogeneous nucleation such as a new phase forming at grain boundaries in a polycrystalline material. In this case, as was discussed in Chapter 6, the shape of the particle at the grain boundary is just two of the spherical caps shown in Figure 7.24 back to back. So the areas, volumes, and total interfacial energies are somewhat different but lead to exactly the same result: $\Delta G_{het} = \Delta G_{hom} f(\theta)$ the only difference being that, now

$$f(\theta) = \frac{\left(2 - 3\cos\theta + \cos^3\theta\right)}{2}$$

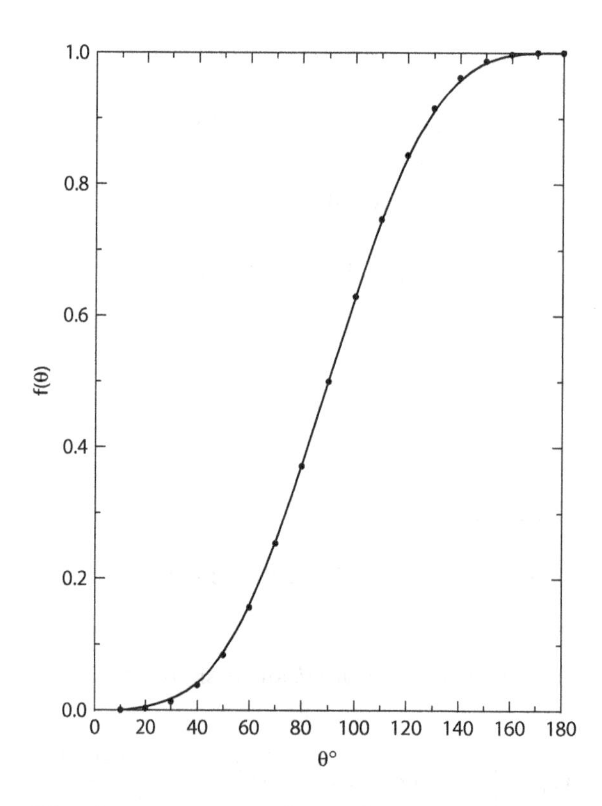

FIGURE 7.25 f(θ) versus θ for a heterogeneous nucleus with a wetting angle of θ as in Figure 7.24.

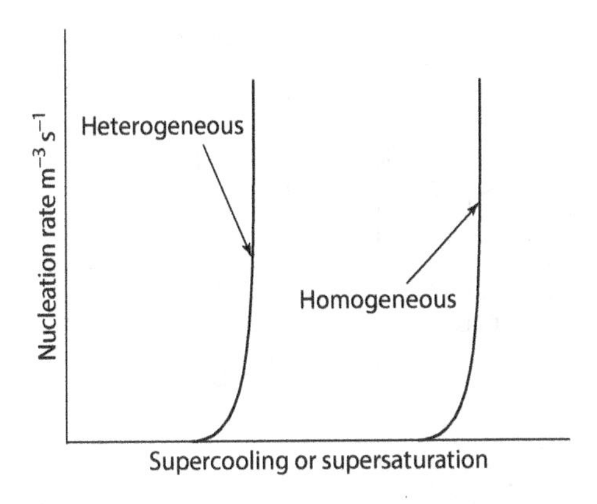

FIGURE 7.26 Effect of heterogeneous nucleation on the nucleation rate.

where θ is one-half of the dihedral angle, φ, as shown in Figure 6.16 of Chapter 6 and can only vary from 0° to 90° and at 90° there is no wetting of the grain boundary and homogeneous nucleation proceeds. In most real systems, there is at least partial wetting of the grain boundary by second phase particles and precipitation of a new phase in polycrystalline solids typically takes place more readily at grain boundaries.

7.10 DISPLACIVE TRANSFORMATIONS

7.10.1 General Concepts

In a typical displacive transformation, a crystalline solid transforms to a crystalline solid of exactly the same composition but with a different crystal structure simply by shifting atoms, ions, or molecules to a different position within a unit cell, distorting it, and changing the cell's crystallography as shown schematically in Figure 7.27. Such a transformation does not require movement of atoms, ions, or molecules by diffusion and therefore *cannot* be prevented by cooling rapidly through the phase transition temperature since there is nothing to restrict the small atom or molecule motion within the unit cell. As a result, they are also called *diffusionless transformations.* In fact, displacive transformations are *cooperative* transformations—sometimes called *military transformations*—in that atoms in neighboring unit cells essentially all move together mutually reinforcing their tendency to move causing the interface between the transformed and untransformed phase moving as fast as the speed of sound (Balluffi et al. 2005). These transformations are also frequently referred to as *martensitic* transformations after the FCC to body-centered tetragonal (BCT) martensite transformation that occurs at low temperatures in steels that gives a very high hardness to steel.*

One feature of diffusionless transitions is the twinning† that occurs in the transformed phase to minimize the strain energy developed due to the difference in crystallography and volume between the two phases: Figure 7.28 illustrates this. Figure 7.29 is a transmission electron micrograph (Chiang 1988) of a grain of $BaTiO_3$ with 0.05 mole additive of (Bi_2O_3 + Nb_2O_5) that form a solid solution with the $BaTiO_3$ and prevent the cubic to tetragonal phase change, discussed below. In this figure, the additives have diffused from the grain boundary into the $BaTiO_3$ grain around its edges keeping it cubic at room temperature. However, the center of the grain is still almost pure $BaTiO_3$ and has undergone the cubic to tetragonal phase transition with the attendant twinning. The other features in the micrograph are stacking faults (determined by changing the diffraction conditions in the microscope and observing the changes in contrast) that consist of a missing plane in a sequence ABC|BCABC...

* Martensite is named after the German metallurgist Adolf Martens, 1850–1914.

† One way to represent a twin is the plane stacking sequence in a cubic-close-packed structure. Instead of ABCABCABC... stacking it becomes ABC|BCABCABC|BCABC...where the vertical lines represent the twin boundaries.

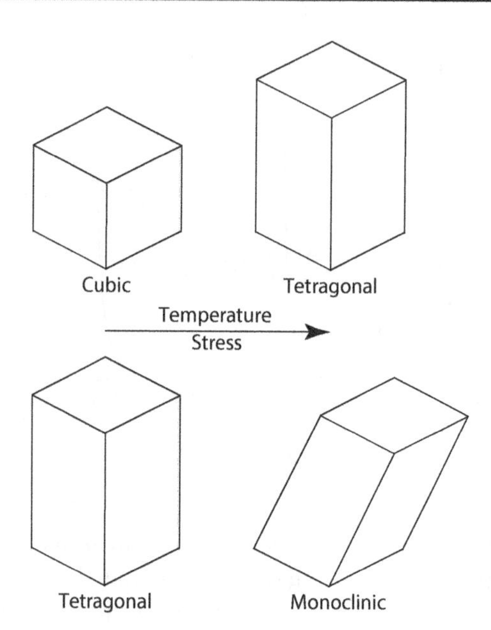

FIGURE 7.27 Schematic of diffusionless or displacive phase transformations: cubic ⇔ tetragonal ⇔ monoclinic structures.

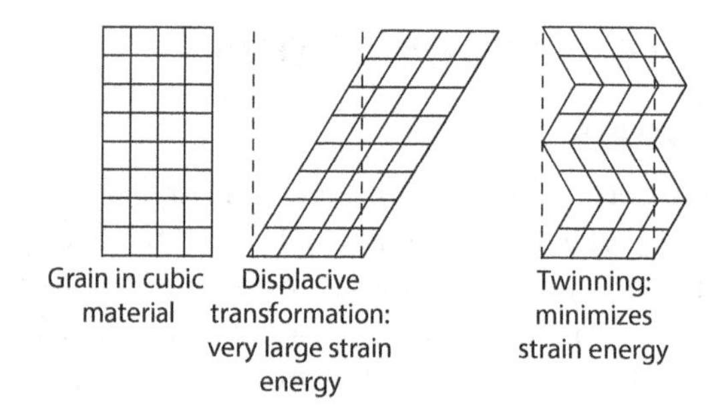

FIGURE 7.28 Cubic material undergoing a displacive transformation. To minimize the strain energy associated with the transformation, the transformed grain forms twins.

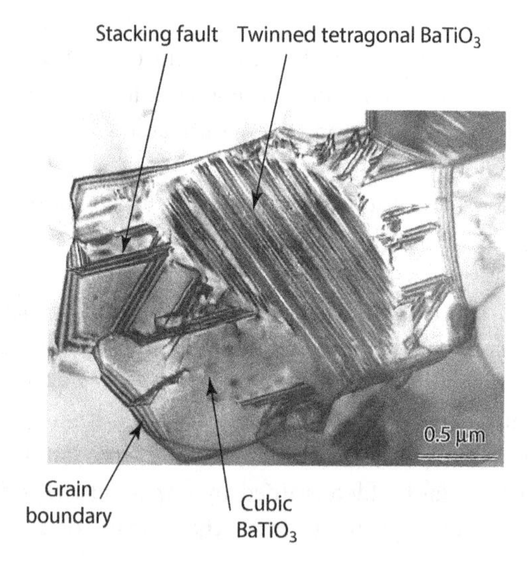

FIGURE 7.29 Transmission electron micrograph of a *core–shell* structure grain in Bi_2O_3 plus Nb_2O_5-doped $BaTiO_3$. The dopants are concentrated near the grain boundary and render the material cubic. The dopants have not had time to diffuse into the center of the grain leaving it essentially pure $BaTiO_3$ allowing it to undergo the cubic-tetragonal phase transition with subsequent twinning. (Chiang 1988)

Because of the rapid propagation of the phase interphases in displacive transformations their kinetics are of limited interest from a practical point of view. Of far greater interest are the detailed crystallographic changes between the two phases, the resultant stresses developed in the two phases, how the transformation is nucleated, how the final microstructure develops, and how the properties—usually mechanical—are affected by the transformation. Most of these issues are outside of the main focus of this book. Nevertheless, some examples of displacive transformations are presented because of their technological importance.

7.10.2 Zirconium Oxide

Important examples of displacive polymorphic phase transformations include the transitions that occur in pure zirconium oxide, ZrO_2:

$$\text{monoclinic} \underset{}{\overset{1170°\text{C}}{\rightleftharpoons}} \text{tetragonal} \underset{}{\overset{2370°\text{C}}{\rightleftharpoons}} \text{cubic}$$

Figure 7.30 shows the fluorite cubic structure of the high temperature cubic polymorph of ZrO_2 in which the Zr^{+4} ion is surrounded by 8 oxygen ions. In reality, the Zr^{+4} ion is really too small to be in eightfold coordination by oxygen and it *rattles around* inside the eightfold oxygen cube. The radius of the Zr^{+4} ion is 87 pm and that of oxygen is 132 pm giving a cation/anion radius ratio of about 0.66 which is smaller than the minimum ratio of 0.732 necessary to stabilize eightfold coordination. Nevertheless, at the very high temperatures where the cubic phase is stable, there is enough thermal vibrational energy to keep the mean Zr^{+4} ion position centered in the oxygen cube. However, as the temperature is lowered, the zirconium ion would prefer to position itself closer to four oxygens on the face of the cube, hence the transformation to tetragonal zirconia. At still lower temperatures, a monoclinic structure is more stable in which the oxygen cube becomes even more distorted to attempt to satisfy the needs for the Zr^{+4} ion to have a lower coordination number than eight. The tetragonal to monoclinic transformation includes about a 3% increase in volume. Therefore, *pure* polycrystalline zirconia, sintered at some elevated temperature—say 1750°C—on cooling will essentially fall apart at this transition temperature due to this large volume change that causes cracking at grain boundaries and grain separation. Because ZrO_2 has a high melting point—about 2680°C—it would make a good high temperature refractory for furnace insulation, and so on. Unfortunately, because of this low temperature phase transition, pure ZrO_2 cannot be used as a thermal insulator. However, it was found almost a century ago that, by forming an alloy of ZrO_2 with MgO, CaO, Y_2O_3, and others, the high temperature cubic phase could be *stabilized* to low temperatures and actually maintained all the way to room temperature.

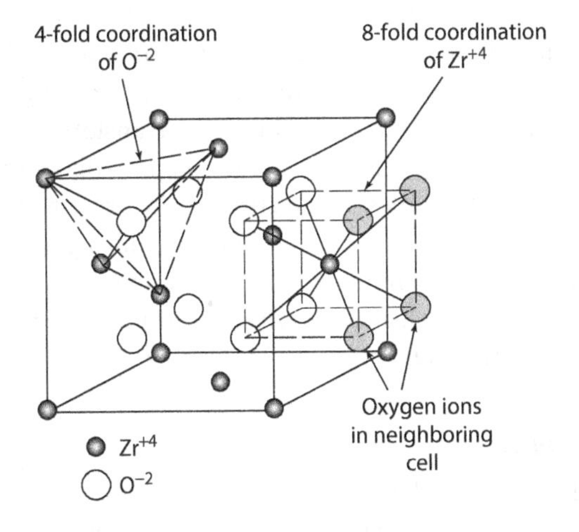

FIGURE 7.30 The cubic structure of ZrO_2, fluorite (CaF_2), showing how the Zr^{+4} ions in the faces are eightfold coordinated by 8 oxygen ions, 4 within the cell and 4 in the neighboring cell. Note the eightfold site in the center of the fluorite cell is not occupied by a Zr^{+4} ion. The ions are drawn to scale. (Emsley 1998)

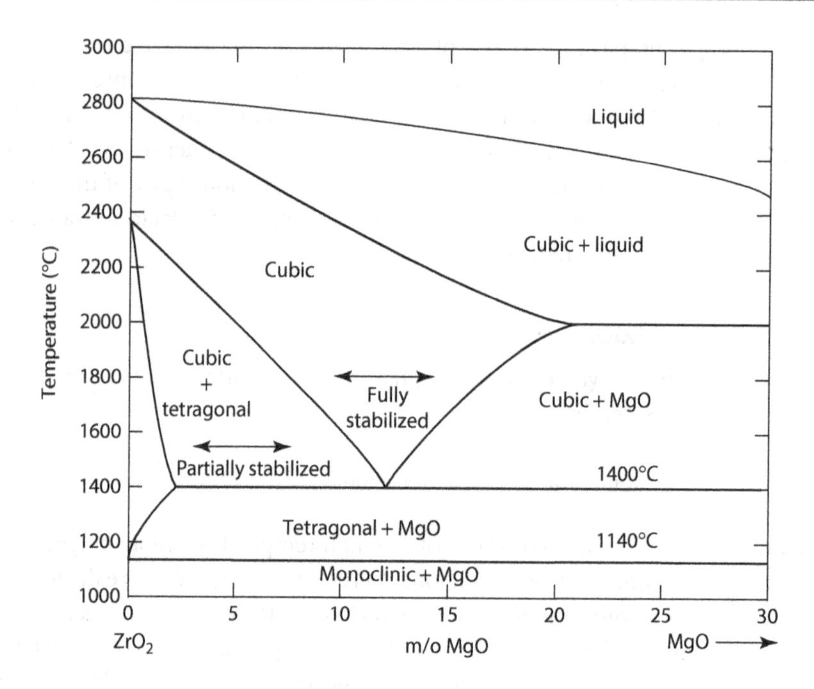

FIGURE 7.31 Part of the MgO-ZrO_2 phase diagram. (After Roth 1995)

Figure 7.31 based on data in the literature (Roth 1995) shows that ZrO_2 with about 12 m/o MgO fired—densified—at about 1750°C is entirely cubic. Hence, such solid solutions are called *stabilized zirconia* and if the material is 100% cubic it is called *fully stabilized zirconia.* The presence of these other elements in solid solution, Mg^{2+}, if they are randomly distributed in the crystal lattice, gives the crystal a high degree of symmetry, a cubic structure. Therefore, in order to transform from the cubic—high symmetry—solid solutions, the added cations must diffuse into positions that give *all* the unit cells either tetragonal or monoclinic symmetry since in displacive transformations, unit cells transform together. Since moving the randomly dispersed cation alloying elements around is a diffusion process, it is possible to retain the cubic structure down to room temperature without transformation—hence, the existence of cubic zirconia gemstones. If the material is not fully stabilized by adding smaller amounts of alloying compounds—say 5 m/o MgO in Figure 7.31—and cooled under proper conditions to room temperature, a two-phase mixture of cubic and tetragonal zirconia is obtained. If the tetragonal grains or crystals are small enough, the tetragonal phase is inhibited from transforming to monoclinic by the surrounding cubic phase that prevents the tetragonal to monoclinic volume expansion. This material is called *partially stabilized zirconia* (*PSZ*) and it has a very high fracture strength for ceramics—on the order of 10^9 Pa. High strength occurs because crack propagation releases the compressive stress on the untransformed tetragonal zirconia allowing it to expand and transform to monoclinic, the expansion partially closes the crack, and increases the fracture resistance—fracture toughness—and strength. In order to achieve high strengths, the grain size of the tetragonal phase must be carefully controlled during firing and cooling. If the grain size is too large, it will simply transform on cooling providing no strengthening. If the tetragonal grain size is too small, cracks will not relieve enough stress to allow the transformation, also providing no strengthening (Green et al. 1989).

7.10.3 Barium Titanate

Another important set of displacive phase transformations are those in barium titanate, $BaTiO_3$, which goes through the following series of transformations (Kingery et al. 1976):

$$\text{rhombohedral} \xrightleftharpoons{-90^\circ\text{C}} \text{orthorhombic} \xrightleftharpoons{0^\circ\text{C}} \text{tetragonal} \xrightleftharpoons{130^\circ\text{C}} \text{cubic}$$

Note that the symmetry of the stable phase increases as the temperature increases similar to ZrO_2 and is generally true for all solids. The high temperature cubic form has the perovskite crystal structure with the barium ions at the cube corners, the oxygen ions on the cube faces, and the Ti^{+4} ion in the center of the octahedron formed by the six oxygen ions on the cube faces, Figure 7.32. In this case, the size of the octahedron is determined by close packing of both the large oxygen and barium ions and the octahedron is a little too large for the Ti^{+4} ion to touch all the neighboring oxygen ions so it too *rattles around* inside the oxygen octahedron until the temperature becomes low enough for it to be bound closer to one or more of the oxygen ions. As a result, cooling from elevated temperatures, the cubic phase transforms to tetragonal at 130°C simply due to a shift in the Ti^{+4} ion from the center of its oxygen octahedron toward one of the cube faces distorting the crystal from cubic to tetragonal. This produces a built-in electric dipole that gives $BaTiO_3$ its very large relative dielectric permittivity—6000—making it useful as a dielectric in capacitors. In addition, cooling through this transition temperature in an electric field, the materials are *poled*—aligning the electric dipoles—making the material *piezoelectric*—stress produces a voltage and an applied voltage produces a strain—enabling many applications including sonar and precise small voltage-controlled movements as required in the scanning tunneling microscope, STM.

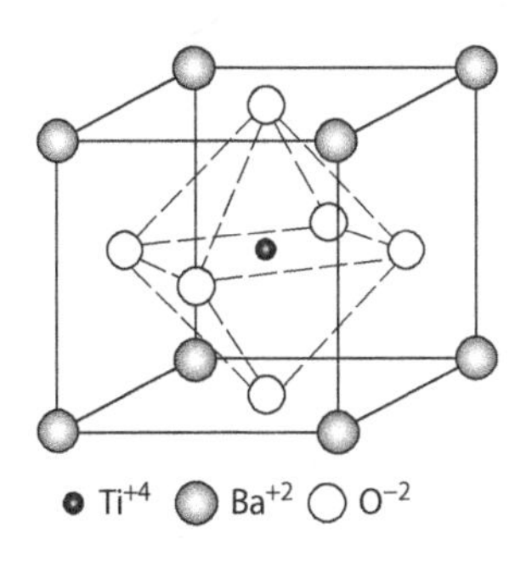

FIGURE 7.32 The perovskite crystal structure of barium titanate, $BaTiO_3$. The ions are drawn to scale. (Emsley 1998)

7.10.4 Martensite in Steels

Pure iron goes through two phase transformations: $\text{BCC}, \alpha \xrightleftharpoons{912^\circ\text{C}} \text{FCC}, \gamma \xrightleftharpoons{1394^\circ\text{C}} \text{FCC}, \delta$ (Brandes and Brook 1992). Of course, a most important effect of carbon on the phase transformation of steels—iron containing carbon—is the displacive transformation to form martensite upon rapid cooling iron–carbon alloys from the cubic austenite* to prevent the diffusion-controlled decomposition of austenite to ferrite and Fe_3C, cementite. "The way in which this transformation occurs, however, is a complex process and even today the transformation mechanism, at least in steels, is not properly understood" (Porter et al. 2009, 369). However, this much *is* known, the metastable FCC austenite undergoes a displacive transformation to BCT martensite that produces a large amount of strain in the steel that greatly increases the hardness: Figure 7.33 (Bain 1924). In principle, the FCC austenite could transform all the way to BCC ferrite via the *Bain strain*: 20% contraction along the c-axis of the BCT cell and 12% expansion along the a-directions. However, the presence of carbon in solid solution in octahedral interstitial site restricts the complete transformation to the BCC cubic phase. The result of the transformation is then a BCT phase whose lattice parameters depend on the carbon content of the alloy, Figure 7.34 (Krauss 1990; Reed-Hill and Abbaschian 1992). Note that 4 a/o carbon corresponds to only about one carbon atom per 12.5 BCT unit cells yet it has a very large influence on the transformation! The temperature below the eutectoid temperature in Figures 7.4 and 7.5 where martensite begins to form, the M_s temperature, also depends on the carbon content, decreasing as the carbon content increases implying that the Gibbs energy difference between the austenite and ferrite phase needs to get larger to overcome the strain energy produced in the transformation. Microscopically, the martensite forms randomly and in some cases consists of plates consisting of extremely small twins. As the temperature is lowered, more plates form until martensite formation stops. Usually, there is some *retained* austenite that does not transform and the amount increases with carbon content. In addition, the detailed microstructure of the martensite formed also depends on the carbon content of the steel. The details of the martensite transformation in steels has been studied extensively mainly with respect to crystallography, martensite nucleation, effect of carbon content and alloying elements, the effects on microstructure, all with a focus on optimizing the mechanical properties of steels for a large number of specific applications (Krauss 1990). For more details about the process—there are many—the literature should be consulted (Krauss 1990; Reed-Hill and Abbaschian 1992; Porter et al. 2009).

* Named after the English metallurgist William Chandler Roberts-Austen, 1843–1902.

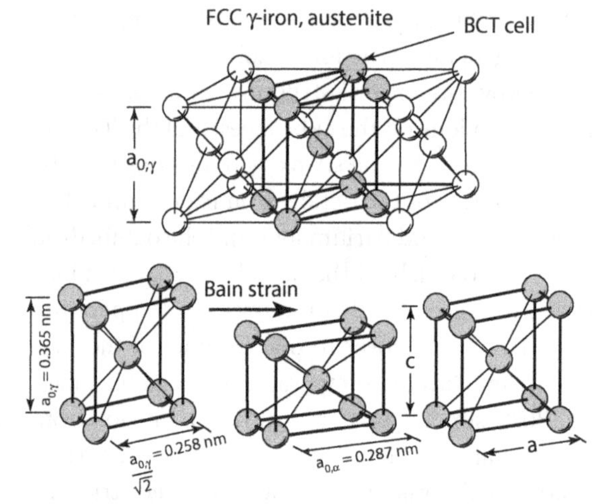

FIGURE 7.33 The Bain model of FCC austenite transforming to BCT martensite in steels, Fe-C alloys. In principle, pure iron should be able to transform from FCC to BCC by about a 20% negative strain in the z-direction and plus 12% in the x- and y-directions relative to the isolated BCT cell at the bottom left: the Bain strain. However, the presence of carbon in octahedral interstices in austenite only allows the transformation to proceed partially to BCT martensite at the bottom right. (Bain 1924)

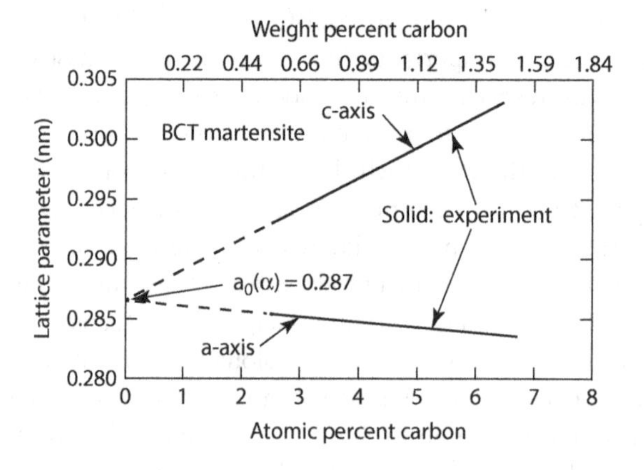

FIGURE 7.34 The variation of the c- and a-axes of BCT martensite as a function of carbon content in steels. Note that they extrapolate to the lattice parameter of BCC iron, ferrite. (Data from Krauss 1989)

Frequently, all displacive transformations are referred to as *martensite or martensitic transformations* (Buchanan and Park 1997) but others would restrict this terminology to displacive transformations in which strain energy plays a dominant role (diffusionless transformations).

7.10.5 Shape Memory Alloys

Shape memory alloys are important examples of displacive transformations. Important shape memory metals are the nitinol alloys of roughly 50 m/o Ni-50 m/o Ti or NiTi—hence the name since they were discovered at the U.S. Naval Ordinance Laboratory in Maryland. The crystal structure of NiTi is CsCl, four Ni at the corners of a cube and the Ti in the center, in its high temperature—$T \approx 60°C$—*austenitic* phase and becomes tetragonal with about an 8% volume change going through the martensite transformation start temperature, M_s, near or slightly above room temperature (Brandes and Brook 1992). Again, like all materials that show a displacive transformation, *the strain associated with the transformation in these materials is frequently accommodated in a*

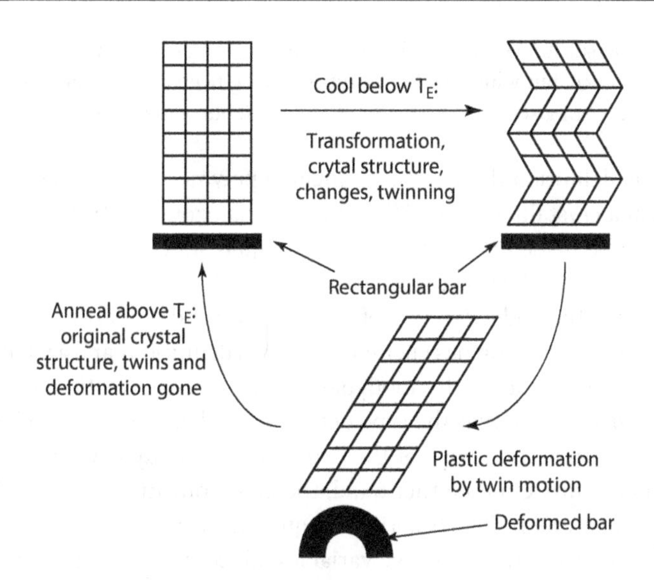

FIGURE 7.35 Schematic showing how the strain accompanying a displacive transformation can be minimized by twinning, and how the twinned material can be plastically deformed by twin reorientation. On reheating above T_E, the twins and deformation go away and the deformed bar returns to its original shape: the *shape memory* effect.

polycrystalline material by the formation of twins as sketched in Figure 7.35. Application of a stress changes the orientation of the twinned crystals producing plasticity or a strain. In these twinned materials, the plastic deformation is caused simply by movement of twin boundaries or reorientation of twins. Therefore, heating back into the austenite region, each crystal again becomes cubic and the strain introduced in the martensitic phase disappears and the material assumes its original shape. Hence, the name: *shape memory alloys.* On the other hand, if these materials are deformed slightly above the phase transformation temperature, then the stress can induce the phase transformation and a great deal of elastic strain—up to 10%—can be induced entirely by the alignment of the transformed crystals in the direction of the stress. On removal of the stress, the structure reverts to the cubic phase and the strain returns to zero. Such materials are called *superelastic* or *pseudoelastic.* Materials that exhibit displacive transformations are also classed under so-called *ferroic materials.* Ferroic crystals may be defined as *crystals that possess two or more orientation states, or domains, and under a suitable driving force such as stress, electric field, or magnetic field the domain walls will move switching the crystal from one domain state to another.* Materials that switch in an electric field are called *ferroelectrics*, those that switch under a stress are called *ferroelastic*, and those that switch under a magnetic field, not surprisingly, are called *ferromagnetic.* These properties and fields can be inter-related in interesting and complex ways depending on the symmetry of the crystals and the directional dependencies of their properties* (Newnham 2005). It should be noted that both ferroelectric and ferroelastic materials actually undergo *phase transitions* to different crystal structures and the *domains*—crystal regions with the same orientation of properties—are actually crystal twins. Most ferromagnetic materials do not undergo phase transitions and their domains merely consist of regions with different orientations of the aligned magnetic moments of the atoms or ions in the solid.

7.11 GLASS TRANSITION

7.11.1 Introduction to the Glass Transition

It is appropriate to say something about the glass transformation particularly since attempts to explain it have gone on for over 400 years. Morey (1938, 34) defined a glass as

* If you enjoy linear algebra, then you will appreciate detailed investigations of the relationship between the structure, properties, and transformation of these materials.

> A glass is an inorganic substance in a condition which is continuous with, and analogous to, the liquid state of that substance, but which, as the result of having been cooled from a fused condition, has attained so high a degree of viscosity as to be for all practical purposes, rigid.

Well yes and no! This definition does not even include polymer glasses and most polymers are in the glassy rather than a crystalline state. What does "...analogous to, the liquid state..." mean? And glasses do not have to be made only from a fused or liquid state. They can be deposited from the vapor phase by some type of PVD process as was seen with the crystallization of amorphous silicon. During the last 100 years, the understanding of the structure and properties of materials has considerably improved, the glass transition has elicited numerous definitions and a wide variety of explanations, including whether or not it is a thermodynamic phase transformation or something different. For example, a *first-order transition* is the type of transition that most of this chapter is focused on; namely, there is a discontinuous change of the extensive thermodynamic variables U, V, H, and S at the transition temperature. On the other hand, the glass transition has often been referred to as a *second-order transition*, one in which derivative quantities such as thermal expansion and heat capacity are discontinuous but the extensive variables are continuous. There is some debate in the literature whether the glass transition is a true second-order transition: "...the glass-rubber transition is not a real thermodynamic phase transition, neither a first nor a second order transition... At the very best the T_g-transition may be seen as a quasi-second-order transition but certainly not as a first-order one" (Van Krevelen 1997, 130). Nevertheless, the IUPAC (International Union of Pure and Applied Chemistry) defines it as a second-order transition (McNaught and Wilkinson 2006, 360). One of the better literature definitions is (Glass transition 2016):

> Thus, the liquid-glass transition is not a transition between states of thermodynamic equilibrium. It is widely believed that the true equilibrium state is always crystalline. Glass is believed to exist in kinetically locked state, and its entropy, density, and so on, depend on the thermal history. Therefore, the glass transition is primarily a dynamic phenomenon.

Despite the variations in the definition, the glass transformation plays an extremely important role in materials technology even if its materials science is somewhat debatable. Metals (by rapid cooling), covalently bonded elements (Si, Ge, and chalcogenides: S, Se, Te), inorganic compounds (silicates, phosphates, borates, halides), and polymers, all show glass transitions (Varshneya 1994). Wherever you are, it doesn't take much looking around to find something made with either an inorganic glass or a glassy polymer. Figure 7.36 shows the volume versus temperature for a material that when cooled from the liquid state can either crystallize at its melting point if cooled slowly or, if cooled sufficiently rapidly,

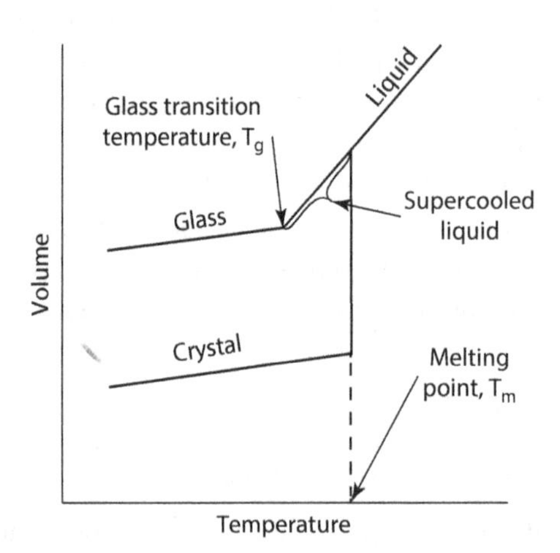

FIGURE 7.36 Volume versus temperature for a material that forms a glass at the glass transition temperature, T_g, or becomes 100% crystalline at the melting point T_m. Note that volume expansion versus temperature for the glass and crystal are virtually the same but that for the liquid is much larger reflecting the additional motion of the atoms, ions, or molecules in the liquid state compared to the solid state.

will form a glass instead. Note the discontinuity in the derivative of the volume, dV/dT, the volume thermal expansion coefficient at the glass transition temperature. The terms *slow* and *rapid* depend on the material. For example, to crystallize a liquid silicate melt to prevent glass formation, the cooling rate might be measured in degrees Centigrade per hour or per day while for metallic glasses, cooling rates on the order of 10^6–10^8 °C/s are necessary. Polymers require cooling rates somewhat in between these extremes. If the cooling rate is sufficiently rapid, then the liquid can be *supercooled* below the melting point. Figure 7.36 illustrates two extremes where a material either crystallizes completely at the melting point or goes completely to a glass at the glass transition temperature. This is typically the behavior of silicate glasses. On the other hand, many polymers are partly crystalline and partly glass. As a result, the V-T curve on cooling can show behavior that is a combination of these two extremes. *However, for purposes of distinguishing between crystallization and glass formation, Figure 7.36 is the most appropriate.*

7.11.2 Thermal Expansion

As the glass transition can be defined in terms of a discontinuity in thermal expansion, it is worthwhile examining the fundamental processes producing expansion. In a monatomic gas, such as argon, only translational motion of the atoms is possible and they travel at speeds of several hundred m/s, about 1000 mph, as was seen in Chapter 5, and travel only about 0.1 μm before colliding with another atom. This is the basis for diffusion of atoms in gases as will be seen later in Chapter 9. Note that all of the thermal energy in the gas—the enthalpy—is in this translational motion and the value of the heat capacity reflects that heat added to a monatomic gas contributes only to higher translational speeds. Note that for one mole of an ideal gas, pV = RT, so the volume thermal expansion coefficient is just the inverse of the temperature,

$$\alpha_g = \frac{1}{V}\frac{dV}{dT} = \frac{d(RT/p)}{(RT/p)dT} = \frac{(R/p)}{(RT/p)}\frac{dT}{dT} = \frac{1}{T} \tag{7.33}$$

which is 3.33×10^{-3} K^{-1} at room temperature. This extra volume reflects the increasing translational velocity of the gas atoms.

In contrast, in a solid, the atoms, molecules, or ions are fixed in position and merely vibrate on their lattice sites—at room temperature this vibration frequency, as is shown in Chapter 9 is on the order of 10^{13} Hz—as sketched in Figure 7.37 where the interatomic bonds essentially act like springs against which the atoms vibrate. The heat content or enthalpy of a solid is entirely contained in the kinetic and potential energy of these atomic vibrations. As the heat content and temperature is increased, the vibration amplitude increases giving rise to the thermal expansion of a solid as shown in Figure 7.38. The potential energy curves versus distance between atoms, ions, or molecules in solids are not perfectly symmetrical, so the *average* interatomic distance increases as the amplitude of vibrations or temperature increases. Figure 7.38 greatly exaggerates the effect of temperature on the average position of the atoms. Typically, the minimum in the potential energy versus distance curve, the equilibrium distance, is several electron volts negative, the bonding energy of the solid. The gas constant in eV/K is

$$\frac{R}{N_A e} = \frac{8.314\,\text{J/mol}\cdot\text{K}}{(6.022\times10^{23}\,\text{atom/mol})(1.602\times10^{-19}\,\text{J/eV})} = 8.61\times10^{-5}\,\text{eV/K}$$

so even at 1000°C the thermal energy is only about 0.11 eV above the minimum in the potential energy curve, hardly noticeable in Figure 7.38. The thermal expansion is related to the mean interatomic distance as a function of temperature, and therefore to the asymmetry of the potential energy curve with distance. The more symmetrical the bonding curves, such as those for covalently bonded materials, the smaller the expansion

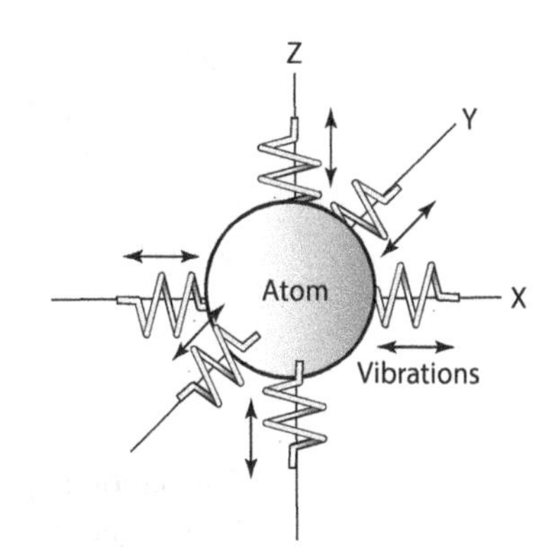

FIGURE 7.37 Schematic showing an atom in a solid vibrating in three directions with the interatomic bonds depicted as springs. Each vibration contributes k_BT to the heat capacity or $3k_BT$ for each atom or ion.

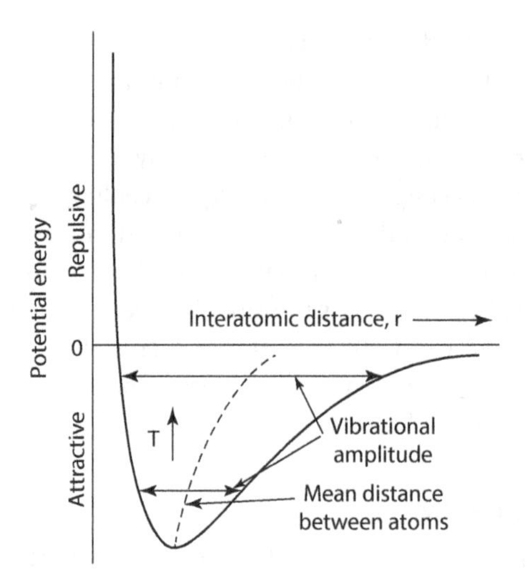

FIGURE 7.38 Potential energy versus interatomic distance in a solid. An asymmetric bonding curve exists for all types of bonding: ionic, covalent, and van der Waals. Because of the asymmetry, as the temperature (energy) increases the amplitude of the atomic vibration increases and its mean position—dashed curve—between the attractive and repulsive parts of the potential increases generating thermal expansion. The effect of temperature is greatly exaggerated in this drawing in that the minimum in the potential is several eV negative—the interatomic bond—and thermal energies are small, only on the order of a tenth of an eV for 1000°C.

coefficient. Also, as the bond strength increases the deeper and more symmetrical the potential well becomes. Therefore, the melting point goes up and the thermal expansion coefficient decreases. For example, for the elements C (diamond), Si, and Ge all having the diamond cubic crystal structure with melting points of 4700 K (under pressure), 1687 K, and 1211 K, their thermal expansion coefficients are $1.18 \times 10^{-6}\ K^{-1}$, $2.6 \times 10^{-6}\ K^{-1}$, and $5.8 \times 10^{-6}\ K^{-1}$, respectively (Haynes 2013). On the other hand, a more asymmetric bonding curve as in an ionic bond leads to higher expansion coefficients. For example, at 300 K, the linear thermal expansion coefficient of reasonably ionic MgO with a melting point of 3100 K, is about $\alpha_l = 12 \times 10^{-6}\ K^{-1}$ which makes the volume expansion coefficient, α_V (see Appendix)

$$\alpha_V = \frac{1}{V}\frac{dV}{dT} \simeq 3\alpha_L = 36 \times 10^{-6}\,K^{-1}$$

about two orders of magnitude smaller than that of a gas. And—as mentioned in Section 5.2.1—each energy-absorbing mode of an atom contributes $1/2\,k_BT$ to the total energy of an atom. A vibration has both kinetic and potential energy each of which contribute $1/2\,k_BT$ for a total energy per vibration of k_BT. But for each atom, there are three directions—or modes—of vibrations so the total energy contribution to a vibrating atom is $3k_BT$. For one Avogadro's number of atoms, $E_{vib} = 3N_Ak_BT = 3RT$. So the heat capacity for a monatomic solid is simply,

$$\overline{C}_P \cong \overline{C}_V \cong \frac{dE_{vib}}{dT} = 3R = 24.94\,J/mol \cdot K. \tag{7.34}$$

This ignores quantum effects and the coupling of atoms that the Einstein and Debye theories of heat capacity take into account and that explain heat capacities at low temperatures. However, at high temperatures, Equation 7.34 is a good approximation. One thing to be careful about is that the molar heat capacity for metals is about 3R since there is only one Avogadro's number of atoms per mole. However, for compounds as shown in Chapter 1, there are more atoms or ions than one N_A per mole. For example, for NaCl, there are 2 N_A per mole so its heat capacity is about $6R \cong 50$ J/mol-K. Similarly, for Al_2O_3 there are 5 N_A ions per mole so, $\overline{C}_P \cong 125$ J/mol-K.

Liquid behavior is between that of a solid and a gas in many ways. As shown in Figure 7.36, on melting, for most solids,* there is an expansion in going from the solid to the liquid state—typically around 10 v/o (volume percent) or more. For example, for copper at its melting point of $T_{mp} = 1083°C$, the density of liquid copper is, $\rho(Cu, l) = 8.00\ g/cm^3$ while that of solid copper is $\rho(Cu, s) = 8.96\ g/cm^3$ (Haynes 2013). As a result, there is about 10% more volume in the liquid than in the solid and now the atoms can move with some translational velocity as well as vibrating back and forth against their interatomic bonds. Furthermore, in a polymer, there can be large and different sections of a polymer molecule that are free to move with increased volume without moving the entire molecule. Therefore, atoms or molecules in a liquid have both *translational and vibrational* motions that don't exist in a crystalline solid, motions that contribute to heat capacity, enthalpy, and thermal expansion coefficient. For example, the heat of fusion at melting is essentially the energy necessary to give the atoms of the solid the translational energy that they have in a liquid. So the volume of a liquid is determined not only by the volumes of the atoms, ions, or molecules of the liquid but also by a certain amount of *free volume* that gives them extra freedom of movement compared to a crystalline solid.

7.11.3 Short-Range and Long-Range Order

Because of the translational motion in a liquid, the atomic structure of a liquid is also somewhere between that of a solid and a gas. A solid has both *short-range order and long-range order*. For example, in solid Cu, which has an FCC lattice, each of the Cu atoms is surrounded by 12 nearest neighbor Cu atoms—all the same distance away—which is the *short-range order*. However, a copper crystal also has *long-range order* in that—if it is a perfect crystal—there is another Cu atom at one lattice parameter—cell edge—away, 10 lattice parameters away, or even a million lattice parameters away traveling along a cell edge in a <100> direction. In contrast, liquid Cu has only the same *short-range order* that exists in the Cu crystal; namely, each Cu is surrounded—on the average—by 12 nearest neighbor Cu atoms. However, there is no *long-range order* in that if one travels in any direction in the liquid, the chance of hitting the center of another Cu atom at 2, 10, or 100 interatomic distances away is highly unlikely. Furthermore, the atom positions are constantly changing in a liquid as well. Figure 7.39 gives an example of SiO_2 that is made up of SiO_4^{4-} tetrahedra bonded at corners to give a fairly open structure in crystals, liquids, and glasses and shows the difference between the ordered array of tetrahedra in a crystal compared to the disorder that exists in both liquids in glasses.† The hexagonal

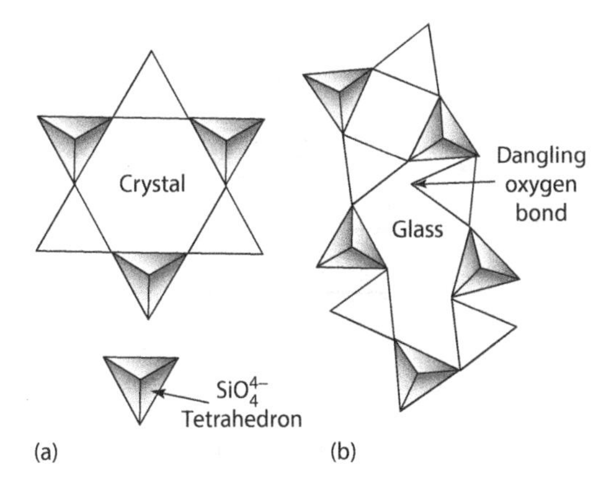

FIGURE 7.39 Comparison between the ordered arrangement of (a) SiO_4^{4-} tetrahedra in crystalline silicates (or PO_4^{3-} in phosphates) with (b) the more random arrangement of the tetrahedra in a silicate glass. Note that the short range order in both consists of four oxygen ions at fixed distances from the silicon ion but the uniform Si–Si distance and 180° angle in the crystal vary in the glass because of the random angular rotation of the tetrahedra around their corner bonds.

* Si and H_2O being notable exceptions because of their directed covalent and hydrogen bonds, respectively.

† This is actually a more accurate representation of phosphate crystals, liquids, and glasses in that the PO_4^{3-} tetrahedra sharing corners have only one unbonded corner oxygen to form sheets and chains.

array is presumed to extend infinitely in three dimensions with additional tetrahedra sharing the currently unshared corners. The short-range order in this case is that each silicon is surrounded by four oxygen ions and each oxygen by two silicon atoms at the same distance. Neighboring silicon–silicon distances in the crystal are all the same, but this distance varies in the liquid or glass because of the ability of the tetrahedra to rotate. Also, note the existence of a broken or dangling bond. In pure SiO_2, at high temperatures, thermal energy generates a number of these broken bonds decreasing the continuity of the three-dimensional network of SiO_4^{4-} tetrahedral and generating smaller groups of interconnected tetrahedra that constantly form and reform into different configurations. It is the breaking of the network into smaller groups that generates the higher thermal expansion in the liquid and decreases the viscosity—the ability of these groups to move past each other under stress.* Note in Figure 7.39, that idealized crystal structure consists of a three-dimensional array on interconnected hexagonal groupings of six tetrahedra. While in the glass or liquid state in Figure 7.39 there is a group of four and a group of eight interconnected tetrahedra. In order for this liquid or glass to crystallize, there must be bond breaking and rearrangement of these tetrahedra to form the ordered crystalline array. As the melting point is approached from above, the number of broken bonds decreases and the viscosity of the liquid continually increases. In polymers, short-range order exists within a chain segment. But if there is any rotational freedom around the bonds, then any long range order along the molecule length disappears to say nothing of the essentially independent orientation from one polymer molecule to each of its neighbors precluding long-range order.

7.11.4 Viscosity

The viscosity of both glasses and liquids is critically important in flow behavior and the glass transition. Figure 7.40 shows how the viscosity of a liquid is defined—and frequently measured. This figure shows two parallel plates with a fluid of thickness y between them and a shear stress applied between the plates of τ† causing a gradient in the velocity of the liquid from v = 0 at the bottom, stationary plate, to a maximum value, v, at the top plate. The shear strain rate, $\dot{\gamma}$, is defined as the gradient of the velocity with distance or,

$$\dot{\gamma} = \frac{dv}{dy} = \frac{dx/dt}{y} = \frac{(dx/y)}{dt} = \frac{d\gamma}{dt} \quad (7.35)$$

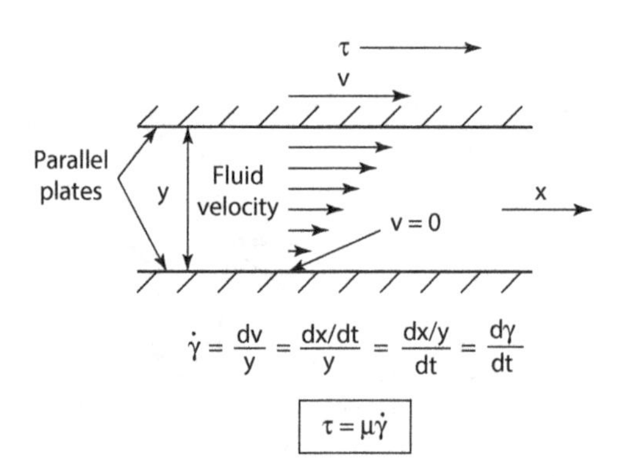

FIGURE 7.40 The shear motion of a moving plate relative to a stationary one with a fluid of thickness y in between showing how the viscosity of the fluid is defined by the shear strain rate, $\dot{\gamma}$, and the shear stress, τ.

* In silicate glasses, sodium oxide, Na_2O, calcium oxide, CaO, and other oxides are added to break up the continuous three-dimensional network of pure SiO_2 as will be shown in Chapter 9.

† The symbol τ has already been used for the *relaxation time* in Chapter 2. Unfortunately, in the literature both the relaxation time and the shear stress are given this symbol. Hopefully, the contexts in which they appear will not make their use confusing. However, the relaxation time for stress relaxation will be used shortly and here a different notation for the relaxation time will be used to hopefully minimize confusion.

and the viscosity, μ,* is defined as the resistance to flow, or

$$\tau = \mu\dot{\gamma}. \quad (7.36)$$

The units of viscosity are Units(μ) = Pa-s. However, in the older literature the cgs version of viscosity is used and in the cgs system, Units(μ) = P for poise: 1 P = 0.1 Pa-S. Water, for example, has a viscosity of around one centipoise or $\mu = 10^{-3}$ Pa-s. The viscosity for forming inorganic glass objects, *the working point* (Varshneya 1994), is about 10^3 Pa-s around the same as viscous syrup and the viscosity for the injection molding of polymers is about 100 Pa-s (Morton-Jones 1989).

7.11.5 The Glass Transition Temperature, T_g

The formation of a glass takes place in the following way. As the temperature of a liquid is decreased, atomic and molecular *translational*—and some group migrational and rotational—motion is less energetic and becomes more difficult until, at the melting point in a simple monatomic material such as copper, the atoms can easily arrange themselves into the long-range ordered structure of the crystal since the atoms can move easily and do not need to go very far. In contrast, in polymers and inorganic materials the larger molecules and atomic groups in the liquid state require several coordinated motions and, perhaps, bond breaking and reforming to orient them properly to generate the long-range order of the crystal. This takes time and atomic or molecular translational and other motions—diffusion and/or viscous flow. If the material is cooled sufficiently below the normal melting temperature—*a supercooled liquid*—the temperature may become so low that these translational motions essentially stop and the lack of long-range order of the liquid is solidified in place at the *glass transition* temperature, T_g (Figure 7.41). Now the only thermal motions experienced by the atoms and molecules in the glass are the same thermal vibrations of the atoms that occur in the crystal. Therefore, the thermal expansion coefficient of the glass is essentially the same as the crystal and less than that of the liquid since there is no longer any translation or diffusional motion. As a result, the glass no longer flows like a liquid and becomes rigid and exhibits properties of a solid such as hardness, elastic modulus, and heat capacity in addition to the thermal expansion coefficient. Figure 7.41 shows that the slopes of the volume versus temperature curves—the volume thermal expansion coefficients—of the crystal and the glass are essentially equal and both less than that of the liquid: $\alpha_{glass} \cong \alpha_{crystal} < \alpha_{liquid}$.

Therefore, a good definition of a glass is: ***a material that has the structure of a liquid—short range order but no long range order—and the properties of a solid—elastic modulus, hardness, thermal expansion coefficient, heat capacity, etc*** And at the expense of being repetitious, that is why the volume versus temperature plots for the crystalline solid and the glass are essentially the same, or parallel, in Figure 7.41 since the only mechanism that produces volume expansion—thermal vibrations of atoms on stationary positions—is the same in both. While in the liquid state, in addition to thermal vibrations, the molecules in polymers and the large atomic groups in inorganic crystals have translational, rotational, and complex vibrational motions that increase with temperature increasing the volume at a faster rate than atomic vibrations alone.

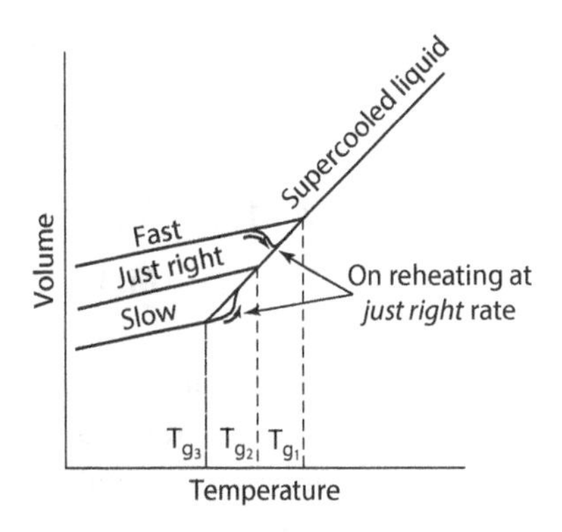

FIGURE 7.41 Schematic showing how the cooling rate affects the glass transition temperature: the fastest cooling rate gives T_{g1} at the largest glass volume; the slowest cooling rate give T_{g3} and the smallest glass volume; and the middle cooling rate give T_{g2} and the middle glass volume. Reheating all three glasses at the same rate as the middle cooling results in different behavior of each glass. The V-T curve of the glass heated and cooled at the same rate, on reheating follows the cooling curve closely. However, the fastest cooled glass, at the slower reheating rate tries to equilibrate its structure with the slower rate and, near the glass transition temperature, it actually shrinks to try to follow the V-T of the slower reheating rate. For the more slowly cooled glass, the opposite occurs: it tries to maintain its structure—volume—above the original T_{g3}.

* The symbol η is frequently for viscosity and is the symbol recommended by the IUPAC (McNaught and Wilkinson, compilers 2006). However, here η is already being used as number per unit volume and μ is frequently used for viscosity.

Another important feature of the glass transition temperature is that *it is not a material property* like the melting point or the boiling point that are fundamentals of a first-order phase transition. In contrast, the glass transition temperature depends on the rate of cooling as shown in Figure 7.41: the faster the rate of cooling, the higher the glass transformation temperature but the glass transition temperature change with cooling rate is small. For all glass-forming materials, the glass transformation temperature occurs when the glass viscosity reaches about $\mu \cong 10^{12}$ Pa-s. Figure 7.41 implies that glasses of identical composition, cooled at different rates with different glass transition temperatures, have different densities related to their volumes at the glass transition temperatures. Optical glasses used as lenses will have spatially different indices of refraction if cooled at different rates, which, of course is undesirable. As a result, such glasses are heated for long times just below the glass transition temperature and slowly cooled to insure uniform density. On the other hand, *tempered (or toughened) glass*, is a normal silicate glass that is rapidly cooled that puts the surface in compression when the entire part has cooled to room temperature increasing the strength of the glass (Jones 1956; Doyle 1979; Pfaender 1983; Varshneya 1994). Note also in Figure 7.41 when all three of the glasses are heated at the same rate as the intermediate cooling rate, the rapidly cooled glass shrinks as it reaches the glass transition temperature and actually gives it a *negative* thermal expansion in this region. At temperatures this close to the glass transition temperature, the glass has enough thermal energy for structural rearrangement and tries to have the same degree of order and volume as the glass cooled at that just right rate. In contrast, the glass cooled at the slowest rate, will actually try to maintain its greater degree of order until above its original glass transition temperature and then when its viscosity is low enough, it rapidly expands to give the supercooled liquid structure appropriate to the intermediate cooling rate. These changes and their rates are all functions of the glass viscosity as a function of temperature. Therefore, engineers in the glass industry working with silicate glasses have designated key viscosity points (Kingery et al. 1976; Varshneya 1994):

Working point, where the glass is shaped	10^3 Pa-s
Softening point, where the glass gets soft	$10^{6.65}$ Pa-s
Annealing point, where cooling-induced stresses relax	$10^{12.4}$ Pa-s
Strain point, below which nonuniform strain cannot form	$10^{13.5}$ Pa-s

and the *working range* is the viscosity interval between the *working point* and the *softening point.*

7.11.6 Viscoelastic Behavior

7.11.6.1 Material Differences

The intent here is not to discuss the many and varied mechanical properties of materials, but only those that relate to material behavior at temperatures near the glass transition temperature that lead to an improved understanding of T_g. Therefore, the focus is on glassy polymers and inorganic glass-forming materials. Because of the different histories of development and processing requirements, particularly temperature, for these two classes of materials; testing procedures; and the properties measured are frequently different and difficult to compare. Furthermore, for many inorganic glasses, particularly oxides, the details of the various structural units in the melt and in the glass phase are not that well understood. The viscosity of inorganic glasses and the effects of additives, particularly in silicate glasses, is the primary property for the production of glass objects. A large volume of literature on the empirical effects of additives and temperature on viscosity has been developed over many years (Morey 1938) without a detailed understanding of the structural units in the glass that determine properties. In contrast, for polymers both their viscous and elastic (hence *viscoelastic*) behaviors are important for a number of reasons. It is important is to understand the effects of all the possible pre-determined structural variables of polymers such as chain length, chain composition, branching, cross-linking, chain and branch stiffness, chain and branch length, and so on, on the glass transition and mechanical properties (Scherer 1986; Ward and Hadley 1993; Sperling 2006). So for polymers, the structural elements in the glass are known and the major effort is to relate these structural elements to properties such as viscosity as a function of temperature and the glass transition temperature. The glass transition temperature for many polymers is within roughly 100° of room temperature: high density polyethylene, −90°C; polypropylene, −18°C; poly(vinyl chloride), 87°C; and polycarbonate, 150°C (Callister and Rethwisch 2009). Even elastomers in which the polymer chains are cross-linked but can move freely past each other at room temperature can undergo a glass transition at low temperatures; for example, polybutadiene, $T_g \cong -90°C$ (Askeland and Phulé 2003). These low temperatures allow the use of *in situ* experimental techniques on polymers that are difficult to conduct at the high temperatures necessary for most inorganic glasses. One of these techniques involves the application of either a periodic stress or strain to the polymer and observation of its dynamic modulus and internal energy loss. Performing these experiments as a function of temperature and frequency can generate a great deal of information about the effects of chain structure on polymer melts and glasses. For example, polymers above their glass transformation temperatures exhibit several different regions for the shear modulus, G, as a function of temperature as shown in Figure 7.42. Most noncross-linked linear amorphous polymers exhibit a *rubber-like* region just above the glass transition temperature where inhibition to translational molecular motion is caused by a polymer chain that may be sufficiently tangled with other molecules that it takes time to untangle. For short times, this entanglement acts almost like a cross-linking covalent bond of the polymer molecules. As the temperature approaches the glass transition temperature from above, the polymer becomes stiffer or, more accurately, the greater is its elastic modulus. As a result, polymers exhibit both elastic and viscous behavior. For temperatures well above the glass transition temperature, thermal energy overcomes these *tangles* between the molecules and they flow or move past each other and the modulus drops rapidly as shown in Figure 7.42. As the chain length increases, the temperature where this drop occurs increases. This is in contrast to the real elastomers in which permanent cross-links retard flow at high temperatures (Figure 7.43). However, even for elastomers, it becomes more difficult for the molecules to move as the glass transition temperature is approached. This rubbery region above the glass transition temperature has only recently been observed for the first time in inorganic phosphate glasses that have individual chains of connected tetrahedra.* These experiments on lithium metaphosphate glass, in the neighborhood of 200°C, based on $LiPO_3$, in which PO_4^{3-} tetrahedra form long chains by bonding with other tetrahedra at two corners (not too dissimilar to the silicate tetrahedra sharing corners in Figures 7.9 and 7.39) with the Li^+ ion forming ionic bonds between the chains. With substitutions of other larger monovalent and weaker bonding ions such

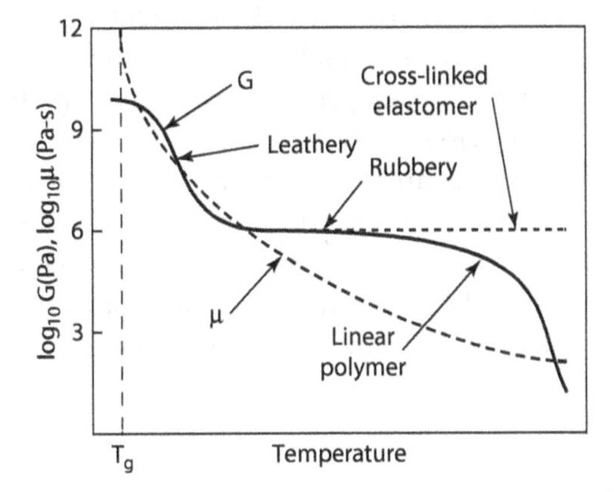

FIGURE 7.42 Temperature dependence of the shear modulus, G, and the viscosity, μ, for a linear polymer above the glass transition temperature emphasizing the *rubbery* region and contrasting it with a cross-linked elastomer. The temperature dependence of the viscosity roughly corresponds to $\mu = \mu_0 \exp(Q/RT)$ with Q and μ_0 chosen so that the magnitude of the viscosity change is about that expected over the temperature range of about 150°C. The viscosity probably varies in a more complex fashion than shown except near the glass transition temperature.

* Note that in a PO_4^{3-} tetrahedron, one of the oxygen atoms has a double bond to the phosphorus with the other three oxygen ions cancelling the remaining plus three charge on the phosphorus giving a net charge of minus three to the tetrahedron.

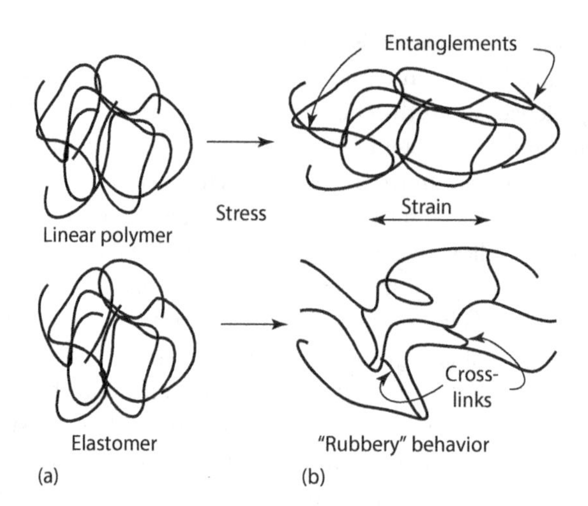

FIGURE 7.43 (a) Entanglements between polymer chains lead to rubbery behavior of linear polymers. (b) Real elastomers have covalently bonded cross-links between chains that allow molecules to slide past each other between the cross-links that limit the amount of sliding or elastic deformation.

as Na^+, K^+, and Cs^+ it is only the smaller Li^+ ions that form the more distant *pinning* points between chains behaving similar to the tangles in linear polymers. As a result, recoverable elastic strains on the order of 30% are observed (Inaba et al. 2015). These results suggest that similar behavior might be found in metasilicate glasses based on the tetrahedral chain structure of metasilicate crystals such as $MgSiO_3$, but at higher temperatures. Other candidates that might show similar behavior are the chalcogenide glasses, many of which consist of isolated chains of atoms. A potential implication of such a controlled chain structure in a silicate glass is that fibers based on such glasses might have higher strengths, all else being equal, due to the orientation of the chains during viscous drawing of the fibers.

7.11.6.2 Some Viscoelastic Behavior

A common mechanical property test, particularly for metals, is the tensile test in which a metal sample is elongated at a constant applied tensile strain rate, $\dot{\varepsilon}_a$, and the stress and strain are measured and used to define the mechanical properties of the metal such as elastic modulus, yield stress, tensile stress, plastic strain, and fracture stress. Similar constant strain rate tests are also used for crystalline ceramics to determine the elastic modulus, the fracture stress, and the fracture toughness. In glasses, above the glass transition temperature, the main properties of interest are the shear modulus, G, Units(G) = Pa, and the viscosity, μ, Units(μ) = Pa-s, as shown in Figure 7.42.* Below T_g, the main properties of interest for glass are the shear modulus and the fracture strength. There are several ways to express the relationship between the properties of interest and the shear strain rate, $\dot{\gamma} = d\gamma/dt$ (Scherer 1986; Ward and Hadley 1993). However, for a constant applied strain rate, $\dot{\gamma}_{applied}$, the type of test commonly used to determine the mechanical properties of metals and ceramics, general behavior can be obtained from the simple general relation,

$$\dot{\gamma}_{applied} = \dot{\gamma}_{elastic} + \dot{\gamma}_{plastic} \tag{7.37}$$

where $\dot{\gamma}_{applied} = \dot{\gamma} =$ constant is the applied strain rate. Now, $\tau = G\gamma_{elastic}$ so $\dot{\gamma}_{elastic} = (1/G)(d\tau/dt)$ where τ = shear stress defined in Figure 7.40 with Units(τ) = Pa. Finally, $\tau = \mu\dot{\gamma}_{plastic}$. These substitutions into Equation 7.37, gives the following nonhomogeneous differential equation:

$$\dot{\gamma} = \frac{1}{G}\frac{d\tau}{dt} + \frac{\tau}{\mu} \tag{7.38}$$

* Finding viscosity versus temperature data for polymers is difficult and its behavior with temperature can be complex. The viscosity data in this figure were calculated with $\mu = \mu_0 \exp(Q/RT)$ but with a much larger Q than would be expected for polymers in order to cover the range of viscosity values shown.

which is easy to solve by standard integrations when $\dot{\gamma}$ = constant. Rewriting this as

$$\frac{d\tau}{(\dot{\gamma} - \tau/\mu)} = Gdt$$

and integrating,

$$-\mu \ln(\dot{\gamma} - \tau/\mu) = Gt + C$$

where C is an integration constant. Rearranging and taking the exponent of both sides of this equation gives

$$\dot{\gamma} - \frac{\tau}{\mu} = C'e^{-\frac{G}{\mu}t} = C'e^{-\frac{t}{t_{rel}}}$$

where C′ is another constant and $t_{rel} = \mu/G$ is the relaxation time and when t = 0, τ = 0, gives $C' = \dot{\gamma}$ so the final solution is

$$\tau = \mu\dot{\gamma}\left(1 - e^{-\frac{t}{t_{rel}}}\right). \quad (7.39)$$

Figure 7.44 shows the stress–strain behavior near the glass transition of Figure 7.42 for a fixed value of G and different values of the viscosity. Figure 7.45 shows similar data but for both varying shear modulus and viscosity, which is a more realistic model near T_g. For both of these figures, the constant strain rate was 0.001 s^{-1}, which is typical for a constant strain rate test and the resulting total strains are small (Askeland and Phulé 2003). This strain rate is much lower than typically used in forming polymeric and inorganic glasses where strain rates about 10^3 s^{-1} are more common and the viscosity and the shear modulus may vary with shear rate. Figure 7.46 shows a relaxation phenomenon typical of viscoelastic behavior. In this case, the applied strain rate was stopped at 15 s where the total stain is held at γ = 0.015 and the stress relaxes to zero with a relaxation time of t_{rel} = 4 seconds.

Many of the mechanical properties of polymers above the glass transition temperature are performed not with a constant strain rate (or stress rate) but one that is periodic, that is, $\dot{\gamma}_{applied} = A \sin \omega t$, as a function of both temperature and frequency, ω. As a result, the solution to Equation 7.37 becomes more difficult (although not much more, but requires solution techniques beyond what are necessary here) but this technique provides detailed information about structural elements in the glass and their relaxations. This experimental technique is frequently referred to as the measurement of

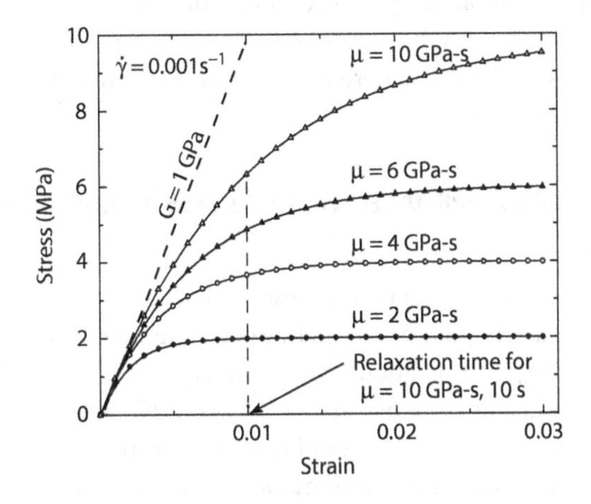

FIGURE 7.44 Viscoelastic stress–strain behavior for a polymer with a fixed value of the shear modulus G = 1 GPa and four different viscosities that give rise to different relaxation times. The relaxation time of 10 seconds is shown for the material that has a viscosity of μ = 10 GPa-s.

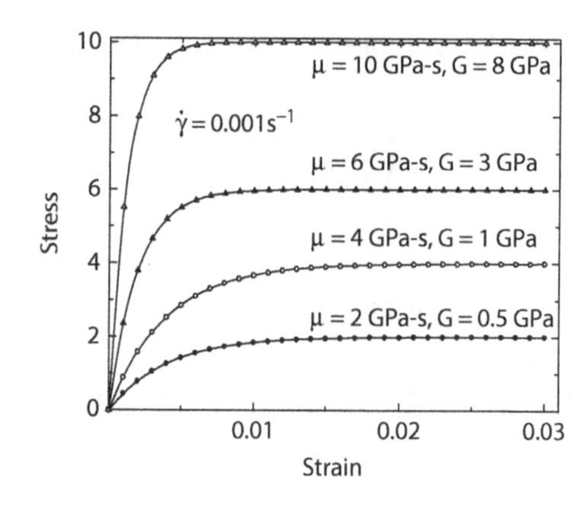

FIGURE 7.45 This figure is similar to Figure 7.44 except in this case, each stress–strain curve has a different value of the shear modulus and viscosity, which is more realistic behavior particularly as the glass transition temperature is approached from higher temperatures.

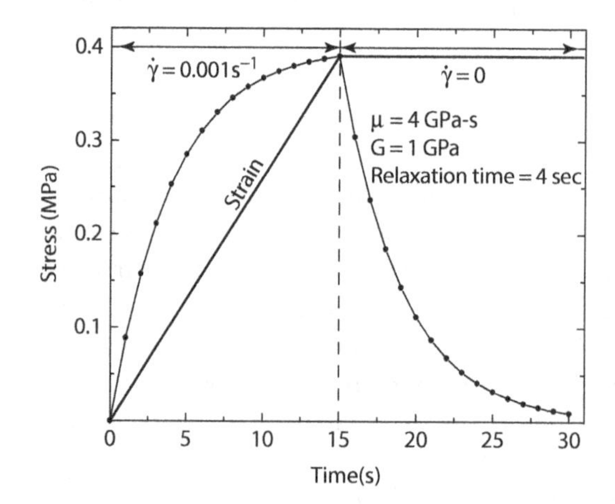

FIGURE 7.46 Example of stress relaxation for a viscoelastic material in a constant strain rate test in which the applied strain rate was dropped to zero at 15 seconds. The figure shows the constant strain and the decreasing stress with time with a relaxation time of 4 seconds after the applied constant strain rate was stopped.

internal friction and gives correlations between frequency and temperature that help to determine the structural features of the glass that produce a specific relaxation. One of the reasons that the technique has not been applied more frequently to inorganic glasses is that it is relatively easy experimentally near room temperature but becomes considerably more difficult as the temperature gets higher, particularly in excess of 500°C, where it would have the most interest for inorganic glasses, particularly silicates.

7.11.7 A Free Volume Approach to the Glass Transition

7.11.7.1 Free Volume

A somewhat different approach to the glass transition involves a concept known as *free volume*. As was shown earlier, the increase in volume on melting of many metals is on the order of about 10%. For ionic compounds, this increase is about 20% (Kingery 1959). The volume change on melting of polymers is less certain but has been estimated between 2% and 10% (Sperling 2006). This *free volume* is associated with the increased degrees of freedom for the molecules, atoms, and ions in the liquid to perform translational and other larger modes of motion than in the solid state. As a result, the volume of a liquid can be given as

$$V = V_a + V_f$$

where:

V_a is the volume of atoms, ions, and molecules
V_f is the free volume.

The volume thermal expansion of liquids is typically larger than that of solids, and so the free volume increases more rapidly with temperature than that required of vibrational motion in the solid state.

7.11.7.2 Free Volume and T_g

As the temperature is decreased, the amount of free volume decreases because the extra motion of the molecules or atoms gets smaller. There are many, but not terribly different, approaches to relating the free volume to the glass transition temperature. In all cases, it is assumed that the changing configurations of the molecules cannot keep up with the rate of the decreasing temperature, $\beta = dT/dt \cong \Delta T/\Delta t$ for a constant rate, thereby *freezing in* an amount of free volume. Figure 7.47 shows the free volume of a liquid as a function of temperature with the free volume no longer changing at the glass transition temperature. As mentioned before, the glass transition is still not completely understood. However, a *plausibility* argument, developed by Varshneya, based on the free volume and modeled on the glass transition in inorganic glasses gives some valuable insight into the transition (Varshneya 1994).

7.11.7.3 Volume Viscoelastic Behavior

In Section 7.11.6, it was shown that the shear strain rate could be expressed as

$$\dot{\gamma}_{applied} = \dot{\gamma}_{elastic} + \dot{\gamma}_{plastic} \tag{7.40}$$

that leads to the result

$$\tau = \mu\dot{\gamma}\left(1 - e^{-\frac{t}{t_{rel}}}\right) \tag{7.41}$$

where:

$t_{rel} = \mu/G$
μ is the viscosity
G is the shear modulus
τ is the shear stress
$\dot{\gamma}$ is the applied strain rate.

A similar equation to Equation 7.40 can be written for the volume strain rates $\dot{\xi}_i$ where $\dot{\xi}_i = (1/V_i)(dV_i/dt)$ so

$$\dot{\xi}_{applied} = \dot{\xi}_{plastic} + \dot{\xi}_{elastic} \tag{7.42}$$

with a similar result,

$$P = \eta\dot{\xi}\left(1 - e^{-\frac{t}{t_{rel}}}\right) \tag{7.43}$$

where:

$t_{rel} = \mu/K = \mu\kappa$
η is the *volume* viscosity $\cong \mu$
K is the bulk modulus
κ is the compressibility $= 1/K$
P is the pressure
$\dot{\xi}$ is the applied volume strain rate.

The main point is that the volume viscous relaxation time is given by $t_{rel} = \mu\kappa$ where $\kappa = (1/V)(dV/dP) = d\xi_{elastic}/dP$ with Units$(\kappa) = Pa^{-1}$.

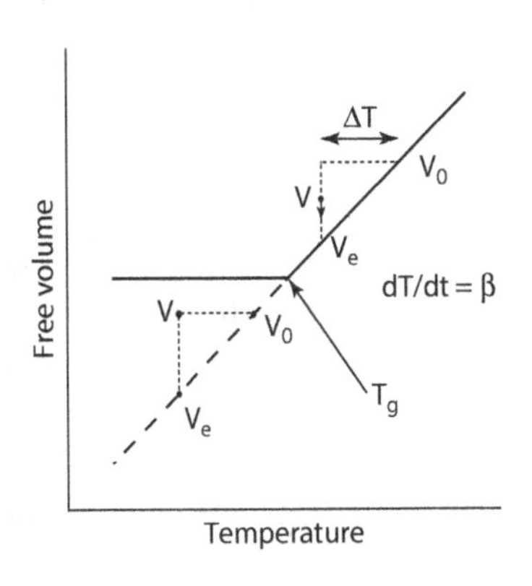

FIGURE 7.47 Model for the relaxation of free volume as a function of temperature. The solid diagonal line and its dashed extension represent the equilibrium free volume of the liquid as a function of temperature. Above T_g, if the temperature is suddenly changed by an amount ΔT, the liquid will have time to relax to its equilibrium value V_e with the cooling rate, β. However, below T_g, there is insufficient time to relax and the volume does not change. By making ΔT sufficiently small, the glass transition temperature occurs at $\mu \cong 10^{12}$ Pa-s, and the horizontal line is developed where no relaxation occurs and the free volume in the glass does not change with temperature. (After Varshneya 1994)

7.11.7.4 Free Volume Relaxation

Referring to Figure 7.47, suppose that at the V_0 above T_g, the temperature were suddenly dropped by an amount ΔT where the new equilibrium free volume is V_e. The question is, *Can the free volume relax fast enough to keep up with the constant rate of temperature decrease*, $\beta = dT/dt = \Delta T/\Delta t$. This can be modeled with first order kinetics,

$$\frac{dV}{V - V_e} = -\frac{dt}{t_{rel}} \tag{7.44}$$

and integrated to give

$$V - V_e = C'e^{-\frac{\Delta t}{t_{rel}}} \tag{7.45}$$

where Δt simply represents the time from t = 0 when the temperature was changed to current time, t. When t = 0, $V = V_0$ so the solution is

$$f_{unrel} = \frac{V - V_e}{V_0 - V_e} = e^{-\frac{\Delta t}{\mu\kappa}} = e^{-\frac{\Delta T}{\mu\kappa\beta}} \tag{7.46}$$

where f_{unrel} = the fractional *unrelaxed free volume* after an instantaneous temperature change of ΔT. If Equation 7.46 is 1.0 for a given set of parameters, then there is no relaxation and the liquid is now a glass. If Equation 7.46 is zero, then the liquid can completely relax. At the glass transition temperature, T_G, it will be somewhere in between zero and one. Data in the literature used in this model for a soda–lime–silica (Na_2O, CaO, SiO_2) glass of unspecified composition are (Varshneya 1994): $\kappa \cong 10^{-11}\ Pa^{-1}$ and $\mu = 10^{-43}\exp(90{,}000/T)\,Pa\cdot s$. At a cooling rate of $\beta = 10°C/s$ the glass transition temperature of this glass is given to be about 710 K or about 412°C. Therefore, temperatures of 770, 730, 720, 710, and 680 K are chosen to calculate the relaxation with ΔT = 10 K. The results are given in Table 7.2. The results show that, at 770 K, the volume change can keep up with the rate of cooling, while at 680 K, there is no volume relaxation. Notably, at 710 K, the viscosity is $\mu \cong 10^{12}$ Pa-s, the viscosity of the glass transition temperature. The reason that the relaxation rate changes so rapidly with temperature is because of the *double exponential of the relaxation and the viscosity*. From the table, it is clear that if the cooling rate, β, were decreased by a factor of 10 to 1°C/s the exponent in Equation 7.46 is increased by a factor of 10 and will decrease the transition temperature by around 10°. Conversely, an increase in rate by a factor of 10 will increase the glass transition temperature to a similar amount.

To make the model more illustrative, Figure 7.48 shows the relative free volume of the glass described above as a function of the cooling rate. The calculations were performed in the following way. First, obviously the fraction relaxed, f_{rel}, $f_{rel} = 1 - f_{unrel}$, so from Equation 7.46,

$$f_{rel} = \frac{V_0 - V}{V_0 - V_e} = \frac{V - V_0}{V_e - V_0} = 1 - f_{unrel}. \tag{7.47}$$

TABLE 7.2
Volume Relaxation of a Silicate Glass

T(K)	T(°C)	μ(Pa-s)	ΔT/μκβ	f
770	472	5.78×10^7	1.73×10^3	0.00
730	432	3.49×10^{10}	2.87	0.057
720	422	1.94×10^{11}	0.515	0.60
710	412	1.13×10^{12}	8.89×10^{-2}	0.91
680	382	3.02×10^{14}	3.31×10^{-4}	1.00

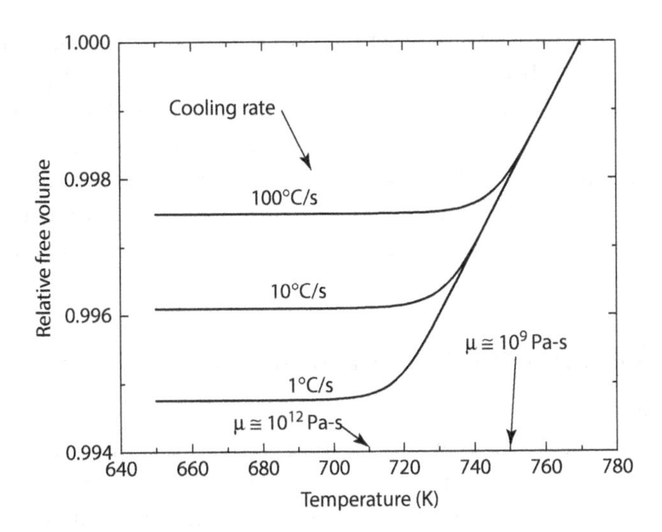

FIGURE 7.48 Glass transition temperature as a function of cooling rate for the model shown in Figure 7.47 with a temperature change of $\Delta T = 2$ K for each relaxation. The data points are not shown since they form essentially continuous curves. The details of the calculation are in the text. Note that the glass transition varies only about 40 K while the cooling rate varies by two orders of magnitude. The calculated viscosities for this glass over this temperature range are also shown.

However, from Figure 7.47, on the free volume versus temperature curve it is obvious that

$$V_e = V_0 - V_0 |\Delta T| \alpha_f$$

where the absolute value of ΔT is taken and α_f, the thermal expansion coefficient of the free volume, $(1/V_f)(dV_f/dT)$. Therefore, Equation 7.47 can be written,

$$V = V_0 \{1 - (1 - f_{unrel}) |\Delta T| \alpha_f\}. \tag{7.48}$$

The relaxation calculations were done at a $|\Delta T| = 2$ K and the free volume, V(T), as a function of temperature was calculated from

$$V(T) = V(T+2)\{1 - (1 - f_{unrel}) |\Delta T| \alpha_f\} \tag{7.49}$$

again, $|\Delta T| = 2$ K and $\alpha_f \cong 10^{-4}$ since $\alpha_f \cong 3\alpha_V \cong 3(3\alpha_L)$ and $\alpha_L \cong 10 \times 10^{-6} K^{-1}$ (Varshneya 1994). That is, the relaxation to a volume V(T) was taken as the relaxation at the volume 2 K higher. This expression was calculated as a function of temperature on a spreadsheet as a function of the cooling rate from 650 to 770 K and plotted in Figure 7.48. Since the actual free volume as a function of temperature requires more empirical data such as density as a function of temperature, relative free volumes were calculated with V(770 K) = 1.00. In contrast to most plots, the calculated data points are not shown in this figure since they essentially form a continuous curve for each of the three plots. Again, this simple plot of free volume versus temperature shows all of the characteristics of the glass transition temperature: a relatively sharp transition from an equilibrium free volume with falling temperature to one that is constant; the faster the cooling rate, the higher the transition temperature; and the glass transition occurs at $\mu \cong 10^{12}$ Pa·s for reasonable cooling rates.

This simple model of free volume relaxation shows clearly why a glass transition occurs, namely, the very strong *double-exponential temperature dependence of the relaxation*. Also, it predicts relatively small changes in the glass transition temperature with orders of magnitude changes in the cooling rate. Finally, it predicts that the glass transition temperature should occur at a viscosity of about 10^{12} Pa-s, which is experimentally found. Again, perhaps not all of the details of the glass transition are understood, but this simple plausibility model predicts the behavior of the glass transition extremely well!

It should be noted that 100°C/s is a rather rapid cooling rate for a glass. If this rate is applied to the surface of a cooling glass, because of the poor thermal conductivity of glasses, the interior of a piece is cooling more slowly. Therefore, the surface becomes rigid well before the interior that continues to shrink until it reaches the glass transition temperature at some lower temperature than the surface. This additional shrinkage of the interior puts the surface under compressive stress thereby increasing the glass strength. This is a commercial process called *thermal tempering* to distinguish it from *chemical tempering* where the composition of the glass is changed at its surface to put the surface under a compressive stress.

7.11.8 Glass Viscosity and Urban Legends

7.11.8.1 The Legends

There is an *urban legend* that exists that suggests that glass—essentially window glass, soda–lime–silica glass, as above—is a supercooled liquid (sic) and therefore flows (very slowly) over long periods of time resulting in changes in thickness and optical properties of window panes! There are two observations that contributed to this legend. First, optical imperfections or striations that exist in old windowpanes are assumed to be flow marks due to the flow of the glass over time since it is just a very viscous liquid. The second version says that if the thickness of an old windowpane is measured, it will be thicker at the bottom than at the top because the glass has flowed over time under its own weight, as sketched in Figure 7.49 (Plumb 1989). Without spending much effort searching, several contradictory articles can be found in the printed literature and on the internet that refute both of these, imply that the data are insufficient to decide their truth, or simply lend some credibility to these legends. They neither attempt to demonstrate why the legends are true or not, nor do most give good definitions of a glass. These articles serve as good examples of why "...because something is in print or online implies that it must be correct..." is not always true and refute the validity of the often-heard comment, "I don't need to know this stuff, I can just look it up!"

7.11.8.2 Dispelling Legend One

In old buildings, glass windows have striations which are actually *flow marks* produced when the glass was poured. Years ago, flat glass for windowpanes was made by pouring the molten glass onto a flat iron or steel sheet to cool and solidify, or in more recent times, by passing it through cold rollers. As a result, the cooling and thickness was very nonuniform and flow marks remained on the surface of—and in—the glass. As a result, distortions are observed when looking through the glass that give the appearance of being caused by continuing flow. An extreme example of this is glass made in the United States during colonial times by just pouring the molten glass onto a cold iron plate as shown in Figure 7.50. This is a window in the *Public House*—an inn and restaurant—in Sturbridge,

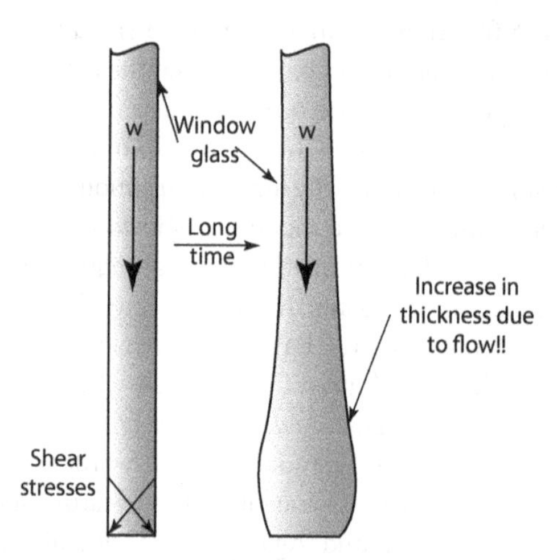

FIGURE 7.49 Sketch of the expected increase in thickness, right, of a window pane deforming under its own weight, w, which the text demonstrates to be unmeasurable.

FIGURE 7.50 Front door windowpane in the over 200-year-old *Public House* in Sturbridge Massachusetts. The pane was formed by pouring molten glass onto a cold iron plate producing the large variations in thickness and flow marks in the glass that result in severe optical distortion. The distorted image is that of a tree outside the window some 10 ft away.

Massachusetts, that is over 200 years old. Such nonuniformities do not occur today because window glass is made by the *float-glass process* developed in 1959 (Doyle 1979) in which the molten and relatively low viscosity glass ($\mu \cong 10^2$ Pa-s), a viscosity similar to that of molasses, at about 1100°C, is poured onto a bath of molten tin where it reaches a uniform thickness and smoothness. The glass is slowly pulled along the molten tin bath with a gradually decreasing temperature until it exits the tin-bath furnace at about 600°C where the viscosity has now risen to about 10^{12} Pa-s (Doremus 1994, p. 100), near the glass transition temperature and the glass no longer flows and is of a uniform thickness and flatness.

7.11.8.3 Dispelling Legend Two

A few rough calculations can quickly disprove this version of the myth. The viscosity of a soda–lime–silica glass (Na_2O–CaO–SiO_2)—window glass—is strongly temperature dependent, as is the viscosity of all silicate and other inorganic and organic glasses. Specifically, the viscosity of soda–lime–silica window glass can be approximated over a wide temperature range by an exponential function of temperature

$$\mu = \mu_0 e^{\frac{Q}{RT}}$$

where again μ_0 = pre-exponential and Q = activation energy. Surprisingly, it is difficult to find data for Q and μ_0 in the literature at low temperatures, so existing data need to be extrapolated to room temperature. With the viscosity data from Section 7.11.7.4, namely, $\mu = 10^{-43} \exp(90{,}000/T)\,\text{Pa}\cdot\text{s}$, at 298 K, room temperature, $\mu \cong 1.5 \times 10^{88}$!!

As was seen in Section 7.11.4, the relaxation time for stress or strain provided by viscous flow is $t_{rel} = \mu/G$. Now Young's modulus for window glass is E ≅ 70 GPa (Kingery et al. 1976) and the shear modulus is given by $G = E/2(1+\nu)$ (McClintock and Argon 1966) and ν = Poisson's ration ≅ 0.25. So, G ≅ 28 GPa and the relaxation time would be,

$$t_{rel} = \mu/G = \frac{1.5\times10^{88}}{28\times10^{9}} \cong 5\times10^{77}\,\text{seconds !!!!}$$

Now the age of the earth is about 4.5 billion years and there are 3 ≅ 10^7 s/year so the age of the earth is ≅ 10^{17} s; 60 orders of magnitude too small to produce any detectable strain!

This calculation assumes that the high temperature data can be extrapolated down to room temperature. This is undoubtedly not entirely correct and literature data suggest that the activation energy for viscosity below the glass transition is about one-half that above, or, at 298 K, room temperature, $\mu = 10^{-43}\exp(45{,}000/T)\,\text{Pa}\cdot\text{s}$ and the viscosity at room temperature could be calculated from this based on the viscosity being 10^{12} Pa-s at the glass transition temperature of about 412°C. However, there are other data in the literature that suggest at room temperature the viscosity for window glass is about 10^{20} Pa-s (Jones 1956). In that case, the relaxation time is

$$t_{rel} = \mu/G = \frac{10^{20}}{28\times10^{9}} \cong 3.6\times10^{9}\ \text{seconds}$$

or about 100 years.

As was seen earlier, the relationship between viscosity, μ; shear stress, τ; and shear strain rate, $\dot{\gamma}$, is $\tau = \mu\dot{\gamma}$. This allows the following more precise calculation. For the sake of concreteness, take a windowpane 4 ft × 4 ft × 1/8 in. (these would be typical window dimensions in the United States) which has a total volume of 4.72×10^3 cm³. The density of this glass is $\rho \cong 2.3$ g/cm³ so the weight of the window is (4.72×10^3 cm³)(2.3 g/cm³) = 1.08×10^4 g = 10.8 kg so that the total force at the bottom is 10.8 kg × 9.8 m²/s = 106.4 N. Now the bottom area is $(12\,\text{in./ft}\times4\,\text{ft}\times2.54\,\text{cm/in.})(0.125\,\text{in.}\times2.54\,\text{cm/in.}) = 38.70\ \text{cm}^2 = 3.87\times10^{-3}\,\text{m}^2$. So the compressive stress at the bottom of the windowpane is σ = (106.4 N) ÷ (3.87×10^{-3} m²) = 2.79×10^4 Pa (~4 psi). The shear stresses will be on the same order of magnitude as the compressive stress, that is, $\tau \cong \sigma/2$ but at 45° to the plane of the window. Therefore, the strain rate can be calculated to be

$$\dot{\gamma} = \frac{d\gamma}{dt} \cong \frac{\Delta x/x}{\Delta t} = \frac{\tau}{\mu} \cong \frac{2.79\times10^{4}\ \text{Pa}}{10^{20}\ \text{Pa}\cdot\text{s}} = 2.7\times10^{-16}\,\text{s}^{-1}$$

where x = thickness of the glass = 0.125 in. = 3.175×10^{-3} m. Assume that the difference in thickness between the top and bottom could be measured to ± 1 μm = Δx (±1 mil = 0.001 in. = 25.4 μm would be more realistic). The time necessary then for this increase in thickness would be

$$\Delta t \cong \frac{\Delta x}{x\dot{\gamma}} = \frac{1\times10^{-6}\ \text{m}}{(3.17\times10^{-3}\,\text{m})(2.7\times10^{-16}\,\text{s}^{-1})} \cong 1.2\times10^{12}\ \text{s.} \tag{7.50}$$

Again, since there are about 3×10^7 s/year this corresponds to about 40,000 years, far longer than most windowpanes have been around. This *urban legend* is clearly just that, a legend that has no basis in fact! In reality, variations in thickness of unused float glass produced today on the order of 140 μm are observed (Dorn et al. 2013). This implies that in order to observe a thickness difference between the top and bottom of a windowpane caused by flow and not a statistical variation, Δx would have to be at least 100 times larger in Equation 7.50 requiring something on the order four million years to be measureable! This is not an accurate model but certainly good enough for a *back of the envelope* calculation sufficiently plausible to send this myth to the *dust-bin of history!*

7.11.9 The Challenger Disaster and the Glass Transition

Figure 7.51 shows the liftoff of the shuttle *Challenger* on its last—and fatal—flight on January 28, 1986. The night of January 27–28, 1986, was an extremely cold night in Florida, freezing or below. The relevant conclusion of the Challenger Accident Report (The Presidential Commission on the Space shuttle Challenger Accident Report 1986) was that a hot gas leak occurred around an elastomer O-ring seal in one of the solid booster rockets and ultimately caused ignition of the main hydrogen and oxygen fuel tanks. There were several factors for the seal failure including poor design. However, the crash investigation team found that the temperature dependence of the elastic behavior of elastomer O-rings probably contributed to the seal failure as suggested by physicist and team

FIGURE 7.51 The *Challenger* space shuttle launch on January 28, 1986, just moments before it was destroyed because of a hot gas leak in one of the O-ring seals in one of the solid booster rockets. Original NASA color photograph converted to grayscale.

member, Richard Feynman. At these low temperatures, the fluorelastomer O-rings on the boosters either became brittle or lost some of their elasticity, entered the *leathery* region of Figure 7.42, and no longer provided a gas-tight seal. This allowed hot gases to leak and burn through eventually leading to ignition of the hydrogen and oxygen fuel tanks and complete destruction of the system. Data on the specific elastomer used for the solid booster O-rings could not be found other than they were fluoroelastomers. These are copolymers of polytetrafluoroethylene—or other fluorinated polymer—with another polymer and can have a glass transition temperature as high as −8°C (17.6°F) (Viton). Lacking modulus data as a function of temperature, it is a valid conjecture that this glass transition is high enough so that at the cold ambient temperatures that morning, near freezing, the elastomer probably became much less elastic contributing to the shuttle's tragic destruction.

7.12 PARTICLE SIZE AND PHASE TRANSITIONS

Research on surface energies of solids has shown that polymorphs that are unstable under normal conditions tend to have lower surface energies than the stable phases (Navrotsky 2011). For example, the surface energy of hematite, α-Fe_2O_3—corundum structure—has a surface energy of 1.9 J/m^2. In contrast the metastable phase of large particles, maghemite, γ-Fe_2O_3—defect spinel structure—has a surface energy of 0.71 J/m^2 (Navrotsky 2011). The implication of this is that a normally metastable phase may become the *stable phase* when the particle size is small since, at constant temperature, pressure, and composition, $\overline{G} = \gamma\overline{S}$ where γ is the surface energy and $\overline{S}$ is the specific surface area per mole, Units($\overline{S}$) = m^2/mol (see Chapter 6, Section 6.9). Figure 7.52 schematically shows how this can occur. At large particle sizes, near $\overline{S} = 0$, the stable phase is alpha, α; the next most stable is beta, β; and the least stable is delta, δ. As the specific surface area increases, because of the differences in surface energies—the slopes of the lines—around 20,000 m^2/mol, beta is the stable phase and at specific surface areas in excess of about 25,000 m^2/mol, the delta phase is stable. There are several implications of this. First, frequently the metastable phases result from the process used to attempt to make the stable phase of a material. For example, in the production of magnetic iron oxides such as spinels, iron oxide, Fe_2O_3, is often made by the decomposition of iron oxalate, $Fe_2(C_2O_4)_3$

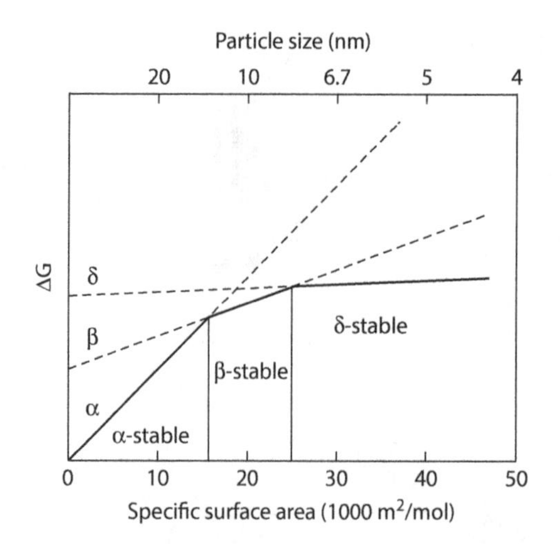

FIGURE 7.52 Schematic drawing illustrating the effect of particle size on the stability of various polymorphs of a given material. With bulk or large particles the alpha phase is stable while with small particle sizes both the beta and delta phases are more stable because of their lower surface energies stabilizing small particles of these *metastable* phases. (Data from Navrotsky 2011)

$$Fe_2(C_2O_4)_3 \xrightarrow{100^\circ C} Fe_2O_3 + 3CO_2(g) + 3CO(g)$$

and can be accomplished by combustion synthesis since the CO can be ignited to provide the heat for the reaction. The resulting Fe_2O_3 produced at this low temperature is small particle size γ-Fe_2O_3 long thought to be the result of the kinetics of decomposition rather than the thermodynamics. That is, it is thought that kinetically it is easier to form the gamma phase rather than the alpha phase at this low temperature. However, if the particle size is small enough, less than about 40 nm ($\bar{S} \cong 5000\,m^2/mol$) then gamma *is* the stable phase because of its lower overall energy because its total surface energy is much lower than that of the alpha phase, hematite. (Navrotsky 2011). Similarly, hydrated surfaces of oxides have lower energies than those of oxides. As a result, the decomposition of goethite, α-FeOOH, to hematite occurs almost 100°C higher for 10 nm particles of goethite compared to bulk particles since it is the stable phase to higher temperatures because of its higher surface energy (Navrotsky 2011). Finally, because of this effect of particle size and relative energies of phases and their region of stability, the equilibrium phase boundaries as a function of $p(O_2)$ and temperature can be moved with small particle sizes, for example,

$$6Fe_2O_3 \xrightleftharpoons{800K} 4Fe_3O_4 + O_2(g)$$

the equilibrium oxygen pressure at 800 K is $p(O_2) \cong 10^{-16}$ bar for bulk particles (Roine 2002) while for 10 nm particles, $p(O_2) \cong 10^{-12}$ bar. And the phase field for FeO, in T-$p(O_2)$ space for iron and oxygen, completely disappears for 10 nm particles! (Navrotsky et al. 2010).

7.13 CHAPTER SUMMARY

This chapter covered many different topics, some in more detail, others in less. The thermodynamics of phase transformations is reviewed. Then the various types of phase transformations are presented focusing on the difference between displacive and reconstructive transformations. Reconstructive transformations in pure materials require the diffusive motion of atoms, ions, or molecules and, therefore, are dependent on time. As a result, most of these transformations occur through a nucleation and growth process that can be controlled by either a surface reaction or diffusion. Phase transformations between gases and solids and gases and liquids are emphasized since the interaction between a gas and a condensed phase interface is easiest to model. In order to do this, some kinetic theory of gases was introduced in order to calculate rates of condensation and growth. Then a model of homogeneous nucleation is introduced. The overall rate of phase transformation by a nucleation and growth process is calculated assuming that a surface reaction controlled the growth process. The model of homogeneous nucleation is then modified for heterogeneous nucleation based on the wetting behavior of a heterogeneous nucleus. The other important class of transformations but probably less well understood, displacive transformations, is illustrated in systems such as stabilized zirconia, $BaTiO_3$, martensite in steels, and shape-memory alloys. However, the kinetics of such transformations are of less interest than their crystallography and resulting lattice strains since the transitions propagate at essentially the speed of sound. Finally, the glass transition is discussed because of its importance for both polymers and inorganic glasses and a definition of a *glass* is presented. A simple free-volume model is then presented that surprisingly lends significant insight into the glass transition: it is a relatively sharp transition because the relaxation of free volume depends on a double exponential with temperature! Then some calculations of viscoelastic behavior of inorganic glasses are made that help to dispel some *urban legends* about glass. Finally, a proximate reason for the space shuttle *Challenger* disaster may very well have been the glass transition in its booster rocket polymer O-ring seals.

APPENDIX

A.1 Volume Thermal Expansion Coefficient

The relation between the volume thermal expansion coefficient, α_V, and the linear thermal expansion coefficient, α_L, is given by $\alpha_V \cong 3\alpha_L$ and can be determined in the following way. The two expansion coefficients are defined as

$$\alpha_L = \frac{1}{L_0}\frac{dL}{dt} \cong \frac{1}{L_0}\frac{\Delta L}{\Delta T}$$

$$\alpha_V = \frac{1}{V_0}\frac{dV}{dt} \cong \frac{1}{V_0}\frac{\Delta V}{\Delta T}.$$

Consider a cube of initial volume V_0 and length L_0 so $V_0 = L_0^3$ then with some temperature increase of ΔT

$$L = L_0 + \Delta L = L_0 + L_0\alpha_L\Delta T = L_0(1+\alpha_L\Delta T)$$

$$V = V_0 + \Delta V = V_0 + V_0\alpha_V\Delta T = V_0(1+\alpha_V\Delta T)$$

but $V = L^3$ so,

$$L^3 = L_0^3(1+\alpha_L\Delta T)^3$$

$$L^3 = L_0^3\left(1+3\alpha_L\Delta T+3(\alpha_L\Delta T)^2+(\alpha_L\Delta T)^3\right)$$

$$V \cong V_0(1+3\alpha_L\Delta T)$$

since the squared and cubed terms are very small. Comparing the equations for V, it is clear that

$$\alpha_V \cong 3\alpha_L.$$

EXERCISES

7.1 The Gibbs energy, G, as a function of temperature for solid, liquid, and gaseous sodium chloride, follow closely the approximate expressions (Roine 2002):

$$G(\text{solid}) = -430.86 - 0.1126 \times T$$

$$G(\text{liquid}) = -361.87 - 0.1988 \times T$$

$$G(\text{gas}) = -244.35 - 0.2792 \times T$$

where Units(G) = kJ/mol and Units(T) = °C and the gas consists of NaCl molecules.

a. Plot these three curves as a function of temperature between 0°C and 2000°C.
b. Calculate the melting and boiling points of NaCl in °C from these data and compare them to literature values. Give the reference for the literature data.
c. Calculate the Gibbs energy (J/mol) for condensation of NaCl vapor to solid NaCl at 500°C.
d. Calculate the equilibrium NaCl vapor pressure at 500°C.

7.2 This problem uses some of the data from Exercise 7.1.

a. The molecular weight and density of solid NaCl are 58.433 g/mol and 2.5 g/cm^3, respectively, calculate the molar volume (m^3/mol) (Haynes 2013).
b. If solid NaCl is in contact with NaCl gas that is 10^4 times the equilibrium partial pressure at 500°C, calculate the Gibbs energy per unit volume (J/m^3) to form a homogeneous nucleus.
c. If the surface energy of solid NaCl from Chapter 6 is 300 mJ/m^2, calculate the critical nucleus radius in meters.
d. Calculate the number of NaCl molecules in this critical nucleus.
e. Calculate the Gibbs energy of the critical nucleus, J.
f. Calculate the number of NaCl molecules/m^3 in the gas.
g. Calculate the concentration of critical nuclei (m^{-3}).
h. If the sticking coefficient is unity, calculate the Θ_0 in the expression for the nucleation rate.
i. Calculate the nucleation rate for NaCl nuclei from the gas phase (m^{-3}s^{-1}) at 500°C with 10^4 times the equilibrium pressure of NaCl gas.

7.3 The major vapor species over solid and liquid selenium, Se, is $Se_2(g)$ so the sublimation or vaporization reaction is 2 Se(s, l) = $Se_2(g)$. The Gibbs energy, G, as a function of temperature for solid, liquid, and gaseous selenium follow closely the approximate expressions (Roine 2002):

$$G(2\text{Se}, \text{solid}) = -23.24 - 0.0970 \times T \text{ kJ} / 2 \text{ moles}$$

$$G(2\text{Se}, \text{liquid}) = -6.20 - 0.1739 \times T \text{ kJ} / 2 \text{ moles}$$

$$G(\text{Se}_2, \text{gas}) = 72.74 - 0.2790 \times T \text{ kJ} / \text{mol}$$

where Units(T) = °C. Note that the values of G for the solid and liquid are for *two moles* since two moles of each form one mole of Se_2 gas.

a. Plot these three curves as a function of temperature between 0°C and 1000°C.
b. Calculate the melting and boiling points of Se in °C from these data and compare them to literature values. Give the reference for the literature data.
c. Calculate the Gibbs energy (J/mol) for condensation of Se vapor to liquid Se at 500°C.
d. Calculate the equilibrium Se vapor pressure at 500°C.

7.4 a. Given the data below (Brandes and Brook 1992) and the melting point of liquid selenium determined in Exercise 7.3, make a plot of volume versus temperature, from 0°C to 500°C for liquid and crystalline Se (see Figure 7.36).

Linear thermal expansion coefficient of solid, $\alpha = 3.7 \times 10^{-6}$ K^{-1}

Solid density at 0°C, ρ(solid) = 4.79 g/cm^3

Liquid density at the melting point, ρ(liq) = 3.989 g/cm^3

Molecular weight of selenium, M = 78.96 g/mol

Change in liquid density with temperature, $d\rho/dT = -1.44 \times 10^{-3}\,\text{g cm}^{-3}\text{K}^{-1}$

b. Calculate the volume expansion coefficient of the solid, K^{-1}.

c. Calculate the volume expansion coefficient of the liquid, K^{-1}.

7.5 The TTT diagram for 1080 steel is Figure 7.5 and shows the times for 1% and 99% transformed.

a. From the 99% transformed curve, make a table of the time for this amount of transformation and the rate of transformation (i.e., 99%/time) versus temperature from 700°C down to 300°C at 50°C intervals.

b. From these data, plot the temperature vertically and the rate of transformation horizontally (%/s) and indicate the equilibrium temperature on your graph.

7.6 A silicate glass is crystallizing following the Johnson–Mehl–Avrami equation, where the fraction crystallized is given by $f_{transformed} = 1 - e^{-At^4}$. The following temperature and times are observed for 50% crystallization: 2 seconds at 600°C, 7 seconds at 500°C, 30 seconds at 700°C, 200 seconds at 750°C, 400 seconds at 300°C, and 300 seconds at 800°C.

a. Calculate the value of "A" for each of these temperatures and times.

b. Plot the Johnson–Mehl–Avrami equation for each of the values of "A" calculated in part "a." *all on the same plot* of fraction transformed versus $\log_{10}$ (time in seconds). Plot at 0.001, 0.01, 0.1, 0.2...0.9, 0.99, and 0.999 transformed to show the shape of the curves.

c. From these data make a table of 0.1, 0.5, and 0.9 transformed as a function of temperature and also include the rate of transformation (s^{-1}) for 50% transformed by taking 50/time(s).

d. From the data, make a plot of temperature vertically and rate of transformation horizontally for the rate calculated for 50% crystallized.

e. Make a TTT diagram for this crystallization by plotting the points for 0.1, 0.5, and 0.9 crystallized with a vertical temperature axis and a horizontal axis of $\log_{10}$ (time, seconds).

7.7 For a certain metal that undergoes a phase transformation from alpha to beta on cooling by a nucleation and growth process at 500°C, given the following data for the nucleation rate Θ (nuclei/m^3-s) and the growth rate of G(m/s) as a function of temperature:

T (°C)	475	400	325
Θ(m^{-3}s^{-1})	5.0×10^{11}	1.0×10^{15}	1.0×10^{16}
G(m/s)	2.0×10^{-8}	5.0×10^{-8}	1.00×10^{-10}

a. On the same graph, calculate and plot the fraction transformed as a function of time ($\log_{10}$ [time, seconds] scale) from the Johnson–Mehl–Avrami equation for each of these temperatures with a minimum of 10 points on each curve.

b. From the data and plots in "a" above, estimate the times for 5%, 50%, and 95% transformed and plot a TTT diagram for this transformation for 5%, 50%, and 95% transformed with temperature vertical and time horizontal ($\log_{10}$ [time, seconds] scale). Include and show the equilibrium temperature.

7.8 Glycerin (glycerol), $C_3H_5(OH)_3$, has a melting point of 17.8°C but a fairly high viscosity at that temperature that allows it to be cooled to form a glass. The following data exist on the viscosity of glycerin (Haynes 2013): −42°C, μ = 6710 Pa-s; −25°C, μ = 262 Pa-s; −15.4°C, μ = 66.5 Pa-s; −4.2°C, μ = 14.9 Pa-s; 6°C, μ = 6.26 Pa-s; 15°C, μ = 2.33 Pa-s; and 25°C, μ = 0.954 Pa-s

a. Plot these data on a semilog plot versus 100/T (K^{-1}) and determine Q and μ_0 in for the $\mu = \mu_0 \exp(Q/RT)$ viscosity (use a *least squares* fit).
b. Calculate the glass transition temperature, °C, for glycerin; that is, where $\mu \cong 10^{12}$ Pa-s.

7.9 A soda–lime–silica window glass (0.15 w/o Na_2O, 0.15 w/o CaO, 0.70 w/o SiO_2) has the following values of viscosity ($\log_{10}\mu$(Pa-s)) and temperature (°C): 9.3, 600; 7.2, 700; 5.4, 800; 4.5, 900; 3.9, 1000; 2.9, 1100; 2.0 1200; 1.7, 1300; and 1.2, 1400 (several sources).
a. Plot these data on a semilog plot versus 1000/T (K^{-1}) and determine Q and μ_0 in for $\mu = \mu_0 \exp(Q/RT)$ the viscosity (use a *least-squares* fit).
b. Calculate the temperature at which the viscosity is 10^{12} Pa-s, °C, for this silicate glass.

7.10 From the data for viscosity calculated in 7.9, make a plot of the relative free volume from 50 K above to 50 K below the temperature at which $\mu \cong 10^{12}$ Pa-s at a cooling rate of 0.1 K/s: a single plot like that if Figure 7.48 and use the same parameters as in Section 7.11.7 in the text. Take the free volume at the +50 K temperature to be 1.00.

REFERENCES

American Society for Metals. 1973. *Atlas of Isothermal Transformation Diagrams.* Metals Park, OH: American Society for Metals.

Askeland, D. R. and P. P. Phulé. 2003. *The Science and Engineering of Materials,* 4th ed. Pacific Grove, CA: Brooks/Cole-Thomson Learning.

Bain, E. C. 1924. The Nature of Martensite. *Trans. AIME, Min. Metall.* 70(Paper No. 1299-S): 1–11.

Balluffi, R. W., S. M. Allen, and W. C. Carter. 2005. *Kinetics of Materials.* New York: Wiley.

Brandes, E. A. and G. B. Brook. 1992. *Smithells Metals Reference Book,* 7th ed. Oxford: Butterworth-Heinemann.

Buchanan, R. C. and T. Park. 1997. *Materials Crystal Chemistry.* New York: Marcel Dekker.

Callister, W. D., Jr. and D. G. Rethwisch. 2009. *Fundamentals of Materials Science and Engineering: An Integrated Approach,* 3rd ed. New York: Wiley.

Chiang, S.-K. 1988. *Microstructure Development in Temperature-Stable $BaTiO_3$.* PhD Dissertation. Columbus, OH: Ohio State University.

Davidovits. P. et al. 2004. Mass Accommodation Coefficient of Water Vapor on Liquid Water. *Geophys. Res. Lett.* 31(L22111): 1–4.

"Diffusionless transformation." *Wikipedia, The Free Encyclopedia,* last modified May 5, 2016. https://en.wikipedia.org/w/index.php?title=Diffusionless_transformatin&oldid=718711113. (accessed July 13, 2016.)

Doremus, R. H. 1994. *Glass Science,* 2nd ed. New York: Wiley, p. 100.

Dorn, H. et al. 2013. Unusual Variation of Thickness Within a Pane of Annealed Float Glass. *Can. Soc. Forensic Sci. J.* 46(3): 166–169.

Doyle, P. J., ed. 1979. *Glass Making Today.* Redhill, UK: Portcullis Press.

Eitel, W. 1954. *The Physical Chemistry of the Silicates.* Chicago, IL: University of Chicago Press.

Emsley, J. 1998. *The Elements,* 3rd ed. Oxford: Oxford University Press.

Emsley, J. 2001. *Nature's Building Blocks.* Oxford: Oxford University Press.

"French invasion of Russia." *Wikipedia, The Free Encyclopedia,* last modified June 24, 2016. https://en.wikipedia.org/w/index.php?title=French_invasion_of_Russa&oldid=726728034. (accessed July 13, 2016.)

Glass Transition. Wikipedia Encyclopedia. http://en.wikipedia.org/wiki/Glass_transition.

"Glass transition." *Wikipedia, The Free Encyclopedia,* last modified July 7, 2016. https://en.wikipedia.org/w/index.php?title=Glass_tansition*oldid=728786909. (accessed July 13, 2016.)

Green, D. J., R. H. J. Hannink, and M. V. Swain. 1989. *Transformation Toughening of Ceramics.* Boca Raton, FL: CRC Press.

Haynes, W. M. Editor-in-Chief. 2013. *CRC Handbook of Chemistry and Physics,* 94th ed. Boca Raton, FL: CRC Press.

Inaba, S., H. Hosono, and S. Ito. 2015. Entropic shrinkage of an oxide glass. *Nat. Mater.* 14(March 15): 312–317.

Jones, G. O. 1956. *Glass.* New York: Wiley.

Kingery, W. D. 1959. Surface Tension of Some Liquid Oxides and Their Temperature Coefficients. *J. Am. Ceram. Soc.* 45(1): 6–10.

Kingery, W. D., H. K. Bowen, and D. R. Uhlmann 1976. *Introduction to Ceramics,* 2nd ed. New York: Wiley.

Krauss, G. 1990. *Steels: Heat Treatment and Processing Principles.* Metals Park, OH: ASM International.

Lipschutz, S., M. Spiegel, and J. Liu. 2012. *Schaum's Outline of Mathematical Handbook of Formulas and tables,* 4th ed. New York: McGraw-Hill.

McClintock, F. A. and A. S. Argon, eds. 1966. *Mechanical Behavior of Materials.* Reading, MA: Addison-Wesley Publishing.

McNaught, A. D. and A. Wilkinson, compilers. 2006. *IUPAC. Compendium of Chemical Terminology,* 2nd.ed. (the "Gold Book"). London: Blackwell. http://goldbook.iupac.org/G02641.html.
Morey, G. W. 1938. *The Properties of Glass.* New York: Rheinhold Publishing.
Morton-Jones, D. H. 1989. *Polymer Processing.* London: Chapman & Hall.
Navrotsky, A. 2011. Nanoscale Effects on Thermodynamics and Phase Equilibria in Oxide Systems. *ChemPhysChem.* 12:2207–2215.
Navrotsky, A. et al. 2010. Nanophase Transition Metal Oxides Show Large Thermodynamically Driven shifts in Oxidation-Reduction Equilibria. *Science.* 330:199–201.
Newnham, R. E. 2005. *Properties of Materials.* Oxford: Oxford University Press.
Pfaender, H. G. 1983. *Schott Guide to Glass.* New York: Van Nostrand Reinhold.
Plumb, R. C. 1989. Antique Window Panes and the Flow of Supercooled Liquids. *J. Chem. Ed.* 66(12): 994–996.
Porter, D. A., K. E. Easterling, and M. Y. Sherif. 2009. *Phase Transformations in Metals and Alloys,* 3rd ed. Boca Raton, FL: Taylor & Francis Group.
Pound, G. M. 1972. Selected Values of Evaporation and Condensation Coefficients for Simple Substances. *J. Phys. Chem. Ref. Data.* 1: 135–145.
Presidential Commission Report on Space Shuttle Challenger Accident. June 1986. http://science.ksc.nasa.gov/shuttle/missions/51-l/docs/rogers-commission/table-of-contents.html.
Reed-Hill, R. E. and R. Abbaschian. 1992. *Physical Metallurgy Principles,* 3rd ed. Boston, MA: PWS Kent.
Roine, A. 2002. *Outokumpu HSC Chemistry for Windows, Chemical Reaction and Equilibrium Software with Extensive Thermochemical Database,* Version 5.1. Pori, Finland: Outokumpu Research Oy.
Roth, R. S. 1995. *Phase Equilibrium Diagrams, Phase Diagrams for Ceramists,* Volume XI. Washington, DC: National Institute for Standards and Technology.
Roy, B. 2006. *Development of Copper Strontium Oxide and Recrystallization of Hot Wire Chemical Vapor Deposited Amorphous Silicon Films for Solar Cell Applications.* PhD Thesis. Colorado School of Mines, Golden, CO.
Scherer, G. W. 1986. *Relaxation in Glass and Composites.* New York: Wiley.
Silbey, R. J. and R. A. Alberty. 2001. *Physical Chemistry,* 3rd ed. New York: Wiley.
Sosman, R. B. 1965. *The Phases of Silica.* New Brunswick, NJ: Rutgers University Press.
Sperling, L. H. 2006. *Introduction to Physical Polymer Science,* 4th ed. Hoboken, NJ: Wiley.
"Tin pest." *Wikipedia, The Free Encyclopedia,* last modified 13 May 2016. https://en.wikipedia.og/w/index.php?title=Tin_pest&oldid=720004224. (accessed July 13, 2016.)
Van Krevelen, D. W. 1997. *Properties of Polymers,* 3rd ed. Amsterdam, the Netherlands: Elsevier B. V.
Varshneya, A. K. 1994. *Fundamentals of Inorganic Glasses.* Boston, MA: Academic Press.
Video frame from: www.periodictable.ru. (2009 May 2). *Grey tin (tin pest) time-lapse video* [Video file]. Retrieved from https://www.youtube.com/watch?v=FUoVemHuykM.
Viton®. http://www.dupontelastomers.com.
Ward, I. M. and D. W. Hadley. 1993. *An Introduction to the Mechanical Properties of Solid Polymers. Chichester,* England: Wiley.
Wells, A. F. 1984. *Structural Inorganic Chemistry,* 5th ed. Oxford: Oxford University Press.

Section IV
Diffusion with a Constant Diffusion Coefficient

8

Fundamentals of Diffusion

8.1 INTRODUCTION

The process of diffusion—atomic or molecular motion—is central to understanding the kinetics of many processes of interest to the materials scientist or engineer. In fact, the study of diffusion was initially stimulated more by its application to materials than any other field of science or engineering: for example, the heat treatment of steel. The purpose here is to introduce the microscopic mechanisms of diffusion in solids, liquids, and gases, and an elementary rationale given for the differences in diffusion behavior in various materials. The macroscopic differential equations that are used to describe these kinetic processes are also examined. For these, several diffusion geometries and conditions necessary in quantitatively modeling kinetic processes important to materials are presented, each of which requires a somewhat different mathematical treatment. This chapter serves as an introduction only, and both the microscopic and macroscopic aspects of diffusion are expanded in more detail in later chapters. Furthermore, the order in which the next four Chapters 9 through 12 are presented and studied is more a matter of individual choice than one of instructional necessity.

8.2 THE PROCESS OF DIFFUSION

Again, *diffusion* can be defined as *spontaneous intermingling or mixing of atoms, ions, molecules, or particles due to random thermal motion.* This intermingling occurs in the solid, liquid, or gaseous state.

Figure 8.1 schematically shows the *macroscopic* results of the *microscopic or atomistic* processes of diffusion. The diffusion process sketched in Figure 8.1 is pure nitrogen and oxygen of equal amounts initially separated by some kind of barrier, and at $t = 0$, this barrier is removed without introducing convection allowing the gas molecules to diffuse into each other. In this case of nitrogen and hydrogen, as in all diffusion processes, the kinetic energy of the molecules is determined by the temperature. Even at room temperature, gas molecules have rapid thermal motion—speeds on the order of thousands of miles per hour—and they travel only about 0.1 μm before colliding with another gas molecule. So they experience over a billion scattering collisions per second, which leads to their mixing or randomization.

Referring to Figure 8.1, if these are ideal gases, an assumption that will always be made unless otherwise specified, then there is a thermodynamic driving force for the gases to intermingle, the molar ideal Gibbs energy of mixing:

$$\Delta\bar{G}_m^{id} = -T\Delta\bar{S}_m^{id} = RT(X_1 \ln X_1 + X_2 \ln X_x) \tag{8.1}$$

where X_1 and X_2 are the mole fractions of components 1 and 2: nitrogen and oxygen, respectively. Note that the $\Delta\bar{H}^{id}$ for ideal mixing is zero and is a good definition of an ideal solution. This is the thermodynamic equivalent of saying that the molecules are mixing by thermal motion: there is an increase in entropy on mixing. The individual molecules do not know anything about entropy but are mixing because of their random motion. Nevertheless, this can be a convenient way of describing the driving force for diffusion; namely, that the ideal Gibbs energy of mixing is negative. For the interdiffusion of one mole of nitrogen and one mole of oxygen, the ideal Gibbs energy of mixing at 300 K is $\Delta\bar{G}_m^{id}(300\,K) = 8.314 \times 300 \times (\ln 0.5) = -1729\,J/mol$. Not much, but enough.

Before being allowed to diffuse, the gases are kept in separate chambers so that the concentration in each chamber is 100%. At time $t = 0$, they are allowed to begin diffusing into each other. When the two gases have completely mixed, $t = \infty$, and if there were one mole of each, then the final concentration of each is 50 m/o as shown in Figure 8.1. For times $0 \le t \le \infty$, the concentration is a function of distance and time, C(x, t), as shown. For engineers, the interest is: (1) how long is close

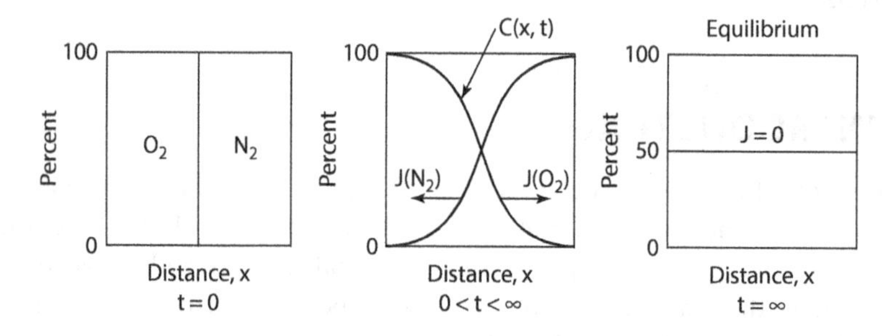

FIGURE 8.1 Schematic diagram showing the mixing of nitrogen, N_2 and oxygen, O_2, by diffusion as a function of time.

enough to $t = \infty$ so the process can be considered complete and (2) what is the concentration as a function of time and distance, C(x, t).

8.3 FICK'S FIRST LAW OF DIFFUSION

Adolf Fick, a German physiologist who was concerned about how fast sodium chloride solutions moved around in the human body, verified the following equation in about 1850, now known as *Fick's first law* of diffusion (Cussler 1997; Narasimhan 2004)

$$J = -D\frac{dC}{dx} \tag{8.2}$$

where:

J is the flux density
D is the *diffusion coefficient* or the *diffusivity*
dC/dx is the concentration gradient.

One consistent set of units is Units(J) = mol/cm^2 s, Units(D) = cm^2/s, and Units(C) = mol/cm^3 so Units(dC/dx) = mol/cm^4 (or the comparable SI units such as Units(D) = m^2/s).* Note that the flux is opposite to the concentration gradient, as is shown in Figure 8.1.

Equation 8.2 is for one dimension but, in general, J is a vector and dC/dx is just one component of the gradient vector, $\vec{\nabla}C$, so, in general, D is a second rank tensor. That is,

$$\vec{J} = -\tilde{D}\vec{\nabla}C$$

where

$$\vec{\nabla}C = \frac{\partial C}{\partial x_1}\vec{i} + \frac{\partial C}{\partial x_2}\vec{j} + \frac{\partial C}{\partial x_3}\vec{k}$$

and

$$\tilde{D} = \begin{bmatrix} D_{11} & D_{12} & D_{13} \\ D_{21} & D_{22} & D_{23} \\ D_{31} & D_{32} & D_{33} \end{bmatrix}$$

so that

$$J_1 = -D_{11}\frac{\partial C}{\partial x_1} - D_{12}\frac{\partial C}{\partial x_2} - D\frac{\partial C}{\partial x_2}$$

with similar equations for the other two directions. However, in all following discussions, it will be assumed that D is scalar and isotropic; that is, it is the same in all directions. This is, of course, actually true and not just a mathematical simplification for gases, liquids, randomly oriented polycrystalline, and amorphous solids that do not have any crystallographic texture or preferred orientation. If considering diffusion in strongly anisotropic solids such as some single crystals, then it might be necessary to consider the varying components of the diffusion coefficient, but is not a concern here.

In the above form, Fick's first law applies *only to thermodynamically ideal solutions*† such as two gases, N_2 and O_2, or two liquids such as ethanol, C_2H_5OH, and water, or solids such as Cu and Ni, or Al_2O_3 and Cr_2O_3. How Fick's first law changes for nonideal solutions, more typically the case for solids, is examined later.

* This is very similar Fourier's law of heat flow, $\dot{q} = -k\,dT/dx$, where $\dot{q}$ is the heat flux, Units($\dot{q}$) = J/cm^2 s = W/cm^2, T is the temperature, and k is the thermal conductivity, Units(k) = W/cm K or W/m K.

† One in which there is no enthalpy on mixing and the Gibbs energy of mixing is given by Equation 8.1.

8.4 VALUES OF DIFFUSION COEFFICIENTS

8.4.1 General

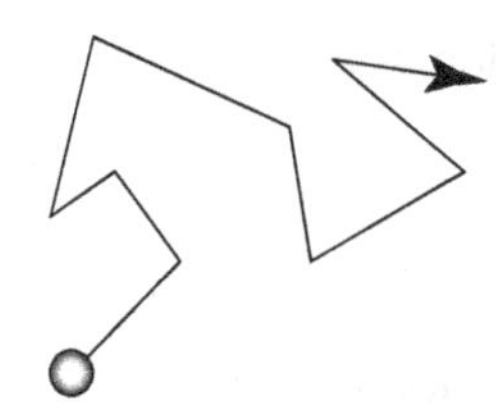

FIGURE 8.2 Random thermal motion of gas atoms or molecules that travel at high speeds and make many collision with other atoms/molecules per second. The distance between collisions, λ, is called the *mean free path.*

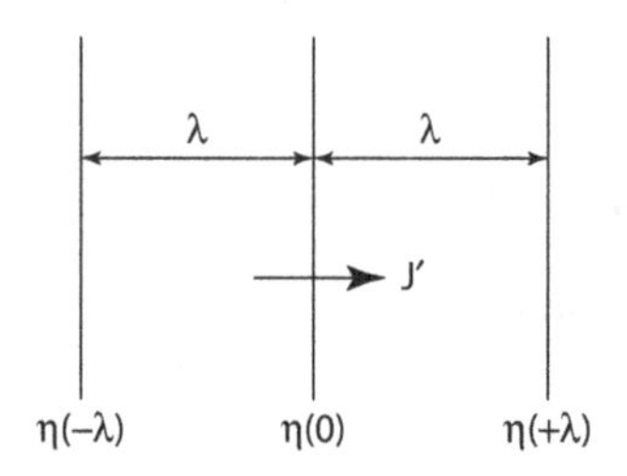

FIGURE 8.3 Geometry used to calculate the diffusion coefficient in a gas.

When the atomistic mechanisms of diffusion are discussed later, the reasons for the following temperature and pressure dependences for diffusion coefficients in gases, solids, and liquids will become more apparent. However, for now, the interest is primarily in how and why the diffusion coefficients in these three phases vary, particularly with temperature, and to get a better idea of the magnitudes for diffusion coefficients in the different states of matter. Values for diffusion coefficients are found in a number of texts on diffusion and other publications (Cussler 1997).

Whether the diffusion or atom motion is in gases, liquids, or solids, it is a random process. Figure 8.2 shows the motion of atoms or molecules in the gas phase. As was mentioned in Chapter 5, footnote to Section 5.2.2, gas atoms move with very high velocities but they do not travel very far—on the order of 0.1 μm—before colliding with another gas atom or molecule, which alters their directions. The mean distance between the collisions is called *the mean free path,* λ, for atoms in a gas. Figure 8.2 shows how the gas atoms move randomly from collision to collision.

From this simple picture of atoms or molecules moving at high speeds in the gas phase with a distance between collisions of λ, a very general relationship that applies to diffusion in all states of matter can be easily obtained. Consider the three planes in Figure 8.3, a mean free path distance apart. Gas atoms at −λ traveling to the right will cross the plane at x = 0 before making a collision, and gas atoms at +λ traveling to the left will also cross the plane at x = 0 before making a collision and changing direction. Assume that there is a concentration gradient that decreases going from left to right as shown in Figure 8.4. If η = number of atoms or molecules/cm^3 with η(−λ) the concentration at x = −λ and η(+λ) the concentration at x = λ, and J′ = atomic or molecular flux, Units(J′) = atoms/cm^2 s, then as was seen in Chapter 5, the flux of atoms crossing a surface can be approximated by*

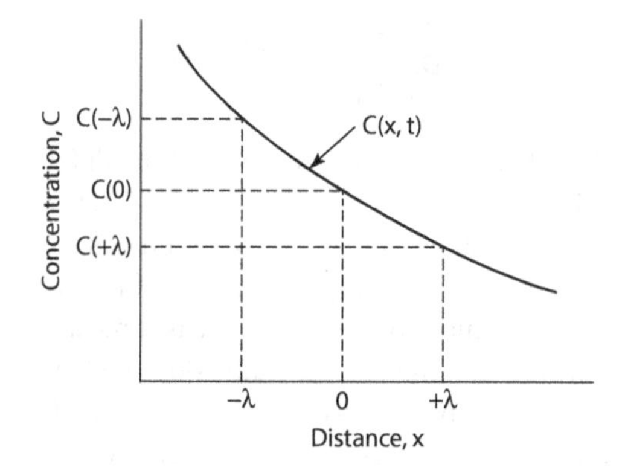

FIGURE 8.4 Concentration of gas atoms versus distance used to calculate the gas diffusion coefficient.

* In Chapter 5, it was shown that the correct values were J′ = ηv/4 and the mean distance was (2/3)λ from the plane of collision. Substitution of these two values into Equation 8.3 gives exactly the same result: the two errors in the approximation just cancel each other.

$$\begin{aligned} J' &= I_{\text{from left}} - I_{\text{from right}} \\ &= \frac{1}{6}v\eta(-\lambda) - \frac{1}{6}v\eta(+\lambda) \\ &= \frac{1}{6}v\left(\eta - \frac{d\eta}{dx}\lambda\right) - \frac{1}{6}\left(\eta + \frac{d\eta}{dx}\lambda\right) \\ J' &= -\frac{2}{6}v\lambda\frac{d\eta}{dx} \end{aligned} \tag{8.3}$$

with the final result

$$J' = -\frac{1}{3}v\lambda\frac{d\eta}{dx}.$$

Dividing both sides by Avogadro's number gives $J = -(1/3)v\lambda(dC/dx)$ and comparing this to Fick's first law shows that

$$D = \frac{1}{3}v\lambda \tag{8.4}$$

and is a very *general result* that can be applied to diffusion not only in gases, but also in solids and liquids. In fact, the difference in values between the diffusion coefficients in solids, liquids, and gases at any given temperature can be reasonably explained by this result.

8.4.2 Diffusion in Gases

The diffusion coefficients for gases have very weak temperature dependences. As will be shown in Chapter 9, for gases,

$$D \cong 10^{-4}\frac{T^{\frac{3}{2}}}{p}\ \text{cm}^2/\text{s} \tag{8.5}$$

where:

Units(p) = bar
Units(T) = K
Note: "≅" is used to denote "approximately equal to."

So for p = 1 bar at 300 K, $D \cong 0.5\ \text{cm}^2/\text{s}$ and at T = 1500 K, $D \cong 6\ \text{cm}^2/\text{s}$. This only about an order of magnitude change over this temperature range that, compared to solids, is a very small change in D. Also, if you need a *back of the envelope* value for the diffusion coefficient of a gas, taking $D \cong 1\ \text{cm}^2/\text{s}$ would be within an order of magnitude for most gases at most temperatures.

Gases have the largest diffusion coefficients because the diffusion or mixing occurs through collisions of gas atoms moving with high thermal velocities on the order of 500 m/s (1125 mph) and the atoms or molecules can move relatively large distances (relative to solids and liquids), in any direction, on the order of 0.1 μm at room temperature, mainly because of the low density of molecules in the gas. The mean free path for diffusion is, in part, determined by the number of atoms or molecules per cm^3. For example, for a gas at room temperature $T \cong 300$ K and at one bar (10^5 Pa) pressure, P,

$$\begin{aligned} \eta &= \frac{n}{V}N_A = \frac{P}{RT}N_A = \frac{10^5}{(8.314)(300)}6.022\times 10^{23} \\ &= 2.41\times 10^{25}\ \text{m}^{-3} = 2.41\times 10^{19}\ \text{cm}^{-3} \end{aligned}$$

where:

Units(η) = atoms/cm^3 or molecules/L, and so on

N_A = Avogadro's number.

This is where the P term comes from in the denominator of Equation 8.5 and why T ends up in the numerator. The other $T^{1/2}$ in the numerator comes from the gas mean speed, $\bar{v} = \sqrt{8RT/\pi M}$. The molecules or atoms in a gas experience some 10^9 collisions per second, which produces rapid mixing and diffusion. If the gas velocity is 5×10^4 cm/s and the mean free path is $\lambda = 3 \times 10^{-5}$ cm, then $D \approx 0.5$ cm/s from the general formula $D = (1/3)v\lambda$ and is consistent with Equation 8.5.

8.4.3 Diffusion in Solids

The situation for solids is very different. In solids, the atoms are very densely packed; take for example copper, with a density of $\rho = 8.92$ g/cm^3 and molecular weight of $M = 63.55$ g/mol*:

$$\eta = \frac{\rho}{M} N_A = \frac{8.92 \text{ g/cm}^3}{63.55 \text{ g/mol}} 6.022 \times 10^{23} = 8.45 \times 10^{22} \text{ cm}^{-3}.$$

Note that in solids the atoms (ions, molecules) are tightly packed, almost like billiard balls, so it makes no difference what type of bonding there is between the atoms in a solid; the number of atoms (ions, molecules) per cm^3 is going to be *something* $\times$ 10^{22} cm^{-3}. The density of solids is some three orders of magnitude higher than the density of gases—in terms of the number of atoms per unit volume—which is one reason that diffusion coefficients in solids are much lower than in gases.

For solids, the diffusion coefficient is strongly temperature dependent and follows an Arrhenius equation; namely,

$$D = D_o e^{-\frac{Q}{RT}} \tag{8.6}$$

where:

D_o is the *pre-exponential*, Units(D_o) = cm^2/s or m^2/s, typically close to $D_o \cong 10^{-3}$ cm^2/s

Q is the *activation energy*, Units(Q) = J/mol.

For many solids, $Q \cong 100$–200 kJ/mol, or more. For example, if Q = 200 kJ/mol, then at room temperature (300 K),

$$D(300\text{K}) = 10^{-3} e^{-\frac{200{,}000}{8.314 \times 300}} = 1.5 \times 10^{-38} \text{ cm}^2/\text{s}.$$

And at 1500 K,

$$D(1500\text{K}) = 10^{-3} e^{-\frac{200{,}000}{8.314 \times 1500}} = 1.08 \times 10^{-10} \text{ cm}^2/\text{s}.$$

This is a difference of *28 orders of magnitude* over the same temperature range where the gas diffusion coefficient varied by only a factor of 10! Clearly, with such a small diffusion coefficient, there would be no evidence for any diffusion at room temperature. However, at 1500 K, the diffusion coefficient is large enough to see changes in microstructure in reasonable periods of time; 10^4 seconds, or about 3 hours. So this is a good *back of the envelope* value for the diffusion coefficient in a solid when property changes occur, $D \cong 10^{-10}$ cm^2/s.

* Or g/g-atom: in either case it is the weight of one Avogadro's number of the atomic or molecular species. In general, no distinction is made between the two: one "mole" of something is one Avogadro's number, $N_A = 6.022 \times 10^{23}$ molecules (or atoms)/mol).

There are two main mechanisms for diffusion in solids: (1) interstitial mechanism (Figure 8.5) and (2) the vacancy mechanism (Figure 8.6). Other mechanisms have been postulated and some do occur but only in special cases and will not be considered here because they are not essential to understanding atomistic diffusion in most solids. The interstitial mechanism occurs for carbon and nitrogen in iron, hydrogen in platinum and palladium, oxygen in SiO_2 glass, H_2O and other gases in polymers. In fact, most diffusion of engineering interest in solid polymers is essentially interstitial diffusion. The vacancy mechanism is important in all close-packed solids and is the mechanism of diffusion of Cu in Cu; Fe in Fe; O^{2-} in MgO; Si, B, and As in Si; Ga and As in GaAs, and so on. Typically, the activation energy for interstitial diffusion is less than that for vacancy diffusion because only the energy for the atom or ion to move or *jump* from one interstitial site to another is necessary. In contrast, for the vacancy mechanism, the atom must jump, but to do that, a neighboring site must be vacant. The concentration of vacancies also depends exponentially on the temperature and, at a given temperature, there is an *intrinsic* concentration of vacancies that increases the randomness of the system, the entropy, even though it takes energy to create them. At the melting point of many solids, the fraction of lattice sites that are vacant—*also the probability that a single site is vacant*—is on the order of 10^{-4}. As a result, the activation energy for vacancy diffusion is the sum of the energy for the atom to jump as well as the energy to form vacancies. This will be explored in some detail in Chapter 9.

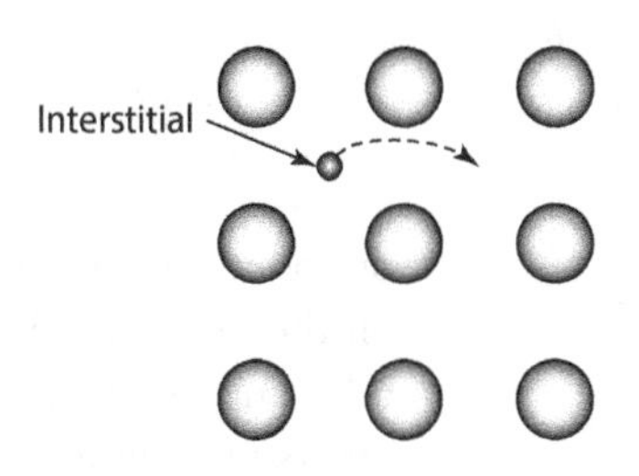

FIGURE 8.5 Interstitial mechanism for diffusion in solids. The small circle represents the interstitial atom and the larger ones are the lattice atoms.

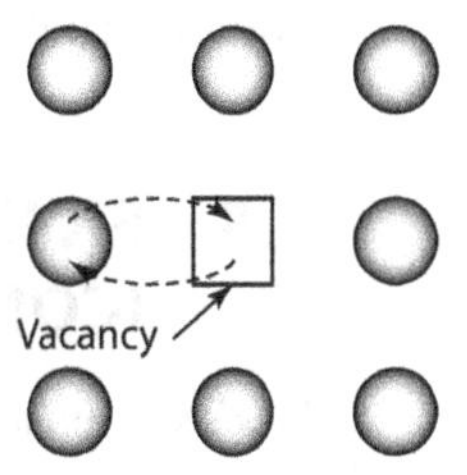

FIGURE 8.6 Vacancy mechanism for diffusion in solids. The "square" in the center represents the vacant site or *vacancy.*

8.4.4 Diffusion in Liquids

Liquids fall between solids and gases in that their diffusion coefficients have stronger temperature dependences than gases but much less so than solids. In fact, diffusion in liquids also follows an exponential temperature dependence, $D = D_o\exp(-Q/RT)$ but the activation energies are on the order of 10–20 kJ/mol. For example, if Q = 10 kJ/mol, at 300 K,

$$D(300\,K) = 10^{-3}e^{-\frac{10,000}{8.314\times 300}} = 1.8\times 10^{-5}\ cm^2/s$$

and at 1500 K,

$$D(1500\,K) = 10^{-3}e^{-\frac{10,000}{8.314\times 1500}} = 4.5\times 10^{-4}\ cm^2/s.$$

This is only a factor of 40 different! This is a little larger than for gases but much smaller than the difference in the diffusion coefficients in a solid over the same temperature range. So a good choice for a *back-of-the-envelope* calculation would be $D \cong 10^{-5} - 10^{-4}$ cm²/s for most liquids.

In liquids, the atoms have kinetic energies and velocities similar to the atoms in gases, on the order of 5×10^4 cm/s but they can only move roughly an interatomic distance (about 0.2 nm) before colliding with another atom or molecule. But they also need space to move. When a crystalline solid melts, it increases its volume about 10% or so: so-called *free volume.* As shown earlier, on melting, for most solids (Si and H_2O being notable exceptions), there is an expansion in going from the solid to the liquid state—typically around 10 v/o (volume percent) or more. For example, for copper at its melting point of $T_{mp} = 1083°C$, the density of liquid copper is, $\rho(Cu, L) = 8.00$ g/cm³ (Brandes and Brook 1992) so that the number of atoms per cm³, η, is

$$\eta(\text{Cu, L}) = \frac{\rho}{M} N_A = \frac{8.0\,\text{g/cc}}{63.55\,\text{g/mol}} 6.022 \times 10^{23}\ \text{atoms/mol} = 7.58 \times 10^{22}\ \text{atoms/cm}^3$$

or about 10% less than calculated above for solid copper. Note that the number density is still on the order of 10^{22} atoms/cm^3. This means that the *vacancy* concentration in a liquid is about 0.1 compared to 10^{-4} in the solid at the melting temperature. This factor alone contributes three orders of magnitude to the diffusion coefficient in the liquid compared to the solid at the melting point. As a result, there is about 10% more volume for a liquid atom to move around in. So, in any given direction, there is about a 10% chance the atom will have the extra volume in this direction. Thus, $1/3v\lambda \approx 1/3 \times (5 \times 10^4) \times (2 \times 10^{-8}) \times 0.1 \approx 3 \times 10^{-5}$ cm^2/s. This is a good approximation to the diffusion coefficients in liquids and explains the roughly five orders of magnitude difference in diffusion coefficients between solids and liquids.

8.5 FICK'S SECOND LAW OF DIFFUSION (CONSERVATION OF MATTER)

Fick's second law of diffusion—really the conservation of matter—is derived for the one-dimensional case in a relatively simple graphic way. The volume of the "box" in Figure 8.7 is $V = Ah$. The number of moles in the box, n, is simply: $n = CAh$, where C is the concentration in mol/cm^3. So the change in the number of moles in the box with time, $\partial n/\partial t$, is just the difference between the flux in the front and the flux out the back:

$$\frac{\partial n}{\partial t} = \frac{\partial C}{\partial t} Ah = AJ(x) - AJ(x+h)$$

$$\frac{\partial C}{\partial t} = -\left\{\frac{J(x+h) - J(x)}{h}\right\}.$$

where the partial derivative with respect to time is being used because the flux is given in terms of distance and the change in flux will be a partial derivative with respect to distance, x. The difference between flux out and flux in is what stays inside the box and increases the concentration. Taking the limit as h goes to zero, the right-hand side of this equation is the definition of the derivative of J; that is,

$$\lim_{h \to 0} \frac{J(x+h) - J(x)}{h} = \frac{\partial J}{\partial x}$$

Therefore,

$$\frac{\partial C}{\partial t} = -\frac{\partial J}{\partial x} = -\frac{\partial}{\partial x}\left\{-D\frac{\partial C}{\partial x}\right\}$$

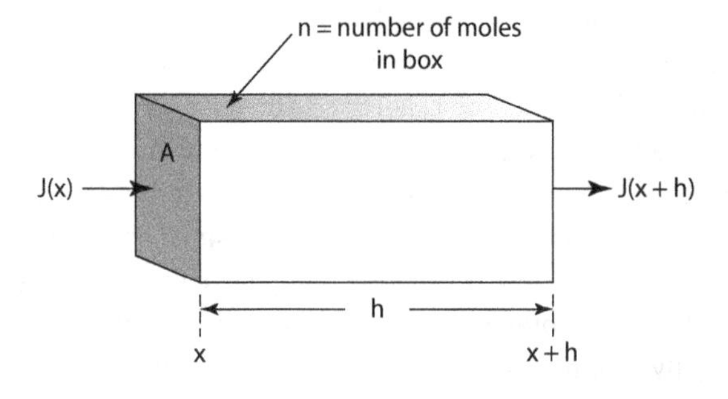

FIGURE 8.7 Schematic used for the derivation of Fick's second law that is essentially the conservation of matter.

and if D is a constant, not a function of x, then

$$\frac{\partial C}{\partial t} = D\frac{\partial^2 C}{\partial x^2}. \quad (8.7)$$

Equation 8.7 is the one-dimensional form of Fick's second law that is an expression of the conservation of mass. This is very similar to Fourier's law for non steady-state heat conduction,

$$\frac{\partial T}{\partial t} = \frac{k}{\rho c}\frac{\partial^2 T}{\partial x^2} \quad (8.8)$$

where:

- T is the temperature
- $k/\rho c$ is the *thermal diffusivity*
- k is the thermal conductivity (Units(k) = W/cm K)
- ρ is the density (Units(ρ) = g/cm^3)
- c is the mass heat capacity (Units(c) = J/g-K) so Units ($k/\rho c$) = cm^2/s: the same as the mass diffusivity.

In many cases, solutions of C(x, t) of Fick's second law as a function of time and temperature are needed to solve many macroscopic diffusion or mass flow problems. The same is true of T(x, t) in Fourier's law to determine heat flow. Because these equations have the same form, they have the same solutions. As a result, solutions to these equations can be found in books on either heat or mass transfer (Carslaw and Jaeger 1959; Crank 1975; Glicksman 2000; Wilkerson 2000).

8.6 FICK'S SECOND LAW REDUX

A more elegant derivation of Fick's second law is as follows for those more mathematically passionate. Considering again the box in Figure 8.7—or any volume of arbitrary shape—the total number of moles in the box (volume), n, is simply:

$$n(t) = \int_V C(x,t)dV$$

so

$$\frac{dn}{dt} = \int_V \frac{\partial C}{\partial t}dV$$

and $\vec{J} = -D\vec{\nabla}C$ assuming that D is a scalar constant. Gauss' divergence theorem is (Kreyszig 1988)

$$\int_S \vec{J}\cdot d\vec{A} = \int_V \vec{\nabla}\cdot\vec{J}dV$$

where $d\vec{A}$ represents the surface normal so that the integral of the flux parallel to the outward-pointing normal to the surface integrated over the surface is the divergence of the flux integrated over the volume. This becomes

$$\frac{dn}{dt} = \int_V \frac{\partial C}{\partial t}dV = -\int_S \vec{J}\cdot d\vec{A} = -\int_V \vec{\nabla}\cdot\vec{J}dV = \int_V D\nabla^2 C dV$$

where ∇^2 is the *Laplacian operator.* The negative sign is necessary because the flux is parallel to the surface normal that points out from the surface of the object. Therefore,

$$\int_V \frac{\partial C}{\partial t}dV = \int_V D\nabla^2 C dV$$

or

$$\int_V \left(\frac{\partial C}{\partial t} - D\nabla^2 C \right) dV = 0$$

or

$$\frac{\partial C}{\partial t} = D\nabla^2 C$$

So in three dimensions, Fick's second law is

$$\frac{\partial C}{\partial t} = D\nabla^2 C \tag{8.9}$$

where

$$\nabla^2 C = \frac{\partial^2 C}{\partial x^2} + \frac{\partial^2 C}{\partial y^2} + \frac{\partial^2 C}{\partial z^2} \tag{8.10}$$

in Cartesian coordinates and

$$\nabla^2 C = \frac{\partial^2 C}{\partial r^2} + \frac{1}{r}\frac{\partial C}{\partial r} = \frac{1}{r}\frac{\partial}{\partial r}\left\{ r\frac{\partial C}{\partial r} \right\} \tag{8.11}$$

in cylindrical or polar coordinates if there is no angular dependence (i.e., there is only circular symmetry and Equation 8.11 is developed in the Appendix). And in spherical coordinates (again no angular dependence; i.e., there is spherical symmetry and also developed in the Appendix)

$$\nabla^2 C = \frac{\partial^2 C}{\partial r^2} + \frac{2}{r}\frac{\partial C}{\partial r} = \frac{1}{r^2}\frac{\partial}{\partial r}\left\{ r^2\frac{\partial C}{\partial r} \right\}. \tag{8.12}$$

In future problems where cylindrical, polar, or spherical coordinates are necessary, only the radial dependence on concentration is considered. Adding angular dependences only adds needless complications at this stage of demonstration of principles.

8.7 GENERAL CONCEPTS ON SOLVING DIFFUSION PROBLEMS

Solutions to Fick's second law are necessary and important in a large number of materials science and engineering problems when C(x, t) is required. The approach to solving Equation 8.7 depends on the type of *boundary conditions* (C(0, t) and C(L, t)) and the *initial condition* (t = 0 usually) that are necessary to solve a given diffusion problem. Because Fick's second law has the second derivative with respect to distance and the first derivative with respect to time, two *boundary conditions* and one *initial condition* are necessary to solve a given diffusion problem.

8.7.1 Infinite Boundary Conditions

For example, a piece of pure Cu and a piece of pure Ni—initial conditions: $[Cu(0 \le x \le \infty, t = 0]^* = 0$ and $[Ni(-\infty \le x \le 0, t = 0)] = 0$—each 10 cm thick are placed in contact with each other at t = 0, as shown in Figure 8.8, and heated to a temperature (maybe around 1000°C) where the interdiffusion coefficient between them $D \cong 10^{-10}$ cm²/s. As will be seen in Chapters 11 and 12, a rough estimate of the interdiffusion distance, L, is $4Dt \cong L^2$ so the extent of the interdiffusion in this *diffusion couple* after about 3 hours (~10^4 seconds) is about $L \cong 20$ μm around the original interface.

* Here "[ ]" represents concentration.

At the far ends of each piece ($x = -10$ cm $\cong -\infty$ and $x = +10$ cm $\cong \infty$), the composition remains pure Cu and pure Ni, respectively. This is an example of a problem that has *infinite boundary conditions* in that the Ni and Cu do not have time to diffuse to the opposite ends of the diffusion couple so $C_{Cu}(\infty, t) = 0$ and $C_{Ni}(-\infty, t) = 0$ remain constant for all reasonable times. Applying the approximation $4Dt \cong L^2$, it would take almost 10,000 years for the Cu and Ni to diffuse to the opposite ends of the 10 cm thick pieces at 1000°C! So infinite boundary conditions apply when *measureable* composition changes never occur at the far ends of *diffusion couple*. The equations describing the interdiffusion do give changes in concentration at the far ends of the couple but these are well below any measureable level and are simply due to the behavior of the mathematical functions.

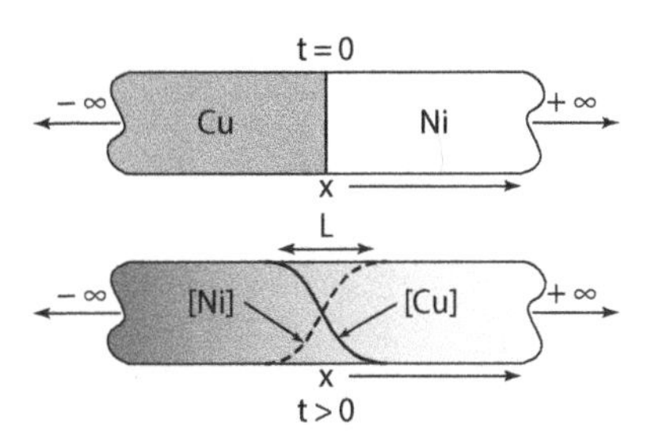

FIGURE 8.8 Pieces of copper, Cu and nickel, Ni, interdiffusing into each other by solid-state diffusion. The size of the pieces is large compared to the distance the atoms of each diffuse, L, during the time allowed for the diffusion to take place. [Cu] is the concentration of copper and [Ni] that of nickel. This is an illustration of *infinite boundary conditions*.

8.7.2 Semi-Infinite Boundary Conditions

As might be expected, semi-infinite boundary conditions are depicted in Figure 8.9 in which there is one end at $x = 0$ and the other end of the diffusion field extends to infinity. This example might represent a piece of pure iron, $C(x, 0) = 0$ (an *initial condition*), into which carbon is being diffused from the surface. The carbon activity (concentration) is fixed at the surface, $C(0, t) = C_S$—a *finite boundary condition*—by controlling the pressures of methane, CH_4, and hydrogen in the gas phase, for example. Then the total depth of diffusion L could be estimated by $4Dt \cong L^2$ again. Again L, the depth of diffusion, would be on the order of micrometers for temperatures around 800°C and never reach very deep into the piece of iron. So $C(\infty, t) = 0$ is the boundary condition at $x = \infty$. These are *semi-infinite boundary conditions* where one boundary condition is given at a fixed position ($x = 0$ in this case) and the other is fixed at $x = \infty$ (or $-\infty$).

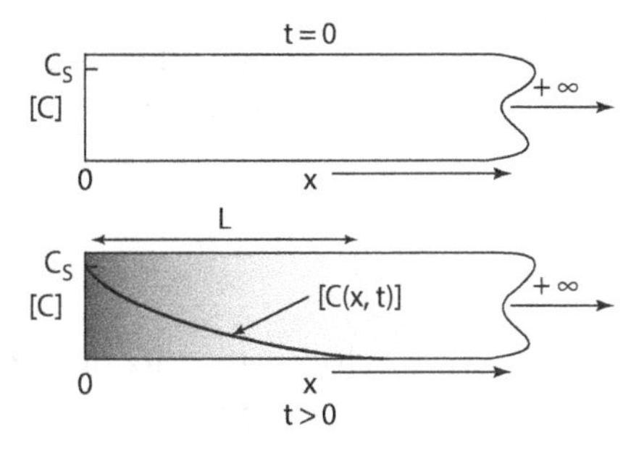

FIGURE 8.9 Carbon diffusing from a surface concentration, C_S, fixed by the surrounding gas composition and thermodynamics, into a piece of iron whose thickness, is large compared "diffusion distance." This is a case of *semi-infinite boundary conditions*.

8.7.3 Finite Boundary Conditions

Finite boundary conditions apply when the diffusion extends over the entire spatial region of interest during the time of diffusion. For example, Figure 8.10 represents a piece of relatively pure iron containing some carbon impurities being heat treated in hydrogen to remove the carbon by reaction with the hydrogen to form methane gas at both surfaces at $x = 0$ and at $x = L$ with the *initial conditions* that $C(x, 0) = C_0$ throughout the piece of iron. Assume that at $t = 0$, the hydrogen is introduced into the hot furnace and the carbon content at both surfaces drops immediately to zero (i.e., $C(0, t) = C(L, t) = 0$) and remain zero for all values of t as shown in the left sketch of Figure 8.10. These are the two *finite boundary conditions* in this case. As diffusion proceeds, the *concentration profiles*, $C(x, t)$, change with time as shown in the middle sketch—actually approach sine curves—until finally at $t = \infty$, the concentration in the sheet has dropped to $C(x, \infty) = 0$ as shown in the right-hand sketch. The approximation $4Dt \cong L^2$ can be used again to estimate how long it will take for the concentration to fall to zero: if $L = 1$ mm and $D = 10^{-8}$ cm²/s, then $t \cong 2 \times 10^5$ seconds or about 70 hours, which is much less than $t = \infty$. Of course, this is only approximate and the partial differential equation really needs to be solved to determine when the carbon concentration has fallen to an acceptable level say, 0.01%, or whatever is necessary for the application of the carbon-free iron sheet: iron cores for special transformers is one application of carbon-free iron.

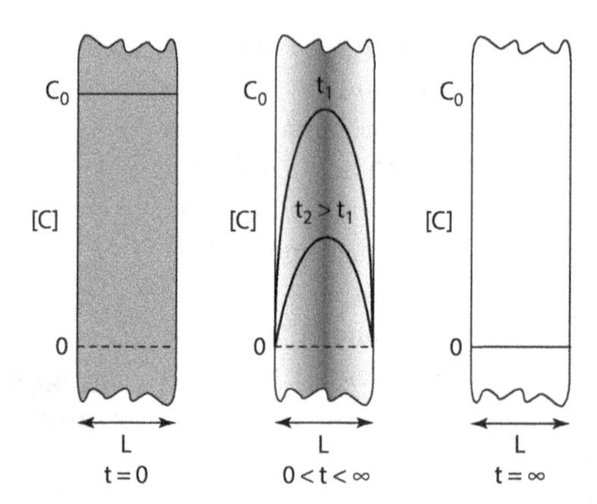

FIGURE 8.10 *Finite boundary conditions* where the concentrations for all times are specified at x = 0 and at x = L. In this case, C(0, t) = C(L, t) = 0.

The interdiffusion of nitrogen and oxygen, shown in Figure 8.1, is also a finite boundary condition problem. However, in this case, the boundary conditions are

$$\frac{\partial[N_2]}{\partial x}(0,t) = \frac{\partial[N_2]}{\partial x}(L,t) = 0.$$

That is, the flux of nitrogen at both ends of the box is zero because $J(N_2) = -D(\partial[N_2]/\partial x)$. The same boundary conditions also apply for oxygen. These boundary conditions basically imply that *neither the nitrogen nor the oxygen can leave the box* of width L so that their fluxes at x = 0 and x = L are zero. These are *derivative boundary conditions but still are finite boundary conditions.*

For finite boundary conditions, the solutions to Fick's second law are of the form,

$$C(x,t) = f(x) + g(x,t)$$

where f(x) is a function of distance only and is either the *equilibrium* concentration or the *steady-state* concentration, and g(x, t) is called the *transient* part of the solution, where g(x, ∞) = 0, meaning that the transient disappears with time. In Figures 8.1 and 8.10, the transient terms are shown in the center sketches. In the right-hand sketch of both figures, the transient has gone to zero and only the *equilibrium* concentrations of C(x, ∞) = 0.5 remain in Figure 8.1 and C(x, ∞) = 0 in Figure 8.10. Note, that when *equilibrium* is reached,

$$J = -D\frac{dC}{dx} = 0$$

which implies that dC/dx = 0. In contrast, however, in the *steady state* ∂C/∂t = 0, which means that

$$\frac{\partial^2 C}{\partial x^2} = 0$$

which can be integrated to give C(x).

It is extremely important to note the difference between *equilibrium* and *steady state*! *Steady state* means that there is a *concentration gradient and a resulting diffusion flux* but the concentration does not change with time, ∂C/∂t = 0. While equilibrium implies that there is no concentration gradient or diffusion flux, so dC/dx = ∂C/∂x = 0. The difference between the two situations will become clearer when finite boundary condition diffusion problems are analyzed in Chapter 12.

8.8 CHAPTER SUMMARY

The purpose of this chapter is to introduce concepts related to modeling diffusion at both the macroscopic and atomistic levels. Fick's first law is introduced for ideal solutions, and the physical implications of the law are discussed for atoms moving in solids, liquids, and gases. The temperature dependences for diffusion coefficients in solids, liquids, and gases are qualitatively summarized as approximate values and are, in fact, very good estimates for diffusion coefficients lacking any other information. Chapter 9 models the atomistic mechanisms of diffusion in gases and solids in detail and justifies the form of the diffusion coefficients given in this chapter. The details of the diffusion in liquids are given in Chapter 13. Fick's second law, conservation of mass, is derived for various coordinate systems. Solutions to Fick's second law, C(x, t), that give the concentration as a function of time and distance, are typically obtained by the materials scientist of engineer with various mathematical techniques to solve the partial differential equation: $\partial C/\partial t = D\partial^2 C/\partial x^2$. These solutions depend strongly on the initial conditions and the type of boundary conditions for the model: infinite, semi-infinite, or finite boundary conditions that are defined in terms of the concentrations at the boundaries of the diffusion zone. Here, only the definitions of these terms are introduced, while Chapters 11 and 12 develop some exact and approximated methods of solving these equations. In between, in Chapter 10, steady-state solutions to the diffusion equations are modeled for a wide variety of materials applications in which diffusion plays an important role.

APPENDIX

A.1 FICK'S SECOND LAW IN POLAR OR CYLINDRICAL COORDINATES

The geometry for developing Fick's second law in cylindrical coordinates is given in Figure A.1. Again, only the partial differential equation for change in the r direction will be considered because for the diffusion models here there is no interest in the angular dependence. In this case, the hollow cylinder has a height L, an inner radius of r, and an outer radius of "r + h" or a wall thickness of h that is considered to be very small. The volume of the cylinder is

$$V_{cyl} \cong 2\pi rLh$$

because h is small. The total number of moles of whatever is diffusing, n, in the cylinder is

$$n = V_{cyl}C \cong 2\pi rLhC$$

where C is the concentration: Units(C) = mol/cm^3. So,

$$\frac{\partial n}{\partial t} = 2\pi rhL\frac{\partial C}{\partial t} \tag{A.1}$$

Also, the change in the number of moles is the difference between the flux in and the flux out:

$$\begin{aligned} \frac{\partial n}{\partial t} &= A(r)J(r) - A(r+h)J(r+h) \\ \frac{\partial n}{\partial t} &= 2\pi rLJ(r) - 2\pi(r+h)LJ(r+h) \end{aligned} \tag{A.2}$$

where:

A(r) is the inner area of the cylinder
A(r + h) is the outer area.

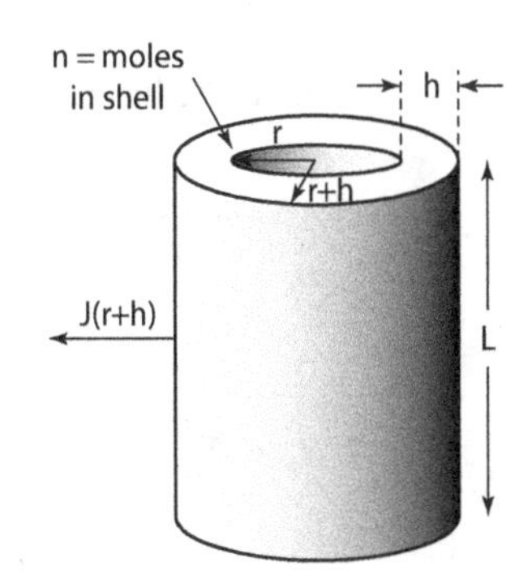

FIGURE A.1 Geometry for deriving Fick's second law in cylindrical coordinates.

Equating Equations A.1 and A.2 gives

$$r\frac{\partial C}{\partial t} = -\left\{\frac{(r+h)J(r+h) - rJ(r)}{h}\right\} \tag{A.3}$$

and

$$\lim_{h\to 0}\left\{\frac{(r+h)J(r+h) - rJ(r)}{h}\right\} = \frac{\partial}{\partial r}\{rJr\} \tag{A.4}$$

where Equation A.4 is the definition of the derivative of rJ(r). Equating Equations A.3 and A.4 and substituting $J = -D(\partial C/\partial r)$ gives

$$r\frac{\partial C}{\partial t} = D\frac{\partial}{\partial r}\left\{r\frac{\partial C}{\partial r}\right\}$$

and the desired result of

$$\frac{\partial C}{\partial t} = \frac{D}{r}\frac{\partial}{\partial r}\left\{r\frac{\partial C}{\partial r}\right\}. \tag{A.5}$$

A.2 Fick's Second Law in Spherical Coordinates

Figure A.2 shows the geometry for the same derivation in spherical coordinates, again not worrying about angular dependence. In this case, the volume considered is a spherical shell of thickness h. Following the same procedure as above,

$$n = V_{shell}C \cong 4\pi r^2 hC$$
$$\frac{\partial n}{\partial t} = 4\pi r^2 h\frac{\partial C}{\partial t} \tag{A.6}$$

Again, the change in the number of moles is the difference between the flux in and the flux out:

$$\frac{\partial n}{\partial t} = A(r)J(r) - A(r+h)J(r+h)$$
$$\frac{\partial n}{\partial t} = -4\pi\{(r+h)^2 J(r+h) - r^2 J(r)\} \tag{A.7}$$

and equating Equations A.6 and A.7 gives

$$r^2\frac{\partial C}{\partial t} = -\left\{\frac{(r+h)^2 J(r+h) - r^2 J(r)}{h}\right\} \tag{A.8}$$

and in the limit as $h \to 0$,

$$\lim_{h\to 0}\left\{\frac{(r+h)^2 J(r+h) - r^2 J(r)}{h}\right\} = \frac{\partial}{\partial r}\{r^2 J(r)\}. \tag{A.9}$$

Equating Equations A.8 and A.9 and substituting $J = -D(\partial C/\partial r)$ gives

$$r^2\frac{\partial C}{\partial t} = D\frac{\partial}{\partial r}\left\{r^2\frac{\partial C}{\partial r}\right\}$$

and the desired result of

$$\frac{\partial C}{\partial t} = \frac{D}{r^2}\frac{\partial}{\partial r}\left\{r^2\frac{\partial C}{\partial r}\right\}. \tag{A.10}$$

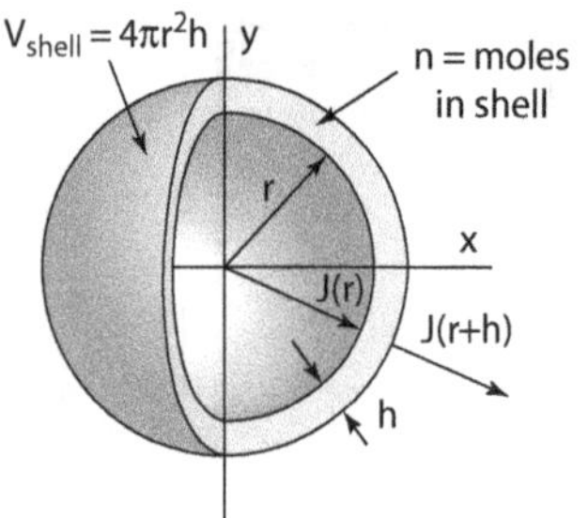

FIGURE A.2 Geometry for deriving Fick's second law in spherical coordinates.

EXERCISES

8.1 The experimentally measured diffusion coefficients as a function of temperature for hydrogen diffusion in SiO_2 glass are given in the table below

a. Plot of $\log_{10}$ D versus 1/T (K^{-1})

Hydrogen Diffusion in SiO_2		
T (°C)	T (K)	D (cm^2/s)
100	373	8.10×10^{-10}
200	473	9.99×10^{-9}
300	573	3.22×10^{-8}
400	673	2.58×10^{-7}
500	773	3.20×10^{-7}
600	873	1.50×10^{-6}
700	973	4.21×10^{-6}
800	1073	4.54×10^{-6}
900	1173	6.85×10^{-6}
1000	1273	7.19×10^{-6}
1100	1373	1.30×10^{-5}
1200	1473	1.68×10^{-5}
1300	1573	3.32×10^{-5}
1400	1673	2.56×10^{-5}
1500	1773	2.85×10^{-5}

b. From these data, calculate Q (kJ/mol) with a least squares fit to the data

c. From these data, calculate D_0 (cm^2/s)

8.2 The temperature dependence of the diffusion coefficient for the p-type dopant aluminum in silicon single crystal is given by $D = D_0\exp(-Q/RT)$, where D_0 is the pre-exponential term of 0.5 cm^2/s and Q is the activation energy of 289.5 kJ/mol. Plot on a semi-log scale the diffusion coefficient of Al in silicon, $\log_{10}D(cm^2/s)$, from 500°C to 1500°C versus 1/T (K^{-1})

8.3 a. The temperature dependence of the diffusion coefficient for He in polypropylene is given by $D = D_0\exp(-Q/RT)$, where D_0 is the pre-exponential term = 4.1 cm^2/s and Q is the activation energy of 30.5 kJ/mol. Calculate the He diffusion coefficient (cm^2/s) at both 30°C and 100°C

b. Similarly, for the diffusion of H_2O in polyethylene, D_0 is the pre-exponential term of 0.028 cm^2/s and Q is the activation energy of 24.3 kJ/mol. Calculate the H_2O diffusion coefficient (cm^2/s) at both 30°C and 100°C

8.4 Cobalt, melting point 1495°C, oxidizes under certain oxygen pressures to CoO, cobalt monoxide. Given below are the diffusion coefficients for cobalt and oxygen ions in CoO as a function of temperature

Diffusion in Cobalt Oxide		
T (°C)	D(Co) cm^2/s	D(O) cm^2/s
750	4.54E−11	2.07E−19
800	1.03E−10	1.83E−18
850	2.18E−10	1.33E−17
900	4.31E−10	8.13E−17
950	8.08E−10	4.30E−16
1000	1.44E−09	2.00E−15
1050	2.46E−09	8.24E−15
1100	4.04E−09	3.07E−14
1150	6.41E−09	1.04E−13
1200	9.85E−09	3.26E−13

(Continued)

Diffusion in Cobalt Oxide		
T (°C)	D(Co) cm²/s	D(O) cm²/s
1250	1.47E−08	9.46E−13
1300	2.14E−08	2.56E−12
1350	3.05E−08	6.54E−12
1400	4.25E−08	1.58E−11
1450	5.81E−08	3.61E−11

a. Plot the diffusion coefficients (all of the points in the above table) for both cobalt and oxygen on the same semi-log plot of $\log_{10}D$ versus $10^3/T$ (K^{-1})

b. From the plot in (a), calculate D_0 (cm^2/s) and Q (J/mol) for the diffusion of both cobalt and oxygen ions in CoO

c. The diffusion of which of the two ions controls the rate of oxidation of cobalt and explain why

8.5 The experimentally measured diffusion coefficients as a function of temperature for oxygen interstitial diffusion in SiO_2 are given in the table below

Oxygen Diffusion in SiO_2		
T (°C)	T (K)	D (cm²/s)
500	773	5.5E−16
600	873	4.9E−14
700	973	1.2E−12
800	1073	1.3E−11
900	1173	8.6E−11
1000	1273	3.9E−10
1100	1373	1.3E−9
1200	1473	3.6E−9
1300	1573	8.6E−9
1400	1673	1.8E−8
1500	1773	3.4E−8

a. Plot $\log_{10}D$ (cm^2/s) on a semi-log plot versus $1/T$ (K^{-1})

b. From these data, calculate D_0 (cm^2/s) and Q (kJ/mol). (*Note*: use a least squares fit to the data.)

8.6 The temperature dependence for the interstitial diffusion of carbon in iron is given by $D = D_0\exp(-Q/RT)$ where $D_0 = 4 \times 10^{-3}$ cm^2/s and the activation energy Q = 80.1 kJ/mol. Plot $\log_{10}D$ (cm^2/s) for carbon in iron versus $1/T$ (K^{-1}) from 0°C to 1500°C

8.7 The experimentally measured diffusion coefficients as a function of temperature for the diffusion of phosphorus in silicon are given in the table below

Diffusion of P in Si	
Temperature (°C)	D (cm²/s)
500	6.00E−24
600	2.10E−21
700	4.26E−19
800	2.75E−17
900	7.26E−16
1000	2.21E−14
1100	1.42E−13
1200	9.50E−13
1300	7.24E−12
1400	3.15E−11

a. Plot D (cm^2/s) on a semi-log plot (or $\log_{10}D$) versus 1/T (K^{-1})
b. From these data, calculate D_0 (cm^2/s) and Q (kJ/mol). (*Note*: use a least squares fit to the data.)

8.8 The experimentally measured diffusion coefficients as a function of temperature for Ni and Fe vacancy diffusion coefficients in NiO at low iron concentrations are given below.

	Ni in NiO	Fe in NiO
T (°C)	D (cm^2/s)	
500	1.41E−16	2.79E−17
600	3.80E−15	9.49E−16
700	5.21E−14	1.56E−14
800	4.39E−13	1.53E−13
900	2.57E−12	1.01E−12
1000	1.14E−11	4.97E−12
1100	4.06E−11	1.94E−11
1200	1.22E−10	6.29E−11
1300	3.19E−10	1.76E−10
1400	7.42E−10	4.33E−10
1500	1.57E−09	9.67E−10

a. Plot $\log_{10}D$ (cm^2/s) on a semi-log plot versus 1/T (K^{-1}) for both Ni and Fe on the same plot
b. From these data, calculate D_0 (cm^2/s) and Q (kJ/mol) for the diffusion of *both* Ni and Fe in NiO

REFERENCES

Brandes, E. A. and G. B. Brook. 1992. *Smithells Metals Reference Book,* 7th ed. Oxford, UK: Butterworth-Heinemann, Ltd.

Carslaw, H. S. and J. C. Jaeger. 1959. *Conduction of Heat in Solids,* 2nd ed. Oxford, UK: Oxford University Press.

Crank, J. 1975. *The Mathematics of Diffusion,* 2nd ed. Oxford, UK: Clarendon Press.

Cussler, E. L. 1997. *Diffusion, Mass Transfer in Fluid Systems,* 2nd ed. Cambridge, UK: Cambridge University Press.

Glicksman, M. E. 2000. *Diffusion in Solids.* New York: Wiley.

Kreyszig, E. 1988. *Advanced Engineering Mathematics,* 6th ed. New York: Wiley.

Narasimhan, T. N. 2004. Fick's insights on liquid diffusion, *Eos. Trans. Am. Geophys. Union* 85[47] November 2004: 499–501.

Wilkerson, D. S. 2000. *Mass Transport in Solids and Fluids.* Cambridge, UK: Cambridge University Press.

9

Atomistics of Diffusion

9.1 INTRODUCTION

In this chapter, simple atomistic models of the mechanisms of diffusion in solids, gases, and liquids are modeled and related to the macroscopic parameters: *diffusion coefficients—or diffusivities.* The mechanisms in solids and gases are relatively easy to model and give quantitative results consistent with experimentally measured values of diffusion coefficients. For gases, relatively straightforward models give calculated results within 10%–20% of measured values. For solids, it is demonstrated that a simple Arrhenius relationship holds for the diffusion coefficient

$$D = D_0 e^{-\frac{Q}{RT}} \tag{9.1}$$

and plausible arguments about the temperature dependence are made without actually being able to calculate actual values of Q. This would require considerably more sophisticated modeling involving interatomic bonding and other parameters. Nevertheless, the approximate models give values of D_0 that are reasonably close to those measured in systems where diffusion is well-characterized; that is, where the experimental data are pretty good.

Liquids are harder to model and the atomistic concepts fall between those for solids and gases. The Stokes–Einstein model for liquids is readily modeled—except for the fluid dynamics part—and gives reasonable quantitative results for diffusion in liquids. This is discussed later in Chapter 13. However, a plausibility argument will be given in this chapter for the diffusion in liquids.

The atomistic models presented here are relatively simple but include and illustrate all the important principles involved in atomic or molecular diffusive motion. More sophisticated models, of which there are many, take into consideration correlations with other experimental data that reflect more details about interatomic bonding, electronic structure, and molecular size and their interactions. These models have more precise agreement with experiment but without adding many new principles. These detailed models typically involve advanced mathematical and computational techniques and some physics above the level of this book and are good examples of the application of "computational materials science." At this point in understanding diffusion processes, the simple models are all that are necessary to explain why diffusion coefficients are different for solids, liquids, and gases and how they vary with temperature, pressure, and composition.

9.2 ENERGY ABSORPTION IN MOLECULES AND SOLIDS

As discussed in Chapter 5, the kinetic energy in the x-direction, KE_x, for a moving atom is simply

$$KE_x = \frac{1}{2}mv_x^2 = \frac{1}{2}k_BT. \tag{9.2}$$

Again, the velocity vector has three components in the x, y, and z direction, so

$$v^2 = v_x^2 + v_y^2 + v_z^2$$

and the total kinetic energy, KE, is given by

$$\begin{aligned} KE &= \frac{1}{2}mv^2 = \frac{1}{2}mv_x^2 + \frac{1}{2}mv_y^2 + \frac{1}{2}mv_z^2 \\ &= \frac{1}{2}k_BT + \frac{1}{2}k_BT + \frac{1}{2}k_BT = \frac{3}{2}k_BT. \end{aligned} \tag{9.3}$$

The second of these last two equations is a statement of the principle of *equipartition of energy* (Silbey and Alberty 2001), which essentially says that the microscopic modes for absorbing energy all have the same energy, $(1/2)k_BT$, when in thermal equilibrium. For a monatomic gas, the only energy-absorbing modes are translational energies in the three directions leading to $KE = (3/2)k_BT$ in Equation 9.3 and the RMS—root mean square—speed, v_{rms}, of gas molecules is

$$v_{rms} = \sqrt{\frac{3k_BT}{m}} = \sqrt{\frac{3RT}{M}} \tag{9.4}$$

and the mean speed, $\bar{v}$

$$\bar{v} = \sqrt{\frac{8RT}{\pi M}} \tag{9.5}$$

where M is the molecular or atomic weight (kg/mol) (Chapter 5).

For diatomic molecules such as H_2, N_2, O_2, and so on, there are other mechanisms for absorbing energy as indicated in Figure 9.1. These molecules have two degrees of rotational freedom each of which contribute $(1/2)k_BT$ to the total energy of the molecule. In addition, the molecules can change their separation distance and vibrate very much like two masses at the ends of a spring—the chemical bond. For vibrations, or a moving mass on a spring, there are both kinetic energy of the motion and the potential energy of the stretched spring. As a result, the vibrational energy contribution is $(1/2)k_BT$ for each of these energy-absorbing modes or k_BT total for each vibrational mode. These are also examples of the equipartition of energy among the various energy-absorbing modes. So, the total energy of a diatomic molecule would be as follows: energy = kinetic + rotational + vibrational = $(3/2)k_BT + 2 \times (1/2)k_BT + k_BT$ = total of $(7/2)k_BT$ per molecule or (7/2) RT per mole of gas. For molecules with more atoms, the number of vibrational modes increases as the number of atoms increases. However, both the vibrational and rotational modes are quantized and the vibrational modes, in particular, do not contribute much at low temperatures, particular for light or strongly bonded atoms such as H_2.

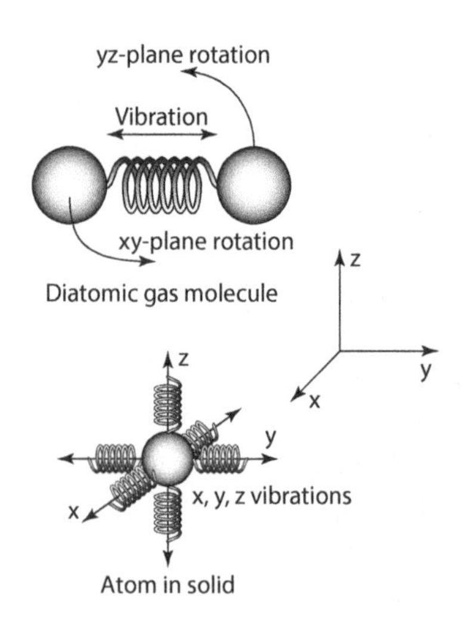

FIGURE 9.1 Diatomic molecule showing both rotational and vibrational modes of energy absorption and an atom in a solid showing three vibrational modes.

The molar heat capacity at constant volume, $\bar{C}_V$, is given by (Chapter 1)

$$\bar{C}_V = \left(\frac{\partial Q}{\partial T}\right)_V = \left(\frac{\partial \bar{U}}{\partial T}\right)_V \tag{9.6}$$

so the molar heat capacity at constant volume for argon should be $\bar{C}_V(Ar) = (3/2)R = 12.47$ J/K and that is what is used in databases (Haynes 2013). For Cl_2 at 1000 K, the experimental value is 29.325 J/K and very close to the theoretical value of (7/2)R is 29.099 J/K (Stull and Prophet 1971).

For solids that are strongly bonded—ionic, metallic, or covalent—neglecting the small contributions that the electrons make to the heat capacity of metals, the only energy-absorbing motions are atomic or ionic vibrations in each of three directions that give rise to a total energy per vibrating atom or ion of $3k_BT$ or a molar heat capacity of $3R = 24.94$ J/K per *Avogadro's number of atoms or ions.* The heat capacity of solids was discussed in Chapter 7. However this latter point is re-emphasized because it is something not always appreciated. As noted in Chapter 7, the *heat capacity per mole* for a compound such as Al_2O_3 depends on the number of atoms/ions in the chemical formula, in this case five. Therefore, the molar heat capacity of copper (with one atom per molecule) should be 24.94 J/K and is experimentally found to be 24.45 J/K at 300 K (Stull and Prophet 1971). However, Al_2O_3 contains $5N_A$ ions per mole, so its molar heat capacity should be $5 \times 24.94 = 124.7$ J/K and is found to be 124.74 at 1000 K (Stull and Prophet 1971). These heat capacities of copper and alumina were chosen at a sufficiently high temperature, so quantum effects are no longer affecting heat capacity. A molar heat capacity at constant volume of 3R for metals is known as the law of Dulong and Petit in older literature (Darken and Gurry 1953).

9.3 INTERSTITIAL DIFFUSION IN SOLIDS

9.3.1 Atom Fluxes

Interstitial diffusion in solids can be simply modeled. In Figure 9.2 interstitial atoms are diffusing with a net flux to the right in a simple cubic lattice.* The spacing between the planes is a, the lattice parameter. Consider planes 1 and 2, where the interstitial atoms lie. As depicted in the figure, the interstitial atoms are vibrating both forward and backward in all three directions. The number of atoms per unit area in plane 1 is n_1 (Units(n) = atoms/cm^2) and the number interstitial atoms in plane 2 is n_2. Then, the number flux density—J', Units(J') = atoms cm^{-2} s^{-1}—of atoms moving from plane 1 → 2 is

$$J'_{1\to 2} = \frac{n_1\Gamma}{6} \tag{9.7}$$

where Γ is the number of jumps from one interstitial site to another per second for an atom, the *jump frequency*, with Units(Γ) = s^{-1} and the 1/6 comes from the fact that the atom can jump in any of the six possible directions. Similarly, the flux of atoms going from 2 → 1 is

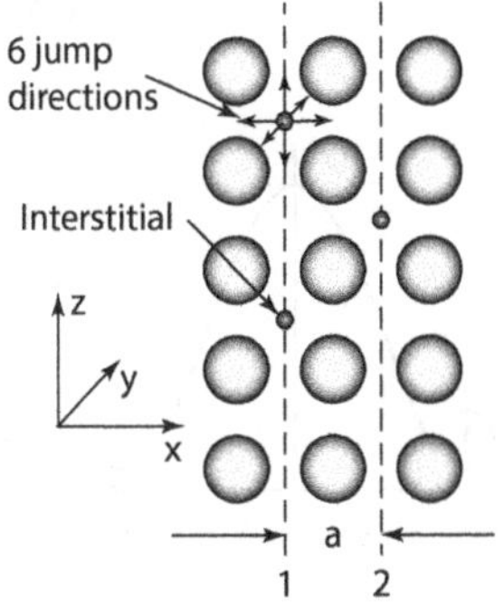

FIGURE 9.2 Model of interstitial atoms diffusing in a simple cubic crystal.

* Polonium is the only element that has a simple cubic lattice at room temperature (transforms to rhombohedral at 36°C) and atmospheric pressure; a radioactive element discovered and named by Madame Curie (Figure 9.3) (Emsley 1998). She was the first person—and one of the few—to be awarded two Nobel prizes in science, one in physics in 1903 with her husband Pierre and Antoine Becquerel for their discovery of radioactivity, which she named. She received a second Nobel Prize in chemistry in 1911 for her discovery of polonium and radium. She called the element polonium after Poland, where she was born Marja Sklodowska in 1867 (Quinn 1995). Sadly, both she and her daughter, Irène Joliot-Curie, died of leukemia caused by radiation, and in the case of Irène, it was thought to be due to an explosion of a vial of polonium some 15 years earlier (Emsley 2001). Polonium 210, $^{210}_{84}Po$, is highly radioactive with a half-life of 138 days (Emsley 1998) and has been suspected as being used in several political assassinations. It is also found in tobacco plants and was thought to be a possible cause of lung cancer from smoking.

FIGURE 9.3 A young Marie Curie. (Wikimedia Commons, Public Domain Photograph, https://commons.wikimedia.org/wiki/File:Marie_Curie_1903.jpg.)

$$J'_{2\to1} = \frac{n_2\Gamma}{6} \tag{9.8}$$

and the net number flux moving from plane 1 to plane 2 is given by

$$J' = J'_{1\to2} - J'_{2\to1} = \frac{n_1\Gamma}{6} - \frac{n_2\Gamma}{6} = \frac{\Gamma}{6}(n_1 - n_2) \tag{9.9}$$

The concentrations, Units(C) = mol/cm^3, are given by $C_1 = n_1/(aN_A)$ and $C_2 = n_2/(aN_A)$, and because the spacing between the planes is very small, a, $C_2 = C_1 + a(dC/dx)$. Putting in the values for the concentrations in Equation 9.9 gives

$$\begin{aligned} J' &= \frac{\Gamma}{6}(n_1 - n_2) = \frac{\Gamma}{6}(aN_AC_1 - aN_AC_2), \\ &= \frac{\Gamma aN_A}{6}(C_1 - C_2) = \frac{\Gamma aN_A}{6}\left(C_1 - \left(C_1 + a\frac{dC}{dx}\right)\right) \\ &= -\frac{\Gamma a^2N_A}{6}\frac{dC}{dx} \end{aligned} \tag{9.10}$$

Dividing Equation 9.10 by Avogadro's number, N_A, the molar flux density, J, is given by

$$J = \frac{J'}{N_A} = -\frac{\Gamma a^2}{6}\frac{dC}{dx}. \tag{9.11}$$

Comparison of Equation 9.11 with Fick's first law, $J = -D(dC/dx)$, the diffusion coefficient, D, is given by

$$D = \frac{\Gamma a^2}{6} \tag{9.12}$$

which is a *general* expression for the diffusion coefficient. In fact, this could be written as

$$D = \frac{\Gamma a^2}{6} = \frac{1}{3}(\Gamma a)\frac{a}{2} = \frac{1}{3}v\lambda \tag{9.13}$$

as was seen earlier for the case of a gas, where v is the speed = Γa, because Units(Γa) = cm/s and $\lambda = a/2$ the mean distance during a jump because $\lambda = (1/a)\int_0^a x\,dx = a^2/(2a) = a/2$. Again, this shows that Equation 9.13 is a very general relation that applies to diffusion in solids, liquids, and gases.

9.3.2 Jump Frequency, Γ

What is the jump frequency, Γ? To jump from plane 1 to plane 2, an interstitial atom has to move between at least four atoms, where it does not fit as well as in the interstitial site. This takes a certain amount of energy, ΔG^*, an activation energy as shown in Figure 9.4. Alternatively, as discussed in Section 4.7.1 during reaction rates, only a certain number of atoms, n^*

$$n \rightleftharpoons n^*;\ \Delta G^* \tag{9.14}$$

on a given plane have enough energy to jump. And from the equilibrium constant for Equation 9.14,

$$n^* = n e^{-\frac{\Delta G^*}{RT}}. \tag{9.15}$$

Because this is the energy necessary to move, rename $\Delta G^* = \Delta G_m$, and ΔG_m is the energy for motion of the interstitial atoms.

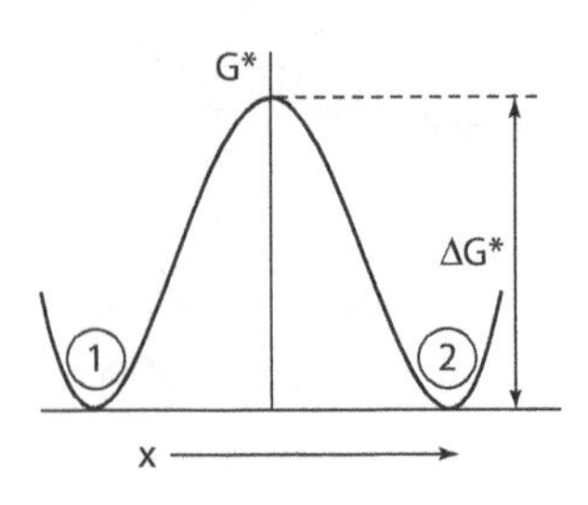

FIGURE 9.4 Activation barrier ΔG^* that must be overcome by an atom moving from plane 1 to plane 2 in a crystal lattice.

The next part of the question that needs to be answered is as follows: "How often do the atoms jump over the barrier into the next interstitial site?" The answer to this is depends on how often they attempt to get over the barrier. The number of attempts per second is taken to be the *vibrational frequency* of the atoms on their sites in the crystal, f, the same vibrations that contribute to the heat capacity. This is reasonable because the atoms on plane 1 that move per second, Γ, is just the number of jump attempts per second, f, times the probability that an atom has enough energy to jump, $\exp(-\Delta G_m/RT)$.

9.3.3 Vibration Frequency, f

Atoms vibrate on their lattice or interstitial sites as shown schematically in Figure 9.5. This leads to the differential equation for the vibrational motion:

$$ma = F \quad \text{or} \quad m\frac{d^2x}{dt^2} = -\beta x \tag{9.16}$$

where β is the spring constant ((2β)/2, because one spring is compressed, whereas the other is extended), Units(β) = N/m. The solution to Equation 9.16 is either $x = x_0\sin(\omega x)$ or $x = x_0\cos(\omega x)$ (Appendix A.1), where the angular frequency $\omega = 2\pi f$ and f is the frequency, Units(f) = cycles/s = Hz. Substitution of either the sine or cosine solution into Equation 9.16 gives

$$-mx_0\omega^2\sin(\omega x) + \beta\sin(\omega x) = 0 \tag{9.17}$$

leading to $\omega^2 = \beta/m$, or, because $f = \omega/(2\pi)$,

$$f = \frac{\omega}{2\pi} = \frac{1}{2\pi}\sqrt{\frac{\beta}{m}}. \tag{9.18}$$

To estimate f, a value for β is needed. This can be approximated from an elastic constant, Young's modulus, E: $\sigma = \varepsilon E$, where σ is the stress (Units(σ) = Pa), ε the strain, and E the Young's modulus (Units(E) = Pa). From Equation 9.16, $F = \beta x$ and dividing both sides by a^2, because x/a is the strain, ε,

$$\frac{F}{a^2} = \frac{\beta}{a}\frac{x}{a} = \sigma = \frac{\beta}{a}\varepsilon = E\varepsilon$$

or

$$\beta = aE \tag{9.19}$$

where a is the interatomic separation.

9.3.3.1 Example: Iron, Fe

For iron, $\eta(\text{Fe}) = (\rho/M)N_A = (7.86\,\text{g/cm}^3/55.847\,\text{g/mol})6.022\times10^{23} = 8.48\times10^{22}\,\text{cm}^{-3} = 8.48\times10^{28}\,\text{m}^{-3}$ is the number of iron atoms per m³. And an approximation to the interatomic spacing is given by $a \cong 1/(\eta^{1/3}(\text{Fe})) = 1/(8.48\times10^{28})^{1/3} = 2.28\times10^{-10}$ m. Young's modulus for iron is $E = 211.4\times10^9$ Pa (Brandes and Brook 1992), so $\beta = aE = (2.28\times10^{-10}\,\text{m})(211.4\times10^9\,\text{N/m}^2) = 48.12$ N/m. The mass per atom of iron is $M/N_A = 55.847\times10^{-3}$ kg/mol/(6.022×10^{23} atoms/mol) $= 9.27\times10^{-26}$ kg/atom. Therefore,

$$f = \frac{1}{2\pi}\sqrt{\frac{\beta}{m}} = \frac{1}{2\pi}\sqrt{\frac{48.12\,\text{N/m}}{9.27\times10^{-26}\,\text{kg}}} = 3.6\times10^{12}\,\text{s}^{-1}\ (\text{Hz}).$$

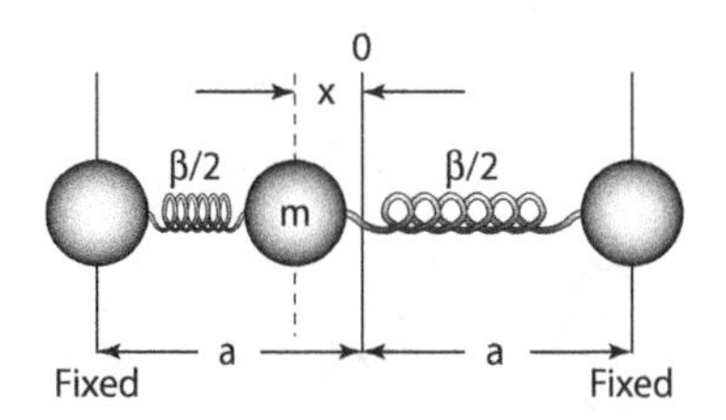

FIGURE 9.5 An atom of mass "m" vibrates horizontally between its two nearest neighbors at a distance a with a spring constant on each side of β/2 and moved from its equilibrium position a distance x. The two adjacent atoms are assumed to not move.

9.3.3.2 Example: Copper, Cu

For copper, $\eta(Cu) = (\rho/M)N_A = (8.92\,g/cm^3 / 63.546\,g/mol) \times 6.022 \times 10^{23} = 8.44 \times 10^{22}\ cm^{-3}$ = 8.44 × $10^{28}\ m^{-3}$ is the number of copper atoms per m^3. Again, an approximation to the interatomic spacing can be given by $a \cong 1/(\eta^{1/3}(Cu)) = 1/(8.44 \times 10^{28})^{1/3} = 2.28 \times 10^{-10}$ m. Young's modulus for copper is E = 129.8 × 10^9 Pa (Brandes and Brook 1992), so $\beta = aE$ = (2.28 × 10^{-10} m)(129.8 × 10^9 N/m^2) = 29.69 N/m. The mass per atom of copper is M/N_A = 63.546 × 10^{-3} kg/mol/(6.022 × 10^{23} atoms/mol) = 1.055 × 10^{-25} kg/atom. Therefore,

$$f = \frac{1}{2\pi}\sqrt{\frac{\beta}{m}} = \frac{1}{2\pi}\sqrt{\frac{29.69\,N/m}{10.55 \times 10^{-26}\,kg}} = 2.53 \times 10^{12}\ s^{-1}\ (Hz)$$

These two calculations suggest that the vibrational frequency is on the order of 3 × 10^{12} Hz.

9.3.3.3 Quantized Vibrations

Another approach to approximating the vibrational frequency is from quantized vibrations. From above, the total energy for a vibrational mode is $E_x = (1/2)k_BT + (1/2)k_BT = k_BT$. Einstein quantized the atomic vibrations and assumed that all the vibration frequencies were the same—as Max Planck had quantized radiation waves—to successfully explain the fact that the heat capacity of solids approached zero as the absolute temperature approached zero. This leads to a quantized energy of vibration: $E = hf = k_B\theta_E$, where θ_E = Einstein temperature—determined experimentally—and h = 6.64 × 10^{-34} Js is *Planck's constant.* Debye modified Einstein's theory of heat capacity noting that the atoms could not vibrate independently, because they are coupled together by interatomic forces (springs), and there must be a series of vibration wavelengths or frequencies up to some maximum frequency, $hf_{max} = k_B\theta_D$, where θ_D is the Debye temperature, as sketched in Figure 9.6. The Debye theory models reality better than Einstein's, and the Debye temperatures of a number of solids have been measured and are available (Gray 1972). So, the maximum vibrational frequencies of atoms in solids can be obtained from the Debye temperatures: θ_D(Fe) = 467 K, θ_D(Cu) = 343 K (Gray 1972), so

$$f_{max}(Fe) = \frac{k_B\theta_D}{h} = \frac{(1.38 \times 10^{-23}\,J/K)(467\,K)}{6.62 \times 10^{-34}\,Js} = 9.74 \times 10^{12}\ Hz$$

and for copper

$$f_{max}(Cu) = \frac{k_B\theta_D}{h} = \frac{(1.38 \times 10^{-23}\,J/K)(343\,K)}{6.62 \times 10^{-34}\,Js} = 7.15 \times 10^{12}\ Hz.$$

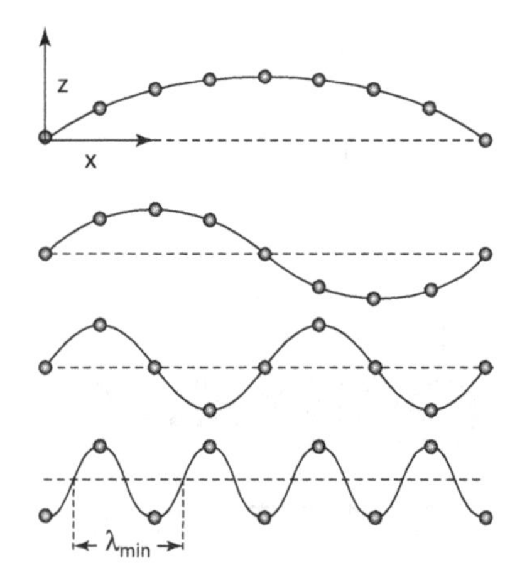

FIGURE 9.6 A few vibration modes of coupled atoms used in the Debye model of heat capacity. The bottom mode has the smallest wavelength, λ_{min}, or the highest frequency.

9.3.3.4 Velocity of Sound

The vibrational frequency can also be calculated from the velocity of sound in a solid, because (Appendix A.2):

$$v = \sqrt{\frac{E}{\rho}} = \lambda f \tag{9.20}$$

where:

ρ is the density (kg/m^3)

λ is the wavelength = 2a (m) (Figure 9.6).

Therefore, for iron

$$f = \frac{1}{2a}\sqrt{\frac{E}{\rho}} = \frac{1}{2\times 2.28\times 10^{-10}\ \text{m}}\sqrt{\frac{211.4\times 10^{9}\ \text{Pa}}{7.86\times 10^{3}\ \text{kg}}} = 1.14\times 10^{13}\ \text{Hz}$$

and for copper

$$f = \frac{1}{2a}\sqrt{\frac{E}{\rho}} = \frac{1}{2\times 2.28\times 10^{-10}\ \text{m}}\sqrt{\frac{129.8\times 10^{9}\ \text{Pa}}{8.92\times 10^{3}\ \text{kg}}} = 8.37\times 10^{12}\ \text{Hz.}$$

On the basis of these three calculations (which in fact are really not independent), a good approximation for the vibrational frequency of atoms in strongly bonded solids would be on the order of $f \cong 10^{13}$ Hz.

9.3.4 Putting It All Together

Combining Equations 9.12, 9.14, and 9.15, the expression for D becomes

$$D = \frac{1}{6}a^2\Gamma = \frac{1}{6}a^2 f e^{-\frac{\Delta G_m}{RT}} = \frac{1}{6}a^2 f e^{\frac{\Delta S_m}{R}} e^{-\frac{\Delta H_m}{RT}} \tag{9.21}$$

where:

ΔS_m is the entropy for motion

ΔH_m is the enthalpy for motion.

ΔS_m is associated with the change in positions of the lattice atoms as the interstitial atom moves through and is not terribly easy to evaluate, but it is expected to be not very significant. For example, it is suggested that $\Delta S_m \cong R$ (Shewmon 1989), so $e^{(\Delta S_m/R)} \cong e^1 = 2.718$. Therefore, for interstitial diffusion,

$$D = D_0 e^{-\frac{Q}{RT}}$$

where

$$D_0 \cong \frac{1}{6}a^2 f e^{\frac{\Delta S_m}{R}} = \frac{1}{6}(2.3\times 10^{-8}\ \text{cm})^2(10^{13}\ \text{s}^{-1})(2.718) = 2.21\times 10^{-3}\ \text{cm}^2/\text{s}$$

and $Q = \Delta H_m$. Table 9.1 gives some experimental data for interstitial diffusion in iron (Shewmon 1989). The agreement between experimental and calculated values of D_0 are surprisingly good given the lack of details in the model. So, for interstitial diffusion, $D \cong 10^{-3}\exp(-\Delta H_m/RT)$ cm²/s, where ΔH_m is the enthalpy of motion of the atoms, molecules, or ions, is a good approximation. Of course, ΔH_m, will vary considerably depending on the material and the atoms or molecules moving because it depends on the crystal structure, the interstitial and lattice atom sizes, and the details of their atomic bonding.

9.3.5 Interstitial Jumps per Second, Velocity

For interstitial motion of carbon in iron, from the data in Table 9.1, Γ, the jumps per second is given by

$$\Gamma = f e^{-\frac{\Delta H_m}{RT}} \cong 10^{13} e^{-\frac{80{,}100}{(8.314)(1000)}} = 6.54\times 10^{8}\ \text{s}^{-1}$$

at 1000 K. The speed is $a\Gamma = 2.3 \times 10^{-8}$ cm $\times$ 6.54×10^8 s^{-1} = 15.04 cm/s! This is pretty fast, yet still much slower than the speeds of atoms in a gas, which are on the order of 5×10^4 cm/s. Also, in the solid interstitials case, the mean free path, $\lambda = a/2 \cong 10^{-8}$ cm, is small, making $D \cong 1/3v\lambda$ on the order of $D \cong 10^{-8}$ to 10^{-7} cm²/s as expected.

TABLE 9.1
Interstitial Diffusion in Iron

Interstitial Atom	D_0 (cm^2/s)	Q (kJ/mol)	D (1000 K) (cm^2/s)
C	4×10^{-3}	80.1	2.6×10^{-7}
N	5×10^{-3}	76.8	4.9×10^{-7}
O	2×10^{-3}	86.0	6.4×10^{-6}

Source: Shewmon, P., *Diffusion in Solids*, 2nd ed, The Minerals, Metals, and Materials Society, Warrendale, sPA, 1989.

9.4 VACANCY MECHANISM OF DIFFUSION IN SOLIDS

9.4.1 Vacancy Diffusion

The vacancy mechanism of diffusion is an important mechanism of atom movement in high-temperature, primarily crystalline, materials and less so in low-temperature materials such as polymers. This is primarily because atomic bonds must be broken to produce vacancies, which requires high temperatures not accessible to most solid polymers. The end of a polymer molecule might be considered a vacancy or the site of a missing atom. However, this does not lead directly to any useful insights into diffusion in polymers, as discussed later. In contrast, inorganic glasses can be treated much like their crystalline counterparts and the concept of a vacant site and other point defects* can be identified. In this case, the concepts of point defects do provide insight into diffusion as well as temperature and additive effects on viscosity and glass transition temperatures. This will also be explored briefly later.

For now, the focus is on crystalline materials where high temperatures are necessary for diffusion to be an important factor in a material's behavior. A vacancy or a vacant lattice site is one in which an atom or ion is missing, as shown again in Figure 9.7. It takes energy to create vacancies, but their presence increases the entropy because of mixing of the vacancies and atoms—the *configurational entropy*—in the crystal lattice and leads to a finite vacancy *equilibrium* concentration at any temperature. As will be more evident later, it is convenient to express the vacancy concentration in terms of the fraction of atomic or lattice sites vacant; that is, [V] = *fraction of lattice sites vacant*, so $[V] = \eta_V/(\eta_V + \eta)$, where η_V is the number of vacancies/cm^3 and η the number of atoms/cm^3, so that $\eta_V + \eta$ = number of lattice sites/cm^3 (or per cubic meter). *Then [V] is also the probability that any single site is vacant.* The model for the diffusion coefficient for vacancy diffusion is very similar to that for interstitial diffusion except that it must be multiplied by [V]; that is, the probability that the neighboring site into which an atom or ion is going to move is vacant; that is,

$$D_{vacancy} = \frac{1}{6}a^2\Gamma[V] = \frac{1}{6}a^2 f e^{-\frac{\Delta G_m}{RT}}[V] = \frac{1}{6}a^2 f e^{\frac{\Delta S_m}{R}} e^{-\frac{\Delta H_m}{RT}}[V] \quad (9.22)$$

The problem then is to determine [V].

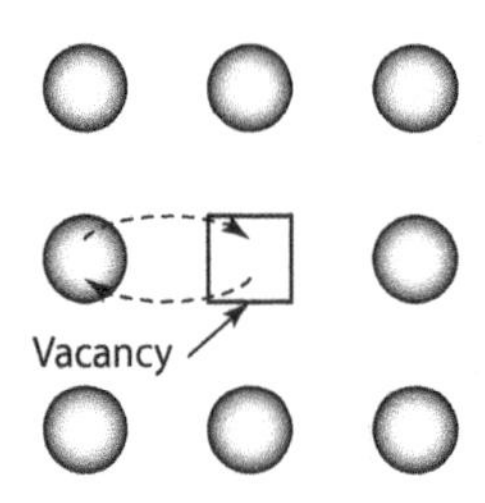

FIGURE 9.7 The vacancy mechanism of diffusion: an atom moves into the vacancy and the vacancy takes the atom's place in the crystal lattice.

9.4.2 Vacancies in Elements (Metals, Ge, Si, etc.): Statistical Approach

Vacant lattice sites (vacancies) and other point defects exist in crystals because they increase the entropy of the system. However, it takes energy to form them largely because of the unsatisfied chemical bonds that are created between the neighboring atoms or ions and the vacancy. The Gibbs energy of

* Vacancies and interstitials are *point defects*, because they have zero dimension in contrast to *dislocations* that have one dimension, *grain boundaries* and other boundaries that have two dimensions, and *pores* and *second phases* that have three dimensions.

a crystal with vacancies compared to one without—that is, a perfect crystal—G(n) with n = number of vacancies and N = number of atoms—(it is more convenient to use numbers per unit volume with no loss in generality)—is given by the following equation with the configurational entropy term of Boltzmann, $S = k_B \ln W$ (Chiang et al. 1997):

$$G(n) = n g_V - T\Delta S_{m,\,config}$$
$$G(n) = n g_V - k_B T \ln W \qquad (9.23)$$
$$G(n) = n g_V - k_B T \ln\left(\frac{(N+n)!}{N!\,n!}\right)$$

where g_V is the Gibbs energy to form a single vacancy and the entropy of mixing is given by (Chapter 1)

$$\Delta S_{m,\,config} = k_B \ln W = k_B \ln\left(\frac{(N+n)!}{N!n!}\right) \qquad (9.24)$$

where W is just the number of ways that N atoms and n vacancies can be arranged on N + n total sites, or it is the number of possible physical arrangements of the system.* Stirling's approximation is $\ln(n!) \cong n \ln n - n$, Chapter 1. So, applying Stirling's approximation in Equation 9.23, G(n) becomes,

$$G(n) \cong n g_V - k_B T[(N+n)\ln(N+n) - (N+n) - N\ln N + N - n\ln n + n]. \qquad (9.25)$$

Differentiating G(n) with respect to n to minimize the total Gibbs energy as a function of the number of vacancies present—the minimum in energy in Figure 9.8—results in

$$\left(\frac{\partial G}{\partial n}\right)_{p,T,N} = \mu_V = 0 = g_V + \ln(N+n) + 1 - \ln(n-1)$$

and rearranging gives

$$0 = g_V - k_B T \ln\left(\frac{N+n}{n}\right). \qquad (9.26)$$

Note that minimizing the Gibbs energy of a crystal with vacancies is tantamount to *setting the chemical potential of vacancies, μ_V, in equilibrium at a given temperature and pressure to zero.*

The final result is

$$[V] = \frac{n}{N+n} = e^{-\frac{g_V}{k_B T}} = e^{-\frac{\Delta G_V}{RT}} \qquad (9.27)$$

where:

$g_V = e_V + p v_V - T s_V$
e_V is the energy to form a single vacancy
v_V is the volume of a vacancy
s_V is the entropy associated with the formation of a single vacancy†
ΔG_V is the equivalent term for a mole of vacancies.

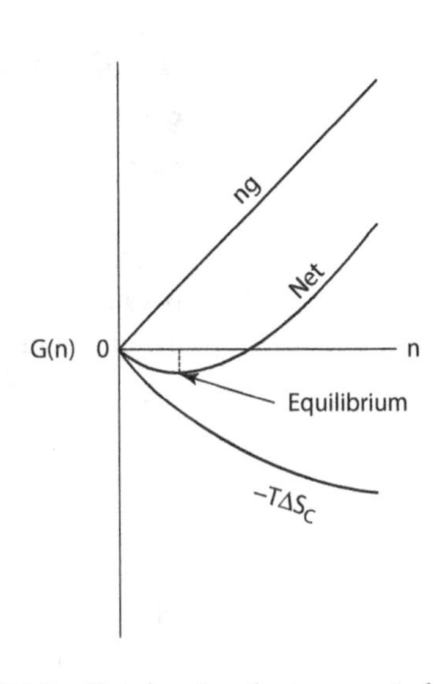

FIGURE 9.8 Plot showing the increase in free energy, ng, with an increase in the number of vacancies, n, the decrease in free energy due to the entropy of mixing—the configurational entropy, ΔS_C—and the resulting net energy that leads to a equilibrium concentration of vacancies at the minimum of the free energy.

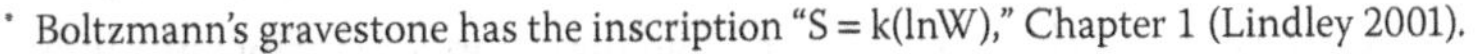

* Boltzmann's gravestone has the inscription "S = k(lnW)," Chapter 1 (Lindley 2001).

† This entropy term relates mainly to the formation of new local lattice vibration frequencies in the neighborhood of a vacancy and the entropy associated with the introduction of these new modes. It is thought to be small and usually neglected.

Frequently in the literature, the approximation is made that $(N + n) \cong N$, but there is no need or reason to do this. However, in calculations, the number of vacancies compared with the number of atoms or lattice sites is usually very small and can indeed be neglected. Nevertheless, it is important to note that Equation 9.27 is *exactly correct* and $[V] = n/(N + n)$ *is the **fraction** of vacancies relative to the total of lattice sites, N + n, occupied by N atoms and n vacancies.* This is emphasized because this fraction of vacancies, no units, is frequently confused in the literature with the actual concentration, number per cm^3, of vacancies. This is incorrect!

9.4.3 Regular Solution Approach

This same result can be obtained by finding the minimum in the energy for vacancies in a crystal with a regular solution (Reed-Hill and Abaaschian 1992), where for a regular solution (Darken and Gurry 1953)

$$\Delta G_{soln} = n_V \Delta G_V - T\Delta S_{ideal} = n_V \Delta G_V + RT(n_V \ln X_V + n_N \ln X_N)$$

where:

- ΔG_{soln} is the total Gibbs energy of the solution
- ΔG_V is the energy difference between the perfect crystal and one with a mole of vacancies not counting the entropy of mixing term
- n_V is the moles of vacancies
- n_N is the moles of atoms
- X_i is the mole fraction.

If both the top and bottom of the two terms on the right-hand side of the equation are multiplied by Avogadro's number, N_A,

$$\Delta G'_{soln} = n_V \Delta G_V - T\Delta S_{ideal} = ng_V + k_B T(n \ln X_V + N \ln X_N) \quad (9.28)$$

with $X_V = n/(N + n)$, $X_N = N/(N + n)$. Substituting these values into Equation 9.28 gives Equation 9.25, which leads to the same result as before, Equation 9.27, upon differentiation.

9.4.4 Vacancy Concentrations

9.4.4.1 Fractions Not Concentrations

Obviously, if both the numerator and denominator of the left-hand side of Equation 9.27 are divided by some volume, V, $\eta_V = n/V$ and $\eta = N/V$ will now be the number of vacancies and the number atoms per unit volume, respectively, and $\eta_V + \eta$ the total number of lattice sites per unit volume. Because electrical conductivity, diffusion, and other properties depend on the number of vacancies per unit volume, the number of vacancies per cm^3 or per m^3 are logical choices. As a result, the "concentration" of vacancies, [V], is from above,

$$[V] = \frac{\eta_V}{\eta_V + \eta} = e^{-\frac{g_V}{k_B T}} = e^{-\frac{\Delta G_V}{RT}} \quad (9.29)$$

where [V] is the ***fraction** of lattice sites* (bold added for emphasis) that are vacant: as such, it is *unit less* and not really a "concentration." As mentioned in Section 9.4.1, equivalently, [V] is the *probability that any given single site* is vacant.

Now $\Delta \bar{G}_V = \Delta \bar{H}_V - T\Delta \bar{S}_V = \Delta \bar{U}_V + P\Delta \bar{V}_V - T\Delta \bar{S}_V$ for a mole of vacancies and the entropy term is usually small as is the $P\Delta \bar{V}$ term and both can be ignored in calculating the concentration of vacancies. Of course, if there are pressure gradients that give rise to vacancy concentration gradients that lead to mass transport during sintering or creep, then the $P\Delta \bar{V}$ term is important. However, even then, the overall concentration of vacancies does not necessarily change very much. So,

$\Delta\bar{G}_V \cong \Delta\bar{H}_V \cong \Delta\bar{U}_V$ with values usually on the order of several electron volts (1 eV = 1.602×10^{-19} J ≌ 96,500 J/mol and $k_B = 1.38 \times 10^{-23}$ J/atom/1.602×10^{-19} J/eV = 8.61×10^{-5} eV/atom) and even for large pressure values, for example, 700 mPa (100,000 psi) for a molar volume around 10 cm³/mol ≡ 10^{-5} m³/mol, $P\Delta\bar{V}_V \cong 700 \times 10^6 \times 10^{-5} = 7000$ J/mol $\ll$ 100 kJ/mol $\cong \Delta\bar{U}_V$ and can be neglected. At atmospheric pressures, where P = 10^5 Pa, the $P\Delta\bar{V}$ term is on the order of 1 J/mol and certainly negligible in determining vacancy concentrations. Note that these energies are frequently given the subscript f to designate *energy of formation*; for example, ΔH_f. Table 9.2 gives some values of the enthalpy of formation of vacancies in metals.

9.4.4.2 Example: Copper at Its Melting Point, 1083°C

For Cu, $\Delta H_V = 1.03$ eV (Borg and Dienes 1988), so at its melting point of 1083°C

$$[V] = \exp(-1.03 \times 96{,}500/(8.314 \times (1083 + 273))) = 1.48 \times 10^{-4}$$

This implies that $\eta \gg \eta_V$, so $\eta_V + \eta \cong \eta$, which is the number of copper atoms per unit volume, which can be calculated from the molar volume,

$$\eta = \frac{\rho}{M} \times N_A = \frac{8.92 \text{ g/cm}^3}{63.55 \text{ g/mol}} \times 6.022 \times 10^{23} \text{ atoms/mol} = 8.45 \times 10^{22} \text{ atoms/cm}^3$$

which means that $\eta_V = 1.48 \times 10^{-4} \times 8.45 \times 10^{22} = 1.25 \times 10^{19}$ vacancies/cm³. Note again that for all solids because the atoms and ions more or less pack like billiard balls, the concentration of atoms, ions, or (simple) molecules in most solids in something times 10^{22} cm^{-3}.

TABLE 9.2
Enthalpies of Vacancy Formation in Pure Metals and Elements

Metal	Lattice	ΔH_f (eV)	Melting T (°C)	References
Al	FCC	0.68, 0.87, 0.76	660	1,2,3
Ag	FCC	1.12, 0.99, 1.09	961	1,2,3
Au	FCC	0.89, 0.87, 0.96	1063	1,2,3
Cu	FCC	1.29, 1.03, 1.17	1083	1,2,3
Fe	FCC	1.4	1536 (910–1400)	1
Ni	FCC	1.78	1453	1
Pt	FCC	1.32, 1.49, 1.2	1552	1,2,3
Pd	FCC	1.85	1552	1
Pb	FCC	0.57, 0.5, 0.53	327	1,2,3
Mo	BCC	3.0, 2.3	2620	1,2
Nb	BCC	2.65, 2.0	2467	1,2
Ta	BCC	2.8	3015	1
V	BCC	2.1	1902	1
W	BCC	4.0, 3.3	3400	1,2
Fe	BCC	1.6, 1.5	T < 910	1,2
Na	BCC	0.04, 0.48,[c] 0.34	98	2,5,5
Si	Diamond	2.4	1410	4
Ge	Diamond	2.2	937	4

References for the table: 1, Shewmon (1989); 2, Borg and Dienes (1988); 3, Henderson (1972); 4, Mayer and Lau (1990); 5, Catlow and Mackrodt (1982). Another reference gives the same data as Shewmon (1989) (Glicksman 2000). c = calculated.

9.4.4.3 Example: Silicon at Its Melting Point, 1410°C

For Si, $\Delta H_V = 2.4$ eV (Mayer and Lau 1990), at its melting point of 1410°C,

$$[V] = \exp(-2.4 \times 96{,}500/(8.314 \times (1410 + 273))) = 6.48 \times 10^{-8}$$

and

$$\eta = \frac{\rho}{M} \times N_A = \frac{2.32 \text{ g/cm}^3}{28.08 \text{ g/mol}} \times 6.022 \times 10^{23} \text{ atoms/mol} = 4.98 \times 10^{22} \text{ atoms/cm}^3$$

so that $\eta_V = 6.48 \times 10^{-8} \times 4.98 \times 10^{22} = 3.22 \times 10^{15}$ vacancies/cm³.

Note that the energy to form a vacancy is considerably higher in silicon than in copper, indicating that the bonds between atoms in silicon are significantly stronger than those in copper, which is also reflected in the higher melting point of silicon.

9.4.5 Vacancies, Quasichemical Approach: Point Defect Chemistry

9.4.5.1 In an Element

All of the above discussion is fine and works well for simple monatomic systems such as Cu and Si, where there is only one crystal site to consider and the interaction between lattice and electronic defects does not have to be taken into consideration. However, when considering more complex materials, for example, $Y_3Fe_5O_{12}$ with the possibility of additives in solid solution that can affect the concentration of vacancies (or interstitials) or introduce electronic defects, the statistical treatment becomes unwieldy and unattractive. The statistical approach can be used for more complex solids but becomes increasingly algebraically messy, impractical, and unnecessary particularly when electronic point defects such as electrons accompany the formation of lattice point defects, and/or impurity or atmosphere effects are introduced (Readey 1966a,b). In contrast, the quasichemical approach is very straightforward, simple, and absolutely correct!

Furthermore, there is interest in the concentrations of all of the *point defects*—interstitials, vacancies, impurity atoms, electrons, and so on—present and their interactions and how they affect the properties of the material.

Because the interest is only in equilibrium concentrations, some type of chemical equilibrium approach should be equivalent to the statistical method. Indeed, *the exact same result, Equation 9.29, can be obtained by a simple quasichemical approach or writing a chemical equation for the formation of point defects*, and then writing the equilibrium constant for the equation or reaction. For example, for the formation of vacancies in copper, a quasichemical equation is

$$\text{perfect crystal} = \text{null} = 0 \rightleftharpoons V_{Cu};\ \Delta\bar{G}_V \tag{9.30}$$

where $\Delta\bar{G}_V$ is the same as that given earlier—the energy necessary to form *a mole* of vacancies in a perfect crystal. As a result, the Gibbs energy is barred: $\Delta\bar{G}_V$.* So, the "concentration" of vacancies—more precisely, the *fraction of lattice sites vacant*—is obtained from the equilibrium constant for this reaction, Equation 9.30:

$$[V_{Cu}] = K_e = e^{-\frac{\Delta G_V}{RT}}. \tag{9.31}$$

Obviously, Equations 9.29 and 9.31 are the same with the caveat in the quasichemical approach that the vacancy concentrations, $[V_{Cu}]$, must be defined as *fraction of lattice sites that are vacant* to be

* However, to reduce the notation, in the following discussions and calculations on point defects, the bars are left off the molar quantities for defects and is more consistent with the literature.

consistent with the statistical approach. The left-hand side of Equation 9.30 is zero and is appropriate because the perfect crystal generates vacancies and there needs to be nothing on the left-hand side of the equation to balance it. It is a balanced equation because there is no mass or charge on either side!* However, the objection could be raised that "the lattice sites are not balanced, there are vacant sites on the right but none on the left." True! Equation 9.30 might be better written as

$$Cu_{Cu} \rightleftharpoons V_{Cu} + Cu_{surface};\ \Delta G_V. \tag{9.32}$$

That is, a copper atom on a copper site in the crystal, Cu_{Cu}, moves to the surface, $Cu_{surface}$, leaving behind a vacant copper site, V_{Cu}, resulting in an equilibrium constant of

$$\frac{[V_{Cu}][Cu_{surface}]}{[Cu_{Cu}]} = K_e = e^{-\frac{\Delta G_V}{RT}} \tag{9.33}$$

where:

$[V_{Cu}]$ is the fraction of copper sites vacant
$[Cu_{Cu}]$ is the fraction of copper sites in the crystal occupied by a copper atom
$[Cu_{surface}]$ is the fraction of surface sites occupied by copper.

For most real situations where $\Delta G_V \cong 1$ eV or greater, as it is for the majority of elements in Table 9.2, the vacancy site fraction is small as shown for copper and silicon. Because $[V_{Cu}] + [Cu_{Cu}] = 1$, then $[Cu_{Cu}] \cong 1$ for small vacancy concentrations. Similarly, the fraction of surface sites does not change, they are all occupied, so $[Cu_{surface}] = 1$ and Equation 9.33 becomes the same as Equation 9.31. However, if for some reason $\Delta \bar{G}_V$ were small, then Equation 9.33 would be necessary. Nevertheless, for real materials, the concentration of vacancies is small and Equations 9.30 and 9.31 are quite appropriate.

9.4.5.2 Very Small $\Delta \bar{G}_V$

For example, if, ΔG_V were zero, then $[V_{Cu}] = [Cu_{Cu}]$ or $\eta_V = \eta_{Cu}$, and $[V_{Cu}] = [Cu_{Cu}] = 1/2$. Note that in Table 9.2, one of the values for the enthalpy of vacancy formation in sodium is 0.04 eV, which is very small. So, from Equation 9.33

$$\frac{[V_{Na}]}{[Na_{Na}]} = \frac{[V_{Na}]}{1-[V_{Na}]} = K_e \tag{9.34}$$

because $[V_{Na}]$ is not small compared to 1.0 because ΔH_f is so small. Rearranging Equation 9.34 gives $[V_{Na}] = K_e/(1+K_e)$ and at 300 K,

$$[V_{Na}] = \frac{K_e}{1+K_e} = \frac{e^{-\frac{\Delta H_f}{RT}}}{1+e^{-\frac{\Delta H_f}{RT}}} = \frac{e^{-\frac{(0.04)(96{,}500)}{(8.314)(300)}}}{1+e^{-\frac{(0.04)(96{,}500)}{(8.314)(300)}}} \cong 0.176!$$

Such a high vacancy concentration should be easily measured by measuring the density of sodium and comparing to the theoretical or x-ray density of pure sodium. In fact, this is a technique that has been used to measure the vacancy concentration of solids if the vacancy concentration is high enough and the technique to measure the density is precise enough. The theoretical or x-ray density of a crystalline solid is given by

$$\rho = \frac{\sum_i n_i A_i}{N_A V_{cell}}$$

* This is a good reason to call this a *quasichemical* approach.

where:

n_i is the number of atoms of species i per unit cell
A_i is the atomic weight of species i
N_A is the Avogadro's number
V_{cell} is the volume of the unit cell.

Take for example sodium, which has a BCC unit cell of two sodium atoms per cell, an atomic weight of A = 22.99 g/mol, and a lattice parameter of $a = 4.2906 \times 10^{-8}$ cm (Gray 1972). Therefore, the theoretical or x-ray density of sodium is

$$\rho\,(\mathrm{g/cm^3}) = \frac{2(22.99)}{(6.022\times10^{23})(4.2906\times10^{-8})^3} = 0.9667\ \mathrm{g/cm^3}.$$

A literature value for the density of sodium is $\rho = 0.9712$ g/cm³ (Gray 1972). The calculated vacancy concentration is quite large and the calculation of the density must be modified because the number of atoms per unit cell is not 2.0 but $2.0 \times (1 - [V_{Na}])$. So, the density with vacancies should be

$$\rho\,(\mathrm{g/cm^3}) = \frac{2(1-0.176)(22.99)}{(6.022\times10^{23})(4.2906\times10^{-8})^3} = 0.7965\ \mathrm{g/cm^3}$$

which is much smaller than the value for the observed density of sodium, which is very close to the x-ray value assuming that there were no vacant sodium sites in the crystal! What is the problem? A likely culprit is the value for the formation of vacancies in sodium, $\Delta H_f = 0.04$ eV, is just too small and probably was tabulated incorrectly in the literature.* It might be expected that the enthalpy for vacancy formation would scale with the melting point of a solid because both are determined by the strength of the interatomic bonds (Shewmon 1989). For BCC iron, ΔH_V s(Fe) = 1.4 eV from Table 9.2. The melting point of iron is 1809 K, whereas that of sodium is 371 K. For iron, $\Delta H_V/T_m = 7.74 \times 10^{-4}$ eV/K. Multiplying this by the melting point of sodium gives $\Delta H_V(Na) \cong 0.287$ eV, about a factor of 9 greater than the one literature value of 0.04 and closer to the other tabulated values. With this value of the enthalpy to form vacancies in sodium, the vacancy concentration at 300 K would be

$$[V_{Na}] = e^{-\frac{\Delta H_f}{RT}} = e^{-\frac{(0.287)(96{,}500)}{(8.314)(300)}} \cong 1.5 \times 10^{-5}$$

which is too small to detect in the density and similar to vacancy concentrations in copper and silicon at their melting points. So from these data and calculations, it could be concluded that the value given in the literature $\Delta H_f(Na) \cong 0.04$ eV is probably incorrect. In any event, this calculation for sodium is a good review of theoretical density calculations, the effect of vacancy concentrations on densities, and how large vacancy concentrations need to be handled, Equation 9.34.

9.4.6 Charges of Point Defects in Compounds and Kröger–Vink Notation

9.4.6.1 Rationale

The quasichemical approach embodied in Equation 9.30 is particularly convenient for compounds such as very strongly ionically bonded solids.† In NaCl or MgO, creating just sodium (magnesium) or chlorine (oxygen) vacancies without the other is essentially impossible because of electrical charge imbalance. In compounds, whether ionically bonded or not, vacancies must be created on both anion and cation sites together—or all sublattices (ternary, etc. compounds)—for the crystal to remain electrically neutral

* Yes, this actually happens sometimes implying that "... just because it is published doesn't mean it's correct!"

† The approach also works perfectly well for metallic or covalently bonded compounds. To get completely correct results, all that needs to be done is to treat these compounds as if they were ionic and associate a formal charge to each element in the compound. Take GaAs for example. If gallium is taken to be Ga^{3+} and As^{3-}, then point defect chemistry can easily and correctly be applied to point defects and dopants in gallium arsenide.

(also to maintain "cation" and "anion" site balance). In other words, an electrically neutral "molecule" or "formula unit"—NaCl or MgO—must be removed from the crystal. The key to developing point defect chemistry and defect concentrations in compounds is to have a suitable method to account for charge on the defects and balance these charges to determine the concentrations of defects.

9.4.6.2 Kröger–Vink Notation

In the 1940s and 1950s, scientists at the Phillips-Eindhoven laboratory in Holland were developing phosphors for cathode ray tubes, CRTs. Phosphors are usually some inorganic compound that is "doped" (small additions) with one or more other compounds and emit light at a desired wavelength when irradiated with electrons, x-rays, or light. For example, a major breakthrough promoting the development of color television many years ago was the discovery of Eu_2O_3-doped YVO_4 (YVO_4:Eu), which gives off a bright red light under electron bombardment in a CRT. A bright red phosphor combined with existing green and blue phosphors allowed accurate color rendition on the TV screen. Because the additives and host lattice ions frequently have different charges—which is important in light emission—Kröger and Vink (1956) developed a convenient notation to consider the charge on the different ions in the phosphor that was far superior to all other notations in use at the time and is now quite universally accepted. Prior to this, each branch of engineering and science—or individual—had its own, different, notation for defects that made it difficult if not impossible for workers in different fields to understand the results of others working in a related field on similar materials. Unfortunately, although widely accepted*, the meaning of the K–V notation is still frequently misinterpreted even in many textbooks. As a result, some confusion still exists today on how to apply K–V notation and defect chemistry to calculate concentrations of point defects and their effect on properties such as diffusion and electrical conductivity. Being able to calculate the concentrations—in number per unit volume—is extremely important in predicting the influence of point defects on properties.

9.4.6.3 Charge Relative to the Perfect Crystal

Figure 9.9 shows the essence of K–V notation that takes into consideration the charge on a point defect. Here vacancies in MgO are used as examples. If an Mg^{2+} ion is removed from a perfect crystal of MgO (rock-salt structure), the initially uncharged crystal must now have a net 2− charge. As a result, the vacant magnesium site can be represented as V''_{Mg}, where the V indicates a vacancy, the subscript Mg indicates the lattice site, and the two apostrophes represent a 2− charge: each

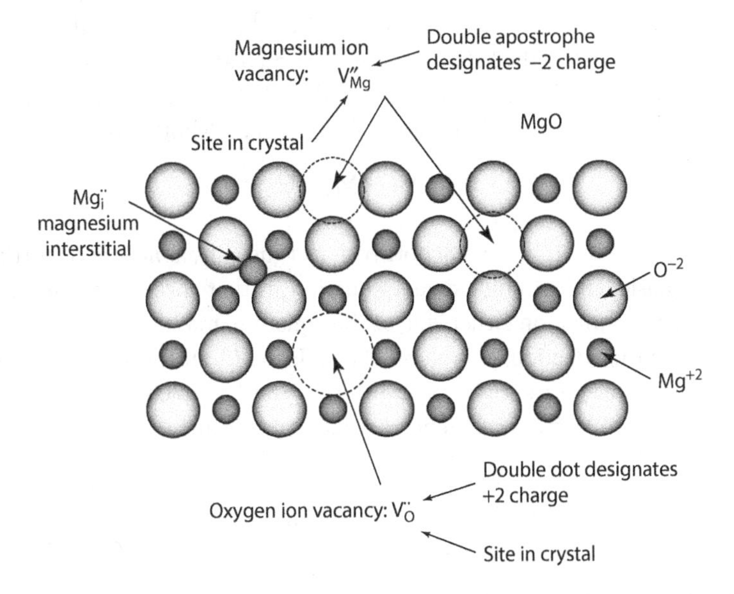

FIGURE 9.9 Drawing of a magnesium oxide crystal showing vacancies and interstitials: a Schottky defect and a Frenkel defect and how the Kröger–Vink notation applies to these defects.

* K–V notation is now the accepted point defect notation by the International Union of Pure and Applied Chemistry (Connelly et al. 2005).

apostrophe representing a single negative charge. This is more than just a "bookkeeping" method to indicate charge because the vacant site really does have a net negative charge. Figure 9.9 shows the vacant Mg^{2+} site is surrounded by six O^{2-} ions, four in the plane and one above and one below, whose charges are not balanced, giving the site a 2− charge relative to the perfect crystal. Similarly, a vacant oxygen site is designated by $V_O^{\bullet\bullet}$, where the double-dot superscript represents a 2+ charge relative to the perfect lattice: each dot representing a single positive charge. Again, Figure 9.9 shows that the $V_O^{\bullet\bullet}$ really is surrounded by six cations whose charges are not satisfied giving the site a net positive charge. In fact, in crystals, the positively charged vacancy can capture an electron that will orbit the positive charge in stable energy states just as an electron has stable states around a hydrogen nucleus. Electron transitions between these stable states give rise to color absorption bands, and the trapped electron plus vacancy is known as an F-center (F-for "color" from the German word for color, Farben) (Shulman and Compton 1962).

9.4.7 Intrinsic Point Defects in Compounds

9.4.7.1 Intrinsic and Extrinsic Defects

Intrinsic point defects are those whose concentrations depend on the properties of the given crystal itself. As a result, any two very pure crystals of the same material at the same temperature and pressure will have the same point defect concentration: for example, copper, where $[V_{Cu}] = K_e = \exp(-\Delta G_V/RT)$, Equation 9.31, the vacancy site fraction depends only on the energy to form vacancies in copper, ΔG_V, whose value depends on the strength of the atomic bonds in copper. On the contrary, in most *compounds*, impurities or additives can be present that can control the concentrations of point defects that vary with the additive concentration. Such defect concentrations that are controlled by something from "outside" the material and are called *extrinsic* defects. There are several types of intrinsic point defects in solids. But for now, the focus is on intrinsic Schottky and Frenkel defects that are considered here in some detail.

9.4.7.2 Schottky Defects

Because vacancies in compounds are charged, cation and anion vacancies cannot be created independently: they can be only created in electrically neutral groups; that is, "formula groups" or "molecules" of vacancies. An electrically neutral "molecule" of *vacancies* is called a *Schottky defect* and the equilibrium constant is called the *Schottky product* named after Walter Schottky, who developed the concept while working at Siemens corporation in the 1920s and 1930s (Hoddeson et al. 1992). In NaCl, for example, a point defect equation for the formation of Schottky defect is given by

$$0 \rightleftharpoons V'_{Na} + V^{\bullet}_{Cl};\ \Delta G_S. \tag{9.35}$$

where ΔG_S is the Gibbs energy to form a Schottky defect (usually, a *mole* of Schottky defects). This is a perfectly valid quasichemical point defect equation in that three equation-balancing criteria are satisfied: (1) there is no mass on either side of the equation, mass is balanced; (2) there is no net charge on either side of the equation, charge is balanced; and (3) the number of sodium sites equals the number of chlorine sites on the right as required by the 1:1 stoichiometry of Na^+ and Cl^- ions in NaCl.* ΔG_S is the energy to form the Schottky defect or vacancy pair. Writing the equilibrium constant for Equation 9.35 gives

$$[V'_{Na}][V^{\bullet}_{Cl}] = e^{-\frac{\Delta G_S}{RT}} \tag{9.36}$$

* Actually, the number sodium and chlorine sites are balanced because Equation 9.35 similar to Equation 9.32 would be $Na^{+}_{Na} + Cl^{-}_{Cl} \rightleftharpoons NaCl_{surface} + V'_{Na} + V^{\bullet}_{Cl}$, where there are both sodium and chlorine sites on both sides of the equation. Requiring a stoichiometric relationship between the number of cation and anion sites on the right-hand side of Equation 9.35 is equivalent to this more complete equation and is correct as long as the number of Schottky defects is small so that the site fractions on the left-hand side are essentially 1: that is, $[Na^{+}_{Na}] = [Cl^{-}_{Cl}] \cong 1$.

where $[V'_{Na}]$ and $[V^{\bullet}_{Cl}]$ are the *fraction of sodium sites vacant and fraction of chlorine sites vacant,* respectively. This is where confusion often arises: these concentrations are *site fractions* and, therefore, *have no units*! This is what the statistical approach, Equation 9.29, gives. In the literature, these *concentrations* (site fractions) are frequently given units such as "mol/m^3," which is *absolutely incorrect,* they are fractions! Making this error, calculations of *real* concentrations (mol/m^3) are usually incorrect and can be wrong by a factor of 2 or 3. On the other hand, when combining lattice or atomic point defects in equations with electronic defects such as electrons, treating these as "mol/m^3" or some other concentration units, can lead to errors of several orders of magnitude when calculating the *actual* defect concentrations per unit volume! Appendix A.4 expands on this. This incorrect usage occurs frequently in a rather large number of books and published papers that perpetuate and lend legitimacy to this error and make point defect chemistry more confusing than it need be. To emphasize this, it is worth repeating that $[V'_{Na}]$ and $[V^{\bullet}_{Cl}]$ are the *fraction of sodium sites vacant and fraction of chlorine sites vacant,* respectively, and *therefore do not have units.*

If the number of NaCl Schottky defects per unit volume is $\eta_{V_{NaCl}}$, then, obviously, $\eta_{V'_{Na}} = \eta_{V^{\bullet}_{Cl}} = \eta_{V_{NaCl}}$ are the number of sodium and chlorine vacancies per unit volume respectively. Because the concentrations of vacancies are usually small in strongly bonded solids—on the order of 10^{-4} at the melting point—as was the case for copper and silicon, the number of NaCl sites per cm^3 is given approximately by* $\eta_{NaCl} = (\rho/M)N_A$ where ρ = density of NaCl = 2.165 (g/cm^3), M = molecular weight NaCl = 58.44 g/mol (Haynes 2013), so

$$\eta(\text{NaCl}) = \frac{\rho}{M}N_A = \frac{2.165}{58.44} \times 6.022 \times 10^{23} = 2.23 \times 10^{22}\ \text{cm}^{-3}$$

and the number of sodium and chlorine lattice sites per cm^3 is just $\eta_{Na} = \eta_{Cl} = \eta_{NaCl}$, which has the advantage of showing the charges on the vacancies because there is one site for each ion for each "molecule" of NaCl. So,

$$[V'_{Na}] = \frac{\eta_V(\text{Na})}{\eta(\text{Na})} = [V^{\circ}_{Cl}] = \frac{\eta_V(\text{Cl})}{\eta(\text{Cl})} = [V_{NaCl}] = \frac{\eta_V(\text{NaCl})}{\eta(\text{NaCl})}.$$

That is, for the Schottky product in NaCl, the site fraction of sodium and chlorine vacancies are equal: $[V'_{Na}] = [V^{\bullet}_{Cl}]$, as they should be.

Another approach, that is more general and applicable to all types of point defect equations, is to consider *electrical neutrality* more formally. That is, the number of positive charges *per unit volume* must be equal to the number of negative charges per unit volume. For a Schottky defect in NaCl, this means:

$$\begin{gathered}\text{Positive/cm}^3 = \text{Negative/cm}^3 \\ (z_+)(e)\eta_{V^{\bullet}_{Cl}} = (z_-)(e)\eta_{V'_{Na}}\end{gathered} \tag{9.37}$$

where:

e is the electron charge, 1.602×10^{-19} C

z_+ and z_- are the number of electron charges on the positively and negatively charged vacancies, respectively.

---

* To enhance readability here and what follows, equivalent notation is used: for example, $\eta(\text{NaCl}) = \eta_{NaCl}$ both mean the number of NaCl molecules cm^{-3}. If the concentration of vacancies was not small, then as shown above for a metal, then the concentration of sodium vacancies, for example, is:

$$[V'_{Na}] = \frac{\eta_V(\text{Na})}{\eta_V(\text{Na}) + \eta_{Na}(\text{Na})} \quad \text{and} \quad [Na_{Na}] = \frac{\eta_{Na}(\text{Na})}{\eta_V(\text{Na}) + \eta_{Na}(\text{Na})} = \frac{\{\eta_V(\text{Na}) + \eta_{Na}(\text{Na})\} - \eta_V(\text{Na})}{\eta_V(\text{Na}) + \eta_{Na}(\text{Na})} = 1 - [V'_{Na}]$$

and similarly for the fraction of chlorine sites occupied by chlorine and chlorine vacancies.

For NaCl, $z_+ = z_- = 1$ and so,

$$\eta_{V_{Cl}^{\bullet}} = \eta_{V_{Na}'} \tag{9.38}$$

and dividing by η_{NaCl}

$$\frac{\eta_{V_{Cl}^{\bullet}}}{\eta_{NaCl}} = \frac{\eta_{V_{Na}'}}{\eta_{NaCl}}$$

$$\frac{\eta_{V_{Cl}^{\bullet}}}{\eta_{Cl}} = \frac{\eta_{V_{Na}'}}{\eta_{Na}} \tag{9.39}$$

$$[V_{Cl}^{\bullet}] = [V_{Na}']$$

as before. Therefore, from the equilibrium constant, the Schottky product, Equation 9.36 becomes

$$[V_{Na}'][V_{Cl}^{\bullet}] = e^{-\frac{\Delta G_S}{RT}}$$

$$[V_{Na}']^2 = [V_{Cl}^{\bullet}]^2 = e^{-\frac{\Delta G_S}{RT}} \tag{9.40}$$

$$[V_{Na}'] = [V_{Cl}^{\bullet}] = e^{-\frac{\Delta G_S}{2RT}}$$

where $\Delta G_S/2$ is the *average energy* to form a vacancy. Table 9.3 gives some Schottky defect formation energies.

TABLE 9.3
Schottky Defect Formation Energies in Compounds

Compound	Structure	ΔH_S (eV)	Melting T (°C)	References
NaBr	"	1.72, 1.68	747	2,6
NaCl	"	2.3, 2.3, 2.3, 2.02	801	1,2,3,5,6
KCl	"	2.3, 2.3, 2.6, 2.3	771	1,2,3,5,6
KBr	"	2.4, 2.4, 2.53, 2.27[c]	734	1,2,6,7
CsI	CsCl	1.9, 2.14	632	1,6
CsCl	CsCl	1.86	646	3,6
LiF	NaCl	2.5, 2.5, 2.5, 2.68	848	1,2,3,5,6
LiCl	"	2.12	610	5,6
LiI	"	1.34	469	5,6
CaF_2	CaF_2	5.5	1418	3,4
BeO	Wurtzite, ZnS	6, 6	2578	3,4
MgO	NaCl	7.7, 6, 7.5[c]	2825	3,4,7
CaO	"	6, 6	2613	3,4
BaO	"	3.4	1973	3,7
MnO	"	4.6	1842	3,7
UO_2	"	6.4	2847	3,4
TiO_2	Rutile	5.2	1843	3
Al_2O_3	α-Al_2O_3	23, 25.7[c]	2050	3,7
$MgAl_2O_4$	Spinel	29.1[c]	2105	3,7

References for the table: Crystal structures (Wells 1984); melting points (Haynes 2013); 1, Hayes and Stoneham (1985); 2, Henderson (1972); 3, Chiang et al. (1997); 4, Kingery et al. (1976); 5, Glicksman (2000); 6, Gray (1972); 7, Catlow and Mackrodt (1982). c = calculated.

To write an equation for the formation of both Na and Cl vacancies and the equilibrium constant as above assumes *that the vacancies are not associated into a "molecule" of vacancies on neighboring lattice points.* Quite the contrary, it assumes that they are *dissociated and each cation and anion vacancy is free to mix randomly with the ions on the cation and anion lattices,* respectively. In reality, the vacancies are charged and attract each and can associate. An equilibrium for this association can be written, but is not discussed here to keep things simple.

9.4.7.3 Example: Vacancy Concentrations in NaCl at Its Melting Point

$\Delta\bar{G}_S \cong \Delta\bar{H}_S = 2.3$ eV × 96,472 J/eV = 221,890 J/mol (Chiang et al. 1997). So, at the melting point of NaCl of T = 801°C = 1074 K,

$$[V'_{Na}] = [V^{\bullet}_{Cl}] = e^{-\frac{221{,}890}{2(8.314)(1074)}} = 4.02\times10^{-6}$$

roughly four vacancies for every million sites, which does not seem like very many. However, $_V$(Na) = η_V(Cl) = $[V'_{Na}]\eta$(NaCl) = 4.02 × 10^{-6} × 2.23 × 10^{22} cm^{-3} = 8.96 × 10^{16} cm^{-3}, which now seems to be a reasonable number relative to the number of total sites. Note that for both copper and NaCl, the energy to form a single vacancy is on the order of 1 eV and their melting temperatures are not too different: T_m(Cu) = 1083°C and T_m(NaCl) = 801°C.

9.4.7.4 Schottky Defects in Al_2O_3

What about a more complex compound such as Al_2O_3? The same principles hold true in that an electrically neutral "molecule" of vacancies must be formed, $2\,V'''_{Al}$ and $3V^{\bullet\bullet}_O$, and the quasichemical equation for the formation of the vacancies is

$$0 \rightleftharpoons 2V'''_{Al} + 3V^{\bullet\bullet}_O;\ \Delta G_S \tag{9.41}$$

with the Schottky product given by

$$[V'''_{Al}]^2[V^{\bullet\bullet}_O]^3 = e^{-\frac{\Delta G_S}{RT}} \tag{9.42}$$

where the ΔG_S in this case is the energy to form 1 mole of "Al_2O_3" vacancies. So, if $\eta_V(Al_2O_3)$ is the number of "Al_2O_3 vacancies" per unit volume, then η_V(Al) = $2\eta_V(Al_2O_3)$ and η_V(O) = $3\eta_V(Al_2O_3)$ because there are 2Al and 3O per Al_2O_3 "molecule." Then, the number of Al_2O_3 "molecule" sites per cm^3 is again, $\eta(Al_2O_3) = (\rho(Al_2O_3)/M(Al_2O_3))N_A$, where ρ is the density (g/cm^3) and M is the molecular weight (g/mol) of Al_2O_3. Here, η(Al) = $2\eta(Al_2O_3)$ and η(O) = 3 $\eta(Al_2O_3)$, so

$$\begin{aligned} [V'''_{Al}] &= \frac{\eta_V(Al)}{\eta(Al)} = \frac{2\eta_v(Al_2O_3)}{2\eta(Al_2O_3)} = \frac{\eta_v(Al_2O_3)}{\eta(Al_2O_3)} \\ [V^{\bullet\bullet}_O] &= \frac{\eta_V(O)}{\eta(O)} = \frac{3\eta_v(Al_2O_3)}{3\eta(Al_2O_3)} = \frac{\eta_v(Al_2O_3)}{\eta(Al_2O_3)} \end{aligned} \tag{9.43}$$

therefore, $[V'''_{Al}] = [V^{\bullet\bullet}_O]$; that is, the vacancy site fractions are equal in Al_2O_3 just as they were in NaCl.

Similarly, application of *electrical neutrality* in this case is

$$(2e)\eta_{V^{\bullet\bullet}_O} = (3e)\eta_{V'''_{Al}}$$

and dividing by $\eta_{Al_2O_3}$

$$\frac{(2e)\eta_{V_O^{\bullet\bullet}}}{\eta_{Al_2O_3}} = \frac{(3e)\eta_{V_{Al}'''}}{\eta_{Al_2O_3}}$$

$$\frac{(2e)\eta_{V_O^{\bullet\bullet}}}{\frac{\eta_O}{3}} = \frac{(3e)\eta_{V_{Al}'''}}{\frac{\eta_{Al}}{2}}$$

$$\frac{(6e)\eta_{V_O^{\bullet\bullet}}}{\eta_O} = \frac{(6e)\eta_{V_{Al}'''}}{\eta_{Al}}$$

$$[V_O^{\bullet\bullet}] = [V_{Al}''']$$

again as before for NaCl.

This is true for all Schottky defects in binary compounds: *The site fractions of the cation and anion vacancies are equal for Schottky defects.* However, the *numbers* of vacancies per unit volume must be different to balance the charge difference, but their site fractions are the same. (Note: the charge has already been taken care of in the formula, e.g., Al_2O_3, so charge is counted twice and just cancels out.) So, the equilibrium constant in Equation 9.42 becomes

$$[V_{Al}''']^2[V_O^{\bullet\bullet}]^3 = [V_{Al}''']^5 = [V_O^{\bullet\bullet}]^5 = e^{-\frac{\Delta G_S}{RT}}$$

$$[V_{Al}'''] = [V_O^{\bullet\bullet}] = e^{-\frac{\Delta G_S}{5RT}}. \tag{9.44}$$

Again, the energy to form either aluminum or oxygen vacancies is the "average" energy per vacancy, $\Delta G_S/5$.

9.4.7.5 Example: Vacancy Concentrations in Al_2O_3 at Its Melting Point

$\Delta G_S \cong \Delta H_S \cong 22$ eV × 96,472 J/eV = 2,122,400 J/mol. At the melting point of Al_2O_3 of T = 2050°C = 2323 K,

$$[V_{Al}'''] = [V_O^{\bullet\bullet}] = e^{-\frac{2,122,400}{5(8.314)(2323)}} = 2.85 \times 10^{-10}$$

Now, $\eta(Al_2O_3) = (\rho/M)N_A = (4.00/101.96) \times 6.022 \times 10^{23} = 2.36 \times 10^{22}$ cm^{-3}, so $\eta_V(Al) = [V_{Al}''']\eta(Al) = [V_{Al}''']\, 2\eta(Al_2O_3) = 2.85 \times 10^{-10} \times 2 \times 2.36 \times 10^{22}$ cm^{-3} = 1.35×10^{13} cm^{-3}. For $\eta_V(O) = [V_O^{\bullet\bullet}]\eta(O) = [V_O^{\bullet\bullet}]\, 3\eta(Al_2O_3) = 2.85 \times 10^{-10} \times 3 \times 2.36 \times 10^{22}$ cm^{-3} = 2.02×10^{13} cm^{-3} and the vacancy concentrations per cm^3 are in a ratio of 2:3 as required by the formula for Al_2O_3.

Note that the energy to form a single vacancy in aluminum oxide is in the neighborhood of about 4 eV, which is considerably larger than that for NaCl. This is not too surprising considering the higher charges on Al^{3+} and O^{2-} ions compared with Na^+ and Cl^- produce stronger bonding as reflected in the melting points: 801°C for NaCl compared with 2050°C for Al_2O_3. However, from a practical point of view, the *intrinsic vacancy site fraction* at the melting point of alumina is only about 0.3 parts per billion! This means that to have a chance to observe intrinsic behavior in this material, the starting purity of the material must be greater than nine-9s pure*—semiconductor-grade material—otherwise, the impurities present are likely to influence the vacancy concentrations as will be discussed below.

One final point about compounds. There are many important compounds that contain more than a single cation such as $Y_3Fe_5O_{12}$, $MgAl_2O_4$, $NiFe_2O_4$, and so on. Again for such ternary

* Nine-9s pure is 99.9999999% pure!

compounds, a Schottky defect is defined a "molecule" or formula unit of vacancies and a Schottky product can be written. Take $MgAl_2O_4$ for example:

$$0 \rightleftharpoons V''_{Mg} + 2V'''_{Al} + 4V^{\bullet\bullet}_{O}; \Delta G_S$$

and

$$[V''_{Mg}][V'''_{Al}]^2[V^{\bullet\bullet}_{O}]^4 = K_S = e^{-\frac{\Delta G_S}{RT}} \tag{9.45}$$

as was the case for binary compounds. Just like binary compounds, the site fraction of vacancies are equal, that is, $[V''] = [V'''_{Al}] = [V^{\bullet\bullet}_{O}]$.

9.4.7.6 Frenkel Defects

Another type of intrinsic defect that is important in diffusion in solids is the *Frenkel defect* postulated by Jacob Frenkel in the 1920s while at the Leningrad Institute for Electrical Engineering in Russia (Hoddeson et al. 1992). A Frenkel defect is a vacancy and an intestinal pair. As such, they can occur in either elemental solids or compounds. It is difficult to force large lattice atoms or ions into small interstitial sites. However, they are particularly important in certain compounds that have either large interstitial sites such as the fluorite crystal structure or small ions such as the silver ion in the silver halides, where the interstitial silver plays an important role in the—rapidly becoming extinct—film photographic process. Figure 9.9 shows a magnesium Frenkel defect in MgO consisting of an interstitial magnesium ion and a vacant magnesium site. This Frenkel defect can be written as

$$Mg_{Mg} \rightleftharpoons Mg^{\bullet\bullet}_{i} + V''_{Mg} : \Delta G_F \tag{9.46}$$

where ΔG_F is the energy to form the Frenkel defect, the vacancy–interstitial pair, and the subscript i indicates an interstitial site. Again, for this point defect equation, mass and charge are balanced. To give a site balance, the equation would have to be written:

$$Mg_{Mg} + I \rightleftharpoons Mg^{\bullet\bullet}_{i} + V''_{Mg} : \Delta G_F \tag{9.47}$$

where I is an interstitial site. But as was the case when the vacancy concentration was small, $[I] \cong 1$ when the concentration of Frenkel defects is small; that is, very few interstitial sites are occupied, so Equation 9.46 works. Note that the Mg_{Mg} has no charge relative to the perfect crystal in K-V notation, although it actually refers to a Mg^{2+} ion on a magnesium site. On the contrary, the Mg^{2+} on the interstitial site has a 2+ charge relative to the perfect crystal because the empty interstitial site has no charge. So, the equilibrium constant can be written as

$$\frac{[Mg^{\bullet\bullet}_{i}][V''_{Mg}]}{[Mg_{Mg}]} \cong \frac{[Mg^{\bullet\bullet}_{i}][V''_{Mg}]}{1} = [Mg^{\bullet\bullet}_{i}][V''_{Mg}] = K_e = e^{-\frac{\Delta G_F}{RT}} \tag{9.48}$$

where $[Mg^{\bullet\bullet}_{i}]$ is the *fraction of interstitial sites occupied.* To continue to calculate the site fractions or the concentrations (number/cm^3), information about the number of interstitial sites relative to the number of lattice sites is needed. This means that the specific interstitial site in the crystal that the magnesium goes into must be known. For example, MgO has the rock-salt structure with cubic close packing of oxygen ions. There are *four* oxygen sites per cell, all occupied with O^{2-} ions and *four* octahedral cation sites per cell, all occupied with Mg^{2+} ions. The largest unoccupied site is the tetrahedral site inside the corners of the MgO unit cell, none of which are occupied in the perfect crystal. However, there are eight tetrahedral sites per cell. So, if the $Mg^{\bullet\bullet}_{i}$ goes into tetrahedral sites, the number of interstitial sites per unit volume is just twice the number of magnesium sites or $\eta_i = 2\eta_{Mg}$. Electrical neutrality is now

$$(2)(e)\eta_{Mg^{\bullet\bullet}_{i}} = (2)(e)\eta_{V''_{Mg}} \tag{9.49}$$

and cancelling and dividing each side by the number of magnesium sites per unit volume,

$$\frac{\eta_{Mg_i^{\bullet\bullet}}}{\eta_{Mg}} = \frac{\eta_{V_{Mg}''}}{\eta_{Mg}}$$
$$\frac{2\eta_{Mg_i^{\bullet\bullet}}}{\eta_i} = \frac{\eta_{V_{Mg}''}}{\eta_{Mg}} \tag{9.50}$$
$$2[Mg_i^{\bullet\bullet}] = [V_{Mg}'']$$

where the last equation states that the fraction of interstitial sites filled is half of the fraction of vacant lattice sites because there are twice as many interstitial sites per unit volume. Putting this result into Equation 9.48 gives

$$2\left[M_i^{\bullet\bullet}\right]^2 = K_e$$
$$\left[M_i^{\bullet\bullet}\right] = \frac{1}{\sqrt{2}} e^{-\frac{\Delta G_F}{2RT}} \tag{9.51}$$

or equivalently, $[V_{Mg}''] = \sqrt{2}\, e^{-\frac{\Delta G_F}{2RT}}$.

Table 9.4 gives some literature values of the energy to form Frenkel defects in compounds.

9.4.8 Back to the Vacancy Diffusion Coefficient

Substitution for the concentration of vacancies as the probability that a neighboring site is vacant, [V], in something like copper, from Equation 9.22

$$D = \frac{1}{6} a^2 f e^{\frac{\Delta S_m}{R}} e^{-\frac{\Delta H_m}{RT}} [V]$$
$$= \frac{1}{6} a^2 f e^{\frac{\Delta S_m}{R}} e^{-\frac{\Delta H_m}{RT}} e^{\frac{\Delta S_V}{R}} e^{-\frac{\Delta H_V}{RT}}$$
$$= \frac{1}{6} a^2 f e^{\frac{\Delta S_m + \Delta S_V}{R}} e^{-\frac{\Delta H_m + \Delta H_V}{RT}} \tag{9.52}$$
$$= D_0 e^{-\frac{Q}{RT}}$$

TABLE 9.4
Frenkel Defects in Compounds

Compound	Type	Structure	ΔH_F (eV)	Melting T (°C)	References
AgCl	Cation	NaCl	1.1, 1.4,[c] 1.44	455	1,2,3
AgBr	Cation	NaCl	1.13	430	3
ZnO	Anion	ZnS	2.51[c]	1974	2
Li_2O	Cation	Antifluorite	2.28,[c] 2.78	1438	2,3
MgF_2	Anion	TiO_2, rutile	3.12[c]	1263	2
CaF2	Anion	CaF_2	2.5, 2.75,[c] 2.8	1418	1,2,3
	Cation		7		1
SrF_2	Anion	CaF_2	2.38,[c] 2.5, 2.3	1477	2,2,3
BaF_2	Anion	CaF_2	1.98,[c] 1.91, 1.86	1368	2,2,3
UO_2	Anion	CaF_2	5.1, 5.47[c]	2847	1,2
	Cation		9.5		1
Na_2S	Cation	Antifluorite	1.77	1172	3

References: Crystal structures (Wells 1984); melting points (Haynes 2013); 1, Kingery et al. (1976); 2, Catlow and Mackrodt (1982); 3 Gray (1972).

[c] indicates a calculated value.

so $D_0 = (1/6)a^2 f \exp(-(\Delta S_m + \Delta S_V)/R)$ and $Q = \Delta H_m + \Delta H_V$ for vacancy diffusion; that is, *the activation energy for vacancy diffusion is the sum of the energy for atomic motion plus the energy to form vacancies*. In the literature, ΔH_V is typically called the *enthalpy of formation* and sometimes written as ΔH_f and, similarly, $\Delta S_V = \Delta S_f$. Note, that for compounds, ΔH_f is the *average energy of defect formation*, which was shown above to be $\Delta H_S/2$ for NaCl and $\Delta H_S/5$ for Al_2O_3, where ΔH_S is the enthalpy to form the Schottky defect: the "molecule" of vacancies.

In materials such as copper, iron, NaCl, Al_2O_3, and so on, vacancy diffusion is the main mechanism by which the atoms or ions in the crystals move about and diffuse. This is the process of *self-diffusion* and the diffusion coefficient of copper in copper is the *self-diffusion coefficient of copper* and similarly, the diffusion coefficient of sodium in sodium chloride is the *self-diffusion coefficient of sodium in sodium chloride*, and so on.

9.4.8.1 Example : Diffusion in Cu

Considering self-diffusion in copper, as calculated above, the vacancy concentration (site fraction) of copper at its melting point of 1083°C is 1.48×10^{-4} and the $\Delta H_f = 1.03$ eV $= 1.03 \times 96{,}472 = 99{,}366$ J/mol. In Table 9.4, the activation energy for Cu diffusion in copper, Q = 210,900 J/mol. Therefore, $\Delta H_m = Q - \Delta H_f = 210{,}900 - 99{,}366 = 111{,}534$ J/mol. Also, from this table, $D_0 = 0.1$ cm²/s, so the self-diffusion coefficient of copper in copper at its melting point is

$$D = D_0 e^{-\frac{Q}{RT}} = 0.1 \times e^{-\frac{210{,}900}{(8.314)(1356)}} = 7.51 \times 10^{-10}\ \text{cm}^2/\text{s}$$

9.4.8.2 Example: Diffusion in NaCl

For self-diffusion in NaCl, as calculated above, the vacant sodium ion concentration in NaCl at its melting point of 801°C is 4.02×10^{-6} and $\Delta H_S/2 = 221{,}890/2 = 110{,}945$ J/mol. From Table 9.5 of diffusion coefficients in compounds, Q = 202,600 J/mol so $\Delta H_m = Q - \Delta H_{S/2} = 202{,}600 - 110{,}945 = 91{,}655$ J/mol. In this case, $D_0 = 18$ cm²/s, so at its melting point, the self-diffusion coefficient of Na in NaCl is

$$D = D_0 e^{-\frac{Q}{RT}} = 18.0 \times e^{-\frac{202{,}600}{(8.314)(1074)}} = 2.52 \times 10^{-9}\ \text{cm}^2/\text{s}$$

Also, in Table 9.5, for the diffusion of Cl in NaCl, Q = 199.7 kJ/mol so $\Delta H_m = Q - \Delta H_S/2 = 88{,}755$ J/mol. Here, $D_0 = 2.2$ cm²/s, so at the melting point, the diffusion of Cl in NaCl is

$$D = D_0 e^{-\frac{Q}{RT}} = 2.2 \times e^{-\frac{199{,}700}{(8.314)(1074)}} = 4.26 \times 10^{-10}\ \text{cm}^2/\text{s}$$

which is about a factor of 5 smaller than the diffusion coefficient for sodium at the same temperature.

9.4.8.3 Example: Diffusion in Al_2O_3

As a final example, consider diffusion in aluminum oxide, Al_2O_3. To emphasize the difference between the cation and anion diffusion coefficients, Figure 9.10 schematically shows the diffusion coefficients for aluminum and oxygen in aluminum oxide: the activation energies for the diffusion of the cation and anion are different and, typically, the diffusion coefficient for the anion is usually smaller because it is the larger ion and has more difficulty in moving. But for both, $Q = \Delta H_m + (\Delta H_S/n)$, where n is the number of atoms in the "molecule. For example, in aluminum oxide, Al_2O_3, the activation energies for diffusion of aluminum and oxygen are, respectively,

$$Q(\text{Al}) = \Delta H_m(\text{Al}) + \frac{\Delta H_S}{5}$$

and

$$Q(\text{O}) = \Delta H_m(\text{O}) + \frac{\Delta H_S}{5}.$$

TABLE 9.5
Diffusion Data

Material	D_o (cm²/s)	Q (kJ/mol)	T_{mp} (°C)	T (°C)	D at T (cm²/s)	Type	Source
			Liquids				
Na	1.10×10^{-3}	10.25	98	98	3.96×10^{-5}	Liquid	1
Pb	2.37×10^{-4}	13.04	327	327	1.74×10^{-5}	Liquid	1
Ag	5.80×10^{-4}	32.47	961	961	2.45×10^{-5}	Liquid	1
H_2O	5.21×10^{-2}	19.37	0	35	2.70×10^{-5}	Liquid	3
			Metals				
Na	0.72	48.1	98	98	1.22×10^{-7}	Vacancy	1
Cu	0.10	210.9	1083	1083	7.51×10^{-10}	Vacancy	1
Ag	0.67	189.1	961	961	6.63×10^{-9}	Vacancy	1
Au	0.09	176.1	1063	1063	1.17×10^{-8}	Vacancy	1
Pt	0.33	284.9	1552	1552	2.31×10^{-9}	Vacancy	1
Mo	0.13	437.2	2610	2610	1.56×10^{-9}	Vacancy	1
Mo	0.13	437.2	2610	1000	1.49×10^{-19}	Vacancy	1
Zn in Cu	0.34	190.8	1083	1000	5.04×10^{-9}	Vacancy	1
Fe(fcc)	0.49	284.1	1538	1000	1.08×10^{-12}	Vacancy	1
Fe(bcc)	1.39	240.2	1538	1000	1.93×10^{-10}	Vacancy	1
C in fcc Fe	0.20	142.3	1538	1000	2.90×10^{-7}	Interstitial	1
C in bcc Fe	0.02	84.1	1538	1000	7.08×10^{-6}	Interstitial	1
N in fcc Fe	0.001	74.1	1538	1000	9.11×10^{-7}	Interstitial	1
			Semiconductors				
Si	1.50×10^{3}	484.3	1414	1414	1.51×10^{-12}	Vacancy	2
B in Si	0.76	333.8	1414	1414	3.51×10^{-11}	Vacancy	2
As in Si	12.00	390.7	1414	1414	9.58×10^{-12}	Vacancy	2
P in Si	3.85	353.1	1414	1414	4.49×10^{-11}	Vacancy	2
Al in Si	0.50	289.4	1414	1414	5.47×10^{-10}	Vacancy	2
Ge	18.50	296.2	937	937	3.02×10^{-12}	Vacancy	2
Ga in GaAs	2.90×10^{8}	578.8	1238	1238	2.84×10^{-12}	Vacancy	2
As in GaAs	0.70	308.7	1238	1238	1.49×10^{-11}	Vacancy	2
			Ceramics				
Ag in AgBr	57	93.6	432	432	6.62×10^{-6}	Interstitial	6
Br in AgBr	3.00×10^{4}	186.2	432	432	4.79×10^{-10}	Vacancy	6
Na in NaCl	18	202.6	801	801	2.52×10^{-9}	Vacancy	6
Cl in NaCl	2.2	199.7	801	801	4.26×10^{-10}	Vacancy	6
Mg in MgO	0.25	330.54	2850	2850	7.40×10^{-07}	Vacancy	6
O in MgO	4.3×10^{-5}	343.5	2850	2850	7.73×10^{-11}	Vacancy	6
Al in Al_2O_3	28	477	2050	2050	5.26×10^{-10}	Vacancy	6
O in Al_2O_3	2.90×10^{3}	636	2050	2050	1.45×10^{-11}	Vacancy	6
Ni in NiO	4.40×10^{-4}	184.9	1957	1957	2.31×10^{-8}	Vacancy	6
O in NiO	6.20×10^{-4}	240.6	1957	1957	1.67×10^{-9}	Vacancy	6
Ti in TiO_2	6.20×10^{-2}	256.9	1843	1843	2.49×10^{-8}	Vacancy	6
O in TiO_2	2.00×10^{-3}	252.04	1843	1843	1.06×10^{-9}	Vacancy	6
Cu in Cu_2O	4.36×10^{-2}	151.04	1244	1244	2.56×10^{-7}	Vacancy	6
O in Cu_2O	6.50×10^{-3}	164.43	1244	1244	1.31×10^{-8}	Vacancy	6
Na in $Na_2O{\cdot}11Al_2O_3$	2.71×10^{-4}	16.5	2000	2000	1.13×10^{-4}	Interstitial	7
	2.71×10^{-4}	16.5	"	300	8.49×10^{-6}	Interstitial	
O in $0.15CaO–0.85ZrO_2$	6.20×10^{-2}	138.3	2550	1000	1.31×10^{-7}	Vacancy	7
Na in SiO_2-glass	6.9	125.4	1475 T_g	1000	4.93×10^{-5}	Interstitial	7
H_2 in SiO_2-glass	5.65×10^{-4}	43.03	1475 T_g	1000	9.69×10^{-6}	Interstitial	2

(Continued)

TABLE 9.5 (*Continued*)
Diffusion Data

Material	D_o (cm^2/s)	Q (kJ/mol)	T_{mp} (°C)	T (°C)	D at T (cm^2/s)	Type	Source
O_2 in SiO_2-glass	2.70×10^{-4}	111.9	1475 T_g	1000	6.91×10^{-9}	Interstitial	2
O in SiO_2-glass	1.20×10^{-4}	295	1475 T_g	1000	9.42×10^{-17}	Vacancy	2
			Polymers				
He in polypropylene	4.1	30.5	−18T_g	27	2.00×10^{-5}	Interstitial	4
H_2 in "	2.4	34.7	−18T_g	27	2.18×10^{-6}	Interstitial	4
CO_2 in polybutadiene	0.24	30.5	−85T_g	27	1.17×10^{-6}	Interstitial	4
H_2 in "	0.053	21.3	−85T_g	27	1.04×10^{-5}	Interstitial	4
N_2 in "	0.22	30.1	−85T_g	27	1.26×10^{-6}	Interstitial	4
O_2 in "	0.15	28.5	−85T_g	27	1.64×10^{-6}	Interstitial	4
CO_2 in butyl rubber	36	50.2	−73T_g	27	6.54×10^{-8}	Interstitial	4
He in "	0.015	24.3	−73T_g	27	8.81×10^{-7}	Interstitial	4
H_2 in "	1.36	33.9	−73T_g	27	1.70×10^{-6}	Interstitial	4
N_2 in "	34	50.6	−73T_g	27	5.26×10^{-8}	Interstitial	4
O_2 in "	43	49.8	−73T_g	27	9.17×10^{-08}	Interstitial	4
Polystyrene-self			100T_g	135	1.2×10^{-15}	Reptation	5
Poly(methyl methacrylate)-self			105T_g	135	6.9×10^{-17}	Reptation	5

References for Table: Melting points (Haynes 2013); glass transition temperatures, T_g (Sperling 2006), (Callister and Rethwisch 2009). 1, Brandes and Brook (1992); 2, Mayer and Lau (1990); 3, Robinson and Stokes (2002); 4, Barrer (1941); 5, Sperling (2006); 6, Gray (1972); 7, Kinergy et al. (1976).

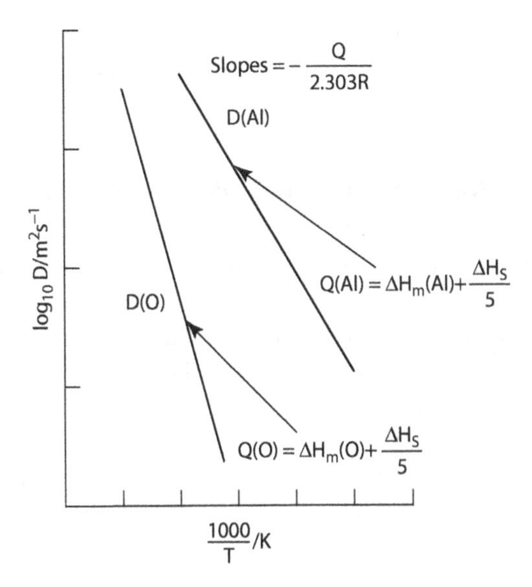

FIGURE 9.10 Representation of the intrinsic vacancy diffusion coefficients for aluminum and oxygen in aluminum oxide. The activation energy for diffusion consists of the activation energy for motion and the energy for formation of vacancies: in this case, the latter is the energy to form a Schottky defect divided by 5—the average energy of formation per vacancy.

9.4.9 Additional Comments About Vacancy Diffusion

9.4.9.1 Vacancy Jumps and Vacancy Diffusion Coefficient

As was noted above, the model for the diffusion coefficient for a vacancy mechanism is not much different from that for the interstitial mechanism except that the expression for interstitial diffusion must be multiplied by the site fraction of vacancies, which is again, Equation 9.22:

$$D = \frac{1}{6}a^2\Gamma[V] = \frac{1}{6}a^2 f e^{-\frac{\Delta G_m}{RT}}[V] = \frac{1}{6}a^2 f e^{\frac{\Delta S_m}{R}} e^{-\frac{\Delta H_m}{RT}}[V]$$

That means that the number of jumps per second for a vacancy, Γ, will be on the same order as that for interstitials, namely, about 10^8 jumps per second, and their speed will be similar to interstitials, on the order of cm/s. However, for the atoms in the lattice, for example, Cu, the number of jumps per second is about a factor of 10^{-4} jumps less—essentially that for vacancies times the site fraction of vacancies—and the speed will also be reduced by a similar amount.

The above expression for vacancy diffusion, Equation 9.22, could be interpreted as

$$D = \frac{1}{6}a^2\Gamma[V] = D_V[V]$$

where D_V is *the diffusion coefficient for vacancies*. This could be carried further

$$D = D_V[V] \cong D_V \frac{\eta_V}{\eta}$$

where:

η_V is the number of vacancies per cm^3

η is the number of atoms per cm^3, so that

$$D\eta = D_V\eta_V \quad (9.53)$$

That is, the diffusion coefficient for atoms times the number of atoms per cm^3 is equal to the vacancy diffusion coefficient times the number of vacancies per cm^3.

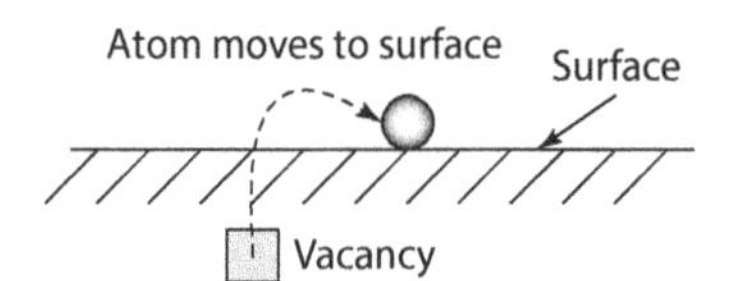

FIGURE 9.11 Depiction of a surface acting as a "source" (or sink) for vacancies. Other two-dimensional defects such as grain and interface boundaries can behave in a similar way to act as sources or sinks for vacancies.

9.4.9.2 Sources and Sinks for Vacancies

To equilibrate with changing temperatures, and so on, the concentration of vacancies needs to change. Vacancies can be created at vacancy *sources* and destroyed at vacancy *sinks*. As shown in Figure 9.11, the surface certainly can be a source or sink for vacancies but so can dislocations and grain boundaries—as these can be considered to be dislocation arrays—as sketched in Figure 9.12. Vacancies created or destroyed at the end of an edge dislocation gives rise to *dislocation climb*—important during creep—as opposed to dislocation *glide*, which is responsible for plastic deformation in crystalline solids.

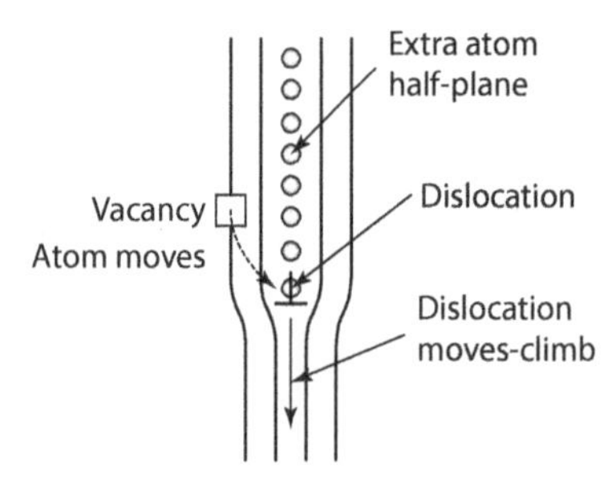

FIGURE 9.12 How a dislocation can act as a source or sink for vacancies. An atom leaves a lattice position and attaches itself to the end of an edge dislocation half plane leaving behind a vacancy in the lattice. The dislocation line moves downward, dislocation climb, and in doing so creates a vacancy; therefore, it is a vacancy source. Conversely, an atom can move from a dislocation filling a vacancy: again the dislocation "climbs" and a vacancy disappears, so the dislocation is acting as a vacancy sink.

9.4.10 Extrinsic Diffusion

In a discussion of vacancy diffusion, the vacancies are *intrinsic vacancies* and the resulting diffusion coefficients are *intrinsic diffusion coefficients*. These are *intrinsic* because they only depend on the temperature and the properties of the pure crystal—internal or intrinsic to the crystal; primarily, the type and strength of the interatomic bonding. On the contrary, something that is both a benefit and a problem is that impurities or additives can alter the diffusion coefficient in compounds. This is a benefit if adding something increases or decreases a diffusion coefficient in a desired and controlled way. On the contrary, it is a major problem when trying to measure intrinsic diffusion coefficients in compounds where impurities—frequently at only the part per million (ppm) level—control the vacancy concentration (see the calculation for Al_2O_3 in Section 9.4.7.4).

A classic example of extrinsic diffusion is the addition of $CdCl_2$ to NaCl. For small concentrations, $CdCl_2$ forms a solid solution with NaCl with Cd^{2+} ions going onto sodium sites and chlorine ions going onto chlorine sites. Consider the following quasichemical point defect equation for the dissolution of $CdCl_2$ into NaCl.

$$CdCl_2 \xrightleftharpoons{NaCl} Cd^{\bullet}_{Na} + 2Cl^{X}_{Cl} + V'_{Na};\ \Delta G_{sol} \quad (9.54)$$

where ΔG_{sol} is the Gibbs energy for the reaction.* The equation has both backward and forward directions: forward = dissolution and backward = precipitation. The equilibrium constant for this reaction can be written as (because $[Cl^{X}_{Cl}] \cong 1$)

$$\frac{\left[Cd^{\bullet}_{Na}\right]\left[V'_{Na}\right]}{a_{CdCl_2}} = K_e = e^{-\frac{\Delta G_{sol}}{RT}} \quad (9.55)$$

where a_{CdCl_2} is the thermodynamic activity of $CdCl_2$. A point defect equation such as Equation 9.54 must again have (1) mass balance (both sides of the equation must have the same mass: $CdCl_2$ on both sides here), (2) charge balance (both sides must have the same charge: zero here), and (3) site balance (on the right-hand side of the equation, the number of cation and anion sites involved in the *host lattice* (NaCl) must be in the same ratio of anions and cations in the compound: 2 cation:2 anion in this case.† The Cd^{2+} replacing a Na^+ ion has a net positive charge relative to the perfect crystal, hence the 1+ charge notation, which must be balanced by a negatively charged sodium ion vacancy, V'_{Na}.‡ The electrical neutrality requirement becomes

$$\text{Positive charge/cm}^3 = \text{Negative charge/cm}^3$$

$$(1)(e)(\eta_{Cd^{\bullet}_{Na}}) = (1)(e)(\eta_{V'_{Na}})$$

so that for every cadmium ion, Cd^{2+}, that goes into solid solution, a sodium vacancy, V'_{Na}, is created. If the concentration of vacancies created by the $CdCl_2$ is much greater than the intrinsic concentration, then the $[V'_{Na}]$ is *independent* of temperature and depends only on the concentration of $CdCl_2$, $[CdCl_2]$, in solid solution, and $[V'_{Na}] \cong [Cd^{\bullet}_{Na}]$ to maintain electrical neutrality. The vacancy concentrations now depend on something other than the *intrinsic* properties of NaCl but depend on something from the outside, or *extrinsic* to the NaCl. Therefore, vacancy concentrations controlled by additives or impurities are called *extrinsic vacancies or extrinsic point defects* and the diffusion and diffusion coefficients that result are called *extrinsic diffusion and extrinsic diffusion coefficients*, respectively. Figure 9.13 schematically shows the results of a classic experiment on the effects of additives such as $CdCl_2$ on the diffusion of sodium in sodium chloride showing both the *intrinsic*—$Q = \Delta H_m + \Delta H_S/2$—and the extrinsic region—$Q = \Delta H_m$—which depends on the level of $CdCl_2$ additive, or *doping*. The extrinsic region exists at the lower temperatures where the extrinsic vacancy concentration is greater than the intrinsic vacancy concentration. At higher temperatures, because of the exponential temperature dependence of the intrinsic vacancy concentration, the

* A result of Equation 9.54 is that the formation of the solid solution of x-moles of $CdCl_2$ in NaCl can be written as $xCdCl_2 + (1-2x)NaCl = Na_{1-2x}Cd_xCl$ or $Na_{1-2x}Cd_x(V_{Na'})_xCl$.

† Note that Equation 9.54 could also be written as $CdCl_2 + 2V'_{Na} + 2V^{\bullet}_{Cl} \xrightleftharpoons{NaCl} Cd^{\bullet}_{Na} + V'_{Na} + 2Cl^{X}_{Cl}$.

‡ Cl^- going on Cl^- sites are electrically neutral and the neutrality is frequently designated with a superscript X, which is really not necessary. Also, it is useful, as shown in Equation 9.54 to indicate the host crystal—NaCl over the arrow.

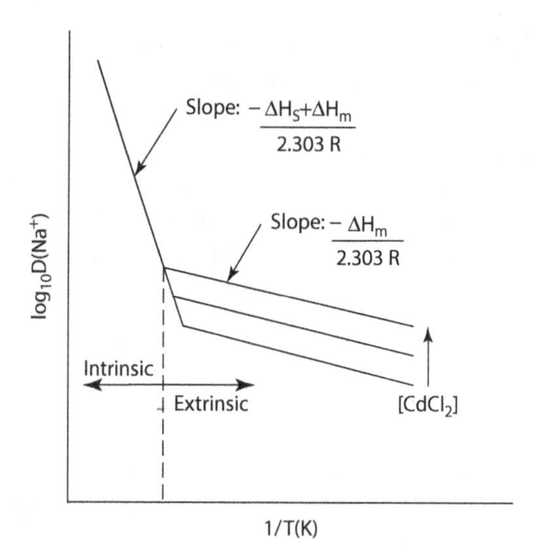

FIGURE 9.13 Idealized plot of the diffusion of sodium in sodium chloride with additions of $CdCl_2$ that create sodium ion vacancies giving rise to extrinsic diffusion at low temperatures. As the concentration of $CdCl_2$, $[CdCl_2]$, increases, so does the $[V'_{Na}]$ and the sodium diffusion coefficient.

intrinsic vacancy concentration will exceed the extrinsic concentration and the slope of the $\log_{10}D$ versus 1/T curve will increase at higher temperatures.

Note that if the $a_{CdCl_2} = 1$, then Equation 9.55 gives the solubility of $CdCl_2$ in NaCl as a function of the temperature. It should also be noted, that a different, but perfectly valid, point defect reaction for the solution of $CdCl_2$ into NaCl can be written as

$$CdCl_2 \xrightarrow{NaCl} Cd^{\bullet}_{Na} + Cl^{x}_{Cl} + Cl'_{i};\ \Delta G_{sol} \qquad (9.56)$$

with the extra chlorine ion going into an interstitial position to balance the charge of the cadmium on the sodium site. The reality is that both reactions, Equations 9.54 and 9.56 will take place simultaneously, and both vacancies and interstitials will be produced. However, the important reaction is the one that takes place to a greater extent. In this case, it is very difficult to "squeeze" an extra chlorine ion into the NaCl unit cell. As a result, the ΔG_{sol} for the reaction in Equation 9.56 will be much larger than for the formation of the sodium vacancies and the concentration of interstitials will be much less than the concentration of vacancies. However, for a real material, only an experiment would show which of the two reactions dominates. The diffusion data for $CdCl_2$ in NaCl strongly suggests that the cadmium ion's charge is *compensated* by the formation of sodium vacancies rather than chlorine interstitials.

Extrinsic behavior in compounds is much more important than just academic interest. First, as was shown above for aluminum oxide at its melting point, the concentration of intrinsic vacancies is only a few tenths of a part per billion! To observe intrinsic diffusion in Al_2O_3 then the impurity concentration must be less than this or the material must be purer than nine-9s pure! This is about the purity of semiconductor-grade silicon, which has a considerably lower melting point and is much easier to purify. As a result, property measurements on Al_2O_3 that depend on point defect concentrations, such as diffusion, will almost always be measuring extrinsic rather than intrinsic behavior.

On the contrary, in Chapter 7, it was shown that cubic ZrO_2 could be made the stable phase down to room temperature by additives such as CaO and Y_2O_3 that can go into solid solution up to about 20 m/o (Figure 7.31). One of the reasons that these additives stabilize the cubic structure is, in addition to the additive cation, vacancies are created that also help make the phase transformation from cubic to tetragonal to monoclinic more difficult. For example, if Y_2O_3, yttrium oxide, is added to ZrO_2, oxygen vacancies are created:

$$Y_2O_3 \xrightleftharpoons{ZrO_2} 2Y'_{Zr} + 3O^{X}_{O} + V^{\bullet\bullet}_{O}. \qquad (9.57)$$

With some 10 or 20 m/o Y_2O_3 in solid solution, the vacant oxygen concentration is on the order of $n_{V_O^{\bullet\bullet}} \cong 10^{21}\ cc^{-1}$, many orders of magnitude higher than the intrinsic vacancy concentration. So the oxygen vacancy diffusion coefficient is very high even at modest temperatures (Table 9.5). As will be seen later in Chapter 13, the oxygen ion electrical conductivity is directly proportional to the oxygen ion diffusion coefficient and such heavily doped ZrO_2 has a high oxygen ion electrical conductivity at relatively low temperatures. As a result, it has been used as an oxygen sensor in automobiles for many years because of its high ionic conductivity. Similarly, it serves as the solid electrolyte in SOFCs, ***solid oxide fuel cells*** that are being considered for power generation, energy storage, and automobile propulsion.

9.4.11 Nonstoichiometry

9.4.11.1 General Concept

Most of the time, compounds in phase diagrams are shown as essentially vertical lines and do not show any solid solution between the end members as shown schematically in Figure 9.14. The main reason for this is that in compounds, solid solutions would require the formation of extrinsic point defects, which takes considerable energy and so the solubility is limited. As noted, Equation 9.56 represents the solubility of $CdCl_2$ in NaCl if the activity of $CdCl_2$, $a_{CdCl_2} = 1$ (or some other constant value). However, as shown in the inset in Figure 9.14, particularly at higher temperatures for a general compound AB, the exact 1:1 stoichiometry of A:B does not hold and the stoichiometry of AB can vary between AB_{1-x} to AB_{1+y}, where $x \neq y$ in general. Now x and y are generally small, on the order of 10^{-6} or 10^{-4} (and may be much larger in some materials), but can have a significant effect on properties such as diffusion because the *nonstoichiometry* is taken care of by the creation of point defects, frequently vacancies. For an ionic compound with A^{2+} and B^{2-},

$$2A_A + B(g) \overset{AB}{\rightleftharpoons} 2A_A^{\bullet} + B_B^X + V_A''; \Delta G_R \tag{9.58}$$

where ΔG_R is the energy for the reaction, mainly the energy to create a vacancy. The equilibrium constant for this reaction is

$$\frac{\left[A_A^{\bullet}\right]^2\left[V_A''\right]}{p(B)} = K_e = e^{-\frac{\Delta G_R}{RT}} \tag{9.59}$$

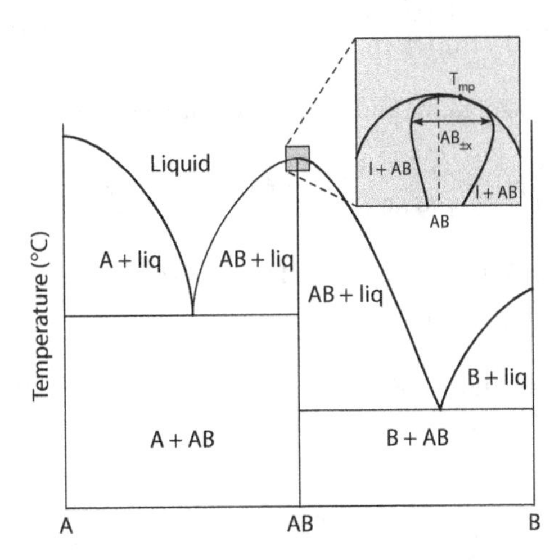

FIGURE 9.14 A–B phase diagram with a single compound AB that exhibits virtually no solid solubility of either excess A or B. However, as shown by magnified region in the insert, there is in reality a range of solid solubility $AB_{\pm x}$ or nonstoichiometry that might be small—$x \sim 10^{-4}$—but introduces defect concentrations greater than intrinsic concentrations that can control defect-dependent parameters such as the diffusion coefficient.

because $\left[A_A\right] \cong 1$ and $\left[B_B^X\right] \cong 1$. If the gas pressure, p(B), the temperature, T, and the energy for the reaction, ΔG_R, are known, then Equation 9.59 can be solved for the point defect concentrations, which may or may not be higher than the intrinsic concentrations. In fact, in a real material, intrinsic defects, impurities or additives, and nonstoichiometry all need to be considered together to determine the defect concentrations that will, in turn, determine the values of the material parameters such as the diffusion coefficient.

9.4.11.2 Example: Nonstoichiometry in Fe_2O_3

To make this more concrete, consider the loss of oxygen from Fe_2O_3 to make it nonstoichiometric, Fe_2O_{3-x}, where the missing oxygen implies vacant oxygen sites, for example,*

$$2Fe_{Fe} + O_O \rightleftharpoons 2Fe'_{Fe} + V_O^{\bullet\bullet} + \frac{1}{2}O_2(g);\ \Delta G_R \tag{9.60}$$

Equation 9.60 says that some Fe^{+3} ions (Fe_{Fe}) are being reduced to Fe^{2+} ions (Fe'_{Fe}) with the loss of oxygen and the formation of oxygen vacancies. An equilibrium constant can be written for Equation 9.60, namely

$$\frac{\left[Fe'_{Fe}\right]^2\left[V_O^{\bullet\bullet}\right]p_{O_2}^{1/2}}{\left[Fe_{Fe}\right]^2} = K_e = e^{-\frac{\Delta G_R}{RT}} \tag{9.61}$$

If the concentration of vacancies is small, then $[Fe_{Fe}] \cong 1$ and the electrical neutrality condition requires that

$$\eta_{Fe'_{Fe}} = 2\eta_{V_O^{\bullet\bullet}}$$

and $\eta_{Fe} = 2\eta_{Fe_2O_3}$ and $\eta_O = 3\eta_{Fe_2O_3}$ resulting in $[Fe'_{Fe}] = 3[V^{\bullet\bullet}]$, so Equation 9.61 becomes

$$9\left[V_O^{\bullet\bullet}\right]^2\left[V_O^{\bullet\bullet}\right]p_{O_2}^{1/2} = K_e = e^{-\frac{\Delta G_R}{RT}}$$

and finally,

$$\left[V^{\bullet\bullet}\right] = \left(\frac{1}{9}\right)^{\frac{1}{3}} p_{O_2}^{-(1/6)} e^{-\frac{\Delta G_R}{3RT}}. \tag{9.62}$$

Suppose $\Delta G_R = 3\,\text{eV/mol} \cong 3 \times 96{,}500\ \text{J/eV} = 289{,}500\ \text{J/mol}$ and $p_{O_2} = 10^{-6}$ bar then at $T = 1200°C$

$$\left[V_O^{\bullet\bullet}\right] = \left(\frac{1}{9}\right)^{\frac{1}{3}}\left(\frac{1}{10^{-6}}\right)^{\frac{1}{6}} e^{-\frac{289{,}500}{(3\times 8.314\times 1473)}} = 1.82\times 10^{-3}$$

For Fe_2O_3, the density is $\rho = 5.24\ \text{g/cm}^3$ and the molecular weight $M = 159.69$ g/mol. Therefore,

$$\eta_{Fe_2O_3} = \eta = \frac{\rho}{M}N_A = \frac{5.24\ \text{g/cm}^3}{159.69\ \text{g/mol}} \times 6.022\times 10^{23}\ \text{mol}^{-1} = 1.98\times 10^{22}\ \text{cm}^{-3}$$

and $\eta_O = 3\eta = 3\times 1.98\times 10^{22} = 5.93\times 10^{22}\ \text{cm}^{-3}$ and $\eta_{Fe} = 2\eta = 2\times 1.98\times 10^{22} = 3.95\times 10^{22}\ \text{cm}^{-3}$. So the number of vacant oxygen ions per cm^3 is

$$\eta_{V_O^{\bullet\bullet}} = \left[V_O^{\bullet\bullet}\right]\eta_O = \left(1.82\times 10^{-3}\right)\left(5.93\times 10^{22}\right) = 1.08\times 10^{20}\ \text{cm}^{-3}$$

Now, for Fe_2O_{3-x},

$$\frac{\eta_O}{\eta_{Fe}} = \frac{3-x}{2} = \frac{\eta_O - \eta_{V_O^{\bullet\bullet}}}{(2/3)\eta_O} = \frac{3-3\left[V_O^{\bullet\bullet}\right]}{2}$$

* If the temperature is high enough, the electron is ionized off of the Fe'_{Fe}, so Equation 9.60 could be written $O_O^X \rightleftharpoons V_O^{\bullet\bullet} + 2e' + 1/2\,O_2(g);\ \Delta G_R$. See Appendix A.9 for more details.

so $x = 3[V_O^{\bullet\bullet}]$ or the formula is $Fe_2O_{3(1-[V_O^{\bullet\bullet}])}$ not surprisingly. So for this example, the composition is $Fe_2O_{3-0.0055}$ or $Fe_2O_{2.99454}$.*

9.5 OTHER DIFFUSION PROCESSES AND SOLIDS

9.5.1 Inorganic Glasses

There are many different kinds of inorganic glasses such as chalcogenides, phosphates, borates, and silicates, but silicate glasses embody by far the greatest number of applications for these materials: just look out the window! All of these materials readily form glasses because of their very open structures with oxygen polyhedral sharing corners as seen in Chapter 7, for example Figure 7.39, which leads a very open glass or crystal structure. Again, to make the discussion more concrete, silicate glasses are examined. Crystalline SiO_2 melts at 1705°C and has a glass transition temperature of about 1475 K. However, with such a high T_g, very high temperatures are needed to be able to get the pure silica glass viscosity low enough to work the glass and fabricate all the kinds of things that are made from glass. The viscosity decreases with temperature because oxygen–silicon–oxygen bonds are being broken by thermal energy decreasing the connectivity of the three-dimensional silica network. Fortunately, the silica network can be more easily broken up with the addition of additives such as Na_2O and CaO and other alkali and alkaline earth oxides. This can be understood as the creation of extrinsic point defects:

$$2Na_2O \xrightarrow{SiO_2} 4Na_i^{\bullet} + V_{Si}^{''''} \tag{9.63}$$

where the sodium ions go into the large interstitial sites in the SiO_4^{4-} tetrahedral network and are *compensated* by a vacant silicon site with a net charge of 4−. So, the four oxygens, which were associated with the tetrahedron that now has a silicon vacancy, are now *dangling bonds or dangling oxygens* because they are no longer joining two tetrahedra at their corners: the network is being broken up (Figure 9.15).† As a result, at a given temperature, SiO_2, which has appreciable concentrations—10 wt.%

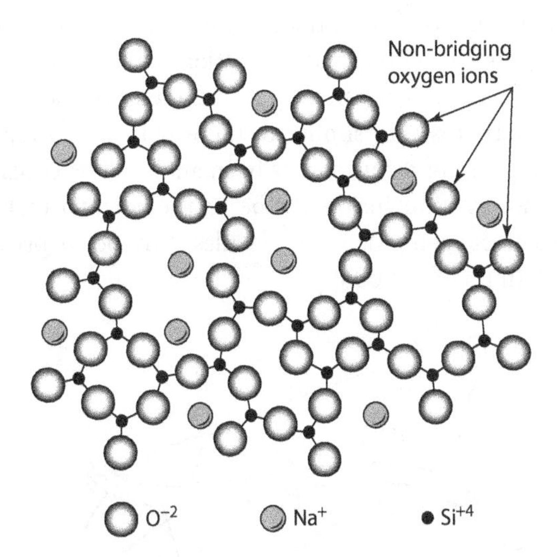

FIGURE 9.15 Schematic structure of a Na_2O–SiO_2 glass that shows the Na^+ interstitial ions that electrically compensate for silicon vacancies that result in nonbridging oxygen ions.

---

* This could also be obtained by the following solid solution reaction:

$$2xFeO + (1-x)Fe_2O_3 \rightleftharpoons Fe_2O_{3-x}$$

since Fe'_{Fe} is an Fe^{2+} ion on an Fe^{3+} site in the Fe_2O_3 crystal. Note that the extra electron on the Fe^{+2} ion can easily move to a neighboring Fe^{+3} ion, and so on, through the crystal giving an electrical current and electronic conductivity. This model of electronic conductivity is the so-called hopping electron or small polaron model of electronic conductivity in nonstoichiometric compounds.

† In reality, Figure 9.15 represents the structure of amorphous B_2O_3 better than it does SiO_2 because boron is coordinated by three oxygen ions in a plane, whereas silicon is coordinated by four oxygen ions at the corners of a tetrahedron.

or more—of *modifiers*, alkali and alkaline earth oxides, has a much lower viscosity than pure silica glass, resulting in a much lower glass transition temperature—below 600°C. Also, because of the open structure, interstitial diffusion in silicate glasses would be expected and many gases diffuse in them as interstitials (Table 9.5). Similarly, the alkali ions such as Na^+ are interstitials as well and have a large diffusion coefficient characteristic of interstitial diffusion.

9.5.2 Diffusion in Polymers

Figure 9.16 shows the two mechanisms by which diffusion can take place in polymers. The first is the familiar interstitial diffusion of gases through the relatively open structure of an amorphous polymer, either above or below the glass transition temperature. Therefore, it is very similar to interstitial diffusion in inorganic glasses but has the advantage that there is little interaction between the diffusing gas and the nonpolar polymer. As a result, interstitial diffusion is relatively fast and is very important commercially because polymers are used as *gas separation membranes* to separate different gases by their rates of diffusion (Kesting and Fritsche 1993; Mulder 1996). Such membranes are used to separate impurity gases (e.g., H_2S) from methane pumped from natural gas wells and CO_2 from biogas. In this application, the degree of separation really depends on the product of the gas diffusion coefficient D and its solubility in the polymer, S, that is, $P = DS$, where P is called the *permeability* and is the important parameter for this application. Gas permeation by diffusion will be discussed in more detail in Chapter 10. Although there are abundant data on the permeability of different polymers to different gases, the permeabilities are generally *not* separated into the solubility and diffusion coefficients (Mark 2007). As a result, interstitial diffusion coefficients for different gases in different polymers are not always readily obtainable but, with some searching, they can be found (Van Krevelen 1997).

What about the *self-diffusion* of the polymer molecules? It is not easy for the individual polymer molecules to move when entangled with other polymer molecules in a polymer melt or solid. As schematically shown in Figure 9.16, the individual polymer molecules are restrained from moving transverse to their long axis because of the neighboring molecules. However, they can move along their length or long axis by the thermal generation and movement of small "defects" along their length. When a defect of length L has moved to the end of the molecule, the molecule has moved a distance L. This motion is very similar to the way in which reptiles move and, hence, is called *reptation*. This mechanism of diffusion has been analyzed by considering the motion of the polymer molecule of n-monomer units inside a "tube" in the polymer melt about 5 nm in diameter, roughly the amount of space between polymer molecules. Not too surprisingly, the result is similar to Equation 9.12; namely (Balluffi et al. 2005)

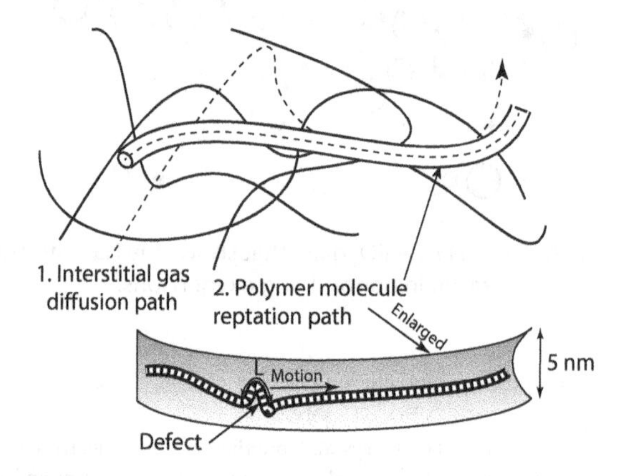

FIGURE 9.16 Diffusion in polymers showing (1) how small gas atoms can diffuse interstitially between the polymer molecules and (2) polymer diffusion along the length of the molecule by the process of *reptation*, where the polymer molecules move along their length inside a tubular volume by the motion of kink-like defects of length L.

$$D \cong \frac{1}{6} f \left(\frac{L}{n} \right)^2 \tag{9.64}$$

where:

f is the frequency related to the defect motion

L/n is a fractional length of the polymer chain similar to the jump distance in Equation 9.12.

Equation 9.64 indicates that the diffusion coefficient is inversely proportional to the square of the length or molecular weight of the polymer chain. As a result, diffusion by reptation gets rather insignificant with large molecular weight polymers. Table 9.5 gives data for reptation diffusion for two polymers normalized to a molecular weight of 150,000 g/mol (Sperling 2006). Clearly, these diffusion coefficients at these low temperatures will not lead to any observable composition changes in reasonable periods of time and this process of reptation is not very important for composition changes and phase transitions. On the contrary, these small movements of polymer molecules can have important effects on dielectric and mechanical relaxation processes.

9.5.3 Diffusion in a Variety of Solids

Table 9.5 gives data for diffusion in a variety of solids collected from a number of sources. This compilation is not the result of an exhaustive search nor have the data been evaluated to present what appear to be best values of the parameters from available literature data. Rather, the information is presented to show how the discussion about diffusion mechanisms leads to actual diffusion coefficients and activation energies for different materials. First, the separation into the various material categories is somewhat artificial because, for example, "semiconductors" really refers to a property—electrical conductivity—that is, the conductivity is somewhere in between that of a good conductor and that of an insulator. Nevertheless, this method of classification of materials is rather common and various materials are recognized as being in one of these categories. A very good compilation on diffusion in solid and liquid metals is readily available and used as a source for data in this table (Brandes and Brook 1992). Similarly, there is a good compilation of diffusion data for semiconductors and liquids (and gases) (Haynes 2013). There are other sources for information on semiconductors as well as alkali halides and some oxides (Gray 1972). There is a continuing series of publications that provides current abstracts on diffusion data for ceramics without any compilation or evaluation of the data (Fisher 2011). An extension of this is the publication *Defect and Diffusion Forum* published by Trans Tech Publications (http://www.ttp.net). Probably the most extensive publication of data on solids is in the *Landolt-Börnstein Tables* and *Springer Materials* published by Springer (http:\\www.materials.springer.com).

As was shown for aluminum oxide, a material that has a high melting point, the energy to form intrinsic defects is also large so the concentration of defects at almost all temperatures is very small. As a result, measured diffusion coefficients can be influenced by small concentration of impurities, making it difficult to separate extrinsic and intrinsic behavior. Therefore, *computational materials science* techniques are frequently used to calculate defect formation energies and diffusion parameters (Catlow and Mackrodt 1982). These types of calculations have a long history (Mott and Gurney 1948), and, with the computational techniques and computing power available today, calculated parameters can be more reliable than those measured in experiments. Several of the parameters for the high-temperature materials in this table are actually calculated values.

Table 9.5 shows that interstitial diffusion is typically on the order of $D \cong 10^{-6}$ cm^2/s regardless of the material, but at different temperatures, of course. On the contrary, vacancy diffusion coefficients are on the order of $D \cong 10^{-10} - 10^{-12}$ cm^2/s at the melting point, regardless of the material. Two unusual cases with high diffusion coefficients are that of oxygen in 15 w/o CaO in solid solution in ZrO_2 and beta alumina, $Na_2O \cdot 11Al_2O_3$. Calcia-doped zirconia, $Zr_{0.85}Ca_{0.15}O_{1.85}$, was shown to have a very large concentration of oxygen ion vacancies and, therefore, is expected to have a high oxygen diffusion coefficient. Beta alumina has a hexagonal crystal structure with all of the sodium ions residing in planes perpendicular to the hexagonal axis with only about half of the possible sodium sites occupied. Therefore, the sodium ions can move very easily and the diffusion can be considered as either interstitial with a large interstitial concentration or as vacancy diffusion with about half of the sites vacant. Both of these materials behave as solid electrolytes and are being considered for power generation and electrical energy storage.

It is worth noting that the difference between interstitial and self-diffusion coefficients is best illustrated in a solid such as iron or steel. Because of its industrial importance in the heat treatment of steels, the diffusion coefficient of C, an interstitial, and the self-diffusion coefficients of iron are reasonably well characterized. From Table 9.5, at 1000°C, the diffusion coefficient of the interstitial carbon is 2.90×10^{-7} cm^2/s, whereas that of iron via a vacancy mechanism is 1.93×10^{-10} cm^2/s, the difference being mainly in the energy necessary to form vacancies. So the diffusion of carbon in iron is much faster than the diffusion of iron, and during the heat treatment of steel, it is the diffusion of carbon that controls the changes in composition and microstructure, which, in turn, determine the mechanical properties of the steel.

9.5.4 Surface and Grain Boundary Diffusion

Dislocations, grain boundaries, and surfaces are possible paths for rapid diffusion compared to diffusion through the crystal lattice or amorphous structure in a solid (Shewmon 1989). Dislocations and grain boundaries are likely paths for more rapid diffusion because the structure in and around them is distorted and atoms have more room to move from place to place. As a result, atoms or ions that move via a vacancy mechanism might find it easier to diffuse along these paths more so than interstitial atoms or ions. Diffusion along dislocations is called *dislocation pipe diffusion* or simply *pipe diffusion.* On surfaces, there is no hindrance to atom or ion motion normal to the surface, so, here too, more rapid diffusion might be expected. Such diffusion processes are difficult to model and perhaps even more difficult to unequivocally determine. One of the reasons for this can be analyzed with Figure 9.17, which models a single grain in a polycrystalline material surrounded, of course, by grain boundaries.* In this figure, assume that the material is copper and zinc is being diffused from the vapor into the copper via a vacancy mechanism. The zinc atoms may indeed move faster down the grain boundaries into the copper. How important this is to the overall composition as a function of distance and time into the bulk of the copper depends on the relative fluxes of zinc diffusing down the grain boundaries compared to the flux through the crystal. Consider the ratio of these total diffusion fluxes:

$$\frac{\text{Flux}_{gb}}{\text{Flux}_g} = \frac{A_{gb}J_{gb}}{A_gJ_g} = \frac{(\pi d\,\delta/2)D_{gb}}{(\pi d^2/4)D_g}\frac{dC}{dx} \tag{9.65}$$

where:

the subscript "gb" refers to the grain boundary and "g" to the grain
A is the area
D is the diffusion coefficient
δ is the thickness of the grain boundary
d is the grain size or diameter
x is the direction perpendicular to the surface.

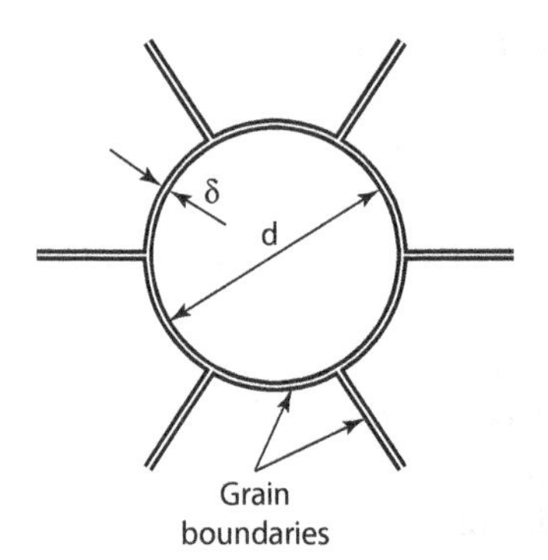

FIGURE 9.17 An idealized circular grain of diameter "d" in a polycrystalline material with grain boundaries that have a width for diffusion of "δ." The importance of grain boundary diffusion on the change of concentration normal to the surface depends on the ratio of δ:d—if small, not important, if large, it is.

The grain boundary thickness, δ, is divided by 2 because only half of the material diffusing down it competes with the diffusion down the grain. Equation 9.65 becomes

$$\frac{\text{Flux}_{gb}}{\text{Flux}_g} = \frac{2\delta D_{gb}}{dD_g}. \tag{9.66}$$

Most of the parameters in Equation 9.66 are essentially constant in a given material with the exception of the grain size d. As a result, the amount of zinc diffused into a material along the grain boundaries relative to that in the grains

* In a real polycrystalline material, the grain would not be circular but more like a hexagon if the grain boundary energies were isotropic and the edges of the grain would be straight and not curved. A circular grain is assumed to make the approximations easier.

will depend on the grain size. Assume that the grain boundary diffusion coefficient is about the same as for interstitials or $D_{gb} \cong 10^{-6}$ cm²/s; the width of grain boundary is slightly larger than the interatomic separation in the crystal, $\delta \cong 3\times10^{-8}$ cm. From Table 9.5, the diffusion coefficient of zinc into copper at 1000°C is $D_g \cong 5 \times 10^{-9}$ cm²s. With these values, Equation 9.66 becomes

$$\frac{Flux_{gb}}{Flux_g} = \frac{2(3\times10^{-8})(10^{-6})}{d(5\times10^{-9})} = \frac{1.2\times10^{-5}}{d}. \tag{9.67}$$

For copper, a grain size of 10–100 μm (10^{-3} to 10^{-2} cm) or larger would be realistic for a typical polycrystalline copper. So the grain boundary contribution to diffusion would be only 0.1%–1% of that in the bulk and would have little effect on the zinc composition as a function of distance into the material. On the contrary, if this were nano-grain size copper, say $d \cong 10$ nm $\cong 10^{-7}$ cm, then the grain boundary diffusion would dominate and determine the composition as a function of distance and time. A little more is said about grain boundary diffusion in Chapter 11 when measurement of diffusion coefficients is discussed.

One final note about grain boundary diffusion is that care must be taken to clearly define what the "grain boundary" consists of: is it really an array of dislocations, a structural rearrangement at the interface of two grains to accommodate the difference in their relative crystallographic orientations, or more something more complex? For example, impurities in compounds really do not like to go into solid solution because vacancies or other kinds of extrinsic defects are created, for example, Equation 9.54, which takes energy so their solubility is limited. As a result, impurities would rather form a second phase, perhaps continuous at grain boundaries. For example, SiO_2 is a common impurity in most polycrystalline oxides and it can react with the host oxide or other impurities to form a silicate glass that wets the grain boundaries. In such a material "grain boundary diffusion" may actually represent interstitial diffusion in a silicate glass that can be much faster than in the crystalline host.

9.6 DIFFUSION IN GASES

9.6.1 The Diffusion Coefficient

In gases, gas atoms are moving at high velocities and collide with other atoms changing their paths and mix by these collisions (Figure 9.18). As was seen in Chapter 8, the diffusion coefficient for a gas can be written in the very general form:

$$D = \frac{1}{3}v\lambda \tag{9.68}$$

where:

v is the molecular or atomic speed

λ is the mean free path, or the distance between collisions.

It was shown in Chapter 7 that the RMS speed of molecules in a gas is given by

$$v = v_{rms} = \sqrt{\frac{3RT}{M}} \tag{9.69}$$

and the mean speed by

$$\bar{v} = \sqrt{\frac{8}{\pi}\frac{RT}{M}} \tag{9.70}$$

where:

R is the gas constant = 8.314 J/mol K

T is the temperature, K

M is the molecular weight, kg/mol.

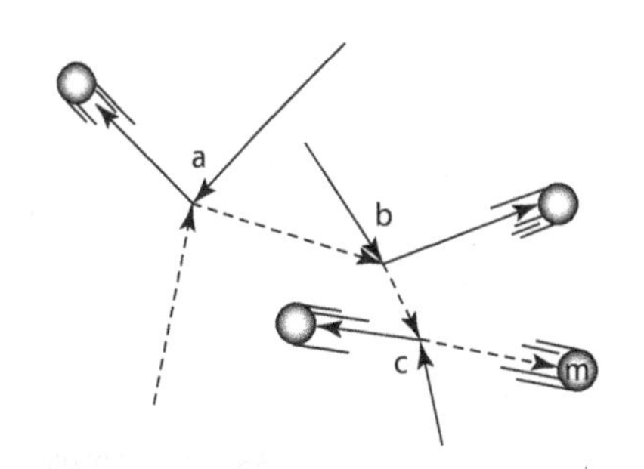

FIGURE 9.18 Schematic diagram showing a few collisions of atom of mass "m" colliding with three other gas molecules, at a, b, and c, that change the direction of the atom and produces "random thermal mixing," diffusion, of the atoms.

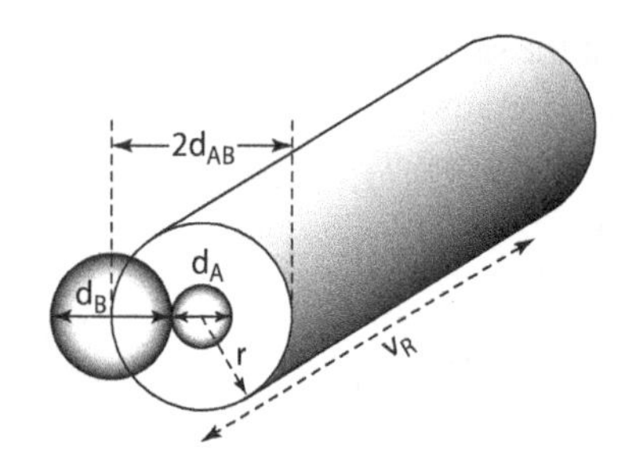

FIGURE 9.19 Geometry used to calculate the number of collisions per unit time that a gas atom traveling at a relative velocity of v_R and diameter d_A makes with other gas atoms of diameter d_B. If a B atom has its center within a distance of d_{AB} of A, then it will collide with A, so the collision volume swept out by A per unit time is volume $= \pi d_{AB}^2 v_R$.

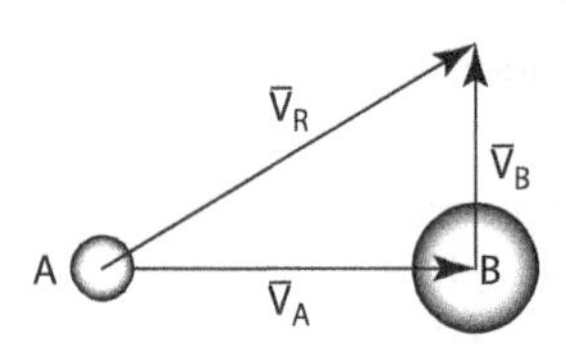

FIGURE 9.20 Relative gas atom velocities. On the average, the velocities of A and B will be perpendicular to each other in a collision, so $v_R^2 = v_A^2 + v_B^2$.

In gas diffusion, the atoms are moving at high velocities and the distance they travel before a collision, the mean free path = λ, is determined by how far apart the atoms are in the gas phase—or on the number of atoms per unit volume. To model the diffusion coefficient in a gas, the mean free path must be determined.

9.6.2 Mean Free Path in a Gas

To calculate the mean free path of a gas atom, the corners of the circuitous path in Figure 9.18 can be neglected and the path straightened as shown in Figure 9.19, where a cylinder of gas of length v_R is considered. To be completely general, it is convenient to consider the *interdiffusion* of two gases, A and B, so their relative speed, v_R, needs to be considered. Their respective velocities are vectors, $\vec{v}_R = \vec{v}_A + \vec{v}_B$, but as shown in Figure 9.20, on the average, two atoms will collide on paths that are normal to each other so

$$v_R = \sqrt{v_A^2 + v_B^2}. \tag{9.71}$$

For *self-diffusion* in the gas, A = B, the relative velocity is simply $v_R = \sqrt{2}\, v_A$. So the distance an atom of A will move in one second relative to atoms of B, which it is going to collide with is the length of the column of gas in Figure 9.19, v_R. Now all the B atoms that are at least halfway into that column will collide with *A* as shown in Figure 9.19. So, the collision distance and radius r of the column are simply

$$r = \frac{d_A + d_B}{2} = d_{AB} \tag{9.72}$$

where:

d_A and d_B are the "collision diameters"—essentially the atomic diameters—of atoms A and B, respectively

d_{AB} is simply the mean diameter of A and B.

So, the number of collisions per second, Z, that an atom A makes with B atoms is simply the volume of the cylinder times the number of B atoms per unit volume, η_B, in the cylinder. That is

$$Z = \pi d_{AB}^2 v_R \eta_B \tag{9.73}$$

and for self-diffusion, $v_R = \sqrt{2}\, v_A$,

$$Z = \pi d_{AB}^2 \sqrt{2} v_A \eta_A.$$

Finally, the mean free path, λ, is the distance traveled per second, v_A, divided by the number of collisions per second, Z:

$$\lambda = \frac{v_A}{Z} = \frac{1}{\sqrt{2}\pi d_{AB}^2 \eta_A} \tag{9.74}$$

As an example, consider oxygen at 1 bar pressure and 300 K. For oxygen, $d_{O_2} = 3.467 \times 10^{-8}$ cm (Geankoplis 1972), and

$$\eta = \frac{N_A P}{RT} = \frac{6.022 \times 10^{23} \times 1}{(83.14)(300)} = 2.41 \times 10^{19}\ \text{cm}^{-3}$$

The mean speed of the oxygen molecules is

$$v_{O_2} = \sqrt{\frac{8}{\pi}\frac{RT}{M}} = \sqrt{\frac{8(8.314)(300)}{\pi(32\times10^{-3})}}$$

$$= 4.46\times10^2 \text{ m/s} = 4.46\times10^4 \text{ cm/s}$$

so that the number of collisions per second is

$$Z = \pi d_{AB}^2 \sqrt{2} v_A \eta_A = \pi\sqrt{2}\left(3.467\times10^{-8}\right)^2\left(4.46\times10^4\right)\left(2.41\times10^{19}\right)$$

$$Z = 5.74\times10^9 \text{ collisions/s.}$$

Notice that this is about the same as the jumps per second for atoms in solids where the temperature is high enough for observable diffusion: for example, for C in iron, $\Gamma \cong 10^9$ jumps/s at 1000°C. The mean free path for oxygen is

$$\lambda = \frac{1}{\sqrt{2}\pi\left(3.467\times10^{-8}\right)^2\left(2.41\times10^{19}\right)}$$

$$\lambda = 7.77\times10^{-6} \text{ cm} = 0.0777\times10^{-4} \text{ cm} \simeq 0.1\ \mu\text{m}$$

which means that the jump—collision—distance in gases is almost 1000 times that of a solid. As a result, diffusion coefficients in gases are—because of the 10^4 faster speed and 10^3 greater jump distance—about 10^7 times higher than interstitial diffusion in solids or on the order 1 cm²/s. As will be seen, this is a good approximation.

9.6.3 The Gas Diffusion Coefficient

For interdiffusion between two gases A and B, again, $D_{AB} = (1/3)v_R\lambda$. So,

$$v_R = \sqrt{v_A^2 + v_V^2}$$

$$= \left(\frac{8}{\pi}\frac{RT}{M_A} + \frac{8}{\pi}\frac{RT}{M_B}\right)^{\frac{1}{2}} \tag{9.75}$$

$$v_R = \left(\frac{8}{\pi}RT\right)^{\frac{1}{2}}\left(\frac{1}{M_A} + \frac{1}{M_B}\right)^{\frac{1}{2}}$$

and the mean free path is given by

$$\lambda = \frac{1}{\sqrt{2}\pi d_{AB}^2 \eta}$$

$$= \frac{1}{\sqrt{2}\pi d_{AB}^2 (PN_A/RT)} \tag{9.76}$$

$$\lambda = \frac{RT}{\sqrt{2}\pi N_A}\frac{1}{d_{AB}^2 P}$$

Putting Equation 9.75 in for v_R, Equation 9.76 for λ, and combining all the constants into the first factor:

$$D_{AB} = \frac{1}{3}v_R\lambda = \left\{\frac{1}{3}\left(\frac{8}{\pi}R\right)^{\frac{1}{2}}\frac{(R)}{\sqrt{2}\pi N_A}\right\}\left(\frac{1}{M_A} + \frac{1}{M_B}\right)^{\frac{1}{2}}\frac{T^{\frac{3}{2}}}{d_{AB}^2 P}$$

Evaluation of the constant term gives

$$\left\{\frac{1}{3}\left(\frac{8}{\pi}R\right)^{\frac{1}{2}}\frac{(R)}{\sqrt{2}\pi N_A}\right\}=\frac{1}{3}\left(\frac{8}{\pi}8.314\right)^{\frac{1}{2}}\frac{\left(\frac{83.14}{10^6}\right)}{\sqrt{2}\pi\left(6.022\times10^{23}\right)}$$

$$=4.765\times10^{-29}\left(\frac{kg^{1/2}\ bar\ m^4}{mol^{1/2}\ K^{3/2}\ s}\right)$$

if the Units(M_A) and Units(M_B) are kg/mol, Units(d_{AB}) = m, Units(T) = K, and Units(P) = bar, then the Units(D_{AB}) = m^2/s. So, the final result is,

$$D_{AB}=4.765\times10^{-29}\left(\frac{1}{M_A}+\frac{1}{M_B}\right)^{1/2}\frac{T^{3/2}}{d_{AB}^2P}\ m^2/s$$

If the more convenient set of units is chosen: Units(M_i) = g/mol, Units(d_{AB}) = 10^{-8} cm or Ångstroms, and Units(P) still bar, then multiplying the above by $(10^3\ g/kg)^{1/2}(10^{20}\ \mathring{A}^2/m^2)(10^4\ cm^2/m^2)$ to get a much more convenient set of units to work with

$$D_{AB}=1.507\times10^{-3}\left(\frac{1}{M_A}+\frac{1}{M_B}\right)^{\frac{1}{2}}\frac{T^{3/2}}{d_{AB}^2P}\ cm^2/s. \tag{9.77}$$

9.6.3.1 Example 1: He and Ar Interdiffusion at 298 K

d(Ar) = 3.542 × 10^{-8} cm, d(He) = 2.551 × 10^{-8} cm (Geankoplis 1972), M(Ar) = 39.9 g/mol, M(He) = 4.0 g/mol. Therefore, d_{AB} = (3.542 + 2.551)/2 = 3.00 × 10^{-8} cm, so

$$D_{He-Ar}=1.507\times10^{-3}\left(\frac{1}{4.0}+\frac{1}{39.9}\right)^{\frac{1}{2}}\frac{(298)^{3/2}}{(3.00)^2(1)}$$

$$D_{He-Ar}=0.452\ cm^2/s$$

which can be compared with a measured value of D = 0.729 cm^2/s (Cussler 1997; Haynes 2013) and is different by less than a factor of 2. Note that the diffusion coefficient for helium and argon can be written as

$$D_{He-Ar}=0.87\times10^{-4}\frac{T^{3/2}}{P}$$

which is why the approximation for diffusion in gases is roughly $D\simeq10^{-4}(T^{3/2}/P)$.

9.6.3.2 Example 2: H_2O in N_2 at 298 K and 1 bar

d(H_2O) = 2.641 and d(N_2) = 3.798 (Geankoplis 1972), M(H_2O) = 18.0, M(N_2) = 28.01, so

$$D_{H_2O-N_2}=1.507\times10^{-3}\left(\frac{1}{28.01}+\frac{1}{18}\right)^{\frac{1}{2}}\frac{(298)^{3/2}}{(3.22)^2(1)}$$

which gives $D_{H_2O-N_2}=0.226\ cm^2/s$ compared to a measured valued of D = 0.242 cm^2/s (Haynes 2013).

9.6.4 The Chapman–Enskog Equation

The preceding derivation for the gas diffusion coefficient is based on what is known as the *hard sphere model* because it assumes that there is no interaction between gas atoms or molecules until they touch as shown in Figure 9.21a: the atoms or molecules essentially interact as billiard balls. In reality, gas atoms or molecules will, at a minimum, experience van der Walls attractive forces and electron cloud repulsive forces and are "squishier" than billiard balls and lead to a potential energy

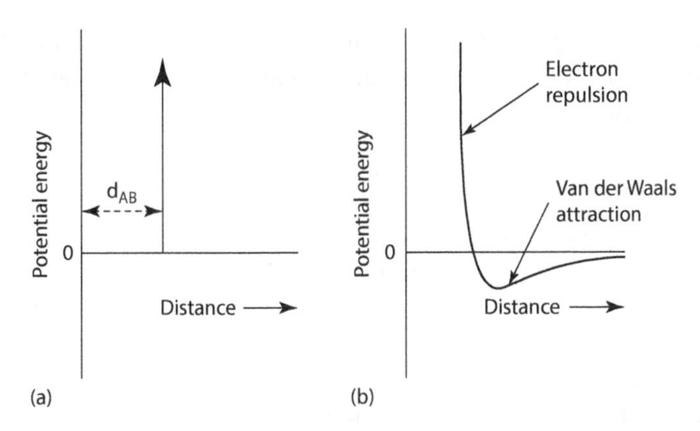

FIGURE 9.21 (a) Potential energy versus distance for a "hard sphere" model for atoms and molecules colliding in the gas phase. (b) Potential energy versus distance curve for a combined van der Waals attractive and a power dependence of the electron repulsion at small separation distances as an attempt to model gas collisions more precisely with a more realistic interatomic potential energy.

versus distance curves that look more similar to that shown in Figure 9.21b. Taking into consideration such a potential energy curve leads to the so-called "Chapman–Enskog" equation for diffusion in gases (Geankoplis 1972; Cussler 1997; Bird et al. 2002),

$$D_{AB} = 1.86 \times 10^{-3} \left(\frac{1}{M_A} + \frac{1}{M_B} \right)^{\frac{1}{2}} \frac{T^{3/2}}{d_{AB}^2 \Omega_{AB} p} \ \mathrm{cm^2/s} \tag{9.78}$$

where Ω_{AB} is the "collision integral" that accounts for the nonhard sphere interaction of the gas atoms, as do some of the d_{AB} values that are used in the calculation of gas diffusivities. The values of Ω_{AB} are generally between about 0.4 and 2.6 can be evaluated from data given in several references (Geankoplis 1972; Cussler 1997; Bird et al. 2002). Note that there is usually not much difference between the D_{AB} calculated from the simple hard sphere model and that calculated with the Chapman–Enskog result. Diffusion coefficients calculated with Chapman–Enskog typically agree with measured gas diffusion coefficients to about 10%, which is good enough for "highway work." Some of the d_{AB}s that were used in the above calculations were those tabulated in the references for use in this equation. Also, note that for the case of $D_{He\text{-}Ar}$, $\Omega \approx 0.5$ (Geankoplis 1972), which if included in the above calculation, would make the calculated and measured diffusion coefficients much closer. Also, for the case on interdiffusion of H_2O and N_2, $\Omega \approx 1.0$ (Geankoplis 1972), so that the hard sphere calculation and the experimental values are much closer.

The most important point is that the simple hard sphere model, Equation 9.77, gives all the correct temperature, pressure, collision diameter, and molecular weight dependencies that the more realistic—and more complex—model does. The main differences are in the initial constant and the inclusion of the collision integral. For calculation purposes, in future gas diffusion problems, the results of the simple hard sphere model will be completely satisfactory because the Chapman–Enskog equation only applies to nonpolar molecules. It does not strictly even apply to N_2–H_2O interdiffusion because it does not include dipolar effects of the H_2O molecules: another model is necessary for this. This is a perfect example of, "How good do the models have to be to be useful?" With the simple billiard ball model, gas diffusivities can be computed quite accurately and can—and will—be used to quantitatively model reactions controlled by gaseous diffusion and the models compare pretty well with reality. Given the variety of other uncertainties in most models, improving the value of a gas diffusion coefficient by 10% will probably not have a significant effect of the predictability of the model. Is it important to be able to model gaseous diffusion coefficients more accurately? Yes, but the added complexity may not be important to understand, predict, and control an industrial process of interest. Compared with the uncertainty in the activation energies for defect formation and diffusion in solids that can be 10% or larger, the ability to accurately calculate gaseous diffusion coefficients even with the simple model, Equation 9.77, is much better than the ability to calculate solid diffusion coefficients. A difference of 10% in the activation energy can easily lead to a factor of 10 or more in the calculated diffusion coefficient.

9.6.5 Knudsen Diffusion

One final point about gaseous diffusion worth mentioning is so-called *Knudsen diffusion*, which concerns diffusion of gases in porous media where the pore size is smaller than the mean free path in the gas phase. Knudsen diffusion is important in many applications in materials. For example, during the calcining of materials such as carbonates to give oxides, the evolving CO_2 gas must diffuse through the porous outer layer of the oxide. As in all cases, the diffusion coefficient for Knudsen diffusion can also be written, $D = (1/3)v\lambda$. But in Knudsen diffusion, the mean free path is the distance that a gas atom or molecule can diffuse before it hits the wall of the pore as shown in Figure 9.22. In this case, the mean free path is essentially equal to the pore diameter, d, so that the Knudsen diffusion coefficient, D_K becomes

$$D_K = \frac{1}{3}vd. \tag{9.79}$$

At room temperature and one atmosphere the mean free path in a gas is on the order of 0.1 μm. This means that the pore size would have to be very small for pure Knudsen diffusion. Generally, it is more realistic to consider that both pore and gas collisions might be taking place simultaneously so that the total collision rate, Z_{total} is given by $Z_{total} = Z_{gas} + Z_{pore}$. Dividing the total number of collisions by the molecular speed, v, putting in the definition of the mean free path, λ, and finally dividing by 1/3v gives

$$\frac{Z_{total}}{v} = \frac{Z_{gas}}{v} + \frac{Z_{pore}}{v}$$

$$\frac{1}{\lambda_t} = \frac{1}{\lambda_g} + \frac{1}{d}$$

$$\frac{1}{(1/3)v\lambda_t} = \frac{1}{(1/3)v\lambda_g} + \frac{1}{(1/3)vd}$$

where the last equation is

$$\frac{1}{D} = \frac{1}{D_{gas}} + \frac{1}{D_K} \tag{9.80}$$

which is nothing more than the reiteration of the principle that "when there are series steps, the slower or smaller one controls the rate." Here, the smaller of the gas and Knudsen diffusion coefficients determines the overall value of D.

As an example, take the case of N_2 at STP diffusing in a porous material with 10-nm diameter pores. Here, D_{gas} is given by

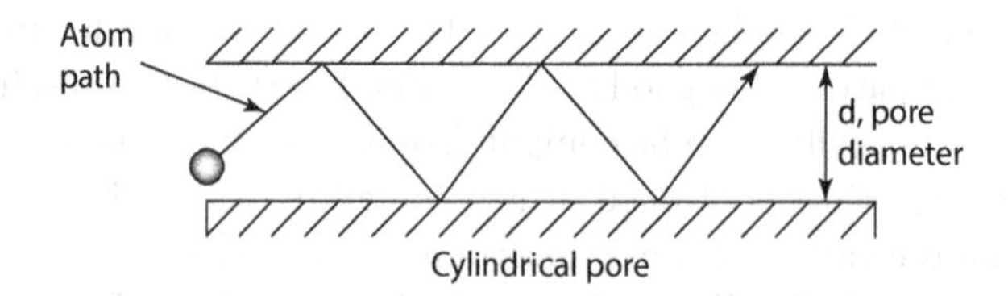

FIGURE 9.22 Schematic drawing illustrating the principle of Knudsen diffusion that can occur with gases diffusing in porous material. In pure Knudsen diffusion, the mean distance between collisions of the atoms or molecules in the gas phase, the mean free path, is considerably greater than the diameter, d, of the pores. As a result, gas atoms strike the walls of the pore before they collide with other gas atoms and their mean free paths are now determined by the pore diameter.

$$D_{N_2} = 1.507\times10^{-3}\left(\frac{2}{28.01}\right)^{\frac{1}{2}}\frac{(298)^{3/2}}{(3.798)^2(1)}$$

$$D_{N_2} = 0.144 \text{ cm}^2/\text{s}$$

$$v = \sqrt{\frac{8}{\pi}\frac{RT}{M}} = \sqrt{\frac{8\times8.314\times298}{\pi\times28.01\times10^{-3}}}\times10^2 = 4.74\times10^4 \text{ cm/s}$$

so

$$D_K = \frac{1}{3}vd = \frac{1}{3}\left(4.74\times10^4\right)\left(10\times10^{-7}\right) = 1.58\times10^{-2} \text{ cm}^2/\text{s}$$

and from Equation 9.80

$$D = \frac{D_K D_{N_2}}{D_K + D_{N_2}} = \frac{1.58\times10^{-2}\times0.144}{\left(1.58\times10^{-2}\right)+(0.144)} = 1.42\times10^{-2} \text{ cm}^2/\text{s}^*$$

$D_K = (1/3)vd = (d/3)\sqrt{(8/\pi)(RT/M)}$ demonstrates that the diffusion of a gas through a porous material depends inversely on $\sqrt{M}$. So if a mixture of gases with different masses are passed through long small diameter tubes, those with the lighter mass will emerge first. This is the principle of uranium isotope separation by *gaseous diffusion.* $^{238}UF_6$ and $^{235}UF_6$ gases are prepared from the uranium oxide ore and separated by Knudsen diffusion through many miles of corrosion-resistant porous material with the lighter gas, $^{235}UF_6$ coming out first. By repeated cycling, higher and higher enrichments can be obtained. Gaseous diffusion plants have been displaced by centrifuge facilities for the enrichment of uranium isotopes for a number of reasons, not the least of which, they can be made much smaller—and harder to detect—for a given output.

9.7 DIFFUSION IN LIQUIDS

The atomistics of diffusion in liquids is between that for gases and solids but more similar to diffusion in gases, as might be expected, because at the critical temperature, the structure of the liquid and gas are indistinguishable. Therefore, the general equation for diffusion is applied to a liquid; namely, Equation 9.68

$$D = \frac{1}{3}v\lambda$$

where:

- v is the molecular speed
- λ is the mean free path.

The mean speed for gas molecules is given by Equation 9.70

$$\bar{v} = \sqrt{\frac{8}{\pi}\frac{RT}{M}}$$

and the mean free path is roughly the distance between molecules in the liquid,

$$\lambda \cong \frac{1}{\eta^{1/3}}$$

* Strictly speaking, for a 10-nm pore, the molecular diameter, on the order of 0.4 nm is a sizeable fraction of the pore diameter and this approximate calculation does not take this into consideration.

where η is the number of atoms or molecules per unit volume. However, much of the space between molecules is actually taken up by the volume of the molecules themselves. Therefore, a better approximation to λ is

$$\lambda \cong \frac{1}{\eta^{1/3}} - d$$

where d is the diameter of the liquid molecule. Take water as an example,

$$\eta = \frac{\rho}{M} N_A \cong \frac{1}{18}\left(6.022\times10^{23}\right) = 3.35\times10^{22}\ \text{cm}^{-3}$$

and $d_{H_2O} \cong 0.276$ nm is the diameter of the water molecule (Wells 1984), so

$$\lambda \cong \frac{1}{\eta^{1/3}} - d_{H_2O} = \frac{1}{\left(3.35\times10^{22}\right)^{1/2}} - 2.76\times10^{-8}\text{cm} = (3.10-2.76)\times10^{-8} = 0.34\times10^{-8}\ \text{cm}$$

and

$$\bar{v} = \sqrt{\frac{8}{\pi}\frac{8.314\times300}{18\times10^{-3}}}\times100\ \text{cm/m} = 5.94\times10^{4}\ \text{cm/s}.$$

Therefore, for water at 300 K, $D \cong (1/3)(5.94\times10^4\ \text{cm/s})(0.34\times10^{-8}\ \text{cm}) = 6.7\times10^{-5}\ \text{cm}^2/\text{s}$. The literature value for diffusion in water at 300 K is $D \cong 2.4 \times 10^{-5}\ \text{cm}^2/\text{s}$ (Robinson and Stokes 2002). This is less than a factor of 3 difference, so this is a pretty good plausibility argument. A better approach to the diffusion in liquids that can be applied more generally is given in Chapter 13.

9.8 CHAPTER SUMMARY

This chapter covers the atomistics of diffusion in solids, gases, and an introduction to diffusion in liquids. The model of interstitial diffusion in solids is developed first showing the origin of D_0 and Q in the diffusion coefficient $D = D_0 \exp(-Q/RT)$. The preexponential term, D_0, is shown to be $D_0 = (1/6)a^2 f \exp(-\Delta S_m/R)$, where a is the jump distance ≅ the lattice parameter and ΔS_m is the entropy for interstitial motion and usually involves the movement of neighboring atoms allowing the interstitial atom to move. For interstitial diffusion, the activation energy for diffusion, Q, includes only the energy for the interstitial to move, ΔH_m. The jump frequency, f, is essentially the vibration frequency of atoms on their lattice sites and several approaches are used to show that $f \cong 10^{13}$ Hz. Vacancy diffusion is essentially the same as interstitial diffusion but multiplied by the *site fraction* of vacancies, [V], which is first determined in an elemental material by a statistical approach and found to be exponentially dependent on the energy to form a vacancy, ΔH_f. Therefore, the activation energy for vacancy diffusion is $Q = \Delta H_m + \Delta H_f$ and is much higher than the activation energy for interstitial diffusion leading to vacancy diffusion coefficients smaller than interstitial diffusion coefficients in a given material. A quasichemical approach or *point defect chemistry* is shown to give the same results as the statistical approach and is extremely useful when considering vacancies and interstitials in compounds. The Kröger–Vink notation for defects in compounds is used to balance charges on the crystal lattices. Schottky and Frenkel defects are introduced as *intrinsic* defects in compounds. Then *extrinsic* lattice defects produced by impurities, dopants, and atmospheric effects are shown as well as how the atmosphere can change the *nonstoichiometry* and point defect concentrations in compounds. A major point made, often repeated, that care must be used to distinguish the number of defects per unit volume, for example, the number of vacant oxygen ions per unit volume, $\eta_{V_O^{\bullet\bullet}}$, and the fraction of oxygen sites that are vacant, $[V_O^{\bullet\bullet}]$, which has *no units*. It is the site fractions that must be used in chemical equilibria because they have no units and the equilibrium constant also has none. And because they are fractions, *they have no charge*. There is significant confusion in the literature about using these different terms in point defect chemistry and can lead to errors of many orders of magnitude in the calculated number of charged species per unit volume of interest in calculating properties. As stated before, literature must be viewed with a critical eye! Point defect chemistry is shown to apply to diffusion in glasses as well. Diffusion in polymers is briefly discussed where the main interest in these materials is interstitial diffusion. Finally, a few comments are made about grain boundary and surface diffusion.

A rather simple model of atoms colliding is developed for diffusion in gases that gives quite precise values for gaseous diffusion coefficients. This model is applied to diffusion in porous media, *Knudsen diffusion*, and demonstrates that small pore sizes can limit the mean free path of gas atoms and molecules and determine the value of the diffusion coefficient.

Finally, a plausibility argument is given for diffusion in liquids similar to the argument for diffusion in gases that predicts reasonable values for diffusion coefficients. An alternative approach that is more commonly used to model diffusion in liquids is left to Chapter 13.

APPENDIX

A.1 Solution to $x'' + \omega^2 x = 0$

For a mass m vibrating on a spring with a force constant, β, the equation of motion is

$$F = ma = m\frac{d^2x}{dt^2} = -\beta x$$
$$\frac{d^2x}{dt^2} + \frac{\beta}{m}x = \frac{d^2x}{dt^2} + \omega^2 x \tag{A.1}$$

where $\omega^2 = \beta/m$. Equation A.1 is a linear ordinary differential equation with constant coefficients and the standard method for solving such equations is to assume that the solutions are of the form $x = e^{\lambda t}$ (Kreyszig 1988), which gives for the “characteristic equation,”

$$\lambda^2 e^{\lambda t} + \omega^2 e^{\lambda t} = 0$$
$$\lambda^2 + \omega^2\lambda = 0$$
$$(\lambda - iw)(\lambda + i\omega) = 0$$

which has roots of $\lambda = \pm i\omega$, where $i = \sqrt{-1}$. The general solution is

$$\begin{aligned} x &= c_1 e^{i\omega t} + c_2 e^{-i\omega t} \\ &= c_1(\cos\omega t + i\sin\omega t) + c_2(\cos\omega t - i\sin\omega t) \\ &= (c_1 + c_2)\cos\omega t + (ic_1 - ic_2)\sin\omega t \\ x &= A\cos\omega t + B\sin\omega t \end{aligned} \tag{A.2}$$

because $e^{ix} = \cos x + i\sin x$ (which is easily seen from the infinite series of the three functions). In other words, the solution to the mass on a spring equation is sinusoidal motion and whether it is a cosine or sine solution or a combination depends on the conditions at $t = 0$. The angular frequency is $\omega = \sqrt{\beta/m}$, rad/s, and the frequency, f, Hz (s^{-1}) is given by

$$f = \frac{\omega}{2\pi} = \frac{1}{2\pi}\sqrt{\frac{\beta}{m}}. \tag{A.3}$$

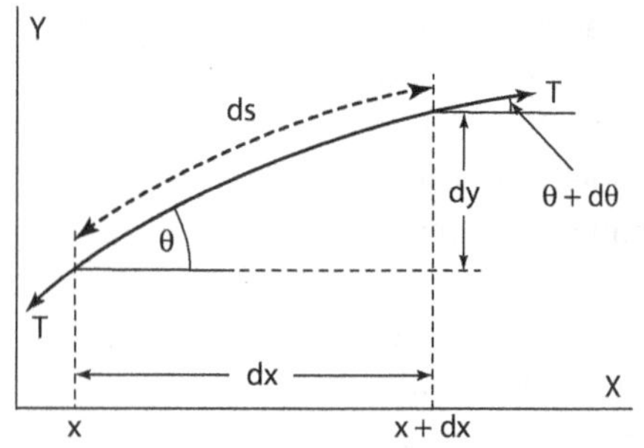

A.2 Vibrations of String

The following derivation of the vibration of a string is covered in many books where different approximations are made (Pearson 1966; French 1971; Main 1984; Elmore and Heald 1985; Pain 1999) but all leading to the same result, the wave equation for a vibrating string.

Figure A.1 shows a tiny part of vertically displaced string with arc length ds with mass per unit length, μ (kg/m) under a tension, T (Newton). Now from Figure A.1, the arc length is given by

FIGURE A.1 Geometry to determine the transverse vibrations of a string.

$$(ds)^2 \cong (dx)^2 + (dy)^2 \tag{A.4}$$

which can be manipulated in the following way

$$(ds)^2 \cong (dx)^2\left[1+\left(\frac{dy}{dx}\right)^2\right]$$

so

$$ds \cong \sqrt{1+\left(\frac{dy}{dx}\right)^2}\,dx \tag{A.5}$$

which is the formula for arc length, ds. If the string is not displaced too far in the y direction so that slope $dy/dx \cong 0$, then $ds \cong dx \cong \Delta x$ so the mass of the string of length "ds" is $dm = \mu ds \cong \mu\Delta x$. The forces in the x and y directions are

$$\begin{aligned} F_x &= T\cos(\theta+\Delta\theta) - T\cos(\theta) \\ F_y &= T\sin(\theta+\Delta\theta) - T\sin(\theta). \end{aligned} \tag{A.6}$$

For θ and Δθ very small, $\cos(\theta) \cong \cos(\theta + \Delta\theta) \cong 1$ so $F_x \cong 0$. Similarly,

$$\begin{aligned} F_y &\cong T\sin(\theta+\Delta\theta) - T\sin(\theta) \\ &\cong Tx(\theta+\Delta\theta) - Tx(\theta) \cong T\Delta\theta \end{aligned} \tag{A.7}$$

Now

$$\tan\theta = \frac{\partial y}{\partial x} \tag{A.8}$$

so

$$\begin{aligned} \frac{1}{\partial x}(\tan\theta) &= \frac{1}{\partial x}\left(\frac{\partial y}{\partial x}\right) \\ \sec^2\theta\frac{\partial\theta}{\partial x} &\cong \frac{\Delta\theta}{\Delta x} = \frac{\partial^2 y}{\partial x^2} \end{aligned} \tag{A.9}$$

because $\sec\theta = 1/\cos\theta \cong 1$ when θ is small as it is in this case and partial derivatives with respect to x are being used because the partial derivative with respect to t will be necessary for the acceleration. Therefore,

$$(dm)a_y = F_y$$

$$(\mu\Delta x)\frac{\partial^2 y}{\partial t^2} = T\Delta\theta = T\frac{\partial^2 y}{\partial x^2}\Delta x \tag{A.10}$$

where a_y is the acceleration in the y direction (m/s²). Rearranging Equation A.10 gives the desired result:

$$\frac{\partial^2 y}{\partial t^2} = \frac{1}{v^2}\frac{\partial^2 y}{\partial x^2} \tag{A.11}$$

where the wave velocity, v, is given by $v = \sqrt{T/u}$ m/s because Units(T/u) = (m/s)².

A.3 Longitudinal Vibrations in a Solid

What follows is very similar to what was done above to develop Equation A.11 for the transverse vibrations of a string. The difference is that, here, longitudinal vibrations along the x-direction in Figure A.2 are considered. This results in the displacement of a section of the solid, Δx, by an amount w in the x-direction. As was true for the vibrating spring, many of the references cited in Appendix A.2 also derive the equation for longitudinal vibrations in a solid and, again, with a variety of different approximations all leading to the same result.

In the case of longitudinal vibrations, the piece of solid of length Δx is not only being displaced an amount w, but it is also being stretched in the x-direction by an amount Δw by the opposing forces in the x-direction, F_{neg} and F_{pos}. Then the average strain over Δx is Δw/Δx and the average stress, σ, is given by $\sigma = E(\Delta w/\Delta x)$ and at any particular point, $\sigma = F/A = E(\partial w/\partial x)$, where A is the cross-sectional area of the bar. Now,

$$\sigma(x + \Delta x) = \sigma(x) + \frac{\partial \sigma}{\partial x}\Delta x \tag{A.12}$$

so

$$F_{neg} = A\sigma(x) = AE\frac{\partial w}{\partial x}$$

$$F_{pos} = A\sigma(x) + A\frac{\partial \sigma}{\partial x}\Delta x = AE\frac{\partial w}{\partial x} + AE\frac{\partial^2 w}{\partial x^2}\Delta x \tag{A.13}$$

or

$$F_{net} = F_{pos} - F_{neg} = AE\frac{\partial^2 w}{\partial x^2}\Delta x. \tag{A.14}$$

The difference between the forces leads to the acceleration of the piece of material Δx long that has mass, $m = A\rho\Delta x$, where ρ is the density (kg/m³), that is,

$$ma = (A\rho\Delta x)\frac{\partial^2 w}{\partial t^2} = F_{net} = AE\frac{\partial^2 w}{\partial x^2}\Delta x \tag{A.15}$$

and cancelling and rearranging gives the desired result:

$$\frac{\partial^2 w}{\partial x^2} = \frac{\rho}{E}\frac{\partial^2 w}{\partial t^2} = \frac{1}{v^2}\frac{\partial^2 w}{\partial t^2} \tag{A.16}$$

which is the wave equation for longitudinal waves in a solid with velocity $v = \sqrt{E/\rho}$ because Units(E/ρ) = Pa/(kg/m³) = (m/s)².

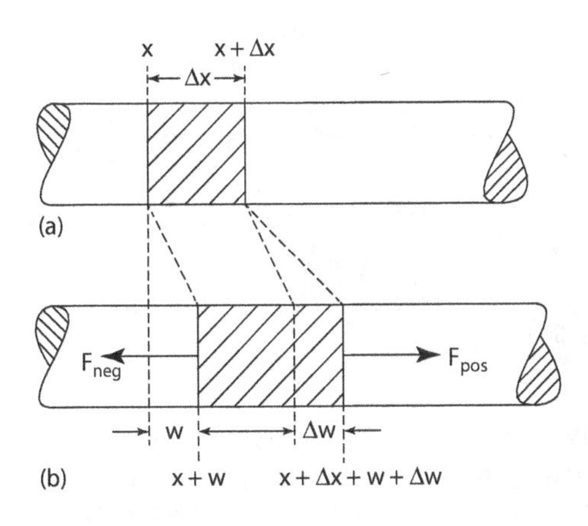

FIGURE A.2 Geometry to determine the longitudinal vibrations of a rod.

A.4 Electronic Defects

A.4.1 Rationale

An extensive review of point defect chemistry including electronic defects is available from many sources (Kröger 1964; Tilley 1987; Barsoum 1997; Chiang et al. 1997; Smyth 2000; Maier 2004), but each of these references uses a different but similar notation with frequent confusion between concentrations (numbers/cm^3) and site fractions. In some cases, the notation $[V_O^{\bullet\bullet}]$ is defined as a site fraction in one place then, later, is defined as a concentration, number per cm^3 that has been defined here as $\eta_{V_O^{\bullet\bullet}}$. This is one of the reasons that η was defined as concentration with Units(η) = number/cm^3. Although this leads to a slightly more complex notation, it does not lead to the confusion that exists in the literature. Another option used in the literature is, for example, that the Schottky defect is defined as

$$\left[V_{Mg}''\right]\left[V_O^{\bullet\bullet}\right] = K_S = \eta_{Mg}\eta_O e^{-\frac{\Delta G_S}{RT}}$$

and $[V_{Mg}''] = \eta_{V_{Mg}''}$ and $[V_O^{\bullet\bullet}] = \eta_{V_O^{\bullet\bullet}}$ are concentrations per cm^3. As long as this convention is used consistently, the resulting point defect equations should give the correct results sought: namely, the concentrations per cm^3 as a function of temperature, partial pressures, and dopant concentrations. However, this gets a little messy with something like, say Fe_2O_3:

$$0 \rightleftharpoons 2V_{Fe}''' + 3V_O^{\bullet\bullet},$$

where the convention used here gives for the Schottky product

$$\left[V_{Fe}'''\right]^2\left[V_O^{\bullet\bullet}\right]^3 = K_S = e^{-\frac{\Delta G_S}{RT}}$$

and the "[]" indicate *site fraction*. If "[]" is used as concentration then the equivalent equation would have to be:

$$\left[V_{Fe}'''\right]^2\left[V_O^{\bullet\bullet}\right]^3 = K_S = \eta_{Fe}^2\eta_O^3 e^{-\frac{\Delta G_S}{RT}} = 108\eta_{Fe_2O_3}^5 e^{-\frac{\Delta G_S}{RT}}$$

Confusing! Another variation is the use of the notation "$V_O^{\bullet\bullet}$" for *both* the vacant oxygen species in an equation *and* the number of vacant oxygen per unit volume! Finally, sometimes the "$[V_O^{\bullet\bullet}]$" notation is used to represent a site fraction and then used in electrical neutrality expressions, which is clearly incorrect and will lead to incorrect answers: another example of "...just because it's published does not mean it's correct..." So when adopting material from these and similar resources, make sure that a consistent point defect chemistry is used and be aware that the literature must be used with caution and, perhaps, corrections.

The addition of electronic defects into point defect chemistry is modeled to show how easy and consistent it is to include them and gives concentrations that are the same as given by a more solid state physics or statistical approach. Furthermore, the confusion between concentrations and site fractions described above can lead to large errors in calculated concentrations that are important for electrical conductivity in particular. So it is important to have a consistent model of electronic defects along with lattice defects to help decipher the confusion that exists in the literature.

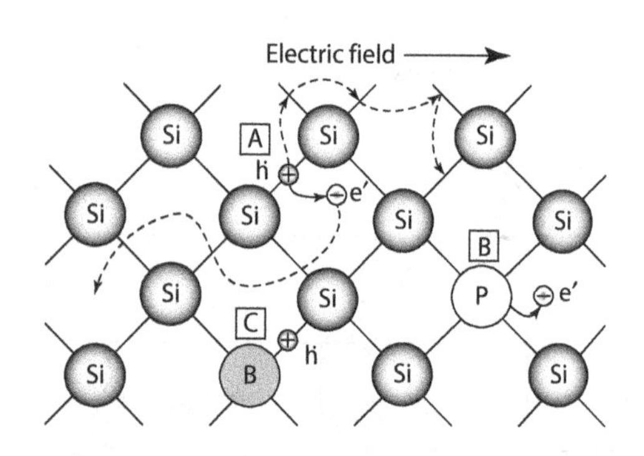

FIGURE A.3 Schematic drawing of s silicon crystal showing (A) the generation of electron hole pairs by thermal energy; (B) phosphorus dopant acting as an electron donor; and (C) boron acting as an electron acceptor.

A.4.2 Intrinsic Electronic Defects

Figure A.3 uses a silicon crystal as an example with intrinsic electronic defects, free electrons and missing electrons in covalent bonds, *holes*, at point A, and extrinsic electronic point defects at points B and C introduced by the *dopants* phosphorus and boron. First, thermal energy may be sufficient to knock electrons out of bonds (covalent or ionic) leaving behind an empty electron site that has a charge of plus one relative to the perfect crystal, called a *hole* and designated $h^{\bullet}$ in Kröger–Vink (K–V) notation. The electron of course is negative and designated e'. So, similar to the Schottky product, the ionization of a bond can be represented as

$$0 \rightleftharpoons e' + h^{\bullet};\ E_g \qquad \text{(A.17)}$$

where E_g is the energy to form the electron–hole pair. As shown in the figure, under the influence of an electric field to the right, the electron will move to the left and contribute to electronic conductivity. Similarly, electrons in neighboring covalent bonds will move to the left and into the hole. This means that the hole moves to the right in the direction of the field, so it acts like a positively charged electron as far as electrical conductivity is concerned. This is why it is given a positive charge. An equilibrium constant can be written for Equation A.17; namely,

$$\left[e'\right]\left[h^{\bullet}\right] = K_i = e^{-\frac{E_g}{RT}} \qquad \text{(A.18)}$$

where $[e']$ is the *fraction of electron energy states* occupied $[e'] = n/N_C$, where n is the number of electrons per cm^3 and N_C is the number of electron energy states per cm^3. Similarly, $[h^{\bullet}] = p/N_V$, where p is the number of holes per cm^3 and N_V is the number of hole energy states per cm^3. For perfectly free electrons and holes—an empty box with just walls (there are no ions present to scatter electrons and the electrons do not collide)—$N_C = N_V = 2.51 \times 10^{19}\ cm^{-3}$, whereas in real materials that have atoms or ions, they are not the same but usually have not largely different values. These values for the *density of states* occur because the motion of both electrons and holes behave as waves that are quantized. The Heisenberg uncertainty principle states that the uncertainty in the position, Δx, times the uncertainty in the momentum, Δmv, is given by

$$\Delta x\, \Delta(mv) = \frac{h}{2\pi}$$

where h is the Planck's constant $= 6.626 \times 10^{-34}$ Js. If the electron or hole is in a cubic piece of silicon that has a length $= L$, then $\Delta x = L$ because the electron must be in the silicon, and each electron energy state must be separated by a momentum difference of $(\Delta mv) = h/(2\pi L)$ or a kinetic energy difference of

$$\Delta E = \frac{1}{2}\frac{(\Delta mv)^2}{m} = \frac{h^2}{8\pi^2 mL^2}$$

and limits the number of states per unit volume. Completing the calculation of the total number of states in three dimensions gives $N_C = 2.51 \times 10^{19}\ cm^{-3}$ (Mayer and Lau 1990). So, Equation A.18 becomes

$$\left(\frac{n}{N_C}\right)\left(\frac{p}{N_V}\right) = e^{-\frac{E_g}{RT}}$$

and electrical neutrality requires that $n = p$ (always true for *intrinsic* material) and

$$np = N_C N_V e^{-\frac{E_g}{RT}}$$

so

$$n = p = \left(N_C N_V\right)^{\frac{1}{2}} e^{-\frac{E_g}{2RT}}. \qquad \text{(A.19)}$$

For silicon, $E_g = 1.1$ eV so at room temperature, 300 K (Mayer and Lau 1990),

$$n = p = 2.51\times10^{19}\ e^{-\frac{1.1\times 96{,}500}{2\times 8.314\times 300}} = 1.44\times10^{10}\ \text{cm}^{-3}$$

and makes pure silicon a pretty good insulator at room temperature!

A.4.3 Extrinsic Electronic Defects

A.4.3.1 Phosphorous and Boron in Silicon

Extrinsic electronic defects are produced by additives—dopants—or impurities in the crystal. For silicon, two common dopants are phosphorus, P, and boron, B. Phosphorous has five outer electrons and uses four of them to make four covalent bonds with neighboring silicon atoms in silicon as shown in Figure A.3. The fifth electron is loosely bound and easily ionized to give a free electron. Similarly, boron has only three outer electrons to make three covalent bonds with neighboring silicon atoms leaving one of the four bonds missing an electron, a hole. It too can be easily ionized to move through the crystal and contribute to electrical conductivity. Dopants that *donate* electrons to conduct are called *donor* dopants and those that generate holes are called *acceptor* dopants because they accept electrons from other covalent bonds to give electrical conductivity.

To demonstrate the principles of extrinsic electronic defects, consider phosphorus as a dopant. The equation for the P atom to donate an electron is

$$P_{Si}^{X} \rightleftharpoons P_{Si}^{\bullet} + e'; E_D$$

where E_D is the energy to ionize the electron. The equilibrium constant for this reaction is,

$$\frac{\left[P_{Si}^{\bullet}\right]\left[e'\right]}{\left[P_{Si}^{X}\right]} = K_D = e^{-\frac{E_D}{RT}} \tag{A.20}$$

and putting the expressions for the site fractions,

$$\frac{\left(\eta_{P_{Si}^{\bullet}}/\eta\right)\left(n/N_C\right)}{\left(\eta_{P_{Si}^{X}}/\eta\right)} = \frac{\left(\eta_{P_{Si}^{\bullet}}\right)(n)}{\left(\eta_{P_{Si}^{X}}\right)N_C} = K_D \tag{A.21}$$

where $\eta = \eta_{Si}$ is the number silicon sites per cm³. Now the total number of phosphorus donors per cm³ is the number of ionized and unionized: $\eta_{P_{Si}} = \eta_{P_{Si}^{X}} + \eta_{P_{Si}^{\bullet}}$. The total electrical neutrality requires that $\eta_{P_{Si}^{\bullet}} + p = n$, but assuming that the hole concentration is going to be small and can be ignored, $\eta_{P_{Si}^{\bullet}} \cong n$. Therefore, Equation A.21 becomes

$$\frac{n^2}{\eta_{P_{Si}} - n} = N_C K_D$$

giving the quadratic equation

$$n^2 + nN_C K_D - N_C K_D \eta_{P_{Si}} = 0$$

the solution of which is

$$n = \frac{-N_C K_D \pm \sqrt{(N_C K_D)^2 + 4N_C K_D \eta_{P_{Si}}}}{2}$$

$$n = \frac{-N_C K_D + N_C K_D\sqrt{1 + 4(\eta_{P_{Si}}/N_C K_D)}}{2}$$

The solutions to this quadratic equation at two extremes are as follows: first, if $4(\eta_{P_{Si}}/(N_C K_D)) \gg 1$ then,

$$n \cong (N_C \eta_{P_{Si}})^{\frac{1}{2}} e^{-\frac{E_D}{2RT}} \tag{A.22}$$

exactly the same as obtained from statistical approaches. Equation A.22 is essentially the case when E_D is large. The second extreme is when $4(\eta_{P_{Si}}/(N_C K_D)) \ll 1$, then the solution becomes*

$$n = \frac{-N_C K_D + N_C K_D \left(1 + 2(\eta_{P_{Si}}/(N_C K_D))\right)}{2}$$

$$n \cong \eta_{P_{Si}} \tag{A.23}$$

or all of the phosphorus atoms are ionized implying that E_D is small. For phosphorus in silicon, E_D = 0.045 eV (Mayer and Lau 1990) and for a total phosphorus concentration of $\eta_{P_{Si}} = 10^{16}$ cm^{-3} gives,

$$4\frac{\eta_{P_{Si}}}{N_C K_D} = 4\frac{10^{16}}{2.51\times 10^{19}\left(\exp(-0.045\times 96{,}500/8.314\times 300)\right)} \cong 10^{-2} \ll 1$$

so Equation A.23 should apply and all of the donors are ionized, $n \cong \eta_{P_{Si}} = 10^{16}$ cm^{-3}. To get a more accurate number, the quadratic equation would have to be solved. Figure A.4 shows the electron concentration as a function of the energy to ionize the donor, E_D. This Figure clearly shows that all of the donor will be ionized up to about a value of $E_D \cong 0.1$ eV, regardless what the host material is.† Silicon doped with 10^{16} phosphorus atoms per cm^3 would have an electronic conductivity about six orders of magnitude higher than intrinsic silicon where $n = p = 1.44 \times 10^{10}$ cm^{-3}. Also, because $np = 2.07\times 10^{20}$, $p = 1.88 \times 10^4$! It is clear why the electrons, in this case, are called *majority carriers* and the holes *minority carriers* because there is a factor of 10^{12} difference in their concentrations! Boron, also with a small ionization energy of $E_A = 0.054$ eV for holes (Mayer and Lau 1990), gives

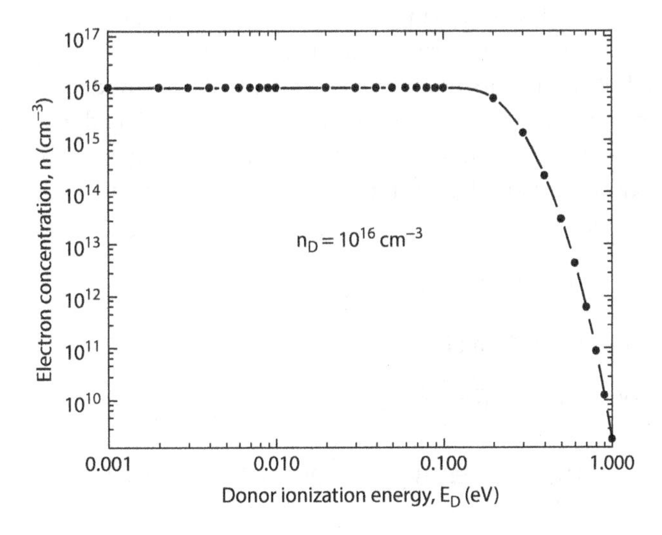

FIGURE A.4 The effect of the donor ionization energy on the electron concentration for a donor concentration of 10^{16} donors/cm^3.

* $(1+x)^n \cong 1 + x/n$ for x small: that is, only the first two terms in the infinite series are important.

† Donors that are completely ionized and follow Equation A.23 are called *shallow donors* while those that have larger ionization energies and follow Equation A.22 are called *deep donors*.

similar results but for p-type material. It is this ability to dope silicon and other solids to change their electrical properties that have made this the "silicon age."

A.9.4.3.2 Electronic Doping of Compounds Fluorine-doped and indium-doped tin oxide (ITO) are used as transparent conducting oxides (TCOs) for a number of applications including transparent contacts for solar cells, heatable de-icing coating for aircraft windshields and other applications. For pure SnO_2, $E_g \cong 4$ eV so the intrinsic electron and hole concentrations are quite small and visible light does not have sufficient energy to create an electron hole pair.* So, it is transparent in visible light and a pretty good insulator. However, it can be doped with fluorine to make a sufficiently good conductor for transparent contacts and de-icing. The doping of SnO_2 with fluorine can be written as

$$SnF_4 \underset{}{\overset{SnO_2}{\rightleftarrows}} Sn_{Sn}^X + 2F_O^{\bullet} + 2e' + F_2(g); \Delta G_D \tag{A.24}$$

where:

$F_O^{\bullet}$ represents the F^- ion on O^{-2} site and is a n-type dopant

ΔG_D is the energy to produce the four point defects: two fluorine on oxygen and two electrons.

Also, their concentrations depend on the fluorine pressure.† But this raises a concern, because in Section 9.4.10, such dopants introduced lattice defects. In fact, the solid solution of SnF_4 into SnO_2 could be written with the formation of vacant tin sites:

$$SnF_4 \underset{}{\overset{SnO_2}{\rightleftarrows}} Sn_{Sn}^X + 4F_O^{\bullet} + V_{Sn}^{''''}; \Delta G_V. \tag{A.25}$$

This is also a valid point defect formation reaction. Both Equations A.24 and A.25 occur to some extent but which concentration, vacancies or holes, dominates depends on the relative values of the energies, ΔG_D and ΔG_V. Because fluorine-doped tin oxide is a good electronic conductor, the reaction in Equation A.24 must at least produce enough electrons for good conductivity.

A.9.4.4 Kröger–Vink or Brouwer Diagrams

There have been many studies of dopants on the electrical conductivity and defect chemistry of different compounds. However, very few of them cover the entire range of possible defect equilibria and not all of the results confirm these simple defect models. Therefore, rather than using data on real materials, a model of an imaginary oxide AO is doped with another oxide, B_2O_3 is developed. Note that this is not boron oxide but just some general oxide with a cation B^{3+}. Assume that Schottky defects are the predominant lattice defects in AO. So, the complete point defect chemistry in AO is

Schottky defect: $0 \rightleftharpoons V_A'' + V_O^{\bullet\bullet}; K_S$

Electron-hole pairs $0 \rightleftharpoons e' + h^{\bullet}; K_i$

Nonstoichiometry $O_O^X \rightleftharpoons V_O^{\bullet\bullet} + 2e' + \frac{1}{2}O_2(g); K_R$

Doping‡ $B_2O_3 \overset{AO}{\rightleftarrows} 2B_A^{\bullet} + 2e' + 2O_O^X + \frac{1}{2}p_{O_2}(g); K_D$

---

* Light energy is given by $E = hf = hc/\lambda \cong (6.625 \times 10^{-34}\ \text{J s})(3 \times 10^8\ \text{m/s}) / (1.602 \times 10^{-19}\ \text{J/eV}) \times 10^6\ \mu\text{m/m} = 1.24\ \text{eV}\ \mu\text{m}$. The lower limit of visible light is about 0.43 μm corresponding to an energy of 2.9 eV. So Eg must be greater than 2.9 eV, so that light is not absorbed by the production of electron–hole pairs.

† This equation really is the sum of two equations: one to put the SnF_2 into solid solution and another to ionize the fluorine to produce the electrons.

‡ Only one of the possible doping reactions is necessary, the formation of vacancies will take care of itself.

that lead to the following equations

$$\eta_{V_A''}\eta_{V_O^{\cdot\cdot}} = \eta^2 K_S$$
$$np = N_C N_V K_i$$
$$\eta_{V_O^{\cdot\cdot}} n^2 p_{O_2}^{\frac{1}{2}} = \eta N_C^2 K_R \tag{A.26}$$
$$\eta_{B_A^{\cdot}}^2 n^2 p_{O_2}^{\frac{1}{2}} = \eta^2 N_C^2 a_{B_2O_3} K_D$$

where $\eta = \eta_{AO}$, along with the complete electrical neutrality equation,

$$n + 2\eta_{V_A''} = p + 2\eta_{V_O^{\cdot\cdot}} + \eta_{B_A^{\cdot}}. \tag{A.27}$$

This gives five unknowns and five equations assuming temperature and oxygen pressure are constant, so the nonlinear equations could be solved simultaneously with perhaps more than a little effort. But of greater interest is to solve the equations for all of the concentrations as a function of oxygen pressure. This generates a Kröger–Vink diagram or the so-called *Brouwer diagram*, an approximation to the K–V diagram in which logarithm of all the concentrations are assumed to be linear with the oxygen pressure (Kröger 1964) that can be more easily constructed. To do this, it is first assumed that certain defects dominate in different regions of the diagram. Then the oxygen dependencies of all of the defects are calculated from Equations A.26 (Kröger 1964). This diagram for this system is shown in Figure A.5 with four different regions in the diagram and the concentrations determined from the set of equations in Equation A.26:

Region 1: reduction, $n \cong 2\eta_{V_O^{\cdot\cdot}} \propto p_{O_2}^{-1/6}$ so $p, \eta_{V_A''}$ and $a_{B_2O_3} \propto p_{O_2}^{+1/6}$
Region 2: donor doping, $n \cong \eta_{B_A^{\cdot}}$ = constant so p = constant also, and $\eta_{V_O^{\cdot\cdot}} \propto p_{O_2}^{-1/2}$ from K_R and $\eta_{V_A''}$ and $a_{B_2O_3} \propto p_{O_2}^{1/2}$
Region 3: vacancy compensation,* $2\eta_{V_A''} \cong \eta_{B_A^{\cdot}}$ = constant so $\eta_{V_O^{\cdot\cdot}}$ and $a_{B_2O_3}$ are also constant and because $n \propto p_{O_2}^{-1/4}$ and therefore, $p \propto p_{O_2}^{+1/4}$
Region 4: oxidation, $p \cong \eta_{V_A''} \propto p_{O_2}^{+1/6}$ as well as $a_{B_2O_3} \propto p_{O_2}^{+1/6}$ while n and $\eta_{V_O^{\cdot\cdot}} \propto p_{O_2}^{-1/6}$

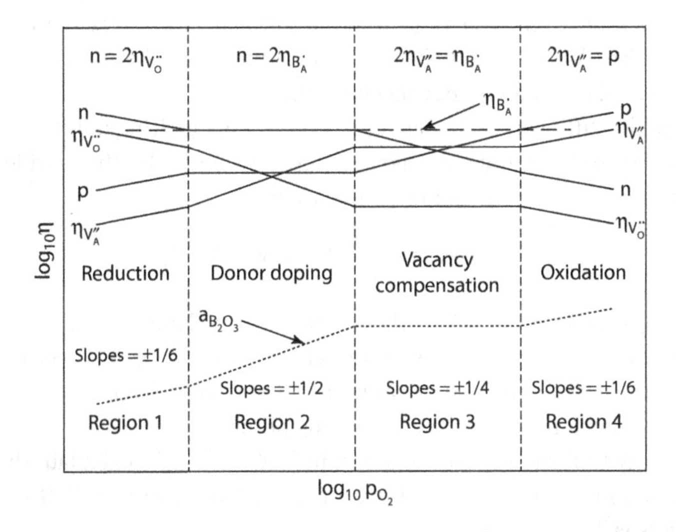

FIGURE A.5 Brouwer approximation to the Kröger–Vink diagram of a compound AO with a dopant B_2O_3 that gives the point defect concentrations as a function of oxygen pressure. Note that this represents compounds other than oxides such as GaAs, CdTe, and so on, where the p_{O_2} is replaced by p_{As_2}, p_{Te_2}, and so on.

* When vacancies balance the charge on the dopant the dopant charge is *compensated* by the vacancy charge. When the dopant charge is balanced by the hole or electron charge it is called *electronic compensation.*

What the diagram shows is that, a given temperature, the point defect concentrations depend on both the oxygen pressure and the dopant concentration. Furthermore, the dopant will act as either an electron donor or produce vacancies depending on the partial pressure of oxygen. It should be noted that it is assumed that the concentration of B_2O_3 is well below its solubility, so that its activity varies with oxygen pressure. If the activity were to exceed 1.0, B_2O_3 would precipitate, more with increasing oxygen pressure, while the activity remains constant. As a result, the concentration of the dopant also changes with oxygen pressure. Needless to say, this complication makes the drawing of that diagram much more interesting!

Three things are worthy of note. First, the entire range of oxygen pressures necessary to experimentally construct a complete K–V is usually not accessible. For example, low oxygen pressures can be achieved with $H_2O - H_2$ mixtures. For example, at 500°C, for $H_2O{:}H_2 \cong 10^{-3}$ which is easily achievable, the $p_{O_2} \cong 10^{-34}$ bar (Roine 2002), whereas pressures greater that a few bar are hard to achieve. Similarly, CO_2:CO gas mixtures can also achieve low oxygen pressures. Experimentally, this means that most data will be obtained toward the left hand, or reducing end of the K–V diagram. For example, for the AB diagram, this may only cover region 2, where B_2O_3 is acting as a donor dopant. Second, the concentrations of all of the point defect concentrations are rather compressed near the top of the diagram to get all of the other information into the graph. But, for example, if region 2 covered the experimental range of $10^{-34} < p_{O_2} < 1$ bar then because of $p_{O_2}^{\pm 1/2}$ pressure dependencies, the vacant oxygen and vacant cation concentrations vary by 10^{17} over this oxygen pressure range while the electron concentration remains constant! Third, the linear approximation is not valid near the transition from one region to another. In this case, the Equations A.26 and A.27 really do have to be solved simultaneously.

EXERCISES

9.1 Would it be appropriate to say that [V] = mole fraction? Why or why not?

9.2 a. The density of silver, Ag, is 10.5 g/cm³ and its atomic weight is 107.87 g/g atom. Calculate the number of Ag lattice sites per cm³.

b. Calculate the molar volume (cm³/mol) of Ag.

c. The energy to form lattice vacancies in silver is $\Delta H_f = 105$ kJ/mol. Calculate the fraction of vacant lattice sites in Ag at its melting point of 961°C.

d. Calculate the number of vacant Ag sites per cm³ at its melting point.

e. For diffusion of silver in silver (silver "self-diffusion"), $D_o = 0.67$ cm²/s and Q = 189.1 kJ/mol. Calculate ΔH_m (kJ/mol), the enthalpy for motion of Ag atoms in Ag in which the diffusion takes place by a vacancy mechanism.

f. Calculate the diffusion coefficient of Ag in Ag at its melting point.

9.3 a. Suppose that sodium and chlorine vacancies attract each other and form nearest neighbor pairs according to the following equation:

$$V'_{Na} + V^{\bullet}_{Cl} \rightleftharpoons \left(V'_{Na} V^{\bullet}_{Cl}\right); \Delta G_p \cong \Delta H_p$$

where $\Delta G_p \cong \Delta H_p = -0.2$ eV is the pairing energy. Note that it is negative because the vacancies attract each because of the charge difference. Neglecting possible entropy effects because the pairs can orient in different crystal directions, calculate the number of vacancy pairs/cm³ in NaCl at its melting point of 801°C.

b. If the energy to form a Schottky defect in NaCl is 2.3 eV, calculate the concentration of paired and unpaired defects at 100°C intervals from 100°C to 800°C. Plot these data as $\log_{10}$[] versus T (°C).

9.4 Silicon replaced germanium, Ge, as the semiconductor material of choice about 50 years ago because it is easier to grow a good stable and insulating oxide, SiO_2, on silicon while the oxide on Ge, GeO_2, is not nearly as good. However, Ge has better semiconducting properties than silicon and there are some new devices in which the oxide on silicon has to be replaced with some other oxide. Therefore, Ge has the potential of making a "comeback" as a semiconductor material.

a. The density of Ge is 5.35 g/cm^3 and its atomic weight is 72.59 g/g atom. Calculate the number of Ge lattice sites per cm^3.
b. Calculate the molar volume (cm^3/mol) of Ge.
c. The energy to form lattice vacancies in germanium is $\Delta H_f = 1.73$ eV/vacancy. Calculate the fraction of vacant lattice sites in Ge at its melting point of 937°C.
d. Calculate the number of vacant Ge sites per cm^3 at its melting point.
e. For diffusion of germanium in germanium (germanium "self-diffusion"), $D_o = 18.50$ cm^2/s and Q = 296.2 kJ/mol. Calculate ΔH_m (in both kJ/mol and eV/atom), the enthalpy for motion of Ge atoms in Ge in which the diffusion takes place by a vacancy mechanism.
f. Calculate the diffusion coefficient of Ge in Ge at its melting point.

9.5 a. The density of TiO_2, rutile (titanium dioxide), is 4.26 g/cm^3 and its molecular weight is 79.88 g/mol. Calculate the number of titanium and oxygen sites per cm^3 in TiO_2.
b. Calculate the fraction of vacant oxygen ion sites at the melting point of TiO_2, 1850°C, if the energy to form a Schottky defect is 5.2 eV per "molecule" of vacancies.
c. Calculate the concentration of oxygen vacancies (cm^{-3}) at the melting point of TiO_2.

9.6 Show that the vacancy site fractions for a Schottky defect in $MgAl_2O_4$ are equal.

9.7 Show that Equation 9.62 follows from Equation 9.61.

9.8 An oxide compound, AO, undergoes nonstoichiometry at certain oxygen partial pressures by the reaction:

$$2A_A + O_O = \frac{1}{2}O_2(g) + V_O^{\bullet\bullet} + 2A_A'; \Delta G_R = 4.0\ \text{eV}$$

Of course, the A_A' ion is an A^{2+} ion reduced to an A^+ ion in Kröger–Vink notation. Also, for a Schottky defect in AO:

$$0 = V_A'' + V_O^{\bullet\bullet}; \Delta G_S = 2.1\ \text{eV}.$$

Because of three possible defects, the total electrical neutrality condition is

$$2\eta_{V_O^{\bullet\bullet}} = 2\eta_{V_A''} + \eta_{A_A'}$$

But because $\eta_A = \eta_O$, so $2[V_O^{\bullet\bullet}] = 2[V_A''] + [A_A']$. This implies that two concentration ranges are possible, the stoichiometric range where $[V_O^{\bullet\bullet}] \cong [V_A'']$ and the $[A_A']$ can be neglected and a nonstoichiometric range where $2[V_O^{\bullet\bullet}] \cong [A_A']$ and $[V_A'']$ can be neglected.

a. Calculate the vacancy concentration in the stoichiometry range where $[V_O^{\bullet\bullet}] \cong [V_A'']$ at 1000°C.
b. Calculate $[A_A']$ in the stoichiometric range when $p_{O_2} = 10^4$ bar at 1000°C.
c. From parts (a) and (b), calculate the nonstoichiometry at $p_{O_2} = 10^4$ bar at 1000°C; i.e. x in AO_{1-x}.
d. In the nonstoichiometry range, $2[V_O^{\bullet\bullet}] \cong [A_A']$, calculate $[V_O^{\bullet\bullet}]$ and x at $p(O_2) = 10^{-14}$ bar and 1000°C.
e. Calculate all of the vacancy concentrations, x, and the oxygen pressure when $2[V_A''] \cong [A_A']$.

9.9 a. Calculate the RMS speeds (cm/s) of He and Ar gases at 1000°C. M(He) = 4.002 g/mol, M(Ar) = 39.948 g/mol.
b. Calculate η (atoms/cm3) at 1000°C and 1 atmosphere pressure.
c. Calculate the mean free path, λ (cm), at 1000°C and 1 atmosphere pressure if d(He) = 256 pm and d(Ar) = 348 pm.
d. Calculate $D_{He\text{-}Ar}$ (cm^2/s) at 1000°C and 1 atmosphere pressure.

9.10 The general form for the gas diffusion coefficient is $D = A(1/M)^{1/2}(T^{3/2}/d^2p)$ where A is a constant, Units(M) = g/mol, Units(T) = K, Units(d) = Å (0.1 nm) and Units(p) = atm. Calculate the value of the "A" and give its units if Units(D) = cm2/s.

9.11 The mean speed of gas atoms is given by

$$v = \sqrt{\frac{8RT}{\pi M}}$$

where v is the speed and M is the molecular weight.

a. Calculate the mean speed (cm/s) for hydrogen molecules at 1000°C.
b. The mean free path, λ, for atoms in a gas is given by $\lambda = 1/(\sqrt{2}\pi d^2 \eta)$ Where d is the molecular diameter, d = 282.7 pm and η is the number of molecules per unit volume. Calculate the number of hydrogen molecules per cm^3 at 1-torr pressure and 1000°C.
c. Calculate the mean free path (cm) for hydrogen at 1000°C and 1-torr pressure.
d. The gas diffusion coefficient is given by $D = (1/3)v\lambda$. Calculate the gas diffusion coefficient (cm^2/s) for hydrogen at 1000°C and p = 1 torr.

9.12 Assuming that Schottky defects are the only lattice defects, construct a K–V (Brouwer) diagram for the point defects (electrons, holes, and vacancies) in a pure compound AO with the following parameters: Eg = 3 eV, $\Delta G_V = 3$ eV, $\Delta G_R = 3$ eV, density = 5.0 g/cm^3, and molecular weight of M = 80 g/mol. Use a range of oxygen pressures so that the diagram will extend from inside the reduction region to inside the oxidation region. Include the numerical values for the concentrations and pressures on the axes.

REFERENCES

Balluffi, R. W., S. M. Allen, and W. C. Carter. 2005. *Kinetics of Materials.* New York: John Wiley & Sons.
Barrer, R. M. 1941. *Diffusion in and Through Solids.* New York: MacMillan.
Barsoum, M. 1997. *Fundamentals of Ceramics.* New York: McGraw-Hill.
Bird, R. B., W. E. Stewart, and E. N. Lightfoot. 2002. *Transport Phenomena,* 2nd ed. New York: John Wiley & Sons.
Borg, R. J. and G. J. Dienes. 1988. *Solid State Diffusion.* New York: Academic Press.
Brandes, E. A. and G. B. Brook. 1992. *Smithells Metals Reference Book,* 7th ed. Oxford, UK: Butterworth-Heinemann, Ltd.
Callister, W. D., Jr. and D. G. Rethwisch. 2009. *Fundamentals of Materials Science and Engineering,* 3rd ed. New York: John Wiley & Sons.
Catlow, C. R. A. and W. C. Mackrodt. 1982. *Computer Simulation of Solids.* Berlin, Germany: Springer-Verlag.
Chiang, Y.-M. C., D. Birnie, III, and W. D. Kingery. 1997. *Physical Ceramics.* New York: John Wiley & Sons.
Connelly, N. G., T. Damhus, R. Hartshorn, and A. Hutton. 2005. IR-11.4 Point Defect (Kröger-Vink) Notation. *Nomenclature of Inorganic Chemistry. Redbook of the International Union of Pure and Applied Chemistry.* Cambridge: The Royal Society of Chemistry, 239–241.
Cussler, E. L. 1997. *Diffusion,* 2nd ed. Cambridge, UK: Cambridge University Press.
Darken, L. S. and R. W. Gurry. 1953. *Physical Chemistry of Metals.* New York: McGraw-Hill Book Company.
Elmore, W. C. and M. A. Heald. 1985. *Physics of Waves.* New York: Dover Publications, Inc.
Emsley, J. 1998. *The Elements,* 3rd ed. Oxford, UK: Oxford University Press.
Emsley, J. 2001. *Nature's Building Blocks.* Oxford, UK: Oxford University Press.
Fisher, D. J., ed. 2011. *Defects and Diffusion in Ceramics XII.* Zurich, Switzerland: Trans Tech Publications.
French, A. P. 1971. *Vibrations and Waves.* New York: W. W. Norton & Co.
Geankoplis, C. J. 1972. *Mass Transport Phenomena.* Columbus, OH: Christie J. Geankoplis.
Glicksman, M. E. 2000. *Diffusion in Solids.* New York: John Wiley & Sons.
Gray, D. E., editor-in-chief. 1972. *American Institute of Physics Handbook,* 3rd ed. New York: McGraw-Hill.
Hayes, W. and A. M. Stoneham. 1985. *Defects and Defect Processes in Nonmetallic Solids.* New York: John Wiley & Sons.
Haynes, W. M., editor-in-chief. 2013. *Handbook of Chemistry and Physics,* 94th ed. Boca Raton, FL: CRC Press.
Henderson, B. 1972. *Crystalline Solids.* New York: Crane, Russak & Co.
Hoddeson, L., E. Braun, J. Teichmann, and S. Weart. 1992. *Out of the Crystal Maze.* New York: Oxford.
Kesting, R. E. and A. K. Fritzsche. 1993. *Polymeric Gas Separation Membranes.* New York: John Wiley & Sons.
Kingery, W. D., H. K. Bowen, and D. R. Uhlmann. 1976. *Introduction to Ceramics,* 2nd ed. New York: John Wiley & Sons.
Kreyszig, E. 1988. *Advanced Engineering Mathematics,* 6th ed. New York: Wiley.
Kröger, F. A. 1964. *The Chemistry of Imperfect Crystals.* Amsterdam, the Netherlands: North Holland Publishing Co.
Kröger, F. A. and V. J. Vink. 1956. Relations between the concentrations of imperfections in crystalline solids. In *Solid State Physics,* Vol. 3. F. Seitz and D. Turnbull, eds. New York: Academic Press, 307–435.
Lindley, D. 2001. *Boltzmann's Atom.* New York: The Free Press.
Maier, J. 2004. *Physical Chemistry of Ionic Materials.* Chichester, YK: John Wiley & Sons, Ltd.

Main, I. G. 1984. *Vibrations and Waves in Physics,* 2nd ed. Cambridge, UK: Cambridge University Press.

Mark, J. E., ed. 2007. *Physical Properties of Polymers Handbook,* 2nd ed. New York: Springer Science and Business Media, LLC.

Mayer, J. W. and S. S. Lau. 1990. *Electronic Materials Science.* New York: MacMillan.

Mott, N. F. and R. W. Gurney. 1948. *Electronic Processes in Ionic Solids.* New York: Dover.

Mulder, M. 1996. *Basic Principles of Membrane Technology,* 2nd ed. Dordrecht, the Netherlands: Kluwer Academic Publishers.

Pain, H. J. 1999. *The Physics of Waves and Vibrations,* 5th ed. Chichester, UK: John Wiley & Sons, Ltd.

Pearson, J. M. 1966. *A Theory of Waves.* Boston, MA: Allyn & Bacon, Inc.

Quinn, S. 1995. *Marie Curie.* New York: Simon & Schuster.

Readey, D. W. 1966a. Mass transport and sintering in impure ionic solids. *Journal of the American Ceramic Society.* 49(7): 366–369.

Readey, D. W. 1966b. Chemical potentials and initial sintering in pure metals and ionic compounds. *Journal of Applied Physics.* 37: 2309–2312.

Reed-Hill, R. E. and R. Abaaschian. 1992. *Physical Metallurgy Principles,* 3rd ed. Boston, MA: PWS Kent.

Robinson, R. A. and R. H. Stokes. 2002. *Electrolyte Solutions,* 2nd ed. Mineola, NY: Dover.

Roine, A. 2002. Outokumpu HSC Chemistry for Windows, Chemical Reaction and Equilibrium Software with Extensive Thermochemical Database, Version 5.1. Pori, Finland: Outokumpu Research Oy.

Shewmon, P. 1989. *Diffusion in Solids,* 2nd ed. Warrendale, PA: The Minerals, Metals, and Materials Society.

Shulman, J. H. and W. D. Compton. 1962. *Color Centers in Solids.* Oxford, UK: Pergamon Press.

Silbey, R. J. and R. A. Alberty. 2001. *Physical Chemistry,* 3rd ed. New York: John Wiley & Sons.

Smyth, D. M. 2000. *The Defect Chemistry of Oxides.* Oxford, UK: Oxford University Press.

Sperling, L. H. 2006. *Introduction to Polymer Physical Chemistry,* 4th ed. New York: John Wiley & Sons.

Stull, D. R. and H. Prophet. 1971. *JANAF Thermochemical Tables,* 2nd ed. NSRDS-NBS 37. Washington, DC: U.S. Government Printing Office.

Tilley, R. J. D. 1987. *Defect Crystal Chemistry.* New York: Chapman & Hall.

Van Krevelen, D. W. 1997. *Properties of Polymers,* 3rd ed. Amsterdam, the Netherlands: Elsevier B. V.

Wells, A. F. 1984. *Structural Inorganic Chemistry,* 5th ed. Oxford, UK: Oxford University Press.

10

Steady-State Diffusion

10.1 INTRODUCTION

Many kinetic processes in materials are controlled by steady-state, or near-steady-state, diffusion. This chapter models several steady-state diffusion problems important in a wide variety of materials science and engineering situations to illustrate, if nothing else, the breadth of application of the models to many areas of materials science and engineering and the various principles involved in their modeling. As emphasized in Chapter 8, the *steady state* for diffusion means that the concentration does not change with time, yet there still is a diffusive flux of material. Again, this is in contrast to *equilibrium* in which, not only are the concentrations constant with time, but there is no flux because there is no concentration or Gibbs energy gradient. For the models developed in this chapter, many second-order effects such as boundary motion and the attendant fluid flow or the flow induced by the diffusion flux are ignored. These are small perturbations on the results and can be neglected in a first-order approximation. Different geometries, different materials, and different processes are emphasized with the intent of demonstrating the wide-ranging applicability of steady-state diffusion models in materials kinetics. Furthermore, to obtain quantitative results, the values of various parameters must be obtained. In most cases, exact values for the parameters of interest are not readily available and some reasonable approximations must be made. If the model gives a reasonable quantitative fit to an experimental or industrial process, then the approximations are good enough.

10.2 GAS DIFFUSION THROUGH SOLIDS: CARTESIAN COORDINATES

Gas diffusion through solids is a relatively simple and useful example of steady-state diffusion that has many commercial applications. Usually, gases diffuse as interstitial molecules in polymers and inorganic glasses and interstitial atoms in metals. Consider a sheet of thickness L, which is small compared to the other two dimensions; that is, the size of the sheet is large compared to the thickness—a window pane or a sheet of poly(methyl methacrylate)—Plexiglas®—for example. Figure 10.1 shows gas pressures of $p_0(x = 0)$ and $p_L(x = L)$ that determine the concentration of the diffusing species in the solid—hydrogen in palladium, water in polyvinyl chloride, oxygen in SiO_2—where $C(0) = C_0$ and $C(L) = C_L$, which are the two necessary boundary conditions for solving the steady-state diffusion equation. Also shown are some *transient* values of the concentration, C(x, t), that finally lead to the steady-state concentration. The flux or flux density J (Units (J) = mol/cm^2 s) as a function of time is sketched in Figure 10.2. The flux through the solid becomes constant when the steady state is reached. For now, the interest is only on the steady state: what happens during the transient is left for Chapter 11. To determine the flux of gas through this plate, it is necessary to begin with the steady-state version of Fick's second law in one dimension:

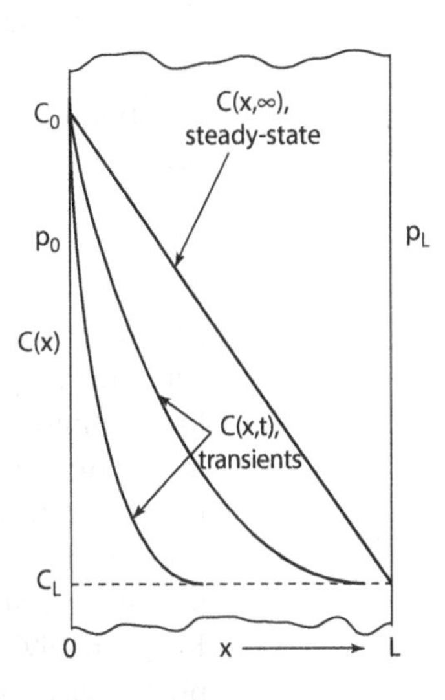

FIGURE 10.1 Concentrations determined by diffusion through a cell wall with an internal concentration of C_0 determined by an internal pressure, p_0, and C_L determined by an external pressure, p_L. Two transients C(x, t) are shown as well as the steady-state concentration, C(x, ∞), when dC/dt = 0.

$$\frac{\partial C}{\partial t} = D\frac{\partial^2 C}{\partial x^2} = 0 \qquad (10.1)$$

or

$$\frac{d^2C}{dx^2} = 0$$

which is now an ordinary differential equation because the change in concentration with time is no longer being considered. This is integrated twice to give $C(x) = Ax + B$; for $x = 0$, $C_0 = B$ and for $x = L$, $A = (C_L - C_0)/L$ so the solution to the differential equation with the two fixed boundary conditions is

$$C(x) = \frac{(C_L - C_0)}{L}x + C_0 \qquad (10.2)$$

which clearly satisfies the two boundary conditions and is the linear function of x—which no longer changes with time—shown in Figure 10.1. So the steady-state flux, J, is given by

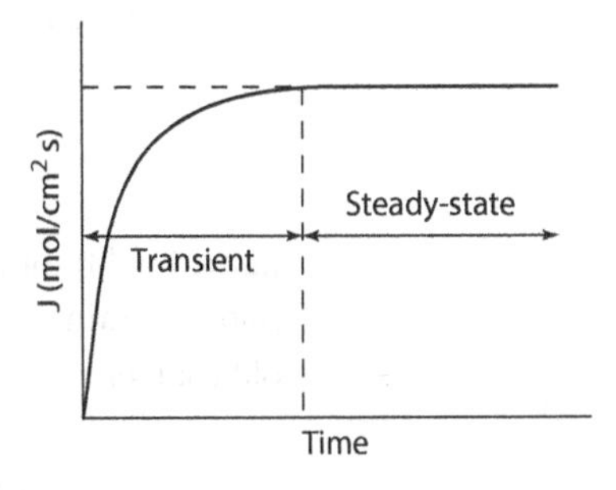

FIGURE 10.2 Flux through the wall in Figure 10.1 as a function of time showing the increasing flux during the transient period and a constant flux during the steady state.

$$J = -D\frac{dC}{dx} = -D\frac{(C_L - C_0)}{L} = D\frac{(C_0 - C_L)}{L} \tag{10.3}$$

which is a positive outward flux because $C_0 > C_L$.

10.3 GAS DIFFUSION THROUGH A POLYMER

10.3.1 Permeability and Solubility

Many polymers are used as separation membranes to separate gases in mixtures or to remove impurities from gases (Kesting and Fritzsche 1993). In the case of gases in glasses—polymeric or inorganic—the gases diffuse as molecules and their concentration depends of the gas pressure (lowercase p will be used for pressure to avoid confusion with the permeability coefficient, P, defined below) as

$$C = pS \tag{10.4}$$

where:

C is the concentration
p is the pressure
S is the solubility constant.

Equation 10.4 is a form of *Henry's law* of solubility (Silbey and Alberty 2001). The flux, Equation 10.3, becomes

$$J = DS\frac{(p_0 - p_L)}{L} \tag{10.5}$$

and the product of D and S is called the *permeability* or *permeability coefficient*, P. If a flux is measured with a pressure difference across the plate or membrane, then what is measured is P and by this simple diffusion arrangement, the values of D and S cannot be separated. Unfortunately, in the literature, S and D are rarely separated, particularly for polymers, and only P is reported (Mark 2007). Nevertheless, significant amounts of data are available, in a limited number of sources, for S and D in polymers as well as correlations between them via the Lennard–Jones potential used for the Chapman–Enskog equation for gas diffusion seen in Chapter 9 (Van Krevelen 1997). More data are available on gas solubilities for metals than other materials simply because the solubility of gases in metals is important for a number of applications other than just transport through the material (Brandes and Brook 1992). In any event, P is defined as (Sperling 2006)

$$P = \frac{(\text{Quantity of gas}) \times (\text{Sheet thickness})}{(\text{Area}) \times (\text{Time}) \times (\text{Pressure drop})}$$

and is frequently given in the units (Cussler 1997)

$$\text{Units}(P) = \frac{10^{-10}\ \text{cm}^3\ \text{gas (STP)(cm thickness)}}{(\text{cm}^2\ \text{membrane area})(\text{cm Hg pressure})\,\text{s}} = 1\ \text{Barrer}$$

named in honor of R. M. Barrer who did research on diffusion and permeability in solids and whose book contains a considerable amount of still useful data, even though much of it is almost 100 years old (Barrer 1941). The preferred set of units today are (Sperling 2006)

$$\text{Units}(P) = \frac{\text{cm}^3\ \text{gas (273 K}, 10^5\ \text{Pa)}(\text{cm thickness})}{(\text{cm}^2\ \text{area})\text{s}(\text{Pa})}$$

so the conversion factor is

$$\text{Units(P)} = \frac{76\,\text{cm Hg}}{1.013\times10^5\ \text{Pa}}\times10^{-10}\times1\ \text{Barrer}$$
$$\text{Units(P)} = 7.50\times10^{-14}\ \text{Barrer}. \tag{10.6}$$

For example, one literature source gives the permeability of oxygen in low-density polyethylene as 2.9 Barrer (Mulder 1996) (note that some references use "Barrer" and others "barrer"), while another gives $P = 2.2 \times 10^{-13}\,cm^2/Pa\ s$ (Sperling 2006), consistent with the conversion factor in Equation 10.6. If the interest is in Units (J) = mol/cm^2 s, and Units (D) = cm^2/s, Units (C) = mol/cm^3, and Units (S) = mol/cm^3 atm or mol/cm^3 bar, more useful units of P would seem to be

$$\text{Units(P)} = \text{Units(D)}\times\text{Units(S)} = \left(\frac{\text{cm}^2}{\text{s}}\right)\left(\frac{\text{mol}}{\text{cm}^3\ \text{atm}}\right) = \frac{\text{mol}}{\text{cm atm s}}.$$

For dissolution of many gases as molecules in amorphous polymers and inorganic glasses, a good empirical approximation for much of the experimental data in the literature is (Barrer 1941; Geankoplis 2003)

$$S \cong 0.05\frac{\text{m}^3\ \text{(gas solute)}}{\text{m}^3\ \text{(solid solvent)atm}}. \tag{10.7}$$

Because pV = RT = (0.08205 L atm/mol K)(273 K), 1 mole of gas = 22.4 L/mol =22,400 cm^3/mol at STP (Standard Temperature and Pressure: p = 1 atm and T = 0°C = 273.15 K),* so

$$S \cong 0.05\frac{\text{m}^3\ \text{(gas solute)}}{\text{m}^3\ \text{(solid solvent)atm}}\times\frac{1}{(22{,}400\ \text{cm}^3/\text{mol})}$$
$$S = 2.23\times10^{-6}\ \text{mol/cm}^3\ \text{atm}\times\frac{1}{1.01325\times10^5\ \text{Pa/atm}} = 2.2\times10^{-11}\ \text{mol/cm}^3\ \text{Pa}. \tag{10.8}$$

Neoprene is one of the first synthetic rubbers and is made by polymerizing chloroprene, 2-chloro-1,3-butadiene, $CH_2{=}CCl{-}CH{=}CH_2$, and is used for a wide variety of flexible tubing. It will be used to illustrate steady-state diffusion simply because separate data could be found for both the diffusion and solubility of hydrogen (Barrer 1941).

10.3.2 Cylindrical Coordinates

Consider a long tube of neoprene schematically shown in Figure 10.3. In Appendix A.1, the steady-state concentration as a function of radial distance for diffusion in cylindrical coordinates

$$\frac{\partial C}{\partial t} = \frac{D}{r}\frac{\partial}{\partial r}\left(r\frac{\partial C}{\partial r}\right) \tag{10.9}$$

namely,

$$\frac{d}{dr}\left(r\frac{dC}{dr}\right) = 0 \tag{10.10}$$

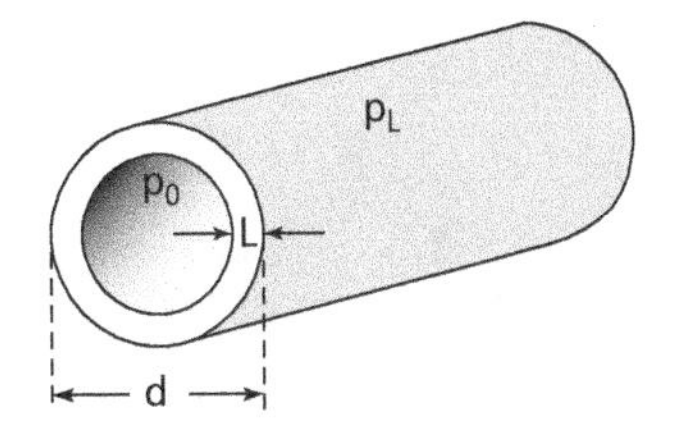

FIGURE 10.3 Dimensions for diffusion through a cylindrical tube wall with an external diameter of d and a wall thickness of L with an internal gas pressure of p_0 and an external pressure of p_L.

* NIST (National Institute for Science and Technology) uses p = 1 atm for STP, while IUPC (International Union of Pure and Applied Chemistry) uses p = 1 bar for STP. It should also be noted that frequently data are not readily available for the gas–solid system of particular interest and approximations or extrapolations of other data must be made to quantitatively model the process. This is true not only for these gas–solid systems but for kinetic processes in materials in general and it seems to be universally true for the particular process and system of *your* interest.

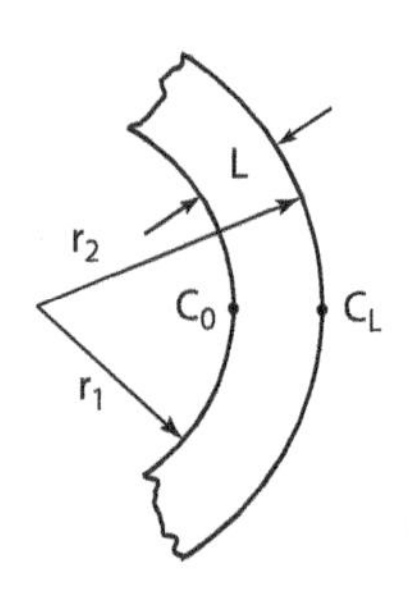

FIGURE 10.4 Dimensions and concentrations for cylindrical coordinates.

is integrated to give

$$C(r) = \frac{(C_0 - C_L)}{\ln(r_1/r_2)} \ln\left(\frac{r}{r_2}\right) + C_L \tag{10.11}$$

where the relevant dimensions and concentrations are shown in Figure 10.4. Equation 10.11 is plotted in Figure 10.5 for $r_1/r_2 = 2/3$ along with a linear approximation to the concentration versus radius for diffusion through a cylindrical tube wall.

10.3.3 Thin-Wall Result

If the tube wall thickness, L, is small enough compared to the diameter, d, then cylindrical coordinates can be ignored and the solution for a flat wall obtained above, Equation 10.5, can be applied—what is called *the thin-wall solution.* Assume that the hydrogen gas pressure in the tube is 1 atm, $p_0 = 1$ atm, the external hydrogen pressure is zero, $p_L = 0$, the wall thickness L = 1 mm, and the temperature is about room temperature, T = 300 K. As shown in Appendix A.1, if the tube radius is greater than about 5 mm, the thin-wall solution will overestimate the flux by about 10%. This may or may not be satisfactory for this application: Is the value of the permeability known to better than 10%? For diffusion of H_2 though neoprene, the literature gives $D_0 = 9.0$ cm²/s and Q = 38,200 J/mol so the diffusion coefficient of hydrogen through neoprene is given by (Barrer 1941)

$$D = D_0 e^{\frac{-Q}{RT}} = 9.0 e^{\frac{38,204}{(8.314)(300)}} = 1.64 \times 10^{-6}\ \text{cm}^2/\text{s}$$

so $P = SD \cong (2.23 \times 10^{-6}$ mol/cm³ atm)$(1.64 \times 10^{-6}$ cm²/s) $= 3.59 \times 10^{-12}$ mol/cm atm s (because the value of S is very close to that in Equation 10.8 (Barrer 1941)), which could be converted to "barrers" if necessary. Therefore, the flux, J, is

$$J = \frac{DSp_0}{L} = \frac{Pp_0}{L} = 3.59 \times 10^{-12} \frac{1}{0.1} = 3.59 \times 10^{-11}\ \text{mol/cm}^2\ \text{s}.$$

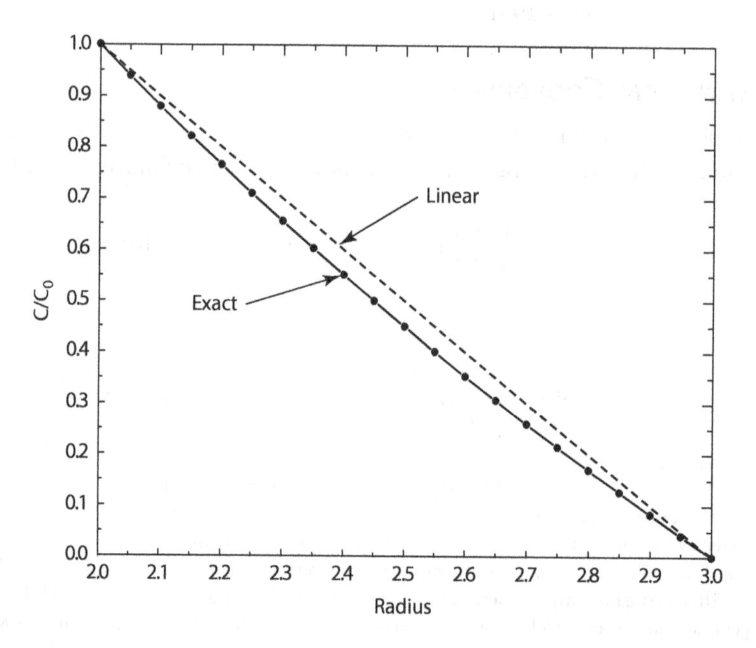

FIGURE 10.5 Calculated concentration as a function of radius through a cylindrical wall with an outer to inner radius ratio of 3:2 along with the linear *thin-wall* approximation.

This value of the flux might be good or bad depending on what the application is. If it is to purify hydrogen by diffusion through the membrane, this might be too low and a different polymer or tube thickness might be necessary. On the other hand, if hydrogen were being piped into a furnace with this tubing, there might be a concern about hydrogen leaking into the atmosphere from the tubing. However, the flux may be low enough so leakage is not a problem. So the significance of the result depends on what question the engineer is trying to answer.

10.4 GAS DIFFUSION THROUGH METALS: CYLINDRICAL COORDINATES

For metals, it is easier to find separate solubility and diffusion data because gas solubility and diffusion in metals are important in many applications in addition to membranes. Hydrogen diffusion through a palladium tube is one example. Hydrogen diffuses relatively easily through palladium while other gases such as water vapor and oxygen do not. In addition, Pd does not form a surface oxide that could be a barrier to hydrogen reaching the Pd metal and going into solid solution. As a result, diffusion of hydrogen through Pd is a system of choice to purify hydrogen that may be made by electrolysis of water, or by some other process, that introduces gas impurities. High-purity hydrogen is essential when it is to be used in applications such as semiconductor device processing where purity is critical.

For simplicity, assume that hydrogen is inside a Pd tube, shown in Figure 10.3, that has an outer diameter of d and a wall thickness of L. Assume that the thickness of the wall is very small compared to the diameter of the tube so that thin-wall solution can be used. Again, Fick's second law in Cartesian coordinates in one dimension applies. However, in the case of metals, molecules such as H_2, O_2, and N_2 dissociate into *atoms* through interaction with the free electrons in the metal, diffusing as individual atoms rather than the molecules. As a result, their dissolution follows *Sievert's law* (Gaskell 1992)

$$H_2\,(g) = 2\,H\,(\text{solid soln});\ \Delta G^\circ \tag{10.12}$$

so that

$$\frac{a^2(H)}{p(H_2)} = K = e^{\frac{-\Delta G^\circ}{RT}}$$

where a(H) is the thermodynamic activity of hydrogen in solid Pd. Assuming an ideal solution so that a(H) = X(H), where X(H) is the mole fraction of hydrogen, then the mole fraction in solution is given by

$$X(H) = K^{1/2}p^{1/2}(H_2).$$

For hydrogen in palladium, $K = 6.25 \times 10^{-4}$ at 477°C (Brandes and Brook 1992)—750 K—and for the sake of concreteness, assume that $p_0(H_2) = 2$ atm and $p_L(H_2) = 1$ atm—Figure 10.3—so that the pure hydrogen is delivered at one atmosphere after being purified by passing through the Pd tube. So, $X_L(H) = (6.25e^{-4})^{1/2}(1)^{1/2} = 0.025$ and $X_0 = 2^{1/2}X_L = 0.0353$, which are both small enough to be neglected compared to the concentration of Pd when calculating the mole fraction.* Now,

$$[Pd] = \frac{\rho\,(g/cm^3)}{M\,(g/mol)} = \frac{12.02}{106.4} = 0.112\ \text{mol/cm}^3$$

where:

ρ is the density

M is the molecular (atomic) weight of palladium (Haynes 2013).

* $X(H) = \frac{\eta(H)}{\eta(H) + \eta(Pd)} \cong \frac{\eta(H)}{\eta(Pd)}$, when $\eta(H) \ll \eta(Pd)$.

Therefore, $C_L(H) = (0.112)(0.025) = 2.80 \times 10^{-3}$ mol/cm³. And similarly for $C_0(H) = (0.112)(2)^{1/2}X_L(H) = 3.96 \times 10^{-3}$ mol/cm³. The diffusion coefficient of hydrogen in Pd, $D_0 = 2.9 \times 10^{-3}$ cm²/s and Q = 22.2 kJ/mol (Brandes and Brook 1992) so D(750K) is

$$D(750\,K) = D_0 e^{\frac{-Q}{RT}} = 2.9\times10^{-3}\, e^{\frac{-22{,}200}{(8.314)(750)}} = 8.25\times10^{-5}\ \text{cm}^2/\text{s}$$

which is *very* large diffusion coefficient for a solid—and is actually quite similar to the diffusion coefficient in liquids. If the thickness of the Pd tube wall is 0.1 mm, then the flux of H_2—which is just one-half of the H flux—is

$$J(H_2) = \frac{1}{2}J(H) = \frac{D}{2}\frac{(C_0 - C_L)}{L}$$

$$= \frac{8.25\times10^{-5}}{2}\frac{(3.96\times10^{-3} - 2.80\times10^{-3})}{0.01}$$

$$= 4.78\times10^{-6}\ \text{mol/cm}^2\,\text{s}.$$

To produce 1 mol of pure hydrogen per minute, an area of $1/(4.78 \times 10^{-6} \times 60) = 3.48 \times 10^3$ cm² of tubing would be necessary. If the tubing had a diameter of d = 0.5 cm, then the total length of tubing necessary would be $l = 3.48 \times 10^3/(\pi \times 0.5) = 2.21 \times 10^3$ cm or 22.1 m, an easily manageable length of tubing.

10.5 DIFFUSION OF HYDROGEN FROM A NUCLEAR FUSION SPHERE: SPHERICAL COORDINATES

10.5.1 Introduction

Another interesting example of steady-state—actually *quasi-steady-state*—diffusion is see how fast hydrogen diffuses out of either a silicate glass or a polymer sphere filled with deuterium and tritium that is to be used as a *laser fusion* target. Here, spherical coordinates are appropriate and this is really a *quasi-steady-state* solution because the pressure inside the sphere will change with time. However, the assumption will be made—a very valid assumption—that the transient effects in the sphere wall are quite fast and the concentration profile through the wall of the sphere quickly becomes the steady-state profile for whatever the gas pressure inside the sphere happens to be at a given time.

10.5.2 Nuclear Fusion

Attempts to harness the energy developed during fusion of hydrogen nuclei for the peaceful generation of electricity has been an ongoing endeavor for over 60 years. This is the same process that produces the energy in stars and thermonuclear weapons. For fusion to occur, hydrogen nuclei must almost contact each other: about 10^{-13} cm apart. However, there is electrostatic energy between the charges on the nuclei that must be overcome, as shown in Figure 10.6, given by

$$E = \frac{e^2}{4\pi\varepsilon_0 r} \tag{10.13}$$

where:

ε_0 is the permittivity of free space, $\varepsilon_0 = 8.85 \times 10^{-12}$ F/m (C²/J/m)*
e is the charge on the electron, $e = 1.602 \times 10^{-19}$ C
r is the radius of the hydrogen nucleus ≅ 10^{-13} cm (Emsley 1998).

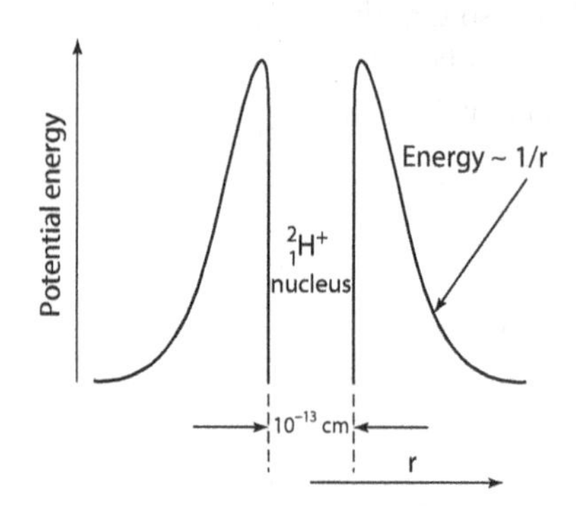

FIGURE 10.6 Potential energy versus radial separation distance for two hydrogen nuclei undergoing nuclear fusion.

* Units (charge) = coulombs designated as C.

The potential energy in Figure 10.6 is rounded near the nucleus because of the attraction of the nuclear strong force. If the potential were to be overcome by accelerating two hydrogen ions at each other, their kinetic energy, $KE = 3/2k_BT$, would have to equal or exceed the electrostatic energy, which means that the hydrogen atoms would have to be heated to such a temperature that

$$\frac{3}{2}k_BT = \frac{e^2}{4\pi\varepsilon_0 r} \tag{10.14}$$

$$T = \frac{2}{3}\frac{e^2}{4\pi\varepsilon_0 r k_B} = \frac{2}{3}\frac{\left(1.602\times10^{-19}\,C\right)^2}{(4)(3.1416)\left(8.85\times10^{-12}\,C^2/J/m\right)\left(1\times10^{-15}\,m\right)\left(1.38\times10^{-23}\,J/K\right)} = 1.1\times10^{10}\,K$$

which is a little tough to achieve! However, this does not take into account the strong nuclear attractive forces that start to operate at these distances and lower and round the energy barrier near the nucleus to look something like Figure 10.6. To increase the nuclear attractive force—which increases with the number of nucleons in the nucleus—deuterium, ${}_1^2H$, and tritium, ${}_1^3H$, are used, yet the ions still must be heated or accelerated to the equivalent of hundreds of millions of degrees to overcome the electrostatic potential energy barrier to enable fusion.

The interest in fusion, or course, is that the energy released, $E = \Delta mc^2$, where Δm is the mass difference between the reactant and product nuclei, is very large:

$${}_1^3H + {}_1^2H \rightarrow {}_2^4He + {}_0^1n + 17.6\,MeV. \tag{10.15}$$

To put this quantity of energy in perspective, fusing 1 mol of deuterium and 1 mol of tritium, a total of 5 g of fuel would produce $E = (17.6 \times 10^6$ eV/atom)$(1.602 \times 10^{-19}$ J/eV)$(6.022 \times 10^{23}$ atoms/mol) $= 1.7 \times 10^{12}$ J. A 1 GW electrical generation power plant—typical size of a nuclear- or coal-fired plant—produces $E = (1 \times 10^9$ J/s)(24 h/d)(3600 s/h) $= 8.64 \times 10^{13}$ J of electricity per day. Because 1 ton of coal $= 2.13 \times 10^{10}$ J/ton, this would require $(8.64 \times 10^{13}$ J)/$(2.13 \times 10^{10}$ J/ton) = 4000 tons of coal. However, the efficiency of generating electricity from burning coal is only about 30%, so closer to 10,000 tons of coal per day are needed, while producing about 40,000 tons of CO_2 a day, a less-than-desirable byproduct. Because a typical U.S. railroad coal hopper car holds about 100 tons of coal, a coal train about 100 cars long—the so-called "unit train"—is just one day's supply of coal for a coal-fired power plant. Count the cars next time you get stopped at a rail crossing by one of these coal trains. If this same electricity were generated by fusion with close to 100% efficiency, only about *250 g* $(2.75 \times 10^{-4}$ tons) of deuterium and tritium would be needed instead of 10,000 tons of coal: a factor of 3.6×10^7 less material. This makes storage and transportation much easier along with no climate-altering greenhouse gas produced in the process. The envisioned fusion reactor for the generation of electricity would be surrounded by blanket of molten lithium to absorb the energy of the fusion neutrons and to form additional tritium fuel by the reaction:

$${}_3^6Li + {}_0^1n \rightarrow {}_1^3H + {}_2^4He \tag{10.16}$$

to ensure sustainability of the fuel. The hot liquid lithium ($T_{mp} = 180.5°C$) would then be pumped to a heat exchanger to generate steam, to drive a turbine, and to drive a generator to produce electricity. There is a virtually limitless supply of deuterium in sea water and lithium is a reasonably abundant element—as long as the world's supply is not used up in lithium batteries—so that, if such a tritium breeding reactor could be achieved, there would be sufficient fusion fuel for millions of years of power generation.

The only problem is that a working fusion reactor has yet to be developed—one that generates more fusion energy than the electricity required to generate it. There are two primary designs and several major programs in progress to achieve this. One of these involves *magnetic confinement*

FIGURE 10.7 Photograph of a spherical hydrogen-containing target for nuclear fusion in the National Ignition Facility. (LLNL 2002, https://commons.wikimedia.org/wiki/File:Fusion_microcapsule.jpg.) (Original color photograph changed to grayscale and sharpness and contrast enhanced in Photoshop®.)

of high-energy deuterium and tritium in a high-temperature ionized gas. The other approach is so-*called inertial confinement* or *laser fusion* where a large amount of energy from banks of high-power lasers are focused onto the surface of a tiny glass or polymer sphere containing deuterium and tritium gas (Figure 10.7) (*Science and Technology Review* 2002). The high power during a laser pulse—petawatts (10^{15} W)—impinging on the surface of the sphere causes rapid sublimation of the sphere that produces an equal and opposite force on the hydrogen isotopes and compresses them some 10^4 to 10^5 times forcing their nuclei into contact to ignite the fusion reaction. Note that the solar radiation received by the earth is about 170 PW (Orders of Magnitude (power) 2014). Figure 10.8 shows a schematic of the National Ignition Source at the Lawrence Livermore National Laboratory in California that is attempting to achieve inertial fusion. High-power lasers—a total of 192—generate 20 MJ on the surface of the fusion sphere during a laser pulse on the order of a few nanoseconds; hence 10^{15} W. In the bay that juts out at about the 9 o'clock position in Figure 10.8, two people can be seen standing in the hallway, which gives an indication of the size of the facility—roughly that of about three football fields.

10.5.3 Laser Fusion and Materials

"What," you may ask, "has laser fusion to do with materials kinetics?" Over the several years of development of laser fusion, significant advances in materials have been achieved, bringing the process closer to realization. These include the development of a laser glass that has a small nonlinear index of refraction to prevent self-focusing of the laser beam inside the glass laser components that could destroy them. The glass lasers are silicate glasses that contain Nd_2O_3 in solid solution and the Nd^{+3} ion gives off light—lases—at a wavelength of 1.06 μm. In addition, glass-forming processes needed to be developed to produce the large pieces of glass necessary—meters square. Finally, the infrared wavelength of 1.06 μm interacts too strongly with the ionized gas surrounding the ablating target so large nonlinear optical crystals—potassium dihydrogen phosphate, KDP, KH_2PO_4—had to be grown to double the frequency of the laser light to a wavelength of 0.53 μm, which is not so strongly absorbed by the plasma. Finally, over the years several different materials for the laser fusion sphere (Figure 10.7) containing the deuterium and tritium solids (if sufficiently cooled) or gases have been used including silicate glasses, polymers, and beryllium (Be). Because the interest here is diffusion, the relevant diffusion question regarding laser fusion is, "How long does it take for the hydrogen to diffuse out of the fusion sphere lowering the gas content so a stored sphere can no longer be used?"

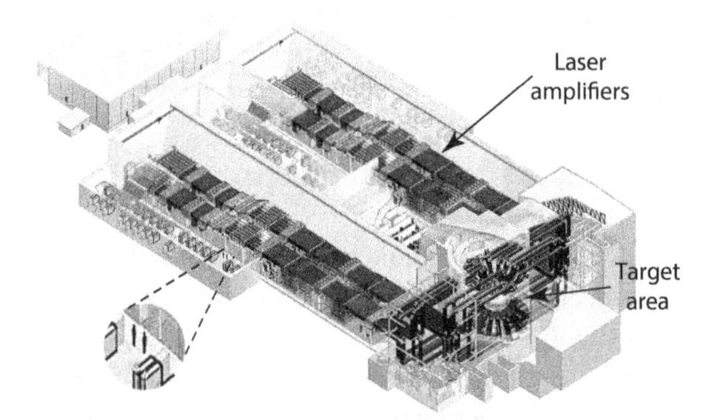

FIGURE 10.8 Schematic drawing of the National Ignition Facility. Inset at left shows two individuals in the hall at about 90° to indicate scale. (LLNL 2002, https://commons.wikimedia.org/wiki/File:Fusion_microcapsule.jpg.) (Original color drawing changed to grayscale and sharpness and contrast enhanced in Photoshop®.)

10.5.4 Hydrogen Diffusion from a Laser Fusion Sphere

10.5.4.1 Model in Spherical Coordinates

In reality, there can never really be a steady state for material diffusing out of a spherical shell, which is true for a closed volume of any shape, as the concentration of the diffusing species inside the sphere is continuously decreasing over time. Yet for now, such complexities can be ignored without much loss of accuracy. For diffusion in spherical coordinates, as was shown in Chapter 8, the mass continuity equation, Fick's second law, is

$$\frac{\partial C}{\partial t} = \frac{D}{r^2}\frac{\partial}{\partial r}\left(r^2\frac{\partial C}{\partial r}\right) \tag{10.17}$$

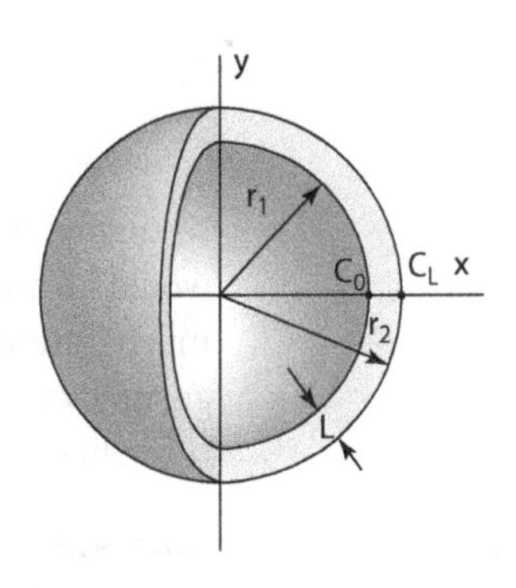

FIGURE 10.9 Spherical shell showing coordinates and concentrations used to calculate concentration profiles and diffusion fluxes in spherical coordinates.

which becomes in the steady state where $\partial C/\partial t = 0$

$$\frac{d}{dr}\left(r^2\frac{dC}{dr}\right) = 0. \tag{10.18}$$

For the dimensions and concentrations shown in Figure 10.9, the solution of Equation 10.18 is (Section A.2)

$$C(r) = -\frac{r_1 r_2}{r}\frac{(C_L - C_0)}{r_2 - r_1} + \frac{(C_L r_2 - C_0 r_1)}{r_2 - r_1} \tag{10.19}$$

which leads to the steady-state diffusion flux through a spherical shell of radius r_2,

$$J(r_2) = D\frac{r_1}{r_2}\frac{(C_0 - C_L)}{r_2 - r_1}. \tag{10.20}$$

The concentration versus radial distance of Equation 10.19 is plotted in Figure 10.10 for $r_1/r_2 = 2/3$ and compared to a linear concentration versus distance used in the thin-wall solution. Note that the ratio of the thin-wall solution flux

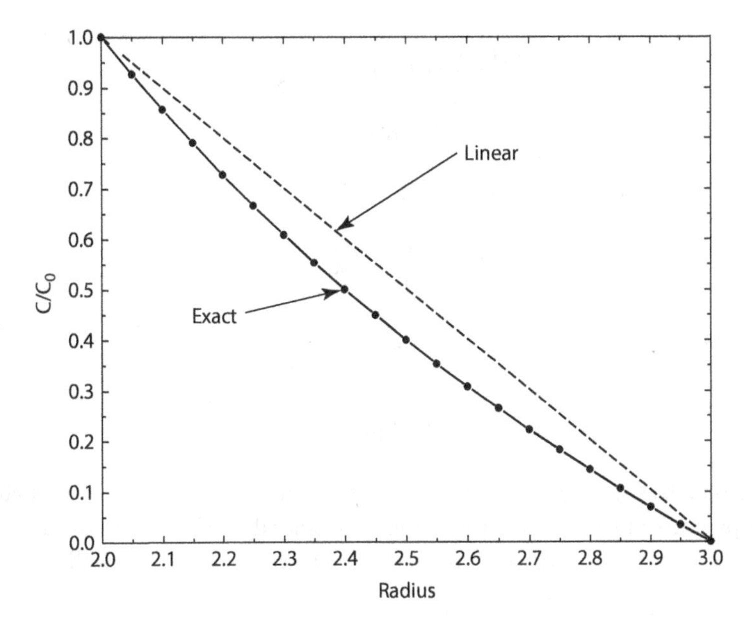

FIGURE 10.10 Calculated concentration versus radial distance through a spherical shell wall of outer to inner radius ratio of 3:2. Also is plotted the linear *thin-wall* solution.

$$J(r_2) = D\frac{(C_0 - C_L)}{r_2 - r_1}$$

to that of the exact solution is just r_1/r_2. So for the curves in Figure 10.10, the exact solution flux would be only 67% that of the linear or thin-wall flux even though there does not appear to be that much difference between the exact solution and the linear approximation in the figure. So to get 90% accuracy with the thin-wall solution for a sphere, the wall thickness would need to be one-tenth of the sphere radius.

10.5.4.2 Polymer Sphere: Approximate

For gases diffusing through polymers, it is the permeability, P, that is the parameter of interest and Equation 10.20 becomes

$$J(r_2) = P\frac{r_1}{r_2}\frac{(p_0 - p_L)}{r_2 - r_1} \tag{10.21}$$

where p_0 and p_L are the internal and external pressures of hydrogen, respectively. Assume that the sphere is made of polystyrene. Unfortunately, neither permeability data nor solubility and diffusion data for hydrogen in polystyrene are readily found—a not untypical situation for polymers. What was found is that the permeability of carbon dioxide, CO_2, in polystyrene is about an order of magnitude lower than it is in natural rubber (Mulder 1996). In Section 10.3, the permeability for hydrogen through rubber was calculated to be $P = 3.59 \times 10^{-12}$ mol/cm atm s. Making the assumption that the behavior of hydrogen is similar to that of carbon dioxide, an estimate for the permeability of hydrogen through polystyrene is $P \cong 3.6 \times 10^{-13}$ mol/cm atm s at room temperature.* Because there is no hydrogen outside the sphere, $P_L = 0$ and $P_0 = P \times p(H_2) = (3.6 \times 10^{-13})(2) = 7.2 \times 10^{-13}$ mol/cm s. The number of moles per second, dn/dt, that leave the sphere is just the area times the flux, $A(r_2)J(r_2) = -AD(dC/dr)_{r_2}$

$$\begin{aligned}\frac{dn}{dt} &= A(r_2)J(r_2) = (4\pi r_2^2)\left(P_0\frac{r_1}{r_2}\frac{1}{r_2 - r_1}\right)\\ &= 4\pi P_0\frac{r_1 r_2}{r_2 - r_1}\\ &= 4\pi(7.2\times10^{-13})\frac{(0.1)(0.075)}{0.025}\\ &= 2.7\times10^{-12}\ \text{mol/s.}\end{aligned}$$

Now the initial number of moles, n, in the sphere at time = 0 is

$$\begin{aligned}n &= V_{\text{sphere}}\frac{p}{RT} = \frac{4}{3}\pi r_1^3\frac{p}{RT}\\ &= \frac{4}{3}\pi(0.075)^3\frac{2}{(82.05)(300)}\\ &= 1.44\times10^{-7}\ \text{mol.}\end{aligned}$$

So the time it takes for the hydrogen pressure to drop to 1 atm is (the 2 in the denominator comes from the 50% drop in pressure or the fact that the number of moles drops by 1/2): $n/(dn/dt) = 1.44 \times 10^{-7}/(2)(2.7 \times 10^{-12}) = 2.7 \times 10^4$ s: or about 8 h and about 30% longer than if the "thin-wall" solution were used.

---

* Again, another example how, in the absence of the appropriate data, the usual situation, some approximations must be made.

Clearly, a polymer with a smaller value of hydrogen permeability needs to be used or the spheres need to be cooled quickly and stored at low temperatures so that D will be sufficiently small that it takes at least a year—about 3×10^7 s—for half of the hydrogen to diffuse out. That means that the permeability would have to be reduced about three orders of magnitude to $P \cong 7.2 \times 10^{-15}$ mol/atm cm s. This model assumes that the C_0 and the concentration profile through the spherical shell quickly change to accommodate the drop in pressure inside the sphere. As will be seen in Chapter 12, this can be estimated from $t \cong L^2/D = (0.025)^2\ \text{cm}^2/(1.64\times10^{-6}\ \text{cm}^2/\text{s}) = 381\,\text{s}$, which is on the order of 1% of the total time for the pressure to drop by one-half and is probably a good assumption that the concentration gradient in the sphere wall is re-established quickly compared to the total time: certainly acceptable given the uncertainties in the other parameters of the problem.

10.5.4.3 Polymer Sphere: Less Approximate

A better approximation can be made by taking into consideration that the pressure drops continuously with time and the rate at which moles leave the inside of the sphere, dn/dt,

$$\frac{dn}{dt} = V_{\text{sphere}}\frac{1}{RT}\frac{dp}{dt} = \frac{4}{3}\pi r_1^3\frac{1}{RT}\frac{dp}{dt}$$

$$= -A_{\text{sphere}}J(r_1) = -\left(4\pi r_1^2\right)DS\frac{r_2}{r_1}\frac{1}{r_2 - r_1}p.$$

Because $L = r_2 - r_1$, this leads to

$$\frac{1}{p}\frac{dp}{dt} = -\frac{3RTDSr_2}{Lr_1^2}$$

and integrating gives

$$\ln\left(\frac{p}{p_0}\right) = -\frac{3RTPr_2}{Lr_1^2}t \qquad (10.22)$$

so the time to drop 50% is calculated to be

$$t = -\frac{Lr_1^2}{3RTDSr_2}\ln\left(\frac{p}{p_0}\right)$$

$$t = -\frac{(0.025\,\text{cm})(0.075\,\text{cm})^2}{3\left(82.05\,\text{cm}^3\,\text{atm/mol K}\right)\left(300\,\text{K}\right)\left(1.64\times10^{-6}\ \text{cm}^2/\text{s}\right)\left(2.26\times10^{-6}\ \text{mol/cm}^3\,\text{atm}\right)\left(0.1\,\text{cm}\right)}\ln\left(0.5\right)$$

$$t = 3561\ \text{s}$$

which is just about 1 h and is about 36% longer than if the pressure drop were not considered. Again, the initial transient is still about 10% of the total time. This calculation also assumes that as the concentration profile through the wall changes with time, its transient is also fast compared to the change in the new steady-state profile as the pressure drops. In any event, the hydrogen still diffuses out too fast and polystyrene spheres need to be cooled to be of value in fusion energy production.

10.5.4.4 Silica Glass Sphere

For hydrogen diffusing in silica glass, it is helpful to use some recent data relating to the properties of SiO_2 important to semiconductor device processing. For example, the diffusion coefficient for hydrogen in glassy SiO_2 is given by $D = D_0\exp(-Q/RT) = 5.65\times10^{-4}\exp(-43.027/RT)\ \text{cm}^2/\text{s}$ (Mayer and Lau 1990), which gives $D\ (298\,\text{K}) = 1.62\times10^{-11}\ \text{cm}^2/\text{s}$ at room temperature of 25°C. Permeability data for hydrogen through fused silica exist, which, along with this diffusion coefficient, give an average value of S that does not vary much between 200°C and 1000°C of $S = 4.1\times10^{-7}\ \text{mol/(cm}^3\,\text{atm)}$. As a result, for hydrogen at room temperature, $P = 6.64 \times 10^{-18}$ mol/cm s atm. From the same expression used for the diffusion of hydrogen from the polystyrene sphere,

$$t = -\frac{Lr_1^2}{3RTDSr_2}\ln\left(\frac{p}{p_0}\right)$$

$$t = -\frac{(0.025\,cm)(0.075\,cm)^2}{3\left(82.05\,cm^3\,atm/(mol\cdot K)\right)\left(298\,K\right)\left(1.62\times10^{-11}\,cm^2/s\right)\left(4.1\times10^{-7}\,mol/(cm^3\,atm)\right)\left(0.1\,cm\right)}\ln(0.5)$$

$$t = 2.0\times10^9\,s$$

Because there about 3×10^7 s/year, and 2×10^9 s are necessary for the hydrogen pressure to drop by one-half, the silica spheres should be satisfactory at room temperature for about 67 years. If loss of hydrogen were the only major consideration, silica would be a much better choice for fusion spheres than polystyrene since they can be stored longer without an unacceptable loss of hydrogen.

10.6 PASSIVE OXIDATION OF SILICON

10.6.1 JUSTIFICATION

Because of its critical importance in the processing of silicon devices, the passive oxidation of silicon has been quite well studied and has led to the Deal–Grove model of silicon oxidation to form SiO_2 (Deal and Grove 1965). As a result, the kinetics of oxidation are well understood and the parameters of the oxidation process have been well characterized (Mayer and Lau 1990).

Two ancillary points are worth noting. First, it is the excellent chemical stability and very good electrical insulating behavior of *native* SiO_2—the oxide that is produced simply by oxidation of the element—compared to the oxides on germanium, GaAs and other semiconductors that have made silicon the chosen material for integrated circuits. This is in spite of the fact that germanium and gallium arsenide are superior semiconductors compared to silicon—Ge was the first semiconductor used for devices and many million Ge devices were made before the introduction of silicon-based components. However, in more recent devices, the oxide thicknesses are 1 nm or less—that is, the gate oxide in MOSFETs—and the native SiO_2 layer is too thin to be a good insulator and new oxides, such as HfO_2, deposited by chemical vapor deposition (CVD), (Chapter 5) or other processes are replacing it. With the need for native SiO_2 oxide diminished, other semiconductors may replace silicon in the future. Nevertheless, because there has been a large capital investment in fabrication facilities based on silicon, it may be some time before another semiconductor replaces silicon in integrated circuits. The other point to be noted is that Andy Grove, co-author of the silicon oxidation model, is a retired former CEO of Intel (Jackson 1997).

The Deal–Grove model takes into consideration all of the series steps associated with the oxidation of silicon to SiO_2:

- Diffusion of oxygen to the O_2–SiO_2 interface
- Diffusion of oxygen through SiO_2
- Reaction of oxygen and silicon at the Si–SiO_2 interface
- Includes the possibility of a pre-existing oxide layer on the surface

The diffusion through the gas phase is usually fast and not rate controlling. For thin oxide layers, the interface reaction at the Si–SiO_2 layer is rate-controlling or the slowest step, while for thicker oxide layers, it is the diffusion of oxygen through the SiO_2 that controls the rate.

10.6.2 OXYGEN DIFFUSION IS RATE CONTROLLING

Here, the assumption is made that it is molecular oxygen diffusion through the SiO_2 layer that controls the rate of oxidation and the other steps in the process are all assumed to be much faster: a good assumption for relatively thick oxide layers. This diffusion model is shown in Figure 10.11 and raises the question, "How is it known that oxygen diffusion through the SiO_2 to the Si–SiO_2 interface controls the rate and not Si diffusion to the O_2–SiO_2 interface?" This has been shown by the double oxidation experiment shown in

Figure 10.12. Oxygen has the following stable isotopes and their relative abundances (Emsley 1998):

- ^{16}O 99.762%
- ^{17}O 0.038%
- ^{18}O 0.200%

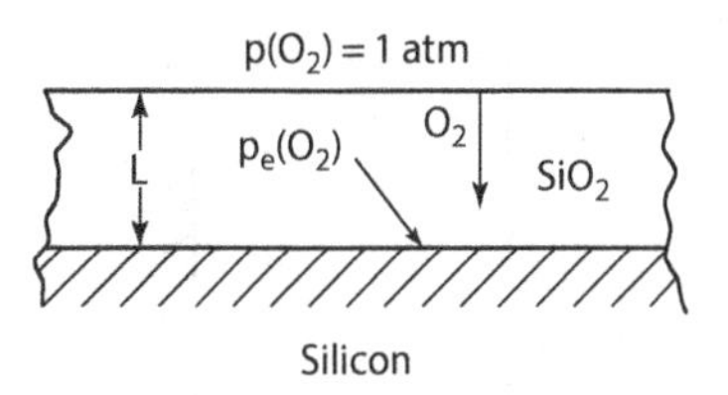

FIGURE 10.11 Schematic of the oxidation of silicon to SiO_2 of thickness L in 1 atm of pure oxygen. The oxygen is shown to diffuse interstitially as O_2 and the equilibrium oxygen pressure at the Si–SiO_2 interface is determined by the reaction: $Si(s) + O_2(g) = SiO_2(s)$.

These isotopes can be readily separated by differences in boiling point or by gas separation membranes. As a result, $^{16}O_2$–$^{18}O_2$ gas mixtures with various $^{18}O_2$ gas enrichments are commercially available. To determine the rate-controlling diffusing species, a double oxidation experiment is carried out at temperatures near 1000°C. Oxidation is first performed in normal oxygen containing almost pure $^{16}O_2$. Then a second oxidation is carried out in an $^{18}O_2$-enriched gas. Then the double oxide layer is analyzed by *SIMS—Secondary Ion Mass Spectrometry*—by sputtering the SiO_2 and forming a hole in the oxide layer—up to a few microns in depth usually. As the hole is sputtered through the oxide, the oxygen isotopes—the secondary ions—are analyzed in a mass spectrometer to determine the [^{18}O] as a function of depth in the layer. A plot of ^{18}O concentration as a function from the SiO_2–gas interface is obtained. If all of the ^{18}O is found at the SiO_2–O_2 interface, then silicon diffused through the oxide layer to form additional oxide. On the other hand, if the ^{18}O is found near the Si–SiO_2 interface—as shown in Figure 10.11—then oxygen diffusing through the oxide layer is rate-controlling. The diffusion of oxygen toward the silicon and the diffusion of silicon away from the silicon are parallel reaction steps. As was shown earlier, the faster of two parallel processes is rate controlling. In this case, it is not surprising that the interstitial diffusion of O_2 would be faster than the—probably—vacancy diffusion of silicon ions because of generally faster interstitial diffusion compared to vacancy diffusion.

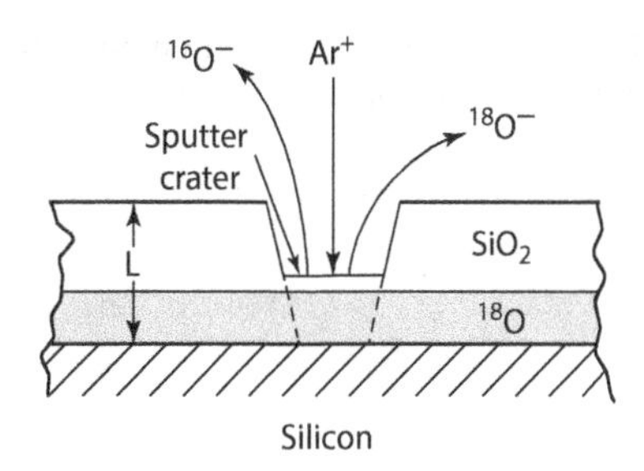

FIGURE 10.12 Schematic drawing showing how the location of the oxygen isotope ^{18}O (gray region) after a two-step oxidation is determined by secondary ion mass spectrometry (SIMS) by sputtering a hole into the oxide with argon ions and measuring the concentrations of ^{16}O and ^{18}O as a function of the sputtering depth.

Experimentally, a concentration profile much like that shown in Figure 10.12 is found and shows oxygen diffusion is rate controlling. In addition, because there is no ^{18}O near the SiO_2–O_2 interface, the data indicate that the oxygen must be diffusing interstitially and it is not exchanging with the oxygen in the SiO_2 network forming the amorphous SiO_2 oxidation product. If it were, the ^{18}O would be found throughout the layer. Figure 10.13 is another depiction of amorphous SiO_2 compared to a crystalline modification, based on the regularity—or lack of it—in the arrangement of the SiO_4^{-4} tetrahedra that share corners forming a three-dimensional structure or network—in the case of the glass.

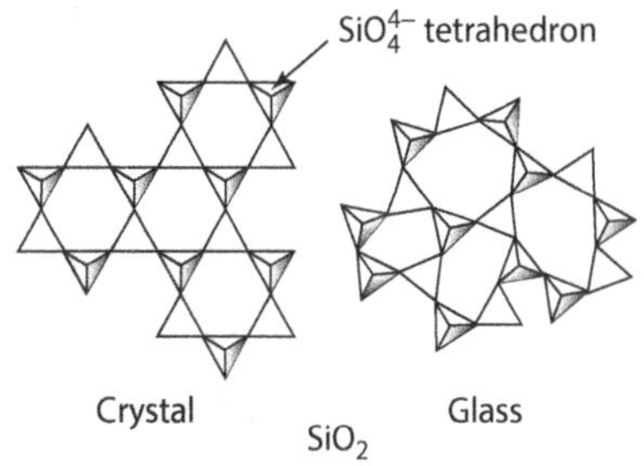

FIGURE 10.13 Another comparison of a SiO_2 crystal with a glass of the same composition, both represented as SiO_4^{4-} tetrahedra attached in three dimensions at corners. The major differences between the two are as follows: the Si–O–Si bond angle in the glass is not always 180° as it is in the crystal, and the rings in the glass do not always contain six tetrahedra as in the crystal.

10.6.3 Thermodynamics of Oxidation

For the sake of concreteness, assume that the temperature of oxidation is 1200°C or 1473 K, which is near the maximum temperature at which silicon oxidation is normally done in practice. The reaction is (Stull and Prophet 1971)

$$\text{Si (s)} + \text{O}_2\text{ (g)} = \text{SiO}_2\text{ (s)}; \quad \Delta G^\circ(1400\text{ K}) = -659{,}813\text{ J/mol} \tag{10.23}$$

so that the equilibrium constant is given by

$$\frac{a(SiO_2)}{a(Si)p(O_2)} = K_p^0 = e^{-\frac{\Delta G^0}{RT}} = e^{\frac{659{,}813}{(8.314)(1400)}} = \frac{1}{2.4\times10^{-25}}$$

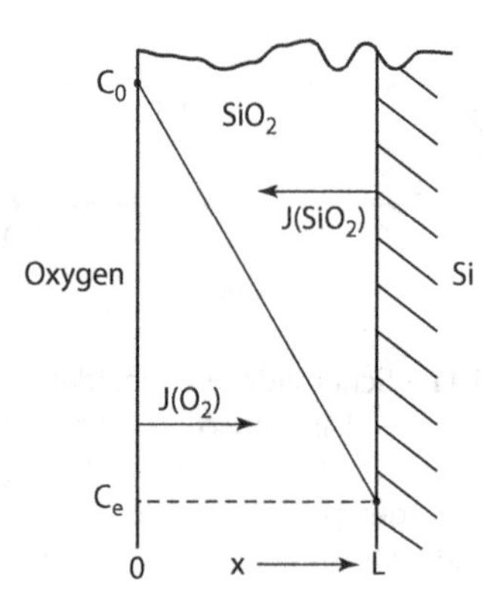

FIGURE 10.14 Geometry, concentrations, and fluxes used for the model of silicon oxidation in oxygen.

At the SiO_2–O_2 interface, if the $p(O_2) = 1$ atm, because $a(SiO_2) = 1$, then $a(Si) = 2.4 \times 10^{-25}$. Conversely, at the Si–$SiO_2$ interface, $a(Si) = 1.0$ and $a(SiO_2) = 1$ so $p_e(O_2) = 2.4 \times 10^{-25}$ atm (Figure 10.11).

10.6.4 Oxidation Kinetics

This is another example where the concentration of the interstitial oxygen is given by $C(O_2) = S\,p(O_2)$, where S is the solubility and $[S] = \text{mol/cm}^3$ atm. Clearly, at the SiO_2–Si interface, the $C_e(O_2) \cong 0$ with such a low $p_e(O_2) = 2.4 \times 10^{-25}$ atm, while at the O_2–SiO_2 interface, $C_0(O_2) = S\,p(O_2) = S \times 1$. Therefore, the diffusion model is shown in Figure 10.14 and $C_e = 0$ and the thickness of the oxide at any given time is L. Therefore, the fluxes of oxygen and SiO_2 become

$$J(O_2) = -\frac{D(O_2)S\,p(O_2)}{L} = -J(SiO_2)$$

$$J(SiO_2) = \frac{\rho(SiO_2)}{M(SiO_2)}\frac{dL}{dt}$$

so

$$\frac{dL}{dt} = \frac{M}{\rho}\frac{DS\,p(O_2)}{L}$$
$$L^2 = \frac{2M}{\rho}DS\,p(O_2)t \tag{10.24}$$

or

$$L^2 = Kt$$

the *parabolic oxidation* kinetics of metals, alloys, compounds, and semiconductors. It is *passive oxidation* because the rate of oxidation, $dL/dt \sim 1/L$, slows down as the oxide layer gets thicker. For SiO_2 at 1400 K, $S = 4.72 \times 10^{-6}$ mol/cm³ atm, and $D_0 = 2.7 \times 10^{-4}$ cm²/s and $Q = 111{,}908$ J/mol (Mayer and Lau 1990), so $D(1400\text{ K}) = 2.7 \times 10^{-4}\exp(-111{,}908/(8.314 \times 1473)) = 2.90 \times 10^{-8}\text{ cm}^2/\text{s}$, a value typical of interstitial diffusion, further confirming the model. So in 3 h ($\cong 10^4$ s) at 1400 K,

$$L^2 = \frac{2M}{\rho}DS\,p(O_2)t$$
$$= 2\frac{(60.08\text{ g/mol})}{(2.2\text{ g/cm}^3)}(2.90 \times 10^{-8}\text{ cm}^2/\text{s})(4.72 \times 10^{-6}\text{ mol/cm}^3\text{ atm})(1\text{atm})(10^4\text{ s})$$
$$L^2 = 7.48 \times 10^{-8}\text{ cm}^2$$

so $L = 2.7$ μm, which is close to what is experimentally found (Mayer and Lau 1990).

10.7 BIOLOGICAL CELL WALLS AND DIFFUSION

10.7.1 The Problem

Another example of steady-state diffusion, again involving spherical coordinates, is the diffusion of simple molecules such as oxygen, carbon dioxide or nitrous oxide through a biological cell wall: specifically, a red blood cell wall where the diffusion is one of the steps involved in the oxidation of blood cell hemoglobin or removal of CO_2 generated by metabolic processes. This example illustrates that

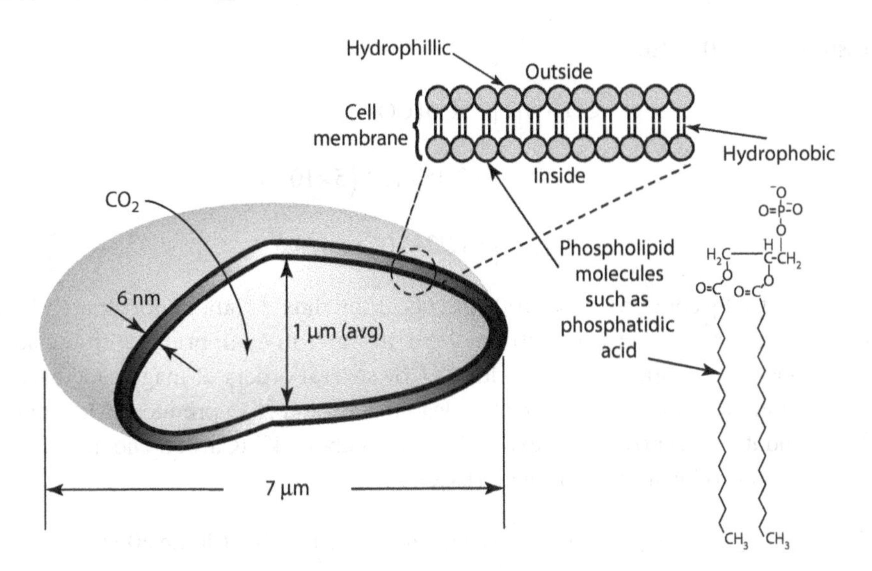

FIGURE 10.15 Model of red blood cell and its phospholipid bilayer consisting, in this case, of phosphatidic acid molecules. About one-half of the membrane thickness, 3 nm, is the hydrophobic part of the membrane.

many biological processes are similar to other processes in materials and, hence, are encompassed in "biomaterials." In addition, the example illustrates some of the difficulties and approximations that must be made for some parameters of the process to make quantitative predictions. More surprisingly, considering the importance of this process in human physiology (Chang 2000), it is not yet completely understood (Endewar et al. 2014) making it even more difficult to obtain useful data to quantitatively model the process.

A cell is surrounded by a phospholipid bilayer—cell membrane—with a hydrophobic core of long $-CH_2-$ chains that only allows diffusion of small, nonpolar molecules—such as carbon dioxide—through it, as sketched in Figure 10.15. Simple, nonpolar molecules, CO_2, O_2, and so on can diffuse through the membrane wall by a dissolution in the wall and diffusion process modeled above. The passage of ions and larger molecules through the cell membrane generally requires more complex processes and pathways such as transfer proteins in the cell wall, processes beyond what can be covered here. For example, the CO_2 that enters a red blood cell reacts with OH^-, with the help of an enzyme, to form HCO_3^- ions that are transported out of the cell via a protein in the cell wall (Lodish 2000). Eventually, the HCO_3^- reaches the lungs where the reverse process takes place and the CO_2 is exhaled. To illustrate another example of steady-state diffusion and finding literature values for the parameters of interest (or estimating them), the simple question that is posed, "How long does it take for the $[CO_2]$ to equilibrate between the inside and outside of red blood cell if the $[CO_2]$ outside is suddenly increased by 20% (maybe due to vigorous exercise)?" This is a simple question but finding accurate values of the necessary parameters to solve the problem is more complex. Nevertheless, some estimates can be made.

10.7.2 Parameters of the Problem

For a red blood cell, the total phospholipid bilayer membrane is about 6 nm thick, while the hydrophobic part is about 3 nm thick (Lodish et al. 2000). Red blood cells are platelets about 7 μm in diameter with an average thickness of about 1 μm (Figure 10.15) (Red blood cell). Therefore, the wall thickness is small compared to the cell radius so that rectangular coordinates can be used (the *thin-wall solution*). Now, normal human blood has enough CO_2 dissolved as CO_2 to be equivalent to about 5 kPa ($\approx 5 \times 10^{-2}$ atm) carbon dioxide gas pressure (Blood). For dissolution of CO_2 in water—a good approximation to blood plasma—at body temperature, $T \cong 37°C$

$$CO_2(g) \rightleftharpoons CO_2(aq); \quad K_e(37°C) = 2.45 \times 10^{-2} \quad \text{(Roine 2002)} \tag{10.25}$$

so at a pressure of 5×10^{-2} atm

$$\left[CO_2(aq)\right] = K_e p(CO_2)$$

$$= 2.45 \times 10^{-2}\left(5 \times 10^{-2}\right)$$

$$\left[CO_2(aq)\right] = 1.23 \times 10^{-3} \text{mol / L.}$$

The atmosphere actually contains about 400 ppm of carbon dioxide (and unfortunately increasing!!) so that Eq. (10.25) gives $[CO_2] = 1.16 \times 10^{-8}$ mol/cm³ in water in equilibrium with the atmosphere. Therefore, our bodies are supersaturated with CO_2 by several orders of magnitude because of the metabolic oxidation processes taking place that lead to CO_2 and H_2O products. Most of this excess CO_2 is exhaled and its concentration in exhaled breath is about 4% (Carbon Dioxide). But dissolved CO_2 reacts with water to form carbonic acid, H_2CO_3:

$$CO_2(aq) + H_2O \rightleftharpoons H_2CO_3(aq); \quad K_e\left(37°C\right) \cong 1.0 \quad \text{(Roine 2002)} \tag{10.26}$$

and because $[H_2CO_3]/[CO_2] = K_e \cong 1.0$, then $[H_2CO_3] \cong [CO_2] \cong 1.23 \times 10^{-3}$ mol/L. More importantly, H_2CO_3 is a weak acid and helps to regulate the pH of blood when H^+ ions are generated by metabolic processes. In this case,

$$H_2CO_3(aq) \rightleftharpoons H^+ + HCO_3^-; K_e\left(37°C\right) = 5.01 \times 10^{-7}. \quad \text{(Roine 2002)} \tag{10.27}$$

The pH of blood is normally around pH = 7.4 (Blood) ($[H^+] = 3.98 \times 10^{-8}$ mol/L) so

$$\left[HCO_3^-\right] = \frac{\left[H_2CO_3(aq)\right]}{\left[H^+\right]} K_e$$

$$= \frac{\left(1.23 \times 10^{-3}\right)}{\left(3.98 \times 10^{-8}\right)}\left(5.01 \times 10^{-7}\right)$$

$$\left[HCO_3^-\right] = 1.54 \times 10^{-2} \text{ mol/L}$$

and the total dissolved CO_2 is

$$\left[CO_2\right]_{\text{Total}} = \left[CO_2(aq)\right] + \left[H_2CO_3\right] + \left[HCO_3^-\right]$$

$$\left[CO_2\right]_{\text{Total}} = 1.23 \times 10^{-3} + 1.23 \times 10^{-3} + 1.54 \times 10^{-2}$$

$$\left[CO_2\right]_{\text{Total}} = 1.79 \times 10^{-2} \text{ mol/L.}$$

so more than 80% of the CO_2 is in the bicarbonate ion and only about 8% as dissolved CO_2 molecules. However, it is only the CO_2 molecules that are of interest because these are the only species in the problem that can diffuse directly through the membrane wall without the help of a transport protein.

10.7.3 The Model

Because spherical coordinates can be neglected in this case (again the *thin-wall solution* is valid), the flux of CO_2 through the cell wall is simply

$$J(CO_2) = \frac{D(C_0 - C_L)}{L}.$$

The concentrations C_0 and C_L at the outside and inside of the cell walls need to be determined. This involves the *partition coefficient*,* K, is defined as

$$K = \frac{[\text{Solute}]_{\text{solvent 2}}}{[\text{Solute}]_{\text{solvent 1}}} \tag{10.28}$$

where solvent 1 in this case could be water—blood—and solvent 2 would be the hydrophobic cell wall, so in this case,

$$K = \frac{[\text{Solute}]_{\text{solvent 2}}}{[\text{Solute}]_{\text{solvent 1}}} = \frac{[CO_2]_{\text{wall}}}{[CO_2]_{H_2O}}.$$

The partition coefficient is used in a variety of applications from medicinal chemistry or drug design to determine how rapidly drugs will penetrate the cell walls in the human body to environmental science and modeling the behavior of organic pollutants. A relatively standard partition coefficient for a given chemical compound is the ratio of its solubility in 1-octanol, $C_8H_{17}OH$, to that in water. Water and octanol are mutually insoluble. Frequently, this partition coefficient is denoted by P or K_{OW} and either log(P) or $\log(K_{OW})$ is reported. Tables of partition coefficients are available for a number of organic compounds (Haynes 2013). For benzene as an example, $\log(K_{OW}) = 2.13$ or benzene is 150 times more soluble in octanol than in water. This is not surprising because the nonpolar benzene would be expected to be more soluble in the nonpolar octanol than polar water. A value for CO_2 was not found in this compilation but, since it is a nonpolar molecule, its K_{OW} would be expected to be high, perhaps on the order of $\log_{10}K_{OW} = 4.0$.

A single value for CO_2 was found $K = 1.6$ (Endewar et al. 2014). Differences in literature values are not uncommon but the variation in measured values is a concern because it makes quantitative modeling of processes, be they drug-related, environmental, or something else, more difficult and imprecise (Renner 2002). Further complicating the issue is that cholesterol molecules in the lipid membrane decrease the solubility in the membrane considerably (Endewar et al. 2014). For the purposes of this model, assume that $K = 4.2$ simply for the purposes of calculation. What to choose for the diffusion coefficient because such data through cell walls are not readily available? Without a better choice, the same D as for diffusion through the rubber wall is used because both walls involve interstitial diffusion among long chain hydrocarbon molecules; that is, $D = 1.64 \times 10^{-6}\ \text{cm}^2/\text{s}$. Now the cell wall has a total thickness of about 6 nm, but only about one-half of that is the hydrophobic part, $L \cong 3$ nm, through which the carbon dioxide must diffuse.

Therefore, the flux density through the wall if there is a 25% difference in concentration is

$$J(CO_2) = \frac{DK\left([CO_2]_{H_2O,0} - [CO_2]_{H_2O,L}\right)}{L}$$

$$= \frac{(1.64\times10^{-6}\ \text{cm}^2/\text{s})(4.2)}{3\times10^{-7}\ \text{cm}}(1.25-1)(1.23\times10^{-5}\ \text{mol/cm}^3)$$

$$J(CO_2) = 7.06\times10^{-5}\ \text{mol/cm}^2\ \text{s}.$$

Approximate the cell shape with a cylindrical cell of $d = 7\ \mu m$ diameter and an average height of $h = 1\ \mu m$ (Figure 10.15). The area of the cell membrane is $2\pi r^2 + 2\pi rh = 9.90 \times 10^{-7}\ \text{cm}^2$ and the cell volume is $\pi r^2 h = 38.5 \times 10^{-12}\ \text{cm}^3$. So the total number of moles of CO_2 that need to be added to the cell is $V(0.25)[CO_2] = (38.5 \times 10^{-12}\ \text{cm}^3)(0.25)(1.23 \times 10^{-5}\ \text{mol/cm}^3) = 1.18 \times 10^{-16}$ mol so the time to equilibrate is

* The term "partition coefficient" is now considered obsolete and "partition constant," "partition ratio," or "distribution ratio" are now the more preferred terms (Partition Coefficient).

$$t = \frac{n}{JA} = \frac{V(0.25)[CO_2(a)]}{JA}$$

$$= \frac{1.18\times10^{-16}\ \text{mol}}{(7.06\times10^{-5}\ \text{mol/cm}^2\,\text{s})(9.90\times10^{-7}\ \text{cm}^2)}$$

$$t = 1.7\times10^{-6}\ \text{s!}$$

This is considerably faster than a uniform concentration can be established within the cell with a liquid diffusion coefficient $D \sim 10^{-5}$ cm²/s estimated by $Dt \cong L^2$. $L = h = 1 \times 10^{-4}$ cm in this case, so $t \cong (1 \times 10^{-4}\ \text{cm})^2/10^{-5}\ \text{cm}^2/\text{s} \cong 1 \times 10^{-3}$ s: three orders of magnitude slower! So equilibration of CO_2 will be controlled by liquid diffusion outside and inside the cell rather than diffusion through the cell membrane and the exact solution to this is a finite boundary problem solution to Fick's second law.

10.8 CVD OF Si FROM SiHCL$_3$: DIFFUSION CONTROL

10.8.1 Introduction

This again will be an approximate model because one gaseous species is being consumed and two others are being generated which produces a net flow and flux in the gas phase in addition to the motion of the interface due to the deposition of silicon. All of these affect the concentration profile and steady-state diffusion. However, in this particular case, these are weak second-order effects and certainly do not affect the quantitative results more than the uncertainty in the other parameters of the process. Therefore, these second-order effects are neglected.

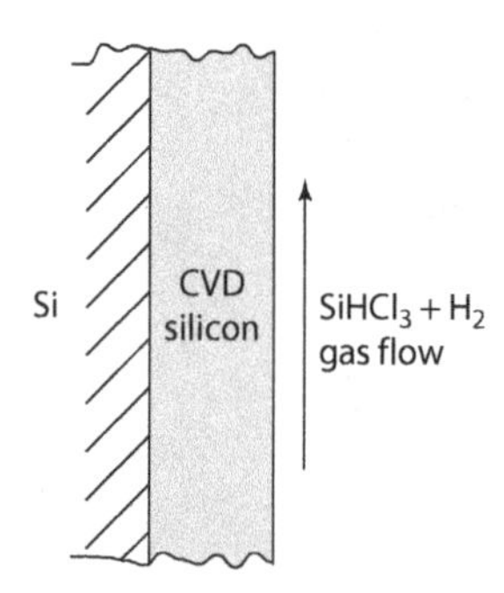

FIGURE 10.16 Model for the chemical vapor deposition of silicon on a silicon single crystal in a flow reactor with trichlorosilane, $SiHCl_3$, and hydrogen as reactants.

10.8.2 Diffusion versus Reaction Control

Earlier, the CVD of silicon from $SiHCl_3$, trichlorosilane, was investigated assuming that the rate was determined entirely by surface reaction and that gas diffusion was infinitely fast. Now, the opposite approach is taken: the surface reaction is assumed to be infinitely fast and diffusion is the rate-controlling step. This implies that the gas species are in equilibrium at the solid surface. The schematic in Figure 10.16 shows the deposition of CVD silicon in a flowing gas stream of H_2 carrier gas and $SiHCl_3$ vapor in a reactor. Gas flowing over a solid surface has a velocity of zero at the surface and increases out into the flowing gas stream. The thickness of the region in which the gas speed changes from zero to the stream velocity is called the gas *boundary layer, the thickness of which,* δ, depends on the geometry, gas properties, and the velocity of flow. With equilibrium at the solid surface and with the gas boundary layer, there will be a concentration gradient from the flowing gas stream to the solid surface that might look something like that depicted in Figure 10.17 where the gas flow is now vertical. Rather than try to calculate the exact gas concentration profile—C(x)—the assumption is made that the gas concentration is linear through the boundary layer and the steady-state diffusion problem to be solved is simplified to that shown by the dotted line in Figure 10.17.

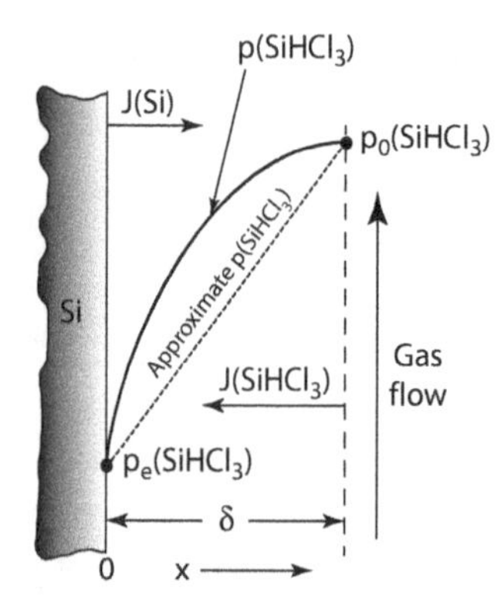

FIGURE 10.17 Schematic showing the effect of gas flow over a flat surface on the concentration of p($SiHCl_3$) as a function of distance away from the surface due to the variation of fluid velocity in the boundary layer at the surface, δ. However, the usual approximation made is that the pressure or concentration is linear over the thickness of the boundary layer as shown by the dashed line.

Again, $\partial C/\partial t = D(\partial^2 C/\partial x^2) = 0$ as usual, and the concentration as a function of distance (positive x as shown in Figure 10.17) is $C(x) = C_e + (C_0 - C_e)(x/\delta)$. So the diffusion flux of $SiHCl_3$ to the surface is

$$J(SiHCl_3) = -D\frac{dC}{dx}$$

$$J(SiHCl_3) = -\frac{D}{RT\delta}\left(p_0(SiHCl_3) - p_e(SiHCl_3)\right) \quad (10.29)$$

and the flux of silicon is just opposite this so

$$J(Si) = \frac{\rho}{M}\frac{dx}{dt} = \frac{D}{RT\delta}(p_0 - p_e) \quad (10.30)$$

combining Equations 10.29 and 10.30 gives for the thickness x of the deposited layer

$$x = \frac{MD}{\rho RT\delta}(p_0 - p_e)t \quad (10.31)$$

where x is the thickness of the silicon deposited as a function of time, t, again a linear rate. Note that this is *identical* to the expression that was obtained for the CVD of silicon by a surface reaction in Chapter 5,

$$x = \frac{Mk}{\rho RT}(p_0 - p_e)t$$

but now the k for the surface reaction rate has been replaced by D/δ.

For a surface reaction being rate-controlling, diffusion was considered to be infinitely fast and, as a result, there were no concentration gradients in the gas phase of either the products or reactants: for diffusion the opposite is true. The surface reaction is assumed to be infinitely fast so surface concentrations are in equilibrium and concentration gradients exist in the gas that drive diffusion in the gas and control the deposition rate.

10.8.3 Thermodynamics

For a diffusion-controlled reaction, at some position the reactants and products are at equilibrium. Therefore, as a starting point for modeling the kinetics of a diffusion-controlled reaction, the equilibrium thermodynamics must be established destroying the myth that kinetic results cannot be obtained from thermodynamics. In this case, the thermodynamics are critical. In Chapter 5, for reaction control, the following conditions were used for the reaction:

$$SiHCl_3(g) + H_2(g) = Si(s) + 3HCl(g) \quad (10.32)$$

and at 1100°C: $\Delta G° = 43{,}280$ J/mol, $p_0(SiHCl_3) = 0.1$ atm, $p_0(H_2) = 1$ atm, and $p_0(HCl) = 0.03$ atm. From these data the equilibrium partial pressure of $SiHCl_3$ is calculated to be $p_e(SiHCl_3) = 1.19 \times 10^{-3}$ atm. This means that about 0.0998 atm of $SiHCl_3$ is consumed and the same amount of hydrogen is generated, while about 3 times this amount of HCl is produced, which is about 10 times the original amount in the gas phase. Therefore, some additional thermodynamic calculations are in order.

Equating the moles of Cl at equilibrium and before the reaction gives

$$3p(SiHCl_3) + p(HCl) = 3p_0(SiHCl_3) + p_0(HCl)$$

where:

- p stands for the equilibrium pressures
- p_0 are the initial pressures.

Of course, pressures rather than concentrations can be used because only ideal gases are being considered and $p \propto$ number of moles. This leads to

$$p(SiHCl_3) = \frac{3p_0(SiHCl_3) + p_0(HCl)}{3} - \frac{p(HCl)}{3}$$

Balancing the moles of H in the same way, the following is obtained:

$$p(SiHCl_3) + p(HCl) + 2p(H_2) = p_0(SiHCl_3) + p_0(HCl) + 2p_0(H_2)$$

and ultimately leads to

$$p(H_2) = p_0(H_2) + \frac{2}{3}p_0(HCl) - \frac{2}{3}p(HCl).$$

Putting these expressions and the initial pressures into the equilibrium constant, $K_e = 2.26 \times 10^{-2}$ and letting p(HCl) = x for simplicity, the following equation needs to be solved for x to calculate the equilibrium partial pressures:

$$\frac{p^3(HCl)}{p(SiHCl_3)p(H_2)} = \frac{x^3}{(0.11-(x/3))(1.02-(2/3)x)} = K_e = 2.26\times10^{-2}.$$

By iteration of x in a spreadsheet program (about 10 iterations), the following equilibrium pressures were obtained: p(HCl) = 0.119094 atm, $p(SiHCl_3)$ = 0.070302 atm, and $p(H_2)$ = 0.940604 atm, which when substituted into the equilibrium constant checks, and the pressures versus distance look similar to those shown in Figure 10.18.

10.8.4 Kinetic Calculation

The parameters used to calculate the rate of deposition are $p_0(H_2)$ = 1 atm, $p_0(SiHCl_3)$ = 0.1 atm, at 1100°C, $p_e(SiHCl_3) = 7.03 \times 10^{-2}$ atm; M(Si) = 28.09 g/mol, ρ(Si) = 2.32 g/cm³ and assume a boundary layer thickness of $\delta \cong 1$ cm.* The diffusion coefficient can be calculated from the approximate kinetic theory of gases, Equation 9.75: (divided by 1.01325 atm/bar to use atmospheres for pressure)

$$D = 1.49\times10^{-3}\left(\frac{1}{M(H_2)} + \frac{1}{M(SiHCl_3)}\right)^{1/2}\frac{T^{3/2}}{d^2p}.$$

The value of $d_{H_2} = 2.827\text{Å}$ is given in the literature (Geankoplis 1972) and that for $SiHCl_3$ can be approximated by $d = 1.18v_b^{1/3}$, where v_b is the molecular volume at the boiling point (Geankoplis 1972). For $CHCl_3$, T_b = 33°C, ρ = 1.331 g/cm³, M = 135.453 g/mol (Haynes 2013). Assuming that the density is about the same at the boiling point, a good assumption because the boiling point is near room temperature, then

$$v_b = \frac{M}{\rho N_A} = \frac{135.453}{(1.331)(6.022\times10^{23})} = 1.68\times10^{-22}\ \text{cm}^3/\text{mol}$$

and $d(SiHCl_3) = 1.18\times(1.68\times10^{-22})^{1/3} = 6.52\times10^{-8}$ cm, so d = (2.827 + 6.52)/2 = 4.67 Å. Therefore, at T = 1373 K, and p = 1 atm so

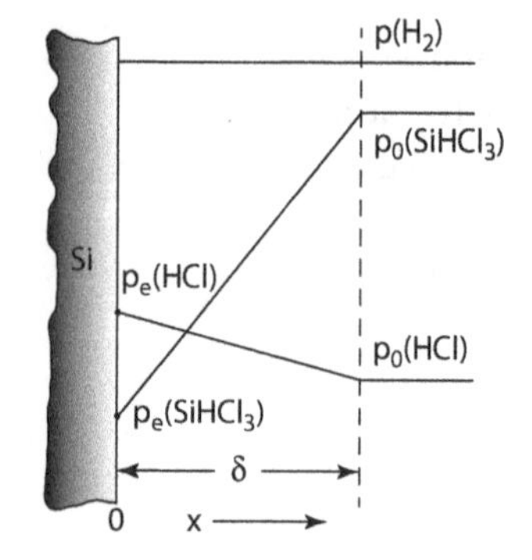

FIGURE 10.18 Schematic of the partial pressures of the reactant and product gases during the deposition of silicon from trichlorosilane controlled by diffusion in the gas. Note that the concentrations at the surface are the equilibrium concentrations for the reaction: $SiHCl_3(g) + H_2(g) = Si(s) + 3HCl(g)$.

* In fluid dynamics, the boundary layer is the region above a surface in a moving fluid stream parallel to the surface over which the gas velocity goes from zero to roughly the stream velocity. It is proportional to the square root of the fluid viscosity divided by the stream velocity. The flow conditions in a gas can easily be adjusted to give a boundary layer of this magnitude.

$$D = 1.49\times10^{-3}\left(\frac{1}{2.00}+\frac{1}{134.45}\right)^{1/2}\frac{1373^{3/2}}{(4.67)^2(1)} = 2.48\,\text{cm}^2/\text{s}.$$

Therefore, the rate of silicon deposition, dx/dt, is

$$\frac{dx}{dt} = \frac{MD}{\rho RT\delta}(p_0 - p_e) = \frac{(28.09)(2.48)}{(2.32)(82.05)(1373)(1)}(0.1 - 7.03\times10^{-2}) = 7.91\times10^{-6}\ \text{cm/s}.$$

This is about a factor of 5 faster than surface reaction rate-controlled deposition calculated in Chapter 5 under almost identical conditions. Therefore, because the reaction-controlled deposition is slower, but not by much, diffusion is the rate-controlling step under these conditions. However, because their rates are sufficiently close, it is worth examining what happens when these rates are nearly the same.

10.9 CVD OF SILICON CONSIDERING BOTH DIFFUSION AND SURFACE REACTION

If the rates of deposition of silicon by CVD of $SiHCl_3$ in hydrogen by diffusion and surface reaction are roughly equal to each other, then the pressure of the $SiHCl_3$ at the surface would be someplace between the two extremes of p_0—required for diffusion being infinitely fast—and p_e—required for surface reaction being infinitely fast. Therefore, the various pressures as a function of distance would look something like that shown in Figure 10.19. The flux equations are now $J_D(Si) = (D/RT\delta)(p_0 - p_S)$ for diffusion and $J_R(Si) = (k/RT)(p_S - p_e)$ for surface reaction, where p_S is the pressure of the $SiHCl_3$ at the surface that is different from both the equilibrium and starting pressures of $SiHCl_3$—actually, in between. A very important concept is that the diffusion flux must equal the reaction flux, $J_D = J_R = J$, *because they are in series.* Now the flux equations can be rearranged in the following way:

$$J_D\frac{RT\delta}{D} = p_0 - p_S \quad\text{and}\quad J_R\frac{RT}{k} = p_S - p_e. \tag{10.33}$$

Adding these two expressions gives

$$J_D\frac{RT\delta}{D} + J_R\frac{RT}{k} = p_0 - p_S + p_S - p_e$$

$$JRT\left(\frac{\delta}{D}+\frac{1}{k}\right) = p_0 - p_e$$

and this last expression can be written as

$$J = \frac{h}{RT}(p_0 - p_e) \tag{10.34}$$

just like the separate expressions for the surface reaction flux and diffusion flux when one is much faster than the other. Here, h is the *mass transfer coefficient* and includes both the effects of surface reaction and diffusion and is given by

$$\frac{1}{h} = \frac{\delta}{D} + \frac{1}{k}. \tag{10.35}$$

So when $k \gg D/\delta$, $h \cong D/\delta$ and when $D/\delta \gg k$, $h = k$, which in words says: "when the surface reaction rate is much greater that the diffusion rate, diffusion controls, and vice versa; that is, the slower of the two series steps controls the rate," which has been seen before.

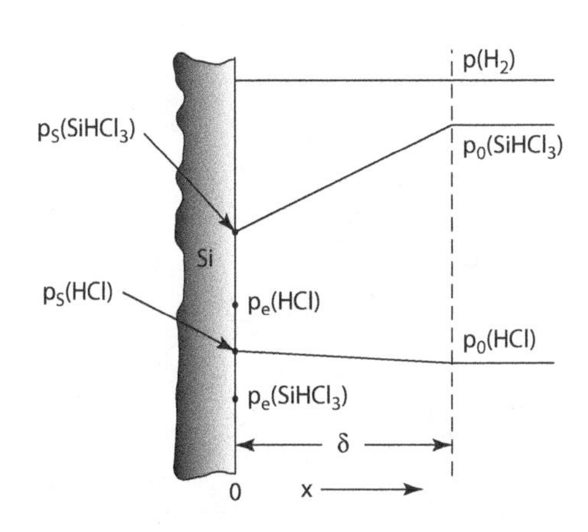

FIGURE 10.19 Gas pressures versus distance from the surface when both diffusion and a surface reaction occur at about the same rate. In this case the surface pressures, p_S, are not the equilibrium pressures, p_e, and, in fact, it is the difference between the surface and equilibrium pressure that drives the surface reaction. The pressure gradients in the gas, of course, cause diffusion.

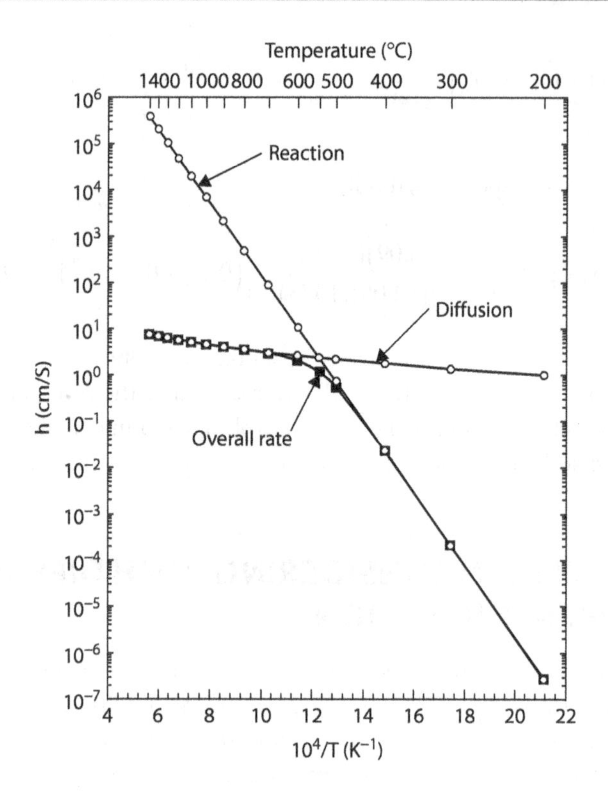

FIGURE 10.20 Plots of the reaction rate constant, k, the diffusion mass transfer, D/δ, and the overall mass transfer coefficient, h, as a function of temperature from actual data for the deposition of silicon from trichlorosilane. The slower process controls h with surface reaction at low temperatures and diffusion at high temperatures as the "overall rate" curve shows.

Figure 10.20 shows the behavior of the mass transfer coefficient—black squares—along with the individual separate behavior of k and D/δ and shows that, at high temperatures, diffusion controls and that, at low temperatures, the surface reaction controls: whichever is the smaller of the two. This is usually the case because the surface reaction is exponentially temperature dependent while the diffusion coefficient only depends on $T^{3/2}$. This is an analytical demonstration that the slower of two series reaction steps indeed controls the overall reaction rate.

10.10 EVAPORATION OF A WATER DROP

10.10.1 The Model

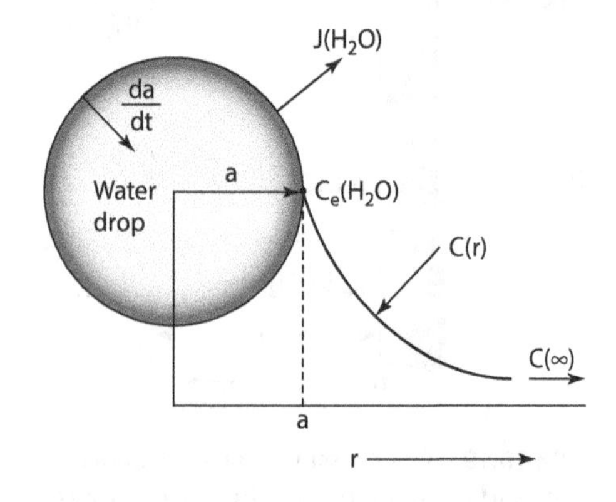

FIGURE 10.21 Schematic of a water drop of radius a evaporating into the atmosphere showing the equilibrium water vapor concentration in the atmosphere at the water–gas interface, $C_e(H_2O)$, and that far from the surface, C(∞), determined by the temperature and relative humidity.

Consider a water drop of radius a in Figure 10.21, evaporating by H_2O diffusion through the gas phase, which is assumed to be air. The equilibrium or saturation pressure of H_2O in the gas is $p_e(H_2O)$ at the water–air interface, and as the radial distance from the center of the sphere, r, goes to infinity, there is a relative humidity in the atmosphere that determines the water vapor pressure, at $r = \infty$, $p_\infty(H_2O)$.

This steady-state* concentration profile in spherical coordinates is determined from

$$\frac{\partial C}{\partial t} = \frac{D}{r^2}\frac{\partial}{\partial r}\left(r^2 \frac{\partial C}{\partial r}\right) = 0$$

* Again, this is assumed to be a *quasi-steady-state* problem in that the concentration profile, C(r), actually does change with time due to gas fluid flow and motion of the sphere–gas interface, but it is assumed that the continuous transient changes are fast compared to the rate of evaporation and the change in sphere radius with time, da/dt, and they can be neglected as is the initial transient necessary to establish the concentration profile.

or

$$\frac{d}{dr}\left(r^2\frac{dC}{dr}\right)=0$$

which gives on the first integration $dC/dr = A/r^2$ and integrating a second time gives the general solution, $C(r) = -(A/r)+B$. When $r = \infty$, $B = C(\infty) = p_\infty(H_2O)/RT$ and when $r = a$, $C(a) = p_e(H_2O)/RT$. Therefore, $C(a) = -(A/a)+C(\infty)$ and $A = -a(C_a - C_\infty)$ so the concentration as a function of r becomes

$$C(r)=\left[C(a)-C(\infty)\right]\frac{a}{r}+C(\infty) \tag{10.36}$$

which checks with the boundary conditions. So the diffusion flux from the drop surface into the atmosphere is

$$J(a)=-D\left(\frac{dC}{dr}\right)_{r=a}=-D\left(-\left[C(a)-C(\infty)\right]\frac{a}{r^2}\right)_{r=a}$$
$$J(a)=\frac{D\left[C(a)-C(\infty)\right]}{a} \tag{10.37}$$

and this must be equal to $(1/A)(dn/dt)$, where A is the drop area $= 4\pi a^2$ and n is the number of moles in the drop which is

$$n=\frac{\rho}{M}V=\frac{\rho}{M}\frac{4}{3}\pi a^3$$

so

$$J(a)=-\frac{1}{A}\frac{dn}{dt}=-\frac{1}{4\pi a^2}\left(\frac{\rho}{M}4\pi a^2\frac{da}{dt}\right)$$
$$J(a)=-\frac{\rho}{M}\frac{da}{dt} \tag{10.38}$$

as usual. The minus sign occurs because there is a net flux out of the sphere so that the sphere is shrinking. Therefore, equating Equations 10.37 and 10.38 at the sphere surface

$$\frac{\rho}{M}\frac{da}{dt}=-D\left[C(a)-C(\infty)\right]\frac{1}{a}\quad\text{or}\quad a\frac{da}{dt}=-\frac{M}{\rho}D\left[C(a)-C(\infty)\right] \tag{10.39}$$

which is integrated to give

$$a^2=a_0^2-\frac{2MD}{\rho RT}\left(p_e(H_2O)-p_\infty(H_2O)\right)t \tag{10.40}$$

which could also be written as

$$\frac{a^2}{a_0^2}=1-\frac{2M}{\rho RT}\left(p_e(H_2O)-p_\infty(H_2O)\right)\frac{Dt}{a_0^2} \tag{10.41}$$

and note the dependence of the particle size on the dimensionless parameter, Dt/a_0^2.

10.10.2 The Rate of Evaporation

For the reaction H_2O (l) = H_2O (g), the equilibrium constant at 30°C is 4.23×10^{-2} (Roine 2002), which means that $p_e(H_2O) = 4.23 \times 10^{-2}$ atm (32.1 torr or mm Hg). The question is, "How long will it take a 2 mm diameter drop to evaporate into air with a relative humidity of 20%, a rather dry climate." First, calculate the gas diffusion coefficient D_{H_2O-air} from*

$$D_{H_2O-air} = 1.49 \times 10^{-3}\left(\frac{1}{M(H_2O)} + \frac{1}{M(air)}\right)^{1/2} \frac{T^{3/2}}{d^2_{H_2O-air}p}.$$

From the literature $d_{H_2O} = 2.641\text{Å}$ and $d_{air} = 3.711$ Å (Cussler 1997) so,

$$d_{H_2O-air} = \frac{d_{H_2O} + d_{air}}{2} = \frac{2.641 + 3.711}{2} = 3.18\ \text{Å}.$$

Therefore, at p = 1 atm,

$$D_{H_2O-air} = 1.491 \times 10^{-3}\left(\frac{1}{18.0} + \frac{1}{28.8}\right)^{1/2} \frac{(303)^{3/2}}{(3.18)^2 1}$$

$$D_{H_2O-air} = 0.234\ \text{cm}^2/\text{s}$$

compared to a measured value of about 0.266 cm²/s (Cussler 1997). Therefore, the time for the sphere to evaporate, $a \to 0$, is given by

$$t = \frac{a_0^2 \rho RT}{2MD} \frac{1}{0.8p_e(H_2O)} = \frac{(0.1)^2(1)(82.05)(303)}{2(18)(0.234)(0.8)(4.23 \times 10^{-2})} = 872\ \text{s} = 14.54\ \text{min}$$

If the relative humidity were 60%, then it would take about twice as long to evaporate or about 30 min.

10.11 DISSOLUTION OF NaCl

10.11.1 The Model

Essentially, the same spherical geometry model applies to other kinetic problems in materials science and engineering. Here, the dissolution of a sphere of pure sodium chloride in water is considered to demonstrate three important points: (1) the same model applies to solid–liquid diffusion-controlled processes; (2) change from molal to molar concentrations; and (3) there is a *discontinuity* of the concentration at the two-phase interface.

The same geometry holds as that in Figure 10.21 only in this case; there will be an equilibrium *aqueous* concentration, C_e, of Na^+ and Cl^- ions at the NaCl–H_2O interface, as shown in Figure 10.22. The dissolution reaction can be written as

$$NaCl_s = Na^+_{aq} + Cl^-_{aq}.$$

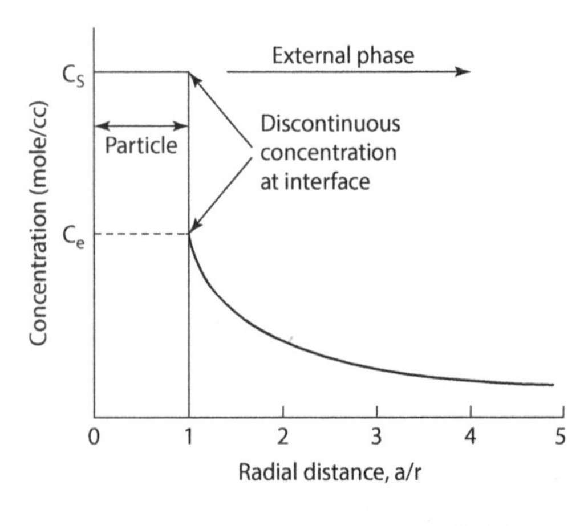

FIGURE 10.22 Concentration versus radial distance from the center of a shrinking sphere to far out into the external phase emphasizing that, regardless of the system and phases present, there is a concentration *discontinuity* at the two-phase interface. At the interface, in the external phase, the concentration is the *equilibrium concentration* between the two phases at the given temperature.

From thermodynamic data, the equilibrium constant for NaCl in H_2O at 30°C is (Roine 2002)

$$\frac{[Na^+][Cl^-]}{a_{NaCl}} = K_e = 39.94. \tag{10.42}$$

The standard state of aqueous species in the database used for this calculation—sodium and chlorine ions—is one *molal*, that is, *1 mol/1000 g*

* See Chapter 9, Equation 9.77 and Section 9.6.3.2.

solution.[*] Therefore, the equilibrium concentrations are molal and dividing by 1000 gives the equilibrium concentration in terms of mol/g. The $[Na^+] = [Cl^-]$ of course if pure NaCl, $a_{NaCl} = 1.0$, is dissolved into pure H_2O. Therefore, the *molal solubility* of NaCl in pure water at 30°C is $[NaCl] = [Na^+] = [Cl^-] = (a_{NaCl}K_e)^{1/2} = \sqrt{39.94} = 6.32$ mol/1000 g of solution, which is the equilibrium concentration, C_e, shown at the interface between the NaCl sphere and the solution in Figure 10.22.

10.11.2 Calculation of Molar Concentration

In general, it is more convenient and consistent with other concentrations in this book to work with mol/cm^3. There are two ways to approach converting $C_e = 6.32$ mol/1000 g of solution to mol/cm^3. The first is to look up the density of NaCl–H_2O solutions in the literature. Tables of this exist but only go up to about 5 molal (Haynes 2013). Nevertheless, extrapolating these data to 6.32 mol/1000 g gives a specific volume of 0.83 cm^3/g so 1000 g of solution would be about 830 cm^3. So the concentration in mol/cm^3 would be

$$[NaCl] = \frac{6.32 \text{ mol}/1000\text{g}}{830 \text{ cm}^3/1000\text{g}} = 7.61 \times 10^{-3} \text{ mol/cm}^3 \tag{10.43}$$

Another approach would assume an ideal solution, so the volume of the of the 1000 g would be

$$V = n_{NaCl}\bar{V}_{NaCl} + n_{H_2O}\bar{V}_{H_2O}$$

where:

n_i are the number of moles

$\bar{V}_i$ are the molar volumes as a function of concentration respectively of NaCl and H_2O, which may not exist in the literature.

However, the solution can be assumed to be ideal and that the molar volumes are those of the pure liquids at 30°C. Now M(NaCl) = 58.443 g/mol so in the saturated 6.32 molal solution, there are 6.32 mol × 58.433 g/mol = 369.3 g of NaCl leaving 630.7 g of H_2O. So $n_{H_2O} = 630.7/18 = 35.04$ mol. The molar volume of H_2O is simply

$$\bar{V}_{H_2O} = \frac{M \text{ g/mol}}{\rho \text{ g/cm}^3} = \frac{18}{1} = 18 \text{ cm}^3/\text{mol}.$$

The density of molten NaCl is given in the literature as (Brandes and Brook 1992)

$$\rho_{NaCl}(\text{liquid}) = 1.911 - 0.543 \times 10^{-3} T\ (°C)$$

which is really only good above the melting point of NaCl. However, it is the only data readily available and so it is extrapolated to 30°C giving

$$\rho_{NaCl}(\text{liquid}) = 1.911 - 0.543 \times 10^{-3}(30) = 1.8947 \text{ g/cm}^3.$$

Note that the density of solid NaCl at 30°C is 2.165 g/cm^3 and the volume of the liquid is about 14% higher than that of the solid as discussed in Chapter 7. The molar volume of NaCl is therefore

$$\bar{V}_{NaCl} = \frac{M \text{ g/mol}}{\rho \text{ g/cm}^3} = \frac{58.443}{1.9847} = 29.45 \text{ cm}^3/\text{mol}$$

[*] The *molar concentration*, M, is when the Units (M) = mol/L of *solution*. Carrying SI units to the extreme, the preferred are Units (M) = mol/dm^3 of solution! Because the interest is usually mol/cm^3 of solution, the molar concentration is the most convenient. However, in many thermodynamic databases, the concentrations of solution species are *molal* concentrations, m, where Units (m) = mol/kg *solution*. For small concentrations, the difference between them is easily within the uncertainty of the thermodynamic data and can often be ignored.

so the ideal molar volume is

$$V_{ideal} = n_{NaCl}\bar{V}_{NaCl} + n_{H_2O}\bar{V}_{H_2O}$$

$$V_{ideal} = 6.32 \times 29.45 + 35.04 \times 18 = 816.8\ cm^3/1000\,g$$

and

$$[NaCl] = \frac{6.32\ mol/1000\ g}{816.8\ cm^3/1000\ g} = 7.73 \times 10^{-3}\ mol/cm^3 \tag{10.44}$$

which is amazingly close to the experimental value in Equation 10.43. It is hard to predict that an NaCl–H_2O solution behaves close to an ideal solution! The point is, if the necessary data cannot be found in the literature, assume an ideal solution. In this case, the average of Equations 10.43 and 10.44 is about $C_e \cong 7.67 \times 10^{-3}\ mol/cm^3$.

The concentration of NaCl in a NaCl crystal is of course $C = \rho/M = 2.165\ g/cm^3/58.44\ g/mol = 3.70 \times 10^{-2}\ mol/cm^3$, also shown in Figure 10.22 (Haynes 2013). Note the *discontinuity* in the NaCl concentration at the NaCl–solution interface: the thermodynamic activity of sodium chloride may be the same in both phases at the interface, but the concentrations are not! If the solution is initially pure water, $C_\infty = 0$, so the equation for the particle size as a function of time becomes essentially the same as Equation 10.41:

$$\frac{a^2}{a_0^2} = 1 - \frac{2M}{\rho} C_e(NaCl)\frac{Dt}{a_0^2}.$$

10.11.3 Ambipolar Diffusion

Data in the literature give for the diffusion coefficients of the two ions in water $D_{Na^+} = 1.334 \times 10^{-5}\ cm^2/s$ and $D_{Cl^-} = 2.032 \times 10^{-5}\ cm^2/s$ (Haynes 2013). Now the two ions cannot diffuse independently because they are charged and must diffuse together to prevent a charge imbalance and resulting electric field so $J(NaCl) = J(Na^+) = J(Cl^-)$. This is so-called *ambipolar* diffusion and more will be said about it in Chapter 13. For now, fluxes of the sodium and chlorine ions are given by

$$J(Na^+) = -D_{Na^+}\frac{C_e}{a}$$

$$J(Cl^-) = -D_{Cl^-}\frac{C_e}{a}.$$

These two equations can be rearranged and added together to give

$$\frac{J(Na^+)}{D_{Na^+}} + \frac{J(Cl^-)}{D_{Cl^-}} = J(NaCl)\left(\frac{1}{D_{Na^+}} + \frac{1}{D_{Cl^-}}\right) = -2\frac{C_e}{a} \tag{10.45}$$

or

$$J(NaCl) = -\frac{2C_e}{a}\frac{1}{\left((1/D_{Na^+}) + (1/D_{Cl^-})\right)}$$

$$J(NaCl) = -\left(\frac{D_{Na^+}D_{Cl^-}}{D_{Na^+} + D_{Cl^-}}\right)\frac{2C_e}{a}$$

$$J(NaCl) = -D_{eff}\frac{C_e}{a}$$

where

$$D_{eff} = \left(\frac{2D_{Na^+}D_{Cl^-}}{D_{Na^+} + D_{Cl^-}} \right) \quad (10.46)$$

or

$$D_{eff} = 2\frac{1.334 \times 2.032}{1.334 + 2.032} \times 10^{-5} = 1.61 \times 10^{-5}\ \text{cm}^2/\text{s}.$$

Clearly, from Equation 10.46, if $D_{Na^+} = D_{Cl^-}$, then $D_{eff} = D_{Na^+} = D_{Cl^-}$, as would be expected. On the other hand, if $D_{Na^+} \ll D_{Cl^-}$, then $D_{eff} \cong 2D_{Na^+}$, and vice versa if $D_{Na^+} \gg D_{Cl^-}$. Neither of these situations exist in this case. However, these last two results show that the faster moving ion essentially "pulls" the slower moving ion giving an effective diffusion coefficient somewhat greater than the slower diffusing ion. In general, $z_c e J(\text{cation}) = z_a e J(\text{anion})$, where $z_c e$ is the charge on the cation and $z_a e$ is the charge on the anion; then $D_{eff} = ((z_c + |z_a|)D_c D_a)/(z_c D_c + |z_a| D_a)$ as obtained by applying Equation 10.45.

The time to dissolve a 0.5 mm diameter salt sphere in water at roughly 30°C is given by Equation 10.41 (again $a \to 0$):

$$t = \frac{a_0^2 \rho}{2MD_{eff}} \frac{1}{C_e} = \frac{(0.025)^2 (2.165)}{2(58.44)(1.61 \times 10^{-5})(7.67 \times 10^{-3})}$$

$$t = 93.8\ \text{s} \cong 1.56\ \text{min}$$

where the density of solid NaCl is ρ = 2.165 g/cm³ and M = 58.44 g/mol (Haynes 2013). Next time you put salt into your beer to augment the head, check to see how long the salt particles are effective. They, of course, act as nuclei for the formation of CO_2 bubbles until they completely dissolve. However, do not shake the glass because that will generate convection currents and decrease the distance that the salt must diffuse and increase the dissolution rate.

10.12 DISSOLUTION OF SPHEROIDIZED CEMENTITE IN AUSTENITE

10.12.1 Introduction

This is another steady-state diffusion problem similar to the previous two in spherical coordinates. In this case, an important solid-solid dissolution process in steel processing is modeled. The photomicrographs that are used come from a publication over 100 years old (Sauveur 1912), long before "materials science and engineering" was even dreamed of, demonstrating how the metallurgy of steel was a major foundation for the field. Furthermore, what would a book about materials kinetics be if it did not include models of steel heat treatment?

Figure 10.23 shows the microstructure of a 0.77 w/o carbon steel, the eutectoid composition, that has been cooled through the eutectoid temperature. As a result, it has a 100% *pearlitic* structure consisting of alternating plates of ferrite, light phase, and cementite, Fe_3C, the dark phase. This typical eutectic and eutectoid microstructure is produced by limited spatial diffusion to produce the two phases each of a different composition than the parent phase. In this steel, this microstructure formed on cooling gamma iron, austenite, through the eutectoid temperature of 727°C, as shown in the Fe-C diagram in Figure 7.4. However, the plate-like structure of the two phases is not thermodynamically stable and, if heat treated at some temperature slightly below the eutectoid temperature, surface tension will cause the Fe_3C platelets to form spheres, or *spheroidize*, which makes the steel easier to machine. Figure 10.24 shows

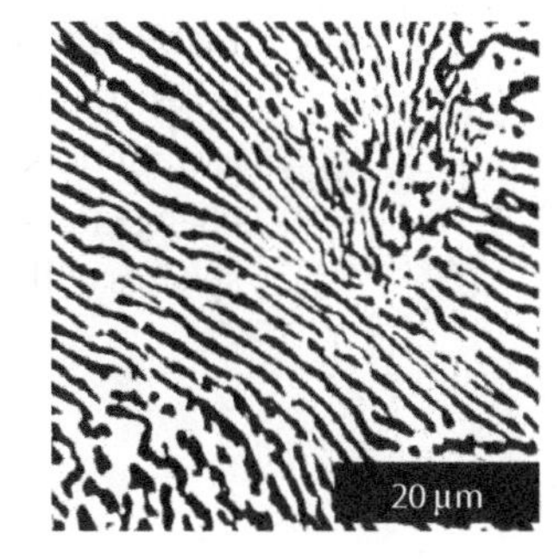

FIGURE 10.23 Eutectoid microstructure of pearlite, Fe_3C and alpha Fe, for a eutectoid steel, 0.77 w/o carbon. (Sauveur 1912). In the public domain because of date of publication. (Contrast and sharpness enhanced in Photoshop®.)

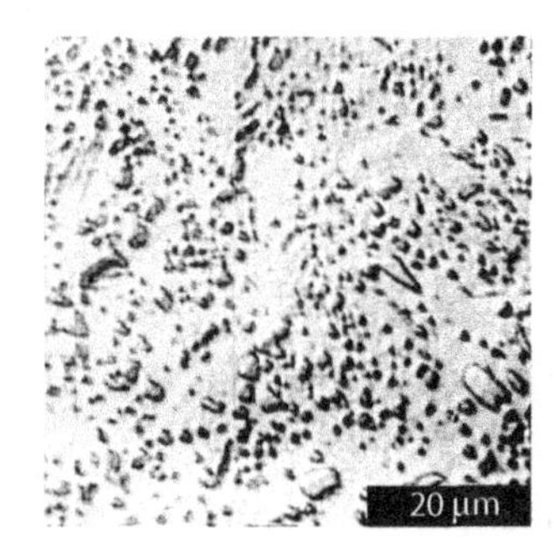

FIGURE 10.24 Spheroidized eutectoid steel. (Sauveur 1912). In the public domain because of date of publication. (Contrast and sharpness enhanced in Photoshop®.)

a spheroidized steel microstructure. The diffusion problem of interest here is: "How long does it take for the spheres of Fe_3C to dissolve if the spheroidized steel is heated into the austenite range, say to 900°C?"

10.12.2 The Model

As can be seen from Figure 10.24, the Fe_3C particles are really not exactly spheres and clearly not all the same size. However, it is assumed that they are close enough to spherical shapes and *monodisperse*—all the same size. Figure 10.24 also shows that the majority of the particles are about 2 μm in diameter with an average initial radius, a_0, of about 1 μm. Again, a steady-state solution in spherical coordinates is used for the dissolution of these particles. Actually, it is another *quasi-steady-state* solution in that C(r) does change with time because of the moving particle–austenite boundary, but it is assumed that this change is much faster than the dissolution rate of the spherical particles so that C(r) is constant. The equivalent to Equation 10.41 for this problem is

$$\frac{a^2}{a_0^2} = 1 - \frac{2M_{Fe_3C}}{\rho_{Fe_3C}}\left(C_e - C_\infty\right)\frac{Dt}{a_0^2}. \tag{10.47}$$

Referring to Figure 10.21, C_e is the concentration of carbon in the austenite at the Fe_3C–austenite interface and C_∞ is the concentration of carbon in the austenite far from the surface of the spherical Fe_3C particle. C_e is easy to determine from the phase diagram in Figure 7.4: at 900°C, the austenite in equilibrium with Fe_3C has about 1.25 w/o carbon. Because the amount of carbon in solid solution is small, for all practical purposes, the total mass per cm^3 is that of Fe and so $\rho \cong \rho(Fe) = 7.86$ g/cm^3. Therefore,

$$C_e(C) = (C\ \text{fraction})\frac{\rho(Fe)}{M(C)} = \left(1.25\times10^{-2}\right)\frac{7.86}{12.01}$$

$$C_e(C) = 8.2\times10^{-3}\ \text{mol/cm}^3.$$

The concentration of carbon in Fe_3C is the same as the number of moles of Fe_3C/cm^3 because there is 1 mol of carbon in each mole of Fe_3C, or (Haynes 2013)

$$C_S = C_{Fe_3C}(C) = \frac{\rho_{Fe_3C}}{M_{Fe_3C}} = \frac{7.694}{179.55} = 4.29\times10^{-2}\ \text{mol/cm}^3.$$

Now the question is, "What should C_∞ be?" Here again, this concentration will change with time because the initial ferrite—and austenite—has essentially no carbon in it. Yet the steel has the eutectoid composition, and when all of the Fe_3C has dissolved, the carbon content of the austenite will be the eutectoid composition or 0.77 w/o carbon. Considering the changing C_∞ concentration with time makes this a more complicated problem—but still solvable. However, the interest here is whether it takes 10 min or 10 h to dissolve the cementite particles rather than determine the exact time—the assumption that all of the particles are of the same size is probably a much bigger assumption than the change in composition with time. Therefore, the composition far from the surface of the dissolving spheres will be assumed to be constant and equal to the eutectoid composition of 0.77 w/o carbon. So,

$$C_\infty(C) = (C\ \text{fraction})\frac{\rho(Fe)}{M(C)} = \left(7.7\times10^{-3}\right)\frac{7.86}{12.01}$$

$$C_\infty(C) = 5.04\times10^{-3}\ \text{mol/cm}^3$$

and the time for dissolution is when a = 0 is, on rearranging Equation 10.47,

$$t = \frac{a_0^2 C_{Fe_3C}}{2D\left(C_e - C_\infty\right)}. \tag{10.48}$$

Now for diffusion of C in FCC iron, austenite, Q = 142.3 kJ/mol and $D_0 = 0.2\ cm^2/s$ from Table 9.5, so

$$D(900°C) = D_0 e^{\frac{Q}{RT}} = 0.2\, e^{\frac{-142{,}300}{8.314 \times 1173}}$$

$$D(900°C) = 9.2 \times 10^{-8}\ cm^2/s.$$

Therefore, the time for dissolution is

$$t = \frac{a_0^2 C(Fe_3C)}{2D(C_e - C_\infty)} = \frac{(1\times 10^{-4})^2 (4.29\times 10^{-2})}{2(9.2\times 10^{-8})(8.2\times 10^{-3} - 5.04\times 10^{-3})}$$

$$t = 0.74\ s.$$

This is pretty quick! If the assumption were made that $C_\infty = 0$, then it would take half as long, which is even faster. Clearly, the assumption about the average particle size plays the biggest role in determining the time for dissolution because the initial particle size is squared, so doubling the initial particle size would increase the time by a factor of 4.

10.13 PRECIPITATION AND TOUGHENING IN ZIRCONIA

10.13.1 Introduction

Precipitation of a second phase from a solid solution has been used as a hardening—strengthening—technique for metals for over 100 years. The aluminum frame that made possible the construction of dirigibles such as the Hindenburg was a precipitation-hardened alloy as is almost all of the metal structure of modern aircraft. Here, the toughening of zirconia ceramics by precipitation is modeled. The purpose is not to compare the model to any particular experimental data but mainly to introduce some considerations involved in precipitation and their relationships to phase equilibria. In addition, the effect of boundary movement on the resulting kinetics presented in Appendix A.3 is applied in this case.

10.13.2 Transformational Toughening

Figure 10.25 shows part of the $MgO–ZrO_2$ phase diagram in Figure 7.31 (Levin et al. 1964). Ceramics are often considered to be "weak" materials because they break before they bend; they have a low fracture toughness. However, partially stabilized ZrO_2 can be made quite tough and crack resistant

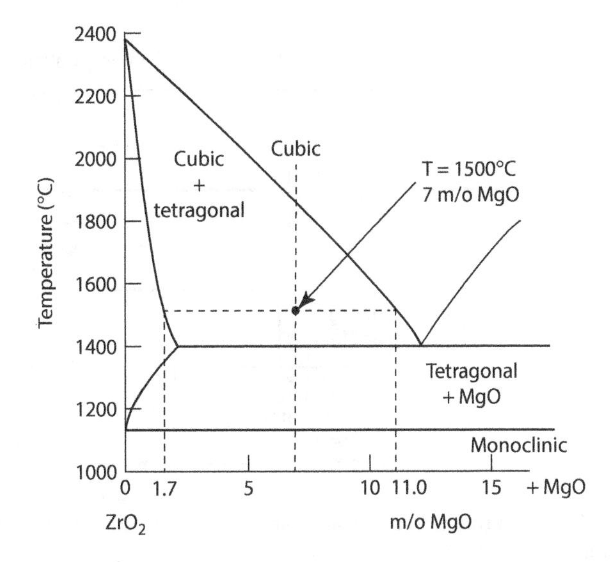

FIGURE 10.25 Part of the $MgO–ZrO_2$ phase diagram showing the equilibrium compositions of the tetragonal phase, 1.7 m/o, and the cubic phase, 11.0 m/o, for a 7 m/o MgO alloy held at 1500°C to precipitate the tetragonal phase.

with strengths up to several GPa (300,000 psi). The reason for the toughening in these materials is the transformation of retained or untransformed tetragonal particles in alloys that have compositions lying in the cubic + tetragonal phase field of Figure 10.25. Tetragonal particles larger than 1 μm or so will transform on cooling to the monoclinic phase with a few percent increase in volume. However, if the tetragonal particles are smaller than 1 μm, they are restrained from transforming by the surrounding cubic phase and can be retained in the tetragonal phase at room temperature. If a crack starts to propagate, some of the constraint of the cubic phase is released and the tetragonal particles transform, expand, and narrow the crack making it more difficult to propagate: increasing the toughness and strength considerably (Green et al. 1989). An ideal way to control the tetragonal particle size is by precipitation of the tetragonal phase from the cubic phase, which is in fact how it is done in industrial practice. This precipitation process is analyzed because it is the opposite of the dissolution process and reinforces the concept of the equilibrium concentration at the two-phase interface. However, in this case, the concentration versus distance profiles during precipitation are somewhat surprising.

10.13.3 The Model

In Figure 10.25, consider an alloy of 7 m/o MgO that has been densified or heat treated at a temperature above the (cubic + tetragonal)-cubic solvus line to form a cubic solid solution, in the neighborhood of 1850°C, also shown in Figure 10.26, at time equal to zero. The alloy is rapidly cooled to 1500°C and held there, while the tetragonal phase precipitates from the cubic solid solution in a nucleation and growth process. Only the growth process is considered here. From the partial phase diagram of Figure 10.25, when the precipitation process is complete—which it may not get to in practice to maximize the toughening—the microstructure will consist of tetragonal precipitate particles containing 1.7 m/o MgO in a matrix of cubic solid solution of 11.0 m/o MgO, as shown in Figure 10.26, when time goes to infinity—equilibrium. At times greater than zero and less than infinity, the middle concentration profile in Figure 10.26, the precipitate particle contains 1.7 m/o MgO, the matrix 7 m/o but the equilibrium concentration at the tetragonal-cubic interface is 11.0 m/o. As a result, in order for the precipitate particle to grow, MgO must diffuse *away* from the particle. In this case, the concentration versus radial distance is

$$C(r) = (C_C - C_0)\frac{a}{r} + C_0 \tag{10.49}$$

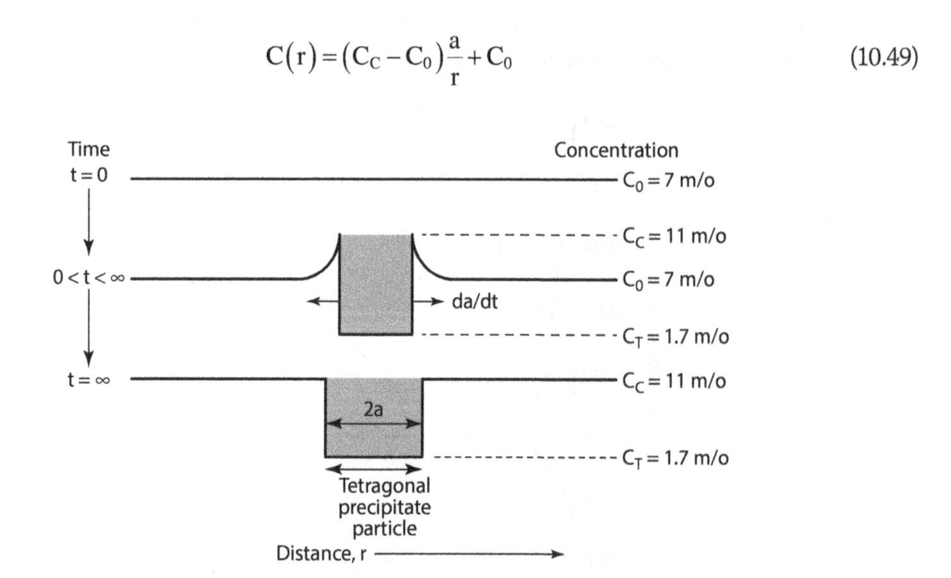

FIGURE 10.26 Compositions as a function of distance and time for precipitation of tetragonal ZrO_2 from a 7 m/o MgO cubic solid solution at 1500°C.

where $C_C = 11$ m/o MgO, the concentration in the equilibrium cubic solid solution, and $C_0 = 7$ m/o MgO, the initial concentration in the cubic phase.* So

$$\left(\frac{dC}{dr}\right)_{r=a} = -\frac{(C_C - C_0)}{a} \tag{10.50}$$

Taking into consideration the motion of the interface boundary discussed in Appendix A.3 (and has been ignored in previous models without justification), gives in this case

$$a^2 = 2D\frac{(C_C - C_0)}{(C_C - C_T)}t \tag{10.51}$$

so it does not make any difference what the units of concentration are, the units cancel so they can be left in mole %.

10.13.4 Calculated Heat-Treat Time

Not surprisingly, the diffusion coefficients of oxygen, magnesium, and zirconium are not known for this particular composition. However, several observations can be made and used to get an approximate diffusion coefficient and heat-treatment time. First, the oxygen diffusion coefficient is much larger than that for the cations because of the large oxygen vacancy concentration produced by MgO in solid solution as was the case for CaO in solid solution, that is,

$$MgO \xrightarrow{ZrO_2} Mg''_{Zr} + O^x_O + V^{\bullet\bullet}_O.$$

Also, as was seen in Section 10.11 on the dissolution of NaCl, when charged species are diffusing, the slower diffusing species controls the rate. Therefore, for the growth of these tetragonal zirconia particles, the diffusion of magnesium and/or zirconium will be rate controlling. Cation diffusion data are not available in the $MgO–ZrO_2$ system but are for the $CaO–ZrO_2$ system (Rhodes and Carter 1966). In this system, $D(Ca^{2+}) = 0.444 \times \exp(-(419{,}240\ J/mol)/RT)\ cm^2/s$ so at 1500°C, $D = 0.444 \times \exp(-(419{,}240)/(8.314)(1{,}773)) = 1.98 \times 10^{-13}\ cm^2/s$.† Assuming that this is the diffusion coefficient that controls the transport of magnesium, then the time it takes at 1500°C to get to a = 0.5 μm is

$$t = \frac{a^2}{2D}\frac{(C_C - C_T)}{(C_C - C_0)} = \frac{(0.5 \times 10^{-4})^2(11.0 - 1.7)}{2(1.98 \times 10^{-13})(11.0 - 7.0)}$$

$$t = 1.47 \times 10^4\ s = 4.08\ h$$

which would be a reasonable time to heat treat in practice because it is neither too short nor too long. In fact, this is entirely in the range of processing conditions used by industry in the manufacturing of transformation-toughened zirconia ceramics (Readey, M. J., pers. comm., December 2015).

* The concentrations here must be in mol/cm³ and all of the given concentrations should be converted to mol/cm³. However, the variation in density with composition is not known but could be approximated by ideal solutions, which they clearly are not, if the lattice parameters of the phases were known as a function of composition, which they also are not. In any event, the units of concentration will cancel and they can be left as mole percent.

† The oxygen diffusion coefficient in this system is about 10^6 times that of the cations (Rhodes and Carter 1966).

There are several assumptions implicit in Equation 10.51. The first is that the concentration of the cubic matrix phase does not change during precipitation and has remained at 7 m/o although the equilibrium value is 11.0 m/o. At equilibrium at 1500°C, the phase diagram shows that the volume fraction of tetragonal will be about $(11-7)/(11-1.7) \cong 0.43$ so for this assumption to be reasonably valid, perhaps no more than 4% of the tetragonal phase has precipitated, about 10% of the equilibrium value. This is probably not a bad assumption as long as the nucleation rate is not extremely fast so that, even though the particle size is small, there are a large number of particles and the fraction of tetragonal phase is much larger. Another assumption is that the density of the solid solutions are the same regardless of the composition, which again is perhaps valid up to about 10%. The addition of MgO to the cubic phase of ZrO_2 will lower its density because Mg^{2+} is lighter than Zr^{4+} and vacant oxygens are formed as well. However, not knowing how the lattice parameter changes with composition the densities and molar volumes really cannot be calculated anyway. Finally, the precipitate-cubic phase boundary moves that distorts the diffusion profile. However, the assumption is made again, that because the growth rate is slow and small, this will not be a large effect and a steady-state concentration profile is assumed. If all of the necessary data are available, then the problem can be solved exactly, albeit not simply. Nevertheless, with all of the assumptions in the given model, the results of the steady-state solution are probably sufficient to quantitatively predict precipitation times in the MgO–ZrO_2 system. The above assumptions will generate small perturbations in the results compared to what the uncertainty in the value for the diffusion coefficient will produce.*

10.13.5 Growth Rate

Taking the square root of both sides of Equation 10.51 gives

$$a = \left[2\frac{(C_C - C_0)}{(C_C - C_T)}\right]^{\frac{1}{2}} \sqrt{Dt}$$

so da/dt can be written as

$$\frac{da}{dt} = \frac{1}{2}\left[2\frac{(C_C - C_0)}{(C_C - C_T)}\right]^{\frac{1}{2}} \sqrt{\frac{D}{t}}$$

where, as before in Chapters 5 and 7, the rate of growth, da/dt, is proportional to a thermodynamic term, $(C_C - C_0)/(C_C - C_T)$, and a kinetic term, $\sqrt{D/t}$. The thermodynamic term is simply the difference between the composition of the alloy in question, C_0 (7 m/o in this example), and the composition of the cubic phase in equilibrium with the tetragonal phase at a given temperature, the solvus. The solvus between the cubic and the cubic + tetragonal regions in the partial phase diagram of Figure 10.25 is essentially linear and is approximately $T = 2400 - 80 \times (\text{m/o MgO})$. So for $C_0 = 7.0$ m/o, the temperature where this composition intersects the solvus line is about 1840°C, and at 1500°C, the composition of the cubic phase is 11.25 m/o, very close to the value of 11 m/o assumed above. So the thermodynamic term is essentially linear and increases as the temperature decreases. In contrast, the kinetic term is strongly dependent on the exponential temperature dependence of the diffusion coefficient and gets smaller rapidly as the temperature decreases because of the large activation energy. As a result, the growth rate reaches a maximum at around 1750°C and decreases with temperature as shown in Figure 10.27.

* The tetragonal particles are not actually spherical. The strain energy between the precipitate and the matrix causes the particles to take on ellipsoidal shapes. In addition, the number density of precipitate particles is rather high, so it is possible that their concentration profiles interact or overlap. Nevertheless, most importantly, the model, with all of its simplifying assumptionsis consistent with experimental results (Green, et al. 1989) and provides the engineer an excellent starting point for developing a technically robust heat-treatment schedule (Readey, M. J., pers. comm., December 2015).

10.14 OSTWALD RIPENING BY DIFFUSION

In Section 6.6 of Chapter 6, referring to Figure 6.10, it was shown that the solubility of a particle of B in a matrix of alpha, with a spherical surface of radius a, $[B_a]$, is higher than that over a plane surface, $[B_\infty]$, because of surface tension:

$$[B_a] = [B_\infty] e^{\frac{2\gamma \bar{V}_B}{aRT}} \tag{10.52}$$

Here:

$\bar{V}_B$ is the molar volume of B

$\gamma = \gamma_{\alpha B}$, or the interfacial energy between B and alpha.

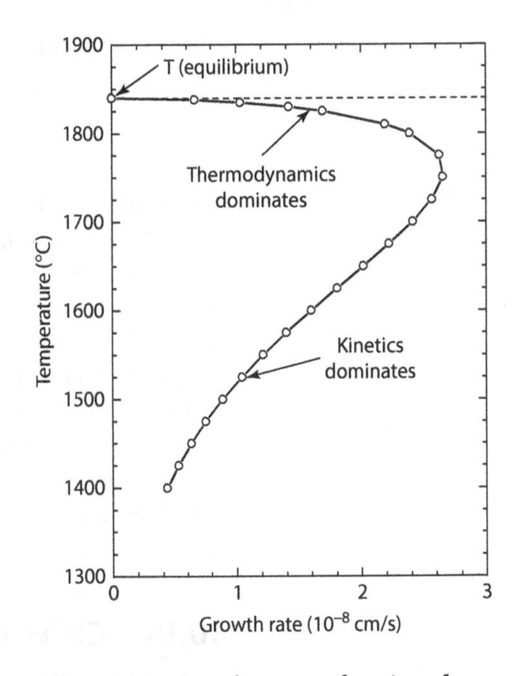

FIGURE 10.27 Growth rate as a function of temperature for a 7 m/o MgO solid solution in the $MgO\text{–}ZrO_2$ system. The overall rate is the product of a *thermodynamic* factor that dominates near the equilibrium temperature and a *kinetic* factor that dominates at low temperatures due to the strong temperature dependence of the diffusion coefficient.

In all of the presentations on dissolution and growth in this chapter, the increased growth or dissolution rate due to surface curvature when the particles are small has been ignored. In these examples, this also would have had a secondary and insignificant effect on the rates. However, what cannot be ignored is that Ostwald ripening or coarsening can be controlled by diffusion as well as by surface reaction as discussed in Chapter 6. In this case, Equation 10.39 becomes

$$\frac{da}{dt} = \frac{D\bar{V}_B}{a}\left([B_{\bar{a}}] - [B_a]\right) \tag{10.53}$$

where $[B_{\bar{a}}]$ is some average concentration in the external phase determined by the average particle size of the size distribution: if $[B_a] > [B_{\bar{a}}]$, the particles shrink, and grow when $[B_a] < [B_{\bar{a}}]$. The difficult part of the problem is, again, finding the particle size distribution and the mean particle size. As in Chapter 6, significant insight into the exact model can be obtained by examining the behavior of a simple two-size distribution rather than a continuous distribution: a number of small particles of radius *a* and some much larger particles so that $[B_{\bar{a}}] \cong [B_\infty]$ so Equation 10.53 becomes

$$\frac{da}{dt} = \frac{D\bar{V}_B}{a}\left([B_\infty] - [B_a]\right)$$

and inserting Equation 10.52

$$\frac{da}{dt} = \frac{D\bar{V}_B}{a}\left([B]_\infty - [B]_\infty \exp^{\frac{2\gamma\bar{V}_B}{aRT}}\right) \cong \frac{D\bar{V}_B[B]_\infty}{a}\left(1 - 1 - \frac{2\gamma\bar{V}_B}{aRT}\right)$$

$$\frac{da}{dt} = -\frac{D\bar{V}_B[B]_\infty}{a}\frac{2\gamma\bar{V}_B}{aRT} \tag{10.54}$$

which leads to

$$a^2\frac{da}{dt} = -\frac{2\bar{V}_B^2 D[B_\infty]\gamma}{RT}t \tag{10.55}$$

and is integrated to give

$$a^3 - a_0^3 = -\frac{6\bar{V}_B^2 D[B]_\infty \gamma}{RT}t = -At. \tag{10.56}$$

where a_0 is the initial particle size. Wagner's result for a particle size distribution is

$$\bar{a}^3 - \bar{a}_0^3 = \frac{8}{9}\frac{\bar{V}_B^2 D[B_\infty]\gamma}{RT}t = A't. \tag{10.57}$$

The results for a two-size distribution, Equation 10.56, and Wagner's solution (Wagner 1961), Equation 10.57, are again notably similar in that the particle size cubed depends linearly on time and all the parameters are exactly the same. The two differences are: his a is $\bar{a}$, the mean particle radius, and he gets a multiplying factor of 8/9 instead of 6. The exact solution takes into account a particular particle size distribution produced by the particle coarsening and that the rate of growth of a particle of a given radius depends on the difference between its solubility and that of the *average* particle size in the distribution. Determining the particle size distribution produced by the coarsening makes the model considerably more complex. The main point of the simple model is that all the other parameters and predicted time dependence are the same: the particle size grows or shrinks roughly as the cube root of time; and the rate is directly proportional to the thermodynamic solubility (over a flat surface), $[B_\infty]$, the interfacial energy, γ, and the diffusion coefficient in the external phase, D.

10.15 CHAPTER SUMMARY

This chapter is committed entirely to steady-state diffusion. It is an important topic, because many processes of interest to the materials engineer can be modeled as, or are, steady-state processes. In principle, the concept of the steady state is relatively straightforward: $dC/dt = 0$ and, therefore, $\nabla^2 C = 0$ from which the concentration of a function of distance, C(x), can be determined. From the concentration profile, C(x), the diffusion flux density can be modeled and the rate of material transported by diffusion can be calculated. As a result, this chapter is largely devoted to investigating steady-state processes of commercial relevance involving different materials, geometries, and processes. To obtain quantitative results, the values of various parameters must be obtained, which for many cases are not available and some reasonable approximations must be made. How reliable the approximations are requires a comparison between the model and the experiment. The first model of interest is diffusion through walls of tubes and spheres for applications ranging from the purification of hydrogen to the *shelve-life* of laser fusion targets. Then the classic model of oxidation of silicon, an extremely important process in the preparation of integrated circuits, is modeled giving quantitative results similar to those used in integrated circuit fabrication facilities, or "silicon foundries." Related to this is the second examination of CVD of silicon; only this time, the process is diffusion controlled. Then the results from both diffusion and surface reaction control are coupled to give an overall mass transfer coefficient that determines the rate of deposition by diffusion at high temperatures and by a surface reaction at low temperatures. Several *quasi-steady-state* phenomena are then examined: the evaporation of a water drop; the dissolution of a NaCl sphere; the dissolution of spheroidized Fe_3C in austenite; and the precipitation kinetics in the $MgO–ZrO_2$ system. These are all quasi-steady-state because the interface between the various phases is moving with time, and to compute an exact result, this moving boundary should really be taken into account, which goes well beyond the phenomena that are modeled here. In reality, if the boundary moves slow enough, the approximate results are often close enough to what actually happens. Finally, the particle coarsening or Ostwald ripening model is developed again, this time with diffusion control.

APPENDIX

A.1 STEADY-STATE SOLUTION FOR CYLINDRICAL COORDINATES

A.1.1 The Model

In both Figures 10.3 and 10.4, it was assumed that the one-dimensional Fick's law in Cartesian coordinates could be used rather than cylindrical coordinates as suggested by the tube geometry. The argument was made that as long as the diameter was large compared to the tube wall thickness, this was a justified assumption. It is instructive to take a look at this more closely by referring to Figure 10.4. Fick's second law in cylindrical coordinates is

$$\frac{\partial C}{\partial t} = \frac{D}{r}\frac{\partial}{\partial r}\left(r\frac{\partial C}{\partial r}\right) \tag{A.1}$$

which becomes for the steady state

$$\frac{d}{dr}\left(r\frac{dC}{dr}\right) = 0. \tag{A.2}$$

Integrating once,

$$\frac{dC}{dr} = \frac{A}{r}$$

and integrating again,

$$C(r) = A \ln r + B \tag{A.3}$$

where A and B must be determined from the boundary conditions: $C(r_1) = C_0$ and $C(r_2) = C_L$. Evaluating A and B,

$$C_0 = A \ln r_1 + B$$

$$C_L = A \ln r_2 + B$$

and subtracting gives

$$C_0 - C_L = A \ln\left(\frac{r_1}{r_2}\right) \text{ or } A = \frac{C_0 - C_L}{\ln(r_1/r_2)}$$

and so

$$B = C_L - A \ln r_2$$

$$= C_L - \frac{(C_0 - C_L)}{\ln(r_1/r_2)} \ln r_2.$$

Substituting these values of the constants A and B into the concentration equation:

$$C(r) = A \ln r + B$$

$$C(r) = \frac{(C_0 - C_L)}{\ln(r_1/r_2)} \ln r + C_L - \frac{(C_0 - C_L)}{\ln(r_1/r_2)} \ln r_2 \tag{A.4}$$

with rearrangement gives the final result:

$$C(r) = \frac{(C_0 - C_L)}{\ln(r_1/r_2)} \ln\left(\frac{r}{r_2}\right) + C_L. \tag{A.5}$$

which checks because $C(r_1) = C_0$ and $C(r_2) = C_L$. This is the steady-state solution for a tube of inner radius = r_1 and an outer radius of $r_2 = r_1 + L$, where L is the tube wall thickness and is plotted in Figure 10.5. Differentiating Equation A.4 gives

$$\frac{dC(r)}{dr} = \frac{(C_0 - C_L)}{r \ln(r_1/r_2)}. \tag{A.6}$$

As might be expected, the fluxes at r_1 and r_2 are different and given by

$$J(r_1) = -D\left(\frac{dC}{dr}\right)_{r_1} = -\frac{D(C_0 - C_L)}{r_1 \ln(r_1/r_2)}$$
$$J(r_2) = -D\left(\frac{dC}{dr}\right)_{r_2} = -\frac{D(C_0 - C_L)}{r_2 \ln(r_1/r_2)}. \quad (A.7)$$

The fluxes are different, of course, because the area on the inside is smaller than the outside area so that the number of mol/cm^2 s will be larger on the inside than on the outside but the number of moles per unit length of the tube—call it J^*, where Units (J^*) = mol/cm s, is independent of position in the tube wall as it should be and can be obtained simply by multiplying the respective fluxes by the inner and outer diameters times π:

$$J^* = 2\pi r_1 J(r_1) = 2\pi r_2 J(r_2) = -\frac{2\pi D(C_0 - C_L)}{\ln(r_1/r_2)}. \quad (A.8)$$

A.1.2 When Can the "Thin-Wall" Approximation Be Made?

First, it needs to be shown that the above concentration versus radius does indeed go to the linear case when the wall thickness is small compared to the tube radii. Let $C_L = 0$ for the sake of simplicity so that Equation A.5 becomes

$$C(r) = C_0 \frac{\ln(r/r_2)}{\ln(r_1/r_2)} \quad (A.9)$$

now $r = r_1 + x = r_2 - L + x$ and $r_1 = r_2 - L$ so Equation A.9 becomes

$$C(x) = \frac{\ln(r_2 - L + x / r_2)}{\ln(r_2 - L / r_2)}.$$

And because $\ln(1 \pm x) \cong \pm x$ for x small,* and $L \ll r_2$,

$$C(x) \cong C_0 \frac{(-(L - x / r_2))}{-(L / r_2)} = C_0\left(1 - \frac{x}{L}\right) \quad (A.10)$$

which is the same result as was obtained for C(x) in the linear—thin wall—case, Equation 10.2. This shows that the steady-state solution in cylindrical coordinates does reduce to the linear case if the wall thickness is small compared to the tube diameter.

When is it valid to make this approximation? The answer to that depends on how accurately do you want to predict the flux? Again, taking $C_L = 0$, the ratio of the flux in cylindrical coordinates (at r_2) to that in the linear coordinates gives

* Note: The approximation $\ln(1 \pm x) \cong \pm x$ comes from the first term in the Maclaurin series for $\ln(1 + x)$:

$$\ln(1+x) = \sum_{n=0}^{\infty} \frac{f^n(0)}{n!} x^n = x - \frac{x^2}{2} + \frac{x^3}{3} - \frac{x^4}{4} \ldots$$

where f^n is the "nth" derivative of the function: $\ln(1 + x)$ in this case.

TABLE A.1
The Effect of the Tube Radii on the Thin-Wall Approximation

r_2	r_1	L	L/r_2	$J(r_2)/J(L)$
3	2	1	0.333	0.822
5	4	1	0.200	0.896
10	9	1	0.100	0.949
10	9.5	0.5	0.050	0.975
10	9.9	0.1	0.010	0.995

$$\frac{J(r_2)}{J(L)} = \frac{-(DC_0/r_2 \ln(r_1/r_2))}{DC_0/L} = \frac{L}{r_2}\frac{1}{\ln\left(1-\frac{L}{r_2}\right)}. \tag{A.11}$$

Table A.1 shows how the ratio of the fluxes varies with the values of L, r_1, and r_2.

So if you can live with 10% accuracy, a 5:1 ratio of outer tube radius to wall thickness would be OK. However, if you need better than 1% accuracy, then a ratio of about 100:1 is necessary.

A.2 Steady-State Solution for Spherical Coordinates

A.2.1 The Model

Similar arguments also hold true for spherical coordinates. In this case, the steady-state solution for Ficks's second law—with the same symbols as used for cylindrical coordinates of Figure 10.4—becomes (Figure 10.9)

$$\frac{d}{d}\left(r^2\frac{dC}{dr}\right) = 0. \tag{A.12}$$

Integrating once,

$$\frac{dC}{dr} = \frac{A}{r^2}$$

and integrating a second time gives

$$C(r) = -\frac{A}{r} + B \tag{A.13}$$

where A and B are again the two integration constants. With the boundary conditions at $C(r_2) = C_L$ and $C(r_1) = C_0$ Equation A.13 becomes

$$C_L = -\frac{A}{r_2} + B \quad \text{and} \quad C_0 = -\frac{A}{r_1} + B$$

$$\text{subtracting,} \quad C_L - C_0 = -\frac{A}{r_2} + \frac{A}{r_1} = \frac{A}{r_1 r_2}(r_2 - r_1) \text{ or } A = r_1 r_2 \frac{(C_L - C_0)}{r_2 - r_1}$$

and

$$B = C_L + \frac{A}{r_2} = C_L + \frac{r_1(C_L - C_0)}{r_2 - r_1}$$

$$= \frac{C_L r_2 - C_L r_1 + C_L r_1 - C_0 r_1}{r_2 - r_1}$$

$$B = \frac{C_L r_2 - C_0 r_1}{r_2 - r_1}.$$

Putting these values for A and B into Equation A.13 gives the concentration as a function of r from the center of the sphere,

$$C(r) = -\frac{r_1 r_2}{r}\frac{(C_L - C_0)}{r_2 - r_1} + \frac{(C_L r_2 - C_0 r_1)}{r_2 - r_1} \tag{A.14}$$

which shows that the concentration through the sphere wall drops with position as 1/r. This is plotted in Figure 10.10 for $C_L = 0$, and $r_1 = 2L$ and $r_2 = 3L$. To make sure, the following checks are made for $C(r_2) = C_L$, $C(r_1) = C_0$, and dC/dr:

$$C(r_1) = \frac{-C_L r_2 + C_0 r_2 + C_L r_2 - C_0 r_1}{r_2 - r_1}$$

$$= C_0 \frac{r_2 - r_1}{r_2 - r_1} = C_0 \quad \therefore \text{checks}$$

$$C(r_2) = \frac{-C_L r_1 + C_0 r_1 + C_L r_2 - C_0 r_1}{r_2 - r_1}$$

$$= C_L \frac{r_2 - r_1}{r_2 - r_1} = C_L \quad \therefore \text{checks}$$

and

$$\frac{dC}{dr} = \frac{r_2 r_1}{r^2}\frac{(C_L - C_0)}{r_2 - r_1} = \frac{A}{r^2} \quad \therefore \text{checks.}$$

So the steady-state solution for C(r) obtained above, Equation A.14 appears to be correct. Therefore, the flux density is given by

$$\begin{aligned} J(r) &= -D\frac{dC(r)}{dr} \\ &= D\frac{r_2 r_1}{r^2}\frac{(C_0 - C_L)}{r_2 - r_1}. \end{aligned} \tag{A.15}$$

And as was the case for the cylindrical coordinates, the total flux, $J^* = AJ$, at any position r is a constant, namely

$$J^* = AJ = (4\pi r^2)\left(\frac{D r_1 r_2 (C_0 - C_L)}{r^2 (r_2 - r_1)}\right)$$

$$= 4\pi D\left(\frac{r_1 r_2 (C_0 - C_L)}{(r_2 - r_1)}\right).$$

A.2.2 When Can the "Thin-Wall" Approximation Be Made?

Another thing that can be done is show that this expression for the concentration goes to that for a flat plate when the thickness of the sphere wall is small compared to the sphere radius, the thin-wall solution. To make life easier, assume that $C_L = 0$, and $r_2 - r_1 = L$ and $r_2 - r = r_1 + x$, so Equation A.14 becomes

$$C(r) = \frac{r_1 r_2}{r}\frac{C_0}{r_2 - r_1} - \frac{C_0 r_1}{r_2 - r_1} = \frac{r_2 r_1 C_0 - r\, r_1 C_0}{rL}$$

$$= \frac{(r_1 + L) r_1 C_0 - r_1 (r_1 + x) C_0}{rL}$$

$$C(r) = \frac{r_1 C_0}{r}\frac{(L - x)}{L}.$$

TABLE A.2
The Effect of the Sphere Radii on the Thin-Wall Approximation

r_2	r_1	L	L/r_2	$J(r_2)/J(L)$
3	2	1	0.333	0.667
5	4	1	0.200	0.800
10	9	1	0.100	0.900
10	9.5	0.5	0.050	0.950
10	9.9	0.1	0.010	0.990

And in the approximation that x and $L \ll r$, then $r \cong r_1 + x \cong r_1$, which leads to the linear approximation as before:

$$C(x) \cong C_0\left(1 - \frac{x}{L}\right) \tag{A.16}$$

and

$$J = -D\frac{dC}{dx} = D\frac{C_0}{L}.$$

Again, how good this approximation is can be obtained by taking the ratio of Equations A.15 and A.16 (Table A.2),

$$\frac{J(r_2)}{J(L)} = \frac{D\dfrac{r_2 r_1}{r_2^2}\dfrac{(C_0 - C_L)}{r_2 - r_1}}{D\dfrac{(C_0 - C_L)}{L}}$$

$$= \frac{r_1}{r_2}.$$

A.3 Moving Boundary Problems

When considering a growing or shrinking phase in some kind of matrix or solution, in addition to the diffusion flux in the solution, there is also a convective flux because of the motion of the phase interface, call it da/dt. In addition, there could be additional motion in the solution if the molar volume of the solute in the solution, $\overline{V}_{soln}$, is different from that in the solute, $\overline{V}_s$. For example, consider a solid sphere of radius a dissolving in a pure solvent. There is a change in the volume of the solution, V_{soln}, given by $V_{soln} = n_{solute}\overline{V}_{soln}$ or $4/3\ \pi r^3 = (C_s 4/3\ \pi a^3)\overline{V}_{soln}$. So

$$v = \frac{dr}{dt} = \frac{a^2}{r^2}C_s\overline{V}_{soln}\frac{da}{dt}$$

is the velocity of the solution due to an increase in volume of the solution relative to the surface of the sphere. Therefore, the total flux, diffusive and convective, relative to the dissolving sphere is given by (Readey and Cooper 1966):

$$J = -D\frac{dC}{dr} + Cv = -D\frac{dC}{dr} + CC_s\overline{V}_{soln}\frac{a^2}{r^2}\frac{da}{dt}$$

at $r = a$

$$C_s\frac{da}{dt} = -D\frac{dC}{dr}(a) + C_e C_s\overline{V}_{soln}\frac{da}{dt}.$$

Of course, this means that the simple steady-state solution no longer applies to the concentration, and obtaining C(r,t) and the rate of dissolution is considerably more complex. This is called a moving boundary or Stefan problem that has been solved (Glicksman 2000) when $\overline{V}_{soln} = \overline{V}_s = 1/C_s$. If it is assumed that the concentration gradient at the interface is the same as that in the nonmoving boundary case, the so-called quasi-static solution is obtained (Glicksman 2000; Readey and Cooper 1966):

$$a\frac{da}{dt} = -D\frac{(C_e - C_0)}{C_s(1 - C_e\overline{V}_{soln})}$$

and if the molar volume of the solute is the same in the solid and the liquid—highly unlikely—then

$$a\frac{da}{dt} = -D\frac{(C_e - C_0)}{(C_s - C_e)}$$

which is frequently found in the literature (Porter et al. 2009).

EXERCISES

10.1 Hydrogen has been considered as the fuel of the future. One way to produce hydrogen is by electrolysis of water with electricity generated by solar cells or wind energy. To be used in a variety of applications, this hydrogen must be purified by removing impurity gases such as O_2, N_2, and H_2O. One way of doing this is by passing hydrogen gas through a polymer diffusion membrane in which hydrogen is soluble and diffuses relatively rapidly through the polymer while the other gases do not. This problem addresses the transport of hydrogen through polybutadiene tubes to purify it. H_2 diffuses as the hydrogen molecule via interstitial diffusion in polymers.

a. For gases in polymers, the solubility or concentration is a function of pressure of the form, C = Sp, where S is the solubility constant, units [S] = mol/cm³ atm and is typically around 2.0×10^{-6} mol/cm³ atm at 300 K. If the $p(H_2)$ on the inside of the tube is 2 atm and $p(H_2) = 1$ atm at the outside of the tube, calculate the concentrations of hydrogen in the inner and outer walls of the polybutadiene tube.

b. Calculate $D(H_2)$ at 27°C when $D_0 = 0.053$ cm²/s and Q = 21.3 kJ/mol.

c. Hydrogen is diffusing through this tube that has an outer diameter of 5 mm and an inner diameter of 3 mm. The cylindrical symmetry cannot be ignored in this case because of the large wall thickness to tube radius ratio. Calculate and plot the concentration, C(r), in mol/cm³ as a function of radius for the steady-state concentration profile through the wall thickness in terms of the inner tube radius, R_1, and the outer tube radius, R_2.

d. From the result in c., calculate dC/dr in mol/cm⁴ at both the inside and outside surface of the tube.

e. Calculate the H_2 flux density, mol/cm² s at both the inside and outside surface of the tube.

f. For this cylindrical diffusion, develop an expression the hydrogen flux per unit length (mol/s cm)of the polybutadiene tube.

g. Calculate how many feet of polymer tube would be needed operating at 27°C and 2 atm H_2 internal pressure to produce 100 ft³/h of H_2 on the outside of the tube.

h. From the result in g., would you suggest any modifications to the tubing? If "No" why not? If "Yes," what might be modified and give some justification for your suggestions?

10.2 An alternative method of generating high purity hydrogen is by passing hydrogen gas through a palladium, Pd, diffusion membrane in which hydrogen is soluble and diffuses rapidly while the other gases do not. This problem addresses the transport of hydrogen through Pd tubes to purify it.

a. Hydrogen is soluble in Pd up to 10 a/o (atomic % as H) at 500°C at a hydrogen pressure of 100 atm (Brandes and Brook 1992). If the density of Pd is $\rho(Pd) = 12.02$ g/cm³ and its molecular weight is M(Pd) = 106.4 g/mol (Haynes 2013), calculate what 10 a/o is in moles of H per cm^3. The hydrogen forms an interstitial solid solution in Pd so that the number of Pd atoms per cm^3 essentially does not change forming the solid solution.

b. The solubility of hydrogen in Pd follows Sievert's law: that is, H_2 (g) = 2H (solid solution) or C(H) = $Sp^{1/2}$. Calculate S so that units [C] are mol/cm^3 when units [p] = atm. Calculate C(H) in mol/cm^3 at 500°C and p = 5 atm.

c. Calculate D(H) at 500°C when $D_0 = 2.9 \times 10^{-3}$ cm^2/s and Q = 22,200 J/mol.

d. Hydrogen is diffusing through a Pd tube with an outer diameter of 30 mils and a wall thickness of 5 mils. Assume that the cylindrical symmetry cannot be ignored again in this case. Calculate and plot the concentration, C(r), in mol/cm^3 as a function of radius for the steady-state concentration profile through the wall thickness in terms of the inner tube radius, R_1, and the outer tube radius, R_2, where it can be assumed that $p_{R_2}(H_2) = 1\,\text{atm}$.

e. From the result in c., calculate dC/dr in mol/cm^4 at both the inside and outside surface of the tube.

f. Calculate the H_2 flux density, mol/cm^2 s at both the inside and outside surface of the tube.

g. For this cylindrical diffusion, develop an expression the hydrogen (Note: H_2) flux per unit length (mol/s cm)of the Pd tube.

h. Calculate how many feet of Pd tube would be needed operating at 500°C and 5 atm internal H_2 pressure to produce 100 ft^3/h of H_2 at STP on the outside of the tube.

i. From the result in h., would you suggest any modifications to the tubing? If "No" why not? If "Yes," what might be modified and give some justification for your suggestions?

10.3 The equilibrium constants for the reaction: KCl (s) = K^+ (soln) + Cl^- (soln) are 19.76 at 100°C and 8.241 at 30°C where the standard state for the solution species is 1 *molal/* (Roine 2002).

a. Calculate the concentration, mol/cm^3, of K^+ and Cl^- in equilibrium with pure KCl at 100°C.

b. Do the same at 30°C.

c. If the density of solid KCl is 1.984 g/cm^3 and its molecular weight M = 74.56 g/mol (Haynes 2013), calculate the concentration, mol/cm^3, of KCl in solid KCl.

d. A Spherical particle of initial radius, a_0, of KCl is dissolving into a large amount of pure water at 30°C. Develop an equation for the steady-state dissolution of a spherical particle of KCl of initial diameter of 2 mm; that is, develop an expression for the radius a as a function of time.

e. Plot the steady-state concentration, mol/cm^3, of KCl in both the solid and liquid as a function of distance, r/a, from the center, of the dissolving sphere out to r/a = 10.

f. Calculate how long it takes to dissolve the 2 mm sphere of KCl if the diffusion coefficient for KCl in H_2O is D = 2.1×10^{-5} cm^2/s at 30°C.

g. Now suppose that a large amount of solution of KCl in equilibrium with pure KCl at 100°C were suddenly cooled to 30°C and that KCl solid particles now grow. Plot the steady-state concentration, mol/cm^3, of KCl in both the solid and liquid as a function of distance, r/a, from the center of the now growing sphere out to r/a = 10.

10.4 a. Salt is often added to a glass of beer to increase the head by forming CO_2 bubbles by heterogeneous nucleation. As these carbon dioxide bubbles rise to the top of the glass, they grow by diffusion of dissolved CO_2 into the bubble. The solubility at room temperature of CO_2 gas is 7.9×10^{-5} mol/cm^3 (Haynes 2013) and because of the pressure in the bottle, the actual amount of gas in solution is five times the solubility (i.e., the solution is supersaturated). If the diffusion coefficient of CO_2 in the liquid is 1×10^{-4} cm^2/s, calculate and plot the CO_2 bubble size as a function of time up to 1 min. Assume that the bubble is growing by diffusion and that your glass is tall enough (a yard) so that the bubble takes that long to get to the top of the glass. Ignore any effects of the motion of the bubble or the motion of the growing bubble–liquid interface and assume steady-state conditions.

b. Plot the CO_2 concentration versus r/a in mol/cm^3 from the center of the bubble out to 10 times the bubble radius.

10.5 Manganese is being deposited by chemical vapor deposition by the following reaction:

$$MnCl_2(g) + H_2(g) \rightleftharpoons Mn(s) + 2\,HCl(g)$$

For the surface reaction, the reaction rate constant, k, is given by $k = k_o \exp(-Q/RT)$ with $k_o = 10^{10}$ cm/s and Q = 150,000 J/mol. For the same reaction controlled by diffusion through the gas phase, the following data are obtained: $D = D_o T^{3/2}/p$ with $D_o = 0.0001$ cm^2 atm/s K$^{3/2}$ and a gas boundary layer thickness of $\delta = 1$ cm.

a. On the same graph plot $\log_{10}h_R$, $\log_{10}h_D$, and $\log_{10}h$ versus $10^4/T$ from 200 to 1500°C, where $h_R = k$ is the mass transfer coefficient for the surface reaction, $h_D = D/\delta$ is the mass transfer coefficient for diffusion through the gas phase, and h is the overall mass transfer coefficient and:

$$\frac{1}{h} = \frac{1}{h_R} + \frac{1}{h_D}$$

b. From the following thermodynamic data, calculate the equilibrium constant for the reaction at 540°C given the following data (Units (T) = K):

$MnCl_2$ (g): $\Delta G^o = -295{,}400 - 263.6\,T$ J/mol

HCl (g) $\Delta G^o = -92{,}300 - 186.8\,T$ J/mol

c. If the initial $MnCl_2$ pressure is 10^{-2} atm and the hydrogen pressure stays constant at essentially $p(H_2) = 1$ atm, calculate the equilibrium pressures of $MnCl_2$ and HCl at 540°C.
d. Calculate h (cm/s) at 540°C.
e. The density and molecular weight of manganese are 7.20 g/cm^3 and 54.938 g/mol, respectively (Haynes 2013). Calculate the deposition rate (mils/h) of Mn at 540°C.
f. Calculate the $MnCl_2$ pressure at the surface of the depositing Mn at 540°C.

10.6 A Hollow amorphous silica (SiO_2 glass) laser fusion sphere 100 μm in diameter is filled with hydrogen gas at a total initial pressure of 1 atm. The diffusion coefficient of hydrogen in SiO_2 is given by $D = D_0 \exp(-Q/RT)$, where $D_0 = 5.65\times10^{-4}$ cm^2/s and Q = 43,027 J/mol. The solubility of hydrogen in the silica is given by $C(H_2) = Sp(H_2)$, where Units (C) = mol/cm^3, Units (p) = atm, and $S = 4.1\times10^{-7}$ mol/cm^3 atm.

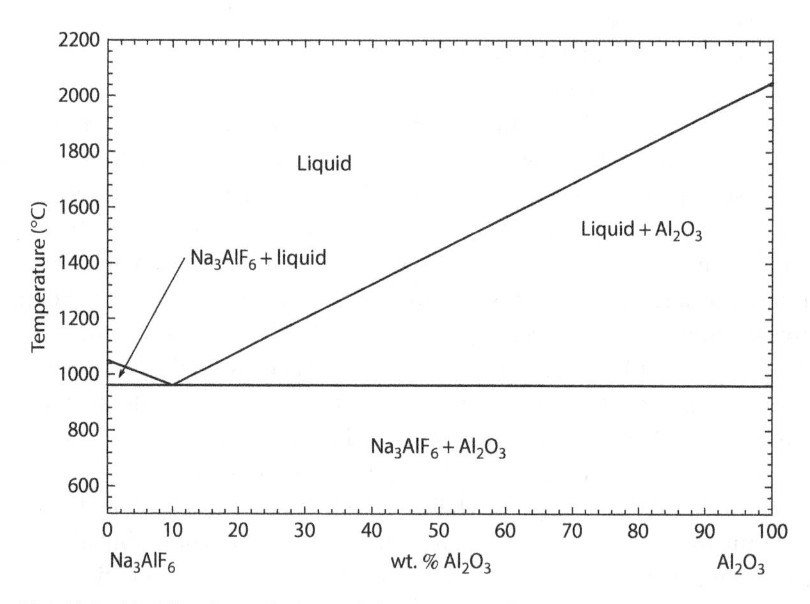

FIGURE E.1 The Al_2O_3–Na_3AlF_6 phase diagram. (After Levin et al. 1964.)

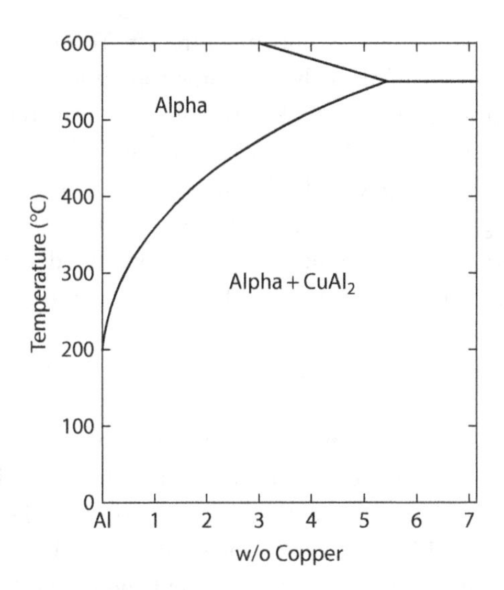

FIGURE E.2 Part of the aluminum-copper phase diagram. (After Brandes and Brook 1992.)

a. From Fick's second law, develop the steady-state model for the concentration as a function of distance, C(x), through the sphere wall assuming that the wall thickness is small relative to the sphere radius: the linear solution. Let C_i = the concentration at the inside of the sphere wall, x = 0, and $p(H_2) = 0$ at the outer wall, x = L.
b. Develop an expression for the H_2 pressure in the sphere as a function of time, S, D, T, L, and sphere diameter, d. Assume only steady state and, because the wall thickness is much less than the sphere radius, ignore spherical coordinates. Take into consideration that C_i will decrease as the pressure drops that leads to an exponential decrease in $p(H_2)$ with time.
c. Calculate the value of D (cm^2/s) at −100°C.
d. Calculate how long it takes (days) for the pressure to drop 10% at −100°C if the sphere has a wall thickness of 10 μm and the hydrogen content of the ambient external atmosphere can be considered to be zero.

10.7 Aluminum metal is made by electrolyzing a cryolite, Na_3AlF_6, melt containing Al_2O_3 dissolved in liquid solution. A schematic of the Na_3AlF_6–Al_2O_3 phase diagram is given in Figure E.1. The overall cryolite melt has 10 w/o Al_2O_3 dissolved. Two hundred micron diameter spherical particles of Al_2O_3 are dumped into the melt for dissolution at 1300°C and the melt is sufficiently large so that the particles do not interact nor do they change the overall concentration of the melt.

a. The density and molecular weight of molten cryolite are 2.084 g/cm^3 and 209.94 g/mol and that of molten Al_2O_3 are 2.89 g/cm^3 and 101.96 g/mol, respectively (Haynes 2013). Molten Na_3AlF_6 and molten Al_2O_3 form an ideal liquid solution. Calculate the Al_2O_3 concentration in mol/cm^3 in the 10 w/o melt far from the dissolving Al_2O_3 and calculate the equilibrium Al_2O_3 concentration (mol/cm^3) in the melt at the solid Al_2O_3–melt interface.
b. Plot the Al_2O_3 concentration (mol/cm^3) from the center of the Al_2O_3 particles to three times the particle radius into the melt. The density of solid Al_2O_3 is 4.0 g/cm^3.
c. Calculate how long it takes the Al_2O_3 particles to dissolve in the cryolite.

10.8 Aluminum alloys containing small amounts of copper (5 w/o or less) are very important alloys as they can be heat treated to precipitate $CuAl_2$ as a second phase, which greatly increases the yield and tensile strengths of the alloy. The relevant part of the aluminum copper phase diagram is given in Figure E.2. These are called "age-hardening" alloys, whose development at the turn of the previous century, enabled the construction of large airships such as the Hindenburg and, today, modern aircraft. To obtain the maximum properties, the particles must be kept small. If they become too large, the strength is not maximized, this is called "over aging." A 4 w/o Cu alloy is "solution" heat treated to single phase alpha at 550°C, quenched, and cooled to 250°C and precipitation heat treated (aging). Assume that the precipitate particles are spherical. The density and molecular weight of aluminum are 2.70 g/cm^3 and 26.98 g/mol, respectively, and assume that the density of the Al solid solution does not change with the small copper concentration. The density and molecular weight of $CuAl_2$ are 4.73 g/cm^3 and 117.51 g/mol and that of copper are 8.92 g/cm^3 and 63.55 g/mol, respectively (Haynes 2013).

a. Calculate the copper concentrations (mol/cm^3) in the original 4 w/o alloy, the alloy in equilibrium with $CuAl_2$ at 250°C, and $CuAl_2$.
b. Plot the concentration of Cu (mol/cm^3) as a function of distance from the center of the growing spherical precipitates out to three times the radius into the alloy.
c. The diffusion coefficient of Cu in Al is given by $D = D_o \exp(-Q/RT)$, where D_o = 0.137 cm^2/s and Q = 123.5 kJ/mol (Brandes and Brook 1992). Calculate D at 250°C.
d. What is the maximum time of heat treatment (hours) at 250°C so the precipitate particles are not larger than 0.2 μm in diameter.
e. Make a plot of $(C_0 - C_e) \times D$ horizontally (which is proportional to the rate of precipitate particle growth) versus temperature vertically from T_e (the temperature at which the precipitate phase becomes stable) down to 250°C and determine the temperature at which the growth rate is a maximum.

10.9 a. With the iron–carbon phase diagram, Figure 7.4, calculate the time it takes for 10 μm diameter spherical cementite particles to dissolve in austenite at 1000°C assuming steady-state diffusion of the carbon in the austenite. Assume that the austenite composition is constant at 0.77 weight% carbon; the density and molecular weights of the cementite, austenite, and carbon are $\rho(Fe_3C)$ = 7.694 g/cm^3, $M(Fe_3C)$ = 179.55 g/mol, ρ(austenite) = 7.86 g/cm^3, M(austenite) = 55.85 g/mol, ρ(C) = 2.2 g/cm^3, and M(C) = 12.01 g/mol; and D_o = 0.20 cm^2/s and Q = 142.3 kJ/mol for carbon diffusion in austenite (Brandes and Brook 1992).

b. Plot the carbon concentration (g atoms/cm^3) as a function of distance from the center of a spherical particle to a distance five times the particle radius from the particle's center at the start of the dissolution process.

10.10 The oxidation rate of silicon can be given by $x^2 = Bt$, where x is the oxide thickness and B is the parabolic rate constant and has the following experimentally determined values in 1 atm of dry oxygen (Mayer and Lau 1990):

Temperature (°C)	B (μ²/h)
1200	0.0450
1100	0.0270
1000	0.0117
920	0.0049
800	0.0011

a. Plot the SiO_2 thickness (nm) as a function of time (0 to 2 h) for each of these temperatures on the same plot.
b. Make a plot of $\log_{10}B$ versus 1/T (K^{-1}) and determine the apparent activation energy for the oxidation process.
c. The diffusion coefficient for O_2 in SiO_2 is given by $D = D_0\exp^{-Q/RT}$ with $D_0 = 2.7 \times 10^{-4}$ cm^2/s and Q = 1.16 eV (Mayer and Lau 1990). How does the calculated activation energy from oxidation data compare with that for diffusion oxygen given in this reference?
d. Assuming that the diffusion coefficient for oxygen is that given in the literature in part c above, calculate the solubility coefficient, S, in $C(O_2) = Sp(O_2)$, where Units (C) = mol/cm^3 and Units (p) = atm at 1000°C.

e. With the value of S from part d (assuming it to be T-independent) and the diffusion coefficient from part c, make a table of T (°C), B (μm²/h) experimental, and B (μm²/h) calculated for the temperatures in the table above.

REFERENCES

Barrer, R. M. 1941. *Diffusion in and Through Solids*. Cambridge, UK: Cambridge University Press.

Blood, 2016. *Wikipedia*. https://en.wikipedia.org/w/index.php?title=Blood&oldid=730458509.

Brandes, E. A. and G. B. Brook. 1992. *Smithells Metals Reference Book*, 7th ed. Oxford, UK: Butterworth-Heinemann, Ltd.

Carbon dioxide, 2016. *Wikipedia*, https://en/wikipedia.org/w/index.php?title=Carbon_dioxide&oldid=730227078.

Chang, R. 2000. *Physical Chemistry for the Chemical and Biological Sciences*. Sausilito, CA: Library Science Books.

Cussler, E. L. 1997. *Diffusion: Mass Transfer in Fluid Systems*, 2nd ed. Cambridge, UK: Cambridge University Press.

Deal, B. E. and A. S. Grove. 1965. General relationship for the thermal oxidation of silicon. *Journal of Applied Physics*. 36: 3770–3778.

Emsley, J. 1998. *The Elements*, 3rd ed. Oxford, UK: Clarendon Press.

Endewar, V., S. Al-Samir, F. Itel, and G. Gros. 2014. How does carbon dioxide permeate cell membranes: A discussion of concepts, results and methods. *Frontiers in Physiology*. 4(January): 1–20.

Gaskell, D. R. 1992. *An Introduction to Transport Phenomena in Materials Engineering*. New York: MacMillan.

Geankoplis, C. J. 1972. *Mass Transport Phenomena*. Columbus, OH: Ohio State University Press.

Geankoplis, C. J. 2003. *Transport Processes and Separation Process Principles*, 4th ed. Upper Saddle River, NJ: Prentice Hall.

Glicksman, M. E. 2000. *Diffusion in Solids*. New York: Wiley.

Green, D. J., R. H. J. Hannink, and M. V. Swain. 1989. *Transformation Toughening of Ceramics*. Boca Raton, FL: CRC Press.

Haynes, W. M., Editor-in-Chief. 2013. *CRC Handbook of Chemistry and Physics*, 94th ed. Boca Raton, FL: CRC Press.

Jackson, T. 1997. *Inside Intel*. New York: Plume Publishing.

Kesting, R. E. and A. K. Fritzsche. 1993. *Polymeric Gas Separation Membranes*. New York: Wiley.

Levin, E. M., C. R. Robbins, and H. F. McMurdie. 1964. *Phase Diagrams for Ceramists*. Columbus, OH: The American Ceramic Society.

Lodish, H., A. Berk, S. L. Zipursky, P. Matsudaira, D. Baltimore, and J. Darnell. 2000. *Molecular Cell Biology*, 4th ed. New York: W. H. Freeman and Company.

Mark, J. E., ed. 2007. *Physical Properties of Polymers Handbook*, 2nd ed. New York: Springer Science and Business Media, LLC.

Mayer, J. W. and S. S. Lau. 1990. *Electronic Materials Science: For Integrated Circuits in Si and GaAs*. New York: MacMillan.

Mulder, M. 1996. *Basic Principles of Membrane Technology*, 2nd ed. Dordrecht, the Netherlands: Kluwer Academic Publishers.

Orders of Magnitude (power). 2016. *Wikipedia*. https://en/wikipedia.org/w/index.php?title=Orders_of_magntude_(power)&oldid=729249444.

Partition Coefficient. 2016. *Wikipedia*. https://en/wikipedia.org/w/index.php?title=Partition_coefficient&oldid=729924295.

Porter, D. A., K. E. Easterling, and M. Y. Sherif. 2009. *Phase Transformations in Metals and Alloys*, 3rd ed. Boca Raton, FL: CRC Press.

Readey, D. W. and A. R. Cooper, Jr. 1966. Molecular Diffusion with a moving boundary and spherical symmetry. *Chemical Engineering Science*. 21: 917–922.

Red blood cell. 2014. *Wikipedia*. https://en.wikipedia.org/w/index.php?title=Red_blood_cell&oldid=729656925.

Renner, R. 2002. The K_{OW} controversy. *Environmental Science and Technology*. November 1: 411A–413A.

Rhodes, W. H. and R. E. Carter. 1966. Cationic self-diffusion in calcia-stabilized zirconia. *Journal of the American Ceramic Society*. 49(5): 244–249.

Roine, A. 2002. *Outokumpu HSC Chemistry for Windows*, Ver. 5.11. Thermodynamic Software Program, Outokumpu Research Oy, Pori, Finland.

Sauveur, A. 1912. *The Metallography of Iron and Steel*. Cambridge, MA: Cambridge University Press.

Silbey, R. J. and R. A. Alberty. 2001. *Physical Chemistry*, 3rd ed. New York: Wiley.

Sperling, L. H. 2006. *Introduction to Physical Polymer Science*, 4th ed. New York: Wiley Interscience.

Stull, D. R. and H. Prophet. 1971. *Janaf Thermochemical Tables*, 2nd ed. Washington, DC: U.S. Government Printing Office.

Van Krevelen, D. W. 1997. *Properties of Polymers*, 3rd ed. Amsterdam, the Netherlands: Elsevier B. V.

Wagner, C. 1961. Theorie der Alterung von Niederschlagen durch Umlosen. *Zeitschrift für Elektrochemie*. 65(7–8): 581–591.

11

Solutions to Fick's Second Law

Infinite and Semi-Infinite Boundary Conditions

11.1 INTRODUCTION

Solutions to Fick's second law, $\partial C/\partial t = D(\partial^2 C/\partial x^2)$, give the concentration as a function of time and distance, C(x, t) (or T(x, t) for the essentially identical heat transfer equation) are critical for evaluating the kinetics of many processes in materials. The intent of this and the following chapter is to introduce and demonstrate various approaches and approximations to some of the solutions of major interest in materials science and engineering without resorting to advanced mathematical technique often used in many books on diffusion; for example, applying a *Laplace transform* to $\partial C/\partial t = D(\partial^2 C/\partial x^2)$ to get an ordinary differential equation; solving the ordinary differential equation; and then applying the inverse transform to get C(x, t). Unfortunately, the determination of the inverse transform is frequently mathematically difficult and often requires additional advanced techniques. Now, tables of Laplace transforms and their inverses are readily available in the literature and can be used to get a solution. Unfortunately, relying on tabulated transforms is basically a *black-box* approach. The intent here is to go beyond simply giving *black-box* solutions, such as $C(x,t) = C_S \text{erfc}\left(x/\sqrt{4Dt}\right)$, for diffusion from a constant surface concentration of C_S, and model these solutions with techniques familiar to someone who has completed only a second or third semester of calculus.

In addition, to keep the mathematics manageable, only solutions to the one-dimensional mass conservation equation $\partial C/\partial t = D(\partial^2 C/\partial x^2)$ are considered in this chapter but are, nevertheless, important in many applications. There are many books devoted to the solution to partial differential equations in general (e.g., Farlow 1982; Snider 2006). Then there are some that are completely, or partially, focused on the solution to the broad range of diffusion or heat transfer equations (Morse and Feshbach 1953; Carslaw and Jaeger 1959; Crank 1975; Glicksman 2000). The intention is not to emulate the rigor and breadth of these works but rather to provide some straightforward and plausible approaches to solutions of a limited number of problems that a materials scientist or engineer might encounter. These and other works in the literature are excellent sources for specific solutions that might be of interest for a given application, provided one is familiar with these solutions and knows how and where to look for them. The goal of this chapter is to provide this familiarity by modeling solutions to Fick's second law for infinite and semi-infinite boundary conditions. Chapter 12 develops models for solutions with finite boundary conditions.

11.2 SOLUTION WITH A DIMENSIONLESS VARIABLE

11.2.1 Introduction

The objective here is first to develop a relationship between the diffusion coefficient and the distance diffused, λ, that leads to $\lambda^2 = 4Dt$ based on a less-than-mathematically rigorous—but intuitive—model. This result suggests that a dimensionless variable of the form $y = x/\sqrt{4Dt}$ might be used to transform the partial differential equation into a solvable ordinary differential equation. From this, it follows that some semi-infinite and infinite boundary condition problems of *real-world* processes can indeed be solved by substitution of the dimensionless variable, $y = x/\sqrt{4Dt}$.

11.2.2 Dimensionless Variable

Figure 11.1 schematically shows the diffusion of boron into silicon from a gaseous diborane, B_2H, source that establishes a constant surface activity or concentration of boron, C_S. The concentration of boron, C(x, t), is a function of both x and t since the boundary conditions are semi-infinite in this case.* The diffusion of boron into the silicon forms a p-n junction 1.0 μm below the surface of a silicon wafer that might be 25 mils†—635 μm—thick, so that semi-infinite boundary conditions clearly apply: 635 μm is essentially infinitely far away from 1 μm and the concentration there will not change during the diffusion. It is assumed that the initial concentration C(x, 0) = 0. This not only makes the problem slightly easier to solve but also is usually true in practice.

The solution to the one-dimensional Fick's second law, $\partial C/\partial t = D(\partial^2 C/\partial x^2)$, gives C(x, t), and the goal here is to find that solution. But first, with a linear approximation to the *diffusion profiles*—C(x, t)—shown in Figure 11.2, a very suggestive result can be obtained that is extremely useful even if the procedure used to obtain the result is less than mathematically rigorous (Mayer and Lau 1990). The procedure is to find a substitute variable that is a function of both time and distance and, when substituted into the partial differential equation, yields an easily solvable ordinary differential equation.

Assume that the C(x, t) profile for some time of diffusion is approximately linear as shown in Figure 11.2 and the boron has diffused to a depth of $x = \lambda$. As usual,

$$J = -D\frac{dC}{dx}$$

and the steady-state solution and the concentration gradient, dC/dx, are given by

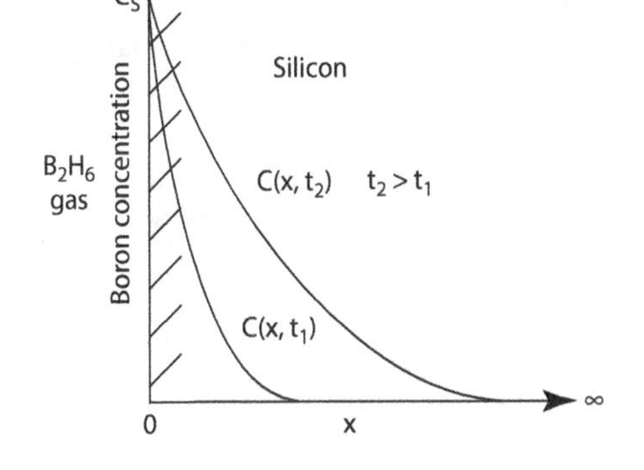

FIGURE 11.1 Schematic showing the C(x, t) of boron into silicon with a fixed surface concentration, determined by the thermodynamics in the gas phase. This is an illustration of diffusion with semi-infinite boundary conditions: $0 \le x \le \infty$.

* As described in Chapter 8, semi-infinite boundary conditions in this case are $C(0, t) = C_S$ and $C(\infty, t) = 0$.

† 1 mil = 0.001 in.

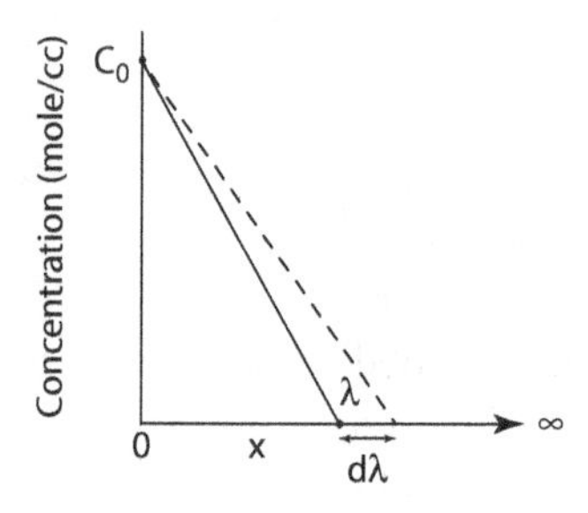

FIGURE 11.2 Linear approximation of diffusion in a semi-infinite medium with a penetration depth of λ in time t and additional penetration of dλ after dt.

$$C(x) = C_S\left(1 - \frac{x}{\lambda}\right)$$

$$\frac{dC}{dx} = -\frac{C_S}{\lambda}$$

where C_S is the surface concentration. Therefore, $J = D(C_0/\lambda)$, as has been seen several times before in Chapter 10. Now the number of moles of boron *per unit length* along the surface of the crystal is just the area, A, of the triangle in Figure 11.2, A = 1/2 $C_0\lambda$. The flux of boron is just the time rate of change of the total amount of boron or the rate of change of the area

$$J = \frac{d}{dt}\left(\frac{C_0\lambda}{2}\right) = \frac{C_0}{2}\frac{d\lambda}{dt} \tag{11.1}$$

equating the two values for the flux,

$$J = D\frac{C_0}{\lambda} = \frac{C_0}{2}\frac{d\lambda}{dt}$$

gives

$$2D = \lambda\frac{d\lambda}{dt}$$

which when integrated gives the desired result

$$\lambda^2 = 4Dt. \tag{11.2}$$

Equation 11.2 basically says that the diffusion distance is proportional to $\sqrt{4Dt}$ and that diffusion time and diffusion distance are not independent. This suggests a single dimensionless variable of (distance²)/(4Dt) or, more precisely, its square root, $y = x/\sqrt{4Dt}$, might be used to solve the partial differential equation. Indeed, if this dimensionless variable is substituted into the partial differential equation of Fick's second law, it will yield an *ordinary differential equation* in y that can be solved by usual straightforward techniques.

11.2.3 Solution of Fick's Second Law by Variable Substitution

It is now demonstrated that the substitution of this dimensionless variable into Fick's second law indeed leads to a solution of

$$\frac{\partial C}{\partial t} = D\frac{\partial^2 C}{\partial x^2}$$

in the form C(x, t). The following contains a number of equations simply because each step is carried out in detail; that is, no *black boxes*—and no new or fancy mathematics! Let the dimensionless variable be $y^2 = x^2/\sqrt{4Dt}$ or $y = x/\sqrt{4Dt}$. Then, substituting first for $\partial C/\partial t$.

$$\frac{\partial C}{\partial t} = \frac{\partial C}{\partial y}\frac{\partial y}{\partial t}$$

$$\frac{\partial y}{\partial t} = \frac{\partial\left(x/\sqrt{4Dt}\right)}{\partial t} = -\frac{1}{2}\frac{xt^{-3/2}}{\sqrt{4D}} = -\frac{1}{2t}\left(\frac{x}{\sqrt{4Dt}}\right) = -\frac{y}{2t}$$

so that

$$\frac{\partial C}{\partial t} = -\frac{y}{2t}\frac{\partial C}{\partial y}. \tag{11.3}$$

Similarly, on the right-hand side of the partial differential equation, substituting for $\partial^2 C/\partial^2 t$,

$$\frac{\partial C}{\partial x} = \frac{\partial C}{\partial y}\frac{\partial y}{\partial x}$$

$$\frac{\partial y}{\partial x} = \frac{\partial\left(x/\sqrt{4Dt}\right)}{\partial x} = \frac{1}{\sqrt{4Dt}}$$

$$\frac{\partial C}{\partial x} = \frac{1}{\sqrt{4Dt}}\frac{\partial C}{\partial y}$$

and differentiating one more time,

$$\frac{\partial^2 C}{\partial x^2} = \frac{\partial}{\partial x}\left(\frac{\partial C}{\partial x}\right) = \frac{\partial}{\partial x}\left(\frac{1}{\sqrt{4Dt}}\frac{\partial C}{\partial y}\right)$$

$$= \frac{\partial}{\partial y}\left(\frac{1}{\sqrt{4Dt}}\frac{\partial C}{\partial y}\right)\frac{\partial y}{\partial x} = \frac{\partial}{\partial y}\left(\frac{1}{\sqrt{4Dt}}\frac{\partial C}{\partial y}\right)\frac{1}{\sqrt{4Dt}}$$

$$= \frac{1}{4Dt}\frac{\partial}{\partial y}\left(\frac{\partial C}{\partial y}\right) = \frac{1}{4Dt}\frac{\partial^2 C}{\partial y^2}$$

so that

$$\frac{\partial^2 C}{\partial x^2} = \frac{1}{4Dt}\frac{\partial^2 C}{\partial y^2}. \tag{11.4}$$

Making the two substitutions of Equations 11.3 and 11.4 for $\partial C/\partial t$ and $\partial^2 C/\partial x^2$ in Fick's second law

$$\frac{\partial C}{\partial t} = D\frac{\partial^2 C}{\partial x^2}$$

$$-\frac{y}{2t}\frac{\partial C}{\partial y} = \frac{D}{4Dt}\frac{\partial^2 C}{\partial y^2}$$

shows that the substitution was successful because the partial differential equation in x and t is now an ordinary differential equation in y,

$$\frac{d^2 C}{dy^2} = -2y\frac{dC}{dy}. \tag{11.5}$$

To solve this ordinary differential equation, let $z = dC/dy$, so

$$\frac{dz}{dy} = -2yz$$

$$\frac{1}{z}\frac{dz}{dy} = -2y$$

which is easily integrated to give

$$\ln z = -y^2 + A'$$

$$z = Ae^{-y^2}.$$

Moreover, integrate z one more time to get C(y)

$$z = \frac{dC}{dy} = Ae^{-y^2}$$

$$C(y) = A\int_0^y e^{-w^2}dw + B$$

where w is just a dummy variable of integration. Putting in $y = x/\sqrt{4Dt}$ leads to the final important result

$$C(x,t) = A\int_0^{x/\sqrt{4Dt}} e^{-w^2}dw + B. \quad (11.6)$$

Therefore, the partial differential equation can be solved as an ordinary differential equation, Equation 11.6, with the dimensionless variable $y = x/\sqrt{4Dt}$, and it gives a specific solution for two integration constants A and B. It should be noted that the partial differential equation was solved simply by making a variable substitution—a *similarity variable*—to give an easily integrated ordinary differential equation: no advanced mathematics beyond calculus! Of course, the right choice of the variable y was critical.

11.3 SEMI-INFINITE BOUNDARY CONDITIONS

11.3.1 Model: Error Functions

As shown in Figure 11.1—diffusion of boron, B, into silicon, for example—the initial conditions in this case are the initial condition C(x, 0) = 0 with the boundary conditions C(0, t) = C_S and C(∞, t) = 0. When x = 0, $x/\sqrt{4Dt} = 0$; therefore, Equation 11.6 is zero and B = C_S. When x = ∞, C = 0; therefore,

$$0 = A\int_0^{\infty} e^{-w^2}dw + C_S$$

$$0 = A\frac{\sqrt{\pi}}{2} + C_S$$

$$A = -\frac{2}{\sqrt{\pi}}C_S$$

where $\int_0^{\infty} e^{-x^2}dx = \sqrt{\pi}/2$, as shown in the Appendix to Chapter 5. Therefore, the solution of Equation 11.6 for these initial and boundary conditions is

$$C(x,t) = C_S\left(1 - \frac{2}{\sqrt{\pi}}\int_0^{x/\sqrt{4Dt}} e^{-w^2}dw\right). \quad (11.7)$$

Now the Gaussian *error function* of y, erf(y), is defined as

$$\mathrm{erf}(y) = \frac{2}{\sqrt{\pi}} \int_0^y e^{-w^2} dw \tag{11.8}$$

which *is just a number* for each value of y, given by $2/\sqrt{\pi}$ times the area under the Gaussian curve, e^{-w^2}, from 0 to y, as shown in Figure 11.3. The error function was defined this way, so that the error function of infinity would equal one, that is, erf (∞) = 1, since $\int_0^\infty e^{-w^2} dw = \sqrt{\pi}/2$. The *complementary error function* of y, erfc(y), is simply 1 − erf(y) and

$$\mathrm{erfc}(y) = 1 - \mathrm{erf}(y) = \frac{2}{\sqrt{\pi}} \int_0^\infty e^{-w^2} dw - \frac{2}{\sqrt{\pi}} \int_0^y e^{-w^2} dw$$

$$\mathrm{erfc}(y) = \frac{2}{\sqrt{\pi}} \int_0^\infty e^{-w^2} dw + \frac{2}{\sqrt{\pi}} \int_y^0 e^{-w^2} dw$$

resulting in

$$\mathrm{erfc}(y) = \frac{2}{\sqrt{\pi}} \int_y^\infty e^{-w^2} dw \tag{11.9}$$

as shown in Figure 11.3. Therefore, the solution to the semi-infinite boundary condition problem of boron diffusing by solid-state diffusion into silicon, Equation 11.7, can be written as follows:

$$C(x,t) = C_S \mathrm{erfc}\left(\frac{x}{\sqrt{4Dt}}\right). \tag{11.10}$$

This same equation applies to the diffusion of carbon from the surface of a piece of iron or steel to increase its hardness; ; or the diffusion of a gas such as hydrogen into a thick piece of polymer ; and the diffusion of ^{18}O into aluminum oxide, Al_2O_3. The latter can be used to measure the diffusion coefficient of oxygen in aluminum oxide by analyzing the ^{18}O content as a function of depth below the surface and comparing it to Equation 11.10. When discussing diffusion in solids, introductory books on materials science and engineering frequently simply introduce Equation 11.10 without any background—the *black-box* approach—and use it to calculate the concentration of carbon in a piece of steel as a function of time and distance. Hopefully, Equation 11.10 is no longer a *black box!*

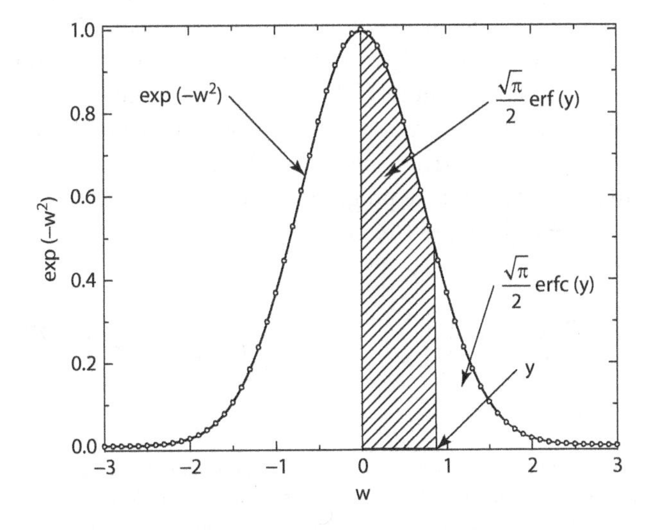

FIGURE 11.3 Plot showing relation of the error function of y, shaded area, and that of the complementary error function to the Gaussian curve.

11.3.2 Diffusion of Boron into Silicon

The diffusion of dopants into silicon at low concentrations, in the order of parts per million, is used to form electronic devices such as diodes and transistors on the surface of single crystal silicon to make integrated circuits. For example, boron could be diffused into the surface of silicon that was doped n-type* with, say phosphorus, to form a p-n junction diode somewhere below the surface. Suppose that the silicon were doped to be an n-type semiconductor with 10^{17} cm^{-3} P atoms. For B diffusion in Si, $D_0 = 0.76$ cm^2/s and Q = 3.46 eV = 333.8 kJ/mol (Mayer and Lau 1990, 210). Take T = 1500 K (1327°C), then $D(B) = 0.76\exp(-333{,}800/(8.314 \times 1500)) = 2.37 \times 10^{-12}\ \text{cm}^2\text{s}^{-1}$. At 1500 K, the equilibrium constant for the reaction $B_2H_6(g) \rightleftharpoons 2\,B(g) + 3\,H_2(g)$ is $K_p = 5.47 \times 10^{-14}$. (Roine 2002) Therefore,

$$\frac{p^2(B)p^3(H_2)}{p(B_2H_6)} = K_p(1500) = 5.47 \times 10^{-14}. \tag{11.11}$$

For silicon, ρ = 2.33 g/cm^3 and M = 28.09 g/mol (Haynes 2013), so η(Si) = the number of Si atoms per cm^3 = (2.33)(6.022 × 10^{23})/(28.09) = 5.0 × 10^{22} cm^{-3}. If $C_s(B) = 10^{19}$ cm^{-3}, then the mole fraction of boron, X(B) needs to be X(B) = 10^{19}/5 × 10^{22} = 2.0 × 10^{-4}. Because the mole fraction is small, it can be taken to be equal to the activity—partial pressure of B(g)—in the diborane–hydrogen atmosphere. To achieve a pressure of p(B) = 2.0 × 10^{-4}, assume a carrier gas of hydrogen having a fixed $p(H_2) = 10^{-2}$ atm The pressure of $p(B_2H_6)$ is then required to be

$$p(B_2H_6) = \frac{p(H_2)^3\,p(B)^2}{K_p}$$

$$= \frac{\left(10^{-2}\right)^3\left(10^{-4}\right)^2}{5.47 \times 10^{-14}}$$

$$p(B_2H_6) = 0.18\ \text{atm}$$

which is experimentally feasible.

Appendix gives some properties of the error function, and there are many places where tables of the error function exist (Abramowitz and Stegun 1965, 310). In addition, there are many software packages and graphing calculators that have both the error function and the complementary error function as built-in functions. So a spreadsheet was used to calculate the concentration of boron as a function of time and distance, $C(x, t)/C_S$ for various times at 1500 K, with the results plotted in Figure 11.4. As can be seen from the figure, if the p-n junction were to be established about 1 μm below the surface, then a total time just short of 1000 seconds would be required for the diffusion.

Of course, the concentration could have been specified at x = 1 μm and the exact time required could have been calculated from the same program. For example, $C/C_S = 0.1$, then by iteration in a spreadsheet of erfc(y) −0.1 = 0, gives y = 1.165 so

$$y = 1.165 = \frac{x}{\sqrt{4Dt}}, \quad t = \frac{1}{4D}\left(\frac{x}{y}\right)^2 = \frac{1}{4(2.37 \times 10^{-12})}\left(\frac{1 \times 10^{-4}}{1.165}\right)^2 = 777\ \text{seconds}$$

* Recall from Chapter 9 that an n-type dopant such as phosphorous has one more electron than silicon, and when it replaces a silicon atom in solid solution, it donates its extra electron to produce n-type electrical conductivity—conduction by a *negative* charge. In contrast, a p-type dopant such as boron has one less outer electron than silicon and accepts an electron producing a net positive charge to give p-type conduction. A p-n junction acts as a diode allowing current to flow in only one direction: one of the many devices critical in integrated circuits.

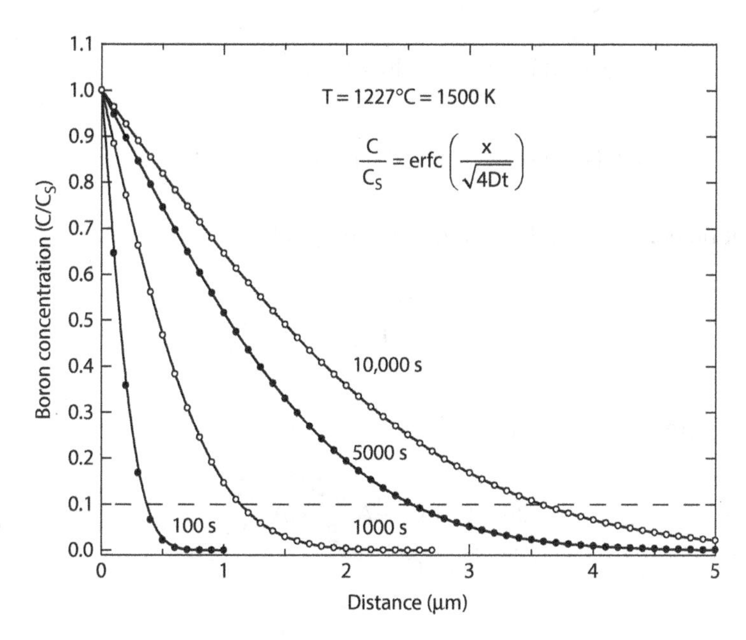

FIGURE 11.4 $C(x, t)/C_S$ for boron diffusing into silicon with surface concentration of $C_S = 10^{20}$ boron atoms cm^{-3} at 1227°C at various times with the formation of a p-n junction at C(x, t) at 10^{19} cm^{-3}.

the exact time necessary for the concentration at x = 1 μm to be $C/C_S = 0.1$. Also note that $x/\sqrt{4Dt} \cong 1$ for this value of 0.1 C_S to be reached. This suggests that an approximate—*back of the envelope*—value for the depth of diffusion in one direction is $x/\sqrt{4Dt} \cong 1$, which was the main point of Section 11.2.2.

11.4 INFINITE BOUNDARY CONDITIONS

11.4.1 Mathematical Model

The general solution that was obtained (see Equation 11.6) can also be used to easily solve infinite boundary condition problems as well. Consider the problem in Figure 11.5 in which the initial concentration left of zero, $-\infty < x < 0$, is C_1 and to the right of zero, $0 < x < \infty$, is C_2. If C_2 were zero, then this would be an example a large piece of silver diffusing into a large piece of gold and vice versa.* This leads to the infinite boundary conditions of $C(-\infty, t) = C_1$ and $C(\infty, t) = C_2$. Therefore,

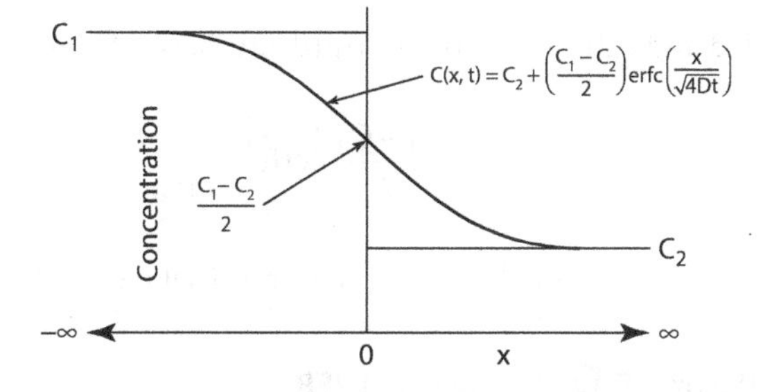

FIGURE 11.5 Depiction of diffusion with infinite boundary conditions for $C(x, 0) = C_1$ for $-\infty \le x \le 0$ and $C(x, 0) = C_2$ for $0 < x \le \infty$ that shows at $x = 0$, $C(0, t) = (C_1 - C_2)/2$ is a constant for all times.

* The same example holds for any other pair of metals, ceramics, or polymers, or liquids for that matter, that form a complete solid or liquid solution and there is no convection allowed for liquids: a good microgravity experiment.

$$
\begin{aligned}
C(\infty,t) &= A\int_0^{\infty} e^{-w^2}dw + B = A\frac{\sqrt{\pi}}{2} + B = C_2 \\
C(-\infty,t) &= A\int_0^{-\infty} e^{-w^2}dw + B = -A\frac{\sqrt{\pi}}{2} + B = C_1
\end{aligned}
\tag{11.12}
$$

because substitution of $z = -w$ in the second equation produces

$$
\int_0^{-\infty} e^{-w^2}dw = -\int_0^{\infty} e^{-z^2}dz = -\frac{\sqrt{\pi}}{2}.
$$

Solving Equation 11.12 for A and B gives

$$
\begin{aligned}
2B &= C_1 + C_2 + A\frac{\sqrt{\pi}}{2} - A\frac{\sqrt{\pi}}{2} \\
B &= \frac{C_1 + C_2}{2}
\end{aligned}
$$

and

$$
\begin{aligned}
C_2 - C_1 &= A\frac{\sqrt{\pi}}{2} + A\frac{\sqrt{\pi}}{2} \\
A &= \frac{2}{\sqrt{\pi}}\left(\frac{C_2 - C_1}{2}\right).
\end{aligned}
$$

Therefore, Equation 11.6 becomes

$$
C(x,t) = \frac{2}{\sqrt{\pi}}\left(\frac{C_2 - C_1}{2}\right)\int_0^{x/\sqrt{4Dt}} e^{-w^2}dw + \left(\frac{C_2 + C_1}{2}\right)
$$

or

$$
C(x,t) = \left(\frac{C_2 - C_1}{2}\right)\mathrm{erf}\left(\frac{x}{\sqrt{4Dt}}\right) + \left(\frac{C_2 + C_1}{2}\right)
\tag{11.13}
$$

which is one form for the solution. If C_1 is both added and subtracted from this equation, the following alternative form for the solution is obtained:

$$
C(x,t) = C_1 + \left(\frac{C_2 - C_1}{2}\right)\left[1 + \mathrm{erf}\left(\frac{x}{\sqrt{4Dt}}\right)\right].
$$

Similarly, if C_2 is both added and subtracted, then a third alternative form of the solution is obtained:

$$
C(x,t) = C_2 + \left(\frac{C_1 - C_2}{2}\right)\mathrm{erfc}\left(\frac{x}{\sqrt{4Dt}}\right).
$$

The reason for giving all three forms for the same solution is that each is found in the literature.

11.4.2 Interdiffusion of Gold and Silver

Gold and silver form a complete solid solution from the melting point of silver, 961°C, to the melting point of gold, 1063°C (Brandes and Brook 1992). From Table 9.5, for silver, $D(Ag) = 0.67\exp(-((189.1\,\mathrm{kJ/mole})/RT))$ and for gold, $D(Au) = 0.09\exp(-((176.1\,\mathrm{kJ/mole})/RT))$ and at 950°C (1223 K), $D(Ag) = 5.61 \times 10^{-9}\ \mathrm{cm^2/s}$ and $D(Au) = 2.35 \times 10^{-9}\ \mathrm{cm^2/s}$. Now these are *self-diffusion coefficients* and the diffusion of silver into gold is really *interdiffusion*, and an

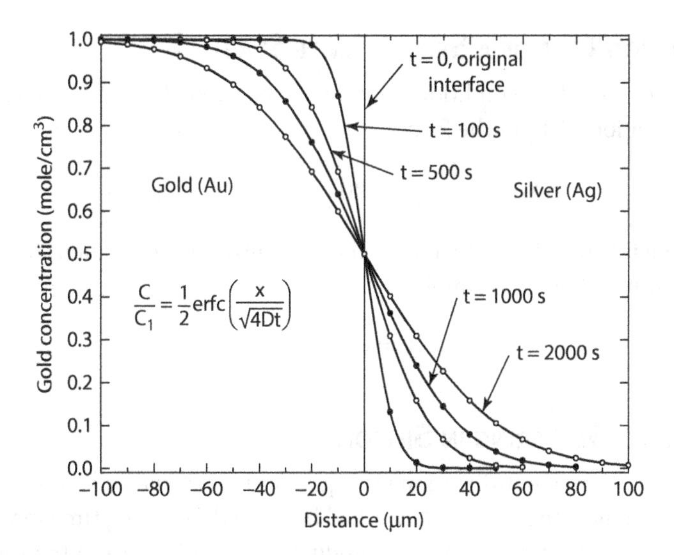

FIGURE 11.6 Interdiffusion of pure silver and gold at 950°C for various times illustrating the infinite boundary condition solution and that the concentration at x = 0, C(0, t) = 0.5 m/o gold. In this case, the self-diffusion coefficients are virtually identical so the infinite boundary condition solution is reasonably accurate. Also note that the diffusion profile for silver diffusing into gold is occurring at the same time, and is the exact mirror opposite of that for gold.

interdiffusion coefficient is necessary and will be discussed later in Chapter 14. For now, these two diffusion coefficients are very close and are essentially the same within the error of the calculation. Therefore, for simplicity, assume that the interdiffusion coefficient is essentially the average of these two values or $D = 3.98 \times 10^{-9}\ cm^2/s$. If, in Figure 11.5, the left of the origin is pure gold and the right of it is pure silver at time equal zero, then if $C_1 = 1$ and $C_2 = 0$, C(x, t) is the concentration of gold as a function of distance and time. Equation 11.13 becomes

$$C(x,t) = \frac{1}{2}\operatorname{erfc}\left(\frac{x}{\sqrt{4Dt}}\right) \tag{11.14}$$

which is plotted in Figure 11.6 for the interdiffusion of gold and silver at T = 950°C for times up to 2000 seconds (a little over one-half hour). Note that the distance of significant concentration change (say up to 10% silver or gold) is much larger for this system then was seen in Section 11.2.2 for diffusion of boron into silicon because of the larger diffusion coefficient here. Nevertheless, again $x/\sqrt{4Dt} = (32\times10^{-4})/\sqrt{4(2.35\times10^{-9})1000} = 1.04 \cong 1$ for the concentration of 0.1 mol/cm³ of silver to be reached: a good measure of *diffusion depth.*

11.5 APPLICATION TO SEMI-INFINITE BOUNDARY CONDITIONS

11.5.1 Constant Surface Concentration

If $C_2 = 0$ (Figure 11.5), then all three solutions, that is, Equation 11.13, become, for *positive values of x*,

$$C(x,t) = \frac{C_1}{2}\left[1 - \operatorname{erf}\left(\frac{x}{\sqrt{4Dt}}\right)\right] = \frac{C_1}{2}\operatorname{erfc}\left(\frac{x}{\sqrt{4Dt}}\right)$$

which is rather interesting because this is the same result that was obtained for the semi-infinite boundary conditions, but in this case $C_1/2$ takes the place of C_S. In other words, the concentration at $C(0, t) = C_1/2$ for all values of time and C(x, t) values for positive values of x are the same as for the fixed-surface concentration, which is now $C_1/2$. This result suggests that infinite boundary condition results can be used to solve other semi-infinite boundary condition problems.

11.5.2 Diffusion Out of a Semi-Infinite Slab

The use of the infinite boundary condition solution can be applied to *diffusion out* of a semi-infinite slab at x = 0. For Equation 11.13 in the form

$$C(x,t) = \left(\frac{C_2 - C_1}{2}\right)\text{erf}\left(\frac{x}{\sqrt{4Dt}}\right) + \left(\frac{C_2 + C_1}{2}\right)$$

if $C_1 = -C_2$ as shown if Figure 11.7, which fixes the concentration at the interface to zero, C(0, t) = 0. The solution for positive x becomes simply

$$C(x,t) = C_2\text{erf}\left(\frac{x}{\sqrt{4Dt}}\right). \tag{11.15}$$

11.5.3 Intrinsic Gettering in Silicon

A very practical application of out-diffusion via Equation 11.15 is *intrinsic gettering* in silicon that is used for integrated circuits. The term *gettering* had been used for a long time in vacuum technology for processes that removed small amounts of residual gas molecules from the vacuum. This terminology was applied to materials processing to describe processes that remove unwanted impurities—particularly those that could interfere with integrated circuit device functions—from solid solution. The undesirable impurities in silicon used for integrated circuits are particularly rapidly diffusing interstitial transition elements such as copper and nickel. These impurities tend to segregate to the stress fields of dislocations* and create highly conducting paths along the dislocation line. If such a dislocation intersects a diode or transistor on the silicon chip, it can short-circuit the device and make it non functional. There are several processes that are used for gettering impurities in silicon, and most of them generate dislocations far from the surface of the wafer where the integrated circuits are being processed: only the top few microns of the wafer surface are used for devices and the more than 600 μm of the rest of the wafer are just there for mechanical support. The basic concept of gettering is to generate dislocations far from the surface that will attract and essentially immobilize the unwanted impurities and keep them away from the devices on the surface.

Intrinsic gettering is a unique process that uses the properties and composition of a grown silicon single crystal wafer to get rid of the impurities (Wolf and Tauber 2000; Campbell 2001). Normal silicon wafers, although very pure, have about 10^{18} oxygen atoms per cm^3 in solid solution—about 20 ppm—because the silicon crystals are grown from molten silicon in silicon dioxide, SiO_2, crucibles. At the elevated temperatures where many of the integrated circuit processing steps are performed, such as the diffusion of dopants to form p-n junctions (Section 11.2.2), residual oxygen precipitates as small SiO_2 particles. These particles, because of a molar volume difference, generate stresses and dislocations in the silicon that can *trap* fast-diffusing impurities. However, it is undesirable to have the impurities near the surface where the integrated circuits are made. Therefore, *intrinsic gettering*

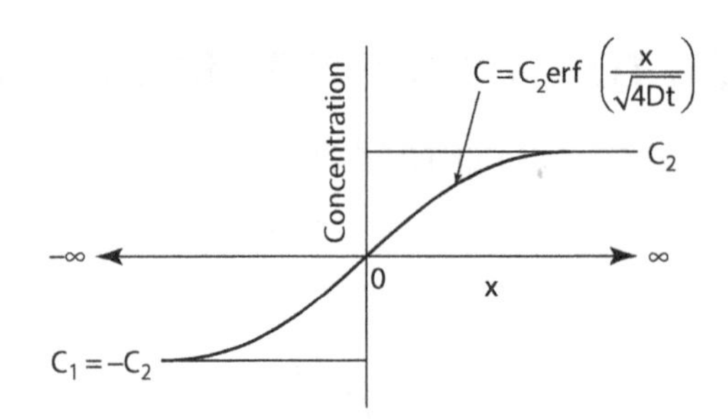

FIGURE 11.7 Infinite boundary condition solution in which $C_1(x, 0) = -C_2(x, 0)$ and the concentration at the original interface remains C(0, t) = 0. This solution can be used for the semi-infinite boundary condition problem where C(0, t) = 0: out-diffusion from a semi-infinite bar.

* This process is called *decorating* the dislocation and allows it to be observed more easily in a transmission electron microscope (TEM).

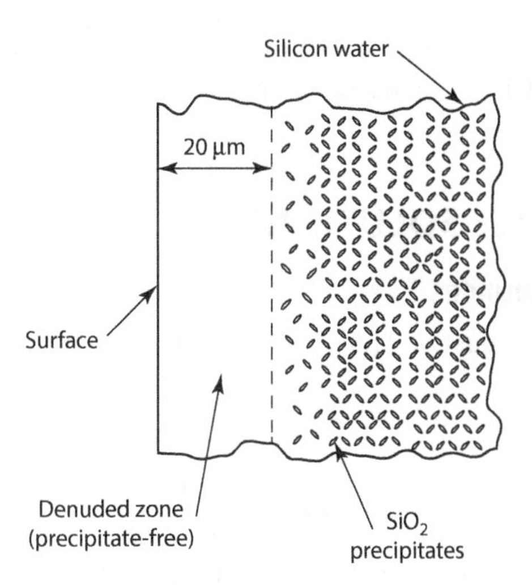

FIGURE 11.8 Schematic microstructure near the surface of a silicon wafer that has undergone the process of *intrinsic gettering*. The wafer is given three heat treatments: one at high temperatures to diffuse oxygen out near the surface; a second at low temperatures to nucleate SiO_2 precipitates in the region still rich in oxygen; and a third high temperature treatment to grow the precipitates, generate dislocations, and trap rapidly diffusion impurities. Precipitation does not occur in the *denuded zone* near the surface where the oxygen content has been reduced below the SiO_2 solubility required for precipitation during the first heat treatment.

involves an initial high temperature heat treatment, about 1200°C, in an inert gas to allow dissolved oxygen near the surface to diffuse out so its concentration will drop below the concentration at which precipitates form, 12 ppm or less (about 6×10^{17} cm^{-3}). Typically, the diffusion time is such that at 1200°C, a zone free of precipitates, the so-called *denuded zone*, is about 20 μm, shown schematically in Figure 11.8. After the surface oxygen removal heat treatment, the wafer is then given a low temperature anneal, about 800°C, to generate a high nucleation rate and a large number of oxide precipitates below the surface. Finally, a third heat treatment near 1000°C is used to grow the nucleated SiO_2 precipitates large enough to generate dislocations needed to trap the transition metal impurities.

For oxygen diffusion in silicon, $D = 7 \times 10^{-2} \exp(-((235{,}460\,\text{J}/\text{mole})/RT))$ (Mayer and Lau 1990) so $D(1200°C) = 3.12 \times 10^{-10}\,\text{cm}^2\text{s}^{-1}$. Figure 11.9 is a plot of Equation 11.15 at 1200°C for

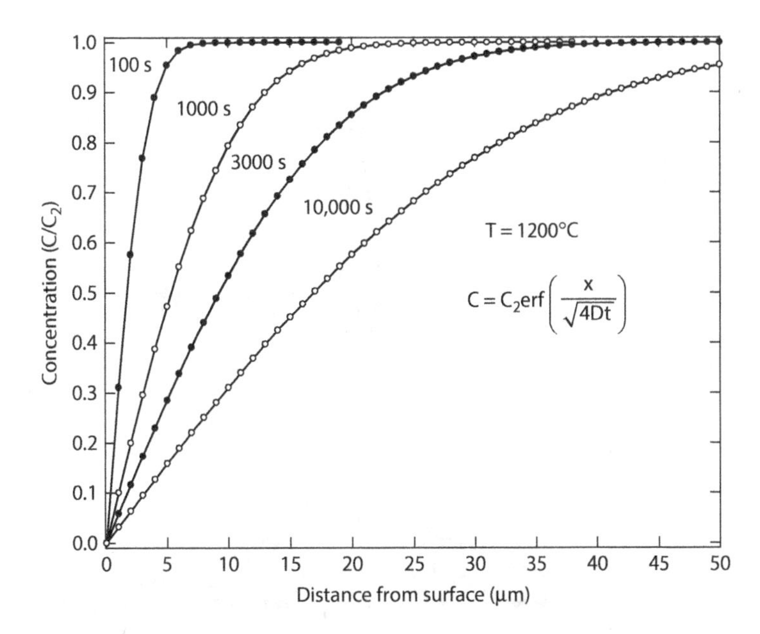

FIGURE 11.9 Diffusion of oxygen from the silicon surface at 1200°C showing that after about 3 hours (10^4 seconds) the oxygen concentration has dropped from an initial concentration of 10^{18} cm^{-3} below about 6×10^{17} cm^{-3} over the first 20 μm or so from the surface. This is sufficiently low so that SiO_2 will not precipitate in this *denuded region* during following heat treatments.

various times showing that the oxygen concentration will drop to about 6 × 10[17] cm[3] at 20 μm depth, as required, in about 10^4 seconds, or about 3 hours.

11.6 FINITE SOURCE SOLUTIONS

11.6.1 Individual Sources

In Appendix, it is shown that Equation 11.6 can be rewritten—with a simple change of variables—to Equation A.8,

$$C(x,t) = \frac{1}{\sqrt{4\pi Dt}} \int_{-\infty}^{\infty} f(x_0) e^{-\frac{(x-x_0)^2}{4Dt}} dx_0 \tag{11.16}$$

where $f(x_0)$ is some concentration function of the dummy variable x_0. It is also shown in Appendix that the function does not necessarily have to be continuous but may consist of a finite number of single source functions, Equation A.10,

$$C(x,t) = \frac{Q(x_0)}{\sqrt{4\pi Dt}} e^{-\frac{(x-x_0)^2}{4Dt}} \tag{11.17}$$

where $Q(x_0)$ is the amount of material (mol/cm²) at x_0, the *source*. Equations 11.16 and 11.17 provide the opportunity to develop many other solutions to Fick's second law for infinite and finite boundary conditions. For example, a single source at x = 0 and Q = 1 is shown as a function of 4Dt, the diffusion parameters, in Figure 11.10. Figure 11.11 shows the resulting concentration as a function of 4Dt for two sources at different positions and different strengths is simply the sum of the concentration profiles for the two sources.

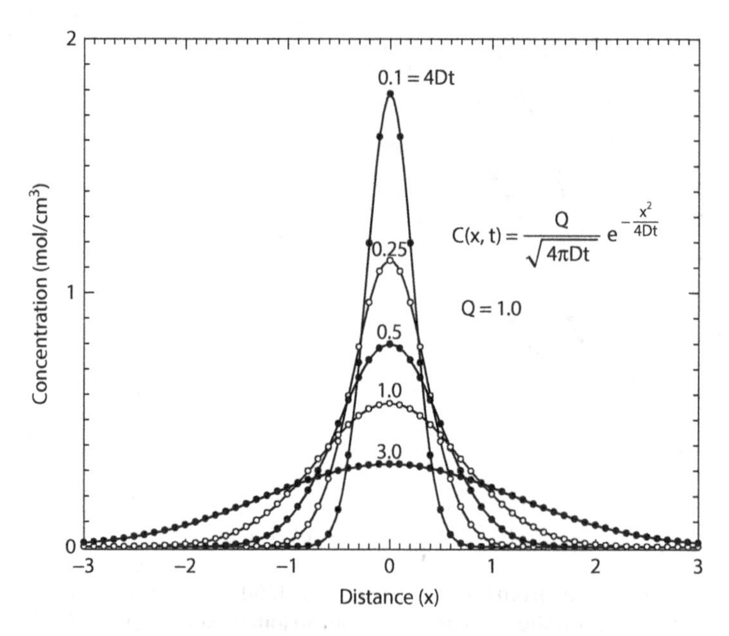

FIGURE 11.10 Diffusion profiles as a function of 4Dt for a single source of strength Q at the origin at t = 0 diffusing with infinite boundary conditions.

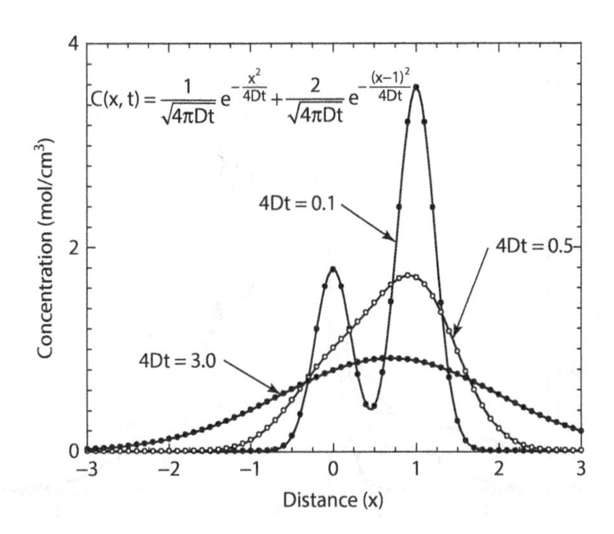

FIGURE 11.11 C(x, t) with infinite boundary conditions from two sources: one at x = 0 of strength Q = 1 and one at x = 1 of strength Q = 2 illustrating how there concentrations blend together.

11.6.2 Extended Sources

Equation 11.16 holds for any type of extended source, f(x), as well as individual sources. For example, consider $f(x) = C_0$ for $-a \le x \le a$. A more complex function of x could be chosen other than a constant concentration, C_0, but this just makes the algebra involving the integrations more time consuming without adding any additional insight. Therefore, for $f(x) = C_0$, Equation 11.16 becomes

$$C(x,t) = \frac{C_0}{\sqrt{4\pi Dt}} \int_{-a}^{a} e^{-\frac{(x-x_0)^2}{4Dt}} dx_0. \tag{11.18}$$

The procedure to put these types of equations into a more familiar and calculable form is just the reverse of that used in the Appendix to obtain Equation 11.16, simply make a variable substitution. An obvious choice is to let $y = (x - x_0)/\sqrt{4Dt}$ so $x_0 = x - \sqrt{4Dt}\, y$ and $dx_0 = -\sqrt{4Dt}\, dy$, and when $x_0 = -a$, $y = (x + a)/\sqrt{4Dt}$ and when $x_0 = a$, $y = (x - a)/\sqrt{4Dt}$ so

$$\frac{C}{C_0} = -\frac{1}{\sqrt{4\pi Dt}} \int_{(x+a)/\sqrt{4Dt}}^{(x-a)/\sqrt{4Dt}} e^{-y^2} \sqrt{4Dt}\, dy.$$

Equation 11.18 is now

$$\frac{C}{C_0} = -\frac{1}{\sqrt{\pi}} \int_{(x+a)/\sqrt{4Dt}}^{0} e^{-y^2} dy - \frac{1}{\sqrt{\pi}} \int_{0}^{(x-a)/\sqrt{4Dt}} e^{-y^2} dy$$

$$\frac{C}{C_0} = \frac{1}{\sqrt{\pi}} \int_{0}^{(x+a)/\sqrt{4Dt}} e^{-y^2} dy - \frac{1}{\sqrt{\pi}} \int_{0}^{(x-a)/\sqrt{4Dt}} e^{-y^2} dy$$

(ugly but no fancy mathematics) which, of course, is

$$\frac{C}{C_0} = \frac{1}{2} \operatorname{erf}\left(\frac{x+a}{\sqrt{4Dt}}\right) - \frac{1}{2} \operatorname{erf}\left(\frac{x-a}{\sqrt{4Dt}}\right) \tag{11.19}$$

which is nothing but a sum of two numbers for specific values of a and $\sqrt{4Dt}$. Figure 11.12 shows the solution for some values of 4Dt. An example where such a solution might apply is for a thin sheet of one metal (e.g., gold) sandwiched between two pieces of silver and heated to get interdiffusion and a *solid-state bond or weld* between the two pieces of silver. Another is an adhesive bond between two polymers or pieces of wood with the diffusion of the solvent (with some of the adhesive) into the pieces being joined.

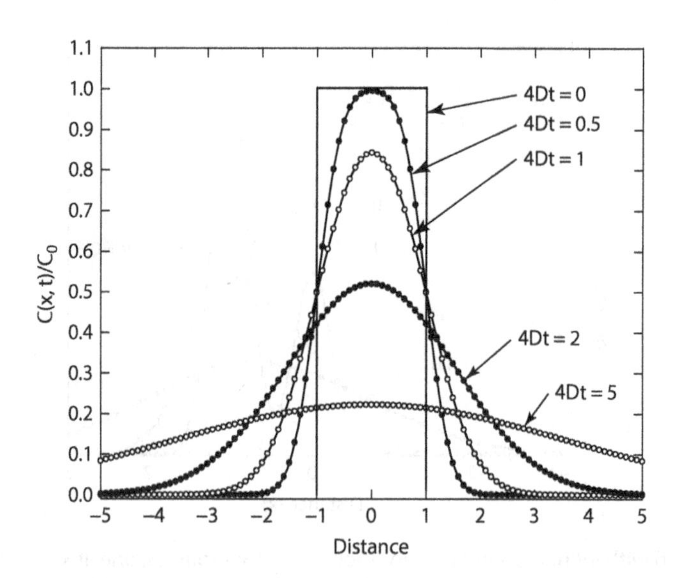

FIGURE 11.12 C(x, t) for infinite boundary conditions for an extended source of $C(x, 0) = C_0$ for $-1 \leq x \leq +1$.

A little mathematical trick is to put Equation 11.19 into a dimensionless variable form by dividing the top and bottom of the parameters in the parentheses by a to give

$$\frac{C}{C_0} = \frac{1}{2}\mathrm{erf}\left(\frac{(x/a)+1}{\sqrt{4(Dt/a^2)}}\right) - \frac{1}{2}\mathrm{erf}\left(\frac{(x/a)-1}{\sqrt{4(Dt/a^2)}}\right).$$

The solution is now in terms of a dimensionless distance variable, x/a, and a dimensionless time variable, $(4Dt/a^2)$.

11.6.3 Method of Images

11.6.3.1 Introduction

For semi-infinite boundary conditions, there are times when the concentration at the boundary of the medium, x = 0, needs to be a fixed value. For example, in Figure 11.11, the influence of the source at x = 1 extends beyond the origin into negative values of x. If this were a semi-infinite medium, clearly the material could not leave the end of the bar at x = 0 but would rather be reflected back to positive x values. The somewhat graphical way to handle this is by the *method of images.*

11.6.3.2 Concentration Confined to Positive Values of x

The top graphic in Figure 11.13 illustrates how this is done by the simple graphical/mathematical trick of applying an *image* source. For example, the shaded portion near the origin is the amount of material from the source at x_0 that is reflected back from the origin. This is the same C(x, t) as the tail of a similar source placed at $-x_0$ because its tail in the positive x direction will add to the source at $x = x_0$. To get the solution, the real source and its image simply have to be added and evaluated for positive values of x. Therefore, the solution for a single source with semi-infinite boundary conditions in which material is not allowed to leave the end of the region at x = 0 is

$$C(x,t) = \frac{1}{\sqrt{4\pi Dt}}e^{-\frac{(x-x_0)^2}{4Dt}} + \frac{1}{\sqrt{4\pi Dt}}e^{-\frac{(x+x_0)^2}{4Dt}} \tag{11.20}$$

where the second term is the image source at $-x_0$. Figure 11.14 shows the resulting concentration as a function of positive x for various values of 4Dt with a source of Q = 1.0 at $x_0 = 1$. Note that, at x = 0, the C(0, t) concentrations approach a constant value or $dC(0,t)/dx = 0$, a *derivative boundary condition* presented in Chapter 8, because there can be no flux in the negative x direction at x = 0.

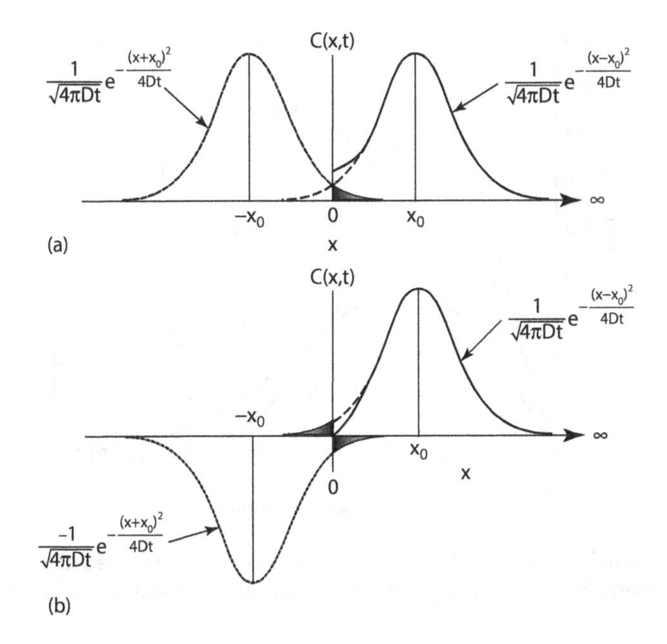

FIGURE 11.13 Illustration of the method of images for semi-infinite boundary conditions with a single source and image. (a) shows how the addition of an image source leads to a derivative boundary condition at x = 0: all of the material is reflected back to positive x at x = 0; (b) shows that with a negative image the concentration C(0, t) = 0: material leaves the surface.

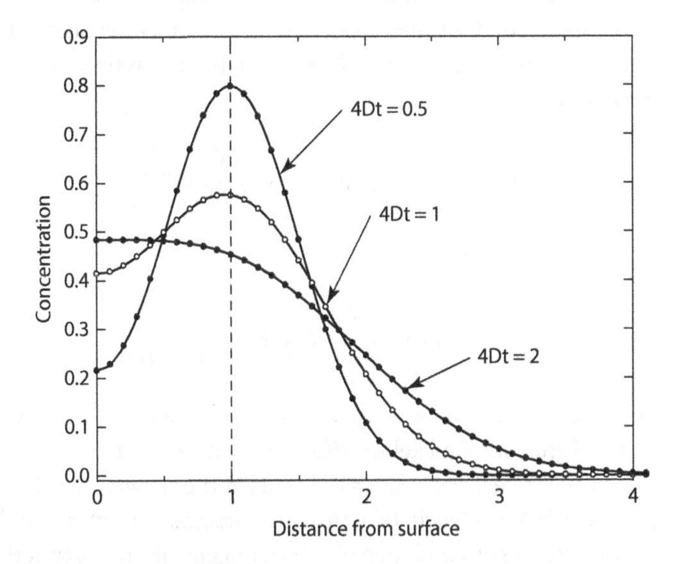

FIGURE 11.14 Semi-infinite boundary conditions with the application of a positive image (at x = −1.0) of a source at x = 1.0 of strength Q = 1 showing that dC/dx(0, t) = 0, or the material diffusing to the left is reflected at the boundary back into the positive direction.

11.6.3.3 C(0, t) = 0

In some applications, the problem requires that the concentration remains zero—or some other fixed value—at the interface of the semi-infinite medium, C(0, t) = 0. The lower graphic in Figure 11.13 shows that this requires a negative image source, so that the concentration for the source at x = 1 is just canceled by the overlapping concentration from the source of Q = −1 at $x = -x_0$. The resulting concentration, C(x, t: t > 0), is shown in Figure 11.15 as a single source at $x_0 = 1$ and Q = 1:

$$C(x,t) = \frac{1}{\sqrt{4\pi Dt}}e^{-\frac{(x-x_0)^2}{4Dt}} - \frac{1}{\sqrt{4\pi Dt}}e^{-\frac{(x+x_0)^2}{4Dt}}. \tag{11.21}$$

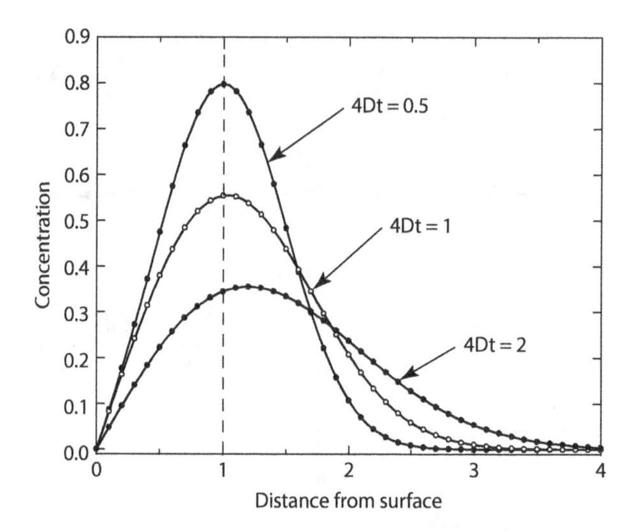

FIGURE 11.15 Semi-infinite boundary conditions with the application of a negative image (at x = −1.0) of a source at x = 1.0 of strength Q = 1 showing that C(0, t) = 0, or the material diffusing to the left leaves the material.

11.6.4 Surface Source and Measurement of D

A very important experimental approach to measuring diffusion coefficients in solids is to place a single source of a radioactive tracer element on the surface of a bar and heat-treat it for a given time. The method of images is very useful for measuring the resultant diffusion coefficient. For a single source in Figure 11.14 placed at the origin, one-half of the material is reflected back to positive values of x and the solution becomes

$$C(x,t) = 2\frac{Q}{\sqrt{4\pi Dt}}e^{-\frac{x^2}{4Dt}} = \frac{Q}{\sqrt{\pi Dt}}e^{-\frac{x^2}{4Dt}} \tag{11.22}$$

and taking the $\log_{10}$ of both sides gives

$$\log_{10} C(x,t) = \log_{10}\left(\frac{Q}{\sqrt{\pi Dt}}\right) - \frac{x^2}{2.303 \times 4Dt} \tag{11.23}$$

which provides a simple, and very commonly used, method of determining diffusion coefficients in solids. As an example, the diffusion of nickel in NiO is to be measured. Deposit a thin surface layer of amount Q of radioactive nickel (e.g., ^{63}Ni, $t_{1/2} = 101$ year) at the surface of a NiO bar then heat for a time (t) at some temperature. When cooled, take thin sections, Δx, and measure the radioactivity in each Δx-thick section as a function of x and then plot the concentration as a function of x^2 on a semi-log plot, Equation 11.23, as shown in Figure 11.16. The diffusion coefficient can be determined from the slope of the line and, once determined, can be used to calculate Q from the intercept. If there are rapid diffusion paths, such as dislocations or grain boundaries, then their contributions will appear with less steep slopes at low concentrations and their diffusion coefficients can also be measured.

11.6.5 Ion Implantation

Ion implantation is a process in which high energy (>0.2 keV and <2 MeV) ions are implanted into a silicon substrate (or any other material) in an ion accelerator to replace doping by diffusion from the surface. Some advantages of ion implantation are: the depth of penetration can be controlled by the ion energy; it can be done at room temperature; and the doping location can be controlled by the pattern in the polymer photoresist material (Wolf and Tauber 2000). However, the concentration profile generated by ion implantation is usually too steep and narrow and must be given a post-deposition *drive-in* diffusion heat treatment. Normally, the concentration profile generated by

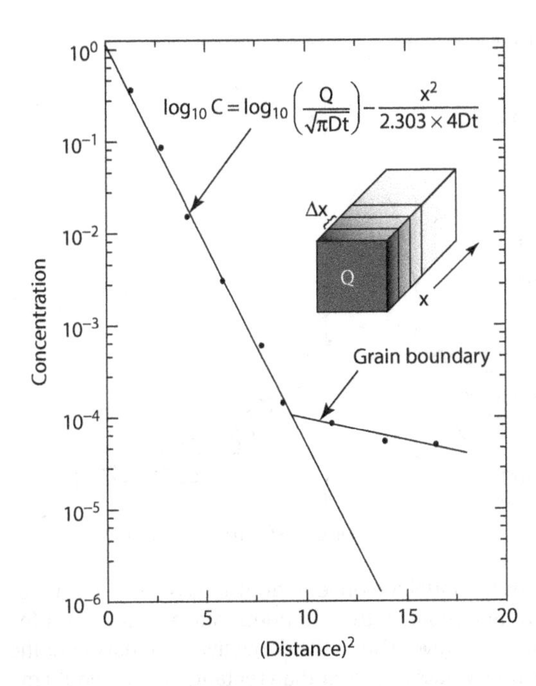

FIGURE 11.16 Schematic illustrating the technique of a surface source to measure diffusion coefficients in solids. The inset shows a solid with an amount Q of radioactive isotope on the surface and allowed to diffuse into the bar. After a given time, the bar is sectioned parallel to the surface into slices of thickness Δx and the radioactivity (concentration) measured and plotted (the data points) on a semilog plot of concentration versus depth, x^2. From both the slope and intercept the diffusion coefficient at that temperature can be calculated. If there are fast diffusion paths present such as grain boundaries or dislocations, these will appear at low concentrations as shown.

ion implantation is Gaussian with some mean penetration depth on the order of a few tenths of a micrometer. After a relatively short diffusion time, the shape of the original as-implanted profile does not have much effect on the concentration profile, C(x, t). To simplify the problem, yet get reasonably accurate values, Figure 11.17 shows concentration profiles as a function of 4Dt starting with the rectangular profile at 4Dt = 0 as the as-deposited ion implantation profile. Then with the method of images described in Section 11.6.2.2, the concentration profiles are given by Equation 11.16:

$$C(x,t) = \frac{C_0}{\sqrt{4\pi Dt}} \int_{-\infty}^{\infty} e^{-\frac{(x-x_0)^2}{4Dt}} dx_0$$

$$C(x,t) = \frac{C_0}{\sqrt{4\pi Dt}} \int_{-b}^{-a} e^{-\frac{(x+x_0)^2}{4Dt}} dx_0 + \frac{C_0}{\sqrt{4\pi Dt}} \int_{a}^{b} e^{-\frac{(x-x_0)^2}{4Dt}} dx_0$$

where the first term in the second equation is simply the image of the second term. Substitution of $y = (x - x_0)/\sqrt{4Dt}$ and $dx_0 = -\sqrt{4Dt}\, dy$ results in four terms, two each for the source and its image,

$$C(x,t) - \frac{1}{\sqrt{\pi}} \int_{(x+b)/\sqrt{4Dt}}^{0} e^{-y^2} dy - \frac{1}{\sqrt{\pi}} \int_{0}^{(x+a)/\sqrt{4Dt}} e^{-y^2} dy - \frac{1}{\sqrt{\pi}} \int_{(x+a)/\sqrt{4Dt}}^{0} e^{-y^2} dy - \frac{1}{\sqrt{\pi}} \int_{0}^{(x+b)/\sqrt{4Dt}} e^{-y^2} dy$$

and from the definition of erf(z)

$$C(x,t) = \frac{1}{2}\text{erf}\left(\frac{x+b}{\sqrt{4Dt}}\right) - \frac{1}{2}\text{erf}\left(\frac{x+a}{\sqrt{4Dt}}\right) + \frac{1}{2}\text{erf}\left(\frac{x-a}{\sqrt{4Dt}}\right) - \frac{1}{2}\text{erf}\left(\frac{x-b}{\sqrt{4Dt}}\right). \quad (11.24)$$

Admittedly, Equation 11.24 looks pretty ugly! But a, b, and $\sqrt{4Dt}$ are known and the Equation 11.24 is simply a sum of four numbers and easily numerically evaluated. These are plotted in Figure 11.17 for

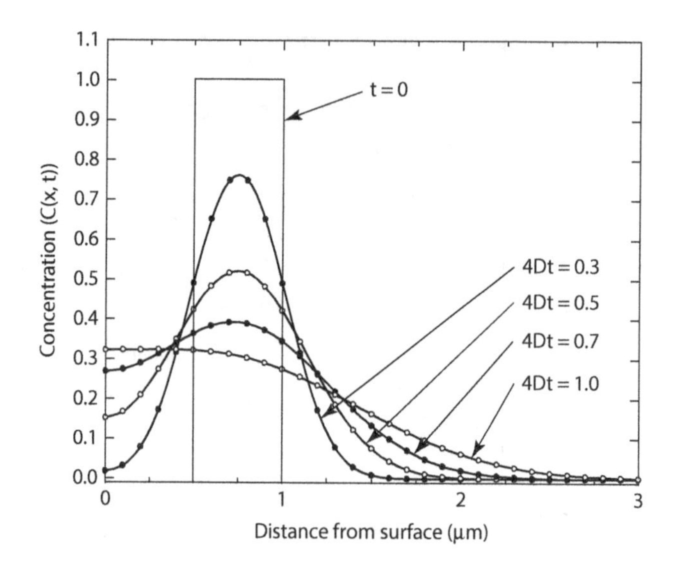

FIGURE 11.17 Use of images to model the drive-in diffusion concentration profiles after ion implantation. The implanted concentration profile is the extended source of $C(x, 0) = 1$ for $0.5 \le x \le 1.0$ and there is a similar extended image source that gives rise to the derivative boundary condition $dC/dx(0) = 0$. An actual implanted profile would be more Gaussian rather than rectangular but would make little difference by the time $4Dt = 1.0$ where the concentration profile is rather flat and ideal for forming a p-n junction.

$a = 0.5\ \mu m$, $b = 1\ \mu m$, and various values of 4Dt. Again, at $4Dt \cong 1$, the diffusion profile is reasonably flat for over 1 μm making it ideal for forming a p-n junction.

11.7 FINITE SOURCE AND RANDOM WALK

11.7.1 Introduction

The equation for a single source at x_0 permits an easy entrance to more fundamental concepts concerning diffusion. Frequently, some of these more fundamental concepts discussed are introduced earlier as the basis for diffusion and much of what has been discussed up to this point in Chapters 8 through 11 follows later. The approach presented here is a more *phenomenological* route in that the movement of atoms in crystals and gases is more easily visualized than the more mathematical, abstract route. However, steps can now easily be retraced to examine more basic ideas that, hopefully, lead to a fuller understanding of the *random thermal motion* described by diffusion.

11.7.2 Mean Displacement

Equation 11.17 for a single source at x_0 is,

$$C(x,t) = \frac{Q(x_0)}{\sqrt{4\pi Dt}} e^{-\frac{(x-x_0)^2}{4Dt}}. \tag{11.25}$$

Instead of concentration, if the interest is in the number of atoms per unit volume, $\eta(x,t)$ then

$$\eta(x,t) = \frac{N(x_0)}{\sqrt{4\pi Dt}} e^{-\frac{(x-x_0)^2}{4Dt}} \tag{11.26}$$

where $N(x_0)$ = the number of atoms/cm^2 at x_0. To make life simpler, assume that the atoms are all initially positioned at $x_0 = 0$, that is, at the origin. Then the probability of finding an atom at some position and time, p(x, t) would simply be

$$p(x,t) = \frac{\eta(x,t)}{N} = \frac{1}{\sqrt{4\pi Dt}} e^{-\frac{x^2}{4Dt}}. \quad (11.27)$$

And the mean position from the origin, $\bar{x}$, for the N atoms is simply,

$$\bar{x} = \frac{1}{\sqrt{4\pi Dt}} \int_{-\infty}^{\infty} x e^{-\frac{x^2}{4Dt}} dx.$$

In addition, making the usual substitution of $y = x/\sqrt{4Dt}$, the integral becomes

$$\sqrt{\frac{4Dt}{\pi}} \int_{-\infty}^{\infty} ye^{-y^2} dy = -\sqrt{\frac{Dt}{\pi}} \left. e^{-y^2} \right|_{-\infty}^{\infty} = 0 \quad (11.28)$$

which is no big surprise because Equation 11.27 is an even function of x.* Equation 11.28 says that the mean position of an atom is zero, which is also not surprising because the atoms can move either in the positive or in the negative x-direction and on the average, there will be as many to the left of the origin as there is to the right.

11.7.3 Mean Square Distance in One Dimension

On the other hand, the *mean square distance* covered by atoms moving only in the positive and negative x-directions is given by

$$\langle x^2 \rangle = \frac{1}{\sqrt{4\pi Dt}} \int_{-\infty}^{\infty} x^2 e^{-\frac{x^2}{4Dt}} dx \quad (11.29)$$

where $\langle \; \rangle$ means the *average of* and again making the usual substitution $y = x/\sqrt{4Dt}$ this becomes

$$\langle x^2 \rangle = \frac{4Dt}{\sqrt{\pi}} \int_{-\infty}^{\infty} y^2 e^{-y^2} dy = \frac{8Dt}{\sqrt{\pi}} \int_{0}^{\infty} y^2 e^{-y^2} dy \quad (11.30)$$

because it is an *even function* of y. To evaluate the integral, let $y^2 = z$, then

$$\int_0^{\infty} y^2 e^{-y^2} dy = \frac{1}{2} \int_0^{\infty} z^{1/2} e^{-z} dz = \frac{1}{2} \int_0^{\infty} z^{((3/1)-1)} e^{-z} dz \text{ so}$$

$$\int_0^{\infty} y^2 e^{-y^2} dy = \frac{1}{2} \Gamma\left(\frac{3}{2}\right) = \frac{1}{2}\left(\frac{\pi}{2}\right) = \frac{\pi}{4} \quad (11.31)$$

where $\Gamma(3/2)$ is the *gamma function* of 3/2 where the gamma function is just a number and is defined by (see Appendix A.3)

$$\Gamma(\alpha) = \int_0^{\infty} x^{\alpha-1} e^{-x} dx.$$

Therefore,

$$\Gamma\left(\frac{3}{2}\right) = \int_0^{\infty} x^{((3/2)-1)} e^{-x} dx = \frac{\pi}{2}.$$

So combining Equations 11.30 and 11.31, the mean square displacement is related to the diffusion coefficient and is given by

$$\langle x^2 \rangle = 2Dt. \quad (11.32)$$

* An *even function* is one for which f(−x) = f(x) such as cos(x). An odd function is one for which f(−x) = −f(x) such as sin(x).

11.7.4 Mean Square Distance in Three Dimensions

In three dimensions, the displacement of an atom from the origin is a vector $\vec{R}$ that can be written as

$$\vec{R} = x\vec{i} + y\vec{j} + z\vec{k}$$

and because $R^2 = x^2 + y^2 + z^2$, then

$$\langle R^2 \rangle = \langle x^2 \rangle + \langle y^2 \rangle + \langle z^2 \rangle.$$

And in the discussion in Section 11.7.3, there was no distinction made about the three directions x, y, and z, so the argument there holds for the other two mean square distances as well, so that

$$\begin{aligned} \langle R^2 \rangle &= 2Dt + 2Dt + 2Dt \\ \langle R^2 \rangle &= 6Dt. \end{aligned} \tag{11.33}$$

11.7.5 Random Walk and Diffusion

11.7.5.1 Example with 15 Steps

Consider the two-dimensional lattice of Figure 11.18. After 15 random atom jumps* of equal distance a in the ±x or ±y directions, a *random walk*, it has moved to position $\vec{R}_{15}$ from the origin, and this is the vector sum of all of the 15 individual atom jumps,

$$\vec{R}_{15} = \vec{r}_1 + \vec{r}_2 + \cdots \vec{r}_{15}$$

where the length of each $\vec{r}_i$ is always a. Then the square of the displacement, R_{15}^2, is just the dot product of the vector with itself

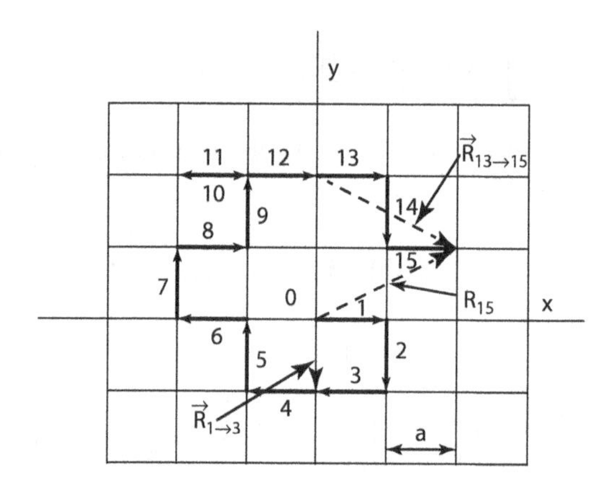

FIGURE 11.18 Two-dimensional random walk on a square grid of 15 steps each of distance *a*. R_{15} is the resulting distance vector of the 15 steps from the origin. $\vec{R}_{1\to3} = \vec{R}_{13\to15}$ are two shorter three-step paths.

* In Section 9.3.5, it was shown that interstitial atoms can take about 10^8 jumps per second, so this random walk of 15 steps takes place in less than a microsecond!

$$\vec{R}_{15}\cdot\vec{R}_{15} = (\vec{r}_1+\vec{r}_2+\cdots\vec{r}_{15})\cdot(\vec{r}_1+\vec{r}_2+\cdots\vec{r}_{15})$$
$$= \vec{r}_1\cdot\vec{r}_1+\vec{r}_1\cdot\vec{r}_2+\cdots\vec{r}_1\cdot\vec{r}_{15}$$
$$+\vec{r}_2\cdot\vec{r}_1+\vec{r}_2\cdot\vec{r}_2+\cdots\vec{r}_2\cdot\vec{r}_{15} \quad (11.34)$$
$$\vdots$$
$$+\vec{r}_{15}\cdot\vec{r}_1+\vec{r}_{15}\cdot\vec{r}_2+\cdots\vec{r}_{15}\cdot\vec{r}_{15}$$

Equation 11.34 is a sum of 225 terms! However, this can be simplified by grouping terms

$$\vec{R}_{15}\cdot\vec{R}_{15} = (\vec{r}_1\cdot\vec{r}_1+\vec{r}_2\cdot\vec{r}_2+\vec{r}_3\cdot\vec{r}_3\cdots+\vec{r}_{15}\cdot\vec{r}_{15})+(\underline{\vec{r}_1\cdot\vec{r}_2}+\vec{r}_1\cdot\vec{r}_3+\vec{r}_1\cdot\vec{r}_4\cdots+\vec{r}_1\cdot\vec{r}_{15})$$
$$+(\underline{\vec{r}_2\cdot\vec{r}_1}+\vec{r}_2\cdot\vec{r}_3+\vec{r}_2\cdot\vec{r}_4\cdots+\vec{r}_2\cdot\vec{r}_{15})\cdots+(\vec{r}_{15}\cdot\vec{r}_1+\vec{r}_{15}\cdot\vec{r}_2+\vec{r}_{15}\cdot\vec{r}_3\cdots+\vec{r}_{15}\cdot\vec{r}_{14}). \quad (11.35)$$

Grouping the diagonal terms separately in the first sum, it follows that (if you spend the time and look *very* carefully at Equation 11.34 [Shewmon 1989]):

$$\vec{R}_{15}\cdot\vec{R}_{15} = \sum_{i=1}^{15}\vec{r}_i\cdot\vec{r}_i + 2\sum_{j=1}^{14}\sum_{i=1}^{15-j}\vec{r}_i\cdot\vec{r}_{i+j}. \quad (11.36)$$

The first term in Equation 11.36 is obvious, while the second term is harder to see. However, the 2 in front of the second term occurs since there are two dot products for each term in the sum that are the same, that is, $\vec{r}_1\cdot\vec{r}_2$ and $\vec{r}_2\cdot\vec{r}_1$ underlined in Equation 11.35. The dot product can be written as

$$\vec{r}_1\cdot\vec{r}_2 = |\vec{r}_1||\vec{r}_2|\cos\theta_{1,2}$$

where $|\vec{r}_i|$ are the magnitudes of each of the vectors, $|\vec{r}_i| = a$ in this case, and $\theta_{i,j}$ is the angle between any two vectors. Therefore, Equation 11.36 can be written:

$$R_{15}^2 = 15a^2 + 2a^2\sum_{j=1}^{14}\sum_{i=1}^{15-j}\cos\theta_{i,i+j}.$$

For the random walk of Figure 11.18, $\cos\theta_{i,j} = 0, 1, -1$, because the vectors are all at right angles, are parallel, or are antiparallel, respectively. Note that in Figure 11.18, $R_{15}^2 \neq 0$, and by inspection it is $R_{15}^2 = x^2 + y^2 = (2a)^2 + a^2 = 5a^2$. This could be shown by taking all 225 of the dot products above—not an exciting project!

11.7.5.2 Example with Three Steps

However, the principle can be demonstrated by taking a shorter path of only three vectors, say $\vec{R}_{1\to3} = (\vec{r}_1+\vec{r}_2+\vec{r}_3)$, so that

$$(\vec{R}_{1\to3})\cdot(\vec{R}_{1\to3}) = (\vec{r}_1+\vec{r}_2+\vec{r}_3)\cdot(\vec{r}_1+\vec{r}_2+\vec{r}_3)$$
$$= \vec{r}_1\cdot\vec{r}_1+\vec{r}_1\cdot\vec{r}_2+\vec{r}_1\cdot\vec{r}_3+\vec{r}_2\cdot\vec{r}_1+\vec{r}_2\cdot\vec{r}_2+\vec{r}_2\cdot\vec{r}_3+\vec{r}_3\cdot\vec{r}_1+\vec{r}_3\cdot\vec{r}_2+\vec{r}_3\cdot\vec{r}_3$$
$$= (\vec{r}_1\cdot\vec{r}_1+\vec{r}_2\cdot\vec{r}_2+\vec{r}_3\cdot\vec{r}_3)+(\vec{r}_1\cdot\vec{r}_2+\vec{r}_1\cdot\vec{r}_3+\vec{r}_2\cdot\vec{r}_1+\vec{r}_2\cdot\vec{r}_3+\vec{r}_3\cdot\vec{r}_1+\vec{r}_3\cdot\vec{r}_2)$$
$$= (\vec{r}_1\cdot\vec{r}_1+\vec{r}_2\cdot\vec{r}_2+\vec{r}_3\cdot\vec{r}_3)+2(\vec{r}_1\cdot\vec{r}_2+\vec{r}_1\cdot\vec{r}_3+\vec{r}_2\cdot\vec{r}_3)$$
$$= 3a^2+2a^2(\cos\theta_{1,2}+\cos\theta_{1,3}+\cos\theta_{2,3})$$
$$(\vec{R}_{1\to3})^2 = 3a^2+2a^2(0-1+0) = a^2.$$

Because $\theta_{1,3} = 180°$, $\cos\theta_{1,3} = -1$ and the other two angles are 90° so that their cosines are zero. Note that in Figure 11.18, the length of the vector $\vec{R}_{1\to3}$ is indeed $|\vec{R}_{1\to3}| = a$ and points in the −y direction from the origin to where vectors 3 and 4 connect.

11.7.5.3 General Result

For any general value of n, Equation 11.36 becomes

$$\vec{R}_n \cdot \vec{R}_n = \sum_{i=1}^{n} \vec{r}_i \cdot \vec{r}_i + 2\sum_{j=1}^{n-1}\sum_{i=1}^{n-j} \vec{r}_i \cdot \vec{r}_{i+j} \quad (11.37)$$

and following the procedures above, Equation 11.37 can be written

$$R_n^2 = n a^2 \left(1 + \frac{2}{n}\sum_{j=1}^{n-1}\sum_{i=1}^{n-j} \cos\theta_{i,i+j}\right) \quad (11.38)$$

where R_n is the path taken by a single diffusing atom. To get the mean position of all the diffusing atoms, the average of Equation 11.38 must be taken so,

$$\langle R_n^2 \rangle = n a^2 \left(1 + \frac{2}{n}\left\langle \sum_{j=1}^{n-1}\sum_{i=1}^{n-j} \cos\theta_{i,i+j} \right\rangle\right) \quad (11.39)$$

where the brackets $\langle\ \rangle$ designate the *average* value. As was shown, some of the cosine values may be positive and some may be negative, and on the average, they will cancel out when summed over a large value of n jumps and the double sum in Equation 11.39 equals zero! Therefore, in general, the *mean square distance*, $\langle R_n^2 \rangle$, in a random walk is just

$$\langle R_n^2 \rangle = n a^2. \quad (11.40)$$

Both the 15-step and the 3-step examples in Sections 11.7.5.1 and 11.7.5.2 are off by a factor of 5 simply because only a very small number of steps were taken in each case.

11.7.6 Diffusion Coefficient

The mean distance between jumps is a as is shown on the square lattice in Figure 11.18, then $\langle R^2 \rangle = n a^2$, where n is the number of jumps necessary to get to position $\vec{R}$, Equation 11.40. Therefore, Equation 11.33 can be rewritten as

$$\langle R^2 \rangle = n a^2 = 6Dt$$

or

$$D = \frac{1}{6}\frac{n}{t}a^2$$
$$D = \frac{1}{6}\Gamma a^2 \quad (11.41)$$

where Γ is the number of jumps per second. Equation 11.41 is the same as Equation 9.11 found in Section 9.3 for interstitial diffusion in solids. However, here Equation 11.41 has been developed from a statistical argument rather that the more concrete concept of atoms moving in a cubic lattice.

11.8 CHAPTER SUMMARY

The solution to Fick's second law of diffusion (conservation of mass), $\partial C/\partial t = D(\partial^2 C/\partial t^2)$, with a constant diffusion coefficient (and its companion equation for heat transfer) is important in many processes in materials science and engineering. With a *similarity variable*, $x^2/4Dt$, the partial differential equation is transformed into an ordinary differential equation that is easily integrated and leads to *error function* solutions for C(x, t) that can be used to solve several infinite and semi-infinite boundary value problems. In the Appendix, it is shown how these solutions are really sums or integrals of source functions that can themselves be used to solve a number of other interesting problems. The *method of images* can be used particularly for semi-infinite boundary problems with fixed surface concentrations or derivative boundary conditions. One such solution, the single surface source, is widely used to measure diffusion coefficients in solids. The concept of extended sources and their result in error function solutions by a simple transformation of variables are useful in solving diffusion problems such as post-deposition diffusion after *ion implantation* in semiconductor processing. Finally, the use of single source to obtain a relationship between the mean square displacement and the diffusion coefficient is introduced as an approach to diffusion as a random walk process via the rear door: the principles of a random walk are usually discussed first and shown how they lead to a diffusion coefficient. The approach taken here was to discuss earlier the more phenomenological processes of diffusion being atom motion and then use these results to show the relationship between the diffusion coefficient and the more esoteric concept of a random walk.

APPENDIX

A.1 Source Function

It was shown in Section 11.4.1 that the solution for the infinite boundary conditions of Figure 11.5 can be written as

$$C(x,t) = C_1 + \left(\frac{C_2 - C_1}{2}\right)\left[1 + \operatorname{erf}\left(\frac{x}{\sqrt{4Dt}}\right)\right]$$

when $C_1 = 0$, then this solution becomes

$$C(x,t) = \frac{C_2}{2}\left\{1 + \frac{2}{\sqrt{\pi}}\int_0^{x/\sqrt{4Dt}} e^{-y^2}dy\right\}$$

$$C(x,t) = \frac{C_2}{2}\left\{1 + \frac{2}{\sqrt{\pi}}\int_0^{x/\sqrt{4Dt}} e^{-y^2}dy\right\} \quad \text{(A.1)}$$

writing out the error function in terms of the integral. The first term in the brackets, 1, can be written as

$$1 = \operatorname{erf}(\infty) = \frac{2}{\sqrt{\pi}}\int_0^{\infty} e^{-z^2}dz. \quad \text{(A.2)}$$

Let $y = -z$ so that $dz = -dy$ and when $z = 0$, $y = 0$ and when $z = \infty$, $y = -\infty$ and substituting these into Equation A.2 and putting it back into Equation A.1 gives

$$C(x,t) = \frac{C_2}{2}\left\{-\frac{2}{\sqrt{\pi}}\int_0^{-\infty} e^{-y^2}dy + \frac{2}{\sqrt{\pi}}\int_0^{x/\sqrt{4Dt}} e^{-y^2}dy\right\}$$

and *flipping* the limits on the first integral that changes its sign and adding:

$$C(x,t) = \frac{C_2}{\sqrt{\pi}} \int_{-\infty}^{x/\sqrt{4Dt}} e^{-y^2} dy. \tag{A.3}$$

Another change of variables in Equation A.3 gives a powerful result that naturally leads to important solutions for infinite and semi-infinite boundary condition diffusional mass transfer problems. Specifically, let $y = (x - x_0)/\sqrt{4Dt}$ then $dy = -(1/\sqrt{4Dt})dx_0$ and when $y = (x/\sqrt{4Dt}) = (x/\sqrt{4Dt}) - (x_0/\sqrt{4Dt})$ or $x_0 = 0$ and when $y = -\infty$ then $x_0 = \infty$, so Equation A.3 becomes

$$C(x,t) = -\frac{C_2}{\sqrt{4\pi Dt}} \int_{\infty}^{0} e^{-\frac{(x-x_0)^2}{4Dt}} dx_0 = \frac{1}{\sqrt{4\pi Dt}} \int_{0}^{\infty} C_2 e^{-\frac{(x-x_0)^2}{4Dt}} dx_0. \tag{A.4}$$

In reality, C_2 in Equation A.4 is really $C_2(x_0)$, that is, it is a function of position x_0, but it was initially assumed to be constant for $x_0 > 0$ and a constant equal to zero for $x_0 < 0$, but certainly it does not have to be a constant. In fact, C_2 could be replaced by some concentration that is a function of x_0: note that x_0 is just a dummy variable in the integral. The limits of integration for Equation A.4 really extend from $-\infty < x_0 < \infty$ because this equation was obtained assuming that $C = 0$ for $x_0 < 0$. Therefore, a more *general form* of Equation A.4 is really

$$C(x,t) = \frac{1}{\sqrt{4\pi Dt}} \int_{-\infty}^{\infty} f(x_0) e^{-\frac{(x-x_0)^2}{4Dt}} dx_0. \tag{A.5}$$

And, of course, $f(x_0)$ does not even have to be a continuous function of x_0. For example, suppose that there is a single source of amount $Q(x_0)/\Delta x_0 = f(x_0)$ where Units[Q] = moles/cm^2. Equation A.5 now becomes

$$C(x,t) = \frac{1}{\sqrt{4\pi Dt}} \frac{1}{\Delta x_0} \int_{x_0}^{x_0+\Delta x_0} Q(x_0) e^{-\frac{(x-x_0)^2}{4Dt}} dx_0. \tag{A.6}$$

In the limit as $\Delta x_0 \to 0$, this is the definition of the derivative of the integral, so in the limit of a point source, $Q(x_0)$, Equation A.6 becomes

$$C(x,t) = \frac{Q(x_0)}{\sqrt{4\pi Dt}} e^{-\frac{(x-x_0)^2}{4Dt}} \tag{A.7}$$

which is the *source function* or *Green's function* for the solution of infinite and semi-infinite boundary condition diffusion (and heat transfer) problems of interest. Furthermore, Equation A.7 predicts how a quantity of material Q at x_0 and $t = 0$ spreads over distance, x, and time as shown in Figure 11.10.

A.2 Notes on the Error Function

The (Gaussian) error function, erf(x), is a definite integral for *x*, that is, it is just a number that must be looked up in tables or calculated approximately just like sine and cosine functions. The error function, erf(x) is given by

$$\text{erf}(x) = \frac{2}{\sqrt{\pi}} \int_0^x e^{-y^2} dy \tag{A.8}$$

and must be evaluated numerically for different values of x. One way of doing this is by using term-by-term integration of the series expansion of $\exp(-y^2)$:

$$e^{-y^2} = 1 - y^2 + \frac{y^4}{2!} - \frac{y^6}{3!} + \frac{y^8}{4!} - \frac{y^{10}}{5!} + \cdots$$

namely

$$\mathrm{erf}(x) = \frac{2}{\sqrt{\pi}}\int_0^x e^{-y^2}dy = \frac{2}{\sqrt{\pi}}\left(x - \frac{x^3}{3} + \frac{x^5}{10} - \frac{x^7}{42} + \frac{x^9}{216} - \frac{x^{11}}{1320} + \cdots\right). \quad \text{(A.9)}$$

The error function is graphically depicted by the shaded area under the curve in the Figure 11.3. The *complementary error function*, $\mathrm{erfc}(x) = 1 - \mathrm{erf}(x)$, is depicted to the right of the shaded area in the figure. The $\sqrt{\pi}/2$ factor comes about because, as was shown in Chapter 5,

$$\int_{-\infty}^{\infty} e^{-x^2}dx = 2\int_0^{\infty} e^{-x^2}dx = \sqrt{\pi}$$

and the definition of the error function.

The error function has the following properties:

$$\mathrm{erf}(0) = 0 \qquad \mathrm{erf}(\infty) = 1 \qquad \mathrm{erf}(-x) = -\,\mathrm{erf}(x)$$

The error function is found in many mathematics handbooks, in many software packages, and on graphing calculators. In addition, a good approximation to the error function is given by (Lasaga 1998)

$$\mathrm{erf}(x) \simeq \left(1 - \exp(-(4x^2/\pi))\right)^{1/2} \quad \text{(A.10)}$$

and if x is small, then since $e^x \simeq 1 + x$, $\mathrm{erf}(x) \simeq (2/\sqrt{\pi})x$, which could be obtained from the first term in the infinite series expansion, Equation A.9, as well. Equation A.10 is a very good approximation and is plotted in Table A.1 along with the error between the approximation and the value of erf(x) and erfc(x) calculated with a spreadsheet. Other approximations tables of the error function are available in the literature as well (Abramowitz and Stegun 1965). Also note that when $(x/\sqrt{4Dt}) \simeq 1 \simeq (x^2/4Dt)$ the concentration is about 15% of the interface concentration for diffusion with semi-infinite or infinite boundary conditions.

A.3 Gamma Function

A.3.1 Introduction

The gamma function or the generalized factorial function of α, $\Gamma(\alpha)$ appears frequently when integrating things related to the normal distribution and similar functions. However, it is just a definite integral and simply represents a number and is not particularly mathematically complex. The gamma function of α, a real variable and not necessarily an integer, is given by

$$\Gamma(\alpha) = \int_0^{\infty} x^{\alpha-1}e^{-x}dx \quad \text{(A.11)}$$

which looks more formidable than it really is.

A.3.2 Recursion Formula

Now, $\Gamma(\alpha + 1)$ can easily be integrated by parts as follows:

$$\Gamma(\alpha+1) = \int_0^{\infty} x^{\alpha}e^{-x}dx$$

$$= -x^{\alpha}e^{-x}\Big|_0^{\infty} + \alpha\int_0^{\infty} x^{\alpha-1}e^{-x}.$$

TABLE A.1
Error Function

x	erf(x)	Approx.	Error	erfc(x)
0	0	0	0	1
0.05	0.056372	0.056374	−3.31E-05	0.943628
0.10	0.112463	0.112479	−1.45E-04	0.887537
0.15	0.167996	0.168051	−3.28E-04	0.832004
0.20	0.222703	0.222832	−5.80E-04	0.777297
0.25	0.276326	0.276573	−8.94E-04	0.723674
0.30	0.328627	0.329042	−1.26E-03	0.671373
0.35	0.379382	0.380020	−1.68E-03	0.620618
0.40	0.428392	0.429309	−2.14E-03	0.571608
0.45	0.475482	0.476731	−2.63E-03	0.524518
0.50	0.520500	0.522131	−3.13E-03	0.479500
0.55	0.563323	0.565378	−3.65E-03	0.436677
0.60	0.603856	0.606368	−4.16E-03	0.396144
0.65	0.642029	0.645019	−4.66E-03	0.357971
0.70	0.677801	0.681278	−5.13E-03	0.322199
0.75	0.711155	0.715115	−5.57E-03	0.288845
0.80	0.742101	0.746526	−5.96E-03	0.257899
0.85	0.770668	0.775528	−6.31E-03	0.229332
0.90	0.796908	0.802161	−6.59E-03	0.203092
0.95	0.820891	0.826483	−6.81E-03	0.179109
1.00	0.842701	0.848571	−6.97E-03	0.157299
1.2	0.910314	0.916591	−6.90E-03	0.089686
1.3	0.934008	0.940064	−6.48E-03	0.065992
1.4	0.952285	0.957887	−5.88E-03	0.047715
1.5	0.966105	0.971084	−5.15E-03	0.033895
1.6	0.976348	0.980607	−4.36E-03	0.023652
1.7	0.983790	0.987303	−3.57E-03	0.016210
1.8	0.989091	0.991887	−2.83E-03	0.010909
1.9	0.992790	0.994943	−2.17E-03	0.007210
2	0.995322	0.996925	−1.61E-03	0.004678
2.1	0.997021	0.998177	−1.16E-03	0.002979
2.2	0.998137	0.998946	−8.10E-04	0.001863
2.3	0.998857	0.999406	−5.50E-04	0.001143
2.4	0.999311	0.999673	−3.62E-04	0.000689
2.5	0.999593	0.999825	−2.32E-04	0.000407
2.6	0.999764	0.999909	−1.45E-04	0.000236
2.7	0.999866	0.999953	−8.78E-05	0.000134
2.8	0.999925	0.999977	−5.19E-05	0.000075
2.9	0.999959	0.999989	−2.99E-05	0.000041
3	0.999978	0.999995	−1.68E-05	0.000022

The first term is clearly zero when $x = 0$. However, at $x = \infty$, it becomes indeterminate. Rewriting this first term as $-x^{\alpha}/e^{x}$ then when $x = \infty$ the ratio is ∞/∞ which is still indeterminate. However, by L'Hôpital's rule, which states (Edwards and Penney 1986)

$$\lim_{x\to\infty}\frac{f(x)}{g(x)} = \lim_{x\to\infty}\frac{f'(x)}{g'(x)}$$

that is, if the ratio of the two functions f and g is indeterminate, then take their respective derivatives f′ and g′ and see if the ratio is now determinate. If not, continue taking derivatives until it becomes so. In this case above, suppose $\alpha = 2.7$ for the sake of concreteness, then taking derivatives of the numerator and denominator until $x^{0.3}$ ends up in the denominator,

$$\lim_{x\to\infty}\frac{f(x)}{g(x)} = \lim_{x\to\infty} -\frac{x^{2.7}}{e^x} = \lim_{x\to\infty} -\frac{2.7x^{1.7}}{e^x}$$

$$= \lim_{x\to\infty} -\frac{(2.7)(1.7)x^{0.7}}{e^x} = \lim_{x\to\infty} -\frac{(2.7)(1.7)(0.7)}{x^{0.3}e^x}$$

$$= 0$$

which goes to zero and shows that the first term in the integration above does go to zero at $x = \infty$. As a result,

$$\Gamma(\alpha+1) = \int_0^\infty x^\alpha e^{-x}dx = \alpha\int_0^\infty x^{\alpha-1}e^{-x}dx$$

$$\Gamma(\alpha+1) = \alpha\Gamma(\alpha) \tag{A.12}$$

is a *recursion formula* that means if there is a value for $\Gamma(\alpha)$ then values of Γ can be determined for values of α that differ by unity.

A.3.3 Some Values for the Gamma Function

Because,

$$\Gamma(1) = \int_0^\infty e^{-x}dx = 1$$

then

$$\Gamma(2) = \Gamma(1+1) = 1\cdot\Gamma(1) = 1$$

$$\Gamma(3) = \Gamma(2+1) = 2\cdot\Gamma 2 = 2\cdot 1 = 2!$$

$$\Gamma(4) = \Gamma(3+1) = 3\cdot\Gamma(3) = 3\cdot 2\cdot 1 = 3!$$

and so forth. This is why the gamma function is sometimes called the *generalized factorial function*. For kinetics, one-half order gamma functions are of interest, such as

$$\Gamma(1/2) = \int_0^\infty x^{1/2-1}e^{-x}dx = \int_0^\infty x^{-1/2}e^{-x}dx$$

This can be easily integrated by making the substitution, $x = y^2$, so $dx = 2y\,dy$ and $x^{1/2} = y$ so

$$\Gamma(1/2) = \int_0^\infty x^{-1/2}e^{-x}dx = 2\int_0^\infty e^{-y^2}dy = 2\left(\frac{\sqrt{\pi}}{2}\right)$$

so

$$\Gamma(1/2) = \sqrt{\pi}. \tag{A.13}$$

Therefore, with the recursion formula, other values of $\Gamma(n/2)$ can easily be obtained.

$$\Gamma(3/2) = \Gamma(1+1/2) = \frac{1}{2}\Gamma(1/2) = \frac{\sqrt{\pi}}{2}$$

$$\Gamma(5/2) = \Gamma(1+3/2) = \frac{3}{2}\Gamma(3/2) = \frac{3\sqrt{\pi}}{4}$$

$$\Gamma(7/2) = \Gamma(1+5/2) = \frac{5}{2}\Gamma(5/2) = \frac{15\sqrt{\pi}}{8}$$

and so on.

EXERCISES

11.1 Show that $C(x,t) = (1/\sqrt{4\pi Dt})e^{-(x^2/4Dt)}$
Is a solution to

$$\frac{\partial C}{\partial t} = D\frac{\partial^2 C}{\partial x^2}.$$

Hint: Show that the left-hand side of Fick's second law is equal to the right-hand side when substituting the solution.

11.2 Make a plot of $C(x,t) = (1/\sqrt{4\pi Dt})e^{-(x^2/4Dt)}$ for $-100 < x < 100$ μm (with a data point every 10 μm) for the following values of Dt: 1×10^{-6}, 2×10^{-6}, 5×10^{-6}, and 1×10^{-5} cm². Put all of the plots on the same graph and normalize all of the values, so that C(max) = 1.0 for $Dt = 1 \times 10^{-6}$.

11.3 a. Given the boundary conditions $C(-\infty, t) = C_1$ and $C(\infty, t) = C_2$ give the appropriate solution to Fick's second law for these boundary conditions.

b. Make plots of C(x, t) with initial conditions of $C(x, 0) = 1$, for $-100\ \mu m \le x \le 0$ and $C(x, 0) = 0$ for $x > 0$, for $4Dt = 5 \times 10^{-4}$ cm², $4Dt = 5 \times 10^{-5}$ cm² and for $4Dt = 5 \times 10^{-6}$ cm² from $x = -100$ to $x = 100$ μm. Put all three plots on the same graph with data points every 5 μm.

11.4 The solution to Fick's second law for a semi-infinite bar (i.e., $0 \le x \le \infty$) is given by

$$C(x,t) = C_0 + (C_S - C_0)\,\mathrm{erfc}\left(\frac{x}{\sqrt{4Dt}}\right)$$

where C_0 is the initial concentration in the bar (i.e., $C(x, 0) = C_0$) and C_s is a constant surface concentration (i.e., $C_s = C(0, t)$). This solution is the appropriate solution to use when studying carburizing of steel or forming a p-n junction in silicon by diffusion. For example, a piece of 1018 steel (0.18 w/o C) is being carburized at 1000°C with a constant surface concentration of 0.8 w/o carbon.

a. If the density and molecular weight of iron are $\rho = 7.87$ g/cm³ and M = 55.845 g/mol and those of carbon 2.2 g/cm³ and 12.011 g/mol, respectively, calculate the atomic percent carbon in 1018 steel.

b. The diffusion coefficient for carbon in iron is given by $D = D_0 e^{-(Q/RT)}$ and $D_0 = 0.20$ cm²/s and Q = 142.3 kJ/mol. Calculate the diffusion coefficient of carbon in iron at 1000°C.

c. Calculate the time necessary to obtain a concentration of 0.4 w/o carbon 1 mm below the surface.

d. Calculate and plot on the same graph, the carbon concentration (w/o) at 0.1, 0.5, and 1.0 times the time calculated in part c. for $0 < x < 1$ cm with data points at every 0.5 mm.

11.5 In another application, it is desired to remove carbon from the surface of this 1018 steel by annealing it in hydrogen at 1000°C, so that the surface concentration of carbon in the iron goes to zero.

a. Give the appropriate solution to Fick's second law for this situation.
b. Plot the carbon concentration as a function of distance for $0 < x < 1$ cm for $4Dt = 0.1$, 0.2, 0.5, and 1.0 all on the same graph with data points every 0.05 mm.

11.6 You have just been hired by Intel to work in their semiconductor fabrication facility (FAB). Your job is to set up a boron diffusion furnace for making MOSFET (metal oxide semiconductor field effect transistors) devices on silicon very-large-scale integration (VLSI) chips.

a. From the following data, calculate what the BCl_3 pressure should be at 1127°C in a furnace that contains essentially a 50–50 mixture of hydrogen and HCl gas at a total pressure of one atmosphere, so that the boron concentration at the surface of the silicon wafer is 10^{19} boron atoms per cm^3. Assume an ideal solid solution of B in Si. The atomic weight and density of silicon are 28.09 g/mol and 2.33 g/cm^3, respectively. *Hint*: The BCl_3 pressure will be low enough, so that it should not affect the total pressure.

$$\Delta G^0 = -80{,}609 \text{ cal/mol at } 1127°C \text{ for } BCl_3 \text{ gas}$$

$$\Delta G^o = -24{,}534 \text{ cal/mol at } 1127°C \text{ for HCl gas}$$

b. The data for boron diffusion in silicon are $D_0 = 0.76$ cm^2/s and $Q = 333.8$ kJ/mol. Calculate the diffusion coefficient for B in Si at 1127°C.
c. Calculate, from the error function solution for this problem, the time it takes for the boron concentration to reach 10^{15} cm^{-3} at 1 μm below the surface of the silicon at 1127°C. The silicon is initially pure.
d. Plot the concentration of boron for $0 < x < 2$ μm with data points every 0.1 μm.

11.7 Three single diffusion sources of strength, $Q = 1$ are initially at $x = -1$, $x = 0$, and $x = 1$ in an infinite medium.

a. Develop an expression for $C(x, t)$ with these three sources.
b. Plot $C(x, t)$ for these three sources between $-4 < x < 4$ for values of $4Dt = 0.1$, 0.5, 1.0, and 2.0 all on the same graph with data points at every $\Delta x = 0.1$.

11.8 There are two single diffusion sources, one of strength $Q = -1$ at $x = -1$ and one of strength $Q = 1$ at $x = 1$.

a. Develop an expression for $C(x, t)$ with these two sources.
b. Plot $C(x, t)$ for these two sources between $-4 < x < 4$ for values of $4Dt = 0.1$, 0.5, 1.0, and 2.0 all on the same graph with data points at every $\Delta x = 0.1$.

11.9 There are two extended diffusion sources, one of strength $Q = -1$ between $x = -1$ and $x = 0$ and one of strength $Q = 1$ between $x = 0$ and $x = 1$.

a. Develop an expression for $C(x, t)$ in terms of error functions with these two sources.
b. Plot $C(x, t)$ for these two sources between $-4 < x < 4$ for values of $4Dt = 0.1$, 0.5, 1.0, and 2.0 all on the same graph with data points at every $\Delta x = 0.1$.

11.10 A piece of newly formed poly(vinyl chloride) (PVC) ($T_g = 360$ K), is exposed to air at room temperature, 25°C and one atmosphere pressure. The solubility and diffusion data are from the literature (Van Krevelen 1997).

a. The solubility parameters for N_2 and O_2 are $\log_{10}S(N_2) = -6.69$ and $\log_{10}S(O_2) = -6.33$ with $\text{Units}(S) = cm^3(STP)/cm^3 \cdot Pa$. Calculate the solubilities of nitrogen and oxygen in mol/cm^3 at *one atmosphere* pressure.
b. The diffusion coefficients for nitrogen and oxygen in PVC at room temperature are $D(N_2) = 3.98 \times 10^{-9}\ cm^2/s$ and $D(O_2) = 1.26 \times 10^{-8}\ cm^2/s$. Calculate and plot the concentrations $C(x, t)$ of both nitrogen and oxygen for times of 1, 12, and 24 hours, all on the same graph, out to an appropriate distance where the highest concentration at that point has dropped to 10% of its surface value. Include at least 20 data points on each curve.

11.11 Another piece of PVC has been exposed to an atmosphere of pure CO_2 at one atmosphere at 25°C. The solubility and diffusion data below are from the literature (Van Krevelen 1997).

a. For CO_2 in PVC at 25°C and one atmosphere CO_2, $\log_{10} S(CO_2) = -5.448$. Calculate the concentration of CO_2 at the surface of this PVC in mol/cm^3.

b. The diffusion coefficient of CO_2 in PVC is of the usual form, $D = D_0 e^{-(Q/RT)}$, with $Q = 50$ kJ/mol and $D_0 = 80$ cm^2/s. Calculate $D(CO_2)$ at 25°C.

c. The dissolved CO_2 is to be removed by allowing the PVC to stand in CO_2-free air until it diffuses out. Assume that the initial CO_2 concentration is constant from the surface to a depth of 10 μm: $C(x, 0) = S$ for $0 < x < 10$ μm and $C(x, 0) = 0$ for 10 μm $< x < \infty$, that is, semi-infinite boundary conditions. Develop an expression in terms of error functions for $C(x, t)$ for $0 < x < \infty$.

d. Calculate the time, t_{max}, it takes for the *maximum* CO_2 concentration in the PVC to reach 10% of its initial value.

e. Plot $C(x, t)$ for $t = 0$, $t = 0.1\ t_{max}$, $t = 0.5\ t_{max}$, and $t = t_{max}$ out to a distance twice as far as the x location of t_{max} all on the same graph. Include at least 20 data points in each plot.

REFERENCES

Abramowitz, M. and I. A. Stegun. 1965. *Handbook of Mathematical Functions.* New York: Dover.

Brandes, E. A. and G. B. Brook. 1992. *Smithells Metals Reference Book,* 7th ed., Oxford, UK: Butterworth-Heinemann.

Campbell, S. A. 2001. *The Science and Engineering of Microelectronic Fabrication,* 2nd ed. New York: Oxford University Press.

Carslaw, H. S. and J. C. Jaeger. 1959. *Conduction of Heat in Solids,* 2nd ed. Oxford: Oxford University Press.

Crank, J. 1975. *The Mathematics of Diffusion,* 2nd ed. Oxford: Oxford University Press.

Edwards, C. H., Jr. and D. E. Penney. 1986. *Calculus and Analytical Geometry,* 2nd. ed. Englewood Cliffs, NJ: Prentice Hall.

Farlow, S. J. 1982. *Partial Differential Equations for Scientists and Engineers.* New York: Wiley.

Glicksman, M. E. 2000. *Diffusion in Solids.* New York: Wiley.

Haynes, W. M., Editor-in-Chief. 2013. *CRC Handbook of Chemistry and Physics,* 94th ed. Boca Raton, FL: CRC Publishing.

Lasaga, A. C. 1998. *Kinetic Theory in the Earth Sciences.* Princeton, NJ: Princeton University Press.

Mayer, J. W. and S. S. Lau. 1990. *Electronic Materials Science: For Integrated Circuits in Si and GaAs.* New York: MacMillan.

Morse, P. M. and H. Feshbach. 1953. *Methods of Theoretical Physics,* Vols. I and II. New York: McGraw-Hill.

Roine, A. 2002. *Outokumpu HSC Chemistry for Windows, Ver. 5.11,* thermodynamic software program, Outokumpu Research Oy, P.O. Box 60, FIN-28102 PORI, Finland.

Shewmon, P. 1989. *Diffusion in Solids,* 2nd ed. Warrendale, PA: The Minerals, Metals, and Materials Society.

Snider, A. D. 2006. *Partial Differential Equations, Sources and Solutions.* Mineola, NY: Dover.

Van Krevelen, D. W. 1997. *Properties of Polymers,* 3rd ed. Amsterdam, the Netherlands: Elsevier B. V.

Wolf, S. and R. N. Tauber. 2000. *Silicon Processing for the VLSI Era,* Volume 1-Process Technology, 2nd ed. Sunset Beach, CA: Lattice Press.

12

Diffusion with Finite Boundary Conditions

12.1 INTRODUCTION

In Chapter 8, the differences between infinite and finite boundary conditions were introduced, and Chapter 11 considered macroscopic diffusion kinetics for infinite and semi-infinite boundary conditions. In this chapter, the emphasis shifts to diffusion kinetics with finite boundaries. One important difference between infinite and finite boundary conditions is that finite boundary condition models will eventually reach either steady-state or equilibrium. In contrast, infinite and semi-infinite boundary conditions preclude every reaching constant fluxes or concentrations with time. As a result, the techniques for solving these two different classes of macroscopic diffusion kinetic problems are usually different. Nevertheless, some finite boundary problem solutions can frequently be modeled with the same techniques used for infinite boundary conditions. However, such routes to solutions may be more complex than techniques designed specifically for the finite case. On the other hand, the infinite solutions *always apply* to finite boundary problems when the extent of diffusion is small compared to the finite dimensions of the problem. In some cases, a combination of both types of solutions to a finite boundary problem can give results sufficiently close to the actual solution and gives satisfactory results with far less mathematical complexity and computation. As always, real examples of materials processing and behavior are used to illustrate the principles involved.

12.2 CORING IN A CAST ALLOY

12.2.1 Introduction

As before, the goal here is not to solve all of the various problems of interest in nonsteady-state diffusion in materials science and engineering but, rather, to provide some insight into how these problems are solved and some implications from the resulting model. The focus again is on the one-dimensional Fick's second law in rectangular or Cartesian coordinates. So the problem to solve is then—for the *umpteenth* time:

$$\frac{\partial C}{\partial t} = D\frac{\partial^2 C}{\partial x^2}$$

with initial conditions, C(x, 0), and *finite boundary conditions*, usually C(0, t) and C(L, t), In addition, as pointed out in Chapter 8, the solution to Fick's second law for finite boundary conditions consists of two terms: (1) a steady-state or an equilibrium term and (2) a transient term that disappears when time goes to infinity. For many diffusion problems in materials science and engineering, it turns out that the starting or initial conditions require mathematical techniques that lead to analytical solutions containing infinite Fourier series in the transient term. However, there are some important problems in which the real initial conditions can be closely approximated by some simpler function that produces solutions that do not require infinite series. One example is the problem of *coring* in a cast metal alloy and its removal by a *homogenization* heat treatment. Unfortunately, the casting of a solid from a liquid of mixed composition occurs almost exclusively for metals, so the application of the principles to other materials is somewhat restricted. Nevertheless, in addition to valuable insight into the solution of finite boundary diffusion processes, it also relates processing to phase diagrams to help emphasize the importance of phase diagrams and their interpretation in materials science and engineering.

12.2.2 What Is Coring?

The origin of coring is illustrated in Figure 12.1, which gives a phase diagram with an alloy of composition x being cooled from the melt. The first solid precipitates when the temperature reaches T_1 (where the composition x intersects the liquidus) and has the composition 1 in equilibrium with the liquid of composition 2. Since this is an equilibrium diagram, if the alloy could be cooled infinitely slowly, the average composition of the solid would follow the composition of the solidus line in the diagram. However, this is usually impractical and more realistic cooling rates do not allow adequate time for solid-state diffusion to equalize the composition throughout the solid. This gives an *average* solid composition that follows a curve similar to 1–6 in the figure. For example, at T_2, the freezing solid has composition 3 in equilibrium with liquid of composition 4. The newly frozen solid is much richer in component B than is the average composition of the solid. At temperature T_3, the chosen composition should be completely solid, but this is not the case since the average composition is deficient in component B. As a result, freezing continues down to T_4 where the average solid composition is now the same as the original alloy, x, at point 6. However, the composition of the solid freezing at this temperature

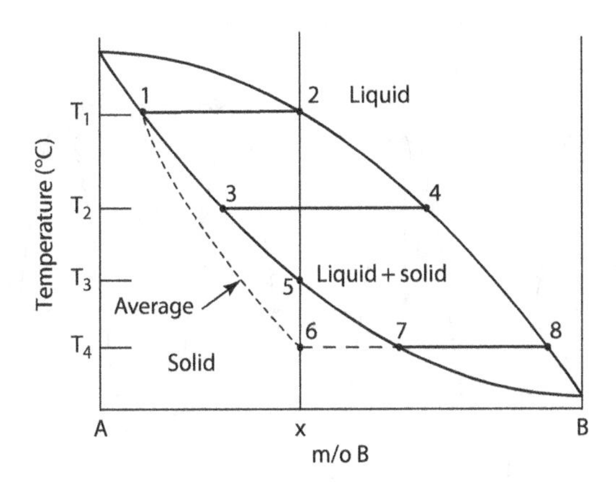

FIGURE 12.1 Part of single solid solution phase diagram showing freezing a liquid of composition x faster than solid-state diffusion can maintain a uniform equilibrium solid concentration. At T_1, and point 2, liquid of composition x is in equilibrium with the first solid to freeze of composition 1. At T_2, the solid of composition 3 that is freezing is now in equilibrium with liquid of composition 4. However, the average solid composition, the dashed *solidus*, has a variation in composition between composition 1 and composition 3. The liquid should be all solidified at T_3 and composition 5. However, the average composition is much less than this and does not all solidify until it reaches T_4 where the solid freezing has composition 7, considerably higher than the average solid composition. In principle, the composition may vary from composition 1, the first solid to freeze, to composition 7, producing a *cored* compositional structure.

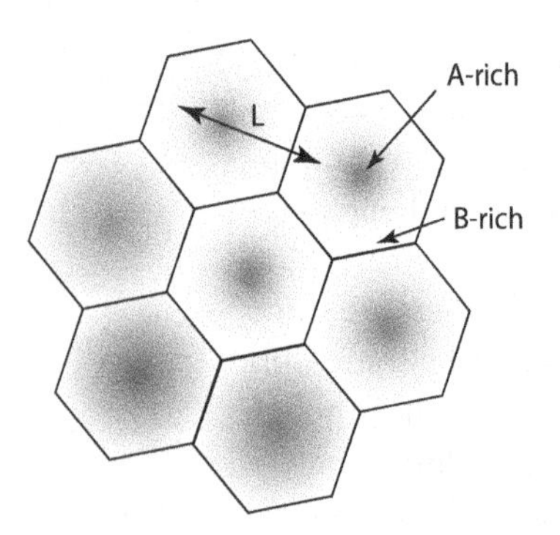

FIGURE 12.2 Schematic cored microstructure that shows several grains that have a compositional variation from the center—the first to freeze—rich in component A as shown in the phase diagram in Figure 12.1, out to the grain boundaries—the last to freeze—which are much richer in component B. The spatial variation in composition is essentially the size of the grains; namely, L as shown in the sketch.

has composition 7, far richer in B. This implies that when the alloy is completely solidified, there will be compositional variations in the solid depending when it solidified as schematically shown in Figure 12.2. Principally, such spatial concentration variations can happen if the cast crystalline solid is a metal, ceramic, or polymer. In reality, however, it is far more important for metals given that almost all metals are cast from a melt at one point in their processing history, which is not the case for ceramics and polymers. Furthermore, it is a particular problem for a metal since these compositional variations give rise to different electrochemical potentials in aqueous solutions that can generate localized corrosion, in addition to the spatial variation in mechanical properties, such as hardness. To remove or minimize these compositional variations, a *homogenization* heat treatment is given the alloy to allow solid-state diffusion to level the concentration to the uniform composition x. Therefore, the problem becomes, "How long should the heat-treatment be at a specific temperature?"

12.2.3 Initial Composition Model

Coring varies reasonably periodically in space with a period roughly equal the grain size, L, as indicated in Figure 12.2. As a result, the as-cast spatial composition variation can be most easily modeled by a sine or cosine function as shown in Figure 12.3. Call the average composition $x = C_0$ so the initial condition at the start of the homogenization heat treatment is

$$C(x,0) = C_0 + \beta \sin\left(\frac{2\pi x}{L}\right) \tag{12.1}$$

where β is the magnitude of the initial composition variation in Figure 12.3.

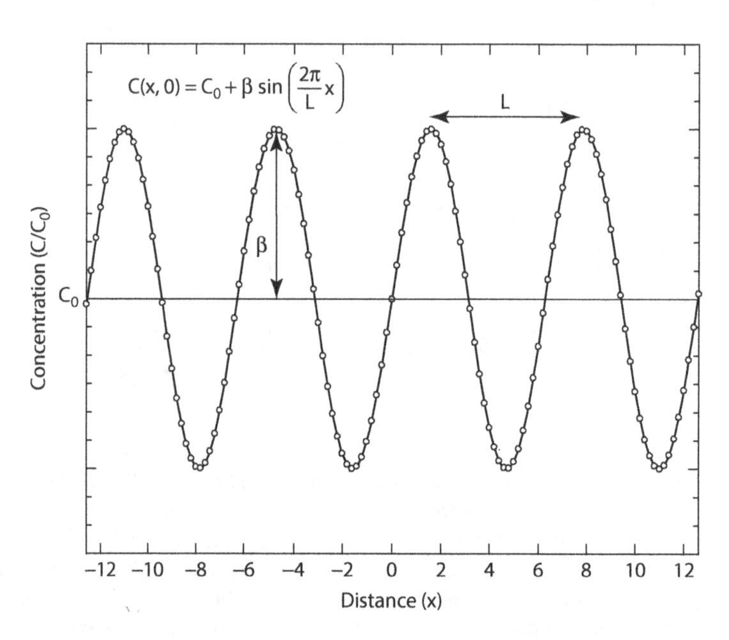

FIGURE 12.3 A sinusoidal *model* of the compositional spatial variation produced during coring as depicted in Figures 12.1 and 12.2. Beta (β) is the maximum positive and negative composition deviation from the mean composition of the solid (C_0) ($L = 2\pi$).

12.2.4 Infinite Boundary Condition Solution

If the sinusoidal composition variation extends to infinity in all directions then the concentration as a function of time and distance is given by the infinite boundary solution from Chapter 11:

$$C(x,t)=\frac{1}{\sqrt{4\pi Dt}}\int_{-\infty}^{\infty} f(x_0)\,e^{-\frac{(x-x_0)^2}{4Dt}}\,dx_0 \tag{12.2}$$

where $f(x_0)$ is just Equation 12.1 or,

$$C(x,t)=\frac{1}{\sqrt{4\pi Dt}}\int_{-\infty}^{\infty}\left[C_0+\beta\sin\left(\frac{2\pi x_0}{L}\right)\right]e^{-\frac{(x-x_0)^2}{4Dt}}\,dx_C \tag{12.3}$$

which can be integrated* to give

$$C(x,t)=C_0+\beta\sin\left(\frac{2\pi x}{L}\right)e^{-\frac{4\pi^2 Dt}{L^2}}. \tag{12.4}$$

Note that as $t \to 0$, Equation 12.1, the initial condition is recovered and when $t \to \infty$, $C(x, \infty) = C_0$: the composition is uniform.

This homogenization problem can also be considered a *finite boundary* problem if the values of x are restricted to $0 \le x \le L$, where L is the grain size since the composition (and solution, Equation 12.4) are periodic in L (Figure 12.4). As is shown later, Equation 12.4 is also the solution of a finite boundary problem by the techniques developed for solving infinite and semi-infinite boundary condition problems. Note that as was pointed out in Chapter 8, finite boundary problem solutions typically are of the form

$$C(x,t)=f(x)+g(x,t) \tag{12.5}$$

where:

$f(x)$ is an *equilibrium* or *steady-state* concentration

$g(x, t)$ is a *transient* term, so that $g(x, \infty) = 0$.

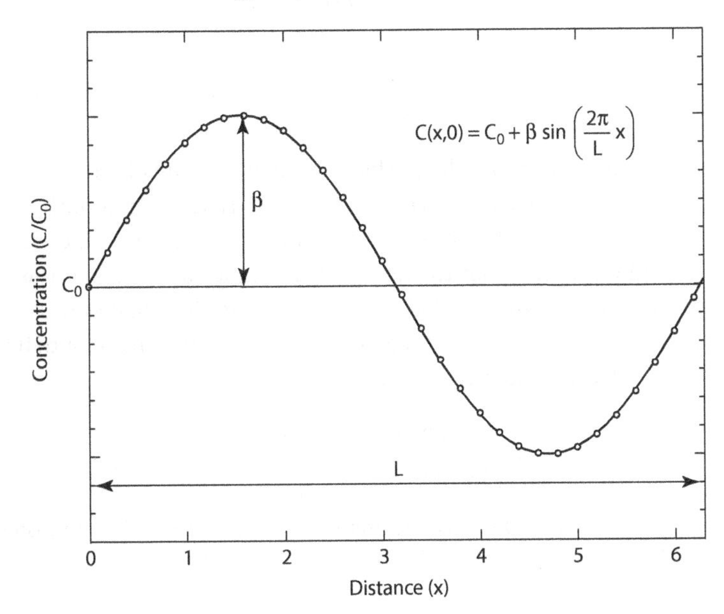

FIGURE 12.4 A single sine cycle from the composition variation of Figure 4.3 used to model the homogenization annealing process used to minimize or eliminate the coring of a cast alloy ($L = 2\pi$).

* The integration requires only simple trigonometric functions and integrations but is a bit lengthy and has been relegated to Section A.1.

In this case, in Equation 12.4, C_0 is an *equilibrium* term, since dC/dx = 0 when time goes to infinity, and the other term is a *transient* since it disappears as time goes to infinity.

Now there are many finite boundary problems for which the solution techniques for infinite and semi-infinite boundary solutions can be used. However, generating the solutions is frequently more complex than techniques designed specifically to solve finite boundary condition problems. Nevertheless, the solutions for infinite and semi-infinite boundary conditions can be applied to finite boundary problems as long as the diffusion times are sufficiently short that diffusion does not extend entirely across the finite region. Examples of this are developed later. In fact, some pretty good approximations to solutions of finite boundary problems can also be made by some *back of the envelope* applications of a combination of the two types of solutions.

Another important point to note in Equation 12.4 is that the transient part of the solution is a product of a function dependent only on x, call it $X(x) = \beta\sin(2\pi x/L)$ and one only dependent on t, $T(t) = \exp(-4\pi^2 Dt/L^2)$; in other words, the independent variables, x and t, are separated into their own distinct functions. This strongly suggests that the general solution of the transient part of finite boundary problems might be solvable by assuming a solution of the form: C(x, t) = X(x)T(t).

12.3 CORING BY SEPARATION OF VARIABLES

12.3.1 Separation of Variables

In fact, a general method to solve finite boundary problems is by *separation of variables* where the solution is assumed to be of the following form: $C(x,t) = X(x)T(t)$; that is, it is the product of a function dependent only on x, X(x), and one dependent only on t, T(t). If this X(x)T(t) can be substituted into the partial differential equation and a solution can be found then the assumption that C(x, t) is of this form is indeed correct. By doing this,

$$\begin{aligned} \frac{\partial C}{\partial t} &= D\frac{\partial^2 C}{\partial x^2} \\ X\frac{\partial T}{\partial t} &= DT\frac{\partial^2 X}{\partial x^2} \\ \frac{1}{DT}\frac{\partial T}{\partial t} &= \frac{1}{X}\frac{\partial^2 X}{\partial x^2}. \end{aligned} \tag{12.6}$$

In the last line of Equation 12.6, the left-hand side is a function only of t and the right-hand side a function only of x. These are independent variables, x and t, in that either or both of them can be varied independently. For example, t can be held constant and x varied at will and vice versa. However, in doing so, Equation 12.6 requires that the left- and the right-hand sides of the equation remain equal. The only way that this is possible is that both the left- and right-hand sides of the equation are constant—the same constant—which, to reduce future algebra, is usually set it to be a constant such as $-\alpha^2$. The last line of Equation 12.6 then becomes

$$\frac{1}{DT}\frac{\partial T}{\partial t} = \frac{1}{X}\frac{\partial^2 X}{\partial x^2} = -\alpha^2. \tag{12.7}$$

There are now two *ordinary* differential equations to solve: one in t, T(t), and one in x, X(x). The equations to be solved are

$$\frac{1}{T}\frac{dT}{dt} = -D\alpha^2 \quad \text{and} \quad \frac{d^2X}{dx^2} + \alpha^2 X = 0$$

and the solutions to these familiar ordinary differential equations have been given before and are

$$T(t) = A'e^{-\alpha^2 Dt} \quad \text{and} \quad X(x) = B'\sin\alpha x + E'\cos\alpha x$$

so

$$C(x,t) = X(x)T(t) = (B\sin\alpha x + E\cos\alpha x)e^{-\alpha^2 Dt} \tag{12.8}$$

where A′, B′, E′, B, and E are simply integration constants that need to be evaluated from the initial and boundary conditions.

12.3.2 Applied to Coring

This homogenization problem can also be considered a *finite boundary* problem if the values of x are restricted to $0 \le x \le L$ because the composition (and solution, Equation 12.4) are periodic in L (Figure 12.4). The easiest way to get a solution to Fick's second law finite boundary condition problems is to make the boundary conditions *homogeneous** (make them equal to zero); that is $C(0, t) = C(L, t) = 0$. To accomplish this for the coring problem, simply define $C^* = C - C_0$, so that the boundary conditions are $C^*(0, t) = C^*(L, t) = 0$ and

$$C^*(x,0) = C(x,0) - C_0 = \beta\sin\left(\frac{2\pi x}{L}\right). \tag{12.9}$$

is now the new initial condition. Obviously, Fick's second law becomes $\partial C^*/\partial t = D(\partial^2 C^*/\partial x^2)$, which is solved with the above two boundary conditions and single initial condition. The complete solution is obtained by just adding C_0 to the solution for $C^*(x, t)$, since $C(x,t) = C^*(x,t) + C_0$.

For the coring example, $C^*(x,t) = X(x)T(t) = (B\sin\alpha x + E\cos\alpha x)\exp(-\alpha^2 Dt)$ and the initial condition at $t = 0$ is $C^*(x,0) = \beta\sin(2\pi x/L)$ implying that $E = 0$, $B = \beta$, and $\alpha = 2n\pi/L$ where n is any integer.† To satisfy the initial condition here, $n = 1$, so that the solution for C^* is

$$C^*(x,t) = \beta\sin\left(\frac{2\pi x}{L}\right)e^{-\frac{4\pi^2 Dt}{L^2}}$$

And adding C_0 the final result is

$$C(x,t) = C_0 + \beta\sin\left(\frac{2\pi x}{L}\right)e^{-\frac{4\pi^2 Dt}{L^2}} \tag{12.10}$$

which (thankfully) is the same as Equation 12.4.

Again, the solution consists of an *equilibrium* term, C_0, plus a *transient term*, the second term, which goes to zero as $t \to \infty$. Figure 12.5 shows the behavior of the solution with $\beta/C_0 = 1.0$. Notice that when $4Dt/L^2 = 0.05$ the peak in the concentration has been reduced about to about 60% of its maximum value and for $4Dt/L^2 = 0.5$, the maximum deviation from the equilibrium concentration (at $x/L = 0.25$ and 0.75) is only 0.7%, which is probably good enough in engineering practice. Clearly, from a practical engineering standpoint, coring can be removed in a homogenization heat treatment in much shorter times than the infinite time that the mathematics requires.

12.3.2.1 Example: Homogenization Anneal of Cu–Sn Alloy

From the literature, (Brandes and Brook 1992) Sn is soluble in Cu up to about 15 w/o (weight %) at about 723°C. For diffusion of Sn in Cu, $D_0 = 0.06\ \text{cm}^2/\text{s}$ and $Q = 180{,}000$ J/mol (Brandes and Brook 1992) so at 723°C, $D = 0.06 \times \exp[-180{,}000/(8.314)(1000)] = 2.4 \times 10^{-11}\ \text{cm}^2\text{s}^{-1}$. Therefore, to reduce the concentration inhomogeneity in an alpha solid solution with less than 15 w/o Sn down to about 1% of its original value ($4Dt \cong 0.5 \times L^2$, Figure 12.5) in about 3 hours—10^4 seconds (a practical heat-treating time)—the spatial variation in composition that could be minimized would be $L \cong \sqrt{8Dt} = \sqrt{8(2.4 \times 10^{-11})(10^4)} = 1.38 \times 10^{-3}$ cm, about 14 μm, not terribly large but on the order of a reasonable grain size and spatial extent of coring. (Reed-Hill and Abbaschian 1992) Therefore, spatial inhomogeneities of this magnitude can be essentially eliminated by heat treating at modest temperatures for a few hours.

* *Homogeneous boundary conditions* are when the boundary conditions are zero.

† This result that $\sin(2n\pi x/L)$ is a solution for any integer value of n is useful a little later.

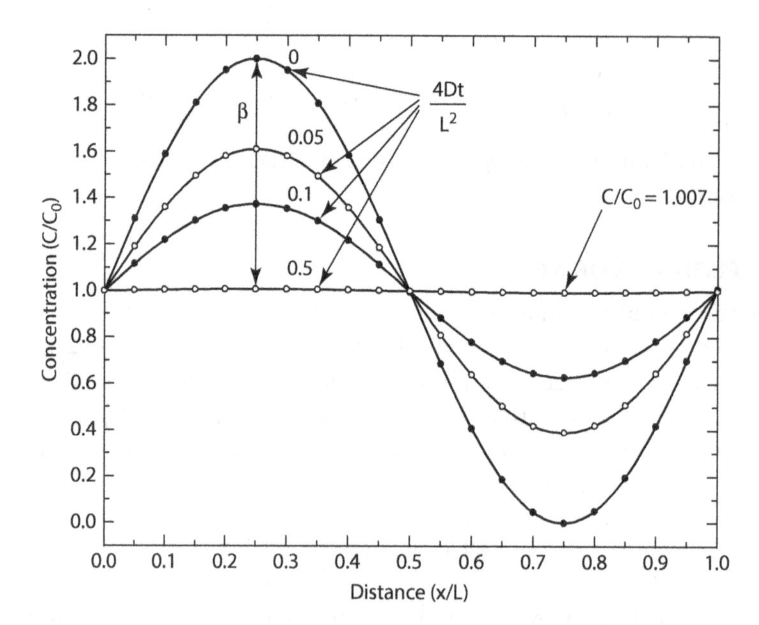

FIGURE 12.5 The decay in compositional inhomogeneity as a function of the dimensionless parameter, $4Dt/L^2$, showing that the inhomogeneity is essentially gone by $4Dt/L^2 \cong 0.5$: Equations 12.4 and 12.10. At the maximum concentrations at $x/L = 0.25$ and 0.75, at $4Dt/L^2 \cong 0.5$, $C/C_0 = 1.007$, less than 1% from a uniform composition of C_0.

12.4 DIFFUSION OUT OF A SHEET OF FINITE THICKNESS

12.4.1 Introduction

Diffusion out of and into a sheet of finite thickness, L, is a widely occurring problem in materials processing. For example, the removal of interstitial impurity atoms such as carbon from iron is important in producing magnetic iron that can be both easily machined and magnetized for special applications such as lens pole pieces in electron microscopes. Conversely, diffusion of carbon into sheet iron or steel can increase the hardness and strength of the sheet. Other applications include the removal of solvent from a cast polymer sheet or the drying of a cast polymer-bonded ceramic sheet for microelectronic circuit packaging after firing at high temperature. These all are examples of the same diffusion problem.

The relative complexity of the solution to this problem depends critically on the initial (or final) concentration of the diffusing species in the sheet. For example, if the material to be removed has a uniform initial concentration, C_0, throughout the thickness of the slab or sheet at $t = 0$, as shown in Figure 12.6, and all of the material diffusing needs to be removed via diffusion, $C(x, \infty) = 0$, the exact mathematical solution requires an infinite sum of terms. In reality, however, usually only a small number of terms of the solution are necessary to get extremely accurate results. Moreover, the nearly exact solution can be easily evaluated with commercial spreadsheet software as shown in the next few sections. On the other hand, approximation of the precise initial condition by a simpler one provides a great deal of insight into the kinetics of the problem. There might be some differences between the approximate and exact solution depending on what is being calculated yet the results still provide a very satisfactory *back of the envelope* practical answer for many engineering problems. For example, Figure 12.6 shows a particular C(x, t) that looks very much like a sine curve. If the initial concentration can be approximated by such a curve, then the solution is indeed easy to solve as was the case with coring. Also, as in coring, both infinite and semi-infinite solution techniques can be applied. Certainly, for an initial concentration of C_0, semi-infinite solutions

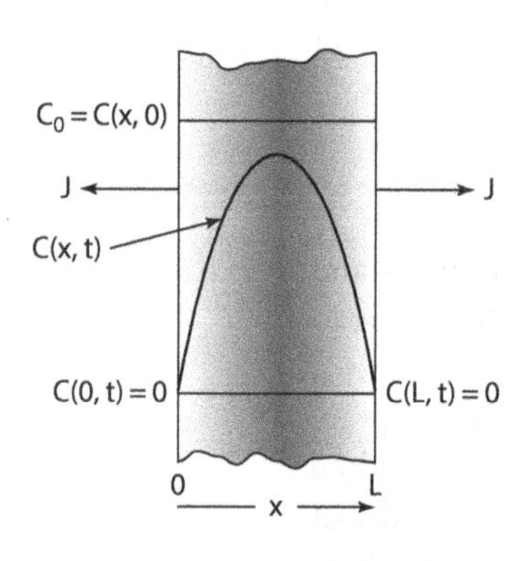

FIGURE 12.6 Model of a sheet of thickness L with an initial concentration of C_0 whose surface concentrations are zero at time zero, $C(0, t) = C(L, t) = 0$. After a certain time, t, the concentration profiles approach that of the single C(x, t) shown: essentially a sine curve.

will apply for dimensionless time parameter, $4Dt/L^2$, sufficiently small that the concentration profiles produced on either side of the sheet (i.e., at x = 0 and at x = L) do not significantly overlap. Even in cases where the overlap is large, useful results still can be obtained.

12.4.2 First Approximation: Initial Condition Is $C(x, 0) = C_0 \sin(\pi x/L)$

First, the simple case where the initial concentration is a specific value of C(x, t) in Figure 12.6 that can be approximated by a single sine function,

$$C(x,0) = C_0 \sin\left(\frac{\pi x}{L}\right) \tag{12.11}$$

which implies that $B = C_0$ and $\alpha = \pi/L$ in the general solution Equation 12.8 above. Such a situation can really exist in practice if a sheet were partially dried or degassed, resulting in the diffusing species having a similar concentration profile. The separation of variables solution for this simple initial condition is then:

$$C(x,t) = C_0 e^{-\frac{\pi^2}{L^2}Dt} \sin\left(\frac{\pi x}{L}\right) \tag{12.12}$$

and is plotted in Figure 12.7 for various values of Dt/L^2. Note, that, the maximum concentration has dropped to about 10% of its initial value when $Dt/L^2 \cong 0.25$ or when $4Dt/L^2 \cong 1$ which is consistent with infinite and semi-infinite boundary conditions. As a result, a good approximate, *back of the envelope*, estimate of the time it takes for such a diffusion process to go to near completion (at least 90%) is when $4Dt/L^2 \cong 1$. When $Dt/L^2 = 1$, the maximum concentration has dropped to about $10^{-5} C_0$, which is such a low value that this is an overestimate of the time needed for the reaction to go to completion. Also note that the above solution is the sum of both an equilibrium term, $C(x, \infty) = 0$ and the transient term of Equation 12.12.

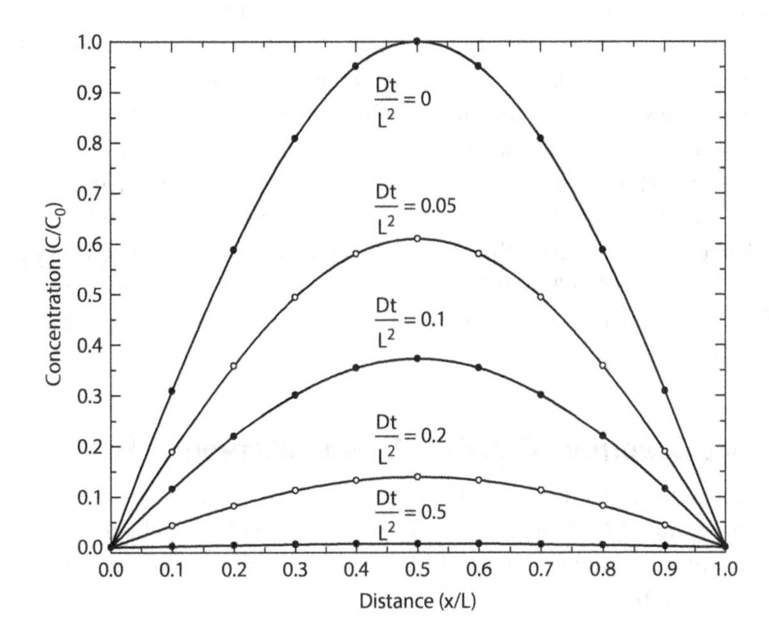

FIGURE 12.7 The decay of concentration within a sheet of thickness L with a starting concentration of a simple single sine function. Note that when $Dt/L^2 = 0.5$, the maximum deviation from the equilibrium concentration is at $x/L = 0.5$ and is $C/C_0 = 0.007$, less than 1%.

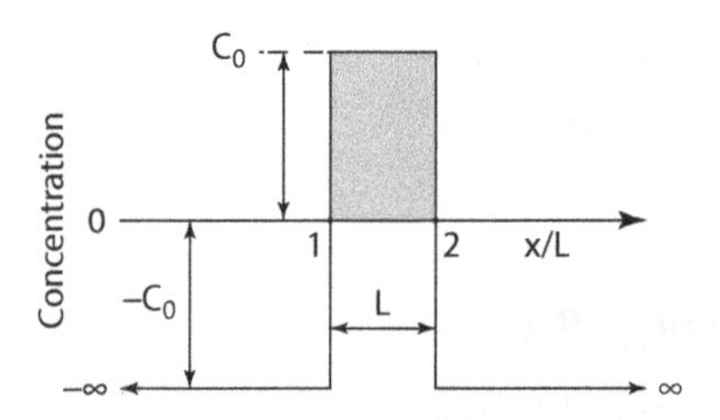

FIGURE 12.8 Concentration versus distance for two semi-infinite boundary conditions that force the concentrations at points 1 and 2 to be zero in the semi-infinite case to see how well it models the finite boundary condition sheet problem.

12.4.3 Initial Condition: C(x, 0) = C_0 with Semi-Infinite Boundary Conditions

As covered in Section 11.5.2, the boundary condition C(0, t) could be obtained for semi-infinite boundary conditions if C(−x, 0) = C(x, 0): that is, the concentration profile to the left of the origin was simply the negative of that for positive x. In this sheet problem, the requirements are that both C(0, t) = 0 and C(L, t) = 0. Certainly, a valid approximation for small values of $4Dt/L^2$, where the diffusion profiles do not overlap, is the assumption that for both x < 0 and x > L, C(x, t) = $-C_0$ as shown in Figure 12.8 (actually in the figure it is for x/L < 1 and x/L > 2 to make the solution more general). The solution to this problem is to put the proper values for $f(x_0)$ into Equation 12.2

$$C(x,t)=\frac{1}{\sqrt{4\pi Dt}}\int_{-\infty}^{\infty} f(x_0)\,e^{-\frac{(x-x_0)^2}{4Dt}}\,dx_0.$$

so it becomes (Exercise 12.1)

$$C(x,t)=-\frac{C_0}{\sqrt{4\pi Dt}}\int_{-\infty}^{L} e^{-\frac{(x-x_0)^2}{4Dt}}\,dx_0+\frac{C_0}{\sqrt{4\pi Dt}}\int_{L}^{2L} e^{-\frac{(x-x_0)^2}{4Dt}}\,dx_0 \\ -\frac{C_0}{\sqrt{4\pi Dt}}\int_{2L}^{\infty} e^{-\frac{(x-x_0)^2}{4Dt}}\,dx_0. \tag{12.13}$$

Application of the usual procedures of substitution of $y=(x-x_0)/\sqrt{4Dt}$, changing the limits, rearranging the integrals, and expressing them as error functions, Equation 12.13 leads to (Exercise 12.1):

$$\frac{C(x,t)}{C_0}=\operatorname{erf}\left(\frac{x-L}{\sqrt{4Dt}}\right)-\operatorname{erf}\left(\frac{x-2L}{\sqrt{4Dt}}\right)-1 \tag{12.14}$$

which is plotted in Figure 12.9. Only the concentrations in the gray area are of interest, the rest are not. The test of how good the solutions are is how close they are to zero at the boundaries, that is, C(L, t) = C(2L, t) = 0. As indicated on Figure 12.9, the concentrations at these points are within 1% of zero up to $4Dt/L^2 = 0.3$, which is a surprisingly long dimensionless time. Also note how the concentration profiles resemble sine curves, which are the solutions obtained with the separation of variables technique for an initial condition of Equation 12.11 $C(x,0)=C_0\sin(\pi x/L)$. This is explored further a little later. However, at larger dimensionless time values, for example, $4Dt/L^2 = 1$, Figure 12.9 shows that the boundary conditions clearly are not satisfied and eventually all of the concentrations will be negative—a mathematically correct but physically not a very practical result. Also note that beyond about $4Dt/L^2 = 0.1$ the concentration profiles become more and more asymmetric around C = 0. This suggests that maybe a better approximation might be the use of image sources as in Section 11.6.3.

12.4.4 Initial Condition: C(x, 0) = C_0 with Method of Images

In this approximation to try to minimize the large volume of negative initial concentration of the approximation in Section 12.4.3, negative images (of finite volume) of the rectangular area are applied at each side of it as shown in Figure 12.10. With the same procedure used in Section 12.4.4, the result is (Exercise 12.2):

$$\frac{C(x,t)}{C_0}=\operatorname{erf}\left(\frac{x-L}{\sqrt{4Dt}}\right)-\operatorname{erf}\left(\frac{x-2L}{\sqrt{4Dt}}\right)-0.5\times\operatorname{erf}\left(\frac{x}{\sqrt{4Dt}}\right)+0.5\times\operatorname{erf}\left(\frac{x-3L}{\sqrt{4Dt}}\right). \tag{12.15}$$

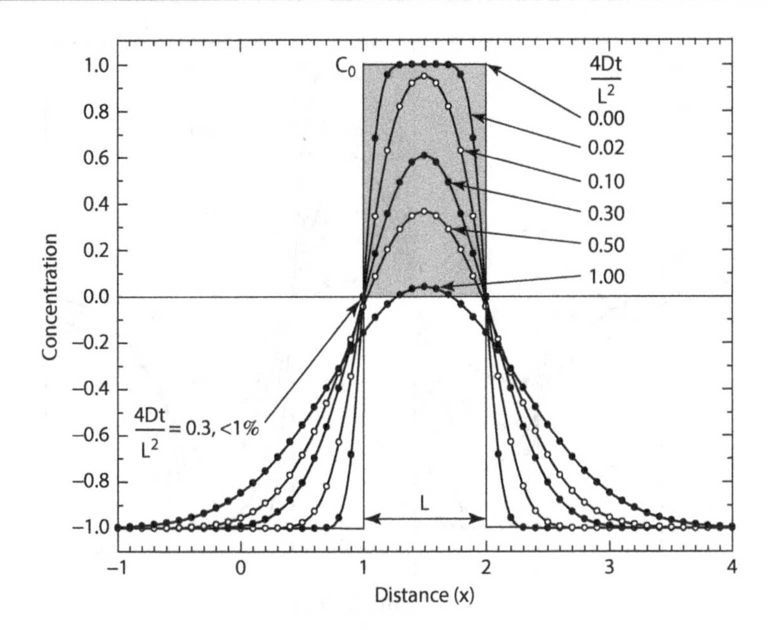

FIGURE 12.9 Concentrations as a function of distance and Dt/L^2 for the semi-infinite boundary condition model of Figure 12.8 for the sheet of thickness L, the gray area. The critical points are x = 1 and x = 2 where the concentration must remain at zero. At $Dt/L^2 = 0.3$, the concentrations are within 1% of C = 0, which means the solution is pretty good up to this time. As can be seen for the curves of $Dt/L^2 = 0.5$ and 1.0, the fit at zero becomes worse, as expected.

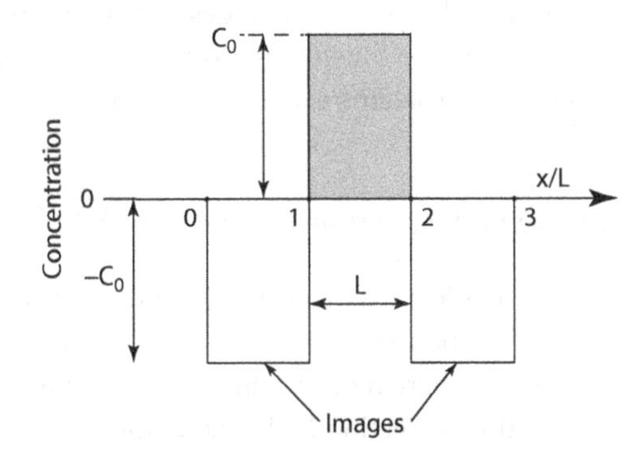

FIGURE 12.10 Concentration versus distance for two semi-infinite boundary conditions with two images that try to force the concentrations at points 1 and 2 to be zero to see how well it models the finite boundary condition sheet problem.

The results are plotted in Figure 12.11. There are a few things worth noting. First, the algebra is getting longer and the number of terms in the approximate solution is larger. Second, the approximation seems to be slightly better up to about $4Dt/L^2 = 0.5$ where the concentrations are only about 2% from zero: somewhat better than the previous approximation. Also note that the concentration profiles are more symmetrical between the gray (actual) sheet and the two images, at least up to about $4Dt/L^2 = 0.3$. However, material is still diffusing to plus and minus infinity at x = 3L and x = 0, respectively. This strongly suggests that if a two additional positive images where placed between −L and 0 and 3L and 4L, the approximation would get better. In fact, to get an exact solution, an infinite number of terms are necessary and such solutions can be found in the classic literature on diffusion and heat transfer (Carslaw and Jaeger 1959).

12.4.5 Initial Condition: $C(x, 0) = C_0$ with Separation of Variables

The technique of using solutions appropriate for infinite and semi-infinite boundary conditions gives surprisingly accurate solutions to a finite boundary problem up to significant values of $4Dt/L^2$.

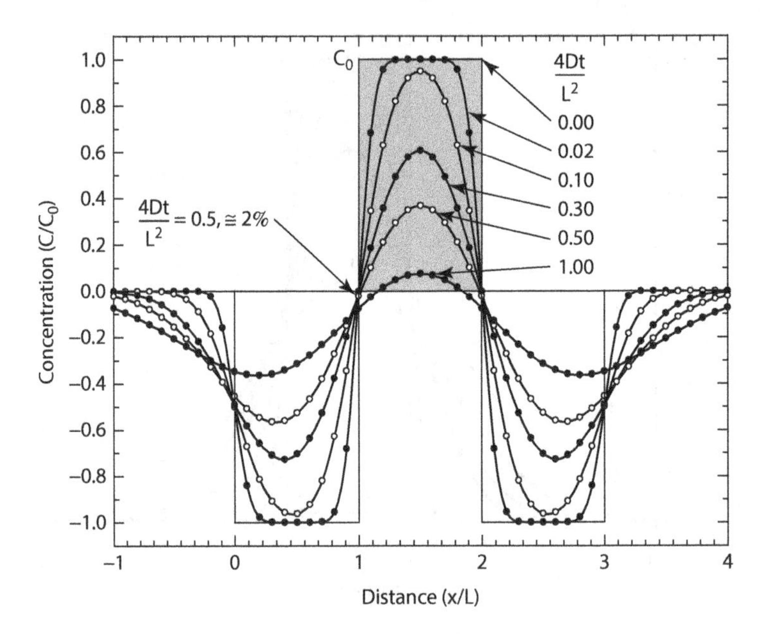

FIGURE 12.11 Concentrations as a function of distance and Dt/L^2 for the semi-infinite boundary condition model with images of Figure 12.10 for the sheet of thickness L the gray area. The critical points are x = 1 and x = 2 where the concentration must remain at zero. At $Dt/L^2 = 0.5$, the concentrations are within 2% of C = 0, which means the solution is only slightly improved over the solution in Figure 12.9. As can be seen for $Dt/L^2 = 1.0$, the fit at zero is not good.

However, they can only give an exact solution for all values of time if an infinite series of terms is used. As a result, a more standard technique for solving finite boundary problems is used. The general technique to solve finite boundary problems exactly is by separation of variables as mentioned in Equation 12.8, namely,

$$C(x,t) = X(x)T(t) = (B \sin \alpha x + E \cos \alpha x)e^{-\alpha^2 Dt}.$$

Again for diffusion out of finite sheet of thickness L, the initial condition is $C(x, 0) = C_0$ with the boundary conditions $C(0, t) = C(L, t) = 0$. The boundary condition at x = L suggests that the solution must consist of terms of the type $\sin(n\pi x/L)$, where n can be any integer and not just n = 1 as was assumed for the initial condition of a sine function in Section 12.4.2, Equation 12.11. This is both good news and bad news. The good news is that an *infinite*, Fourier, series can be created to satisfy most initial conditions, such as $C(x, 0) = C_0$, and most others (e.g., Glicksman 2000). The bad news is that evaluating the coefficients in the Fourier series is often messy as is the final form of the equation and its evaluation.

The theory of Fourier series to represent various functions will not be considered in detail here. However, the rationale for the necessity of an infinite series of terms necessary to satisfy the initial conditions was suggested at the end of Section 12.4.4, and the general method of obtaining a Fourier series solution is presented. Since there an infinite number of similar terms that are solution to the partial differential equation, they can be used to get the exact solution. So instead of a single solution to the partial differential equation, there is the nth solution, or nth term in the infinite series given by

$$C_n(x,t) = A_n e^{-\frac{n^2\pi^2}{L^2}Dt} \sin\left(\frac{n\pi x}{L}\right) \tag{12.16}$$

and the complete solution is simply the sum of these terms,

$$C(x,t) = \sum_{n=1}^{\infty} A_n e^{-\frac{n^2\pi^2}{L^2}Dt} \sin\left(\frac{n\pi x}{L}\right) \tag{12.17}$$

where the A_n coefficients are determined from the initial conditions of a constant, $C(x, 0) = C_0$ by (e.g., Kreyszig 2011)

$$A_n = \frac{2}{L}\int_0^L C(x,0)\sin\left(\frac{n\pi x}{L}\right)dx$$

$$= -\frac{2}{L}C_0\frac{L}{n\pi}(\cos n\pi - \cos 0)$$

$$A_n = \frac{4C_0}{n\pi} \quad \text{for n odd and 0 for n even.}$$

It is not very attractive to have a series with zeroes for all of the even terms, so substituting $2m + 1 = n$ and then summing from $m = 0$ to $m = \infty$ sums over only odd numbers. Therefore, the result is

$$C(x,t) = \frac{4C_0}{\pi}\sum_{m=0}^{\infty}\frac{1}{(2m+1)}e^{-\frac{(2m+1)^2\pi^2}{L^2}Dt}\sin\left(\frac{(2m+1)\pi x}{L}\right) \tag{12.18}$$

and Equation 12.18 *is the exact solution* to the slab problem. At $t = 0$, this becomes

$$\frac{C_0}{C_0} = 1 = \frac{4}{\pi}\sum_{m=0}^{\infty}\frac{1}{(2m+1)}\sin\left(\frac{(2m+1)\pi x}{L}\right) \tag{12.19}$$

and the infinite sine series, the Fourier series, is just equal to $\pi/4$ at $x = L/2$. Take the first five terms at $x = L/2$,

$$\frac{4}{\pi}\left(1 - \frac{1}{3} + \frac{1}{5} - \frac{1}{7} + \frac{1}{9}\cdots\right) = 1.063 \tag{12.20}$$

which is only off by 6% and emphasizes that the number of terms necessary to get a fairly accurate solution does not need to be anywhere near infinite. The requirements become even less stringent for the complete solution, Equation 12.18, because of the rapid decrease in the exponential term from the $-(m + 1)^2\pi^2$ factor in each exponent. The first three terms are

$$C(x,t) = \frac{4C_0}{\pi}\left(\frac{1}{1}e^{-\frac{\pi^2}{L^2}Dt}\sin\left(\frac{\pi x}{L}\right) + \frac{1}{3}e^{-\frac{9\pi^2}{L^2}Dt}\sin\left(\frac{3\pi x}{L}\right) + \frac{1}{5}e^{-\frac{25\pi^2}{L^2}Dt}\sin\left(\frac{5\pi x}{L}\right) + \cdots\right)$$

which at $Dt/L^2 = 0.05$ gives

$$\frac{C(x,t)}{C_0} = \frac{4}{\pi}\left((0.61)\sin\left(\frac{\pi x}{L}\right) + (3.9\times10^{-3})\sin\left(\frac{3\pi x}{L}\right) + (8.78\times10^{-7})\sin\left(\frac{5\pi x}{L}\right) + \cdots\right)$$

for this relatively early diffusion time, the second term in the series is about 0.5% of the first and all the remaining terms beyond the second term on are sufficiently small to be neglected. Note that the first term is close to the simple initial condition sine solution given above in Section 12.4.2.

Figure 12.12 plots this infinite series solution for the first 20 terms—although only 10 would probably be OK—calculated with a spreadsheet. The infinite series solution looks formidable but succumbs to easy evaluation with a computer spreadsheet since only a small number of terms need to be summed to get a good solution. For example, if $m = 20$ and the interval in $x/L = 0.1$ then the spreadsheet contains over 200 entries! However, only one term in the series needs to be entered and then copied to all of the other cells. Figure 12.13 schematically shows how this is done. The x/L values are entered as *columns* and the different m values as *rows*. A single term in the series is entered into a cell [m, x/L], usually the first cell in the upper left-hand corner of the spreadsheet, then copied to all of the other cells. A fixed value of Dt/L^2 is put into a separate cell and used as a parameter. Finally, the last row is a sum of each column to give $\Sigma_m f(Dt/L^2, x/L)$ for each column value of x/L. The row of the sums is then the desired solution for this value of Dt/L^2 as a function of x/L. Then a new value

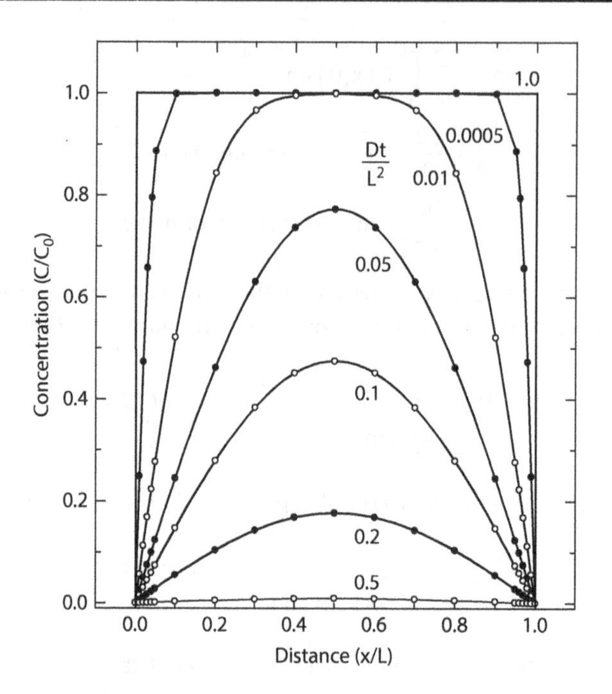

FIGURE 12.12 Twenty-term infinite series solution for diffusion out of a sheet with the initial concentration $C(x, 0) = C_0$. The interval between calculated points near x/L values of zero and one is smaller to get a better curve at small values of Dt/L^2. For values of $Dt/L^2 \geq 0.05$, they are unnecessary.

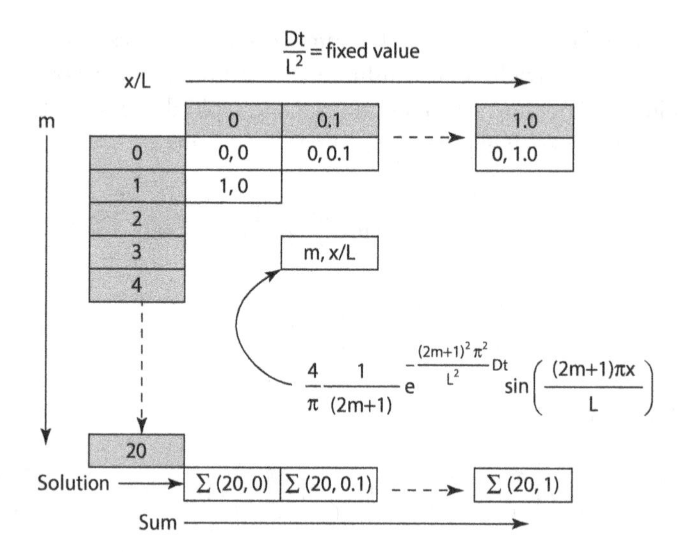

FIGURE 12.13 Schematic showing how a spreadsheet can be used to evaluate the *infinite series* solutions, in this case, for diffusion out of the sheet. Points to note: 1. A fixed value of Dt/L^2 is put into a cell and the series is calculated for that time value. Then Dt/L^2 is changed, and the spreadsheet recalculates the solution. 2. The x/L intervals are placed in the horizontal cells listed at the top. 3. The m in the infinite series are the cells in the rows. Each row represents then a single term in the series. 4. In this case, the sum of the rows is the solution as a function of x/L for this value of Dt/L^2. 5. The term value for m and x/L need only be typed in one cell and then copied throughout the spreadsheet.

of Dt/L^2 is inserted and the spreadsheet recalculates and the next solution is generated in the *sum row.* These are the curves plotted in Figure 12.12.

12.4.6 Comparing Solutions

Figure 12.14 is probably the most important and interesting result of this section. Although the graph is a little cluttered, it contains a lot of data. On this plot are the results of the semi-infinite

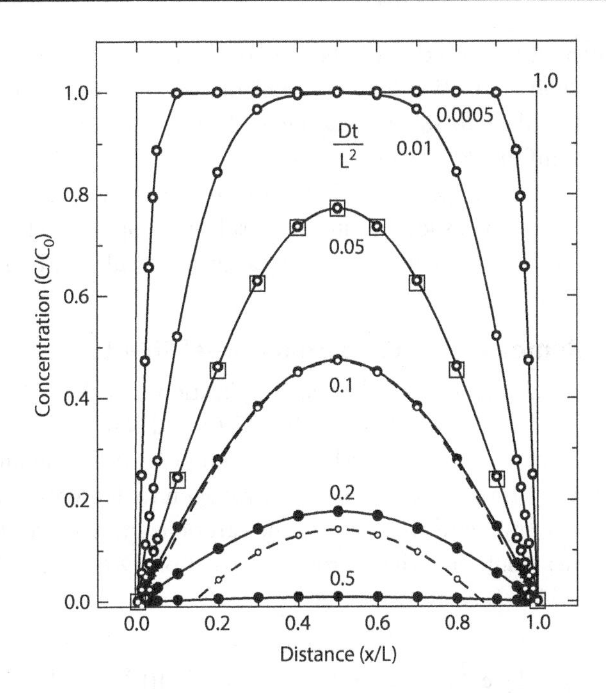

FIGURE 12.14 Comparison of infinite series solution (large black circles), error function solutions (small white circles), and only the *first term* in the infinite series solution at $Dt/L^2 = 0.05$ (squares) for diffusion from a sheet with the initial $C(x, 0) = C_0$ and boundary conditions, $C(0, t) = C(L, t) = 0$.

boundary condition solution, Equation 12.14, Figures 12.8 and 12.9, the curves with the white calculated circles and dashed lines; the infinite series solution, Equation 12.18, the curves with the black circles and solid lines; and a simple sine curve matched at the maximum for the $Dt/L^2 = 0.05$ of the other two curves, the large square points. The surprising result is that the semi-infinite solutions are virtually identical to the more accurate infinite series solution up to almost $Dt/L^2 = 0.1$! Perhaps not as surprising, a single sine function fits very nicely with both solutions at $Dt/L^2 = 0.05$, and will fit even better to the infinite series solution for larger values of time, Dt/L^2. Of course, the semi-infinite B.C. solution gets less accurate for larger values of Dt/L^2 as can be seen by the slight difference between the curves at $Dt/L^2 = 0.1$ and the large separation at $Dt/L^2 = 0.2$. There are two main conclusions from this comparison. First, the more easily calculated semi-infinite boundary condition solution, Equation 12.14, is highly accurate up until somewhere between $0.05 < Dt/L^2 < 0.1$. Then the single sine function solution, Equation 12.12, can be applied. For example, the overall solution can be calculated from Equation 12.14:

$$\frac{C(x,t)}{C_0} = \text{erf}\left(\frac{x-L}{\sqrt{4Dt}}\right) - \text{erf}\left(\frac{x-2L}{\sqrt{4Dt}}\right) - 1$$

up until $Dt/L^2 = 0.5$ and the single sine solution, Equation 12.12 for larger values of Dt/L^2.

$$C(x,t) = C_0 e^{-\frac{\pi^2}{L^2}Dt} \sin\left(\frac{\pi x}{L}\right)$$

Here, the starting value of C/C_0 is $C/C_0 = 0.7723$ the value of at $x/L = 0.5$ at $Dt/L^2 = 0.05$. For example, at $x/L = 0.5$ at $Dt/L^2 = 0.5$, $C/C_0 = 9.15 \times 10^{-3}$. Application of the two solution method of the semi-infinite B.C. solution up $Dt/L^2 = 0.05$ and the single sine solution for the remaining $Dt/L^2 = 0.45$, gives at $x/L = 0.5$, $C/C_0 = 9.10 \times 10^{-3}$, within 0.5% of the infinite series solution! This probably represents more round-off error in the calculations than difference in accuracy, and for all *practical purposes*, the results are the same.

Summarizing the results, for values of $Dt/L^2 \cong 0.5$ or $4Dt/L^2 \cong 2$, such a diffusion reaction has gone to about 99% of completion. In Chapter 11, it was shown that the depth of penetration to get roughly 10% of the surface concentration was $4Dt/L^2 \cong 1$ or $Dt/L^2 \cong 0.25$. In that case, to get 1%, then

$Dt/L^2 \cong 0.5$, which is consistent with the results here for the finite boundary conditions. Therefore, to get a *back of the envelope* calculation for the completion of a finite boundary diffusion problem, $Dt/L^2 \cong 0.25$ ($4Dt/L^2 \cong 1$) will get to 90% completion and $Dt/L^2 \cong 0.5$ ($4Dt/L^2 \cong 2$) will get to within 1% of completion. That is, infinite BCs solutions apply over a surprisingly large range $4Dt/L^2$. Piecing together a few solutions that are accurate within their respective ranges, the results are even better. But, on the other hand, while the exact infinite series solution appears to be complicated, it really isn't, and spreadsheets permit the calculation of very accurate results with only a few terms of the series.

12.4.6.1 Example: Removal of Carbon from a Steel Sheet

As mentioned in Section 12.4.1, for certain electrical applications to achieve the magnetic properties desired, the concentration of interstitial atoms such as C, O, and N need to be decreased to very low levels in iron sheets. To do this, the sheets are heated at temperatures of about 820°C (≅ 1100 K) in hydrogen, so that the interstitials all react with the hydrogen to decrease the interstitial content down to $C \cong 10^{-3}$ w/o $\cong 10^{-5}$ mol/cm³. To decrease the carbon content to this level at the center of a sheet, from either the series solution or the approximate solution, $Dt/L^2 \cong 1.15$. The carbon diffusion coefficient in FCC iron at 1100 K is

$$D = D_0 e^{-\frac{Q}{RT}} = 0.2 \times e^{-\frac{142{,}300}{(8.314\times 1100)}} = 3.5\times 10^{-8}\ \text{cm}^2\text{s}^{-1}.$$

If the sheet is 25 mils thick (1 mil = 0.001 in. ≅ 25 μm), then the time necessary would be

$$t = 1.15\frac{L^2}{D} = 1.15\frac{\left(25\times 25\times 10^{-4}\right)^2}{3.5\times 10^{-8}} = 1.28\times 10^5 \text{ seconds} \cong 36 \text{ hours}$$

which is not an unreasonable time for such a heat treatment.

12.5 OTHER IMPORTANT FINITE BOUNDARY CONDITIONS

12.5.1 Transient in a Membrane and Interdiffusion over Finite Dimensions

Two other finite boundary condition problems are worth looking at since, in one case, the steady-state solution was used to understand diffusion through a wall or membrane in Chapter 10, and it would be nice to compare the transient term with the steady-state solutions. The other model was the interdiffusion between oxygen and hydrogen discussed qualitatively in Chapter 8, as shown in Figure 8.1. It is instructive to get the exact solution to this interdiffusion and see how it compares with the qualitative model and also determine the effect of the parameters on C(x, t).

12.5.2 Diffusion through a Wall or Membrane

For the diffusion through a wall—with an initial concentration of C(x, 0) = 0 and BCs of $C(0, t) = C_S$ and C(L, t) = 0—whose steady-state solution was used for several different diffusion examples earlier: Figure 12.15 shows both the steady-state and the transient infinite series solutions. The infinite series solution to this problem is given by (Section A.3)

$$\frac{C(x,t)}{C_S} = \left(1 - \frac{x}{L}\right) - \frac{2}{\pi}\sum_{n=1}^{\infty}\frac{1}{n}e^{-n^2\pi^2\frac{Dt}{L^2}}\sin\left(\frac{n\pi x}{L}\right). \tag{12.21}$$

Notice, again, that the solution consists of a steady-state term—the first term that should be pretty familiar by now (Chapter 9)—and the transient term. Also note that for t = 0, the transient term is just the Fourier series for the steady-state triangle and that is why it is preceded by a minus sign, so that it is subtracted from the steady-state solution to make the initial concentration at t = 0, C(x, 0) = 0.

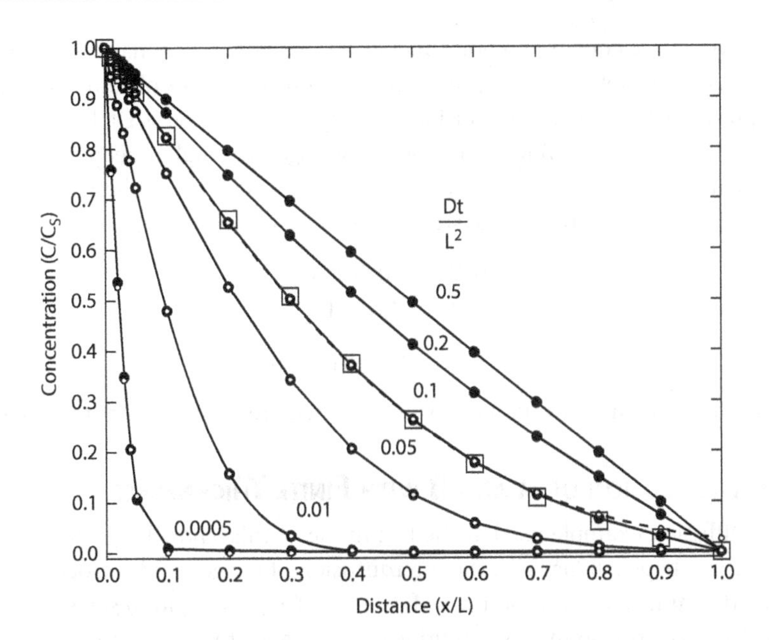

FIGURE 12.15 Comparison of infinite series solution (large black circles), error function solutions (small white circles), and only the first term in the infinite series solution at $Dt/L^2 = 0.05$ (squares) for diffusion through a wall the initial $C(x, 0) = 0$ and boundary conditions, $C(0, t) = C_S$ and $C(L, t) = 0$ so the steady-state solution is $C(x) = C_S(1 - x/L)$.

Also, plotted in Figure 12.15 are the error function solutions for a constant surface concentration, C_S, $C(x,t) = C_S \text{erfc}\left(x/\sqrt{4Dt}\right)$, that was found in Chapter 11 for this semi-infinite boundary condition. The infinite series solutions from Equation 12.21—again a 20-term spreadsheet sum—are the large black circles and are given for all the values of the dimensionless time parameter, Dt/L^2, that are listed. The error function solutions, the smaller white circles and dashed lines, are given only up to $Dt/L^2 = 0.1$ as they diverge quite significantly from the series solution at large values of Dt/L^2 which is not surprising. What is surprising is that the two solutions agree up to such relatively large values of Dt/L^2 again. Notice that the best agreement between the two solutions is between $Dt/L^2 = 0.01$ to 0.1 in this case. For the smallest value, 0.0005 the points do not fall exactly on top of each other and this is most likely due to the error in the 20-term sum (more terms are required for short times) and not due to the error function solution that is most accurate for small times. Up to 0.01, the agreement is most remarkable except for the last few points near $x/L = 1.0$, where now the error function solution is not accurate. At $Dt/L^2 = 0.5$ the concentration has gone to less than 0.5% of the steady-state solution, very similar to sheet diffusion in Section 12.4. Finally, only the first term in the series is included in the solution, Equation 12.21, at $Dt/L^2 = 0.1$ with the calculations plotted as the open squares. This again demonstrates the agreement between all three solutions at least at early times; the 20-term series, the erfc solution, and the single-term solution. Certainly, a good approximation to the solution would be to use

$$C(x,t) = C_S \text{erfc}\left(\frac{x}{\sqrt{4Dt}}\right)$$

up to $Dt/L^2 = 0.1$ and the single-term series solution

$$\frac{C(x,t)}{C_S} = \left(1 - \frac{x}{L}\right) - \frac{2}{\pi} e^{-\pi^2 \times 0.1} \sin\left(\frac{\pi x}{L}\right)$$

$$\frac{C(x,t)}{C_S} = \left(1 - \frac{x}{L}\right) - 0.3887 \times \sin\left(\frac{\pi x}{L}\right)$$

for the remaining time. This would minimize the total amount of computation and can easily be carried out on a handheld calculator. However, calculating the infinites series solution with a spreadsheet is neither time consuming nor difficult. It is easy to take as many terms as desired: even several hundred is not a problem. Finally, since the solution goes 99.5% of completion at $Dt/L^2 = 0.5$, again, a good approximation for the conclusion of the transient period is $Dt/L^2 = 0.5$ or $4Dt/L^2 = 2$.

For example, in Section 10.3.3 for diffusion of hydrogen through a neoprene tube or wall, the diffusion coefficient at 300 K was calculated to be $D = 1.64 \times 10^{-6}\,cm^2/s$. Therefore, for a 1 mm thick tube or wall, the transient time before the steady state is reached is about:

$$t \cong \frac{0.5L^2}{D} = \frac{0.5(0.1)^2}{1.64 \times 10^{-6}}$$

$$t \cong 3.05 \times 10^3 s \cong 51\,\text{minutes}$$

So after about an hour, the flux through the wall of the tube remains constant in the steady-state.

12.5.3 Interdiffusion of A and B with Finite Thicknesses

This is the interdiffusion example used at the beginning of this—now long—discussion about diffusion beginning in Chapter 8, namely, the interdiffusion of oxygen and nitrogen gas, as shown in Figure 8.1. Here, the initial concentration, say of oxygen is $C(x, 0) = C_0$ for $0 \le x \le L/2$ and $C(x, 0) = 0$ for $L/2 \le x \le L$. However, the boundary conditions at $x = 0$ and at $x = L$ are *derivative boundary conditions*, $\partial C(0,t)/\partial x = \partial C(L,t)/\partial x = 0$, which imply that the oxygen must remain inside the volume of length L; there is no flux out. The details for getting the infinite series solution for these derivative boundary conditions are developed in Section A.4:

$$C(x,t) = \frac{C_0}{2} + \frac{2C_0}{\pi}\sum_{n=1}^{\infty}\frac{(-1)^{n+1}}{(2n-1)} e^{-(2n-1)^2\pi^2\frac{Dt}{L^2}} \cos\left(\frac{(2n-1)\pi x}{L}\right) \tag{12.22}$$

where again the first term, $C_0/2$, is the equilibrium term, and the second term is the transient and when $t = 0$, the Fourier series is just the step function, $C(x, 0) = C_0$ for $0 \le x \le L/2$ and $C(x, 0) = 0$ for $L/2 \le x \le L$ (Kreyszig 2011). Again, the solution looks messier than it really is simply to make sure that the sign of each term comes out correct, $(-1)^{n+1}$, and to get rid of all the zero terms, $(2n-1)$ as before. Twenty-term spreadsheet plots for Equation 12.22 for various values of Dt/L^2 are plotted in Figure 12.16 as the larger black circles and solid lines. Notice that the solution goes to about 0.5% of the final solution of $C(x,t)/C_0 = 0.5$ at $Dt/L^2 \cong 0.5$ or $4Dt/L^2 \cong 2$.

The infinite boundary condition solution for these initial conditions is Equation 11.13

$$C(x,t) = \frac{C_0}{2}\text{erfc}\left(\frac{x - L/2}{\sqrt{4Dt}}\right) = \frac{C_0}{2}\text{erfc}\left(\frac{x/L - 0.5}{\sqrt{4Dt/L^2}}\right) \tag{12.23}$$

since $(x - L/2) \to 0$ as $x \to L/2$. The values of Equation 12.23 are plotted up to $Dt/L^2 = 0.05$ in Figure 12.16 as the smaller white circles and the dashed lines. Notice that the agreement between the infinite series solution, Equation 12.22, and the infinite boundary condition solution, Equation 12.23, is not as good as in the last two examples. Once the infinite boundary solution reaches the boundaries, $x = 0$ and $x = L$, the two solutions deviate but are exactly the same up to $Dt/L^2 = 0.01$ and probably up to about $Dt/L^2 \cong 0.025$. Better agreement between the two solutions can be achieved with the use of images in the infinite boundary condition solution to make the slopes at $x = 0$ and $x = L$ equal to zero. But again, to make them agree exactly for all time values, then an infinite series of error functions is necessary.

As in the two previous examples, here again, for $Dt/L^2 \ge 0.05$, only the first term in the series needs to be used and

$$\frac{C(x,t)}{C_0} = \frac{1}{2} + \frac{2}{\pi} e^{-\pi^2\frac{Dt}{L^2}} \cos\left(\frac{\pi x}{L}\right)$$

$$\frac{C(x,t)}{C_0} = \frac{1}{2} + 0.6366 e^{-\pi^2\frac{Dt}{L^2}} \cos\left(\frac{\pi x}{L}\right) \tag{12.24}$$

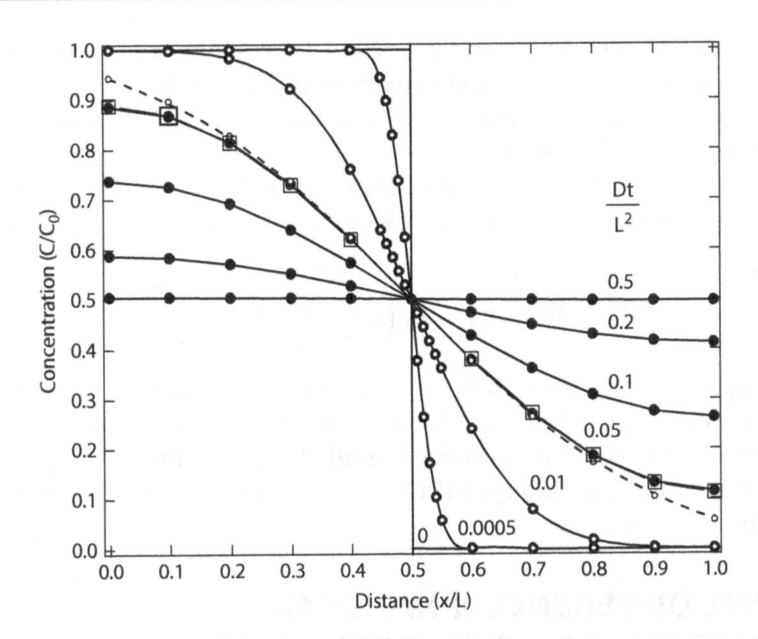

FIGURE 12.16 Comparison of infinite series solution (large black circles), error function solutions (small white circles), and only the first term in the infinite series solution at $Dt/L^2 = 0.05$ (squares) for interdiffusion inside a finite region $0 \le x \le L$ initial $C(x, 0) = C_0$ for $0 \le x \le L/2$ and $C(x, 0) = 0$ for $L/2 \le x \le L$ and boundary conditions, $\partial C/\partial x = 0$ at $C(0, t)$ and $C(L, t)$.

and the values of Equation 12.24 are plotted for $Dt/L^2 = 0.5$ as the large square calculated points in Figure 12.16 again demonstrating that just starting with this single cosine term would give 90% accuracy for the concentrations for times to completion between $Dt/L^2 = 0.05$ to $Dt/L^2 = 0.5$.

For example, for the interdiffusion of nitrogen and oxygen, $D_{N_2-O_2}$ at room temperature is about 0.22 cm^2/s (Cussler 1997). If L = 10 cm, then the time to take equal amounts of oxygen and nitrogen to equilibrate would be

$$t \cong \frac{0.5L^2}{D} = \frac{0.5(10)^2}{0.22}$$

$$t \cong 227\,s = 3.8\,\text{minutes}$$

12.5.3.1 Example: Formation of a NiO–MgO Solid Solution by Reacting MgO and NiO Powders by Solid State Calcination

The results in Section 12.5.3 can provide a very good estimate of how long it might take for a mixture of particles to react. For example, *calcination* is the process used to prepare ceramic oxide powders such as $Y_3F_5O_{12}$ by reacting powders of Fe_2O_3 and Y_2O_3. Or even more simply, two powders such as MgO and NiO could be reacted to form a $Mg_{1-x}Ni_xO$ solid solution. The two powders are mixed and heated to high enough temperatures for interdiffusion, primarily the cations, to take place. Consider the case of forming a 50:50 mole mixture of NiO and MgO.

Assume that the starting powders have a particle size of L/2 = 0.1 μm = 10^{-5} cm and that the diffusion coefficient for the reaction is $D = 6 \times 10^{-14}$ cm^2/s at 1100°C, calculated from the data in Table 9.5. Then the approximate time for calcination would be $4Dt/L^2 \cong 2$

$$t \cong \frac{2L^2}{4D} \cong \frac{2\left(2\times10^{-5}\,\text{cm}\right)^2}{4\left(6\times10^{-14}\,\text{cm}^2\text{s}^{-1}\right)} = 3.13\times10^3\ \text{seconds} \cong 52\ \text{minutes}$$

not an unreasonable length of time and easily achievable in practice. However, if the particle size were 1 μm—10 times larger but still small for powder particles—then the time would be 2 orders of magnitude larger or 5200 minutes or about 3.6 days. If this calcination was being done in a furnace at about 1100°C—about the actual temperature where this is done in practice—then there would be a significant energy savings by using the smaller particle size, so that the furnace would run for a much shorter time. On the other hand, if $D_0 \cong 0.25$ and $Q = 330{,}540$ J/mol, then to calcine the 1 μm particles in the same time—52 minutes—the diffusion coefficient would have to be increased by 2 orders of magnitude to $D = 6 \times 10^{-12}$ cm²/s. Therefore, the temperature must be raised to

$$T(K) = -\frac{Q}{R\ln(D/D_0)} = \frac{-330540}{(8.314)\ln\left((6\times10^{-12})/0.25\right)} = 1626\text{ K} = 1353\text{°C}$$

and is high enough, so that the calcined material will sinter or densify to a very hard mass making it difficult to grind to a fine particle size suitable for final densification—0.1–1 μm. Clearly, the better option in this case is to use the smaller particle size rather than raise the calcining temperature. This example again demonstrates the utility of $4Dt \cong 2L^2$ as an approximation for a diffusion-controlled reaction to go to completion.

12.6 FINITE DIFFERENCE NUMERICAL SOLUTIONS OF FICK'S SECOND LAW

12.6.1 Fick's Second Law in Finite Difference Form

In Chapter 3, Euler's method was used to numerically solve ordinary differential equations by using a difference equation for the derivative, that is,

$$\frac{dC}{dx} \cong \frac{C(x+\Delta x) - C(x)}{\Delta x} \tag{12.25}$$

which is the definition of the derivative in the limit of $\Delta x \to 0$. As was mentioned then, this type of solution is called the *finite difference* technique and how it is applied is covered extensively in the literature (e.g., Lapidus and Pinder 1982; Celia and Gray 1992; Johnson 2009). One attractive feature of this technique is that it lends itself very nicely to solution with a spreadsheet program. As might be expected, partial differential equations can also be solved by a finite difference approach with a spreadsheet if the geometry and boundary conditions are not too complicated. Yet, fairly complex problems can be solved by the technique with relatively simple programs (Kee et al. 2003)*.

Now, Equation 12.25 is the *forward difference* equation for the derivative. A *backward difference* could also be used

$$\frac{dC}{dx} \cong \frac{C(x) - C(x-\Delta x)}{\Delta x} \tag{12.26}$$

with the same result. Fick's second law, $\partial C/\partial t = D(\partial^2 C/\partial x^2)$, a forward difference can be taken for $\partial C/\partial t$, specifically,

$$\frac{\partial C}{\partial t} \cong \frac{C(x,t+\Delta t) - C(x,t)}{\Delta t}. \tag{12.27}$$

Similarly, the first derivative with respect to x can be given by a forward difference,

$$\frac{\partial C}{\partial x} \cong \frac{C(x+\Delta x,t) - C(x,t)}{\Delta x}. \tag{12.28}$$

* The *finite difference* technique is not to be confused with the *finite element* approach, which is very different, more powerful, and can be used to solve many different types of differential equations in not only diffusion and heat flow but also stress and strain analysis and fluid flow. There is also extensive literature on finite element analysis (e.g., Reddy and Gartling 2001; Johnson 2009) with several commercial computer software packages available.

To get the second derivative with respect to x into difference form, it is most convenient to take the backward difference of Equation 12.28 so the differential is symmetric about C(x, t). So this is

$$\frac{\partial^2 C}{\partial x^2} \cong \frac{\left[C(x+\Delta x,t)-C(\{x+\Delta x\}-\Delta x,t)\right]-\left[C(x,t)-C(x-\Delta x,t\right]}{\Delta x \Delta x}$$

$$\frac{\partial^2 C}{\partial x^2} \cong \frac{C(x+\Delta x,t)-2C(x,t)+C(x-\Delta x,t)}{\Delta x^2} \quad (12.29)$$

Putting Equations 12.28 and 12.29 together gives the finite difference equation for the partial differential equation:

$$\frac{C(x,t+\Delta t)-C(x,t)}{\Delta t} \cong D\frac{C(x+\Delta x,t)-2C(x,t)+C(x-\Delta x,t)}{\Delta x^2}$$

and rewriting this into its usual form,

$$C(x,t+\Delta t) \cong C(x,t)+\left(\frac{D\Delta t}{\Delta x^2}\right)\left[C(x+\Delta x,t)-2C(x,t)+C(x-\Delta x,t)\right] \quad (12.30)$$

which is long, but not complicated. Equation 12.30 states that the concentration at a given x at the next time step, Δt, C(x, t + Δt) is just that at the previous time step, C(x, t), plus the parameters of the problem $\left(D\Delta t/\Delta x^2\right)$ times the concentrations at the previous and next distance step, $C(x-\Delta x,t)$ and $C(x+\Delta x,t)$, less twice the previous distance step, C(x, t). The only condition is that the term $\left(D\Delta t/\Delta x^2\right)$ must be small (less than 1/2 in this case [Morton and Mayers 1994]) otherwise the solution oscillates and is easy to see in the solution. Also, the smaller the time step for a given distance step, the closer the solution comes to the analytical solution but requiring more time steps to be taken to get to a given Dt/L^2.

12.6.2 Solving the Finite Difference Equation

The main attractive feature of the finite difference technique is that it lends itself very nicely to a spreadsheet solution. Figure 12.17 shows how the spreadsheet is set up for a solution for diffusion out of slab with boundary conditions C(0, t) = C(L, t) = 0, and the initial condition C(x, 0) = 1.0, the problem solved in Section 12.4. The distance step is horizontal and the time step is vertical.

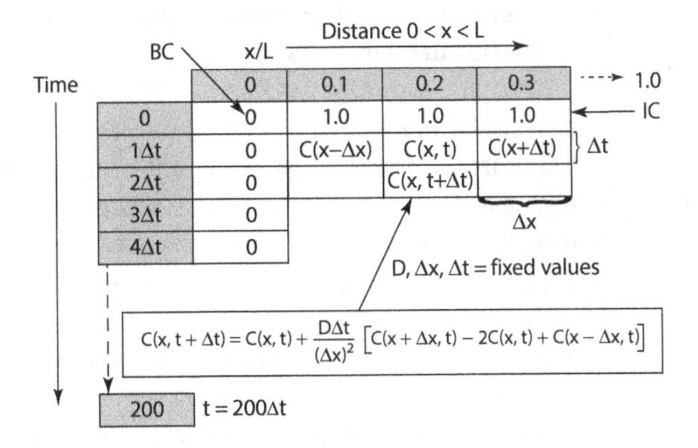

FIGURE 12.17 Schematic showing how a spreadsheet can be used to evaluate a *finite difference* solution, in this case, for diffusion out of a sheet. The boundary conditions, C(0, t) = C(L, t) = 0 are placed in the columns x/L = 0 and x/L = 1.0, respectively. The initial condition, C(x, 0) = 1.0 is entered into row zero. For convenience, the values of D, Δx, and Δt are put into separate cells, so they can be varied separately to see how they affect the solution. Then the finite difference solution is entered into cell (row 1Δt, column x/L = 0.1) and copied to all the remaining cells up t = 200 Δt in this illustration. The values of C(x, t) are automatically calculated at each position.

The boundary condition $C(0, t) = 0$ is the first column and the initial condition is the first row, $C(x, 0) = 1.0$. The values of D, Δx, and Δt can be put into separate cells, so that each is easily changed independently. All that needs to be done is to put the difference equation, Equation 12.30 in the first cell of the $x/L = 0.1$ column and then copy it to all of the other x/L cells in the row. Then the row is copied downward as far as necessary to get the largest value of Dt/L^2 desired: $Dt/L^2 \cong 0.5$ in this case. The solution appears for different values of time vertically and different values of x/L horizontally. Figure 12.18 is a copy of the spreadsheet (up to $x/L = 0.5$) used to calculate diffusion out of a slab of thickness L. In this case $\left(D\Delta t/\Delta x^2\right) = 0.25 < 0.5$ so the solution converges nicely to smooth curves. Since $\Delta x = f\,L$, where f is some fraction of L, this can be written $\left(D\Delta t/\Delta x^2\right) = \left(D\Delta t/f^2L^2\right) < 0.5$. As a result, the value of D necessary to give significant values of Dt/L^2 in a reasonable number of time steps will depend on the value of $D/(f\,L)^2$. In Figure 12.18, $L = 1$ cm and $f = 0.1$ so the diffusion coefficient used is $D = 1.0$ to ensure the inequality of $\left(D\Delta t/\Delta x^2\right) = \left(D\Delta t/f^2L^2\right) < 0.5$ with $\Delta t = 0.0025$. This is a realistic value for the diffusion coefficient for a gas diffusing out of a box. If $D = 10^{-4}\,cm^2/s$, appropriate for a liquid, then the Δt steps would be 25 seconds for this same value of $\left(D\Delta t/\Delta x^2\right) = 0.25 < 0.5$ and the results are the same at any value of Dt/L^2. It should be noted that only four iterations are used when $Dt/L^2 = 0.01$. This is not many iterations in time to smooth out the calculated curves. Nevertheless, Figure 12.19 compares this finite difference solution with the infinite series solution—that also requires a spreadsheet to sum the terms—and clearly, the finite

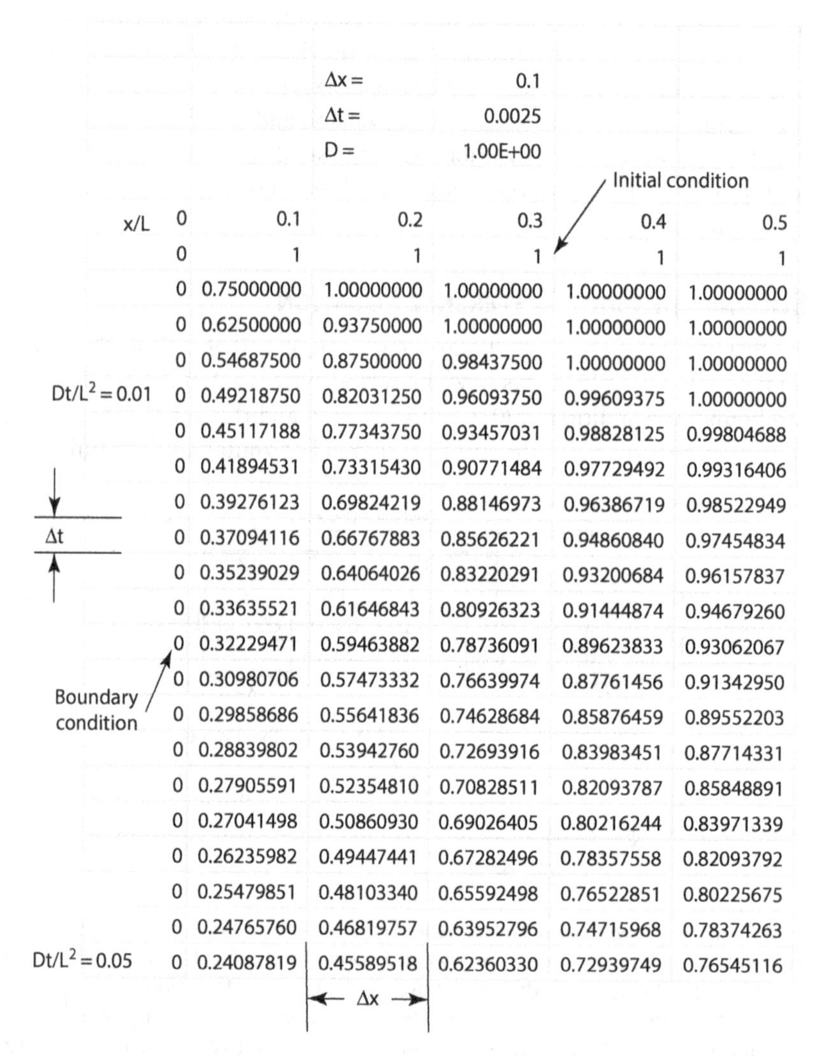

		Δx =		0.1		
		Δt =		0.0025		
		D =		1.00E+00		
x/L	0	0.1	0.2	0.3	0.4	0.5
	0	1	1	1	1	1
	0	0.75000000	1.00000000	1.00000000	1.00000000	1.00000000
	0	0.62500000	0.93750000	1.00000000	1.00000000	1.00000000
	0	0.54687500	0.87500000	0.98437500	1.00000000	1.00000000
$Dt/L^2 = 0.01$	0	0.49218750	0.82031250	0.96093750	0.99609375	1.00000000
	0	0.45117188	0.77343750	0.93457031	0.98828125	0.99804688
	0	0.41894531	0.73315430	0.90771484	0.97729492	0.99316406
	0	0.39276123	0.69824219	0.88146973	0.96386719	0.98522949
Δt	0	0.37094116	0.66767883	0.85626221	0.94860840	0.97454834
	0	0.35239029	0.64064026	0.83220291	0.93200684	0.96157837
	0	0.33635521	0.61646843	0.80926323	0.91444874	0.94679260
	0	0.32229471	0.59463882	0.78736091	0.89623833	0.93062067
	0	0.30980706	0.57473332	0.76639974	0.87761456	0.91342950
Boundary condition	0	0.29858686	0.55641836	0.74628684	0.85876459	0.89552203
	0	0.28839802	0.53942760	0.72693916	0.83983451	0.87714331
	0	0.27905591	0.52354810	0.70828511	0.82093787	0.85848891
	0	0.27041498	0.50860930	0.69026405	0.80216244	0.83971339
	0	0.26235982	0.49447441	0.67282496	0.78357558	0.82093792
	0	0.25479851	0.48103340	0.65592498	0.76522851	0.80225675
	0	0.24765760	0.46819757	0.63952796	0.74715968	0.78374263
$Dt/L^2 = 0.05$	0	0.24087819	0.45589518	0.62360330	0.72939749	0.76545116

FIGURE 12.18 Copy of part of the spreadsheet program for diffusion out of a sheet described in Figure 12.17. The values of $DT/L^2 = 0.01$ and 0.05 are shown for the parameters given at the top of the sheet.

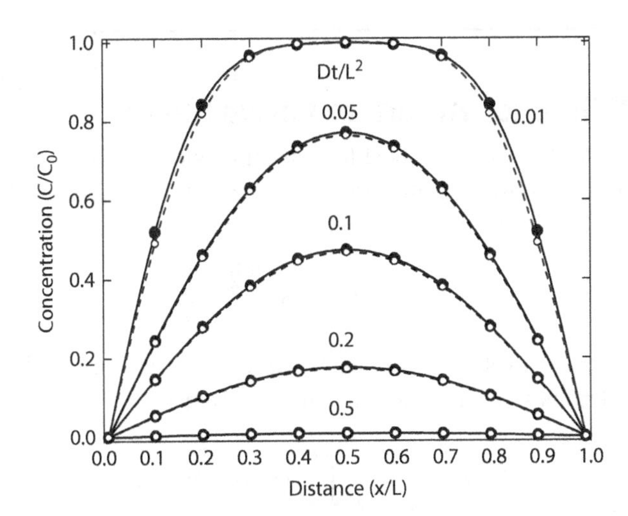

FIGURE 12.19 The spreadsheet finite difference solution, small white data points, compared to the 20-term infinite series solution of Figures 12.12 and 12.14, large black data points. Even after only four time iterations, $Dt/L^2 = 0.01$, the spreadsheet solution is quite close to the infinite series solution and gets closer as time increases.

difference solution is pretty close even at $Dt/L^2 = 0.01$. It could be made closer by making either the time step or the distance step, or both, smaller, which is not a problem. It just requires that more cells be copied.

Of course, a spreadsheet does not have to be used to solve Equation 12.30, but it is extremely convenient way of doing so, particularly with simple boundary geometries. Instead, a simple iterative computer program could easily be written as an alternative for solving the difference equation, Equation 12.30. The main point to emphasize is that, in many cases, numerical techniques can be used to solve Fick's second law with equally good results and sometimes more easily than finding analytical solutions.

12.7 CHAPTER SUMMARY

The main objective of this chapter is to solve kinetic diffusion-controlled problems over finite regions—limited to one dimension—to become familiar with the mathematical techniques typically used to solve these problems; namely, separation of variables that lead to infinite trigonometric or Fourier series when satisfying the boundary and initial conditions. Although these solutions appear formidable at first glance, their evaluation is quite easy. In order to get a very accurate value of $C(x, t)$ only a small number of terms, 10 or 20, are required and can easily be evaluated with a spreadsheet program. These infinite series solutions are compared to error function solutions at short times where the error functions should be more applicable. Surprisingly, the error functions are quite accurate up to values of $Dt/L^2 \cong 0.05$ and are easier to evaluate than the series solutions. Also, when $Dt/L^2 \cong 0.05$, only the first term in the infinite series solution is now satisfactory to accurately represent the solution to the completion of the reaction—within 1%—at $Dt/L^2 \cong 0.5$. Examples used to illustrate these principles are diffusion from a sheet of thickness L with some constant initial concentration; transient diffusion to a steady state though a wall of finite thickness, L; and the interdiffusion of two components over a finite distance L. In each of these as in all finite boundary problems, a good estimate—*back of the envelope calculation*—for the time it takes for the diffusion reaction to go to completion is $Dt/L^2 \cong 0.5$. Finally, the finite difference technique is introduced as a numerical method to solve Fick's second law that lends itself to simple spreadsheet solutions of finite boundary condition problems. Of course, the technique could also be used for infinite and semi-infinite problems as long as the concentrations do not get much above zero far from the interdiffusion zone just like error function solutions for finite boundary conditions.

APPENDIX

A.1 INFINITE BC SOLUTION APPLIED TO CORING MODEL

Figure 12.3 shows the model for the composition variation with distance, *coring*, produced by non-equilibrium freezing of an almost ideal alloy that forms a single solid solution. Specifically, the composition before a homogenization heat treatment can be approximated by

$$C(x,0) = C_0 + \beta \sin\left(\frac{2\pi x}{L}\right). \tag{A.1}$$

In principle, the concentration with time of heat treatment, C(x, t), can be solved by the methods of infinite boundary conditions since the solution can be written as

$$C(x,t) = \frac{1}{\sqrt{4\pi Dt}} \int_{-\infty}^{\infty} f(x_0)\, e^{-\frac{(x-x_0)^2}{4Dt}}\, dx_0$$

where $f(x_0)$ is Equation A.1:

$$C(x,t) = \frac{1}{\sqrt{4\pi Dt}} \int_{-\infty}^{\infty} \left[C_0 + \beta \sin\left(\frac{2\pi x_0}{L}\right)\right] e^{-\frac{(x-x_0)^2}{4Dt}}\, dx_0 \tag{A.2}$$

and all that is left is the integration. The first term in the integration, call it A, involves only C_0 and it is

$$A = \frac{C_0}{\sqrt{4\pi Dt}} \int_{-\infty}^{\infty} e^{-\frac{(x-x_0)^2}{4Dt}}\, dx_0. \tag{A.3}$$

As usual, there needs to be a change in variables, so let $y = (x - x_0)/\sqrt{4Dt}$ and $dx_0 = -\sqrt{4Dt}\,dy$ and when $x_0 = -\infty$, $y = \infty$ and when $x_0 = \infty$, $y = -\infty$. With these substitutions, Equation A.3 becomes

$$A = -\frac{C_0}{\sqrt{4\pi Dt}} \sqrt{4Dt} \int_{\infty}^{-\infty} e^{-y^2} dy$$

$$A = \frac{C_0}{\sqrt{\pi}} \int_{-\infty}^{\infty} e^{-y^2} dy = \frac{C_0}{\sqrt{\pi}} \sqrt{\pi} \tag{A.4}$$

$$A = C_0$$

a reassuring result. The second integral, call it B, is a bit more complicated:

$$B = \frac{1}{\sqrt{4\pi Dt}} \int_{-\infty}^{\infty} \left[\beta \sin\left(\frac{2\pi x_0}{L}\right)\right] e^{-\frac{(x-x_0)^2}{4Dt}}\, dx_0. \tag{A.5}$$

Making the same change in variables gives

$$B = -\frac{\beta}{\sqrt{\pi}} \int_{\infty}^{-\infty} \sin\left[\frac{2\pi}{L}\left(x - \sqrt{4Dt}\,y\right)\right] e^{-y^2} dy. \tag{A.6}$$

It is not obvious that progress is being made but it is. From elementary trigonometry, $\sin(\alpha - \beta) = \sin\alpha\cos\beta - \cos\alpha\sin\beta$ (this can be found in many mathematics reference books and software packages such as Mathematica™). Therefore,

$$\begin{aligned} \sin\left[\frac{2\pi}{L}\left(x - \sqrt{4Dt}\,y\right)\right] &= \sin\left[\frac{2\pi x}{L} - \frac{2\pi}{L}\sqrt{4Dt}\,y\right] \\ &= \sin\left(\frac{2\pi x}{L}\right)\cos\left(\frac{2\pi}{L}\sqrt{4Dt}\,y\right) - \cos\left(\frac{2\pi x}{L}\right)\sin\left(\frac{2\pi}{L}\sqrt{4Dt}\,y\right) \end{aligned} \tag{A.7}$$

and substitution of the second equation in (Equation A.7) into Equation A.6 gives (help!)

$$B = \frac{\beta}{\sqrt{\pi}}\sin\left(\frac{2\pi x}{L}\right)\int_{-\infty}^{\infty}\cos\left(\frac{2\pi}{L}\sqrt{4Dt}\,y\right)e^{-y^2}dy + \frac{\beta}{\sqrt{\pi}}\cos\left(\frac{2\pi x}{L}\right)\int_{-\infty}^{\infty}\sin\left(\frac{2\pi}{L}\sqrt{4Dt}\,y\right)e^{-y^2}dy \quad \text{(A.8)}$$

Now in the second equation in (Equation A.8), the sine function is an *odd* function of y (i.e., f(−y) = −f(y)) so its integration from −∞ to ∞ is zero. (Mathematica™ also gives zero for this integral: the *black-box* approach) So the only integral of interest is

$$\int_{-\infty}^{\infty}\cos\left(\frac{2\pi}{L}\sqrt{4Dt}\,y\right)e^{-y^2}dy \quad \text{(A.9)}$$

which for simplicity can be written as

$$2\int_{0}^{\infty}\cos(ay)e^{-y^2}dy \quad \text{(A.10)}$$

since cos(ay) is an even function (i.e., f(−y) = f(y)) and $a = (2\pi/L)\sqrt{4Dt}$.

Here is one way to integrate Equation A.10. Let I = Equation A.10, then

$$\frac{dI}{da} = -\int_{0}^{\infty}2y\sin(ay)e^{-y^2}dy$$

Integrating by parts $\left(\int f'g = \int fg - \int fg' \text{ and let } f' = -2ye^{-y^2}, g = \sin(ay)\right)$ so $f = e^{-y^2}$ and $g' = a\cos(ay)$ gives

$$\frac{dI}{da} = \left|e^{-y^2}\sin(ay)\right|_0^{\infty} - a\int_0^{\infty}\cos(ay)e^{-y^2}dy$$

$$\frac{dI}{da} = (0-0) - a\frac{I}{2}$$

which is integrated to give,

$$\ln I = -\frac{a^2}{4} + D'$$

$$I = De^{-\frac{a^2}{4}}$$

where D and D′ are constants which can be evaluated at a = 0 from

$$D(a) = 2e^{\frac{a^2}{4}}\int_0^{\infty}\cos(ay)e^{-y^2}dy$$

$$D(0) = 2\int_0^{\infty}e^{-y^2}dy = \sqrt{\pi}$$

Therefore,

$$\int_{-\infty}^{\infty}\cos\left(\frac{2\pi}{L}\sqrt{4Dt}\,y\right)e^{-y^2}dy = \sqrt{\pi}e^{-\frac{a^2}{4}} \quad \text{(A.11)}$$

which is also given by Mathematica™. So Equation A.8 becomes

$$B = \frac{\beta}{\sqrt{\pi}}\sin\left(\frac{2\pi x}{L}\right)\sqrt{\pi}e^{-\frac{\left(\frac{2\pi}{L}\sqrt{4Dt}\right)^2}{4}}$$

$$B = \beta\sin\left(\frac{2\pi x}{L}\right)e^{-\frac{4\pi^2 Dt}{L^2}} \quad \text{(A.12)}$$

Finally, the solution of Equation A.2 by infinite boundary condition methods is obtained by adding Equations A.4 and A.12 to give

$$C(x,t) = C_0 + \beta \sin\left(\frac{2\pi x}{L}\right) e^{-\frac{4\pi^2 Dt}{L^2}}. \tag{A.13}$$

A.2 Fourier Series Solution for Diffusion from a Sheet

A.2.1 Infinite Series Solution

The general separation of variables solution to Fick's second law is

$$C(x,t) = e^{-\alpha^2 Dt}(A \sin \alpha x + B \cos \alpha x). \tag{A.14}$$

For the sheet, the initial condition is $C(x, 0) = C_0$ and the boundary conditions are $C(0, t) = C(L, t) = 0$. With $C(0, t) = 0$ then $B = 0$ in Equation A.14 since the cosine term must be positive at $x = 0$. So the solutions must be of the form

$$C(x,t) = Ae^{-\alpha^2 Dt} \sin \alpha x \tag{A.15}$$

and the other boundary condition, $C(L, t) = 0$ indicates that $\sin \alpha x = 0$, which implies that $\alpha L = n\pi$ with $n = 0, 1, 2, 3, \ldots, n, \ldots, \infty$ or $\alpha = n\pi/L$, so that there are an infinite number of solutions of the form

$$C_n(x,t) = A_n e^{-\frac{n^2\pi^2 Dt}{L^2}} \sin\left(\frac{n\pi x}{L}\right) \tag{A.16}$$

which is convenient since each of these is a solution, then the sum of two or more up to an infinite sum of these terms is also a solution. This means that an infinite series can be used to model virtually any initial condition. That is, the general solution is now

$$C(x,t) = \sum_{n=0}^{\infty} A_n e^{-\frac{n^2\pi^2 Dt}{L^2}} \sin\left(\frac{n\pi x}{L}\right) \tag{A.17}$$

and if the general initial condition is $C(x, 0) = f(x)$, then this becomes

$$f(x) = \sum_{n=0}^{\infty} A_n \sin\left(\frac{n\pi x}{L}\right) \tag{A.18}$$

which is now the *Fourier series* of $f(x)$ and the coefficients A_n are given by (Kreyszig 2011)

$$A_n = \frac{2}{L}\int_0^L f(x) \sin\left(\frac{n\pi x}{L}\right) dx. \tag{A.19}$$

A.2.2 Constant Initial Concentration

For $f(x) = C(x,0) = C_0$

$$\begin{aligned} A_n &= \frac{2C_0}{L}\int_0^L \sin\left(\frac{n\pi x}{L}\right) dx \\ &= \frac{2C_0}{L}\left| -\frac{1}{n\pi/L} \cos\left(\frac{n\pi x}{L}\right)\right|_0^L \\ A_n &= \frac{2C_0}{n\pi}\{1 - \cos(n\pi)\} \end{aligned} \tag{A.20}$$

For n = odd, $A_n = 4C_0/n\pi$ and for n = even, $A_n = 0$. Usually series with many terms equal to zero are rewritten in terms of the dummy variable n to make all the terms take on some value other than zero. In this case, n is replaced by 2m + 1 (which is still an integer) so that

$$A_m = \frac{2C_0}{(2m+1)\pi}\{1-\cos((2m+1)\pi)\}$$
$$A_m = \frac{4C_0}{(2m+1)\pi} \tag{A.21}$$

since the term $\{1-\cos(2m+1)\pi\} = 2$ for any value of m. Therefore, the infinite series solution to the sheet finite boundary problem with the initial concentration of $C(x, 0) = C_0$ and $C(x, \infty) = 0$ is

$$C(x,t) = \frac{4C_0}{\pi}\sum_{n=0}^{\infty}\frac{e^{-(2n+1)^2\pi^2\frac{Dt}{L^2}}}{2n+1}\sin\left[\frac{(2n+1)\pi x}{L}\right]. \tag{A.22}$$

A.2.3 Fourier Series Coefficients

Only the infinite sine series is considered, the procedure for an infinite cosine series is similar. Again, in general, Equation A.18

$$f(x) = \sum_{n=0}^{\infty} A_n \sin\left(\frac{n\pi x}{L}\right)$$

where f(x) is the initial condition at t = 0. Multiply both sides of the equation by $\sin(m\pi x/L)$ and integrate for $0 \le x \le L$,

$$\int_0^L f(x)\sin\left(\frac{m\pi x}{L}\right)dx = \sum_{n=1}^{\infty} A_n \int_0^L \sin\left(\frac{n\pi x}{L}\right)\sin\left(\frac{m\pi x}{L}\right)dx. \tag{A.23}$$

Performing the integration in the summation for m = n first,

$$\int_0^L \sin\left(\frac{n\pi x}{L}\right)\sin\left(\frac{m\pi x}{L}\right)dx$$
$$= \int_0^L \sin^2\left(\frac{n\pi x}{L}\right)dx$$
$$= \int_0^L \left\{\frac{1}{2} - \frac{1}{2}\cos\left(\frac{2n\pi x}{L}\right)\right\}dx$$
$$= \left|\frac{x}{2} - \frac{\sin(2n\pi x/L)}{4n\pi/L}\right|_0^L = \frac{L}{2}$$

Doing the same for m ≠ n,

$$\int_0^L \sin\left(\frac{n\pi x}{L}\right)\sin\left(\frac{m\pi x}{L}\right)dx$$
$$= \frac{1}{2}\int_0^L \left\{\cos\left([n-m]\frac{\pi x}{L}\right) - \cos\left([n+m]\frac{\pi x}{L}\right)\right\}dx$$
$$= \frac{1}{2}\left|\frac{\sin([n-m](\pi x/L))}{[n-m](\pi/L)} - \frac{\sin([n+m](\pi x/L))}{[n+m](\pi/L)}\right|_0^L$$
$$= 0$$

So after these messy, but simple, trigonometric substitutions and integrations, Equation A.23 becomes

$$\int_0^L f(x)\sin\left(\frac{n\pi x}{L}\right)dx = A_n\frac{L}{2}$$

or, Equation A.19

$$A_n = \frac{2}{L}\int_0^L f(x)\sin\left(\frac{n\pi x}{L}\right)dx.$$

A.3 Infinite Series Solution for Diffusion Through a Wall

As pointed out in Chapter 8, in this case, the solution is given by the sum of a steady-state term and a transient term that goes to zero as $t \to \infty$; that is, $C(x, t) = f(x) + g(x, t)$ where $f(x)$ is the steady-state term and $g(x, t)$ is the transient. As was seen in Chapter 10, Figure 10.1, the steady-state solution for diffusion through a wall is

$$C(x) = C_S\left(1-\frac{x}{L}\right)$$

where C_S is the constant concentration on the inside of the wall ($x = 0$) and $C(L) = 0$. Therefore, the solution becomes

$$C(x,t) = C_S\left(1-\frac{x}{L}\right) + \sum_{n=1}^{\infty} A_n e^{-\frac{n^2\pi^2 Dt}{L^2}} \sin\left(\frac{n\pi x}{L}\right). \tag{A.24}$$

For the case when $C(x, 0) = 0$, then

$$0 = C_S\left(1-\frac{x}{L}\right) + \sum_{n=1}^{\infty} A_n \sin\left(\frac{n\pi x}{L}\right)$$

and proceeding as in Equation A.19 to obtain the values for the A_n coefficients

$$A_n = \frac{2}{L}\int_0^L C_S\left(\frac{x}{L}-1\right)\sin\left(\frac{n\pi x}{L}\right)dx$$

$$A_n = -\frac{2C_S}{n\pi}$$

with, again, a simple but lengthy integration leading to the value for A_n. Replacing A_n in Equation A.24 gives the solution for diffusion through a wall of thickness L,

$$\frac{C(x,t)}{C_S} = \left(1-\frac{x}{L}\right) - \frac{2}{\pi}\sum_{n=1}^{\infty}\frac{1}{n} e^{-\frac{n^2\pi^2 Dt}{L^2}} \sin\left(\frac{n\pi x}{L}\right). \tag{A.25}$$

A.4 Infinite Series Solution for Diffusion With Derivative Boundary Conditions

Figure 12.16 shows the initial conditions and solutions for the derivative finite boundary condition problem of

Initial condition: $C = C_0$ for $0 \le x < L/2, C = 0$ for $L/2 \le x < L$
Boundary conditions: $(\partial C/\partial x)(0,t) = 0$ and $(\partial C/\partial x)(L,t) = 0$

As before, the solution must be of the form, Equation A.14,

$$C(x,t) = e^{-\alpha^2 Dt}(A\sin\alpha x + B\cos\alpha x).$$

Taking the first BC,

$$\frac{\partial C}{\partial x}(0,t) = e^{-\alpha^2 Dt}\left(\alpha A\cos(0) - \alpha B\sin(0)\right) = 0$$

which means that A = 0 so the solution must be of the form

$$C(x,t) = e^{-\alpha^2 Dt} B\cos(\alpha x)$$

And with the second boundary condition

$$\frac{\partial C}{\partial x}(L,t) = -\alpha B e^{-\alpha^2 Dt}\sin(\alpha L) = 0.$$

This implies that $\alpha L = n\pi$, n = 0, 1, 2, 3 .... Therefore the nth term in the series, C_n, is given by

$$C_n(x,t) = B_n e^{-\frac{n^2\pi^2 Dt}{L^2}}\cos\left(\frac{n\pi x}{L}\right).$$

Therefore, the solution is

$$C(x,t) = \sum_{n=0}^{\infty} B_n e^{-\frac{n^2\pi^2 Dt}{L^2}}\cos\left(\frac{n\pi x}{L}\right). \tag{A.26}$$

The Fourier coefficients, B_n, are given by (Kreyszig 2011)*

$$B_0 = \frac{1}{L}\int_0^L f(x)dx$$

$$B_n = \frac{2}{L}\int_0^L f(x)\cos\left(\frac{n\pi x}{L}\right)dx$$

for the initial conditions f(x). For the initial conditions in this problem,

$$B_0 = \frac{C_0}{L}\int_0^{L/2} dx = \frac{C_0}{2} \tag{A.27}$$

and

$$B_n = \frac{2C_0}{L}\int_0^{L/3}\cos\left(\frac{n\pi x}{L}\right)dx$$

$$B_n = \frac{2C_0}{L}\left(\frac{L}{n\pi}\right)\sin\left(\frac{n\pi x}{L}\right)\Bigg|_0^{L/2}$$

resulting in,

$$B_n = \frac{2C_0}{n\pi}\sin\left(\frac{n\pi}{2}\right). \tag{A.28}$$

Equation A.28 is kind of messy since for n = 1, 2, 3, 4, 5, ...

$$\frac{1}{n}\sin\left(\frac{n\pi}{2}\right) = 1, 0, -\frac{1}{3}, 0, \frac{1}{5}, 0, -\frac{1}{7}\cdots$$

* These coefficients can easily be determined by the same procedure used in Section A.2.3.

so as before, the zero terms must be eliminated by putting in 2n – 1 for n. Now to get the alternating sign, $(-1)^{n+1}$ must also be included to give

$$\frac{(-1)^{n+1}}{2n-1} = 1, -\frac{1}{3}, \frac{1}{5}, -\frac{1}{7}, \ldots$$

Therefore, this can be substituted for $(1/n)\sin(n\pi/2)$ and get the same numerical values for B_n. Hence B_n becomes

$$B_n = \frac{2C_0(-1)^{(n+1)}}{\pi(2n-1)}$$

with the resulting solution

$$C(x,t) = \frac{C_0}{2} + \frac{2C_0}{\pi}\sum_{n=1}^{\infty}\frac{(-1)^{(n+1)}}{(2n-1)} e^{-\left((2n-1)^2\pi^2\frac{Dt}{L^2}\right)} \cos\left[\frac{(2n-1)\pi x}{L}\right] \quad (A.29)$$

which, thankfully, is the same as the solution in the literature (Crank 1975).

EXERCISES

12.1 a. Show that the initial conditions of Figure 12.8 lead to Equation 12.13.
b. Show that the solution of Equation 12.13 leads to Equation 12.14.

12.2 Show that the solution of the infinite boundary condition problem with the initial concentration profile shown in Figure 12.10 leads to Equation 12.15.

12.3 The diffusion coefficient of carbon in gamma iron is given by $D = D_0 e^{-(Q/RT)}$, where Q = 142.3 kJ/mol and D_0 = 0.2 cm²/s from Chapter 9.

a. Calculate D at 1050°C.
b. If a 30 mil thick sheet of 1018 steel were to have its carbon removed by heat treating in a pure hydrogen atmosphere where

$$C(\text{steel}) + 2H_2(\text{gas}) = CH_4(\text{gas})$$

Estimate how long this will take (hours) at 1050°C.
c. If the density and molecular weight of iron are ρ = 7.87 g/cm³ and M = 55.845 g/mol and those of carbon are ρ = 2.2 g/cm³ and M = 12.011 g/mol, calculate the mole fraction carbon in 1018 steel (0.18 w/o).
d. If the free energy for the above reaction at 1050°C is $\Delta G^o = 50{,}654$ J/mole what is the CH_4/H_2 ratio in the gas phase at so that the activity of carbon is the same in the gas as in the steel. Assume an ideal solution for the solid.
e. Calculate the CH_4/H_2 ratio in the gas to get the mole fraction of carbon down to 10^{-6} in the steel.
f. Calculate D at 950°C.
g. If the temperature were lowered to 950°C, estimate how long it will take to remove all of the carbon, in hours.
h. If the sheet is 1 mm thick, estimate how long it will take in hours to decarburize the sheet at 950°C.

12.4 The single sine function *approximate* solution for this problem (number 12.3) is given by Equation 12.12:

$$C(x,t) = C_0 e^{-\frac{\pi^2}{L^2}Dt} \sin\left(\frac{\pi x}{L}\right)$$

where C_0 is the initial carbon concentration.

a. Calculate the fractional amount of material remaining if this sine function were taken to be the starting point if in the real situation the concentration is uniform, C_0, over the entire thickness. That is, what should the maximum in the sine function be at t = 0 so that the integral over the sine function is C_0L.
b. Develop an expression for the diffusional fluxes out of the sheet at x = 0 and x = L.
c. With this solution, estimate that time that it would take to get the maximum concentration in the sheet down to 10^{-6} w/o (at the maximum) for the 30 mil 1018 steel at 1050°C.
d. If the sheet is 1 mm thick, how estimate how long it will take in hours to decarburize the sheet to this same level at 950°C.

12.5 Another single sine function *approximate* solution for this problem (number 12.3) is taking just the first term in the exact infinite series solution, Equation 12.18:

$$C(x,t) = \frac{4C_0}{\pi} e^{-\frac{\pi^2}{L^2}Dt} \sin\left(\frac{\pi x}{L}\right)$$

where C_0 = the initial carbon concentration.

a. Calculate the maximum concentration at t = 0 for this approximation. (see 12.4a)
b. Calculate the fractional amount of material remaining if this sine function were taken to be the starting point if in the real situation the concentration is uniform, C_0, over the entire thickness.
c. With this solution, estimate that time that it would take to get the maximum concentration in the sheet down to 10^{-6} w/o (at the maximum) for the 30 mil 1018 steel at 1050°C.
d. If the sheet is 1 mm thick, how estimate how long it will take in hours to decarburize the sheet to this same level at 950°C.
e. Calculate the percent differences in the times for parts c and d of this problem and those in problem 12.4.

12.6 In contrast, assume that the initial concentration of the sheet was zero and something was diffusing in. The exact infinite series solution in this case is

$$C(x,t) = C_0 - \frac{4C_0}{\pi} \sum_{m=0}^{\infty} \frac{1}{(2m+1)} e^{-\frac{(2m+1)^2 \pi^2}{L^2}Dt} \sin\left(\frac{(2m+1)\pi x}{L}\right).$$

C_0 is the equilibrium solution and the constant concentration at the interface, the solubility under the specified conditions. The series of course is just C_0 at t = 0 so C(x, 0) = 0, the initial condition. With a spreadsheet (or other computer program), calculate C(x, t) and plot on the same graph for m = 20 terms for values of $Dt/L^2 = 0.01, 0.05, 0.1, 0.2$, and 0.5 at x intervals of 0.1 L from 0 to L.

12.7 Develop a finite difference solution for diffusion into a sheet of thickness L with the initial condition C(x, 0) = 0 and boundary conditions C(0, t) = C_S and C(L, t) = 0. Take $\Delta x = 0.1L$; calculate up to $Dt/L^2 = 0.5$; and plot C(x, t) for $Dt/L^2 = 0.01, 0.05, 0.1, 0.2$, and 0.5 at x/L intervals of 0.1.

12.8 Develop a finite difference solution for diffusion into a sheet of thickness L with the initial condition C(x, 0) = 0 and boundary conditions C(0, t) = C_S and C(L, t) = C_S. Take $\Delta x = 0.1L$; calculate up to $Dt/L^2 = 0.5$; and plot C(x, t) for $Dt/L^2 = 0.001, 0.01, 0.05, 0.1, 0.2$, and 0.5 at x/L intervals of 0.1.

12.9 Develop a finite difference solution for diffusion into a sheet of thickness L with the initial condition C(x, 0) = 0 for $0 \le x/L \le 0.4$, C(0, x) = C_S for $0.4 \le x/L \le 0.6$, and C(0, x) = 0 for $0.6 \le x/L \le 1.0$ with boundary conditions C(0,t) = C(L,t) = C_S. Take $\Delta x = 0.1L$ and calculate and plot C(x,t) for $Dt/L^2 = 0.001, 0.01, 0.02, 0.05, 0.1, 0.2$, and 0.5 at x/L intervals of 0.1L, all on the same graph.

REFERENCES

Brandes, E. A. and G. B. Brook, eds. 1992. *Smithells Metals Reference Book,* 7th ed. Oxford: Butterworth-Heinemann.

Carslaw, H. S. and J. C. Jaeger. 1959. *Conduction of Heat in Solids,* 2nd ed. Oxford: Oxford University Press.

Celia, M. A. and W. G. Gray. 1992. *Numerical Methods for Differential Equations.* Englewood Cliffs, NJ: Prentice Hall.

Crank, J. 1975. *The Mathematics of Diffusion,* 2nd ed. Oxford: Clarendon Press.

Cussler, E. L. 1997. *Diffusion Mass Transfer in Fluid Systems,* 2nd ed. Cambridge: Cambridge University Press.

Glicksman, M. E. 2000. *Diffusion in Solids.* New York: Wiley.

Johnson, C. 2009. *Numerical Solution of Partial Differential Equations by the Finite Element Method.* Mineola, NY: Dover Publications.

Kee, R. J., M. E. Coltrin, and P. Glarsborg. 2003. *Chemically Reacting Flow.* Hoboken, NJ: Wiley.

Kreyszig, E. 2011. *Advanced Engineering Mathematics,* 10th ed. New York: Wiley.

Lapidus, L. and G. F. Pinder. 1982. *Numerical Solution of Partial Differential Equations in Science and Engineering.* New York: Wiley.

Morton, K. W. and D. F. Mayers. 1994. *Numerical Solution of Partial Differential Equations.* Cambridge: Cambridge University Press.

Reddy, J. N. and D. K. Gartling. 2001. *The Finite Element Method in Heat Transfer and Fluid Dynamics.* Boca Raton, FL: CRC Press.

Reed-Hill, R. E. and R. Abbaschian. 1992. *Physical Metallurgy Principles,* 3rd ed. Boston, MA: PWS-Kent Publishing Company.

Section V

Fluxes, Forces, and Interdiffusion

13

Fluxes, Forces, and Diffusion

13.1 INTRODUCTION

Previously, diffusion and the steady-state and transient diffusion models have all assumed that transport occurs in thermodynamically ideal systems. In fact, Fick's laws are based on ideal systems. Unfortunately, the world is far from *ideal* and in many cases transport needs to be considered under nonideal conditions. This does not imply that all of the previous models are invalid. Certainly, under normal conditions, most gases can be considered ideal as can many liquids. And there are many systems that form solid solutions close enough to ideal that can be used to understand the kinetics of important materials science and engineering processes. In reality, examining nonideal systems helps to reinforce understanding of kinetic processes in general, and provides additional tools to investigate more complex systems and processes. In this chapter, some generalizations of transport are considered that have implications beyond kinetic processes and lead naturally to an understanding of *charge transport*—electrical conductivity—in materials as well.

13.2 FLUX DENSITY OF MOVING PARTICLES

It is useful to obtain a fundamental relationship between the number of particles per unit volume η (particles/m^3), moving with a uniform or *drift* velocity, v_d (m/s), and the particle flux density, J' (particles/m^2 s). The resulting general equations can be applied to the modeling of electrical conductivity—both electronic and ionic—the thermal conductivity of solids and liquids, as well as diffusion.

Consider a box of particles shown in Figure 13.1. If these were atoms, molecules, ions, or electrons moving in gases, liquids, or solids, they would be moving with high thermal energies, colliding, and leading to random overall motion. However, if there is some kind of force applied in a given direction, say +x, the particles will all move with some constant net velocity, v_d, in the +x direction. At time $t = L/v_d$, all of the particles in the box, $N_T = \eta L^3$, will have passed through the end of the box with the cross section $A = L^2$. So the particle *number* flux density J' (units: particles/m^2 s) is given by

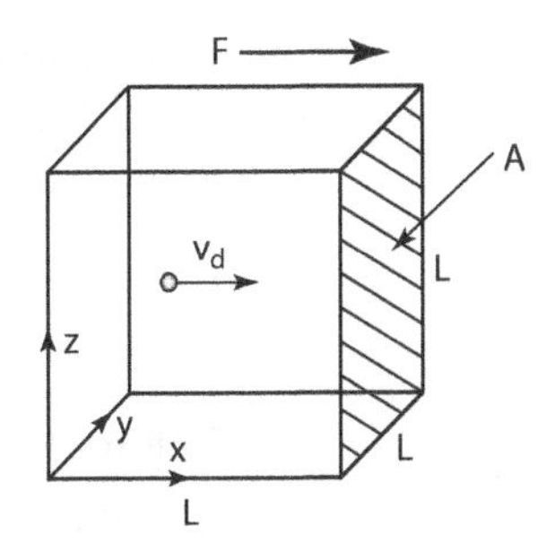

FIGURE 13.1 A particle in a box with sides of length L moving with a constant drift velocity, v_d, in the positive x-direction under an influence of some force, F.

$$J' = \frac{N_T}{At} = \frac{\eta L^3}{L^2(L/v_d)} = \eta v_d. \tag{13.1}$$

By dividing Equation 13.1 by Avogadro's number, N_A, the equivalent expression in terms of the molar flux and the concentration (mol/m^3) is obtained:

$$J = \frac{J'}{N_A} = \frac{\eta}{N_A} v_d = C v_d. \tag{13.2}$$

Thus, not surprisingly, in addition to a flux due to diffusion, there is also one due to flow because of the applied force, and these fluxes just add to get the total molar flux:

$$J = -D\frac{dC}{dx} + C v_d. \tag{13.3}$$

13.3 MOBILITY AND FORCES

13.3.1 Absolute Mobility

Consider a particle (solid, liquid, or gas) settling (or rising) in a liquid under the influence of gravity as shown in Figure 13.2. The particle* will reach a constant drift or *terminal velocity*, v_d, in the (positive or negative) vertical direction when the forces acting on it are balanced. These forces are gravity, F_g, the buoyancy force, F_b, due to the weight of the fluid displaced, and a drag force caused by the viscous forces between the moving particle and the fluid, F_d. As a result, the equation of motion of this particle, F = ma (Newton's first law, one of the few things that one needs to remember in this world), becomes

$$ma = F = F_g + F_b + F_d \tag{13.4}$$

where the resulting acceleration will be in the $-z$ direction if z is the vertical direction and the gravitational force exceeds the drag and buoyancy force. Viscous drag forces occur in many different areas of physics and engineering and they typically are assumed to be proportional to the velocity of the moving particle: for example, $F_d = -\gamma v$. For a particle moving in a fluid in which there is simple laminar flow around the particle, this is indeed the case, and is a good starting point in developing a model for diffusion in liquids. Consider the general result in Equation 13.4 with all the forces

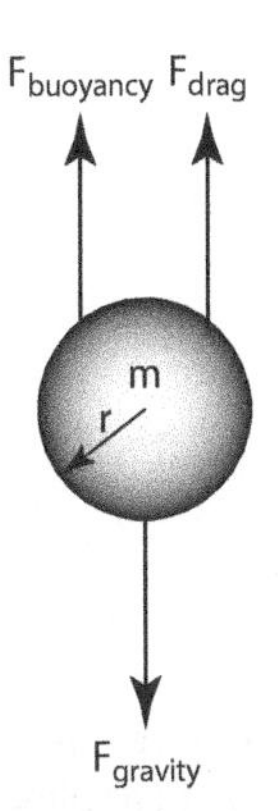

FIGURE 13.2 A particle of radius r and mass m moving vertically in a liquid under the influence of buoyancy, drag, and gravity forces.

* A "particle" in this case *usually* does not refer to the type of particles considered by nuclear physicists. However, it does include things as small as electrons up to actual macroscopic "chunks" of material that are nanometers, or micrometers, or larger in size.

that might be acting on this particle lumped into a single force, F, with the exception of the viscous force. The result is the following simple differential equation for the velocity of the particle as a function of time:

$$ma = m\frac{dv}{dt} = F - \gamma v. \tag{13.5}$$

This too is a very general equation in that F can be any type of net applied force: mechanical, electrical, chemical, and so on. For now, it will be left as a general force. Rearranging Equation 13.5 gives

$$\frac{dv}{F - \gamma v} = \frac{1}{m}dt$$

which is easily integrated by substituting $u = F - \gamma v$ so that $dv - (1/\gamma)du$, resulting in

$$-\frac{1}{\gamma}\ln(F - \gamma v) = \frac{1}{m}t + A$$

and rearranging

$$\ln(F - \gamma v) = -\frac{\gamma}{m}t + A'$$

leading to

$$F - \gamma v = A'' e^{-\frac{\gamma}{m}t}$$

where A, A', A'' are all integration constants. Solving for the velocity with $v = 0$ at $t = 0$, then

$$v = \frac{F}{\gamma}\left(1 - e^{-\frac{\gamma}{m}t}\right) \tag{13.6}$$

Equation 13.6 is plotted in Figure 13.3 and at $t = \infty$, $dv/dt = 0$ and $v = v_d = F/\gamma$.* In any event, the drift velocity can be more conveniently written as

$$v_d = \frac{F}{\gamma} = BF \tag{13.7}$$

where B is called the *absolute mobility* with units of m/s N and $B = 1/\gamma$.

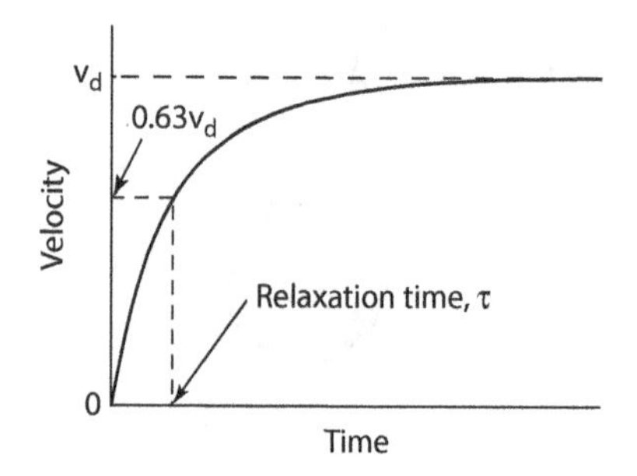

FIGURE 13.3 Velocity versus time for a particle under the influence of an applied force and a drag force that exponentially reaches a steady-state, constant drift velocity, v_d, with a relaxation time of τ.

13.3.2 Relaxation Time—Time to Reach the Drift Velocity

Equation 13.6 has an exponentially decaying term. As in Chapter 2, for exponential decay a relaxation time, τ, can be defined so that when the time has reached τ, the exponential term has dropped to 37% of its final value. So in Equation 13.6, $\tau = (m/\gamma)$ or $\gamma = m/\tau$ and it becomes

$$v = \frac{\tau F}{m}\left(1 - e^{-\frac{t}{\tau}}\right). \tag{13.8}$$

* This could have been obtained directly from Equation 13.5 with $dv/dt = 0$, but Equation 13.6 provides additional insight.

The relaxation time is frequently used as a measure of how long it takes particles to reach the drift velocity, which certainly is important in particle size measurement by Stokes settling.

13.4 STOKES LAW AND PARTICLE SIZE MEASUREMENT

13.4.1 Stokes Law

For particles (or bubbles) in a fluid that have reached the terminal or drift velocity, Equation 13.4 applies and

$$m\frac{dv}{dt} = F_g - F_B - F_d$$
$$0 = F_g - F_B - \gamma v_d. \tag{13.9}$$

For sake of concreteness, assume that the particles are solid spheres of density ρ_s (kg/m^3) settling in a liquid of density ρ_l (kg/m^3) under the force of gravity. Stokes (Gaskell 1992; Clift et al. 1978; Bird et al. 2002) derived that for spherical particles the drag coefficient is given by

$$\gamma = 6\pi r\mu v \tag{13.10}$$

where:

r = radius (m) of the particle
μ = viscosity (Pa s).

Admittedly, this is a "black box" because it is not developed from elementary concepts. But this exception is made here because to model Stokes drag takes substantial computational effort and a familiarity with both fluid dynamics and more general solutions of partial differential equations beyond what is expected here. Qualitatively, the Stokes equation, $\gamma = 6\pi r\mu v$, comes mainly from the viscous force integrated over the surface of the moving spherical particle. Putting Equation 13.10 into Equation 13.9 gives

$$0 = \frac{4}{3}\pi r^3 \rho_s g - \frac{4}{3}\pi r^3 \rho_l g - 6\pi r\mu v_d$$

which is easily solved for v_d, the drift velocity or the velocity of settling of solid particles

$$v_d = \frac{2}{9}\frac{r^2}{\mu}(\rho_s - \rho_l)g \tag{13.11}$$

where g is the acceleration of gravity, 9.8 m/s^2. Equation 13.11 is *Stokes law* of settling and is used to measure particle size and/or a particle size distribution if the densities of the solid and liquid are known, or is used to measure the density of the solid if the particle size is known.

For example, Figure 13.4 shows how a particle size distribution can be measured by settling in a liquid. After a given time of settling, the larger particles have settled more than the smaller particles. From Equation 13.11, the particle size at a given depth is calculated. The density (number) of particles of a given size at a given depth is measured by light or x-ray absorption in the particle suspension. From these data, a particle size distribution is easily generated. Many commercial particle size analyzers operate on this or a similar principle.

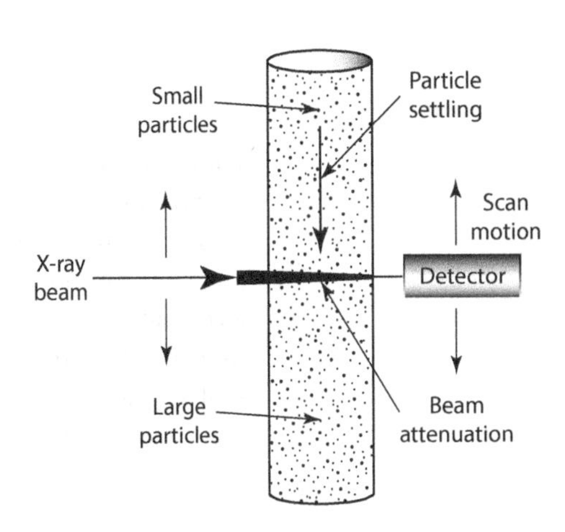

FIGURE 13.4 Schematic representation of a particle size analyzer that uses particle settling in a fluid and x-ray beam attenuation to determine the mass of particles that have settled a certain distance to determine the particle size distribution.

13.4.2 Particle Settling

Consider one micrometer diameter spherical aluminum oxide, Al_2O_3, particles settling in water at 25°C. The density of water is $\rho_l = 0.997$ g/cm³ $\cong 1$ g/cm³ and the viscosity is $\mu = 890 \times 10^{-6}$ Pa s $\cong 10^{-3}$ Pa s and the density of Al_2O_3 $\rho_s = 3.99 \cong 4.0$ g/cm³ (Haynes 2013). The drift velocity is calculated from Equation 13.11

$$v_d = \frac{2}{9}\frac{r^2}{\mu}(\rho_s - \rho_l)g$$

$$v_d = \frac{2}{9}\frac{(0.5\times10^{-6}\text{m})^2}{10^{-3}\text{Pa s}}(4.0\times10^3\text{kg/m}^3 - 1.0\times10^3\text{kg/m}^3)9.8\text{m/s}^2$$

$$v_d = 1.63\times10^{-6}\text{m/s} = 1.63\,\mu\text{m/s}.$$

The relaxation time is calculated from

$$\tau = \frac{m}{\gamma} = \frac{\frac{4}{3}\pi r^3(\rho_s - \rho_l)}{6\pi r\mu}$$

$$\tau = \frac{v_d}{g} = \frac{1.63\times10^{-6}\text{m/s}}{9.8\text{m/s}^2}$$

$$\tau = 1.66\times10^{-7}\text{s}.$$

These calculations show that in 10^4 seconds (about three hours), the particles will have settled about 1.63 cm. Moreover, the relaxation time $\cong 10^{-7}$ seconds is such a small number that the time prior to reaching drift velocity can certainly be neglected.

13.5 ELECTRON MOBILITY

The relationship between mobility and relaxation time obtained in many texts on electronic materials is frequently not transparent: a rather "black box" approach is used or frequently the equation just appears. However, this result is easily obtained from Equation 13.8, which can be written as

$$v_d = \frac{F\tau}{m}. \tag{13.12}$$

For electrons moving in an electric field, E (V/m), the force is given by F = eE, where e = electron charge (1.602×10^{-19} C). When discussing electronic conductivity in solids, the electron mobility (m²/V s), is usually given by the symbol μ_e,* and is related to the relaxation time for electrons to reach a given drift velocity in a given electric field or voltage drop. Of course, if the electrons did not reach a constant drift velocity, the current would continue to increase with time and that is not observed in reality. Electrons collide primarily with atomic vibrations and are scattered and this can be represented as a viscous drag force that leads to Equation 13.12. Substitution of the electrical force acting on an electron into Equation 13.12 gives

$$v_d = \frac{F\tau}{m_e} = \frac{eE\tau}{m_e}$$

* This is essentially the same as the letter used for viscosity although many sources use η for viscosity while here η has been chosen to represent the number of atoms, point defects, particles, and others per unit volume. Because materials science and engineering is *multidisciplinary*, it is necessary to make sure what the notation means in a given situation since the notation typically varies and is standardized—or not—to mean different things in different disciplines.

where m_e is the electron mass (9.11×10^{-34} kg). Because the electron mobility is defined as the *velocity per unit field*, $\mu_e = v_d/E$ or

$$\mu_e = \frac{e\tau}{m_e} \tag{13.13}$$

which appears in most texts on electronic materials. Here, this result follows directly from the model leading to Equations 13.6 and 13.8.

For example, the electron mobility in doped silicon is on the order of 1000 cm^2/V s (Mayer and Lau 1990), which gives for the relaxation time:*

$$\tau = \frac{\mu_e m_e}{e}$$

$$\tau = \frac{(1000\,cm^2/V\,s)(9.11\times10^{-31}\,kg)}{(1.602\times10^{-19}\,C)(10^4\,cm^2/m^2)}$$

$$\tau = 5.7\times10^{-13}\,s$$

a pretty short time!

13.6 ABSOLUTE MOBILITY AND DIFFUSION

A combination of Equations. 13.1 and 13.7 gives

$$J' = \eta v_d = \eta BF. \tag{13.14}$$

In the case of diffusion, there is a *chemical* force per atom or ion undergoing diffusion that can be defined by

$$F_{chem} = -\frac{1}{N_A}\frac{d\bar{G}}{dx} \tag{13.15}$$

where $\bar{G}$ is the partial Gibbs energy (or *chemical potential*, μ) of the diffusing molecules, atoms, or ions in J/mol. The relation between the particle flux density and chemical force is shown in Figure 13.5. Unit analysis of the right-hand side of Equation 13.15 is force per atom or ion or *N/atom* or ion. A combination of Equations 13.14 and 13.15 gives

$$J' = -\frac{\eta}{N_A} B\frac{d\bar{G}}{dx} = -CB\frac{d\bar{G}}{dx} \tag{13.16}$$

where C is the concentration with units of mol/m^3. Recall from thermodynamics in Chapter 1 that

$$\bar{G} = \bar{G}^o + RT\ln a$$

where a is the thermodynamic activity and $\bar{G}^o$ is the Gibbs energy in the standard state.

For an ideal solution, $\bar{G} = \bar{G}^o + RT\ln X$, where X is the mole fraction given by C/C_T where C_T is the total molar concentration for all the species in the solution. For an ideal gas, $C_T = (n/V) = (P/RT)$ and is independent of

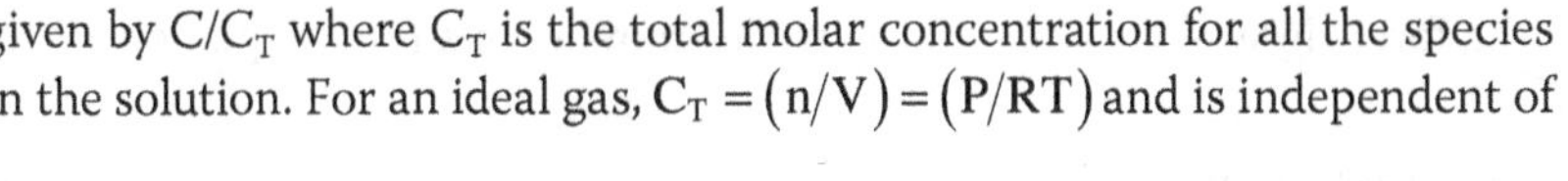

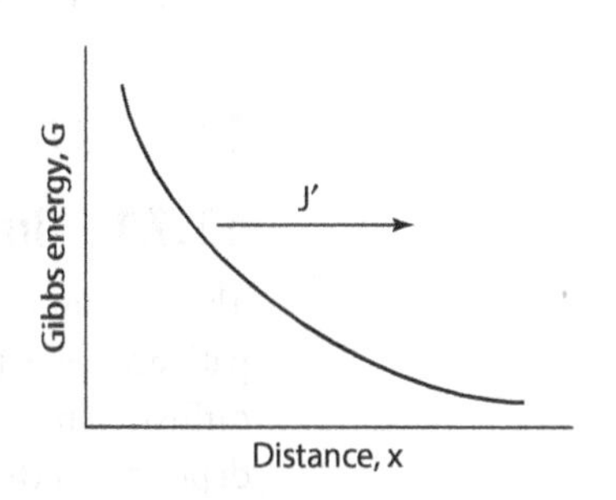

FIGURE 13.5 Schematic drawing showing that the molecular/atomic flux density, J′, is opposite to the gradient in Gibbs energy.

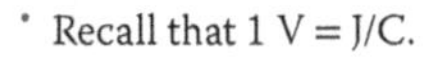

* Recall that 1 V = J/C.

composition—a constant, at a given temperature—and, as a first approximation, it is close to being true for ideal solids and liquids as well.* In any event,

$$\frac{d\bar{G}}{dx} = RT\frac{d}{dx}\ln\frac{C}{C_T} = RT\left(\frac{1}{C}\frac{dC}{dx} - 0\right) = \frac{RT}{C}\frac{dC}{dx} \tag{13.17}$$

for an ideal solution, which put into Equation 13.16 gives

$$J' = -BRT\frac{dC}{dx}$$

or (13.18)

$$J = \frac{J'}{N_A} = -Bk_BT\frac{dC}{dx}$$

where k_B = Boltzmann's constant (1.38×10^{-23} J/atom). Equation 13.18 is a relation between the molar flux and the molar concentration gradient so it is the same as Fick's first law of diffusion, $J = -D(dC/dx)$, which is purely empirical and holds only for thermodynamically ideal systems. Therefore, this must mean that there is a relation between the diffusion coefficient and the absolute mobility; namely,

$$D = Bk_BT \text{ or } B = \frac{D}{k_BT}. \tag{13.19}$$

This is known as the *Einstein* relation, a *particularly important result*, because Equation 13.19 states that the diffusion coefficients for atoms, ions, electrons, molecules, and even dirt or pollen particles are related to their absolute mobilities and vice versa. That means if the diffusion coefficient is known, the velocity and fluxes of any of these species *under any applied force*—mechanical, electrical, chemical, magnetic, and so on—can be determined. Going back to Equation 13.16 with this relation between *B* and *D* gives

$$J' = -C\frac{D}{k_BT}\frac{d\bar{G}}{dx}$$

and dividing by Avogadro's number, N_A, gives the important general result for diffusion in any system:

$$J = -C\frac{D}{RT}\frac{d\bar{G}}{dx}. \tag{13.20}$$

Equation 13.20 is the more general form of Fick's first law of diffusion and can be applied whether the system is thermodynamically ideal or not, and most real condensed matter systems *are not ideal.*

13.7 DIFFUSION IN LIQUIDS

13.7.1 INTRODUCTION

The mechanism of diffusion in liquids was briefly discussed in Chapter 9 in terms of the mean free path and velocities of atoms or molecules in liquids. What follows is the more traditional approach to diffusion in liquids. For atoms, molecules, or particles moving in liquids, the mobility of each of these depends on the viscous drag force derived by Stokes for particles moving in a fluid, Equation 13.10.

* For example, $\bar{V}(NiO) = 11.10$, $\bar{V}(MgO) = 11.26$; $\bar{V}(Ag) = 10.28$, $\bar{V}(Au) = 10.21$, and so on where $\bar{V}$ are molar volumes in cm^3/mol.

Einstein, in explaining *Brownian motion*,* took Equation 13.19 with $B = 1/\gamma$ where, again, $\gamma = 6\pi r\mu v$ and derived the following equation for diffusion in liquids (Einstein 1956):

$$D = \frac{k_B T}{6\pi r\mu}. \tag{13.21}$$

This is the *Stokes–Einstein equation* for diffusion in liquids where *r* is the *hydrodynamic radius* of the diffusing species—atoms, ions, molecules, or actual particles. It was noted earlier in Chapter 8 that diffusion in liquids frequently follows an exponential relationship, $D = D_0 e^{-(Q/RT)}$, and this is consistent with Equation 13.21 because the viscosity is typically temperature dependent and of the form $\mu = \mu_0 e^{(Q/RT)}$, as was seen in Chapter 7; that is, the viscosity decreases as the temperature increases, opposite behavior from that of the diffusion coefficient. As a result, the activation energy for viscosity and the diffusion coefficient for liquids should be the same, and for liquids that do not incorporate large molecules and intermolecular bonding, the value for Q is on the order of 10 kJ/mol.

Polymer diffusion in liquids deserves special attention because of the difficulty in defining the hydrodynamic radius and the variations that can occur depending on the type of polymer and its interaction with the solvent. This is discussed later in a separate section. What follows in the remainder of this section are examples of diffusion in liquids.

13.7.2 Self-Diffusion in Water

In the literature, data exist for the viscosity and diffusion coefficients for pure water over a limited temperature range (Robinson and Stokes 2002) and these data are plotted in Figure 13.6.

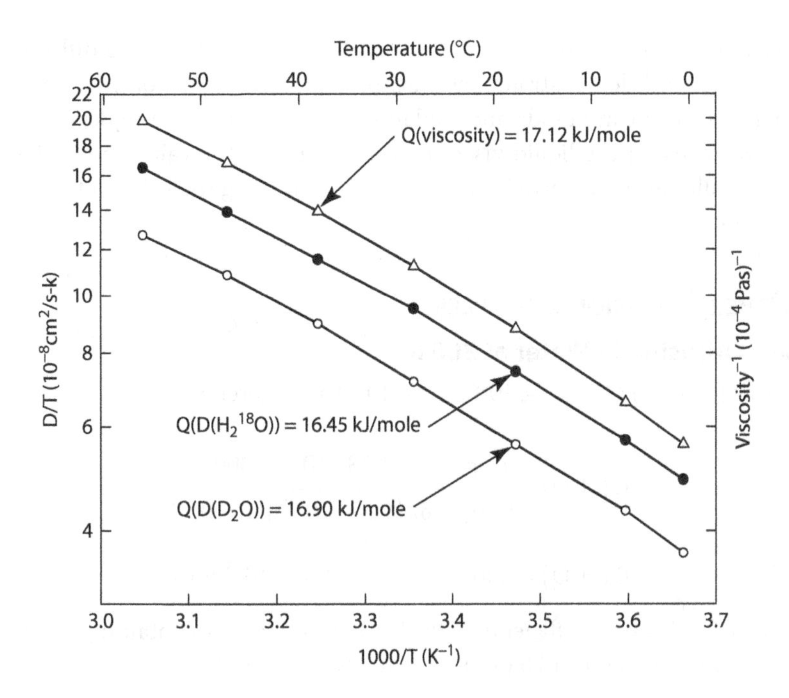

FIGURE 13.6 Water self-diffusion coefficients as a function of temperature measured by both $^2_1H_2O$ and $H_2{}^{18}O$ (Robinson and Stokes 2002) and the viscosity of water versus temperature (Haynes 2013) showing that they all have essentially the same activation energy as predicted by the Stokes–Einstein equation for diffusion in liquids.

* Brownian motion or movement was first described by the botanist Robert Brown, who observed the random movement of pollen and other small particles in liquids. The phenomenon was studied in the late nineteenth century and is now known to be caused by the random motion of the liquid molecules creating random nonuniform forces on the small particles producing their motion. Between about 1906 and 1908, Einstein explained the phenomenon and generalized it to diffusion (Einstein 1956).

The inverses of the viscosity and D/T values are plotted in units so that all three plots have similar values. Note that the activation energies are very close for all three plots, about 17 kJ/mol. Also note that the diffusion coefficients for water with deuterium, D_2O, as the tracer are different from those in which water with oxygen-18 was used, $H_2{}^{18}O$. As a result, for the former, the radius of the water molecule calculated from Equation 13.21 is about 0.113 nm while that for the $H_2{}^{18}O$ diffusion is about 0.086 nm. Both of these values are smaller than those obtained from x-ray measurements, about 0.138 nm (Robinson and Stokes 2002) and 0.132 nm from gaseous diffusion (Geankoplis 1972). So, these calculated values are not too different. An estimate for the size of a molecule or atom is to assume simple cubic packing of the atom as if they were spheres so that

$$d = 2r = \left(\frac{V_b}{N_A}\right)^{1/3} = \left(\frac{V_b}{6.022\times10^{23}}\right)^{1/3} \tag{13.22}$$

$$d = 0.118V_b^{1/3}\ \text{nm}$$

where:

d = molecular diameter

V_b = molar volume at the boiling point (Geankoplis 1972; Bird et al. 2002).

The density of water at its boiling point is 0.95863 (Haynes 2013), so

$$r = \frac{0.118}{2}\left(\frac{M}{\rho}\right)^{1/3} = \frac{0.118}{2}\left(\frac{18.01}{0.95856}\right)^{1/3} = 0.157\ \text{nm}$$

which is larger than the radius of the water molecule calculated from the diffusion coefficients. Nevertheless, these results demonstrate that the Stokes–Einstein equation gives pretty good values for the diffusion coefficient in liquids and explains the weak temperature dependence due to the temperature dependence of the liquid viscosity. It is clear that the value chosen for the radius of the diffusing molecule, atom, or particle is critical in determining the value of the calculated liquid diffusion coefficient.

13.7.3 Other Diffusion Examples

13.7.3.1 Self-Diffusion in Water at 300 K

From Section 13.7.1, $r(H_2O) \cong 0.157 \cong 10^{-9}$ m, $\mu \cong 10^{-3}$ Pa s. Therefore,

$$D(H_2O) = \frac{k_BT}{6\pi r\mu} = \frac{(1.38\times10^{-23})(300)}{6\pi(1.57\times10^{-10})(1\times10^{-3})}$$

$$D(H_2O) = 1.40\times10^{-9}\ \text{m}^2/\text{s} = 1.46\times10^{-5}\ \text{cm}^2/\text{s}$$

compared to tracer diffusion coefficients with deuterium and ^{18}O-containing water molecules of around 2.5×10^{-5} cm^2/s in Figure 13.6 (Robinson and Stokes 2002).

13.7.3.2 Diffusion in Liquid Copper at Its Melting Point

For copper, T_{mp} = 1083°C = 1356 K. The following data were obtained from the literature (Brandes and Brook 1992):

$$D(Cu_l) = D_0e^{-\frac{Q}{RT}} = 1.46\times10^{-3}e^{-\frac{40{,}700}{8.314\times1356}} = 3.95\times10^{-5}\text{cm}^2\text{s}^{-1}$$

$$\mu(Cu_l) = \mu_0e^{-\frac{Q}{RT}} = 0.3\times10^{-3}e^{-\frac{30{,}500}{8.314\times1356}} = 4.5\times10^{-3}\text{Pa s.}$$

Again from the literature, r(Cu) = 1.28×10^{-10} m (Brandes and Brook 1992; Emsley 1998). So, at the melting point of copper, the diffusion coefficient calculated from the Stokes–Einstein equation is

$$D = \frac{\left(1.38\times10^{-23}\right)(1356)}{6\pi\left(1.28\times10^{-10}\right)\left(4.5\times10^{-3}\right)} = 1.72\times10^{-9}\ m^2/s = 1.72\times10^{-5}\ cm^2/s$$

which is a little more than a factor of 2 smaller than D calculated from the experimentally determined values of D_0 and Q, 3.95×10^{-5} cm²/s but, again, reasonably close given the simplicity of the model.

13.7.3.3 Diffusion in Liquid Lead at Its Melting Point

For lead, T_{mp} = 327°C = 600 K. Again, taking data from the literature (Brandes and Brook 1992):

$$D(Pb_l) = D_0 e^{-\frac{Q}{RT}} = 2.37\times10^{-4} e^{-\frac{24,700}{8.314\times600}} = 1.67\times10^{-6}\ cm^2/s$$

$$\mu(Pb_l) = \mu_0 e^{-\frac{Q}{RT}} = 0.4636\times10^{-3} e^{-\frac{8610}{8.314\times600}} = 2.6\times10^{-3}\ Pa\ s.$$

Again from the literature, r(Pb) = 1.75×10^{-10} m (Brandes and Brook 1992; Emsley 1998). So, at the melting point of lead, the diffusion coefficient calculated from the Stokes–Einstein equation is

$$D = \frac{\left(1.38\times10^{-23}\right)(600)}{6\pi\left(1.75\times10^{-10}\right)\left(2.6\times10^{-3}\right)} = 9.65\times10^{-10}\ m^2/s = 9.65\times10^{-6} cm^2/s.$$

In this case, the calculated value from the Stokes–Einstein relation is about a factor of 5, too high compared to the measured diffusion data.

13.7.3.4 Oxygen Diffusion in Water

The above three examples were self-diffusion in a liquid. An example of a solute diffusing in a solvent is oxygen diffusing in water at room temperature, 25°C. From the data used to plot Figure 13.6, the viscosity of water is $\mu = 0.89\times10^{-3}$ Pa s. From tables of collision diameters for gas diffusion (Cussler 1997) $2r(O_2) = 0.3467$ nm so

$$D_{O_2-H_2O} = \frac{k_B T}{6\pi r\mu} = \frac{\left(1.38\times10^{-23}\right)(298)}{6\pi\left(1.73\times10^{-10}\right)\left(0.89\times10^{-3}\right)}$$

$$D_{O_2-H_2O} = 1.42\times10^{-9}\ m^2/s = 1.42\times10^{-5}\ cm^2/s$$

which is only about 70% of the literature value of $D = 2.10\times10^{-5}$ cm²/s (Cussler 1997).

13.7.3.5 Summary

From the above calculations several things are worth noting. First, in liquids including a polar molecular liquid, such as water, around room temperature, and liquid metals, at considerably higher temperatures, the diffusion coefficients are all in the 10^{-4}–10^{-5} cm²/s range. Second, if one had to estimate the value of a diffusion coefficient in a liquid for a 'back of the envelope" calculation, choosing one of these values (e.g., $D \cong 10^{-5}$ cm²/s) would be pretty close. Third, these examples were used because data were relatively available. The same approach could be applied to molten salts and other molecular liquids. However, the data are sometimes harder to locate, whereas the literature holds considerable data on aqueous (Haynes 2013) and metallic liquids (Brandes and Brook1992), respectively. Fourth, although the agreement between measured diffusion coefficients and those calculated is not bad, it is not perfect. As a result, there exist many additional models in the literature to generate more accurate models than the simple Stokes–Einstein relation of Equation 13.21.

13.8 IONIC CONDUCTIVITY AND DIFFUSION

13.8.1 Nernst–Einstein Equation

At the expense of a small diversion, the principles that relate diffusion and mobility can be applied directly to electrical conductivity and diffusion so it is useful to discuss the relationship between the two. Remember that Ohm's law is $V = IR$, where V = voltage (V = J/C), I = current (ampere = C/s), and R = resistance (ohms). And $R = \rho(L/A) = (1/\sigma)(L/A)$, where L and A refer to the length and cross-sectional area of the box with moving particles in Figure 13.1, ρ = resistivity (ohm m) and σ = conductivity (siemens/m = S/m = 1/ohm m). Combining Ohm's law and the definition of resistance and on rearranging terms

$$V = IR = I\frac{1}{\sigma}\frac{L}{A}$$

$$\sigma\frac{V}{L} = \frac{I}{A} \tag{13.23}$$

$$\sigma E = J^*$$

where:

E = electric field (V/m)

J^* = charge flux density (C/m^2 s) = zeJ' where ze is the charge on the moving electron or ion.*

For example, if Mg^{+2} ions were moving, then z = 2, etc. So, combining Equations 13.1, 13.7, and 13.19 plus the force on the moving charges in an electric field, F = zeE, leads to

$$\sigma = \frac{J^*}{E} = ze\frac{J'}{E} = ze\frac{\eta BF}{E} = ze\frac{\eta}{E}\frac{D}{k_B T}zeE$$

$$\sigma = \frac{(ze)^2\eta D}{k_B T} \tag{13.24}$$

which is the *Nernst–Einstein* equation and relates the electrical conductivity due to charge motion to diffusion coefficients. In semiconductor devices, the *diffusion of electrons* is frequently referred to and the electron conductivity is directly proportional to the diffusion coefficient for electrons.

13.8.2 Ionic Conductivity of a 10% Sodium Chloride Solution

For a *back-of-the-envelope* calculation, assume that the density of water at 300 K is essentially 1 g/cm^3 or 10^6 g/m^3. The number of Na^+ ions per m^3 in a 10% solution is approximately

$$\eta = \frac{\rho}{M}N_A \times 0.1 = \frac{1\times10^3\left(\text{kg}/\text{m}^3\right)}{18\times10^{-3}\left(\text{kg}/\text{mole}\right)}6.022\times10^{23}(\text{ions}/\text{mole})\times 0.1$$

$$\eta = 3.5\times10^{27}\,\text{m}^{-3}.$$

In Chapter 10, Section 10.11, the sodium and chlorine ion diffusion coefficients in water found in the literature are $D_{Na^+} = 1.33\times10^{-5}\,\text{cm}^2/\text{s}$ and $D_{Cl^-} = 2.032\times10^{-5}\,\text{cm}^2/\text{s}$, roughly the same as self-diffusion of water at the same temperature. Then, the conductivity for the sodium is

* Or more directly, the electrical conductivity, σ, can be *defined* as the proportionality between the charge flux density, J^*, and the electric field, E.

$$\sigma_{Na^+} = \frac{\left(1.602\times10^{-19}\right)^2\left(3.5\times10^{27}\right)\left(1.33\times10^{-9}\right)}{\left(1.38\times10^{-23}\right)(300)}$$

$$\sigma_{Na^+} = 28.8\ \text{S/m} = 0.28\ \text{S/cm}$$

and that for the chlorine ions

$$\sigma_{Cl^-} = \frac{\left(1.602\times10^{-19}\right)^2\left(3.5\times10^{27}\right)\left(2.032\times10^{-9}\right)}{\left(1.38\times10^{-23}\right)(300)}$$

$$\sigma_{Cl^-} = 44.1\ \text{S/m} = 0.441\ \text{S/cm}$$

to give a total conductivity of

$$\sigma = \sigma_{Na^+} + \sigma_{Cl^-} = 0.288 + 0.441 = 0.729\ \text{S/cm}$$

which makes it a pretty good "semiconductor" at room temperature because its electrical conductivity is between a good conductor like a metal ($\sigma \cong 10^6$ S/cm) and a good insulator such as SiO_2 ($\sigma \cong 10^{-13}$ S/cm). On the other hand, the conduction of electrons by water is essentially zero so it has a very low *electronic* conductivity: an electron conductivity of a good insulator. The combination of high ionic conductivity and low electronic conductivity is necessary for a material to be an *electrolyte* in a battery or fuel cell: ions are easily conducted through the electrolyte between the electrodes, while the electrons must go through an external circuit. This also highlights the confusing word "semiconductor," discussed in Section 9.5.3. The term is commonly interpreted as referring to materials, such as, silicon, germanium, or gallium arsenide, that are used to make electronic devices. However, the term *really* is comparative and refers to how good an electrical conductor a material happens to be: a semiconductor has conductivity somewhere between that of an insulator and a good conductor like a metal. In fact, it would be better to separate materials into just two general classes. (1) Metals, materials that have free electrons to conduct at T = 0 K, that is, they *do not require energy to remove electrons from chemical bonds:* all the free electrons *are* the chemical bond in a metal. (2) Nonmetals, materials that have no free electrons to conduct at T = 0 K because it takes energy to create electron–hole pairs from chemical bonds as described in Chapter 9. Then, the only difference between a semiconductor and an insulator is the amount of energy to create electron–hole pairs: PbS, $E_g \cong 0.25$ eV, a semiconductor; MgO, $E_g \cong 5.5$ eV, an insulator. Furthermore, as shown above, a sodium chloride solution is a good semiconductor as an ionic conductor but is an insulator to electron conduction. Without such *ionic* semiconductors, batteries and fuel cells would not be possible (Xu 2004).

13.8.3 CaO-Doped Zirconium Oxide as a Solid Electrolyte

As was discussed in Chapters 7 and 9, the fluorite crystal structure of ZrO_2 forms extensive solid solutions with other oxides that form oxygen vacancies in ZrO_2, including MgO, CaO, Y_2O_3, etc. A classic example is a CaO-ZrO_2 solid solution having the composition $Ca_{0.15}Zr_{0.85}O_{1.85}$; that is, a 15 m/o (mole percent) solid solution of CaO and ZrO_2. As with Y_2O_3, CaO goes into a solid solution of zirconia with the formation of oxygen vacancies:

$$CaO \xrightleftharpoons{ZrO_2} Ca''_{Zr} + O^x_O + V^{\bullet\bullet}_O \tag{13.25}$$

in the Kröger–Vink notation. Solving the electrical neutrality condition

$$2e\eta_{Ca''_{Zr}} = 2e\eta_{V^{\bullet\bullet}_O} \Rightarrow \frac{\eta_{Ca''_{Zr}}}{\eta_{ZrO_2}} = \frac{\eta_{V^{\bullet\bullet}_O}}{\eta_{ZrO_2}} \Rightarrow \frac{\eta_{Ca''_{Zr}}}{\eta_{Zr}} = \frac{2\eta_{V^{\bullet\bullet}_O}}{\eta_O}$$

because $\eta_{zr} = \eta_{ZrO_2}$ and $\eta_O = 2\eta_{ZrO_2}$ and gives

$$\left[V_O^{\bullet\bullet}\right] = \frac{\eta_{V_O^{\bullet\bullet}}}{\eta_O} = \frac{\left[Ca''_{Zr}\right]}{2} = \frac{0.15}{2} = 7.5\times 10^{-2}$$

where $\eta_{V_O^{\bullet\bullet}}$ are the number of vacant oxygen sites/m^3 and η_O are the number of oxygen ion sites/m^3 in the doped ZrO_2. Data in the literature for the diffusion of oxygen ions in this solid solution are (Kingery et al. 1976)

$$D\left(O^{2-}\right) = D_0 e^{-\frac{Q}{RT}} = 10^{-3} e^{-\frac{81\,kJ/mole}{RT}}\ cm^2/s$$

and give a diffusion coefficient at 1000°C of

$$D\left(O^{2-}\right) = 10^{-3} e^{-\frac{81{,}000}{(8.314)(1273)}} = 4.7\times 10^{-7}\ cm^2/s = 4.7\times 10^{-11}\ m^2/s.$$

To calculate the ionic conductivity of this solid solution at 1000°C it is the number of oxygen ions/m^3, η_O, that must be used for η in Equation 13.24, in this case, because the oxygen ions are doing the conducting. From the literature (Haynes 2013), $\rho(ZrO_2)$ = 5.6 g/cm^3 and $M(ZrO_2)$ = 123.22 g/mol, neglecting the differences between densities of the solid solution and pure ZrO_2, which is a valid first approximation. (Is it? See exercises.)

$$\eta_O = 2\eta_{ZrO_2} = 2\frac{\rho_{ZrO_2}}{M_{ZrO_2}} N_A = 2\frac{\left(5.6\times 10^3 kg/m^3\right)}{(123.22\times 10^{-3} kg/mole)} \times \left(6.022\times 10^{23}\right)$$

$$\eta_O = 5.47\times 10^{28}\ m^{-3}.$$

Substituting these values of D and η into Equation 13.24

$$\sigma_O\left(1000\ °C\right) = \frac{\left(2\times 1.602\times 10^{-19}\right)^2 \left(5.47\times 10^{28}\right)\left(4.7\times 10^{-11}\right)}{\left(1.38\times 10^{-23}\right)(1273)}$$

$$\sigma_O\left(1000\ °C\right) = 15.0\ S/m = 0.15\ S/cm$$

which is a significant ionic conductivity for oxygen in calcia-doped zirconia. Because of its high ionic conductivity, doped zirconia is used as an oxygen-conducting electrolyte in solid oxide fuel cells (SOFCs), in automobile and combustion oxygen sensors, and in oxygen pumps for medical applications. However, the temperature of 1000°C is rather high for many of these applications, and there is a strong desire to reduce the temperature by several hundred degrees to below 600°C while maintaining high ionic conductivity, to simplify the engineering design of the devices. Over the last few decades, considerable progress has been made with various dopant additions to modify the mobility and vacancy concentrations to achieve high conductivities at lower temperatures. Recent materials now demonstrate excellent properties at 600°C or lower, an enabling technology for solid state energy storage devices. Even better would be a good *solid* oxygen ion conductor like doped zirconia with a similar conductivity below 100°C!

A point worth noting with regard to ionic conductivity in something like doped zirconia is that the diffusion coefficient for oxygen can be written as (see Chapter 9)

$$D\left(O^{2-}\right) = \frac{1}{6} f a^2 e^{-\frac{Q_m}{RT}} \left[V_O^{\bullet\bullet}\right].$$

This equation could be written as $D(O^{2-}) = D_{V_O^{\bullet\bullet}}\left[V_O^{\bullet\bullet}\right] = D_{V_O^{\bullet\bullet}}\left(\eta_{V_O^{\bullet\bullet}}/\eta_O\right)$, where $D_{V_O^{\bullet\bullet}}$ is the *vacancy diffusion coefficient* for oxygen vacancies. Replacing $D(O^{2-})$ in Equation 13.24 with this expression, the Nernst–Einstein equation becomes

$$\sigma_O = \frac{(ze)^2 \eta_{V_O^{\bullet\bullet}} D_{V_O^{\bullet\bullet}}}{k_B T} \tag{13.26}$$

implying that the electrical conductivity can be considered to be a result of either the oxygen ions moving or the oxygen vacancies moving. As long as the correct values for concentrations and diffusion coefficients are chosen, the results are the same. If the zirconia were indeed operating as the electrolyte in a cell with an oxygen chemical potential gradient, then the negatively charged oxygen ions would move one way in the cell—from high to low oxygen chemical potential—and the oxygen vacancies would move in the opposite direction, because they are positively charged relative to the crystal lattice.

13.9 COUPLED DIFFUSION IN IONIC SYSTEMS

13.9.1 Electrochemical Potential

In general, for an ionic compound AB, if the ions A and B are both diffusing, there will be an electric field, E, between them due to their different rates of diffusion. As a result, there is both a chemical force and an electrical force acting on each ion that determines its flux. That is, the number flux density of ions, $J' = J/N_A$, which is just the molar flux, J, divided by Avogadro's number is given by a version of the so-called *Nernst–Planck* equation:

$$J' = \eta v_d = \eta BF = \frac{\eta D}{k_B T}\left(F_{chem} + F_{elec}\right) = \frac{\eta D}{k_B T}\left\{-\frac{1}{N_A}\frac{d\bar{G}}{dx} \pm zeE\right\} \tag{13.27}$$

where:

the plus sign is for a cation
the minus sign for the anion
z are the integers α or β for a compound $A_\alpha B_\beta$.

Note that the electrical field is the negative of the gradient in the electrical potential, ϕ, $E = -(d\phi/dx)$ so Equation 13.27 can be written as

$$J' = \frac{\eta D}{k_B T}\left\{-\frac{1}{N_A}\frac{d\bar{G}}{dx} \mp z\frac{\mathcal{F}}{N_A}\frac{d\phi}{dx}\right\} \tag{13.28}$$

where $\mathcal{F}$ is the Faraday = 96,485 C/mol. Equation 13.28 can then be written as

$$J' = -\frac{\eta D}{RT}\frac{d\xi}{dx}$$

where ξ is the *electrochemical potential* and $\xi = \mu \pm z\mathcal{F}\phi$ because $\mu_i = \bar{G}_i$ = the chemical potential of species i (Chiang et al. 1997). To get the coupled equation all that is required is to equate the anion and cation fluxes because the net current is zero. However, this requires a rather significant amount of algebra and so all of the steps are shown in Section A.1 rather than here. The following are the results of systems with charge-coupled diffusion coefficients.

13.9.2 Diffusion in an Electrolyte

In Chapter 10, Section 10.11, the steady-state dissolution of sodium chloride, NaCl, in water was examined and it was found that the coupled *effective diffusion coefficient* for the two ions could be written as

$$D_{eff} = \frac{2D_{Na^+}D_{Cl^-}}{D_{Na^+} + D_{Cl^-}}. \tag{13.29}$$

In Section A.2.1, the general result for NaCl could be written as

$$J_{NaCl} = -\frac{C_{NaCl}}{RT}\left(\frac{D_{Na^+}D_{Cl^-}}{D_{Na^+} + D_{Cl^-}}\right)\left\{\frac{d\bar{G}_{Na^+}}{dx} + \frac{d\bar{G}_{Cl^-}}{dx}\right\} \tag{13.30}$$

and if the solution is dilute, then it will be ideal, and from Equation 13.17 and because $C_{Na^+} = C_{Cl^-} = C_{NaCl}$ then (Equation A.10)

$$J_{NaCl} = -\left(\frac{2D_{Na^+}D_{Cl^-}}{D_{Na^+} + D_{Cl^-}}\right)\frac{dC_{NaCl}}{dx} \tag{13.31}$$

as was found earlier, but this time with the improved justification for Equation 13.30 given in Section A.2. Equation 13.31 shows that if the diffusion coefficients for both ions were the same—which they are not—then $D_{eff} = D_{Na^+} = D_{Cl^-}$, which would be expected intuitively. On the other hand, if one of them was much larger than the other, say $D_{Na^+} \gg D_{Cl^-}$, which it is not, then $D_{eff} = 2D_{Cl^-}$. That is, the overall rate is controlled by the slower diffusing ion but with a diffusion coefficient that is *twice* that of the slower ion. This shows the effect of the electric field coupling between the two diffusing ionic species. The faster diffusing sodium ion creates an electric field that helps to pull the chlorine ion along with it increasing the effective diffusion coefficient.

13.9.3 Diffusion in an Ionic Solid

As shown in Section A.2, the general form of Equation 13.30 for a compound $A_\alpha B_\beta$ is

$$J_{A_\alpha B_\beta} = -\frac{C_{A_\alpha B_\beta}}{RT} D_{eff} \frac{d\bar{G}_{A_\alpha B_\beta}}{dx} \tag{13.32}$$

with D_{eff} now given by (Readey 1966)

$$D_{eff} = \frac{D_A D_B}{\beta D_A + \alpha D_B}. \tag{13.33}$$

For example, suppose solid sodium chloride underwent diffusional creep at some elevated temperature at which a mechanical force was applied to generate the Gibbs energy gradient, $d\bar{G}_{NaCl}/dx$, and the flux of NaCl, J_{NaCl}. In this case, if $D_{Na^+} \gg D_{Cl^-}$, then $D_{eff} = D_{Cl^-}$; that is, the rate of deformation would be *directly proportional* to the diffusion coefficient of the slower diffusing ion and it only takes one Cl^- ion to move a molecule of NaCl!

This coupled diffusion or *ambipolar diffusion* is also important during sintering and other geometry changes such as grain boundary grooving of ionic solids where the chemical potential gradient is produced by the surface tension forces leading to mass transfer. For example, take the case of aluminum oxide, Al_2O_3; if the diffusion coefficients of Al^{3+} and O^{2-} are equal, $D = D_{Al^{3+}} = D_{O^{2-}}$, then Equation 13.33 says that $D_{eff} = D/5$. This is logical because it takes a total of five aluminum plus oxygen ions to move one molecule of Al_2O_3 so the effective diffusion coefficient for the molecule is just one fifth that for the ions. Similarly, if $D_{Al^{3+}} >> D_{O^{2-}}$, then $D_{eff} \cong D_{O^{2-}}/3$ because the aluminum ions move fast, the transfer of oxygen ions controls the rate of aluminum oxide diffusion, and there is a need of three oxygen ions for each Al_2O_3 molecule. So the diffusion coefficient for oxygen divided by three makes sense.

13.10 DIFFUSION OF POLYMERS IN LIQUIDS

13.10.1 Introduction

For the diffusion of polymers in liquids, the situation is more complex because diffusion depends on both the viscosity of the liquid and the effective *hydrodynamic radius* (≅ *radius of gyration*) of the polymer molecule. This in turn depends on the degree of polymerization, the flexibility of the molecules, and their degree of solvation by the solvent (which depends on the polymer–solvent chemistry), among other considerations. These complexities result in many theoretical and empirical variations on the Stokes–Einstein equation and are difficult to generalize in one or more simple expressions. The simple question is, "What to use for the hydrodynamic radius, r, in the Stokes–Einstein equation?" This is an area that has intrigued polymer chemists and physicists for some time and has generated a large number of both theoretical and empirical models. There are also many books that focus on polymers in solution (Molyneux 1983; Des Cloizeaux and Jannink 1990; Fujita 1990; Teraoka 2002; Phillies 2011) to provide some examples. Most books on polymers also dicuss polymers in solution (e.g., Young and Lovell 1991; Rosen 1993; Elias 1997; McCrum et al. 1997; Green 2005) to offer a representative list. The goal is to relate diffusion behavior and viscosity to the molecular weight of the polymer. Because there are many similarities to the behavior of polymers in solution and the random-walk diffusion process discussed in Chapter 11, Section 11.7, the tools necessary to model the diffusion behavior of polymers in liquids are available and can be used.

13.10.2 Molecular Weight Distribution

Contrary to the examples of self-diffusion in liquids or gases or ions in liquids, where in each of these examples a single hydrodynamic radius, r, could be put into (or calculated from) the Stokes–Einstein equation to evaluate a diffusion coefficient, the case for polymers is immediately more complex because polymer molecules do not all have the same chain lengths but show a distribution of lengths or molecular weights* as shown in Figure 13.7 (Callister and Rethwisch 2009). Of course, complicating the matter even further, the mean molecular weight $\langle M \rangle$ can either be a *number average molecular weight* $\langle M_n \rangle$ given by

$$\langle M_n \rangle = \sum_i x_i W_i \tag{13.34}$$

where x_i = the fraction of the polymer in a given length range (e.g., n = 900–1000) and W_i the average molecular weight in the length range (950 × m), or the *weight average molecular weight* $\langle M_w \rangle$ given by

$$\langle M_w \rangle = \sum_i w_i W_i$$

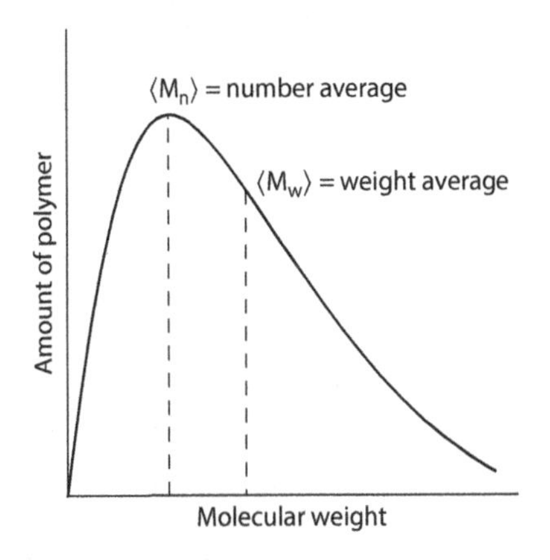

FIGURE 13.7 Schematic molecular weight distribution demonstrating the typical polydisperse (as opposed to monodisperse = one particle size) nature of polymers and that the mean molecular weight by polymer length and polymer weight are usually different.

* Where M = n m (n = number of *repeat units* in a chain and m = molecular weight of the unit): for example, polyethylene, $-[CH_2-CH_2]_n-$, where m = 28 in this case. So, if n = 1000, M = 28,000 g/mol.

where w_i = the fraction of the polymer in a given weight range (e.g., 25,000 → 28,000) and W_i = the average molecular weight in the weight range (26,500 g/mol). As shown in Figure 13.7, these averages are not the same and will only be the same if all of the molecules have the same number of repeating units, *n*.

13.10.3 Solvent Effects

Figure 13.8 shows two examples of the same polymer molecule that has the same *contour length*, L, where L = na, where a = the length of the repeating unit,* but in two different solvents or at two different temperatures. On the left, the polymer is strongly solvated and expands to a large size. The same expansion could be the result of a higher temperature as well. In contrast, the right shows a poorly solvated molecule or one at lower temperature that clearly has a smaller overall spatial extent, however, measured. In the Stokes–Einstein equation, Equation 13.21,

$$D = \frac{k_B T}{\gamma} = \frac{k_B T}{6\pi r \mu}.$$

The denominator really results from the viscous drag on a sphere of radius r and the general drag coefficient, γ, is nonshape specific. As a result, the diffusion coefficient for a given polymer with n segments might be given by (Elias 1997; Green 2005)

$$D = \frac{k_B T}{n\xi} \tag{13.35}$$

where ξ = drag on a single repeating unit of the polymer chain of n repeating units. Because the molecular weight of a polymer chain is given by M = n × m, then the diffusion coefficient is inversely proportional to the molecular weight M, that is, $D \propto M^{-1}$. This assumes that the solvent essentially does not react or interact chemically with any of the polymer segments and just offers hydrodynamic resistance. This is called a *free-draining* polymer. In contrast, the nearly spherical molecules in Figure 13.8 may be either free-draining or not. If there is significant swelling, then the polymer is probably not free-draining because there is obvious strong interaction with the solvent. So, the two polymers in this figure could diffuse essentially as spheres, and because $M = (4/3)\pi\rho r^3$, where ρ = some value of mass density, in these cases, $r \propto M^{(1/3)}$, so $D \propto M^{-(1/3)}$ and some polymers such as tightly coiled polystyrene in cyclohexane show such behavior (Elias 1997). The question is, "What is this mass density, ρ?" As it turns out, the effective radius for a polymer in the Stokes–Einstein equation is very close to the value for the *radius of gyration* of a polymer molecule (Cussler 1997). Because of

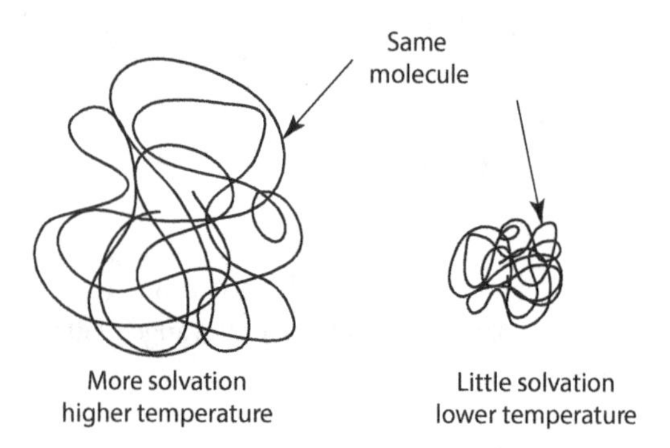

FIGURE 13.8 Schematic drawings of the same polymer molecule and the differences in its spatial extent—tightness of the coils—depending on the temperature and interaction with the liquid in which it is dissolved.

* Frequently, this is denoted by l in the literature. However, a was used as the mean jump distance in diffusion and is used here to emphasize the relation between polymer parameters and diffusion or random walks.

its importance and because it can be measured by other techniques, such as light scattering, a more detailed discussion of the radius of gyration is warranted. However, before doing that, an evaluation of the *end-to-end distance* of a linear polymer chain is useful.

13.10.4 End-to-End Distance of a Freely Jointed Chain

In Figure 13.9, a *freely jointed* polymer of n units, each of length a, and its end-to-end distance, $\vec{R}$, are shown. *Freely jointed* means that each repeating unit can rotate in any direction at any angle from the previous unit. This is typically not true for real polymers but does not change the basic nature of the argument and only slightly affects the final result. So, for a freely jointed chain, the length of the chain, $\vec{R}_n$, is given by

$$\vec{R}_n = \sum_{i=1}^{n} \vec{r}_i \tag{13.36}$$

where the length of the repeating unit vector $\vec{r}_i$ is given by

$$|r_i| = \sqrt{\vec{r}_i \cdot \vec{r}_i} = a$$

and

$$\vec{R}_n \cdot \vec{R}_n = (\vec{r}_1 + \vec{r}_2 + \vec{r}_3 \ldots + \vec{r}_n) \cdot (\vec{r}_1 + \vec{r}_2 + \vec{r}_3 \ldots + \vec{r}_n)$$

$$\vec{R}_n \cdot \vec{R}_n = \sum_{i=1}^{n} \vec{r}_i \cdot \vec{r}_i + 2\sum_{j=1}^{n-1}\sum_{i=1}^{n-1} \vec{r}_i \cdot \vec{r}_{i+j}. \tag{13.37}$$

This is exactly the same result as was obtained for the random walk in diffusion in Chapter 11, Equation 11.37. As was the case for diffusion, the first sum is just na^2 and when the double sum is averaged over all the possible cosine values for the dot product; this term disappears when n is large. Therefore, the mean square of the end-to-end distance for a freely jointed polymer is simply

$$\left\langle R_n^2 \right\rangle = na^2 \tag{13.38}$$

or the root mean square end-to-end distance is

$$\left\langle R_n^2 \right\rangle^{\frac{1}{2}} = n^{\frac{1}{2}} a. \tag{13.39}$$

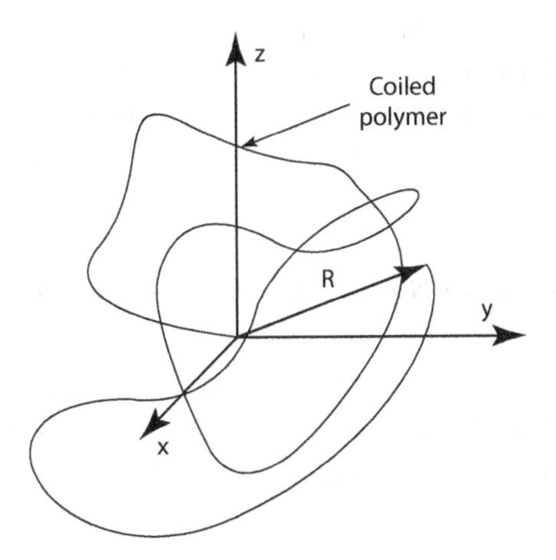

FIGURE 13.9 Coiled polymer showing its end-to-end distance, R, which can be used to gain some information about the spatial extent of a polymer in solution and its relation to the polymer diffusion coefficient in the liquid.

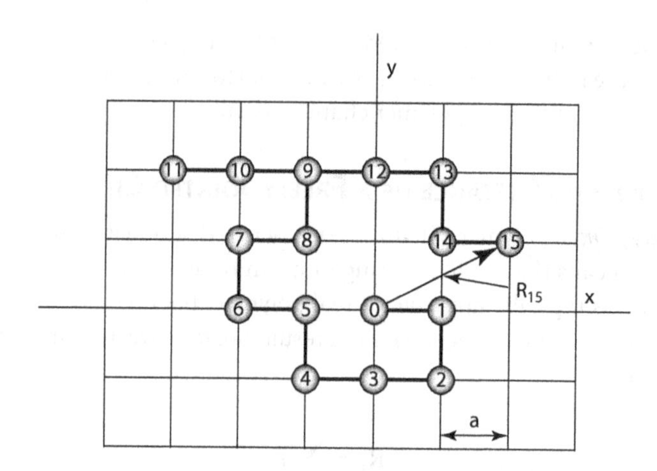

FIGURE 13.10 Two-dimensional polymer chain of n = 15 repeating units of length a showing the end-to-end distance of R_{15}. This figure was drawn to be similar to that of Figure 11.18 to highlight the similarities between polymer statistics in solution and a random walk in diffusion.

These are exactly the same results as obtained for the mean diffusion distance during a random walk of an atom in Chapter 11, Equation 11.40. Figure 13.10 has been drawn to represent a polymer molecule of 15 units in a chain on a square grid in attempt to mimic the random walk of 15 steps given in Figure 11.18. The only difference is that the chain cannot double back on itself as an atom can in a random walk. This occurs in steps 10 and 11 of Figure 11.18. In Figure 13.10, to include the same number of repeating units (steps), 10 and 11 are part of a chain *branch* so the polymer chain depicted in Figure 13.10 is really not a completely linear chain (this will make little difference in the use of this figure later). For the polymer, the vector $\vec{R}_{15}$ is the end-to-end distance calculated in Equation 13.38 and does not equal $15a^2$ but is actually $5a^2$ simply because not enough chain elements, n, have been included to get the cosines between the elements from average to zero.

One implication of Equation 13.39 is the difference in the root mean square end-to-end distance, $\langle R_n^2 \rangle^{\frac{1}{2}}$, and the *contour length* of the polymer, L = na. Assume that a = 300 pm about the length of two C–C bonds (Callister and Rethwisch 2009) and that the single repeating element molecular weight is 30 g/mole; then, if the total molecular weight is 30,000, n = 1000. So $L = 1000 \times \left(300 \times 10^{-12}\,\text{m}\right) = 3.0 \times 10^{-7}\,\text{m} = 300\,\text{nm}$. On the other hand,

$$\langle R_n^2 \rangle^{\frac{1}{2}} = (1000)^{\frac{1}{2}} \times \left(300 \times 10^{-12}\right) = 9.5 \times 10^{-9}\,\text{m} = 9.5\,\text{nm}$$

about a factor of 30 different. In reality, most polymers cannot be stretched into a simple straight line because there are angular restrictions between bonds, such as the 109.5° between C–C bonds in the repeating units of polyethylene, for example. This does introduce changes in the difference between these two parameters of a polymer chain, but not large enough to make a major impact on what follows.

13.10.5 Radius of Gyration and Hydrodynamic Radius

13.10.5.1 Introduction

From mechanics, the *radius of gyration*, R_{gx},* of a body about some axis, for example, the x axis, is the square root of the second moment of the mass divided by the mass, M (or the *moment of inertia, I*, divided by the total mass of the body) (Halliday et al. 1997):

* The notation for the radius of gyration varies: R_g, s, k, etc. Here, R_g will be used because the interpretation of the R_g symbol is rather intuitive.

$$R_{gx}^2 = \frac{\int x^2 dm}{\int dm} = \frac{I_x}{M}. \tag{13.40}$$

For polymer molecules it is more precisely defined as the *root mean square distance* from the elements of the polymer chain to the *center of mass* of the chain, call it r_0,

$$\left\langle R_g^2 \right\rangle^{\frac{1}{2}} = \left(\frac{\int r_0^2 dm}{\int dm} \right)^{\frac{1}{2}}. \tag{13.41}$$

The radius of gyration is very useful because it is very close to the hydrodynamic radius used in the Stokes–Einstein equation. Furthermore, it can be measured independently by light scattering and small angle neutron scattering (Cussler 1997; Green 2005).

13.10.5.2 Mass Distribution and Radius of Gyration

It is generally assumed that the mass of a coiled polymer chain follows some sort of a Gaussian distribution, just as the concentration of a diffusing substance from a single source followed a Gaussian distribution as a function of distance as shown in Chapter 11 (cf. Equations 11.25 and 11.26). In Section 11.7.2 of Chapter 11, this Gaussian source was used to calculate the mean square distance an atom or ion had diffused and the mean distance of a random walk led to the fundamental equation, Equation 11.41,

$$D = \frac{1}{6}\frac{n}{t}a^2 = \frac{1}{6}\Gamma a^2$$

where a = jump distance and G = jumps/s. A similar approach is applied here to obtain a general relation for the radius of gyration of a polymer molecule and its dimensions.

In the literature, the equation for the Gaussian distribution takes on slightly different forms (Young and Lovell 1991; Elias 1997; McCrum et al. 1997), but they all give essentially similar results. For reasons that will be obvious later, here the distribution for one dimension is taken to be

$$\left\langle R_{gx}^2 \right\rangle = \frac{3}{\sqrt{\pi}R} \int_{-\infty}^{\infty} x^2 e^{-\frac{x^2}{\left(\frac{R}{3}\right)^2}} dx \tag{13.42}$$

where $\left\langle R_{gx}{}^2 \right\rangle$ is the mean square distance from the center of mass in the *x*-direction for the mass elements of the polymer and $R = \left\langle R_n^2 \right\rangle^{(1/2)}$ from Equation 13.39 to simplify the notation.

So, Equation 13.42 is solved following the usual procedure of letting $y^2 = 9\left(x^2/R^2\right)$ or $x^2 = \left(R^2/9\right)y^2$ and $x = (R/3)y$ so $dx = (R/3)dy$; making these substitutions into Equation 13.42

$$\left\langle R_{gx}^2 \right\rangle = \frac{1}{\sqrt{\pi}}\frac{3}{R}\frac{R}{3}\int_{-\infty}^{\infty} \frac{R^2 y^2}{9} e^{-y^2} dy$$

$$\left\langle R_{gx}^2 \right\rangle = \frac{R^2}{9\sqrt{\pi}} 2\int_0^{\infty} y^2 e^{-y^2} dy$$

where the 2 in the second equation comes from the fact that the integral in the first equation is even requiring that twice the integral from $0 \to \infty$ equals the integral from $-\infty \to \infty$. Again, as in Chapter 11, the substitution $z = y^2$ is made in the integral so $y = z^{(1/2)}$ and $dy = (1/2)z^{-(1/2)}dz$; thus, the integration becomes

$$\langle R_{gx}^2 \rangle = \frac{R^2}{9\sqrt{\pi}} 2\left(\frac{1}{2}\right)\int_0^\infty z^{\frac{1}{2}} e^{-z} dz$$

$$\langle R_{gx}^2 \rangle = \frac{R^2}{9\sqrt{\pi}} \int_0^\infty z^{\frac{3}{2}-1} e^{-z} dz$$

$$\langle R_{gx}^2 \rangle = \frac{R^2}{9\sqrt{\pi}} \Gamma\left(\frac{3}{2}\right) = \frac{R^2}{9\sqrt{\pi}}\left(\frac{\sqrt{\pi}}{2}\right)$$

with the result

$$\langle R_{gx}^2 \rangle = \frac{R^2}{18}. \tag{13.43}$$

As for the case for diffusion, because $\vec{R}_g$ is really a vector,

$$\langle R_g^2 \rangle = \langle {R_{gx}}^2 \rangle + \langle {R_{gy}}^2 \rangle + \langle {R_{gz}}^2 \rangle = 3\langle {R_{gx}}^2 \rangle. \tag{13.44}$$

Substitution of Equation 13.43 into Equation 13.44 gives

$$\langle R_g^2 \rangle = 3\langle R_{gx}^2 \rangle = \frac{R^2}{6}$$

or

$$6\langle R_g^2 \rangle = \langle R_n^2 \rangle = na^2 \tag{13.45}$$

which can be derived by less intuitive summations (Elias 1997) and frequently quoted without proof or citation. The choice of the distribution function of the mass, Equation 13.42, leads to this result, so this is the correct distribution function. Note that the distribution function is normalized because

$$\frac{3}{\sqrt{\pi}R}\int_{-\infty}^{\infty} e^{-\frac{x^2}{\left(\frac{R}{3}\right)^2}} dx = 1. \tag{13.46}$$

Note the similarity between the radius of gyration, the number of chain elements, n, and the length of the chain elements in Equation 13.45, and the diffusion coefficient and the number of jumps per second and the mean jump distance, a, shown in Chapter 11, Equation 11.41. The importance of Equation 13.45 is that it is a relation between the length and molecular weight of the polymer and its radius of gyration can be used as the hydrodynamic radius in the Stokes–Einstein equation to calculate the diffusion coefficient of a polymer in a liquid! Equation 13.45 can be written as

$$\langle R_g^2 \rangle^{\frac{1}{2}} = \left(\frac{\langle R_n^2 \rangle}{6}\right)^{\frac{1}{2}} = \frac{n^{\frac{1}{2}} a}{\sqrt{6}}$$

because $M \infty n$; then, $D \propto M^{-\frac{1}{2}}$.

13.10.5.3 Polymer Chain with n = 15

Although a chain with only 15 units does not satisfy all of the assumptions in the above model, it is used to illustrate some of the parameters discussed above. Figure 13.11 is the same as Figure 13.10;

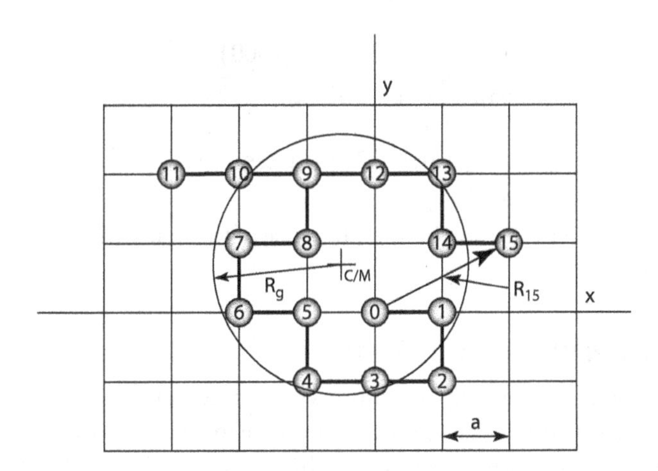

FIGURE 13.11 This is the same as Figure 13.10 but now shows the center of mass (c/m) (x = −0.467a, y = 0.733a) of the polymer molecule and the radius of gyration, R_g, which in this case is given by $R_g = \sqrt{\left(\sum_{i=1}^{15}(r_i - r_{c/m})^2/15\right)} = 1.88a$.

only the center of mass and the radius of gyration have been added. The center of mass for this two-dimensional chain of equal weight elements is

$$\langle x \rangle = \frac{\sum_{i=1}^{15} x_i}{15} = -0.467a$$

$$\langle y \rangle = \frac{\sum_{i=1}^{15} y_i}{15} = 0.733a$$

as shown in the figure. $\langle R_{15}^2 \rangle = 5a^2$ and the radius of gyration is $\langle R_g^2 \rangle^{\frac{1}{2}} = 1.88a$, shown as the circle in Figure 13.11. Note that

$$\frac{\langle R_{15}^2 \rangle}{\langle R_g^2 \rangle} = \frac{5}{3.5} = 1.4 \neq 6$$

indicating again the fact that a chain of 15 repeating units is a poor approximation to an infinite chain to which Equation 13.45 really applies.

13.10.5.4 Example: Diffusion of Poly(Ethylene Glycol) in Water

Poly(ethylene glycol), PEG, also called poly(ethylene oxide), PEO, has the formula $H\text{-}(O\text{-}CH_2\text{-}CH_2)_n$ and is soluble in water. The length of the chain element is 440 pm (Emsley 1998), the weight of the element is 44 g/mol, and a typical average molecular weight for this polymer is about 3400 g/mol. So $n = (3400/44) = 77.3$ repeating units or "mers." However, there are two rotating elements within each unit, a C–C distance of a = 154 pm and C–O of 132 pm (Emsley 1998). So it is probably best to take an average of the two or a = 143 pm,

$$\langle R_g^2 \rangle^{\frac{1}{2}} = \frac{n^{\frac{1}{2}} a}{\sqrt{6}} = \frac{77.3(143\,\text{pm})}{2.45} = 4.51 \times 10^3\,\text{pm} = 5.43\,\text{nm}.$$

In water at room temperature, the diffusion coefficient of this PEG is

$$D = \frac{k_B T}{6\pi\mu r} = \frac{(1.38\times10^{-23})(300)}{6\pi(10^{-3})(5.43\times10^{-9})}$$

$$D = 4.045\times10^{-11}\,m^2s^{-1} = 4.045\times10^{-7}\,cm^2s^{-1}$$

considerably smaller than self-diffusion in water because of the much larger hydrodynamic radius of the PEG compared to that of water.

13.10.6 Diffusion in Polymer Melts

Diffusion in polymer melts is essentially the same as the reptation process of polymer molecules diffusing in a solid polymer. In both cases, a polymer molecule must diffuse through an entanglement—of which it is a member—of the polymer chains. In Chapter 9, Section 9.5.2, Equation 9.62, the diffusion coefficient of a polymer chain by reptation is

$$D \cong \frac{1}{6} f \left(\frac{L}{n}\right)^2 \tag{13.47}$$

where L is the length of the moving defect on the chain that gives rise to diffusion. Again, because $n \infty M$, $D \propto M^{-2}$, or the diffusion coefficient is inversely proportional to the square of the molecular weight, which is what is observed (Green 2005). However, if the Stokes–Einstein equation is used because $\langle R_g^2 \rangle^{1/2} \propto n^{1/2} \propto M^{1/2}$, $D \propto M^{-1/2}$. The difference between these two models is that reptation is for entangled molecules while the Stokes–Einstein result assumes freely moving spheres. In reality, most amorphous polymers are entangled molecules and experimental results suggest that $D \propto M^{-2}$ fitting the reptation model more accurately.

13.11 CHAPTER SUMMARY

There is a general relation between fluxes and forces that leads to the concept of a an *absolute mobility*, the inverse of the drag force, along with the a *drift velocity*, v_d, and a *relaxation time*, τ, to reach the drift velocity under an applied force and a given absolute mobility. These concepts are applied to the motion of solid spherical particles in a fluid with *Stokes law*, $6\pi\mu r$, being the drag force on the particles leading to Stokes settling, Equation 13.11,

$$v_d = \frac{2}{9}\frac{r^2}{\mu}(\rho_s - \rho_l)g$$

which relates the settling velocity, v_d, to the particle radius, r, and the viscosity of the fluid, μ. Particle settling velocity is a convenient method of measuring particle size distributions. Although most texts on electrical properties omit it, a simple model of the electronic mobility of electrons in solids, $\mu_e = (e\tau/m_e)$, is obtained from the absolute mobility, B, viscous drag, and an applied electrical force. The important Einstein relation between the diffusion coefficient and the absolute mobility is established, that is, $D = k_B TB$. This easily leads to the Nernst–Einstein equation that relates ionic conductivity to diffusion: $\sigma = \left((ze)^2 \eta D / k_B T\right)$. Insertion of the Stokes law for 1/B leads to the Stokes–Einstein relation for diffusion in liquids, namely, $D = (k_B T/6\pi\mu r)$, which works well for solutions and liquids where the radius of the diffusing species, r, is well defined. If the diffusing species are ions—that are charged—then a version of the Nernst–Planck equation couples the diffusion of the charged species, Equation 13.27,

$$J' = \frac{\eta D}{k_B T}\left\{-\frac{1}{N_A}\frac{d\bar{G}}{dx} \mp z\frac{\mathcal{F}}{N_A}\frac{d\phi}{dx}\right\}$$

and leads to an effective coefficient for the coupled diffusion. Diffusion of charged species in an electric potential gradient, $d\phi/dx$, also defines an *electrochemical potential*, ξ, $\xi = \bar{G} \pm z\mathcal{F}\phi$, so that Equation 13.27 becomes

$$J' = -\frac{\eta D}{RT}\frac{d\xi}{dx}.$$

For polymer molecules in a solution, defining effective hydrodynamic radii to calculate diffusion coefficients in liquids is less straightforward. For polymers, the best measure for r is the root mean square of the radius of gyration of a polymer molecule, $\langle R_g^2 \rangle^{(1/2)}$, and it can be related to other parameters of the polymer chains such as the average end-to-end distance, the number of repeating units, and the molecular weight.

APPENDIX

A.1 Charge-Coupled Diffusion: Introduction

In electrolytes, the anions and cations cannot diffuse independently because they are charged. As a result of this charge, the faster diffusing ion will exert an electric field, E, and a force, zeE, on the slower diffusing ion with charge ze speeding it up. Similarly, the field will exert a retarding force on the faster diffusing ion slowing it down. Therefore, the resulting diffusion coefficient for sodium chloride, NaCl, diffusing as Na^+ and Cl^- ions in a water solution will be some combination of the diffusion coefficients for the respective ions. This was seen in Chapter 10, Section 10.11, where the dissolution of solid sodium chloride by diffusion was modeled. In addition, for any liquid or solid consisting of ions, such as NaCl, MgO, or $MgAl_2O_3$, the diffusion coefficients of the ions are also coupled by their respective charges that lead to a diffusion coefficient for the *compound* that is a function of the diffusion coefficients of the constituent ions. For example, if a polycrystalline piece of MgO is deformed by diffusional creep—under a mechanical force—at elevated temperature, the effective diffusion coefficient for the deformation of the MgO is a function of both the Mg^{2+} and O^{2-} ions. The purpose here is to model charge-coupled diffusion.

A.2 Fluxes with Two Forces

A.2.1 General Case

Consider the general ionic compound $A_\alpha B_\beta$ whose formula implies that it is composed of ions $A^{\beta+}$ and $B^{\alpha-}$, where α and β are positive integers, for example, Fe_2S_3 consisting of Fe^{3+} and S^{2-} ions. The literature can be a little confusing because the symbols used to designate the charge—here alpha and beta—sometimes include the charge and sometimes do not. Here, they *do not include the charge* and are simply integers as will become clear. In general, if the ions A and B are diffusing, there will be an electric field, E, between them due to their different rates of diffusion. As a result, there is both a chemical force and an electric force acting on each ion that determines its flux. That is, the number flux density of ions, $J' = J/N_A$, which is just the molar flux, J, divided by Avogadro's number is given by

$$J' = \eta v_d = \eta BF = \frac{\eta D}{k_B T}\left(F_{chem} + F_{elec}\right) = \frac{\eta D}{k_B T}\left\{-\frac{1}{N_A}\frac{d\bar{G}}{dx} \pm zeE\right\} \quad (A.1)$$

where the plus sign is for a cation and the minus sign for the anion, and the values of *z* are the integers α or β. Now the charge flux density, J^*, is just $J^* = \pm zeJ'$, and to maintain electrical neutrality, $J_A^* - J_B^* = 0$ or

$$\beta e J'_A = \alpha e J'_B \quad (A.2)$$

and

$$J'_{A_\alpha B_\beta} = \frac{J'_A}{\alpha} = \frac{J'_B}{\beta}. \tag{A.3}$$

for the molecular flux density. The procedure is to write the charge flux density equations for both A and B and obtain an expression for E that can be replaced into either J'_A or J'_B to get the charge-coupled diffusion coefficient. Now (dropping the "bar" on the molar Gibbs energy for the next several equations to simplify the equations),

$$\begin{aligned} J^*_A &= \frac{\beta e \eta_A D_A}{k_B T}\left\{-\frac{1}{N_A}\frac{dG_A}{dx} + \beta e E\right\} \\ J^*_B &= \frac{\alpha e \eta_B D_B}{k_B T}\left\{-\frac{1}{N_A}\frac{dG_B}{dx} - \alpha e E\right\} \end{aligned} \tag{A.4}$$

Equating these two equations because the total charge flux density is zero, and canceling k_B, T, and e,

$$\beta\eta_A D_A\left\{-\frac{1}{N_A}\frac{dG_A}{dx} + \beta e E\right\} = \alpha\eta_B D_B\left\{-\frac{1}{N_A}\frac{dG_B}{dx} - \alpha e E\right\}$$

carrying out the multiplications,

$$-\frac{\beta\eta_A D_A}{N_A}\frac{dG_A}{dx} + \beta^2\eta_A D_A eE = -\frac{\alpha\eta_B D_B}{N_A}\frac{dG_B}{dx} - \alpha^2\eta_B D_B eE$$

and rearranging,

$$\left(\beta^2\eta_A D_A + \alpha^2\eta_B D_B\right)eE = -\frac{\alpha\eta_B D_B}{N_A}\frac{dG_B}{dx} + \frac{\beta\eta_A D_A}{N_A}\frac{dG_A}{dx}$$

and solving for *eE*,

$$eE = \frac{1}{\left(\beta^2\eta_A D_A + \alpha^2\eta_B D_B\right)}\left(-\frac{\alpha\eta_B D_B}{N_A}\frac{dG_B}{dx} + \frac{\beta\eta_A D_A}{N_A}\frac{dG_A}{dx}\right). \tag{A.5}$$

Now, the algebra gets even uglier in that Equation A.5 is substituted into Equation A.1 for either J'_A or J'_B, so

$$J'_A = \frac{\eta_A D_A}{k_B T}\left\{-\frac{1}{N_A}\frac{dG_A}{dx} + \frac{\beta}{\left(\beta^2\eta_A D_A + \alpha^2\eta_B D_B\right)}\left(-\frac{\alpha\eta_B D_B}{N_A}\frac{dG_B}{dx} + \frac{\beta\eta_A D_A}{N_A}\frac{dG_A}{dx}\right)\right\}$$

$$J'_A = \frac{\eta_A D_A}{k_B T\left(\beta^2\eta_A D_A + \alpha^2\eta_B D_B\right)}\left\{-\frac{\left(\beta^2\eta_A D_A + \alpha^2\eta_B D_B\right)}{N_A}\frac{dG_A}{dx} - \frac{\alpha\beta\eta_B D_B}{N_A}\frac{dG_B}{dx} + \frac{\beta^2\eta_A D_A}{N_A}\frac{dG_A}{dx}\right\}$$

$$J'_A = -\frac{\alpha\eta_B\eta_A D_A D_B}{k_B T N_A\left(\beta^2\eta_A D_A + \alpha^2\eta_B D_B\right)}\left\{\alpha\frac{dG_A}{dx} + \beta\frac{dG_B}{dx}\right\}. \tag{A.6}$$

Now, by stoichiometry,

$$\eta_B = \beta\eta_{A_\alpha B_\beta} = \beta\eta$$

$$\eta_A = \alpha\eta_{A_\alpha B_\beta} = \alpha\eta$$

where $\eta = \eta_{A_\alpha B_\beta}$ for short, so Equation A.6 becomes (again using the "bar" for the chemical potential)

$$J'_A = -\frac{\alpha\eta D_A D_B}{RT(\beta D_A + \alpha D_B)}\left\{\alpha\frac{d\bar{G}_A}{dx} + \beta\frac{d\bar{G}_B}{dx}\right\}$$

but because $J_{A_\alpha B_\beta} = (J'_A/\alpha N_A)$ and $C_{A_\alpha B_\beta} = \eta/N_A$,

$$J_{A_\alpha B_\beta} = -\frac{C_{A_\alpha B_\beta}}{RT}\left(\frac{D_A D_B}{\beta D_A + \alpha D_B}\right)\left\{\alpha\frac{d\bar{G}_A}{dx} + \beta\frac{d\bar{G}_B}{dx}\right\} \tag{A.7}$$

where C = concentration of the compound $A_\alpha B_\beta$. Now, because

$$\alpha A + \beta B = A_\alpha B_\beta$$

$$\alpha\bar{G}_A + \beta\bar{G}_B = \bar{G}_{A_\alpha B_\beta}.$$

Equation A.7 could also be written as

$$J_{A_\alpha B_\beta} = -\frac{C_{A_\alpha B_\beta}}{RT} D_{eff} \frac{d\bar{G}_{A_\alpha B_\beta}}{dx} \tag{A.8}$$

where D_{eff} = the "effective" diffusion coefficient,

$$D_{eff} = \frac{D_A D_B}{\beta D_A + \alpha D_B}. \tag{A.9}$$

Equations A.8 and A.9 apply if there *is no concentration gradient* of $A_\alpha B_\beta$ but where there is a *Gibbs energy gradient*, $d\bar{G}_{A_\alpha B_\beta}/dx$, that might be produced by a mechanical stress (creep) or surface curvature (sintering, grain growth, and boundary grooving).

A.2.2 Dissolution of $A_\alpha B_\beta$

If the compound $A_\alpha B_\beta$ is dissolved in a liquid for example (like the previous example of NaCl dissolved in water, Chapter 10, Section 10.11), the concentration of $A_\alpha B_\beta$ varies with distance, $dC_{A_\alpha B_\beta}/dx \neq 0$. For simplicity, consider the solution to be ideal; then Equation A.7,

$$J_{A_\alpha B_\beta} = -\frac{C}{RT}\left(\frac{D_A D_B}{\beta D_A + \alpha D_B}\right)\left\{\alpha\frac{d\bar{G}_A}{dx} + \beta\frac{d\bar{G}_B}{dx}\right\}$$

with $C_A = \alpha C_{A_\alpha B_\beta}$ and $C_B = \beta C_{A_\alpha B_\beta}$ becomes (because $C_i = RT\ln C_i/C_T$, where C_T is the total concentration and assumed constant as a first approximation)

$$J_{A_\alpha B_\beta} = -C_{A_\alpha B_\beta}\left(\frac{D_A D_B}{\beta D_A + \alpha D_B}\right)\left\{\frac{\alpha^2}{\alpha C_{A_\alpha B_\beta}}\frac{dC_{A_\alpha B_\beta}}{dx} + \frac{\beta^2}{\beta C_{A_\alpha B_\beta}}\frac{dC_{A_\alpha B_\beta}}{dx}\right\}$$

$$J_{A_\alpha B_\beta} = -\left(\frac{(\alpha+\beta)D_A D_B}{\beta D_A + \alpha D_B}\right)\left\{\frac{dC_{A_\alpha B_\beta}}{dx}\right\} \tag{A.10}$$

where D_{eff} is now

$$D_{eff} = \left(\frac{(\alpha + \beta) D_A D_B}{\beta D_A + \alpha D_B} \right).$$

Now when $D_A = D_B$, $D_{eff} = D_A = D_B$ as it should. On the other hand, if $D_A >> D_B$, then $D_{eff} = ((\alpha + \beta)/\beta) D_B$. For example, if $CaCl_2$ is dissolved in H_2O and if $D_{Ca^{+2}} >> D_{Cl^-}$, then $D_{eff} = (3/2) D_C$ or the Cl^- ions are being pulled along by the faster diffusing Ca^{+2} ions.

EXERCISES

13.1 One way to separate ore minerals from other minerals is to make use of their settling velocity as determined by the Stokes law. Suppose that the problem is to separate SiO_2 particles from ZnS particles that are ground to roughly the same size, particle diameter $d = 1\ \mu m$. The density of SiO_2 is 2.33 g/cm^3 and that of ZnS is 4.04 g/cm^3 and suppose that they are settled in glycerol–water solution that has a viscosity of $\mu = 50\,mPa$ s and a density of 1.06 g/cm^3 at 25°C (Haynes 2013).

a. Calculate the terminal settling velocity for both of these minerals in the solution at 25°C.

b. Calculate the relaxation times necessary to reach terminal velocity for both of these minerals.

c. Calculate roughly how far apart the two minerals would be in a vertical column of glycerol–water solution after one day of settling.

13.2 Einstein developed the Stokes–Einstein equation for diffusion to explain the motion of small particles in fluids, Brownian motion (Einstein 1956).

a. Calculate the diffusion coefficient of these mineral particles in water at 25°C whose viscosity $\mu = 0.890$ mPa s.

b. If the particles were introduced all together in the water as they began settling (i.e., at time zero, they all were at $x = 0$), approximate how spread out vertically (μm) each of two particle groups would be because of diffusion during settling for one day.

13.3 As part of his theoretical investigations on the Brownian motion of small particles, Einstein, using fluid dynamic arguments, developed the relationship between the viscosity of a fluid with a volume fraction, ϕ, of suspended spherical particles: $\mu/\mu_0 = 1 + 2.5\phi$ (Einstein 1956). From the literature at that time, he found that a 1 wt.% aqueous solution of sugar (sucrose) had a relative viscosity of $\mu/\mu_0 = 1.0245$.

a. Calculate the volume fraction of sugar.

b. If the molecular weight of sucrose $(C_{12}H_{22}O_{11})$ is M = 342.296 g/mol and its density as a solid is 1.5805 g/cm^3, calculate the volume of 1 g of a sugar molecules.

c. Einstein also found in the literature that the specific gravity of a 1 w/o sugar solution in water is 1.00371. If the sugar solution behaves as an ideal solution, calculate (1) what the specific gravity should be and (2) the volume of sugar in the solution.

d. If the volumes of 1 g of sugar are different between (b) and (c), explain this difference: Einstein did!

13.4 In the text, the diffusion coefficients of Na^+ and Cl^- in water are given as $D_{Na^+} = 1.33 \times 10^{-5}\ cm^2/s$ and $D_{Cl^-} = 2.032 \times 10^{-5}\ cm^2/s$ at 25°C. The viscosity of water at 25°C is 0.890×10^{-3} Pa-s. (Haynes 2013).

a. Calculate the sodium and chlorine ion radii from these diffusion coefficients.

b. The radius of the Na^+ ion is given as $r(Na^+) = 98pm$ and that of Cl^- $r(Cl^-) = 181pm$ (Emsley 1998). Compare these radii to those calculated from the Stokes–Einstein equation from the given diffusion coefficients.

c. Any differences between the radii in (a) and (b) are due to solvation of water molecules around the ions. If the radius of a water molecule is roughly 0.120 nm based on the results of Section 13.7.2, calculate the number of water molecules solvating each of the sodium and chlorine ions in water. Hint: Take the volume of the ion calculated from the viscosity and subtract the volume of the bare ion to get the volume of the solvated water molecules.

13.5 If nothing else, this particular exercise demonstrates the number of different references necessary to generate relevant data. Molten salts are good conductors of electricity and the following data are available for molten sodium chloride at 1000°C: $\sigma = 4.16\,S/cm$, $\mu = 0.737 \times 10^{-3}\,Pa\,s$, and $\rho = 1.9911 - 0.543 \times 10^{-3}\,T\,g/cm^3$ where Units(T) = °C (Brandes and Brook 1992).

a. The molecular weight of NaCl is 58.493 g/mol and its density at 25°C is ρ = 2.17 g/cm^3 (Haynes 2013). Its change in volume from 25°C to the melting point, 801°C, is $\Delta V/V = 1.26 \times 10^{-3}$ (Gray 1972). Calculate the density (g/cm^3) of solid sodium chloride at its melting point, the density (g/cm^3) of liquid NaCl at its melting point, and the percent volume change on melting at 801°C.

b. From above, the radius of the Na^+ ion is given as $r(Na^+) = 98\,pm$ and that of Cl^- $r(Cl^-) = 181\,pm$. Assume that these are the radii of the diffusing ions (Emsley 1998). Calculate the diffusion coefficient for Na^+ and Cl^- in molten NaCl.

c. Calculate the ionic conductivity contributions of both the sodium and chloride ions in units of S/cm.

d. Compare the results to a literature value of 4.16 S/cm (Brandes and Brook 1992).

13.6 Potassium ferricyanide, $K_3[Fe(CN_6)]$, is dissolved in pure water at 25°C. The molar ionic conductivities in water at 25 °C are $\sigma(K^+) = 73.5\,S/cm$ and $\sigma\left(\left[Fe(CN)_6\right]^{3-}\right) = 303\,S/cm$ (Burgess 1988).

a. Calculate the diffusion coefficients for these two ions in water at 25°C.

b. Calculate the hydrodynamic radii for these two ions in water at 25°C.

c. Develop an expression for the diffusion coefficient for the dissolving $K_3[Fe(CN_6)]$.

d. Calculate the diffusion coefficient for the dissolving $K_3[Fe(CN_6)]$.

13.7 Poly(vinyl alcohol), PVA, $\left[CH_2CHOH\right]_n$ is water soluble. The carbon–carbon distance is 154 pm.

a. Calculate the "molecular" weight of a mer unit.

b. If the average molecular weight of PVA is 10^5 g/mol, calculate the number of "mer" units in the polymer.

c. Calculate the radius of gyration of this molecule assuming that the chain unit length is the carbon–carbon distance, d = 154 pm.

d. Calculate its diffusion coefficient in water at 25°C, where the viscosity of water is 0.870 mPa.

13.8 Li-polymer batteries use a lithium-ion-conducting electrolyte. To test electrode behavior for such cells, a better liquid electrolyte is used. Aqueous solutions cannot be used because of the reaction of lithium in the anode with water. Therefore, nonaqueous solutions are used. A popular choice for a liquid electrolyte is $LiClO_4$ dissolved in solutions of dimethoxyethane (DME), $C_4H_{12}O_2$, and propylene carbonate (PC, $C_4H_6O_3$). A one molar solution of lithium chlorate at 25°C dissolved in a 60 v/o DMe in PC solution gives a conductivity of 0.015 S/cm. The viscosity of this solution is $\mu = 6 \times 10^{-4}\,Pa\,s$ (Xu 2003).

a. The molar conductivities for Li^+ and ClO_4^- in water are 38.7 S/cm and 67.3 S/cm (Burgess 1988), respectively, where $LiClO_4$ is completely ionized in the solution (Roine 2002). Calculate the hydrodynamic radii of the two ions in water.

b. Similar to problem 4, if the ionic radius lithium is $r(Li^+) = 78\,pm$, calculate the number of solvated water ions.

c. If the conductivities of the two ions are in the same ratio in the organic solvent,
   i. calculate the diffusion coefficients for the two ions,
   ii. calculate the ionic radii for the two ions in the organic solvent.

d. If the difference in conductivity between water and the organic solvent is the incomplete ionization of $LiClO_4$ in the organic liquid, calculate the degree of ionization or dissociation.

e. Give other possible reasons for the differences between the conductivities in water and the organic solvent.

13.9 Pure cubic ZrO_2 has the fluorite (CaF_2) crystal structure, Figure 7.30.

a. The lattice parameter of cubic ZrO_2 is 0.5127 nm (Green, Hannik, and Swain 1989). Calculate the theoretical x-ray density of pure cubic ZrO_2 in g/cm^3.

b. Compare the value of density calculated in a with the value given in the literature of $\rho = 5.6$ g/cm^3. If they are different explain why this might be.

c. Calculate the theoretical density of CaO-stabilized ZrO_2 having the composition of 15 m/o CaO in ZrO_2.

d. Compare the densities calculated in a and c. If there is a difference, can it be ignored in calculating the oxygen ion conductivity? Why or why not?

13.10 A certain liner polymer with a length average molecular weight of 4.4×10^6 g/mol has a viscosity of about 100 Pa s at 200°C.

a. Calculate the number of "mer" units per mean polymer length if the mer molecular weight is 62.5 g/mol.

b. Assuming that there are two rotational chain elements per mer, calculate the radius of gyration of this polymer assuming that the element size is d = 154 pm, the C–C distance.

c. Assuming that the Stokes–Einstein relation holds in this polymer melt, calculate the diffusion coefficient (cm^2/s) for the polymer molecules in this melt.

d. If this were a polymer blend, explain what role diffusion would play in removing compositional inhomogeneities on the order of 0.1 mm in an hour.

REFERENCES

Bird, R. B., W. E. Stewart, and E. N. Lightfoot. 2002. *Transport Phenomena*, 2nd ed. New York: Wiley.

Brandes, E. A. and G. B. Brook, eds. 1992. *Smithells Metals Reference Book*, 7th ed. Oxford: Butterworth-Heinemann.

Burgess, J. 1988. *Ions in Solution:Basic Principles of Chemical Interactions.* New York: Wiley.

Callister, W. D., Jr. and D. G. Rethwisch. 2009. *Fundamentals of Materials Science and Engineering an Integrated Approach*, 3rd ed. New York: John Wiley and Sons.

Chiang, Y.-M. C., D. Birnie, III, and W. D. Kingery. 1997. *Physical Ceramics.* New York: John Wiley and Sons.

Clift, R., J. R. Grace, and M. E. Weber. 1978. *Bubbles, Drops, and Particles.* New York: Academic Press.

Cussler, E. L. 1997. *Diffusion: Mass Transfer in Fluid Systems*, 2nd ed. Cambridge: Cambridge University Press.

Des Cloizeaux, J. and G. Jannink. 1990. *Polymers in Solution Their Modeling and Structure.* Oxford: Clarendon Press.

Einstein, A. 1956. *Investigations on the Theory of the Brownian Movement.* New York: Dover.

Elias, H. -G. 1997. *In Introduction to Polymer Science.* New York: VCH Publishers.

Emsley, J. 1998. *The Elements*, 3rd ed. New York: John Wiley and Sons.

Fujita, H. 1990. *Polymer Solutions.* New York: Elsevier Science.

Gaskell, D. R. 1992. *An Introduction to Transport Phenomena in Materials Engineering.* New York: Macmillan.

Geankoplis, C. J. 1972. *Mass Transport Phenomena.* Columbus, OH: Ohio State University Press.

Gray, D. E., Editor in Chief. 1972. *American Institute of Physics Handbook*, 3rd ed. New York: McGraw-Hill.

Green, P. F. 2005. *Kinetics, Transport, and Structure in Hard and Soft Materials.* Boca Raton FL: CRC Press, Taylor & Francis Group.

Green, D. J., R. H. J. Hannink, and M. V. Swain. 1989. *Transformation Toughening of Ceramics.* Boca Raton FL: CRC Press.

Halliday, D., R. Resnik, and J. Walker. 1997. *Fundamentals of Physics Extended*, 5th ed. New York: John Wiley and Sons.

Haynes, W. M., Editor-in-Chief. 2013. *CRC Handbook of Chemistry and Physics*, 94th ed. Boca Raton FL: CRC Publishing.
Kingery, W. D., H. K. Bowen, and D. R. Uhlmann. 1976. *Introduction to Ceramics*, 2nd ed. New York: John Wiley and Sons.
Mayer, J. W. and S. S. Lau. 1990. *Electronic Materials Science*. New York: Macmillan.
McCrum, N. G., C. Pl. Buckley, and C. B. Bucknall. 1997. *Principles of Polymer Engineering*, 2nd ed. Oxford: Oxford University Press.
Molyneux, P.. 1983. *Water-Soluble Synthetic Polymers: Properties and Behavior, Vol. 1*. Boca Raton, FL: CRC Press.
Phillies, G. D. J. 2011. *Phenomenology of Polymer Solution Dynamics*. Cambridge: Cambridge University Press.
Readey, D. W. 1966. Chemical Potentials and Initial Sintering in Pure Metals and Ionic Compounds. *J. Appl. Phys.* 37(6): 2309–2312.
Robinson, R. A. and R. H. Stokes. 2002. *Electrolyte Solutions*, 2nd ed. New York: Dover.
Roine, A. 2002. *Outokumpu HSC Chemistry for Windows, Chemical Reaction and Equilibrium Software with Extensive Thermochemical Database, Version 5.1*. Pori Finland: Outokumpu Research Oy.
Rosen, S. L. 1993. *Fundamental Principles of Polymeric Materials*, 2nd ed. New York: John Wiley and Sons.
Teraoka, I. 2002. *Polymer Solutions: An Introduction to Physical Properties*. New York: Wiley-Interscience.
Xu, K. 2004. Nonaqueous Liquid Electrolytes for Lithium-Based Rechargeable Batteries. *Chem. Rev.* 104: 4303–4417.
Young, R. J. and P. A. Lovell. 1991. *Introduction to Polymers*, 2nd ed. London: Chapman and Hall.

14

Interdiffusion and Metals

14.1 INTRODUCTION

Most of the previous diffusion modeling largely focused either on a single diffusing species or a single constant diffusion coefficient. In this chapter, *interdiffusion* of two or more atomic or ionic species is examined in detail leading to concentration-dependent diffusion coefficients. This is an important topic because, if the diffusing species have different rates of diffusion, then volume changes can occur that lead to macroscopic flow in addition to the simple random diffusive motion. Interdiffusion plays an important role in the properties and behavior of materials: including phenomena such as the oxidation of metals, osmotic pressure, dialysis, the Kirkendall effect, the swelling of polymers, and reactions to form compounds. As a result, examples are found in metals, inorganic solids, polymers, and biological materials and systems. There have been many studies of these various phenomena, in unrelated scientific and technological disciplines that make the interpretation of the notation and comparison of the literature somewhat difficult. The goal here is not to review and summarize the extensive literature, but to present these topics with a hopefully uniform approach to demonstrate relationships and important technological applications. In fact, this chapter is focused primarily on metals because many of the phenomena involving metals are not only of interest scientifically but also have broad technological applications. As a result, metallic systems have been well studied and phenomena such as oxidation and interdiffusion are well understood. In many of these studies, inert marker motion has led to a better understanding of the phenomenon under investigation.

14.2 OXIDATION OF METALS

14.2.1 Introduction

The oxidation of metals, needless to say, is of great technological and economic importance as well as of interest scientifically. One of the initial things to consider about the oxidation of a given metals is, "Does the oxide provide a coherent protective, or *passive*, layer that reduces the rate of continued oxidation?" This is best summarized by the *Piling–Bedworth ratio*, which is the ratio of the volume of the oxide compared to that of the volume of the metal consumed (West 1986; Kofstad 1988). If the ratio is less than 1, then the oxide layer is under tension and cracks form exposing the surface to further oxidation—the oxide layer does not cover the metal surface completely and is not protective: not *passive*. For the oxidation of magnesium to MgO, the ratio is 0.8 and, as a result, magnesium oxidation does not provide a protective oxide layer. On the other hand, the ratio is greater than 1 for nickel, silicon, iron, and a number of other metals that do form protective layers. For example, the ratio is 1.6 for nickel. However, too large of a positive ratio can lead to compressive stresses in the oxide layer that can also lead to cracking—*spalling* of the oxide—and increase the complexity of the oxidation process. Here, only oxidation with the formation of a protective layer is considered.

Most of the time, the rate of metal oxidation is determined by either the diffusion of the metal out to the surface or oxygen in from the atmosphere as shown in Figure 14.1. In either case, it is diffusion through the oxide layer that determines the rate of oxidation. In fact, much of the understanding of diffusion in oxide compounds has been generated mainly to understand oxidation of metals. It was seen in Chapter 10 that in the oxidation of silicon—or silicon-containing compounds such as SiC, Si_3N_4, and silicides—interstitial diffusion of oxygen through the SiO_2 layer controls the rate of oxidation. In that case, as shown in Figure 14.1, any inert markers* placed at the silicon–oxygen atmosphere would end up at the top of the oxide layer. However, in the oxidation of transition metals such as iron, nickel, and titanium, which is of significant commercial and industrial importance, the diffusing species move via a vacancy mechanism and, for all practical purposes, the composition of the oxide *is constant* throughout the thickness of the oxide layer. In other words, there is no *concentration gradient*—ignoring possible point defect gradients—but there is a large *Gibbs energy gradient* that drives the diffusion-controlled oxidation process. In reality, however, for many of these oxides, there is the small deviation from stoichiometry discussed in Chapter 9 that

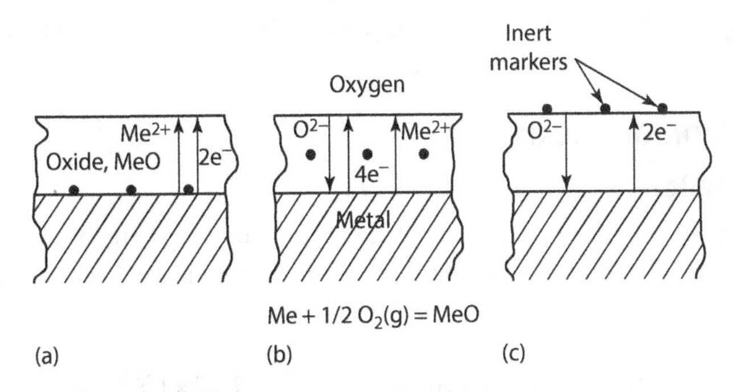

FIGURE 14.1 Possible marker motion during the oxidation of metals. In all cases, the diffusion of electrons is fast compared to that of the ions. (a) The cation, Me^{+2}, diffuses much faster than the oxygen anion, so the markers remain at the metal–oxide interface. (b) The diffusion rates of cations and anions are about the same so the markers end up somewhere inside the oxide layer. (c) The oxygen anion diffusion is much faster than that of the cation, so the markers are on top at the oxide–oxygen interface.

* An inert marker is some nonreactive material such as tungsten wires in copper or zirconia particles in alumina that do not react with the bulk phase at elevated temperatures and serve as a spatial reference point.

produces the requisite electronic charge flux, and this small electron or hole concentration difference will indeed vary from one side of the oxide layer to the other. For example,

$$NiO(s) + x/2O_2(g) \rightleftarrows Ni_{1-x}O(s)$$

$$Fe_2O_3(s) \rightleftarrows Fe_2O_{3-x}(s) + x/2O_2(g)$$

$$TiO_2(s) \rightleftarrows TiO_{2-x}(s) + x/2O_2(g).$$

As was seen in Chapter 10, this concentration difference, 1 – x, is usually quite small (~10^{-3}) compared to the very large Gibbs energy gradient that drives diffusion and mass transport and the oxidation process. In both cases, there must also be a flux of electronic charge or oxidation will not take place according the *Wagner theory of oxidation* (Wagner 1951), which is modeled in the Appendix in terms of the electrical conductivities of the oxide layer. This is in the Appendix to simply reduce the amount of algebra here to minimize the possibility of "losing sight of the forest for the trees." Specifically, for NiO formation the flux of nickel from the *nickel–nickel oxide* interface to the *nickel oxide–oxygen* interface is given by (Wagner 1951)

$$J_{Ni} = -\frac{1}{4\mathfrak{F}^2}\sigma t_i t_e \frac{d\bar{G}_{Ni}}{dx} \tag{14.1}$$

where:

σ = total electrical conductivity = ionic conductivity + electronic conductivity
t_i = the *transference number* for ionic conductivity
t_e = the transference number for the electronic conductivity
$\mathfrak{F}$ = faraday or one mole of electron charges = 96,485 C/mol.

The transference numbers are simply the *fractions* of the total conductivity (current) by either the ions or electrons, $t_i = \sigma_i/\sigma_{total}$. So if $t_e = 0$ and $t_h = 0$—no electronic conduction—then there is no oxidation, which is implied in Figure 14.1.

The principles applied to oxidation can be equally applied to the formation of other compounds or reaction products between a metal and a gas such as sulfides or fluorides. Finally, oxidation can become quite complex if there is more than one oxidation state of the oxide or if an alloy is being oxidized. The modeling of these complexities is best left to studies focused on metals and/or their oxidation and corrosion. Only simple oxidation is covered here to illustrate the general principles involved in a nonideal, diffusion-controlled reaction. These same principles apply to more complex cases but require some modifications. There are many good sources for information on the oxidation of metals that summarize both complex and simple reactions (e.g., West 1986; Kofstad 1988).

14.2.2 Oxidation of Nickel

14.2.2.1 Introduction

For the sake of concreteness, the oxidation of nickel in pure oxygen, $p_{O_2} = 1$ bar, is examined at 1200°C. The melting point of nickel is 1455°C and that of NiO is 1957°C (Haynes 2013). From the diffusion data in Table 9.5,

$$D_{Ni} = D_0 e^{-\frac{Q}{RT}} = 4.4 \times 10^{-4} e^{-\frac{184,900}{(8.314)(1473)}} = 1.22 \times 10^{-10}\ cm^2/s \tag{14.2}$$

$$D_O = D_0 e^{-\frac{Q}{RT}} = 6.2 \times 10^{-4} e^{-\frac{240,600}{(8.314)(1473)}} = 1.82 \times 10^{-12}\ cm^2/s. \tag{14.3}$$

The diffusion coefficient of oxygen is about two orders of magnitude smaller than that of nickel. So for all practical purposes, the oxidation rate is controlled by the faster of the diffusing species, the nickel ion, which is the situation depicted in Figure 14.1a. Any inert markers at the original nickel surface should remain there because all of the diffusion will be from the nickel through the oxide. This, of course, assumes that the transference number for electrons is large enough for oxidation to take place. As pointed out, the electronic conductivity arises because of the nonstoichiometric NiO, which is a fairly good conductor at elevated temperatures due to the formation of point defects by oxidation in oxygen to produce the nonstoichiometric $Ni_{1-x}O$, essentially a solid solution of Ni_2O_3 in NiO:

$$2N_{Ni}^{x} + \frac{1}{2}O_2(g) \rightleftarrows 2Ni_{Ni}^{\bullet} + O_O^{x} + V_{Ni}'' \tag{14.4}$$

where the electronic conductivity comes from the electrons on the Ni^{2+} ions (Ni_{Ni}^{x}) moving to the Ni^{3+} ions ($Ni_{Ni}^{\bullet}$), the so-called *hopping electron conductivity* or *small polaron conductivity*. A literature value of the electrical conductivity of NiO at 1200°C in 1 atm of pure oxygen is $\sigma \cong 1$ S/cm making it a very good *semiconductor* (Smyth 2000). For now, it is assumed that $t_{electronic} \cong 1$ so only the flux density of Ni^{2+} ions needs to be considered (see Exercises).

14.2.2.2 Calculation of the Parabolic Rate Constant

Again invoking the Nernst–Planck equations,

$$\begin{aligned} J_e' &= \frac{\eta_e D_e}{k_B T}\left\{-\frac{1}{N_A}\frac{d\bar{G}_e}{dx} - eE\right\} \\ J_{Ni^{2+}}' &= \frac{\eta_{Ni^{2+}} D_{Ni^{2+}}}{k_B T}\left\{-\frac{1}{N_A}\frac{d\bar{G}_{Ni^{2+}}}{dx} + 2eE\right\} \end{aligned} \tag{14.5}$$

which, after some tedious algebra in the Appendix, if the oxidation is being carried out in pure oxygen at one bar pressure, leads to

$$J_{Ni^{2+}}' = -\frac{\eta_{Ni^{2+}} D_{Ni^{2+}} \eta_{e^-} D_{e^-}}{N_A k_B T\left(4\eta_{Ni^{2+}} D_{Ni^{2+}} + \eta_{e^-} D_{e^-}\right)} \frac{\Delta G_{NiO}^{o}}{L} \tag{14.6}$$

where:

ΔG_{NiO}^{o} is the Gibbs energy of formation of NiO from pure nickel and 1 bar (atmosphere) oxygen
L is the thickness of the oxide layer as shown in Figure 14.2.

Assuming that $t_e \cong 1$ is the same as assuming $4\eta_{Ni^{2+}} D_{Ni^{2+}} \ll \eta_{e^-} D_{e^-}$ because the term in the parentheses in Equation 14.6 is proportional to the total electrical conductivity. Therefore, Equation 14.6 becomes

$$J_{Ni^{2+}}' = -\frac{\eta_{Ni^{2+}} D_{Ni^{2+}}}{RT} \frac{\Delta G_{NiO}^{o}}{L}. \tag{14.7}$$

Again, the left-hand side of this equation needs to be expressed in terms of the change in the oxide thickness with time, dL/dt. Doing this and dividing the number flux density to get the molar flux density of $J_{Ni^{2+}}$,

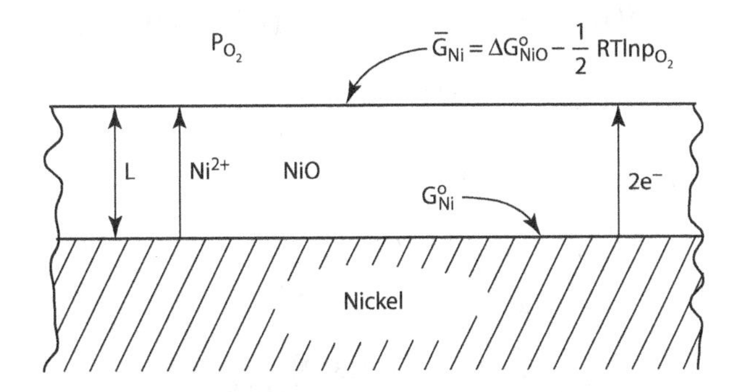

FIGURE 14.2 Schematic for the oxidation of nickel where the nickel diffusion is much faster than that of oxygen. The oxide layer thickness is L and there is a Gibbs energy difference across the oxide layer of $\Delta G_{NiO}^{o} - (1/2)\,RT \ln p_{O_2}$. If the oxidation is in pure oxygen, then it is simply ΔG_{NiO}^{o}.

$$\frac{\rho_{NiO}}{M_{NiO}}\frac{dL}{dt} = J_{NiO} = J_{Ni^{2+}} = \frac{J'_{Ni^{2+}}}{N_A} = -\frac{\eta_{Ni^{2+}} D_{Ni^{2+}}}{N_A RT}\frac{\Delta G^o_{NiO}}{L}$$

$$C_{NiO}\frac{dL}{dt} = -\frac{C_{Ni^{2+}} D_{Ni^{2+}}}{RT}\frac{\Delta G^o_{NiO}}{L}$$

where

$$C_{NiO} = \frac{\rho}{M} = \frac{6.72 \text{ g/cm}^3}{74.692 \text{ g/mol}} = 9.00\times10^{-2} \text{mol/cm}^3 = C_{Ni^{2+}}$$

there is one mole of nickel ions in every mole of NiO. Integration gives the desired result of the oxide thickness as a function of time:

$$L^2 = -\frac{2D_{Ni^{2+}}\Delta G^o_{NiO}}{RT}t = kt \qquad (14.8)$$

again, the parabolic oxidation rate with the rate constant k. The negative sign remains because ΔG^0_{NiO} is negative and, in fact, $\Delta G^o_{NiO}(1200°C) = -107{,}878 \text{ J/mol}$ (Roine 2002).

14.2.2.3 Comparison to Experiment

Data on oxidation of nickel, a very oxidation resistant metal, are summarized in the literature (Kofstad 1988). From these data, it can be estimated that the weight gained by nickel during oxidation is $\Delta W \cong 6$ mg/cm^2 in 10 h at 1200°C. The weight gain is all oxygen so the total thickness of the oxide layer is

$$L = \frac{\Delta W(\text{g/cm}^2)}{\text{mol/cm}^3(A(O)\text{g/mol})}$$

$$L = \frac{6\times10^{-3}}{9\times10^{-2}(16)}$$

where A(O) is the atomic weight of oxygen, so

$$L = 4.17\times10^{-3}\text{cm}.$$

This oxidation was carried out in air so $p_{O_2} \cong 0.21$ bar, so the value for ΔG_{NiO} to be used in Equation 14.8 is

$$\Delta G_{NiO} = \Delta G^o_{NiO} - \frac{1}{2}RT\ln p_{O_2}$$

$$\Delta G_{NiO} = -107{,}878 - \frac{1}{2}(8.314)(1473)\ln(0.21)$$

$$\Delta G_{NiO} = -98{,}322 \text{ J/mole}$$

as shown in Figure 14.2. From Equation 14.8

$$L^2 = -\frac{2D_{Ni^{2+}}\Delta G_{NiO}}{RT}t$$

$$L^2 = -\frac{2(1.22\times10^{-10})(-98{,}322)}{(8.314)(1473)}(10\text{ h}\times3600\text{ s/h})$$

$$L^2 = 7.05\times10^{-5}\text{ cm}^2$$

or

$$L = 8.40 \times 10^{-3}\ \text{cm}$$

roughly about a factor of 2 greater than what is observed. This is close enough to validate the model given all the assumptions, and a factor of 2 error could easily be introduced by the uncertainties in the experimentally derived values: for example, value for the calculated diffusion coefficient can vary by a factor of 2 just by an error of 5% in the activation energy.

14.2.2.4 Other Considerations

How good was the assumption that the transport of electrons is much more rapid than that of the ions so that the transference number for electrons could be assumed to be $t_e \cong 1.0$? This can be checked by calculating the ionic conductivity—in this case, that of Ni^{2+}—with the Nernst–Einstein equation:

$$\sigma_{Ni^{2+}} = \frac{(2e)^2 \eta_{Ni^{2+}} D_{Ni^{2+}}}{k_B T} = \frac{(2\mathcal{F})^2 C_{Ni^{2+}} D_{Ni^{2+}}}{RT} \tag{14.9}$$

where the latter part of the equation is obtained by multiplying both the numerator and denominator by N_A^2 and compare it to that of the electronic conductivity which was found in the literature to be $\sigma \cong 1$ S/cm. Inserting values into Equation 14.9,

$$\sigma_{Ni^{2+}} = \frac{(2 \times 96472)^2 (9.00 \times 10^{-2})(1.22 \times 10^{-10})}{(8.314)(1473)}$$

$$\sigma_{Ni^{2+}} = 3.34 \times 10^{-5}\ \text{S/cm}.$$

The ionic conductivity is almost five orders of magnitude smaller than the electronic conductivity so the assumption that $t_{electronic} \cong 1$ is certainly justified.

One inevitable result of metal oxidation—such as that of nickel where the metal is diffusing through the oxide layer—is the disappearance of the original nickel: it is being used up. One way of looking at this process is illustrated in Figure 14.3. For every nickel atom that enters the oxide as a Ni^{2+} ion and two negative electrons, a vacancy in the nickel is created. This vacancy can be destroyed by diffusing to the end of a dislocation causing dislocation climb, which, in essence, is removing atomic planes from the nickel and replacing them with vacancies. For thin oxide layers on large pieces of metal, this is a satisfactory way of taking care of the volume of the metal that has gone to create the oxide. However, for smaller pieces of metal for which most of the metal ends up as oxide, then ultimately porosity will replace the volume of the metal that has been oxidized as illustrated in Figure 14.4a. If a small sphere is oxidized and the oxide cannot plastically deform as easily as the

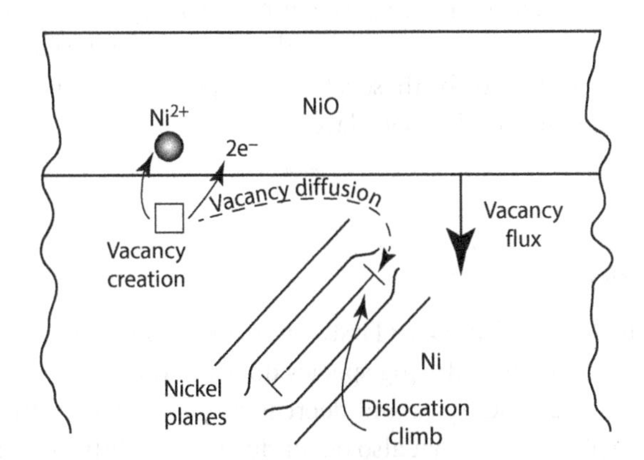

FIGURE 14.3 Schematic that shows the oxidation of a metal such as nickel generates vacancies in the metal that are destroyed at dislocations or at grain boundaries. This means that planes of metal are being removed from the metal and placed into the oxide.

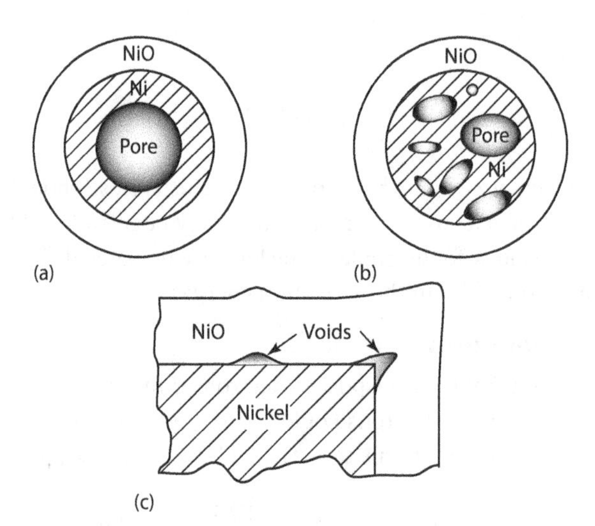

FIGURE 14.4 Mechanisms for the formation of porosity in the metal during oxidation. (a) If the metal and oxide form a completely coherent interface on a small particle, a single pore may form in the center that will grow to meet the oxide surface when the metal is completely oxidized. (b) A more realistic picture in which porosity forms at several points in the oxide. The final result is the same; an empty spherical shell of oxide will be what is left when the oxidation is over. (c) Because of the increase in the volume of the oxide compared to the metal being oxidized, even large pieces of metal can form pores at the metal–oxide interface due to the generation of stresses in the oxide. This is particularly likely at corners and edges.

metal, then a pore or pores will form in the metal. In Figure 14.4a, a single pore is formed in the center and when the oxidation is complete, all that will remain is the shell of oxide with an empty center. Figure 14.4b illustrates a more likely case in which pores form throughout the metal. However, if the oxidation is taken to completion, the net result is the same: the pores will coalesce into one single large pore leaving only the empty oxide shell. Sometimes this porosity is referred to as the "Kirkendall porosity." The Kirkendall porosity, discussed later, is caused by differences in diffusion coefficients in alloys. Naming the porosity that is the result of oxidation the same as that produced by interdiffusion is confusing at a minimum. The mechanisms of the porosity formation are really quite different even though they are both produced by differences in diffusion coefficients of two different atomic or ionic species. Calling this oxidation-induced porosity the "Kirkendall porosity" seems ill-advised.

Another way to form porosity is at the metal–oxide interface as shown in Figure 14.4c. An alternative way of tracing the disappearance of the nickel metal is that the metal–oxide interface simply moves with the disappearing metal. However, this requires that, for a finite-size piece of metal, the oxide must plastically deform to keep up with the shrinking nickel. In some cases, the compressive stresses on a plane surface may become large enough to cause the oxide layer to buckle and debond from the nickel and form a pore at the interface. At a corner, where even greater deformation of the oxide is necessary to take care of shrinking Ni–NiO interfaces, debonding and pores are even more likely to form. In both of these situations, pores form and oxidation stops at these points unless cracks actually form in the oxide layer.

14.3 OSMOSIS

14.3.1 Introduction

What has osmosis have to do with diffusion? First, it is an excellent example of interdiffusion and its consequences when one of the interdiffusing species does not diffuse or has a diffusion coefficient of zero. Also the consequences, increases and decreases in volumes in different regions, and the generation of pressure are phenomena that also occur during interdiffusion in solids. The advantage of examining the traditional process of osmosis is that it is a simple representation of what happens during interdiffusion in solids. Interestingly, Einstein attributed the gradient in osmotic pressure to the force causing Brownian motion and diffusion of particles and molecules, "… a dissolved molecule is differentiated from a suspended body *solely* (his italics) by its dimensions, and it is not apparent why a number of suspended particles should not produce the same osmotic pressure as the same number of molecules" (Einstein 1956, 3).

14.3.2 Definition

To borrow a definition from the literature (Chang 2000, 999),

> Osmosis is the net movement of solvent molecules through a semipermeable membrane from a pure solvent or from a dilute solution to a more concentrated solution.

14.3.3 Semipermeable Membrane

What is a *semipermeable membrane*? It is a thin sheet (usually) of material that has a very definite pore size, as shown in Figure 14.5, such that molecules and ions smaller than the pore size can easily pass through the membrane and anything larger than the pore size cannot. So it is permeable to some chemical species and not others, hence, *semipermeable.* Semipermeable membranes are commercially available with effective pore sizes from 0.1 nm to much larger. The permeability of a membrane is specified by its *molecular weight cutoff*, MWCO, expressed in *daltons* (Haney et al. 2013) where 1 dalton is the *unified atomic mass unit* equal to one-twelfth of a carbon-12 atom (Atomic mass unit). Frequently, these membranes are made from cellulose fibers (see Figure 14.6) and can be made with a wide range of MWCO from 1 to 20,000 daltons and have many applications including the dialysis of blood.

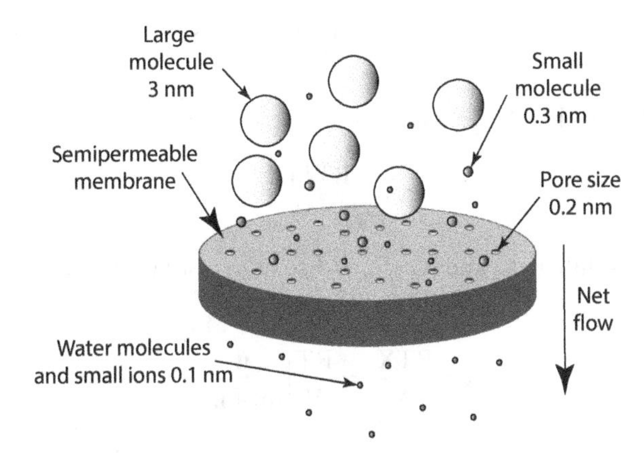

FIGURE 14.5 Schematic showing the operation of a semipermeable membrane in which atoms, ions, molecules, or particles greater than the pore size cannot go through the membrane while those smaller than the pore size can.

Cellulose

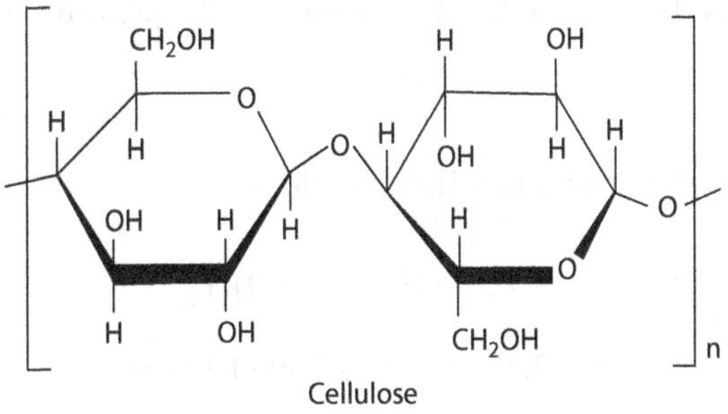

FIGURE 14.6 The structure of cellulose that is frequently the main chemical constituent of semipermeable membranes.

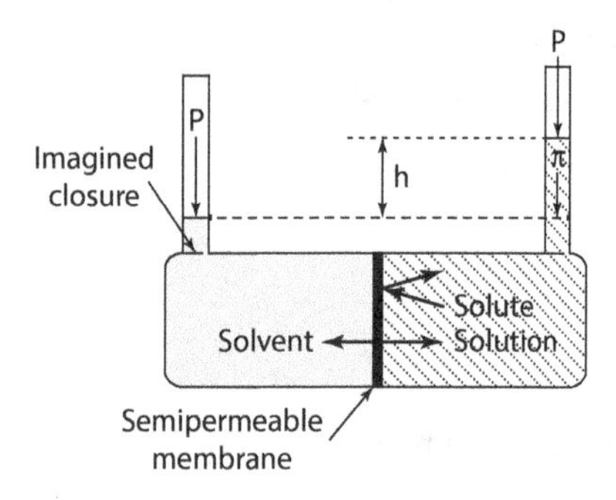

FIGURE 14.7 An apparatus for the measurement of osmotic pressure. Pure solvent is separated from the solution by a semipermeable membrane: permeable to the solvent but not the solute. As a result, solvent diffuses through the membrane into the solution until an osmotic pressure, π, builds up on the solution side of the membrane countering the entropic Gibbs energy between the solvent and the solution. If the vessel were sealed at the bottom of the vertical tubes as indicated, then the pressure would build up on the solution side. If the membrane could move, then it would move toward the faster-diffusing solvent side, a perfect example of marker motion during interdiffusion: the marker moves opposite to the direction of the flux of the faster diffusing species.

14.3.4 Thermodynamics

Figure 14.7 illustrates how the principle of osmosis and osmotic pressure, π, can be observed. Consider a system containing components 1 and 2. Pure solvent, $X_1 = 1$,* is separated from a solution of composition $X_1 < 1$ by a semipermeable membrane that is permeable to the solvent that can freely flow through the membrane in both directions but the solute in the solution cannot. In terms of diffusion, the solvent has a finite diffusion coefficient through the membrane while the diffusion coefficient of the solute through the membrane is zero. As a result, there is a Gibbs energy difference produced by this difference in concentration,

$$\Delta\bar{G} = \bar{G}_{solution} - \bar{G}_{solvent} = G^o_{solvent} + RT\ln X_1 - G^o_{solvent}$$
$$\Delta\bar{G} = RT\ln X_1. \tag{14.10}$$

With this Gibbs energy difference, solvent flows from the pure solvent to the solution increasing its volume. If the apparatus is constructed with a tube for the increase in volume of the solution, as shown in Figure 14.7, the increase in height of the solution in the tube will be such that the pressure that it exerts, π, will just balance the energy difference produced by the solution–solvent interaction. That is, because $d\bar{G} = \bar{V}dp - \bar{S}dT$ and at constant temperature, where $\bar{V}$ = molar volume of the pure solvent. Now, ΔG must be negative since $X_1 < 1$, so

$$\Delta\bar{G} = -\bar{V}\int_P^{P+\pi} dp = -\bar{V}\pi. \tag{14.11}$$

Equating Equations 14.10 and 14.11, gives

$$\pi = -\frac{RT\ln X_1}{\bar{V}}.$$

The mole fraction of the solute is X_2 and $X_1 + X_2 = 1$ so $\ln X_1 = \ln(1 - X_2) \cong -X_2$ because X_2 is usually small, so

$$\pi = \frac{RTX_2}{\bar{V}} = \frac{RT}{\bar{V}}\left(\frac{n_2}{n_1 + n_2}\right)$$

$$\pi \cong \frac{RTn_2}{\bar{V}n_1} = RTC_2 \tag{14.12}$$

where n_1 and n_2 are the number of moles of the solute and solvent, respectively, and $n_2 \ll n_1$ and $V \cong n_1\bar{V}$ is the total volume and C_2 is the concentration of solute in mol/m^3 or something similar. Equation 14.12 is known as the van't Hoff equation for osmotic pressure (Laidler and Meiser 1995).

14.3.5 Example

Suppose that $C_2 = 10^{-3}$ mol/L = 1 mol/m^3. Then, at 300 K,

$$\pi = RTC_2 = (8.314)(300)(1)$$
$$\pi = 2.49\times10^3\ \text{Pa} \cong 2.49\times10^{-2}\ \text{bar}$$

* X_i is the mole fraction of the component i, Chapter 1.

Because 1 atm = 760 mm Hg, if the solvent were water, which has a density of ρ = 1/g/cm^3 while that of mercury is ρ = 13.6 g/cm^3, 1 atm = 1.01325 bar = 1.03 × 10^3 cm H_2O. Or the height to which the water would rise in the right-hand vertical tube in Figure 14.7 would be $h = 1.03\times10^3\,(\text{cm/atm})\times1.01325(\text{atm/bar})\left(2.49\times10^{-2}\ \text{bar}\right) = 25.99\ \text{cm}!$ Note that $p = \rho gh$ so for 1 atm $p = \left(1\times10^3\ \text{kg/m}^3\right)\left(9.81\ \text{m/s}^2\right)\left(10.3\ \text{m}\right) = 1.01\times10^5$ Pa so that 1.03 × 10^3 cm of water is correct.

On the other hand, if the membrane was able to physically move laterally under the applied pressure, the osmotic pressure would cause the membrane in Figure 14.7 to move relative to the two ends of the water column, and keep moving until the pure solvent was gone. This membrane is acting as an *inert marker* relative to the diffusion fluxes of the solvent and the solute where one diffuses much faster than the other: the marker moves toward the faster diffusion species! How fast this marker moves depends on how fast the solvent diffuses through the membrane and into the solution.

14.3.6 Measuring Molecular Weights

Equation 14.12 could be written as

$$\pi = \frac{RTn_2}{\bar{V}n_1} = \frac{RT}{M}\left(\frac{m_2}{V}\right) \tag{14.13}$$

where:

m_2 is the mass of component two so m_2/V is the concentration in kg/m^3
M is the molecular weight.

So the osmotic pressure can be used to measure the molecular weight of a particular substance, including polymers. For example, suppose 1 g of poly(ethylene glycol) is dissolved in 1 L of water, so the concentration is 1 kg/m^3. If the water is raised to a height of 7.26 cm at equilibrium, the molecular weight of the polymer can be calculated from

$$M = \frac{RT}{\pi}\left(\frac{m_2}{V}\right) = \frac{(8.314)(300)}{\left(\frac{7.26}{1.03\times10^3}\text{atm}\left(1.01\times10^5\ \text{Pa/atm}\right)\right)}\left(1\ \text{kg/m}^3\right)$$

$$M = 3.503\ \text{kg/mole.}$$

14.3.7 Desalinization of Seawater

Now consider the case where a pressure, P, is applied to the solute side of Figure 14.7. From Section 14.3.5, if a pressure in excess 2.49 × 10^{-2} atm pressure is applied to the solution side of the membrane, then the pure solvent, water, is forced through the membrane from the solution side to the pure solvent side of the membrane. This process is not too surprisingly called *reverse osmosis*, and it is used to desalinate seawater and provide drinking water for regions near saltwater. However, seawater contains about 3.5 w/o soluble salts, with NaCl as the main constituent at 2.6 w/o (Hecht 1967). The concentration that must be entered into Equation 14.12 is the concentration of *all* of the ionic species in solution. So in a liter of solution there is roughly 35 g of salt. Because most of it is NaCl with M = 58.443 g/mol,

$$C_2 = \left(\frac{35\,\text{g}/1}{58.443\,\text{g/mole}}\times2\right)\times10^3\,1/\text{m}^3$$

$$C_2 = 1.20\times10^3\ \text{mole/m}^3$$

so from Equation 14.12, neglecting small changes in molar volume because of the solution,

$$\pi = RTC_2 = (8.314)(298)\left(1.20\times10^3\right)$$

$$\pi = 2.97\times10^6\,\text{Pa}$$

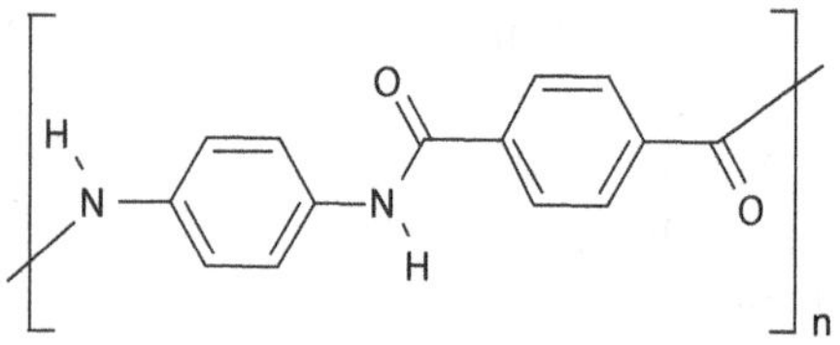

Aramid (aromatic polyamide)

FIGURE 14.8 Typical structure of aramid polymers that are used because of their high strengths for the semipermeable membranes in seawater desalinization plants.

or 2.97×10^6 Pa/1.013×10^5 Pa/atm $\times$ 14.7 psi/atm = 431 psi or 29.3 atm! The practical implication of this result is that desalination by reverse osmosis is an energy intensive process, and in general, industrial-scale desalination plants are almost always located close to a source of electrical generation. In addition, the high pressures necessary to purify seawater by reverse osmosis requires that the membrane materials be quite strong. One of the materials used is made from aramid—poly(aromatic imide)—(see Figure 14.8), fibers, the same family of polymer fibers as the Kevlar™ fibers used in bullet-proof vests. Of course, as the water is removed from the seawater the concentration of dissolved salts increases and the osmotic pressure increases as well. As a result, not only does the reverse osmosis equipment need to be robust enough to withstand the pressures, but also a significant amount of pressure–volume work must be expended to produce fresh water from seawater. Nevertheless, the process is energy-competitive with other desalinization processes such as evaporation and condensation of the water from the sea water salt solution and is widely used throughout the world where freshwater is scarce.[*]

14.3.8 Noteworthy

As mentioned above, the main reason for introducing osmotic pressure and its effects is that some of the phenomena associated with osmosis are very similar—if not identical—to processes that occur during interdiffusion in solids, particularly where the diffusion coefficients of the interdiffusing species are different. For osmosis, the model is that that solvent can diffuse quite readily through the semipermeable membrane while the molecules, ions, or particles (Einstein 1956) cannot: so their diffusion coefficients through the membrane are zero. As a result, the faster (only) diffusing species, the solvent, diffuses through the membrane and *increases the volume* of the solution phase and *depletes the volume* of the solvent on the pure solvent side of the membrane, Figure 14.7. Eventually equilibrium will be reached when the osmotic pressure calculated in Equation 14.12 is reached. By the way, if the two chambers on either side of the membrane in Figure 14.7 are on the order of centimeters in dimensions, because the diffusion coefficient of water and salts in water are on the order of 10^{-5} cm^2 s and $x^2 \cong 4Dt$, times for equilibration will be on the order of 3–8 h. Nevertheless, when equilibrium is reached, the volume on the solution side will have increased considerably while that on the solvent side will have decreased.

Consider the following thought experiment. Suppose both chambers on either side of the membrane were sealed at the intersection of the vertical tubes and chambers and not open to the atmosphere (imagined closure in Figure 14.7). Then, the solvent would move into the solution until there would be a positive pressure in the solution chamber and equal and *negative* pressure on the pure solvent phase. The sum of the absolute values of the pressures would equal the osmotic pressure for the open system. Stated another way, if the vessel were an elastomer—or some other material that would allow free expansion and contraction—then the solution side would expand and the pure solvent side would shrink.

As discussed above, "Suppose that the membrane could move along the tube"? In that case, the membrane—a *marker* indicating macroscopic changes is volume—would move toward the pure solvent side under the increased pressure on the solution side because of the faster diffusion of the

[*] Climate change appears to be generating severe drought conditions throughout many parts of the world and societies may be forced to produce fresh water through desalination of sea water—an energy intensive process regardless of the technology used. This creates a serious dilemma! Unless the desalinization plant is powered with renewable energy, any other source of power will increase atmospheric levels of greenhouse gases that will exacerbate climate change.

solvent into the solution. These same processes occur in solids in which the diffusion coefficients of the interdiffusing species are different.

14.4 INTERDIFFUSION IN METALS AND THE KIRKENDALL EFFECT

14.4.1 Introduction

Interdiffusion in metals and the resulting microstructure effects are extremely important in a number of metallurgical processes, welding for example. A rather interesting and historically important rationale for exploring this topic is that it demonstrates how a single set of experiments can produce a paradigm shift in the understanding of an important area of materials and can be considered *transformational research.* A second reason is that the marker motion and the Kirkendall effect are not isolated to only metals. Finally, how interdiffusion influences the microstructure and behavior of a material or component is illustrated with an industrial application that had significant commercial implications.

14.4.2 History

The history of trying to understand the diffusion of atoms or ions in solids is well over one hundred years old (Tuijn 1997; Cahn 2001). During the 1920s and 1930s, a great deal of effort was made to understand the transport properties—primarily electrical—of ionic and other compounds (Hoddeson et al. 1992) and the concepts of vacancies, Schottky defects, Frenkel defects, and interstitial and vacancy diffusion in compounds were reasonably well-understood by 1940 (Mott and Gurney 1940). In fact, Mott and Gurney calculated energies of vacancy formation from first principles—what today would be called *computational materials science*—probably assisted by only a pencil and paper, and obtained values reasonably close to those existing today. Prior to the 1940s, the preferred mechanisms to explain diffusion of atoms or ions on a crystal lattice were the direct exchange or ring mechanisms, as sketched in Figure 14.9. During the 1940s, a critical series of experiments carried out by Kirkendall and his students (Kirkendall 1942; Smigelskas and Kirkendall 1947) strongly suggested that a vacancy mechanism was responsible for diffusion in metals. After some skepticism-generated spirited discussion, the metallurgical community became convinced of the reality of the vacancy mechanism of diffusion (Nakajima 1997; Sequeira and Amaral 2014). The most convincing set of experiments was the observation of inert marker—Mo wires—movement during interdiffusion between alpha brass (70 w/o Cu–30 w/o Zn) and pure copper that form a continuous solid solution. The markers moved inward toward the brass—markers on opposite side of the brass moved closer together (see Figure 14.10). This marker motion was explained by the more rapid diffusion of zinc compared to copper in the brass, and confirmed by composition measurements across the diffusion couple (Kirkendall 1942; Smigelskas and Kirkendall 1947). If one of the atoms in a substitutional alloy moved faster than the other, this would negate the possibility of the ring and exchange mechanisms that predict the same diffusion coefficient for all constituent atoms. The conclusive paper was presented at a meeting and the comments of the audience are included with the printed paper (Smigelskas and Kirkendall 1947), the most important of which is by Robert F. Mehl a distinguished metallurgist and then head of the Department of Metallurgical Engineering at Carnegie Institute of Technology (now Carnegie

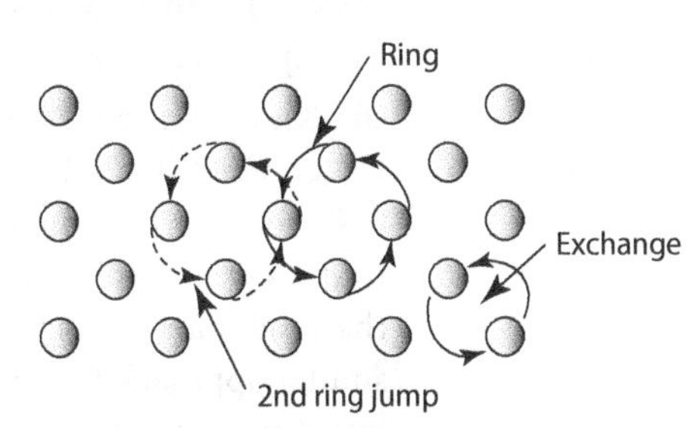

FIGURE 14.9 Ring and direct exchange mechanisms of diffusion that were thought to be the main mechanisms of diffusion in metals before the Kirkendall effect was observed. The Kirkendall effect demonstrated that different atoms could diffuse at different rates while the above two mechanisms require that different atoms must diffuse at the same rate.

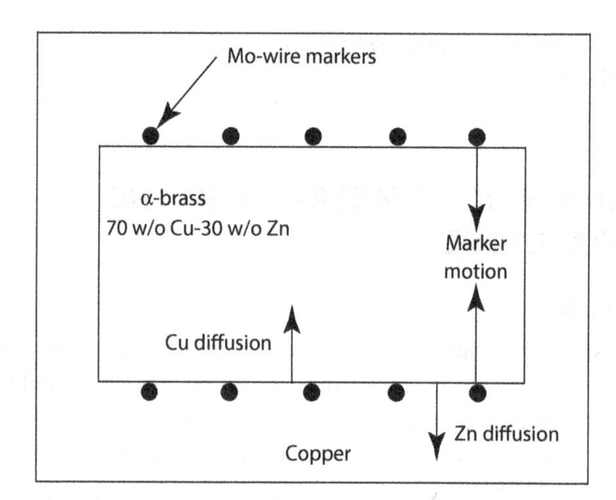

FIGURE 14.10 The classic Kirkendall experiment in which Mo-marker wires were wound around a piece of alpha brass and then plated with pure copper. During heat treatment, the markers moved vertically toward each other showing that zinc diffused faster than copper.

Mellon University) who has a TMS* award named for him and vocal critic of Kirdendall's early results (Smigelskas and Kirkendall 1947, discussion following, 135). Mehl observed,

> ...If atoms in substitutional solid solutions are viewed as diffusing by simple atom interchange, however—perhaps the most likely mechanism—the assumption is obviously justified. Dr. Kirkendall's work calls this assumption into question. If verified, this "Kirkendall effect" would greatly modify not only the treatment of diffusion data but also the theory of the mechanism of diffusion.

About the same time, marker motion was observed in polymers during swelling in a solvent (Hartley 1946) that demonstrated the unequal diffusion rate between polymer and solvent molecules. Theoretical calculations—computational materials science—for the possible mechanisms of diffusion in copper also concluded that the vacancy mechanism was energetically the most favorable diffusion mechanism (Huntington and Seitz 1942). So, by the end of the 1940s, the vacancy mechanism of diffusion in metals had become generally accepted: the convincing observation was marker motion during interdiffusion.

14.4.3 Darken's Analysis

14.4.3.1 Introduction

In 1948, Lawrence Darken from the U.S. Steel Research Laboratory presented an analysis of the Kirkendall effect (Darken 1948) that is often reproduced in texts on diffusion and kinetics (cf. Glicksman 2000; Porter et al. 2009). The goal of Darken's paper was essentially twofold: first, to treat one-dimensional interdiffusion of two components relative to a fixed reference frame and to explain the observation of Kirkendall's marker movement; and second, to include thermodynamic terms in diffusivities. About this same time, Hartley and Crank (1949) used a similar approach to explain Hartley's earlier results on marker motion and swelling of polymers by solvents (Hartley 1946).

Figure 14.11 shows the results of the Kirkendall experiment with two completely isomorphous—complete solid solution, same crystal structures—metals, A and B with inert markers located at the original interface. The initial position of the markers is x_0 and after some time, t, at elevated

* TMS is The Minerals, Metals, and Materials Society.

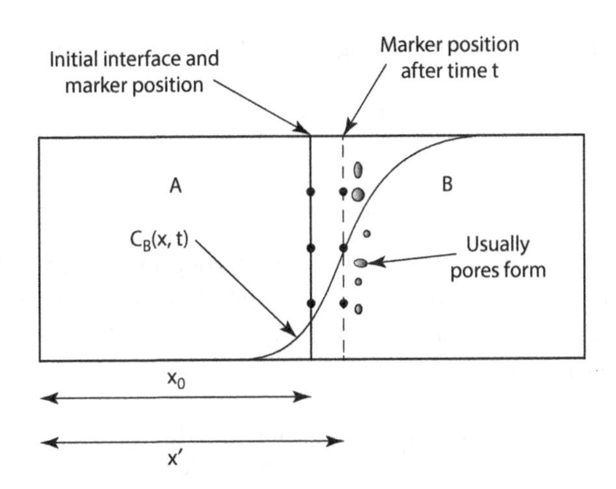

FIGURE 14.11 Results of the Kirkendall experiment. The inert markers initially at the interface between A and B at position x_0 after a given heat treatment time have moved to position x′ relative to the end of the diffusion couple. This indicates that B diffuses faster in A than A in B. As a result, there is a net vacancy flux from A to B that can become supersaturated and precipitate as pores on the B side of the interface.

temperatures A and B have diffused and the markers have moved to position x′. The concentration profile after time t is also shown. It is observed that the concentration at the maker position typically remains constant. In effect, what is happening is that B is diffusing into A faster than A into B. As a result, it is very similar to what happens during osmosis, only in this case, extra planes of alloy are being created to the left of the markers and planes are being destroyed to the right of the markers in Figure 14.11 forcing the markers to move to right. These planes are being created by dislocation climb so vacancies are being created to the left of the markers and being destroyed on the right, both by dislocation climb. As a result, there is a net flux of vacancies from the left to the right compensating for the excess motion of B atoms from the right to the left. There are two additional results of this imbalance in the atom flux. Because extra planes are being squeezed into the lattice to the left of the markers, this region is under compression. Conversely, atom planes are being removed to the right of the markers, so this region is under tension. Finally, porosity is usually found on the side of the markers of the faster diffusing atomic species; the right side of the markers in Figure 14.11. This porosity is generally thought to be generated by supersaturation of vacancies produced by the vacancy flux from the left to the right (Seitz 1953).

14.4.3.2 Marker Motion and Fluxes

From Figure 14.11, the average marker velocity (in the x-direction of course) is simply,

$$v = \frac{x' - x_0}{t}$$

and now diffusion and fluxes must be considered relative to this frame of reference moving with velocity v. The total concentration is assumed to be constant, C_T, and

$$C_T = C_A + C_B \tag{14.14}$$

so

$$\frac{dC_A}{dx} = -\frac{dC_B}{dx}. \tag{14.15}$$

The *diffusion* fluxes with respect to the moving frame of reference, the markers,

$$J_A = -D_A \frac{dC_A}{dx}$$
$$J_B = -D_B \frac{dC_B}{dx} \tag{14.16}$$

where D_A and D_B are called the *intrinsic diffusivities* (Hartley and Crank 1949; Darken and Gurry 1953; Porter et al. 2009). Presumably, they are called *intrinsic* because they are diffusion coefficients relative to a frame of reference moving with the lattice, in case of crystalline materials. The net diffusion flux relative to the markers is

$$J_{net} = J_A + J_B = -\frac{1}{A} C_T \frac{dV}{dt} = -\frac{1}{A} C_T A \frac{dx}{dt}$$
$$J_{net} - C_T v \tag{14.17}$$

the negative sign occurs because the net flux of material to the left forces the markers to move to the right and vice versa: v being the marker velocity. Putting Equations 14.15 through 14.17 together,

$$J_{net} = -C_T v = J_A + J_B = -D_A \frac{dC_A}{dx} - D_B \frac{dC_B}{dx}$$

$$J_{net} = -C_T v = -(D_A - D_B) \frac{dC_A}{dx}$$

or

$$v = \frac{1}{C_T}(D_A - D_B) \frac{dC_A}{dx}$$
$$v = (D_A - D_B) \frac{dX_A}{dx} \tag{14.18}$$

because the mole fraction of A, $X_A = C_A/C_T$. Equation 14.18 is also known as *Darken's first equation.* Also note that the net flux of vacancies, J_v, is just opposite to the net flux of atoms so,

$$J_v = -J_{net} = C_T v = (D_A - D_B) \frac{dC_A}{dx}. \tag{14.19}$$

Figure 14.12 shows these flux relationships as well as the stresses, marker velocity, and vacancy destruction rate, which must go to zero at both ends of the interdiffusion zone.

14.4.3.3 Diffusion Relative to a Fixed Reference Frame

Because the diffusion zone is usually a small fraction of the combined length of A and B, at the ends of the diffusion couple, say x = 0, $dX_A/dx \to 0$ as does the velocity, v (see Figure 14.12). Therefore, it makes sense to measure diffusion relative to the ends of the diffusion couple. However, then in addition to the diffusion fluxes there is the lattice flow of velocity v—a convection term—in the interdiffusion zone. Therefore, the flux of A relative to the fixed reference frame, $\tilde{J}$,

$$\tilde{J}_A = -D_A \frac{dC_A}{dx} + C_A v \tag{14.20}$$

and substituting Equation 14.18 for the velocity,

$$\tilde{J}_A = -D_A \frac{dC_A}{dx} + C_A v = -D_A \frac{dC_A}{dx} + C_A (D_A - D_B) \frac{dX_A}{dx} \tag{14.21}$$

and now a little algebra with Equation 14.21 is necessary. Remembering that C_T is constant,

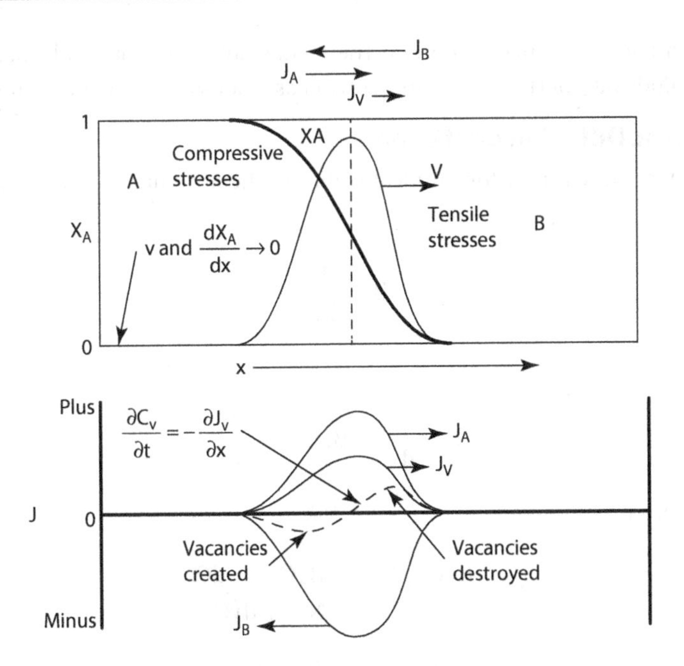

FIGURE 14.12 Further results of the Kirkendall effect as analyzed by Darken. The top figure shows the concentration profile, the marker velocity—that goes to zero at the end of the diffusion couple—and the fluxes of A, B, and vacancies. The bottom figure shows the variation of the three fluxes as well as the rate vacancy creation and destruction as a function of position relative to the concentration profile.

$$\tilde{J}_A = -D_A C_T \frac{d C_A / C_T}{dx} + C_A \frac{C_T}{C_T}(D_A - D_B)\frac{dX_A}{dx}$$

$$\tilde{J}_A = -C_T \left\{ D_A \frac{dX_A}{dx} - X_A D_A \frac{dX_A}{dx} + X_A D_B \frac{dX_A}{dx} \right\}$$

and, of course, $X_A = 1 - X_B$, in the above equation,

$$\tilde{J}_A = -C_T \left\{ D_A \frac{dX_A}{dx} - D_A \frac{dX_A}{dx} + X_B D_A \frac{dX_A}{dx} + X_A D_B \frac{dX_A}{dx} \right\}$$

$$\tilde{J}_A = -C_T \left\{ X_B D_A \frac{dX_A}{dx} + X_A D_B \frac{dX_A}{dx} \right\}$$

$$\tilde{J}_A = -(X_B D_A + X_A D_B)\frac{dC_A}{dx}$$

where the last equation can be written as

$$\tilde{J}_A = -\tilde{D}\frac{dC_A}{dx} \tag{14.22}$$

with $\tilde{D}$ given by

$$\tilde{D} = X_B D_A + X_A D_B \tag{14.23}$$

and Equations 14.22 and 14.23 constitute *Darken's second equation*. This expression for $\tilde{D}$ is the diffusion coefficient that goes into Fick's second law that determines the rate of concentration change with time; specifically,

$$\frac{\partial C_A}{\partial t} = \frac{\partial}{\partial x}\tilde{D}\frac{\partial C_A}{\partial x}. \tag{14.24}$$

Clearly, $\tilde{D}$ is not a constant. Furthermore, if the solid solution is not ideal, then D_A and D_B are a function of the nonideality of the system in question as discussed in the following.

14.4.3.4 Intrinsic Diffusion Coefficients

In Chapter 13, it was shown that for an ideal solution, the absolute mobility, B, and the diffusion coefficient were related by

$$B_i = \frac{D_i}{k_B T}.$$

In general,

$$J_A = -D_A \frac{dC_A}{dx} = B_A C_A \left(-\frac{1}{N_A} \frac{d\bar{G}_A}{dx} \right)$$

which can be rewritten as

$$D_A = \frac{B_A C_A}{N_A} \frac{d\bar{G}_A}{dC_A} = \frac{B_A C_A}{N_A C_T} \frac{d\bar{G}_A}{d(C_A/C_T)}$$

$$D_A = \frac{B_A X_A}{N_A} \frac{d\bar{G}_A}{dX_A} = \frac{B_A}{N_A} \frac{d\bar{G}_A}{d\ln X_A}. \tag{14.25}$$

By definition, $d\bar{G}_A = RTd\ln a_A$, where a is the thermodynamic activity of A, and $a = \gamma_A X_A$, where γ_A = the *activity coefficient*. So,

$$\frac{d\bar{G}_A}{d\ln X_A} = \frac{RTd\ln a_A}{d\ln X_A} = \frac{RTd\ln(X_A \gamma_A)}{d\ln X_A}$$

$$\frac{d\bar{G}_A}{d\ln X_A} = RT\left(\frac{d\ln X_A}{d\ln X_A} + \frac{d\ln \gamma_A}{d\ln X_A} \right)$$

$$\frac{d\bar{G}_A}{d\ln X_A} = RT\left(1 + X_A \frac{d\ln \gamma_A}{dX_A} \right).$$

With the result that

$$D_A = B_A k_B T\left(1 + X_A \frac{d\ln \gamma_A}{dX_A} \right) \tag{14.26}$$

and similarly,

$$D_B = B_B k_B T\left(1 + X_B \frac{d\ln \gamma_B}{dX_B} \right). \tag{14.27}$$

and are Darken's *intrinsic* diffusion coefficients that include thermodynamic nonideality. The term in the brackets is called the *thermodynamic factor*. Of course, if the activity coefficient is constant or unity, these equations revert to Einstein's relation between B and D. Applying some additional thermodynamics, specifically, the *Gibbs–Duhem* equation (Chapter 1),

$$X_A d\bar{G}_A + X_B d\bar{G}_B = 0$$

$$X_A d\ln(X_A \gamma_A) + X_B d\ln(X_B \gamma_B) = 0$$

$$X_A d\ln X_A + X_A d\ln \gamma_A + X_B d\ln X_B + X_B d\ln \gamma_B$$

$$dX_A + X_A d\ln \gamma_A + dX_B + X_B d\ln \gamma_B = 0$$

and, of course, $X_A + X_B = 1$ so $dX_A = -dX_B$; inserting these into the last equation,

$$X_A \frac{d\ln\gamma_A}{dX_A} = X_B \frac{d\ln\gamma_B}{dX_B} \quad (14.28)$$

which shows that the thermodynamic factor is the same for both D_A and D_B.

To get something measureable, values of B_i in Equations 14.26 and 14.27 need to be defined in terms of measureable quantities, which are the tracer* diffusion coefficients, D_A^* and D_B^*. Now for tracers, with the exception of small mass differences between them and the major atomic isotope, there should be no chemical differences so $\gamma_A = \gamma_B = 1$; plus the fact that they are typically used in small concentrations so that γ_i is essentially constant. As a result,

$$D_i^* = k_B T B_i$$

that is, there is no thermodynamic term involved for the tracer diffusion coefficient. As a result, Equations 14.26 and 14.27 immediately become, because of Equation 14.28,

$$D_A = D_A^*\left(1 + X_B \frac{d\ln\gamma_B}{dX_B}\right)$$

$$D_B = D_B^*\left(1 + X_B \frac{d\ln\gamma_B}{dX_B}\right) \quad (14.29)$$

and substitution of Equations 14.29 into Equation 14.23 gives Darken's final result for the interdiffusion coefficient,

$$\tilde{D} = \left(X_B D_A^* + X_A D_B^*\right)\left(1 + X_B \frac{d\ln\gamma_B}{dX_B}\right) \quad (14.30)$$

which is the diffusion coefficient to be substituted into Fick's second law for interdiffusion in a nonideal binary metallic system:

$$\frac{\partial C_A}{\partial t} = \frac{\partial}{\partial x}\tilde{D}\frac{\partial C_A}{\partial x}. \quad (14.31)$$

As Darken points out in his paper, and needs to be stressed because it is not always obvious in the literature, the tracer diffusion coefficients D_A^* and D_B^* *are not the tracer diffusion coefficient in pure A and B*. Darken refers to the D_i^* as *self-diffusivities* which seems to connote tracer diffusion coefficients in the pure material which *they are not* and this term is somewhat misleading. They are the tracer diffusion coefficients of *A* and *B in the alloy of composition X_A and X_B*. As a result, the amount of experimental data necessary to calculate $\tilde{D}$ and $C_A(x,t)$ as a function of composition is rather large. In contrast, $\tilde{D}$ could be determined by a numerical integration of the diffusion profile, determined by Equation 14.31, and compared to an experimental profile. Unfortunately, three parameters need to be determined: D_A^*, D_B^*, and $(d\ln\gamma_B/dX_B)$ as a function of X_B.

14.4.4 Regular Solution Example

14.4.4.1 Diffusion Coefficients

To make the discussion in Section 14.4.3 somewhat more concrete, consider interdiffusion of rhodium and copper (Rh–Cu), for reasons that will become clearer later when the results of the interdiffusion are examined. It should be noted, that this is not a system in which complete data actually exist as a function of composition, but estimates can be made from the limited data that do exist. Table 14.1 lists some properties of interest for rhodium and copper. The phase diagram for Rh–Cu shows phase separation at about 1150°C but with extensive terminal

* A *tracer* is a radioactive isotope of the element in question that can be used to measure the motion—or *trace* the motion—of the nonradioactive element by means of emitted radiation from the radioactive isotope: Chapters 2 and 11.

TABLE 14.1
Parameters for Copper and Rhodium

Parameter	Rhodium	Copper
Melting point (°C)	1962	1083
Crystal structure	FCC (Cu)	FCC (Cu)
Lattice parameter (pm)	380.3	361.5
Thermal expansion (°C^{-1})	8.5×10^{-6}	17.0×10^{-6}
Thermal conductivity (W/m K)	148	397
Atomic weight (g/mol)	102.908	63.546
Density (g/cm^3)	12.4	8.96
Molar volume (cm^3/mol)	8.30	7.09
Diffusion Data		
Cu in pure Cu	Q = 197.8 kJ/mol	D_0 = 0.13 cm^2/s
Rh (trace) in copper	Q = 242.8 kJ/mol	D_0 = 3.3 cm^2/s
Rh in pure Rh*	3.0×10^{-15} cm^2/s	
Cu (trace) in Rh*	1.0×10^{-15} cm^2/s	

Source: Brandes, E. A. and G. B. Brook, eds. 1992. *Smithells Metals Reference Book, 7th ed.* Oxford: Butterworth-Heinemann.
*Estimated based on data for Ir and Pt at 927°C.

solid solution of Cu in Rh and Rh in Cu. (Chakrabarti and Laughlin 1982). Because this is simply an example, assume complete solid solubility at T = 927°C = 1100 K. For Cu in pure Cu at 1100 K, $D = D_0 \exp(-Q/RT) = 0.13 \times \exp(-(197800/(8.314)(1100)) = 5.3 \times 10^{-11}$ cm^2/s and for Rh in Cu $D = D_0 \exp(-Q/RT) = 3.3 \times \exp(-(242800/(8.314)(1100)) = 9.7 \times 10^{-12}$ cm^2/s and for Rh in pure Rh $D \cong 1.0 \times 10^{-15}$ cm^2/s and for Cu in Rh $D \cong 3.0 \times 10^{-15}$ cm^2/s (Brandes and Brook 1992). Call $X_{Cu} = X_A$ and $X_{Rh} = X_B$. Again for concreteness, *assume* that the *self-diffusion coefficient* of Rh is $D^*_{Rh} = 9.7 \times 10^{-12} X_A^2 + 10^{-15}$ cm^2/s and that for copper is $D^*_{Cu} = 5.3 \times 10^{-11} X_A^2 + 3 \times 10^{-15}$ cm^2/s. How these diffusion coefficients actually vary with composition is not known but certainly the diffusion coefficients in almost pure rhodium will be much lower than they are in pure copper. So such a functional variation is not unreasonable.

14.4.4.2 Thermodynamic Factor

If a *regular solution* is assumed, then γ_A, where A refers to copper and B refers to Rh, can be related to the composition by (Darken and Gurry 1953)

$$\ln \gamma_A = \alpha X_B^2 = \alpha (1 - X_A)^2 \tag{14.32}$$

so

$$\frac{d \ln \gamma_A}{dX_A} = -2\alpha (1 - X_A) = -2\alpha X_B$$

and

$$X_A \frac{d \ln \gamma_A}{dX_A} = -2\alpha X_A X_B$$

and the thermodynamic factor becomes

$$\left(1 + X_A \frac{d \ln \gamma_A}{dX_A}\right) = (1 - 2\alpha X_A X_B). \tag{14.33}$$

Choosing $\alpha = 1.5$* leads the values of the activities—$a = \gamma X$—for the two components, A and B, and the deviation from ideality in Figure 14.13. Notice that the activities approach the mole fractions,

* Two phases occur for $\alpha = 2$ but not in Cu-Rh so the choice of $\alpha = 1.5$ seems reasonable (Darken and Gurry 1953).

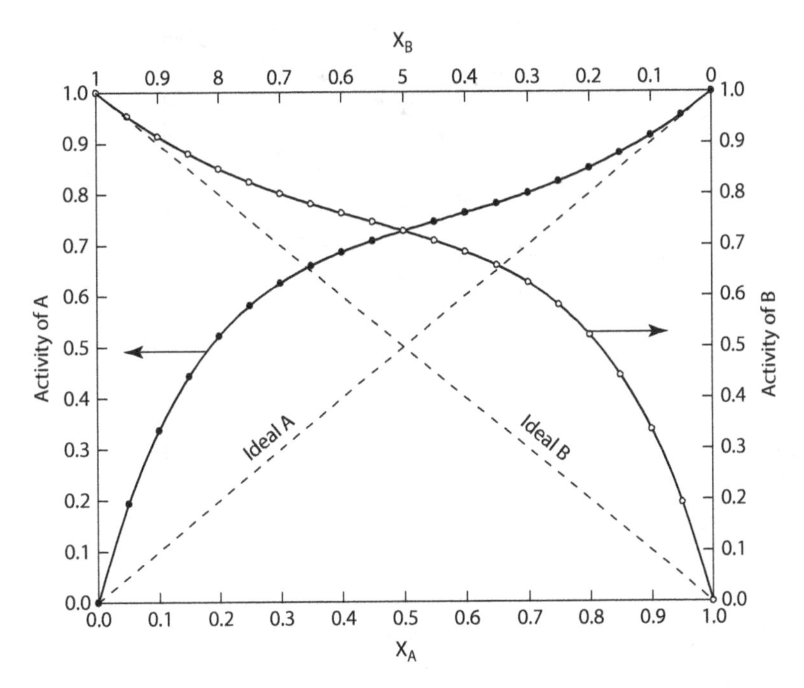

FIGURE 14.13 The deviations from ideal solution behavior—dashed lines—of A and B where $\ln \gamma_A = \alpha X_B^2 = \alpha(1-X_A)^2$ and a similar result for B with $\alpha = 1.5$: a *regular* solution.

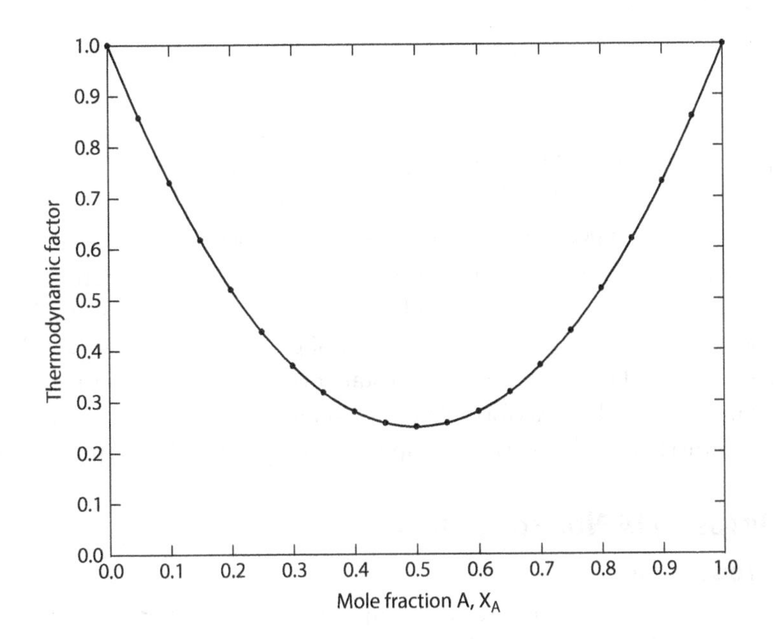

FIGURE 14.14 The *thermodynamic factor* in Darken's interdiffusion equation for the regular solution of Figure 14.13 $\left(1+X_A(d\ln\gamma_A / dX_A)\right)=\left(1-2\alpha X_A X_B\right)$.

Rauolt's law* $a_i \to X_i$ as $X_i \to 1.0$ and that $a_i = kX_i$ as $X_i \to 0$, *Henry's law*,† as expected. (Darken and Gurry 1953). The thermodynamic factor, Equation 14.33 is plotted in Figure 14.14 with $\alpha = 1.5$.

14.4.4.3 Interdiffusion Coefficient

Figure 14.15 gives the diffusion coefficients of interest as a function of the mole fraction of A, X_A. The most interesting thing about this plot is that for mole fractions of A from about 0.15 to 0.8, the

* Chapter 1.

† By definition and as experimentally determined.

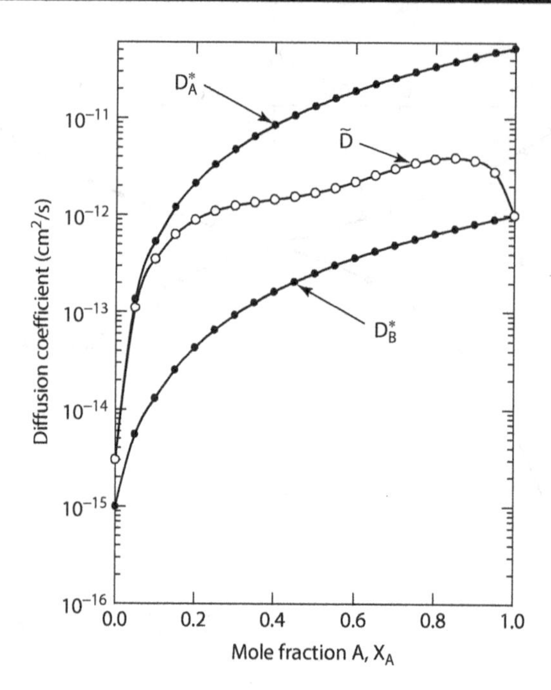

FIGURE 14.15 The tracer diffusion coefficients of A and B and Darken's interdiffusion coefficient based on the data for copper and rhodium in Table 14.1 and some assumptions about how the tracer diffusion coefficients vary with composition and the thermodynamic factor from Figure 14.14. Specifically, $\tilde{D} = \left(X_B D_A^* + X_A D_B^*\right)\left(1 + X_B\left(d\ln\gamma_B / dX_B\right)\right)$.

interdiffusion coefficient, $\tilde{D} \cong 2(\pm 2)\times 10^{-12}$ cm^2/s! As a result, over this range in concentration a *constant diffusion coefficient* could be used without a significant error. However, this depends on what the data are intended for. Also, this result raises several questions: (1) "How important is the thermodynamic factor"? It appears that for the parameters chosen for this calculation it contributes about a factor of three. (2) "What if the D_i^* diffusion coefficients varied differently"? On the other hand, it is expected that they will vary in a similar way with the mole fraction X_A. With three independent variables, there would appear to be a number of possibilities and the fact that the interdiffusion coefficient calculated here is reasonably constant over a large concentration range may simply be an artifact of how the problem was formulated. Therefore, it is important to investigate other features of the interdiffusion model to ensure the approximations are indeed representative of reality.

14.4.5 Diffusion In Nonisomorphous Systems

14.4.5.1 Introduction

The previous discussion focused on diffusion in an isomorphous system, one that has a complete solid solution from A to B requiring that A and B have the same crystal structure and therefore are *isomorphous*. In general, binary systems are generally not isomorphous. Usually, in a binary system as represented by a binary phase diagram, there are more phases than a single solid phase extending from pure A to pure B. Figure 14.16 is a typical example that shows a singular *intermetallic phase*, gamma, γ, in addition to the two terminal solid solutions alpha, α, and beta, β. If a diffusion couple is formed between one mole of A and one mole of B at some temperature T_0 in Figure 14.16, and diffusion is allowed to take place for some time but not so long that A and B completely react, times less than infinity, *all of the phases* in the phase diagram *should* be present. However, the concentration range for one of the phases might be too narrow to be observable or, alternatively, is that a phase cannot nucleate. So, perhaps not all of the phases in the phase diagram are present or obvious in the diffusion couple. For the phase diagram in Figure 14.16, it is assumed that all three

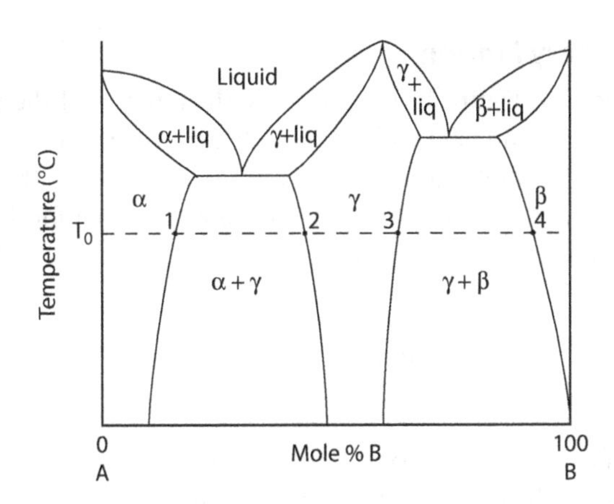

FIGURE 14.16 Hypothetical phase diagram showing two terminal solid solutions, alpha and beta, and an intermediate solid solution, gamma. At some temperature T_0, compositions 1 and 2 are in equilibrium at the interface between alpha and gamma. Similarly, at T_0, compositions 3 and 4 are at equilibrium at the interface between gamma and beta.

phases are present, alpha, gamma, and beta. As the diagram is drawn, if A and B were allowed to completely interdiffuse—time goes to infinity—only the gamma phase would remain at 50 mole % of both A and B.

14.4.5.2 Concentrations and Thermodynamic Activities

Figure 14.17 shows the phase structure of the A-B diffusion couple after some diffusion time. This figure shows four important points that are not always fully appreciated and need to be emphasized. First, as mentioned above, during interdiffusion between A and B, all of the phases in the phase diagram appear in the microstructure.* Second, there are *no two-phase regions* in the microstructure. Two-phase regions occur because the Gibbs energy for two phases is less than that of either of the individual phases over a range of composition and temperature that define the two-phase region: Chapter 1. However, in the *microstructure*, the only places where the phases are in equilibrium are at *the two-phase interfaces*: alpha–gamma and gamma–beta in the top part of Figure 14.17. Third, at these interfaces, 1–2 at the alpha–gamma interface and 3–4 at the gamma–beta interface, the compositions at the interfaces are given by the intersections of the T_0 tie-line in Figure 14.16 with the solvus curves at the points 1, 2, 3, and 4. Consequently, at the two interfaces there are *discontinuities* in the composition–distance profiles: specifically, 1–2 at the alpha–gamma interface and 3–4 at the gamma-beta interface. Fourth, in contrast to composition, the *thermodynamic activity is continuous* across the entire diffusion couple, as shown at the bottom of Figure 14.17. As a result, the activity of *A*—or *B*—is the same in each phase *at the phase interfaces.*

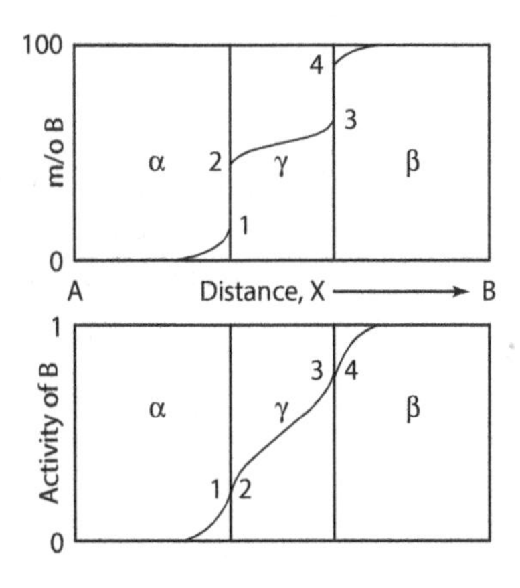

FIGURE 14.17 Diffusion couple between pure A and pure B at some $t < \infty$ where all three phases are present showing the concentration profile of B in the three phases. In the top figure, as pointed out in Figure 14.16, at the interface between alpha and gamma, the equilibrium compositions between the two phases at the interface are 1 and 2. Similarly, 3 and 4 are the equilibrium interfacial concentrations of B between gamma and beta. Note that there are *no two-phase regions* as there are in the phase diagram. The bottom figure shows how the thermodynamic activity of B varies with position and composition. In this case, the activities of B at the two interfaces are the same in both phases comprising each interface because they are at equilibrium at the interfaces so the activities must be the same.

* This is a broad use of the term "microstructure" because it may not be necessary to use a microscope to see the three phases. However, for most crystalline solids, the diffusion coefficients are sufficiently small so that the width or lateral extent of each of the phases is small—micrometers—requiring magnification.

14.4.5.3 Phase Boundary Motion

As interdiffusion of A and B in Figure 14.17 continues, the amount of the gamma phase increases and the amounts of both alpha and beta will decrease. This can only occur by the movement of the alpha–gamma and gamma–beta interfaces. To see what the factors are that determine the motion of these boundaries, consider Figure 14.18 (Porter et al. 2009). In order for the $\alpha-\gamma$ interface to move an amount dx to the left, an amount of B equal to

$$dx\left(C_{B,int}^{\gamma}-C_{B,int}^{\alpha}\right) \tag{14.34}$$

must move into the interface. There is a flux of B to the interface in gamma and a flux away from the interface in alpha. In time dt, the net amount flowing into the interface is

$$\left(J_{\gamma}-J_{\alpha}\right)dt=\left(\left[-\tilde{D}_{\gamma}\frac{dC_{B,int}^{\gamma}}{dx}\right]-\left[-\tilde{D}_{\alpha}\frac{dC_{B,int}^{\alpha}}{dx}\right]\right)dt. \tag{14.35}$$

Equating Equations 14.34 and 14.35,

$$dx\left(C_{B,int}^{\gamma}-C_{B,int}^{\alpha}\right)=\left(\left[-\tilde{D}_{\gamma}\frac{dC_{B,int}^{\gamma}}{dx}\right]-\left[-\tilde{D}_{\alpha}\frac{dC_{B,int}^{\alpha}}{dx}\right]\right)dt$$

and rearranging,

$$v_{\gamma\to\alpha}=\frac{dx}{dt}=\frac{1}{\left(C_{B,int}^{\gamma}-C_{B,int}^{\alpha}\right)}\left(\tilde{D}_{\alpha}\frac{dC_{B,int}^{\alpha}}{dx}-\tilde{D}_{\gamma}\frac{dC_{B,int}^{\gamma}}{dx}\right) \tag{14.36}$$

give the desired result. The way in which the motion was determined, Equations 14.34 and 14.35 give a positive velocity to the *left*, as shown in Figure 14.18. Note that this is the same result obtained in Chapter 10 by a different method that also included the molar volume difference in the two phases.

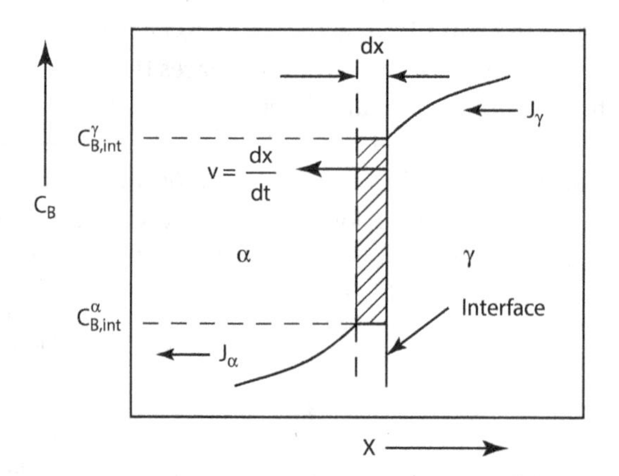

FIGURE 14.18 Schematic of concentration versus distance at an interface used to calculate the velocity of the phase interface, v, between alpha and gamma. The equilibrium concentrations are $C_{B,int}^{\alpha}$ and $C_{B,int}^{\gamma}$ at the alpha and gamma interfaces, and J_α and J_γ are the fluxes of B in alpha and gamma, respectively. (After Porter et al. 2009.)

14.5 CASE STUDY: RHODIUM–COPPER INTERDIFFUSION*

14.5.1 INTRODUCTION

The last section on interdiffusion in metals has a strongly academic flavor in that it is of scientific interest but does it have any practical value? In fact, it has considerable practical value! Certainly, at high temperatures, metals in contact with each other will interdiffuse and form solid solutions or intermetallic phases that can greatly degrade properties critical to the application. For example, solid solutions have lower thermal conductivities than those of pure metals and that could be a problem in a heat transfer application, particularly because the interfacial properties change with time as the amount of solid solution increases at the interface. In addition, Kirkendall porosity may form that also could affect properties such as strength and creep resistance. In contrast, in welding, the temperatures are high and interdiffusion is required to form a strong bond at the weld joint.

In this case study, the effects of unintended interdiffusion leading to failure of a commercial product after some period of time in operation are described: specifically the failure of a rhodium x-ray diffraction tube. Figure 14.19 sketches the pertinent part of the x-ray tube. A thin piece of rhodium metal is attached to the surface of a piece of copper that has been cast onto the rhodium. The copper is water cooled from behind during tube operation to carry away the heat produced by the electron beam current. The rhodium target is irradiated by a beam of electrons of 50 kV energy and about 50 ma total current for a power input of about 2.5 kW (Cullity 1956). The characteristic Rh L_α radiation is excited has a wavelength of $\lambda = 459.7$ pm and is often used for diffraction work with iron-containing (and other elements) materials that the more common copper K_α of wavelength $\lambda = 154$ pm would cause the iron to give off its own fluorescence x-rays prohibiting diffraction studies. Similar tubes are also made with: molybdenum, Mo; cobalt, Co; iron, Fe; and chromium, Cr, targets to provide a variety of characteristic x-rays for diffraction purposes.

14.5.2 TUBE FAILURE

Figure 14.20 shows a cross-section of a failed x-ray target where the rhodium at the surface became sufficiently hot that it melted. As Table 14.1 shows, the melting point of rhodium is 1962°C. Because the back of the copper is water cooled, the temperature here is about 20°C so there is a temperature

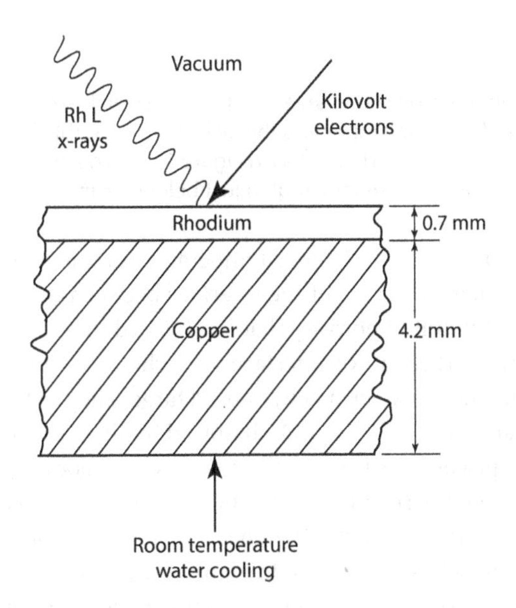

FIGURE 14.19 Schematic of a rhodium x-ray diffraction tube target.

* D. W. Readey, unpublished research.

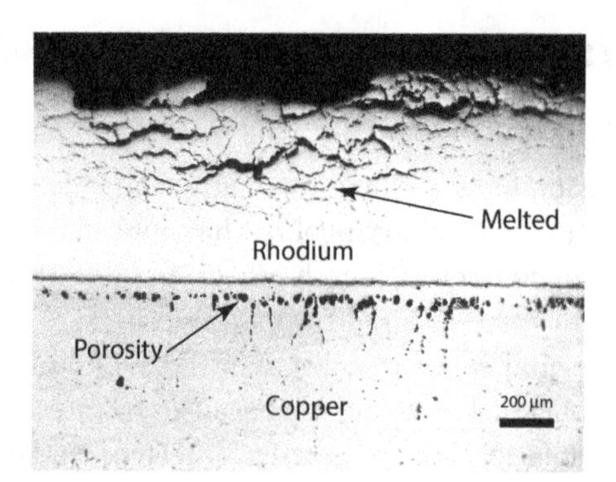

FIGURE 14.20 Photomicrograph of a failed rhodium x-ray target after some unknown time of use. Failure occurs when the rhodium reaches its melting temperature, 1962°C, at the surface. Two distinctive features are the region immediately below the rhodium–copper interface and, below that, a region containing a large volume of pores—the black areas.

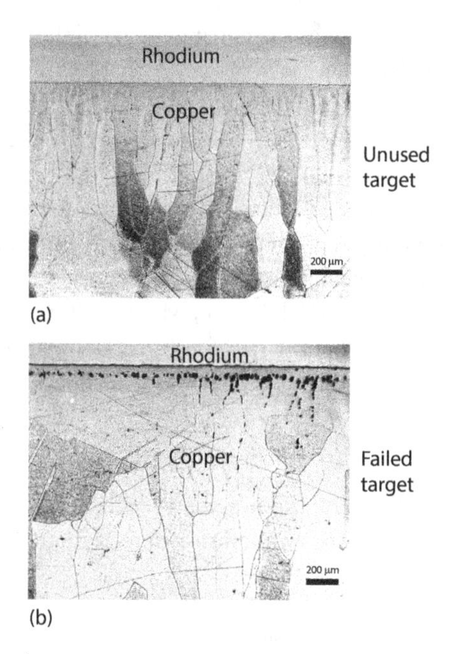

FIGURE 14.21 (a) The top photograph shows an etched—$K_2Cr_2O_7$ etch—cross-section of an unused tube showing the columnar grains of the cast copper and the lack of porosity. (b) The bottom photomicrograph shows the etched version of the same failed target as in Figure 14.20. The pores seem to follow grain boundaries and the etchant appears to etch neither the region just below the interface nor the rhodium.

difference of about 1942°C over about 0.49 cm: a temperature gradient of almost 4000°C/cm! More important, however, is the microstructure of the interface between the rhodium and copper. Most prominent is the very large amount of porosity that has formed in the copper and there is a narrow band of about 60 μm between the Cu–Rh interface and where the porosity begins in the copper. Figure 14.21 shows the difference between a new, unused target and the failed target: both have been etched with a copper etchant, primarily $K_2Cr_2O_7$ dissolved in dilute sulfuric acid. The grain boundaries of the copper of both pieces are easily seen but the etchant does not etch rhodium. Note that the copper grains are elongated perpendicular to the interface a microstructure indicative of the copper having been cast onto the rhodium. The phase diagram for copper–rhodium (Chakrabarti and Laughlin 1982), part of which is reproduced in Figure 14.22, shows phase separation over a wide range of compositions below about 1150°C: above the melting point of copper of 1083°C. Certainly, copper shows no signs of melting but the formation of pores and an apparent interdiffusion layer in the copper at the rhodium–copper interface suggests that the copper just below the interface was operating at a fairly high temperature. The question is "how high and why?"

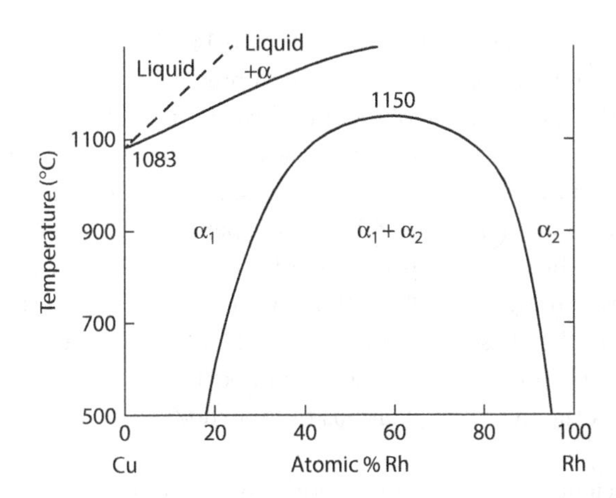

FIGURE 14.22 Pertinent part of the Cu–Rh phase diagram (Chakrabarti and Laughlin 1982). This diagram shows phase separation into two FCC solid solutions, α_1 and α_2 below 1150°C, which is above the melting point of copper at 1083°C.

14.5.3 Evidence for Interdiffusion

Figure 14.23 is a plot of composition versus distance near the interface for both the "new" and "used" (failed) targets. The composition was determined with an electron-beam microanalyzer that uses wavelength dispersion to measure the characteristic x-rays from the material being examined: in this case, Cu K_α and Rh $L_{\alpha,\beta}$. This chemical analysis technique is excellent for detecting small impurity concentrations but lacks the high spatial resolution that can be obtained using energy dispersive analysis in an electron microscope.* For the analysis used here, a resolution of about 2–5 μm is what

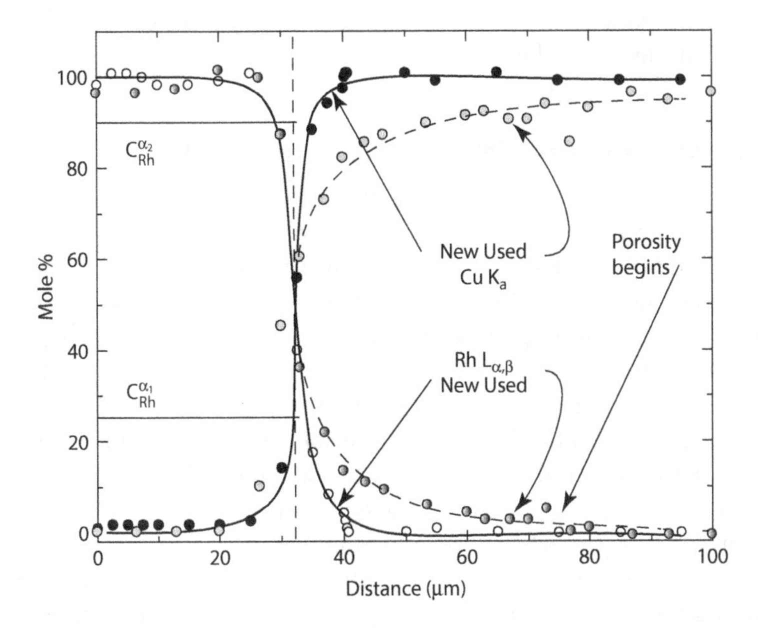

FIGURE 14.23 Concentration of copper (Cu K_α) and rhodium (Rh $L_{\alpha,\beta}$) as a function of distance, as measured by electron probe, for a new—unused—and used x-ray tube target with pure rhodium at the left and copper at the right. The two sets of measurements have been aligned to what appears to be the copper–rhodium interface at about 32 μm. The rhodium concentrations on either side of this interface, $C_{Rh}^{\alpha_1}$ and $C_{Rh}^{\alpha_2}$, from the phase diagram at 800°C are shown. Little interdiffusion is observed on the rhodium side of the interface in either the new or used target. However, significant interdiffusion is observed in the copper of the used target.

* It should be noted that this analysis was carried out in the industrial research laboratory of the tube manufacturer. Although more precise analytical facilities might have been available at a local university or private analytical firm, for proprietary reasons, the analysis was restricted to in-house capabilities applying the best techniques available.

is expected. Based on this resolution, the data of Figure 14.23 suggest that there was little if any interdiffusion in the new target component. Furthermore, the concentration profiles in both the new and failed tube on the rhodium side of the interface are ostensibly identical suggesting little or no copper diffusion into the rhodium. In contrast, the copper side of the failed tube shows considerable interdiffusion between the copper and the rhodium. It is this region between the interface and the porosity in the copper that is not etched by the copper etchant. From Figures 14.20 and 14.21, this region is estimated to be about 60 μm, which is very close to the approximate width of the interdiffusion zone shown in Figure 14.23. Although there is some uncertainty in the exact location of the solvus line(s) in the phase diagram of Figure 14.22 and the interfacial compositions from Figure 14.23, a temperature in this region of at about 800°C is reasonably consistent with both of them. Figure 14.23 shows the interfacial composition of rhodium in the two solid solutions of the phase diagram, α_1 and α_2.

The diffusion coefficients in the rhodium–copper system were used in plotting Figure 14.15 so that they could be compared to the interdiffusion in the failed x-ray tube. Figure 14.15 clearly shows the *intrinsic* diffusion coefficients of copper and rhodium, Equation 14.26, differ by about two orders of magnitude throughout most of the composition range. As a result, copper is the faster diffusing species in the rhodium–copper alloy on the copper side of the interface. As a result, had markers been placed on the copper side of the interface, they would have moved away from the interface and into the copper. Therefore, this seems to be a classic case of the *Kirkendall effect*, which results in the significant porosity shown below the interdiffusion zone in the copper. It should be noted, however, that the interdiffusion zone will also be under a compressive stress because of the increased number of lattice planes in this region causing it to expand—swell—in directions parallel as well as perpendicular to the interface. This results in a tensile stress both in the region directly below the interdiffusion zone and in the rhodium as well. Some of the porosity in the copper may well be due to *creep cavities* due to high-temperature tensile stresses in this region. On the other hand, the diffusion coefficient in rhodium is about two orders of magnitude lower reducing the potential for creep—and cavitation—in the rhodium

14.5.4 Estimate of the Temperature Near the Interface

An approximate—*back of the envelope*—calculation of temperatures when the tube failed can be obtained by approximating the total heat flux through the rhodium and copper in series from the melting point of rhodium, T = 1962°C down to the bottom of the copper in contact with cooling water at about 20°C. This neglects any heat loss by radiation which may be important and that the water may heat up and be at some temperature closer to 100°C. Nevertheless, it is easy to include the interdiffusion layer and this may be important because the thermal conductivity of copper solid solution alloys is significantly lower, k ≡ 30 W/m K, compared to that of copper, k ≡ 399 W/m K (Brandes and Brook 1992). Furthermore, the porosity layer which is on the order of 60 μm thick must have a much lower thermal conductivity than that of copper because just a cursory examination of either Figure 14.20 or 14.21 shows that porosity must contribute about 80% of the cross-sectional area, at least in the regions of highest porosity. So the porous region probably has an effective thermal conductivity very similar to the solid solution layer: k ≅ 30 W/m K. For copper, the thermal conductivity drops to about 350 W/m K at its melting point (Powell et al. 1966). The literature gives the room temperature thermal conductivity of rhodium as 150 W/m K but assuming a value of k(Rh) ≅ 100 W/m K at these high temperatures seems reasonable. Figure 14.24 shows the temperature and thermal flux conditions. In summary,

Rhodium:	$k_1 \cong 100$ W/m K	$\Delta x_1 = 0.7$ mm
Layer:	$k_2 \cong 30$ W/m K	$\Delta x_2 \cong 0.15$ mm
Copper:	$k_3 = 350$ W/m K	$\Delta x_3 = 4.0$ mm

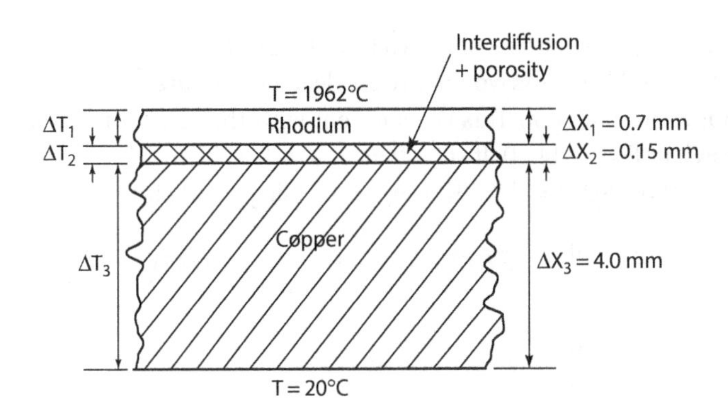

FIGURE 14.24 Model of the x-ray target at failure used to calculate the ΔT_1, ΔT_2, and ΔT_3 temperature differences across the rhodium, interdiffusion and porosity, and copper layers, respectively.

The heat flux, $\dot{q}(\text{watt})$, is given by

$$\dot{q} = -kA\frac{dT}{dx} \tag{14.37}$$

where k is the thermal conductivity (W/m K) and A is the area. In this case, the flux density, $\dot{q}/A$, is a constant through all three layers and the minus sign can be neglected because the direction of the temperature gradient and heat flux density are obviously opposite. Also, the temperature gradient across each layer can be replaced by $\Delta T/\Delta x$ because the gradient across each layer is essentially constant. Therefore,

$$\frac{\dot{q}}{A} = k_1\frac{\Delta T_1}{\Delta x_1} = k_1\frac{\Delta T_2}{\Delta x_2} = k_1\frac{\Delta T_3}{\Delta x_3} \tag{14.38}$$

so proceeding as before with series processes, readily allows calculation of the heat flux density and the temperatures. That is, from Equation 14.38,

$$\begin{aligned} \frac{\dot{q}}{A}\left(\frac{\Delta x_1}{k_1}\right) &= \Delta T_1 \\ \frac{\dot{q}}{A}\left(\frac{\Delta x_2}{k_2}\right) &= \Delta T_2 \\ \frac{\dot{q}}{A}\left(\frac{\Delta x_3}{k_3}\right) &= \Delta T_3 \end{aligned} \tag{14.39}$$

and adding these three equations and equating the sum to the melting point of rhodium, 1962°C, less roughly room temperature, 20°C,

$$\frac{\dot{q}}{A}\left(\frac{\Delta x_1}{k_1} + \frac{\Delta x_2}{k_2} + \frac{\Delta x_3}{k_3}\right) = \Delta T_1 + \Delta T_2 + \Delta T_3 = 1942°\text{C}$$

$$\frac{\dot{q}}{A} = \frac{1942}{\left(\frac{\Delta x_1}{k_1} + \frac{\Delta x_2}{k_2} + \frac{\Delta x_3}{k_3}\right)}. \tag{14.40}$$

Inserting the above values into Equation 14.40,

$$\frac{\dot{q}}{A} = \frac{1942}{\left(\frac{7.0\times10^{-4}}{100} + \frac{1.5\times10^{-4}}{30} + \frac{4\times10^{-3}}{350}\right)}$$

$$\frac{\dot{q}}{A} = 8.29\times10^{7}\ \text{W/m}^2. \tag{14.41}$$

Commercial x-ray tubes operate at tens of kilovolts and tens of milliamps of electron current with a maximum power of about 3 kW. This would give an electron spot area of $A = 3.61\times10^{-5}$ m^2 or about 6.8 mm diameter for a circular spot. This is consistent with the rough diameter of the molten spot on the rhodium shown in Figure 14.20. It should be noted that the maximum radiation heat transfer from the surface (T = 1962 + 274 = 2235 K) is given by (Siegel and Howell 1981)

$$\left(\frac{\dot{q}}{A}\right)_{rad} = \sigma T^4 = 5.67\times10^{-8}\frac{W}{m^2K^4}(2235)^4$$

$$\left(\frac{\dot{q}}{A}\right)_{rad} = 1.41\times10^6\ W/m^2$$

which is about 2% of the heat flux conducted through the target and, certainly for this rough calculation, can be neglected. Combining Equations 14.39 and 14.41,

$$\Delta T_1 = \frac{\dot{q}}{A}\left(\frac{\Delta x_1}{k_1}\right) = 8.29\times10^7\left(\frac{7.0\times10^{-4}}{100}\right) = 580°C$$

$$\Delta T_2 = \frac{\dot{q}}{A}\left(\frac{\Delta x_2}{k_2}\right) = 8.29\times10^7\left(\frac{1.5\times10^{-4}}{30}\right) = 415°C \tag{14.42}$$

$$\Delta T_3 = \frac{\dot{q}}{A}\left(\frac{\Delta x_3}{k_3}\right) = 8.29\times10^7\left(\frac{4.0\times10^{-3}}{350}\right) = 947°C$$

and the sum of the ΔTs is 1942°C as expected. These values seem reasonable except that the temperature in the porous and interdiffusion layer $(\Delta T_2 + \Delta T_3)$ gets considerably above the melting point of copper, 1083°C. From the phase diagram, Figure 14.22 and the composition profile, Figure 14.23, the Rh–Cu interface temperature would seem to be about 800–900°C. Certainly, the maximum temperature must be below 1150°C to have two phases in equilibrium, as Figure 14.22 shows. If either or both of the thermal conductivities of the rhodium and the interdiffusion layers were somewhat larger, then the interface temperature would be lower as well. Because both of these thermal conductivities are estimates, it is not unreasonable that the temperature at the interface is in the range of 800–1000°C.

14.5.5 Other Parameters

14.5.5.1 Interface Motion

Assume that the interface temperature is about 800°C. Around this temperature, the interfacial concentrations do not change drastically as the phase diagram, Figure 14.22, shows. At 800°C, the $C_{Rh}^{\alpha_1} = 25\,m/o$ and $C_{Rh}^{\alpha_2} = 90\,m/o$ from the phase diagram and seem to be consistent with the composition versus distance profile (Figure 14.23), where these two compositions are indicated. Assuming ideal solutions, which they clearly are not, but this also is a negligible effect,

$$\bar{V}_{int}^{\alpha_2} = 0.90\bar{V}_{Rh} + 0.10\bar{V}_{Cu}$$

$$\bar{V}_{int}^{\alpha_2} = 0.90\times8.30 + 0.10\times7.09 = 8.18\ cm^3/mol$$

where $\bar{V}_{int}^{\alpha_2}$ is the molar volume at the solvus in the α_2 solid solution. So

$$C_T^{\alpha_2} = \frac{1}{\bar{V}_{int}^{\alpha_2}} = \frac{1}{8.18} = 0.1223\,mol/cm^3$$

and $C_{Rh}^{\alpha_2} = 0.9C_T^{\alpha_2} = 0.9(0.1223) = 0.110\,mole/cm^3$. Similarly,

$$\bar{V}_{int}^{\alpha_1} = 0.25\bar{V}_{Rh} + 0.75\bar{V}_{Cu}$$

$$\bar{V}_{int}^{\alpha_2} = 0.25\times8.30 + 0.75\times7.09 = 7.39\,cm^3/mol$$

and

$$C_T^{\alpha_1} = \frac{1}{\bar{V}_{int}^{\alpha_1}} = \frac{1}{7.39} = 0.135\,\text{mol/cm}^3$$

and $C_{Rh}^{\alpha_1} = 0.25C_T^{\alpha_1} = 0.25(0.135) = 3.38\times10^{-2}\,\text{mol/cm}^3$. The area under the rhodium composition curve on the copper-rich side of the interface was estimated to be $A \cong 6.91 \times 10^{-5}$ mol/cm^2 of rhodium. This must be the amount diffused from the rhodium side of the interface, so the amount that the interface has moved, Dx, is

$$\Delta x \cong \bar{V}_{Rh} \times A = 8.30\left(6.91\times10^{-5}\right) = 5.7\times10^{-4}\,\text{cm}$$

or 5.7 μm.

14.5.5.2 Time at Temperature

X-ray diffraction tubes are used intermittently rather than continuously so, again, a rough estimate of how long the tube was exposed to high temperatures is required. This can be estimated from amount the interface moved, Δx, calculated above and the rate of motion of an interface during diffusion, Equation 14.36, which for the Rh–Cu interface is

$$v_{Rh-Cu} = \frac{dx}{dt} = \frac{1}{\left(C_{Rh}^{\alpha_2} - C_{Rh}^{\alpha_1}\right)}\left(\tilde{D}_{\alpha_2}\frac{dC_{Rh}^{\alpha_2}}{dx} - \tilde{D}_{\alpha_1}\frac{dC_{Rh}^{\alpha_1}}{dx}\right). \qquad (14.43)$$

Now a few more approximations are necessary. Because there seems to be little diffusion into the rhodium, the first diffusion term in the parentheses can be taken to be zero. From Figure 14.15, $\tilde{D}_{\alpha_1} \cong 10^{-12}\,\text{cm}^2/\text{s}$ and from Figure 14.24, $dC_{Rh}^{\alpha_1}/dx \cong (3.38\times10^{-2})/(10\times10^{-4}) \cong 33.8\,\text{mol cm}^4$ is estimated for the gradient of rhodium at the interface, resulting in

$$\frac{dx}{dt} \cong \frac{1}{\left(C_{Rh}^{\alpha_2} - C_{Rh}^{\alpha_1}\right)}\left(-\tilde{D}_{\alpha_1}\frac{dC_{Rh}^{\alpha_1}}{dx}\right)$$

$$\frac{dx}{dt} \cong \frac{1}{\left(0.110 - 3.38\times10^{-2}\right)}\left(10^{-12}\text{x}33.8\right)$$

$$\frac{dx}{dt} \cong 4.44\times10^{-10}\,\text{cm/s}$$

so

$$\Delta t = \frac{\Delta x}{\frac{dx}{dt}} = \frac{5.7\times10^{-4}}{4.44\times10^{-10}}$$

$$\Delta t \cong 1.29\times10^{6}\,\text{s} \cong 15\ \text{days}.$$

From a practical standpoint, if the x-ray tube were used 8 h/d, 5 d per week, then the life of the tube is only about 9 weeks (about what was actually observed), which is clearly too short to be commercially viable. So in order for the company to be competitive, it was vital to identify a solution. But first it was necessary to validate the idea that the Kirkendall porosity was responsible for the problem.

14.5.6 Heat Treatment of Unused Sample

Based on the data and analysis above, the new and unused sample, the same one shown in Figure 14.21a, was annealed at 790°C for 7 d, 6.1×10^5 s, about one-half the estimated time of operation calculated above. The main goal of the experiment was to validate the hypothesis that, if interdiffusion occurred to about the same extent, whether the Kirkendall porosity would develop similar to that observed in practice. Figure 14.25 shows that indeed both interdiffusion and Kirkendall

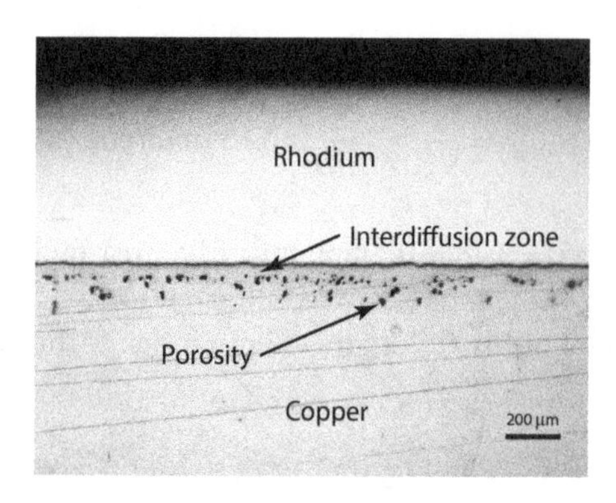

FIGURE 14.25 Unetched photomicrograph of the interface of the unused target after isothermal annealing at 790°C for 7 d—6.1 × 10^5 s. An interdiffusion zone and porosity are generated during this isothermal anneal that appear very similar to the interdiffusion zone and the porosity layers found in the failed tube. This result confirms that the tube failed by interdiffusion between the rhodium and the copper, primarily in the copper, coupled by the generation of the Kirkendall porosity. Together these two regions generate a low thermal conductivity layer that increases in vertical thickness and increasingly lower conductivity—more porosity—with time eventually leading to tube failure.

porosity were found after annealing. The amount of porosity seems less than in the failed sample (see Figure 14.20), and the interdiffusion zone seems somewhat thinner. Nevertheless, these results are strongly suggestive that that x-ray tube failures were caused by interdiffusion; the formation of porosity producing a low thermal conductivity layer; increasing temperature around and above the layer as the tube operated; eventually leading to melting of the rhodium surface. As the temperatures calculated in Equation 14.42, the surface temperature would only be about 1527°C without the low thermal conductivity interdiffusion plus porosity layer. Nevertheless, the temperature at the rhodium–copper interface is still high enough for interdiffusion to occur and Kirkendall porosity to form.

14.5.7 Solution to the Failure Problem

Several potential solutions to prevent tube failure come to mind. The most obvious would be to reduce the power density of the tube. This was not a feasible alternative because the power of the tube had recently been increased to stay competitive with other manufacturers' tubes. In fact, it was after this power increase that failures began to occur. Another obvious potential solution would be to place a diffusion-blocking barrier layer between the rhodium and copper: perhaps a layer of a high melting point metal such as tungsten or molybdenum. However, this adds to the manufacturing cost and would not reduce the temperature of the surface region of the target and may not prevent interdiffusion and failure. Although it seems counterintuitive, a simple solution to the problem is to reduce the thickness of either or both the copper and rhodium layers. For example, if the rhodium layer were reduced to $\Delta x_1 = 0.5$ mm and the copper to $\Delta x_3 = 2$ mm, then from Equation 14.42

$$\Delta T_1 = \frac{\dot{q}}{A}\left(\frac{\Delta x_1}{k_1}\right) = 8.29\times10^7\left(\frac{5.0\times10^{-4}}{100}\right) = 415\,^\circ\text{C}$$

$$\Delta T_3 = \frac{\dot{q}}{A}\left(\frac{\Delta x_3}{k_3}\right) = 8.29\times10^7\left(\frac{2.0\times10^{-3}}{350}\right) = 474\,^\circ\text{C}$$

to give a rhodium–vacuum surface temperature of T = 20 + 415 + 474 = 909°C! And certainly, the temperature near the interface is around 500°C, which is sufficiently low that the interdiffusion coefficient would be lowered by about four or five orders of magnitude extending the tube life to 10^{10} or 10^{11} s or hundreds to thousands of years: at least it should not fail by interdiffusion!

14.6 CHAPTER SUMMARY

This chapter focused on interdiffusion in primarily metallic systems. Oxidation of metals is examined first to demonstrate the use of markers in interdiffusion and how ion and electronic charge fluxes were coupled during oxidation. Osmosis is introduced as an example of interdiffusion between two components when one component essentially does not diffuse. The concept of osmosis is extended to the concept of marker motion during unequal rates of diffusion in a binary system. Finally, the applications of osmosis to both the measurement of molecular weights and the desalinization of seawater are considered. The interdiffusion in a binary metallic system is modeled in some detail including Darken's analysis of the Kirkendall effect: the motion of inert markers by nonequal diffusion rates of the interdiffusion atomic species. The Kirkendall effect has historic significance because it confirmed the existence of vacancy diffusion in metals and practical importance because it is often accompanied by the general of porosity by vacancy condensation that can have deleterious effects on properties. To illustrate this, a real-world case study on the failure of x-ray diffraction tubes by interdiffusion is presented and analyzed in some detail. Many approximations are made in analyzing the failure mode of the tube to illustrate how many technology problems have to be solved even though some of the important data are not available. Nevertheless, this *back of the envelope* approach led to a plausible and at least a semiquantitative model of the tube failure that indicated a direction that could be followed to solve the problem and prevent future failures.

APPENDIX

A.1 WAGNER'S THEORY OF OXIDATION

A.1.1 Introduction

For metal oxidation, as Figure 14.1 shows, the flux of either or both oxygen and metal ions requires a flux of electrons (or holes) for charge neutrality. Therefore, in order for a metal to oxidize by diffusion of ions through the lattice, there must be a counter-current of electrons—electronic defects—for charge compensation. This requires some electronic conductivity in order for the reaction to take place. Take the oxidation of nickel, Ni, to nickel oxide, NiO, by the reaction

$$Ni(s) + 1/2\,O_2(g) \rightarrow NiO(s); \quad \Delta G^o_{NiO}. \tag{A.1}$$

Nickel oxide is a fairly good conductor at elevated temperatures due to the formation of point defects by the oxidation reaction in oxygen to produce the nonstoichiometric $Ni_{1-x}O$, essentially a solid solution of Ni_2O_3 in NiO:

$$2N^x_{Ni} + \frac{1}{2}O_2(g) \rightleftarrows 2Ni^{\bullet}_{Ni} + O^x_O + V''_{Ni} \tag{A.2}$$

where the electronic conductivity arises from the electrons on the Ni^{2+} ions (Ni^x_{Ni}) moving to the Ni^{3+} ions ($Ni^{\bullet}_{NI}$), the so-called *hopping electron conductivity* or *small polaron conductivity*. Even though the deviation from stoichiometry might be small ($x \cong 10^{-4}$), this still provides enough electrons combined with any kind of reasonable mobility to give good electronic transport. As a result, the rate of oxidation is controlled by the diffusion of the ions and usually the metal ion is the faster diffusing species. Because the diffusion of metal ions and oxygen ions are *parallel* processes, it is the faster of the two that dominates, which is usually the metal ion.

A.1.2 Coupled Fluxes: Charge Neutrality

Wagner's model of oxidation is typically given in terms of the relative electronic and ionic conductivity of the oxide layer (West 1986; Kofstad 1988; Maier 2004), and its original development based on conductivities is quite different than the more direct approach used here. The identical result is obtained with the Nernst–Planck relation and it is shown how this leads to the Wagner form of oxidation kinetics.

The model used is NiO, and to reduce the algebra, it is assumed that the diffusion of oxygen is slow compared to that of the nickel ion so only the transport of Ni^{+2} and electrons need to be considered. This does not diminish the main conclusions of the final result but saves a lot of writing with a minor loss in generality. Again, the overall charge flux, J^*, through the oxide layer is zero: that is,

$$J^*_{Ni^{2+}} + J^*_e = 0$$

or

$$2eJ'_{Ni^{2+}} = -eJ'_e = 0 \tag{A.3}$$

where J′ is the number flux density (ions/cm² s). Again, to simplify, for the time being, replace the subscript Ni^{2+} with just N. So the Nernst–Planck equations for the two fluxes are

$$\begin{aligned} J'_e &= \frac{\eta_e D_e}{k_B T}\left\{-\frac{1}{N_A}\frac{d\bar{G}_e}{dx} - eE\right\} \\ J'_N &= \frac{\eta_N D_N}{k_B T}\left\{-\frac{1}{N_A}\frac{d\bar{G}_N}{dx} + 2eE\right\} \end{aligned} \tag{A.4}$$

and replacing these two equations into Equation A.3 (and temporarily dropping the "bar" on $\bar{G}$) gives

$$\frac{\eta_e D_e}{k_B T}\left\{-\frac{1}{N_A}\frac{dG_e}{dx} - eE\right\} = 2\frac{\eta_N D_N}{k_B T}\left\{-\frac{1}{N_A}\frac{dG_N}{dx} + 2eE\right\}.$$

Carrying out the multiplications gives

$$-\frac{\eta_e D_e}{N_A}\frac{dG_e}{dx} - \eta_e D_e eE = -2\frac{\eta_N D_N}{N_A}\frac{dG_N}{dx} + 4\eta_N D_N eE$$

and rearranging gives a value for eE:

$$eE = \frac{1}{(4\eta_N D_N + \eta_e D_e)}\left\{2\frac{\eta_N D_N}{N_A}\frac{dG_N}{dx} - \frac{\eta_e D_e}{N_A}\frac{dG_e}{dx}\right\}. \tag{A.5}$$

Substitution of Equation A.5 for eE into Equation A.4 for the flux density of nickel ions, gives

$$J'_N = \frac{\eta_N D_N}{N_A k_B T(4\eta_N D_N + \eta_e D_e)}\left\{-\frac{dG_N}{dx}(4\eta_N D_N + \eta_e D_e) + 4\eta_N D_N\frac{dG_N}{dx} - 2\eta_e D_e\frac{dG_e}{dx}\right\}$$

$$J'_N = \frac{\eta_N D_N}{N_A k_B T(4\eta_N D_N + \eta_e D_e)}\left\{-4\eta_N D_N\frac{dG_N}{dx} - \eta_e D_e\frac{dG_N}{dx} + 4\eta_N D_N\frac{dG_N}{dx} - 2\eta_e D_e\frac{dG_e}{dx}\right\}$$

giving the final result of

$$J'_{Ni^{2+}} = -\frac{\eta_{Ni^{2+}} D_{Ni^{2+}} \eta_{e^-} D_{e^-}}{N_A k_B T(4\eta_{Ni^{2+}} D_{Ni^{2+}} + \eta_{e^-} D_{e^-})}\left\{\frac{d\bar{G}_{Ni^{2+}}}{dx} + 2\frac{d\bar{G}_{e^-}}{dx}\right\}. \tag{A.6}$$

A.1.3 Thermodynamics

Now, the ionization of nickel can be written as

$$Ni \rightleftarrows N^{2+} + 2e^-$$

and the free energies (partial molar free energies or chemical potentials) for this reaction is

$$\bar{G}_{Ni} = \bar{G}_{Ni^{2+}} + 2\bar{G}_e$$

so the relationship between the derivatives is

$$\frac{d\bar{G}_{Ni}}{dx} = \frac{d\bar{G}_{Ni^{2+}}}{dx} + 2\frac{d\bar{G}_e}{dx}.$$

Therefore, Equation A.6 becomes

$$J'_{Ni^{2+}} = -\frac{\eta_{Ni^{2+}} D_{Ni^{2+}} \eta_{e^-} D_{e^-}}{N_A k_B T\left(4\eta_{Ni^{2+}} D_{Ni^{2+}} + \eta_{e^-} D_{e^-}\right)} \frac{d\bar{G}_{Ni}}{dx}. \tag{A.7}$$

The overall reaction is $Ni(s) + 1/2O_2(g) \rightarrow NiO(s)$, then at both the Ni–NiO and the $NiO–O_2$ interfaces,

$$RT \ln a_{Ni} + \frac{1}{2} RT \ln p_{O_2} = \Delta G^o_{NiO}$$

and if the thickness of the oxide layer is L, then

$$\frac{d\bar{G}_{Ni}}{dx} = \frac{\Delta G^o_{NiO} - \frac{1}{2} RT \ln p_{O_2}}{L}. \tag{A.8}$$

So it would be nice if the oxidation were carried out in pure oxygen at $p_{O_2} = 1$ bar, its standard state, then the gradient in the chemical potential of nickel simply becomes

$$\frac{d\bar{G}_{Ni}}{dx} = \frac{\Delta G^o_{NiO}}{L}. \tag{A.9}$$

A.1.4 Conversion to Wagner's Equation

Wagner (1951) derived expression for the oxidation of metals couched in terms of the electrical and electronic conductivities of the oxides. The approach used in Section 14.3 is the more familiar approach of fluxes and forces to obtain Equation A.7. However, this equation can be easily converted to the more commonly quoted Wagner result by simply making substitutions between diffusion coefficients and conductivities from the Nernst–Einstein equation that was presented in Chapter 13; namely,

$$\sigma_i = \frac{(z_i e)^2 \eta_i D_i}{k_B T}.$$

Therefore, for the electrons,

$$\frac{\eta_e D_e}{k_B T} = \frac{\sigma_e}{e^2} \tag{A.10}$$

and for the Ni^{2+} ions,

$$\frac{4e^2 \eta_{Ni^{2+}} D_{Ni^{2+}}}{k_B T} = \sigma_{Ni^{2+}} = \sigma_i \tag{A.11}$$

where σ_i is the total *ionic conductivity* because the oxygen diffusion and conductivity are assumed to be negligible. Substitution of Equations A.10 and A.11 into Equation A.7 gives

$$J'_{Ni^{2+}} = -\frac{\eta_{Ni^{2+}} D_{Ni^{2+}} \sigma_e}{N_A \left(4e^2 \eta_{Ni^{2+}} D_{Ni^{2+}} + e^2 \eta_e D_e\right)} \frac{d\bar{G}_{Ni}}{dx}$$

$$J'_{Ni^{2+}} = -\frac{\eta_{Ni^{2+}} D_{Ni^{2+}} \sigma_e}{N_A k_B T(\sigma_i + \sigma_e)} \frac{d\bar{G}_{Ni}}{dx}$$

$$J'_{Ni^{2+}} = -\frac{1}{4e^2} \frac{1}{N_A} \sigma_i t_e \frac{d\bar{G}_{Ni}}{dx}$$

where t_i is called the *transference number* and is the fraction of the total current carried by a charged species. So, in this case, $t_e = \sigma_e/(\sigma_e + \sigma_i)$ and $\sigma_i = \sigma t_i$ to get the molar flux of nickel, divide by Avogadro's number, N_A,

$$J_{Ni} = -\frac{1}{4\mathfrak{F}^2} \sigma t_i t_e \frac{d\bar{G}_{Ni}}{dx} \tag{A.12}$$

where $\mathfrak{F}$ = faraday = e × N_A 1.602 × 10^{-19} C/electron × 6.022 × 10^{23} electrons/mol = 96,485 C/mol. Equation A.12 is essentially Wagner's result for oxidation except that it is usually written in terms of the Gibbs energy for the oxygen, $d\bar{G}_{O_2} = (1/2)RTd(\ln p_{O_2}) = -d\bar{G}_{Ni}$.

EXERCISES

14.1 Pure titanium metal is undergoing oxidation in pure oxygen at 1200°C. The diffusion coefficient data Q = 256,000 J/mol and D_0 = 6.2 × 10^{-2} cm²/s for titanium diffusion and Q = 252,040 J/mol and D_0 = 2.0 × 10^{-3} cm²/s for oxygen diffusion. The Gibbs energy for the oxidation reaction: $Ti(s) + O_2(g) = TiO_2(s)$ is $\Delta G^0 = -678.668\,kJ/mol$. Finally, the density and molecular weight of TiO_2 are ρ = 4.17 g/cm³ and M = 79.866 g/mol, respectively. Assume that the diffusion coefficients are sufficiently different that only one matters. Also assume that the TiO_2 is stoichiometric and the stoichiometry does not vary with oxygen pressure. (Actually, the latter is not correct, TiO_2 does become nonstoichiometric by losing oxygen, and so the diffusion coefficients will depend on the oxygen activity and will vary through the oxide layer thickness creating a more difficult problem.)

a. Calculate the diffusion coefficients of Ti^{+4} and O^{2-} in TiO_2 at 1200°C. If the electronic conductivity is high, which of these two ions controls the rate of oxidation?
b. Calculate the titanium and oxygen thermodynamic activities at the Ti-TiO_2 interface.
c. Do the same as in b. at the O_2–TiO_2 interface.
d. Calculate the molar flux density of titanium ions through the TiO_2 for an oxide layer 100 nm thick.
e. Calculate the value of parabolic oxidation rate constant for TiO_2 at 1200°C.
f. Calculate how long it will take to grow an oxide layer 1 μm thick.

14.2 A composite consisting of carbon particles in an MgO matrix is undergoing oxidation at 1200°C in air.

a. The diffusion coefficient data Q = 333,540 J/mol and D_0 = 0.25 cm²/s for magnesium diffusion and Q = 343,500 J/mol and D_0 = 4.3 × 10^{-5} cm²/s for oxygen diffusion. Calculate the diffusion coefficients for Mg^{+2} and O^{-2} diffusion in MgO at 1200°C. Assume that the faster of the two controls the rate of oxidation.
b. The oxidation oxidizes the carbon particles to CO_2 gas. The Gibbs energy of formation for the reaction $C(s) + O_2(g) = CO_2(g)$ is $\Delta G^0 = -396{,}243\,J/mol$ and for the reaction $Mg(l) + 1/2\,O_2(g) = MgO(s)$ is $\Delta G^0 = -438{,}840$ J/mol. Calculate the oxygen pressure at the C–MgO interface when CO_2 first starts to form.
c. Calculate the magnesium thermodynamic activity at the C–MgO interface when CO_2 first starts to form.

d. Calculate the magnesium activity at the O_2–MgO interface.
e. If the carbon particles are 1 μm in diameter, and if the CO_2 cannot escape or diffuse out, calculate the CO_2 pressure in the volume previously occupied by the now completely oxidized carbon particle. Also calculate the oxygen and magnesium activities at this point. Hint: Some tiny bit of carbon remains. The density of carbon is ρ = 2.2 g/cm^3 and its molecular weight is M = 12.011 g/mol.
f. On the other hand, assume that the CO_2 can diffuse out or be released by pores or cracks in the material. There is 5 w/o carbon particles in the MgO (ρ = 3.6 g/cm^3 and M = 40.304 g/mol. Calculate how long it would take to oxidize a layer of the composite 10 μm thick.
g. Develop an expression of the reaction rate constant for this oxidation assuming that the pressure of CO_2 never exceeds 1 atm.
h. Sketch what the microstructure might look like in cross-section from the oxygen–MgO surface down into the material to the unoxidized composite.

14.3 Two grams of poly(vinyl alcohol), $\left[CH_2CHOH\right]_n$, is dissolved in 1 L of water.
a. Calculate the molecular weight of the polymer if the water is raised to a height difference of 11.54 cm at equilibrium in an osmosis apparatus similar to that in Figure 14.7.
b. Calculate the number of chain units (mers) per polymer molecule.
c. Suppose that the osmosis apparatus was a U-shaped tube with an internal diameter of 1 mm and the membrane at the middle of the tube at the bottom of the U. If the membrane were 10 μm thick and the diffusion coefficient for water through the membrane were $D = 10^{-6}\ cm^2/s$, calculate the velocities (cm/s) of the two equal water levels on the top of the left- and right-hand sides of the tubes when the osmosis first begins.
d. Calculate the same velocities when the water level difference in the two sides of the tube has reached 5 cm.

14.4 Consider the apparatus if Figure 14.7. Assume that the solvent is pure water at 25°C and it can diffuse through the membrane, while the solute molecules cannot.
a. Develop an expression for the velocity of the *movable* membrane in terms of the flux of water through the membrane.
b. Assume that the error function solution for semi-infinite BC can be used to get the concentration water at the membrane as function of time:

$$C(x,t) = C_S \operatorname{erfc}\left(\frac{x}{\sqrt{4Dt}}\right).$$

Show that $dC(0,t)/dx = -C_S/\sqrt{\pi Dt}$. Hint: The Leibnitz formula for differentiating a definite integral is

$$\frac{d}{dx}\left(\int_{f_1(x)}^{f_2(x)} g(t)dt\right) = g\left[f_2(x)\right]\frac{df_2(x)}{dx} - g\left[f_1(x)\right]\frac{df_1(x)}{dx}.$$

c. Assume that the water can diffuse rapidly through the membrane and at the membrane–solution interface, x = 0 (the right side of the membrane in Figure 14.7), is pure $H_2O = C_S$, that the solution initial concentration was a 0.1 m/o NaCl solution (99 m/o H_2O), that the interdiffusion coefficient between water and the solution is a constant and $D = 2.5\times10^{-5}\ cm^2/s$, and the length of the solvent chamber is L = 2 cm. Also assume that the error function solution is valid in the solvent chamber for times up until, $C(L,t) = 99.2\ m/o$. Calculate the time it takes for $C(L,t) = 99.2\ m/o\ H_2O$.
d. Develop an expression for the membrane velocity as a function of time and plot the velocity versus time until $C(L,t) = 99.2\ m/o$.
e. Develop an expression for the membrane position as a function of time and plot the position versus time until $C(L,t) = 99.2\ m/o$.

14.5 This exercise is on the Kirkendall effect. Assume that two metals A and B are interdiffusing with B on the left. These two metals diffuse and form an ideal solution so nonideality effects in Darken's equations can be neglected. In fact, the overall concentration is a constant, $C_0 = 0.125$ mol/cm³.

a. The diffusion coefficient of A as a function of mole fraction of B, X_B, is given by $D_A = 10^{-10}(1+9X_B)\text{cm}^2/\text{s}$ and that of B, $D_B = 10^{-11}(1+9X_B)\text{cm}^2/\text{s}$; that is, they are linear with X_B. Calculate and plot D_A, D_B, and the interdiffusion coefficient $\tilde{D}$ as a function of X_B for $0 < X_B < 1.0$.

b. Calculate $\tilde{D}$ for $X_B = 0.5$. Assuming that the interdiffusion coefficient is constant at this value, calculate and plot X_B for $-30\ \mu\text{m} < x < 30\ \mu\text{m}$ with the infinite boundary solution, $C_B = 0.5 \times \text{erfc}\left(x/\sqrt{4Dt}\right)$ for $t = 1000$ s with at least 20 data points.

c. Calculate and plot J_A, J_B, the vacancy flux, J_V, and the velocity of the solid solution at $X_B = 0.5$ relative to the ends of the diffusion couple from the concentration profile and on the same plot as the concentration profile developed in part b. above and get a plot similar to Figure 14.12. Again, use at least 20 data points for each plot.

d. From the data in c. above, calculate the position, x, for the maximum in vacancy destruction.

e. If the vacancies are not destroyed by dislocation climb but rather form pores, calculate the maximum rate of pore formation in cm³ pores/cm³ solid. Assume that a vacancy has the same volume as one of the atoms interdiffusing, about 2×10^{-23}cm³.

14.6 Assume that the solid solution being formed between A and B is not ideal but is actually a regular solution with $\alpha = -0.5$.

a. Calculate and plot the thermodynamic activities of A and B as a function of the mole fraction of A, X_A.

b. Plot the *thermodynamic factor* in Darken's equation as a function of X_A.

c. Plot the interdiffusion coefficient in this nonideal solid solution as a function of X_B with the same data as used in exercise 14.5.

d. Calculate the interdiffusion coefficient $\tilde{D}$ at $X_B = 0.5$ in cm²/s.

e. Comment on the significance of including nonideal behavior on the calculation of $\tilde{D}$ and its value at $X_B = 0.5$.

REFERENCES

"Atomic mass unit." *Wikipedia, The Free Encyclopedia*, http://en.wikipedia.org/w/index.php?title=Atomic_mass_unit&oldid=731029921. (accessed August 11, 2014.)

Brandes, E. A. and G. B. Brook, eds. 1992. *Smithells Metals Reference Book*, 7th ed. Oxford: Butterworth-Heinemann.

Chakrabarti, D. J. and D. E. Laughlin. 1982. The Cu-Rh (Copper-Rhodium) System. *Bulletin of Alloy Phase Diagrams*. 2 (4): 460–462.

Cahn, R. W. 2001. *The Coming of Materials Science*. London: Elsevier Science.

Chang, R. 2000. *Physical Chemistry for the Chemical and Biological Sciences*. Sausalito. CA: University Science Books.

Cullity, B. D. 1956. *Elements of X-Ray Diffraction*. Reading, MA: Addison-Wesley.

Darken, L. S. 1948. Diffusion, Mobility and Their Interrelation through Free Energy in Binary Metallic Systems. *Trans. AIME*. 174: 184–201.

Darken, L. S. and R. W. Gurry. 1953. *Physical Chemistry of Metals*. New York: McGraw-Hill.

Einstein, A. 1956. *Investigations on the Theory of Brownian Motion*. New York: Dover.

Glicksman, M. E. 2000. *Diffusion in Solids*. New York: Wiley.

Haney, P., K. Herting, and S. Smith. 2013. "Separation characteristics of dialysis membranes." http://www.piercenet.com/previews/2013-articles/separation-characteristics-dialysis-membranes/. Accessed August 11, 2014.

Hartley, G. S. 1946. Diffusion and Swelling of High Polymers. Part I. The Swelling and Solution of a High Polymer Solid Considered as a Diffusion Process. *Transactions of the Faraday Society*. 42: B006–B011.

Hartley, G. S. and J. Crank. 1949. Some Fundamental Definitions and Concepts in Diffusion Processes. *Transactions of the Faraday Society*. 45: 801–818.

Haynes, W. M., Editor-in-Chief. 2013. *CRC Handbook of Chemistry and Physics*, 94th ed. Boca Raton, FL: CRC Press.

Hecht, C. E. 1967. Desalination of Water by reverse Osmosis. *Journal of Chemical Education*. 44 (1): 53–54.

Hoddeson, L., E. Braun, J. Teichmann, and S. Weart. 1992. *Out of the Crystal Maze*. New York: Oxford University Press.

Huntington, H. B. and F. Seitz. 1942. Mechanism of Self-Diffusion in Metallic Copper. *Physical Review*. 61(March 1 and 15): 315–325.

Kirkendall, E. O. 1942. Diffusion of Zinc in Alpha Brass. *Transactions AIME*. 147: 104–110.

Kofstad, P. 1988. *High Temperature Corrosion*. Barking, England: Elsevier Applied Science.

Laidler, K. J. and J. H. Meiser. 1995. *Physical Chemistry*, 2nd ed. Boston, MA: Houghton Mifflin.

Maier, J. 2004. *Physical Chemistry of Ionic Materials*. Chichester, England: Wiley.

Mott, N. F. and R. W. Gurney. 1940. *Electronic Processes in Ionic Crystals*. Oxford: Oxford University Press. Note: a second edition was published in 1948 with very minor and clearly identified additions. This was republished as: Mott, N. F. and R. W. Gurney. 1964. *Electronic Processes in Ionic Crystals*, 2nd ed. New York: Dover.

Nakajima, H. 1997. The Discovery and Acceptance of the Kirkendall Effect: The Result of a Short Research Career. *The Journal of The Minerals, Metals & Materials Society*. 49 (6): 15–19.

Porter, D. A., K. E. Easterling, and M. Y. Sherif. 2009. *Phase Transformations in Metals and Alloys*, 3rd ed. Boca Raton FL: Taylor & Francis.

Powell, R. W., C. Y. Ho, and P. E. Liley. 1966. *Thermal Conductivity of Selected Materials*. United States Department of Commerce Report. NSRDS-NBS8. Washington, DC. November 25, 1966.

Roine, A. 2002. *Outokumpu HSC Chemistry for Windows, Ver. 5.11*, thermodynamic software program, Outokumpu Research Oy, PORI, Finland.

Siegel, R. and J. R. Howell. 1981. *Thermal Radiation Heat Transfer*, 2nd ed. Washington, DC: Hemisphere Publishing Co.

Seitz, F. 1953. On the Porosity Observed in the Kirkendall Effect. *Acta Metallurgica*. 1(May): 355–369.

Sequeira, C. A. and L. Amaral. 2014. Role of Kirkendall Effect in Diffusion Processes in Solids. *Transactions Nonferrous Metals Society of China*. 24: 1–11.

Smigelskas, A. D. and E. O. Kirkendall. 1947. Zinc Diffusion in Alpha Brass. *Transactions AIME*. 171: 130–134.

Smyth, D. M. 2000. *The Defect Chemistry of Metal Oxides*. New York: Oxford University Press.

Tuijn, C. 1997. On the History of Models for Solid-State Diffusion. *Defect and Diffusion Forum*. 143–147: 11–20.

Wagner, C. 1951. Diffusion and High Temperature Oxidation of Metals. *Atom Movements*. 153–173. Cleveland OH: Am. Soc. for Metals (now ASM International).

West, J. M. 1986. *Basic Corrosion and Oxidation*, 2nd ed. Chichester, England: Ellis Horwood Ltd.

15

Interdiffusion in Compounds

15.1 PHENOMENA OF INTEREST

Throughout this book, the emphasis is on phenomena that have important consequences on the processing or behavior of materials rather than phenomena that are merely interesting and whose understanding does not necessarily lead to improved processes or materials. For example, in Chapter 14, the Kirkendall effect is covered in some detail. Certainly, the phenomenon has academic and historic implications, but more importantly, it can produce significant property changes as demonstrated by the failure of the rhodium x-ray target. In the same spirit, topics covered in this chapter involve diffusion in more complex systems chemically, thermodynamically, and geometrically. The intent again is not to be all inclusive but to introduce these complexities as illustrations of how similar principles are again applied to important material processes and behavior (Schmalzried 1981). The important phenomena modeled in this chapter include interdiffusion in compounds; the Kirkendall effect in compounds; interdiffusion reactions; the effects of geometry; and finally, inclusion of a volume difference between reactants and products. Oxides are modeled since many of these processes have been analyzed with oxides because of the experimental convenience of fixing the oxygen chemical potential: air, for example. But the same principles apply to any type of compound: sulfides, nitrides, arsenides, borides, intermetallic compounds, and so on.

15.2 INTERDIFFUSION IN NiO–MgO

15.2.1 Several Possible Models

The interdiffusion of MgO and NiO is an example for interdiffusion in compounds. The goal here is not to compare the various models for interdiffusion of NiO and MgO with the limited experimental results (Appel and Pask 1971) but rather to use this system to demonstrate the important features of the different models and the results that each predicts. First, usually an interdiffusion is performed in an atmosphere in equilibrium with the compounds. Usually, the oxygen pressure (the oxygen chemical potential) is fixed and uniform. Certainly, experiments can be devised in which there is an oxygen chemical potential difference from one part of a solid to another. However, the results and analysis of this condition go beyond what is intended here (Yurek and Schmalzried 1975; Schmalzried 1981). With a fixed oxygen chemical potential, there is no *chemical* force (Chapter 13) on the oxygen ions. In one model, *oxygen ions do not move* and only the magnesium and nickel ions interdiffuse in a fixed oxygen ion lattice. In another model, *all three ions* are assumed to be mobile. In these models, the electronic conductivity is considered to be extremely low and charge transport is entirely ionic. A third model assumes fixed oxygens again but with mobile electrons. Of course, there is the fourth model with all three ions *and* electrons mobile. This last situation is the most general case—and algebraically most challenging because of the large number of terms but really introduces neither new or unique concepts nor results. Again, to not miss the forest for the trees, the models with minimal algebraic complexity are covered here. In all of these models, *no new principles are introduced!*

15.2.2 Expected Lack of Oxygen Diffusion

It is usually assumed that in interdiffusion between two compounds such as NiO and MgO, the only consideration is the interdiffusion between the cations (Appel and Pask 1971). The main reason for this is illustrated in Figure 15.1. Specifically, if a Kirkendall experiment were to be carried out on a pure NiO–MgO diffusion couple, the markers at the original interface are not expected to move. There are two plausible reasons for this assumption. First, such an interdiffusion experiment would be carried out in some kind of oxygen-containing atmosphere such as air so that the chemical potential, μ_{O_2}, or partial molar Gibbs energy of oxygen, $\bar{G}_{O_2} = (\partial G/\partial n_{O_2})_{T,p,n_i}$, is the same everywhere, that is, $\nabla\mu_{O_2} = 0$. As a result, there is no chemical potential gradient to drive oxygen diffusion. Second, the number of oxygen ions per unit volume in MgO and NiO are about the same: molar volume MgO, $\bar{V}_{MgO} = 11.26\,\text{cm}^3/\text{mol}$ and molar volume NiO, $\bar{V}_{NiO} = 11.20\,\text{cm}^3/\text{mol}$ (Haynes 2013), which comes pretty close to satisfying the constant volume criterion used in the last chapter. Therefore,

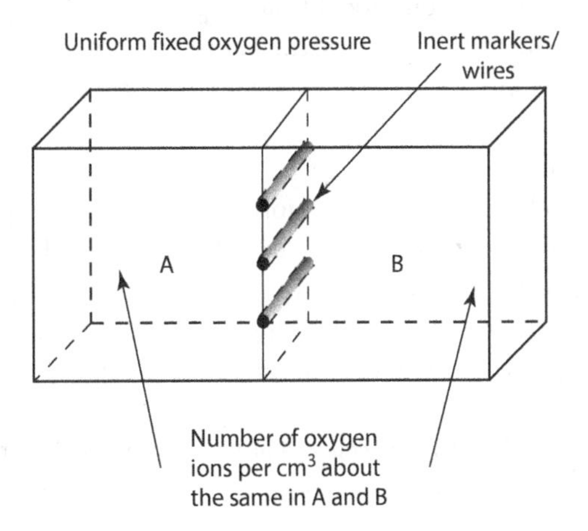

FIGURE 15.1 Two oxide compounds A and B with inert markers at the original interface reacting in a uniform oxygen pressure or chemical potential and assuming that the oxygen ions are immobile.

there should be no tendency for the oxygen to move. As a result, a Kirkendall effect or marker movement would not be expected because neither MgO nor NiO is moving faster than the other past markers. Only the magnesium ions, Mg^{2+}, and the nickel ions, Ni^{2+}, move. This is generally thought to be the case. However, an important factor is ignored in this model that does make a Kirkendall effect possible as is demonstrated below.

In reality, NiO and MgO are a pretty good diffusion couple to try a marker experiment on. The main problem with marker experiments on compounds is that compounds are brittle. If the diffusion couple is made by a solid-state interdiffusion process at some elevated temperature, then on cooling the total thermal expansion difference between the components must be small. For example, if the difference in expansion coefficient were only $\Delta\alpha \cong 1\times10^{-7}\,\mathrm{K}^{-1}$ (about 5%), and Young's modulus $E \cong 250\,\mathrm{MPa}$ (Kingery et al 1976) cooling a diffusion couple from 1200°C would produce stresses in the couple on the order of $\sigma = \varepsilon E \cong \Delta\alpha\Delta TE \cong 1\times10^{-7}(1200)210\times10^{9} \cong 2.52\times10^{7}\,\mathrm{Pa} \cong 36{,}000\,\mathrm{psi}$ close to the fracture strength of these two oxides. Actually, the thermal expansions of NiO and MgO are very close, but both vary with temperature from about 13×10^{-6} to $15\times10^{-6}\,\mathrm{K}^{-1}$ between 200 and 1200°C (Nielsen and Leipold 1963; Nielsen and Leipold 1965). It would be very fortunate if they matched close enough so that fracture would not occur on cooling. Certainly, there are very few compounds whose thermal expansions are this close. This is one reason that there are not many marker experiments on compounds. Metals are much more forgiving because any stresses introduced by thermal expansion differences are easily relieved by plastic deformation. Some plastic deformation can occur in compounds as well, but only at high temperatures and to a limited extent. Disregarding the experimental difficulties with compound interdiffusion, the models are important for their predicted behavior under different assumptions. However, in the interdiffusion experiments with NiO and MgO single crystals, no markers were used (except for pores at the original interface) (Appel and Pask 1971).

In all models, the goal is to develop an expression for the diffusion flux of one of the cations, magnesium for example, in terms of a concentration gradient and an effective or interdiffusion coefficient, $\tilde{D}$, which is a function of measureable tracer diffusion coefficients and, perhaps, the composition. In all the models, an ideal solution is assumed so that the chemical potential for magnesium oxide, μ_{MgO}, is given by

$$\mu_{MgO} = \bar{G}_{MgO} = G^{0}_{MgO} + RT\ln X_{MgO}$$

rather than assuming a regular solution as in Darken's equations for interdiffusion in metals. The regular solution assumption simply adds a small multiplying factor* to the interdiffusion coefficient and adds neither additional insight nor new principles relevant to the interdiffusion models. Furthermore, whether a regular solution model is appropriate for this system is really not known, and from the phase diagram and molar volumes, the NiO–MgO system appears close to ideal. In all of the models generated here, some or all of the following important steps are invoked:

1. Ion flux charge neutrality
2. Thermodynamic plus electric field driving force for each mobile species
3. From Steps 1 and 2, an expression for the electric field term is found
4. Substitution of the electric field term into the number flux density of magnesium, J'_{Mg}†
5. Invoke the Gibbs–Duhem relation to get a single gradient term, $d\bar{G}_{Mg}/dx$
6. Get the molar flux, J_{MgO}, in the form of $J_{MgO} = -\tilde{D}\,dC_{MgO}/dx$

Steps 3–5 involve a lot of algebra but no new science or mathematics. As a result, much of the algebra is left to the Appendix and only the results are presented so as not to supersaturate the models with pages of algebraic equations.

* $\left(1 + X_{MgO}\,d\ln\gamma_{MgO}/dX_{MgO}\right) = \left(1 - 2\alpha X_{MgO}X_{NiO}\right)$ and $\alpha \le 2$: Chapter 14, Section 14.4.4.2, Equation 14.33.

† For all practical purposes $J'_{MgO} = J'_{Mg} = J'_{Mg^{2+}}$ as will be seen and used somewhat interchangeably in the models.

15.2.3 Model with Immobile Oxygen

Here the interdiffusion of Mg^{2+} and Ni^{2+} on a lattice of essentially stationary oxygen ions is considered as shown in Figure 15.2. Again, the very general principle that the charge fluxes must be equal is invoked: Step 1,*

$$\begin{aligned} J^*_{Mg^{2+}} + J^*_{Ni^{2+}} &= 0 \\ 2eJ'_{Mg^{2+}} + 2eJ'_{Ni^{2+}} &= 0 \\ J'_{Mg^{2+}} + J'_{Ni^{2+}} &= 0. \end{aligned} \tag{15.1}$$

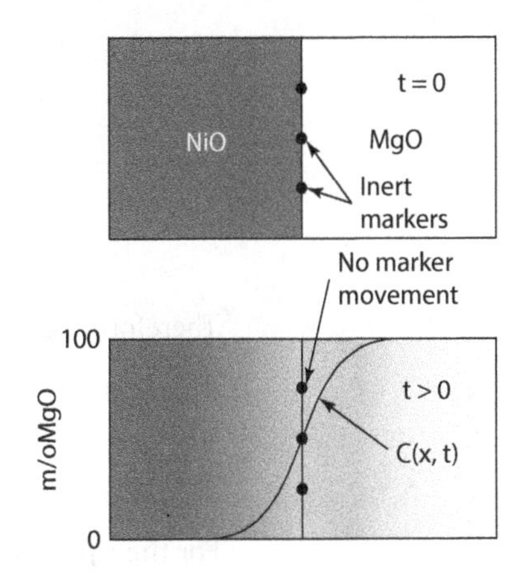

FIGURE 15.2 Interdiffusion between NiO and MgO assuming immobile oxygen ions and inert markers.

Again, the total force on the ions is the chemical force plus the electrical force produced by unequal cation diffusion rates (the overhead "bar" is dropped to minimize symbol complexity, so $\bar{G}_i = G_i$) (Step 2),

$$\begin{aligned} J'_{Mg^{2+}} &= \frac{\eta_{Mg^{2+}} D_{Mg^{2+}}}{k_B T}\left\{-\frac{1}{N_A}\frac{dG_{Mg^{2+}}}{dx} + 2eE\right\} \\ J'_{Ni^{2+}} &= \frac{\eta_{Ni^{2+}} D_{Ni^{2+}}}{k_B T}\left\{-\frac{1}{N_A}\frac{dG_{Ni^{2+}}}{dx} + 2eE\right\}. \end{aligned} \tag{15.2}$$

Substituting Equation 15.2 into Equation 15.1 and solving for the electric field term (Step 3),

$$\frac{\eta_{Mg^{2+}} D_{Mg^{2+}}}{k_B T}\left\{-\frac{1}{N_A}\frac{dG_{Mg^{2+}}}{dx} + 2eE\right\} = -\frac{\eta_{Ni^{2+}} D_{Ni^{2+}}}{k_B T}\left\{-\frac{1}{N_A}\frac{dG_{Ni^{2+}}}{dx} + 2eE\right\}$$

$$\left(\eta_{Mg^{2+}} D_{Mg^{2+}} + \eta_{Ni^{2+}} D_{Ni^{2+}}\right)2eE = \frac{\eta_{Mg^{2+}} D_{Mg^{2+}}}{N_A}\frac{dG_{Mg^{2+}}}{dx} + \frac{\eta_{Ni^{2+}} D_{Ni^{2+}}}{N_A}\frac{dG_{Ni^{2+}}}{dx}$$

$$2eE = \frac{1}{\left(\eta_{Mg^{2+}} D_{Mg^{2+}} + \eta_{Ni^{2+}} D_{Ni^{2+}}\right)}\left\{\frac{\eta_{Mg^{2+}} D_{Mg^{2+}}}{N_A}\frac{dG_{Mg^{2+}}}{dx} + \frac{\eta_{Ni^{2+}} D_{Ni^{2+}}}{N_A}\frac{dG_{Ni^{2+}}}{dx}\right\}. \tag{15.3}$$

Inserting the electric field term, Equation 15.3, into the first of Equations 15.2 (Step 4)

$$J'_{Mg} = \frac{\eta_{Mg^{2+}} D_{Mg^{2+}}}{RT\left(\eta_{Mg^{2+}} D_{Mg^{2+}} + \eta_{Ni^{2+}} D_{Ni^{2-}}\right)}\left\{\begin{array}{l} -\eta_{Mg^{2+}} D_{Mg^{2-}}\dfrac{dG_{Mg^{2+}}}{dx} - \eta_{Ni^{2+}} D_{Ni^{2+}}\dfrac{dG_{Mg^{2+}}}{dx} \\ +\eta_{Mg^{2+}} D_{Mg^{2+}}\dfrac{dG_{Mg^{2+}}}{dx} + \eta_{Ni^{2+}} D_{Ni^{2+}}\dfrac{dG_{Ni^{2+}}}{dx} \end{array}\right\}$$

$$J'_{Mg} = \frac{\eta_{Mg^{2+}} D_{Mg^{2+}} \eta_{Ni^{2+}} D_{Ni^{2+}}}{RT\left(\eta_{Mg^{2+}} D_{Mg^{2+}} + \eta_{Ni^{2+}} D_{Ni^{2+}}\right)}\left\{\frac{dG_{Ni^{2+}}}{dx} - \frac{dG_{Mg^{2+}}}{dx}\right\}. \tag{15.4}$$

From Section A.1, because of the relationships between the chemical potentials of the ions, atoms, and oxides, Equation 15.4 becomes (and now dropping the charge symbols on the η_I and the D_i because the values are the same whether the cations are given charges or not)

$$J'_{Mg} = \frac{\eta_{Mg} D_{Mg} \eta_{Ni} D_{Ni}}{RT\left(\eta_{Mg} D_{Mg} + \eta_{Ni} D_{Ni}\right)}\left\{\frac{dG_{NiO}}{dx} - \frac{dG_{MgO}}{dx}\right\}. \tag{15.5}$$

* Recall that J^* = charge flux density (C/cm² s); J' = number flux density (ions/cm² s); and J = molar flux density (mol/cm² s).

Now, from thermodynamics (see Chapter 1), the Gibbs–Duhem relation is $X_{MgO}dG_{MgO} + X_{NiO}dG_{NiO} = 0$ where X_i = mole fraction. So (Step 5)

$$\frac{dG_{NiO}}{dx} - \frac{dG_{MgO}}{dx} = -\left\{\frac{X_{MgO}}{X_{NiO}}\frac{dG_{MgO}}{dx} + \frac{dG_{MgO}}{dx}\right\} = -\frac{1}{X_{NiO}}\frac{dG_{MgO}}{dx}.$$

Therefore, Equation 15.5 becomes

$$J'_{Mg} = -\frac{\eta_{Mg}D_{Mg}\eta_{Ni}D_{Ni}}{RT\left(\eta_{Mg}D_{Mg} + \eta_{Ni}D_{Ni}\right)}\frac{1}{X_{NiO}}\left\{\frac{dG_{MgO}}{dx}\right\}. \tag{15.6}$$

For the case of an ideal solid solution—which the phase diagram suggests—then

$$G_{MgO} = G^{o}_{MgO} + RT\ln X_{MgO}$$

so

$$\frac{dG_{MgO}}{dx} = \frac{RT}{X_{MgO}}\frac{dX_{MgO}}{dx}.$$

Dividing the numerator and denominator by $\eta = \eta_{Mg} + \eta_{Ni}$ (the total number of cation sites per unit volume, a constant), Equation 15.6 becomes (because $X_{Mg} = X_{MgO}$ and $X_{Ni} = X_{NiO}$ when considering only cations)

$$J'_{Mg} = -\frac{X_{Mg}D_{Mg}X_{Ni}D_{Ni}}{RT\left(X_{Mg}D_{Mg} + X_{Ni}D_{Ni}\right)}\frac{\eta RT}{X_{Ni}X_{Mg}}\left\{\frac{dX_{Mg}}{dx}\right\}$$

$$J'_{Mg} = -\frac{\eta D_{Mg}D_{Ni}}{\left(X_{Mg}D_{Mg} + X_{Ni}D_{Ni}\right)}\left\{\frac{1}{\eta}\frac{d\eta_{Mg}}{dx}\right\}$$

and finally (Step 6)

$$J_{Mg} = \frac{J'_{Mg}}{N_A} = -\frac{D_{Mg}D_{Ni}}{\left(X_{Mg}D_{Mg} + X_{Ni}D_{Ni}\right)}\left\{\frac{dC_{Mg}}{dx}\right\}$$

or

$$J_{Mg} = -\tilde{D}\frac{dC_{Mg}}{dx}$$

where the interdiffusion coefficient, $\tilde{D}$, is given by

$$\tilde{D} = \frac{D_{Mg}D_{Ni}}{\left(X_{Mg}D_{Mg} + X_{Ni}D_{Ni}\right)} \tag{15.7}$$

and, as was the case for interdiffusion in metals, Chapter 14, these diffusion coefficients, D_{Mg} and D_{Ni}, are the *tracer diffusion coefficients at the composition X_{Mg}* (Schmalzried 1981).

15.2.4 Values of $\tilde{D}$

The data for the magnesium and oxygen ion diffusion in pure MgO given in Table 9.5 are as follows: for magnesium, D_0 = 0.25 cm²/s, Q = 330.53 kJ/mol, and for oxygen, $D_0 = 4.3 \times 10^{-5}$ cm²/s and Q = 343.5 kJ/mol. For nickel oxide, the values are $D_0 = 4.4 \times 10^{-4}$ cm²/s and Q = 184.9 kJ/mol for nickel and $D_0 = 6.20 \times 10^{-4}$ cm²/s and Q = 240.6 kJ/mol for oxygen. At 1200°C, the calculated diffusion coefficients for MgO are

$$D_{Mg} = 0.25e^{-\frac{330,540}{(8.314)(1473)}} = 4.74\times10^{-13}\text{cm}^2\text{s}^{-1}$$

$$D_O = 4.3\times10^{-5}e^{-\frac{343,500}{(8.314)(1473)}} = 2.83\times10^{-17}\text{cm}^2\text{s}^{-1}$$

and for NiO

$$D_{Ni} = 4.4\times10^{-4}e^{-\frac{184,900}{(8.314)(1473)}} = 1.22\times10^{-10}\text{cm}^2\text{s}^{-1}$$

$$D_O = 6.2\times10^{-4}e^{-\frac{240,600}{(8.314)(1473)}} = 1.82\times10^{-12}\text{cm}^2\text{s}^{-1}.$$

Notice that oxygen diffusion coefficients are lower than those of the cations in both of the pure materials. An important difference between diffusion in MgO and NiO is the higher melting point of MgO (2825°C for MgO and 1957°C for NiO (Haynes 2013)), hence, the larger activation energies for both cation and anion diffusion. Also, in NiO, there is a deviation from stoichiometry with nickel vacancies, mentioned in Section 14.2.2.1, that gives the nickel ion a higher diffusion coefficient as well. However, the literature suggests that the diffusion coefficient of nickel tracers in MgO at dilute concentrations is very close to that of Mg^{2+} (Kingery et al. 1976). So it would not be surprising that the diffusion coefficient for Mg^{2+} in NiO at low concentrations would not be much different from that of Ni^{2+}. This is expected because both Mg^{2+} and Ni^{2+} diffuse via a vacancy mechanism. At compositions near each of the pure compounds, the vacancy concentration will be essentially that of the pure MgO and pure NiO. However, nickel is heavier than magnesium so that it might be expected that the activation energy for motion of the nickel ion is somewhat higher than that for magnesium giving the nickel ion a lower diffusion coefficient at any given composition. Experimental data are not available for the tracer diffusion coefficients in NiO–MgO alloys so the same approximation made in Section 14.4.4 is applied without any physical or mathematical justification, namely, that the nickel and magnesium tracer diffusion coefficients depend on the square of the mole fraction nickel just to give some functional relation between the diffusion coefficients and composition for purposes of calculation, namely,

$$D^*_{Ni} = 2.0\times10^{-13} + 1.22\times10^{-10}X^2_{Ni}$$

$$D^*_{Mg} = 4.74\times10^{-13} + 3.0\times10^{-10}X^2_{Ni}.$$

These diffusion coefficients are plotted in Figure 15.3 along with $\tilde{D}$, Equation 15.7. Literature data on the MgO–CoO system suggest that $\tilde{D}$ drops much faster with X_{Mg} than this quadratic assumption because the cation vacancy concentration is found to decrease *exponentially* with X_{Mg} in similar CoO–MgO alloys (Schmalzreid 1981). An exponential drop in the interdiffusion coefficient in the NiO–MgO system was also observed for $X_{Ni} \leq 0.5$ and more or less constant for greater NiO concentrations (Appel and Pask 1971). In any event, the individual diffusion coefficients must vary with concentration as does the interdiffusion coefficient. Rather than try to mimic the experimental interdiffusion coefficients, the quadratic variation with X_{Ni} above is used, and Figure 15.3 shows that all three diffusion coefficients, actually the *tracer* diffusion coefficients, vary with composition. The specific functional form of the variation of diffusion coefficients with mole fraction is not important unless a detailed comparison to experimental results* with respect to the unknown tracer diffusion coefficients is desired.

* It should be noted that the experimental results on the NiO–MgO system did not show any marker movement consistent with this model of immobile oxygen ions (Appel and Pask 1971).

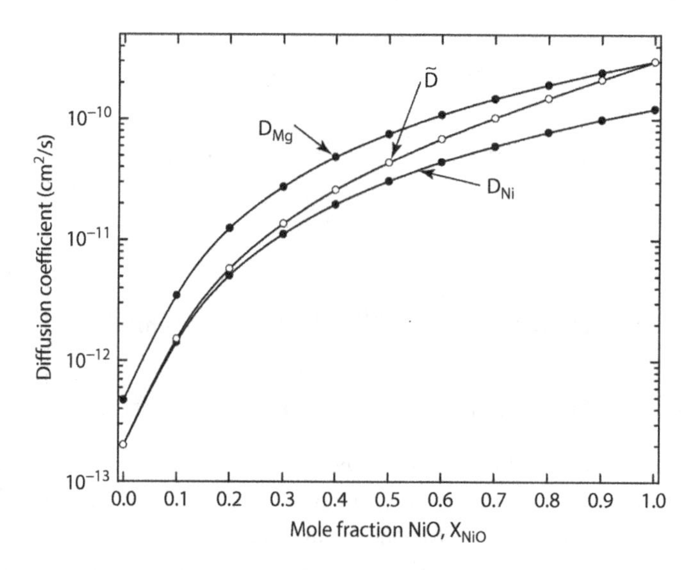

FIGURE 15.3 Calculated diffusion coefficients for nickel and magnesium and the interdiffusion coefficient, $\tilde{D}$, for interdiffusion with immobile oxygen ions. The diffusion coefficients are assumed to vary as X_{NiO}^2.

15.2.5 Interdiffusion with Mobile Oxygen Ions: Kirkendall Effect

15.2.5.1 Expression for Marker Velocity

The electrical neutrality requirement with *mobile oxygen ions* now becomes

$$\begin{aligned} J_{Mg}^* + J_{Ni}^* + J_O^* &= 0 \\ 2eJ'_{Mg} + 2eJ'_{Ni} - 2eJ'_O &= 0 \\ J'_{Mg} + J'_{Ni} - J'_O &= 0. \end{aligned} \tag{15.8}$$

To reduce the physical length of some of the equations, the following briefer notation is used:

$$B_i = \frac{D_i}{k_B T}$$

$$\nabla g_i = \frac{1}{N_A}\frac{dG_i}{dx}$$

where:

i = Mg, Ni, and O and these are the ions Mg^{2+}, Ni^{2+}, and O^{2-}

g_i is the chemical potential or partial molar Gibbs *energy per atom or ion.*

Of course, B = absolute mobility and ∇g_i = thermodynamic force per ion. So the three number flux densities are now

$$\begin{aligned} J'_{Mg} &= \eta_{Mg} B_{Mg}\left(-\nabla g_{Mg} + 2eE\right) \\ J'_{Ni} &= \eta_{Ni} B_{Ni}\left(-\nabla g_{Ni} + 2eE\right) \\ J'_O &= \eta_O B_O\left(-\nabla g_O - 2eE\right). \end{aligned} \tag{15.9}$$

With marker movement, the fluxes relative to a fixed reference frame, J''_i, are (Gopalan and Virkar 1995)

$$J''_{Mg} = J'_{Mg} + \eta_{Mg} v$$
$$J''_{Ni} = J'_{Ni} + \eta_{Ni} v \quad (15.10)$$
$$J''_{O} = J'_{O} + \eta v$$

where η_i = number of ions per cm³ and $\eta = \eta_{Mg} + \eta_{Ni}$ or $1 = X_{Mg} + X_{Ni}$ where X_i = mole fraction and η = constant, independent of composition (constant volume), and v is the marker velocity. Now, with a similar procedure used to develop marker motion in interdiffusing metals, Equation 14.17, the pertinent one, becomes for this system

$$J'_{Mg} + J'_{Ni} = J'_{net} = -v\eta \quad (15.11)$$

so from Equation 15.8

$$J'_{O} = \eta B_O\left(-\nabla\mu_O - 2eE\right) = -\eta v. \quad (15.12)$$

Now, again assuming an ideal solution for the atoms or ions,

$$\mu_O = \mu_O^o + RT\ln\left(\frac{\eta_O}{\eta}\right) \quad (15.13)$$

implies that $\nabla\mu_O = 0$ because $\eta = \eta_O$ = constant; that is, the oxygen ion concentration is everywhere a constant. Therefore, from Equation 15.12

$$-\frac{J'_O}{\eta} = v = 2eEB_O \quad (15.14)$$

or it is the flux of oxygen ions that determines the velocity of the marker motion! In turn, it is determined by the mobility of the oxygen ions and the electric field generated by the motion of the magnesium and nickel cations. What this implies is that oxygen ions move with the faster diffusion cation—for concreteness, assume magnesium (Figure 15.3)—coupled by their electrical charges. This means that MgO moves past the markers forcing them to move in the opposite direction! This is essentially the same situation as in Figure 14.12 for the interdiffusion of metals, but now there is an interdiffusion of the *oxides*. Of course, there is a flux of vacancies in the opposite direction, but now they are vacancy *pairs* or *Schottky defects*! Equation (15.14) means that, if there is a difference in the mobilities of the different ions generating a local electric field, E, as long as the oxygen has some mobility, $B_O > 0$, there *will be* marker movement and a Kirkendall effect during the interdiffusion of compounds! The details of the rest of the algebra applying the six steps listed above are carried out in Section A.2.1 and only the important results are given here.

15.2.5.2 Electric Field Term

As in other cases, the electric field term is obtained from the electrical neutrality equation (15.8) (Equation A.16):

$$2eE = \frac{\left(\eta_{Mg}B_{Mg}\nabla g_{Mg} + \eta_{Ni}B_{Ni}\nabla g_{Ni}\right)}{\left(\eta_{Mg}B_{Mg} + \eta_{Ni}B_{Ni} + \eta_O B_O\right)}. \quad (15.15)$$

15.2.5.3 Velocity Term

Substitution of Equation 15.15 into Equation 15.14 and application of Gibbs–Duhem gives (Equation A.19)

$$v = \frac{\eta_{Mg}B_O\left(B_{Mg} - B_{Ni}\right)}{\left(\eta_{Mg}B_{Mg} + \eta_{Ni}B_{Ni} + \eta_O B_O\right)}\nabla g_{Mg}. \quad (15.16)$$

15.2.5.4 Flux of Magnesium Ions

Also, substitution of Equation 15.15 into the first of Equations 15.9 gives for the number flux of magnesium ions (Equation A.18)

$$J'_{Mg} = -\frac{\eta\eta_{Mg}B_{Mg}(B_{Ni} + B_O)}{(\eta_{Mg}B_{Mg} + \eta_{Ni}B_{Ni} + \eta_O B_O)}\nabla g_{Mg}. \tag{15.17}$$

15.2.5.5 Result

Substituting Equations 15.17 and 15.16 into the first of Equations 15.10, and solving and replacing mobilities with diffusion coefficients gives the desired result (Gopalan and Virkar 1995),

$$J_{Mg} = \frac{J''_{Mg}}{N_A} = -\tilde{D}\frac{dC_{Mg}}{dx}$$

where $\tilde{D}$ is given by (Equation A.20)

$$\tilde{D} = \frac{D_{Mg}D_{Ni} + D_O(X_{Mg}D_{Ni} + X_{Ni}D_{Mg})}{X_{Mg}D_{Mg} + X_{Ni}D_{Ni} + D_O} \tag{15.18}$$

and the marker velocity in terms of the diffusion coefficients is (Equation A.21)

$$v = \frac{D_O(D_{Mg} - D_{Ni})}{(X_{Mg}D_{Mg} + X_{Ni}D_{Ni} + D_O)}\frac{dX_{Mg}}{dx}. \tag{15.19}$$

15.2.5.6 Implications

The diffusion coefficients in Equations 15.18 and 15.19 are all tracer diffusion coefficients at the given composition of X_{Mg}. From Equation 15.18, clearly if $D_O = 0$, then $\tilde{D}$ is the same as in Equation 15.7 where it was assumed that oxygen ions did not move, while Equation 15.19 shows that the marker velocity is zero as it should be. On the other hand, if $D_O >> D_{Ni}, D_{Mg}$, then the marker velocity depends on the difference between D_{Mg} and D_{Ni} and Equation 15.18 becomes

$$\tilde{D} \cong X_{Mg}D_{Ni} + X_{Ni}D_{Mg}.$$

This is Darken's equation for interdiffusion, Chapter 14, Equation 14.23! Therefore, with rapid oxygen diffusion, the interdiffusion is essentially that of *MgO and NiO* and there is a net flux of *Schottky defects*, vacancy pairs, that produces the marker motion! Quantitatively, how large is the marker motion? Assume that interdiffusion has taken place so that X_{Mg} essentially goes from 1.0 → 0 over a distance of 100 μm and that $D_{Mg} \cong 8 \times 10^{-11}$ cm²/s and $D_{Ni} \cong 2 \times 10^{-11}$ cm²/s (Figure 15.3 at $X_{NiO} \cong 0.5$). Then, a very approximate average velocity, $\bar{v}$, would be given by

$$\bar{v} \cong (D_{Mg} - D_{Ni})\frac{\Delta X_{Mg}}{\Delta x} \cong 6\times10^{-11}\times\frac{1}{10^{-2}} \cong 6\times10^{-9}\,\text{cm}\,\text{s}^{-1}.$$

But $4Dt \cong x^2$ so $t \cong x^2/4\tilde{D} \cong (10^{-2})^2/4\times(4\times10^{-11}) \cong 6.25\times10^5\,\text{s} \cong 174\,\text{h}$; hence the distance the markers moved in this time is about $6\times10^{-9}\,\text{cm}\,\text{s}^{-1}\times6.25\times10^5\,\text{s} \cong 3.75\times10^{-3}\,\text{cm} = 37.5\,\mu\text{m}$, about one-third of the interdiffusion zone and easily measurable and clearly important. Also, Kirkendall porosity is likely to form.

But, as shown above in Section 15.2.4, the diffusion coefficients for oxygen in pure MgO and NiO are two to four orders of magnitude lower than the cation diffusion coefficients. This is typical of oxides and other compounds in that the anion diffusion coefficient is usually lower than those of the cations because the anions are larger and it takes more energy to move them than it does the smaller cations. As a result, if the oxygen ions were to diffuse two orders of magnitude slower than the cations, then the equation for the marker velocity, Equation 15.19, becomes approximately

$$v \cong \frac{D_O}{D_{Mg}}\left(D_{Mg} - D_{Ni}\right)\frac{dX_{Mg}}{dx}.$$

So in contrast to the example for a 38 μm marker motion when the oxygen diffusion coefficient is large, it is now reduced to about 0.38 μm over a 100 μm diffusion zone and would be very difficult to detect. The same is true for the interdiffusion coefficient, Equation 15.18: if the oxygen diffusion coefficient is two orders of magnitude lower than those of the cations, the second term in $\tilde{D}$ can be neglected and it reverts to the immobile anion solution, Equation 15.7. Therefore, a major conclusion illustrated by this model is: *when the anion diffusion coefficient is two orders of magnitude or less than the cation diffusion coefficients, for all practical purposes, the anion lattice can be considered to be immobile, the marker velocity is small and can be ignored, and the interdiffusion coefficient is given by Equation 15.7,*

$$\tilde{D} = \frac{D_{Mg}D_{Ni}}{\left(X_{Mg}D_{Mg} + X_{Ni}D_{Ni}\right)}.$$

On the other hand, there are some important systems in which the anion diffusion coefficient is many orders of magnitude greater than the cation diffusion coefficient: specifically ZrO_2 stabilized by other oxides such as CaO, MgO, or Y_2O_3. As discussed in Section 7.10.2, these dopants can form solid solutions with ZrO_2 (and other fluorite structure oxides) up to large mole fractions with the formation of oxygen vacancies:

$$CaO \xrightarrow{ZrO_2} Ca''_{Zr} + O_O^{x} + V_O^{\bullet\bullet}$$

or

$$xCaO + (1-x)ZrO_2 = Zr_{1-x}Ca_xO_{1-x}$$

where x can be on the order of 0.1 or 0.2. As a result, the oxygen diffusion coefficient can be large—$D \cong 10^{-6}$ cm²/s—because of the large oxygen vacancy concentration. However, the cation vacancy concentration is now much smaller through the Schottky product

$$\left[V''''_{Zr}\right]\left[V_O^{\bullet\bullet}\right]^2 = K_S$$

as are the cation (Ca^{2+} and Zr^{4+}) diffusion coefficients (Rhodes and Carter 1966). Therefore, the interdiffusion coefficient in Equation 15.18 and the Kirkendall velocity in Equation 15.19 are still small. Nevertheless, such systems are likely ones in which to observe signs of a Kirkendall effect such as marker movement and porosity.

15.3 INTERDIFFUSION WITH MOBILE ELECTRONS: HIGH ELECTRONIC CONDUCTIVITY

15.3.1 MgO–NiO Interdiffusion with Fixed Oxygen Ions

One very simple—yet correct—approach is as follows. For mobile cations and electrons but fixed oxygen ions, the last equation in Equation 15.1 still holds, namely,

$$J'_{Mg} + J'_{Ni} = 0 \tag{15.20}$$

but with mobile electrons, there is essentially no electric field, so Equation 15.2 is now

$$J'_{Mg} = -\frac{\eta_{Mg^{2+}}D_{Mg^{2+}}}{RT}\frac{dG_{Mg^{2+}}}{dx}$$
$$J'_{Ni} = -\frac{\eta_{Ni^{2+}}D_{Ni^{2+}}}{RT}\frac{dG_{Ni^{2+}}}{dx} \tag{15.21}$$

and combining these two equations gives

$$J'_{Mg} = -\frac{\eta_{Mg^{2+}} D_{Mg^{2+}}}{RT}\frac{dG_{Mg^{2+}}}{dx} = -J'_{Ni} = \frac{\eta_{Ni^{2+}} D_{Ni^{2+}}}{RT}\frac{dG_{Ni^{2+}}}{dx}. \tag{15.22}$$

Again, applying the Gibbs–Duhem equation,

$$X_{Ni^{2+}}\frac{dG_{Ni^{2+}}}{dx} + X_{Mg^{2+}}\frac{dG_{Mg^{2+}}}{dx} = 0$$

Equation 15.22 becomes

$$-\eta_{Mg^{2+}} D_{Mg^{2+}}\frac{dG_{Mg^{2+}}}{dx} = -\eta_{Ni^{2+}} D_{Ni^{2+}}\frac{X_{Mg^{2+}}}{X_{Ni^{2+}}}\frac{dG_{Mg^{2+}}}{dx}$$

and gives the simple result

$$D_{Mg^{2+}} = D_{Ni^{2+}}. \tag{15.23}$$

That is, the tracer diffusion coefficients for the two diffusing cations Mg^{2+} and Ni^{2+} *must be the same* at any given composition. This does not mean that they are independent of composition that they must be (Figure 15.3), but $\tilde{D} = D_{Mg^{2+}} = D_{Ni^{2+}}$. This is not too surprising because if one of the cations diffused faster than the other, then an excess cation concentration would build up and there would be an increase in nonstoichiometry. But, because the oxygen chemical potential is fixed by the atmosphere, so must the cation and anion vacancy concentrations and therefore the stoichiometry, as shown in Chapter 9, Figure A.5. So one cation cannot diffuse faster than the other to ensure that the stoichiometry at a given composition remains constant.

15.3.2 Keeping the Electric Field But with Mobile Electrons

The approach here is similar to that used to obtain the Kirkendall effect where the assumption was mobile oxygen ions. In this model, the electronic conductivity is assumed to be high enough so that an electric field cannot build up. That is, both the electron (or hole) concentration and mobility are high compared to those for the ions. The relevant equations for this model are

$$\begin{aligned} &J^*_{Mg} + J^*_{Ni} + J^*_e = 0 \\ &2eJ'_{Mg} + 2eJ'_{Ni} - eJ'_e = 0 \\ &2J'_{Mg} + 2J'_{Ni} - J'_e = 0 \end{aligned} \tag{15.24}$$

and

$$\begin{aligned} &J'_{Mg} = \eta_{Mg} B_{Mg}\left(-\nabla g_{Mg} + 2eE\right) \\ &J'_{Ni} = \eta_{Ni} B_{Ni}\left(-\nabla g_{Ni} + 2eE\right) \\ &J'_e = \eta_e B_e\left(-\nabla g_e - eE\right). \end{aligned} \tag{15.25}$$

The algebra and other assumptions are given in Section A.2.2, which lead to the general result

$$J'_{Mg} = -\frac{\eta_{Mg} B_{Mg}\left(2\eta B_{Ni} + \eta_e B_e\right)}{\left(2\eta_{Mg} B_{Mg} + 2\eta_{Ni} B_{Ni} + \eta_e B_e\right)}\nabla g_{Mg}. \tag{15.26}$$

If the electronic conductivity is much higher than the ionic conductivity, $\eta_e B_e \gg \eta_{Mg} B_{Mg} \cong \eta_{Ni} B_{Ni}$, then Equation 15.26 reduces to

$$J_{MgO} = J_{Mg} = -D_{Mg}\frac{dC_{Mg}}{dx}$$

the same as Equation 15.23 because the NiO flux is equal and opposite to the MgO flux. Similarly, if the electronic conductivity is low, $\eta_e B_e \ll \eta_{Mg} B_{Mg} \cong \eta_{Ni} B_{Ni}$, then an electric field can be developed with the result

$$J_{Mg} = -\tilde{D}\frac{dC_{Mg}}{dx}$$

where $\tilde{D}$, as obtained in Section 15.2.3, Equation 15.7 where an electric field coupled ion motion, is

$$\tilde{D} = \frac{D_{Mg} D_{Ni}}{\left(X_{Mg} D_{Mg} + X_{Ni} D_{Ni}\right)}.$$

Finally, if $\eta_e B_e \cong \eta_{Mg} B_{Mg} \cong \eta_{Ni} B_{Ni}$, then Equation 15.26 would have to be used to obtain an interdiffusion coefficient, $\tilde{D}$.

15.3.3 Mobile Oxygen Ions and Mobile Electrons

The most comprehensive way to approach this model is similar to the method used earlier in Section 15.2.5, but now

$$J^*_{Mg^{2+}} + J^*_{Ni^{2+}} + J^*_{O^{2-}} + J^*_{e^-} = 0.$$

If the same procedure is followed, the algebra increases by 25%! But a complete and general solution is obtained. However, here the interest is: "Will a Kirkendall effect be observed with high electronic conductivity?" This can be answered with less algebra by a combination of the procedures used in Sections 15.2.5 and 15.3.1, where the electric field is assumed to be zero. Therefore, the model becomes

$$J'_{Mg} + J'_{Ni} - J'_O = 0$$

with

$$J'_{Mg} = -\eta_{Mg} B_{Mg} \nabla g_{Mg}$$

$$J'_{Ni} = -\eta_{Ni} B_{Ni} \nabla g_{Ni}$$

$$J'_O = -\eta_O B_O \nabla g_O$$

and

$$J''_{Mg} = J'_{Mg} + \eta_{Mg} v$$

$$J''_{Ni} = J'_{Ni} + \eta_{Ni} v$$

$$J''_O = J'_O + \eta v$$

where v is the lattice velocity. But again, $J''_O = 0$ so $J'_O = -\eta v = -\eta B_O \nabla g_O$. The algebra is carried out in Section A.2.3 with the results

$$J_{MgO} = -\tilde{D}\frac{dC_{MgO}}{dx}$$

with

$$\tilde{D} = \frac{D_{Mg} D_{Ni} + D_O\left(X_{Ni} D_{Mg} + X_{Mg} D_{Ni}\right)}{D_{Ni} + D_O} \tag{15.27}$$

and

$$v = \frac{D_O\left(D_{Mg} - D_{Ni}\right)}{D_{Ni} + D_O}\frac{dX_{Mg}}{dx}. \tag{15.28}$$

These are essentially the same results as in Sections 15.2.5.5 and 15.3.1. Most importantly, if the oxygen diffusion coefficient is small compared to the cation diffusion coefficients, $\tilde{D} \cong D_{Mg}$ and the marker velocity goes to zero as before. If the oxygen diffusion coefficient is large compared to those of the cations, then Darken's equations are obtained again:

$$\tilde{D} = X_{Ni}D_{Mg} + X_{Mg}D_{Ni}$$

and

$$v = (D_{Mg} - D_{Ni})\frac{dX_{Mg}}{dx}.$$

Therefore, a Kirkendall effect will occur during interdiffusion of compounds as long as the anion diffusion coefficient is sufficiently large regardless of the value of the electronic conductivity. Even the interdiffusion of metallic conductors such as TiB_2 and ZrB_2 can potentially produce a Kirkendall effect.

15.4 INTERDIFFUSION AND SOLID-STATE REACTIONS

15.4.1 Importance

Although there are some important materials that are *solid solutions* produced by the interdiffusion of the constituent compounds, for example, calcia-stabilized zirconia, $Ca_xZr_{1-x}O_{2-x}$, the vast majority of technically important *ternary* compounds are usually formed by solid-state interdiffusion during a calcining reaction.* Examples include $Y_3Fe_5O_{12}$ and $MgFe_2O_4$ for magnetic applications such as radar; $BaTiO_3$ for capacitors and sonar; $LiFePO_4$ and $LiCoO_2$ for fuel cell electrodes; $MgAl_2O_4$ for optical and infrared windows (Ramisetty et al. 2013); and last but not least, the main constituents of Portland cement. Portland cement is the largest volume man-made material by a large margin: in 2013, some four billion metric tons (1 metric ton = 1000 kg) of it were produced, more than one-half in China (USGS 2014). This is about twice the world tonnage production of iron and steel and about an order of magnitude greater than the tonnage of polymers (Review 2012). The main constituents of Portland cement that provide its ability to harden or *set* are $3CaO \cdot SiO_2$† and $2CaO \cdot SiO_2$ along with a minor amounts of $3CaO \cdot Al_2O_3$ and other compounds. They are made by heating clay, $Al_2O_3 \cdot 2SiO_2 \cdot 2H_2O$, and $CaCO_3$ at temperatures up to 1600°C in a *cement kiln* producing a cement *clinker* (large sintered chunk), which is then ground to a fine powder (Bogue 1955; Norton 1974). These compounds are formed by solid-state reactions. It is the reaction of these calcium-rich compounds with water to form an amorphous hydrated calcium silicate solid that bonds to itself and the added sand and silicate filler material in *concrete*.‡ Unfortunately, because of the large amount of Portland cement made, its production alone contributes almost 5% to the total annual world CO_2 gas emissions (Worrell 2001). This is a result of the high firing temperatures requiring large quantities of fuel and the large volume of CO_2 released from the calcium carbonate during the *calcination or calcining* reaction§ (Crowe 2008).

Based on the conclusions reached above in Section 15.2.5, as a first, and good, approximation, the anion sublattice is assumed stationary and compounds are formed simply by interdiffusion of the two cations. Most of the interesting compounds, for example, spinels and titanates, have cations with different charges that make calculations a little longer and more interesting. So, before investigating interdiffusion to form one of these important compounds, the formation of a hypothetical compound ABO_2 from AO and BO is modeled where both cations have the same charge of 2+ simply to remove the slightly complicating factor of two different cation charges. There are several

* Alternatively, but considerably less frequently, they may be formed by precipitation reaction.

† This notation of a chemical compound $3CaO{\cdot}SiO_2$ rather than Ca_3SiO_5 is frequently used in mineralogy. In cement technology it is referred to as C_3S.

‡ Surprisingly, in spite of its technological importance, the details of the kinetics of the hardening or *setting* of the resulting cement powder with water are still not completely understood.

§ In fact, all the solid-state reactions discussed here are frequently referred to as *calcining* reactions because of their similarity to the production of lime, CaO, by the heating of $CaCO_3$ to drive off the CO_2.

compounds with the formula ABO_2 such as $Li^+Ni^{3+}O_2$ that usually consist of monovalent and trivalent cations on alternating {111} planes of the NaCl structure. This distorts the cubic structure along one of the body diagonals to give a rhombohedral, layer-like, structure allowing easy entry and exit for Li^+ ions required of lithium battery electrode material. On the other hand, there seem to be few, if any, real compounds ABO_2 where both cations have a 2+ charge (Wells 1984). Nevertheless, such an imaginary compound is used to illustrate the principles. This is followed by the modeling a more important real-world material, $MgAl_2O_4$ spinel.

15.4.2 Formation of ABO_2 from AO and BO

Figure 15.4 shows a hypothetical phase diagram for AO and BO with a single compound ABO_2 with no measurable variation in composition. At some temperature, T_0, one mole of pure AO is reacting with one mole of pure BO at time t = 0 (Figure 15.5). It is assumed that the AO and BO have the

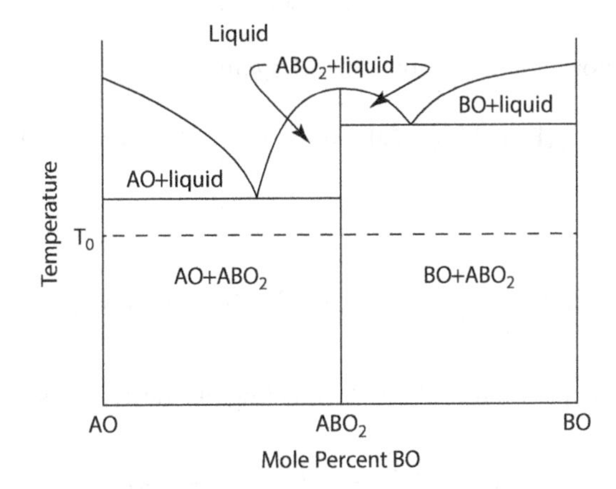

FIGURE 15.4 Hypothetical phase diagram between two oxides AO and BO with a single compound ABO_2 with no evident solid solution in any of the phases and a solid-state reaction temperature of T_0.

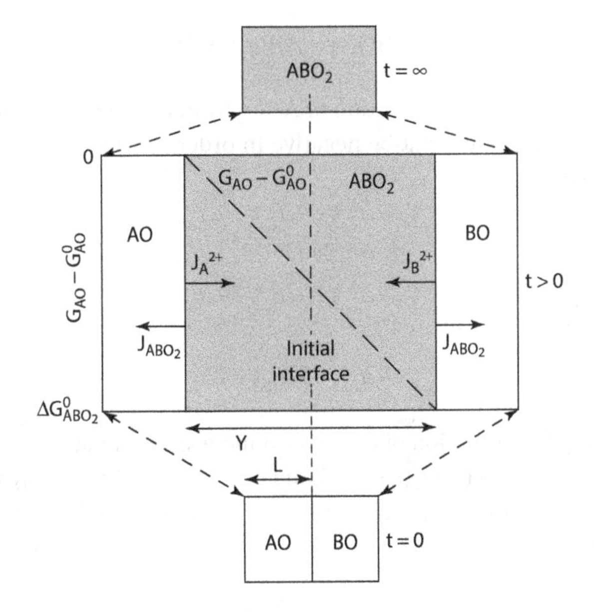

FIGURE 15.5 Microstructure between two equimolar reacting particles of AO and BO: a diffusion couple. The direction of the ion fluxes and the fluxes of ABO_2 are shown. As described in the text, for each mole of ABO_2 produced on the right of the original interface, there is an equal amount produced on the left side. The diagonal dashed line through the ABO_2 regions is the free energy of AO less than in the standard state. It decreases from zero to $\Delta G^0_{ABO_2}$ (which is, of course, negative) across the phase field. The same line for BO would just decrease from zero in the opposite direction. The microstructures are shown for $t = 0$, $0 < t < \infty$, and $t = \infty$, complete reaction.

same molar volume or the number of oxygen ions per unit volume as does the compound ABO_2, so there is no volume change during the reaction, the usual assumption. After some time, $t > 0$, A^{2+} and B^{2+} ions will have interdiffused on the fixed oxygen lattice to produce equal amounts of ABO_2 on the left and right of the initial interface (Figure 15.5). In this case, the charge flux densities of A^{2+} and B^{2+} must be equal as usual,

$$J_A^* + J_B^* = 0$$

$$2eJ'_A + 2eJ'_B = 0$$

$$J'_A + J'_B = 0$$

or

$$J'_A = -J'_B.$$

Consider the shorthand notation of Section 15.2.5.1 again,

$$J'_A = \eta_A B_A\left(-\nabla g_A + 2eE\right) = -J'_B = -\eta_A B_A\left(-\nabla g_A + 2eE\right)$$

and the expression for 2eE,

$$2eE = \frac{1}{\left(\eta_A B_A + \eta_B B_B\right)}\left\{\eta_A B_A \nabla g_A + \eta_B B_B \nabla g_B\right\}.$$

Substituting this into the expression for J'_A and canceling like terms gives

$$J'_A = \frac{\eta_A B_A \eta_B B_B}{\left(\eta_A B_A + \eta_B B_B\right)}\left\{\nabla g_A - \nabla g_B\right\}. \tag{15.29}$$

It is necessary to relate ∇g_A and ∇g_B. Now the overall reaction is

$$AO(s) + BO(s) = ABO_2(s);\ \Delta G^o_{ABO_2} \tag{15.30}$$

where $\Delta G^o_{ABO_2}$ is the Gibbs energy of formation of ABO_2 from the *constituent oxides and not the elements* (Kubaschewski 1972) and it must be negative in order for the reaction to take place. So, the free energies per "molecule" are

$$g_{AO} + g_{BO} = \Delta g^o_{ABO_2}$$

so

$$\nabla g_{AO} + \nabla g_{BO} = 0$$

because the Gibbs energy for formation of the compound is constant at a given temperature. Again, $A + 1/2O_2(g) = AO$ so $A^{2+} - 2e^- + 1/2O_2(g) = AO$ and $g_A - 2g_{e^-} + 1/2g_{O_2} = g_{AO}$ so $\nabla g_A + 2\nabla g_e = \nabla g_{AO}$ because $\nabla g_{O_2} = 0$ as before. Similarly, $\nabla g_B + 2\nabla g_e = \nabla g_{BO}$ so the equivalent of the Gibbs–Duhem equation here is simply

$$\nabla g_A = -\nabla g_B. \tag{15.31}$$

Putting Equation 15.31 into Equation 15.29 and replacing the mobilities with diffusion coefficients,

$$J'_A = -\frac{1}{RT}\frac{\eta_A D_A \eta_B D_B}{\left(\eta_A D_A + \eta_B D_B\right)}\frac{dG_A}{dx}. \tag{15.32}$$

Now $\eta_A = \eta_B = \eta$ so

$$J_A = \frac{J'_A}{N_A} = -\frac{1}{RT}\frac{\eta_A}{N_A}\frac{D_A D_B}{(D_A + D_B)}\frac{dG_A}{dx}$$
$$J_A = -\frac{1}{RT}C_A\frac{D_A D_B}{(D_A + D_B)}\frac{dG_A}{dx} \tag{15.33}$$

where C_A = concentration (mol/cm³). From Equation 15.30, the equilibrium constant for the reaction is given by

$$K_e = \frac{a_{ABO_2}}{a_{AO}\, a_{BO}} = e^{-\frac{\Delta G^o_{ABO_2}}{RT}}. \tag{15.34}$$

So the Gibbs energy of AO at the AO–AO_2 interface, where a_{AO} and a_{ABO_2} are both 1.0, is $G_A = G^o_A$, the Gibbs energy of the standard state because $RT \ln a_{AO} = 0$. Similarly, the Gibbs energy of AO at the BO–ABO_2 interface, where a_{BO} and a_{ABO_2} are both 1, is $G_A = G^o_A + \Delta G^o_{ABO_2}$. If Y is the thickness of the reacted layer as shown in Figure 15.5, then

$$\frac{dG_A}{dx} = \frac{\left(G^0_A + \Delta G^0_{ABO_2}\right) - G^0_A}{Y} = \frac{\Delta G^0_{ABO_2}}{Y} \tag{15.35}$$

and this is negative because ΔG_{ABO_2} is negative. Now,

$$J_{ABO_2} = 2J_A \tag{15.36}$$

because for every new mole created on the right side of ABO_2 by J_A one is also created on the left by the flux of B^{2+}. And this a planar reaction, so as before,

$$\frac{\rho_{ABO_2}}{M_{ABO_2}}\frac{dY}{dt} = C_{ABO_2}\frac{dY}{dt}. \tag{15.37}$$

Combination of Equations 15.33 and 15.35 through 15.37 and recognizing that $C_{ABO_2} = C_{AO} = C_A$ gives the result

$$\frac{1}{Y}\frac{dY}{dt} = -\frac{2}{RT}\frac{D_A D_B}{(D_A + D_B)}\Delta G^o_{ABO_2} \tag{15.38}$$

which integrates to

$$Y^2 = kt \tag{15.39}$$

where the reaction rate constant, k, is given by

$$k = -\frac{4}{RT}\frac{D_A D_B}{(D_A + D_B)}\Delta G^o_{ABO_2}. \tag{15.40}$$

Two things needing emphasis in this result are as follows: first, there are only two tracer diffusion coefficient, D_A and D_B, which, of course, are the diffusion coefficients of A and B in the compound ABO_2 and are constant because the composition is constant across ABO_2; second, if one of the diffusion coefficients is much larger than the other, say $D_A >> D_B$, then $k \propto D_B$ and, not surprisingly, the *slower diffusing species controls* the rate of reaction.

15.4.3 Fraction Reacted

Ideally, the complete reaction—the fraction reacted equals 1—is what is desired but sometimes hard to accomplish. Therefore, of practical importance during solid-state reactions is the fraction reacted as a function of time. For the reaction of AO and BO in Section 15.4.2 with the simple linear model of Figure 15.5, the fraction of AO reacted can be taken to be

$$f_R = \frac{V_0 - V(t)}{V_0} = 1 - \frac{V(t)}{V_0} = 1 - \frac{\left(L - \frac{Y}{2}\right)}{L} = \frac{Y}{2L} \tag{15.41}$$

where:

$V(t)$ = volume unreacted

V_0 = unreacted volume at $t = 0$

L = initial length of the AO half of the diffusion couple. So with Equation 15.40, the fraction reacted becomes

$$f_R^2 = \left(\frac{Y}{2L}\right)^2 = \left(-\frac{\Delta G^0_{ABO_2}}{RT}\right)\left(\frac{D_{eff}t}{L^2}\right) \tag{15.42}$$

where $D_{eff} = D_A D_B / D_A + D_B$. Note that f_R^2 is simply linear with time and depends on the dimensionless Dt/L^2 parameter typical of all diffusion-controlled reactions.

15.4.4 Reaction of MgO and Al_2O_3 to Form $MgAl_2O_4$ Spinel

15.4.4.1 Introduction

A more common solid-state reaction is the formation of a ternary compound from two binary compounds whose cations have different charges such as those presented in Section 15.4.1. The purpose of this model is to show the effect of the different cation valences on the interdiffusion coefficient and reaction rate constant, and where, spatially, the ternary compound is formed. Most of these compounds do not occur in nature—at least not in a sufficiently pure form—and their compositions are often tailored to produce a very specific set of properties dictated by the application such as the magnetic garnet, $Y_{3\text{-}x\text{-}z}Gd_xDy_zFe_{5\text{-}m\text{-}n\text{-}p}Al_mMn_nIn_pO_{12}$, as described in Chapter 1. An important class of ternary compounds is the cubic spinels based on the naturally occurring material, $MgAl_2O_4$, which is colored by impurity ions in solid solution, such as Cr^{3+} or Co^{2+}, and serves as a gemstone. The reaction to form spinels from their constituent oxides has been studied in several systems, $MgAl_2O_4$ and $MgFe_2O_4$ (Carter 1961a) and $NiAl_2O_4$ (Pettit et al. 1966) to give examples. Figure 15.6 shows a simplified phase diagram for the MgO–Al_2O_3 system that does not show any solid solution between the single ternary compound, $MgAl_2O_4$, while diagrams in the literature (Levin et al. 1964, Figure 260) suggest some solid solubility between the spinel and Al_2O_3.* Here, the solid solution is ignored because it complicates the calculation, again without adding new information. It is also small at the temperatures where the reaction is carried out, around 1200°C. The procedure to calculate the reaction rate constant follows the six steps outlined earlier in Section 15.2.2 and the exhaustive algebra—it is only algebra—is ground through in the Appendix and only the results are presented. The assumption is that the spinel phase is growing by the counter diffusion of Mg^{2+} and Al^{3+} ions through the already formed spinel. Again, the oxygen ions are considered immobile and at a fixed concentration. There is essentially no flux of oxygen ions because the concentrations of oxygen in all three phases, MgO, Al_2O_3, and $MgAl_2O_4$, are essentially the same. The molar volume of MgO is (Haynes 2013)

---

* This is probably because Al_2O_3 can exist in a metastable phase, γ-Al_2O_3, which has the spinel structure (Wells 1984).

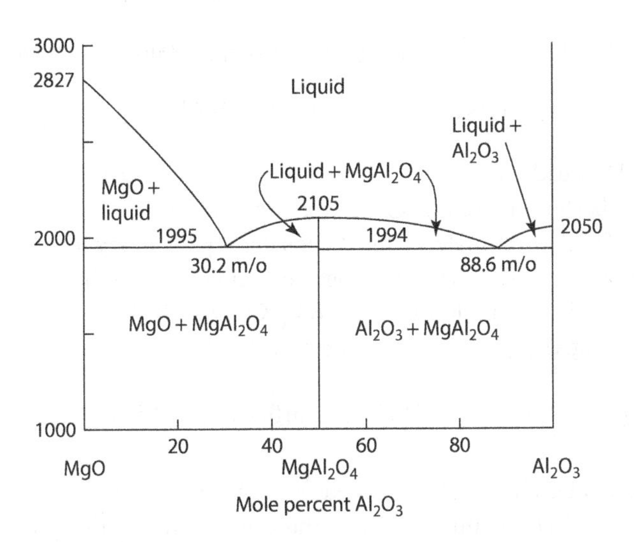

FIGURE 15.6 The MgO–Al_2O_3 system with the single compound, $MgAl_2O_4$ spinel. The temperatures and composition of melting points and eutectics are essentially those given in the literature. However, the literature suggests a significant amount of solid solution at elevated temperatures between $MgAl_2O_4$ and Al_2O_3 that has not been included in this diagram (Levin et al. 1964). The addition of solid solution just complicates the model without adding new information. In any event, the reaction can be assumed to occur at a lower temperature where solid solution is very small.

$$\bar{V}_{MgO} = \frac{M}{\rho} = \frac{40.31\,\text{g/mol}}{3.58\,\text{g/cm}^3} = 11.26\ \text{cm}^3\text{/mol}$$

that of Al_2O_3

$$\bar{V}_{Al_2O_3} = \frac{101.96\,\text{g/mol}}{3.97\,\text{g/cm}^3} = 25.68\,\text{cm}^3\text{/mol}$$

and that of $MgAl_2O_4$

$$\bar{V}_{MgAl_2O_4} = \frac{142.27\,\text{g/mol}}{3.55\,\text{g/cm}^3} = 40.07\,\text{cm}^3\text{/mol.}$$

So the molar volume of $MgAl_2O_4$ is about 8% larger than the combined molar volumes of MgO and Al_2O_3, which is pretty close. Furthermore, the chemical potential of oxygen is the same all over because the reaction is presumably being carried out in an atmosphere of fixed oxygen content, more than likely, air. Therefore, marker motion is precluded in this reaction model. Of course, oxygen diffusion could be included as it was for the interdiffusion of MgO and NiO in Section 15.2.5, making the algebra messier, but giving essentially similar results. So, under the right conditions among the diffusion coefficients, there could be interdiffusion of the oxides and marker motion. Such a model is more complicated than reality warrants and is left to another place and another time.

15.4.4.2 The Reaction

The reaction to form magnesium aluminate spinel from *magnesium and aluminum oxides* is

$$MgO(s) + Al_2O_3(s) \rightarrow MgAl_2O_4(s);\ \Delta G^o_{MgAl_2O_4} \qquad (15.43)$$

where $\Delta G^o_{MgAl_2O_4}$ is the Gibbs energy to form the spinel from the *constituent oxides and not the elements* (Kubaschewski 1972), is a constant throughout the thickness of $MgAl_2O_4$, and must be negative in order for the reaction to take place. For example, the Gibbs energy for the reaction

$$MgO(s) + Al_2O_3(s) = MgAl_2O_4(s)$$

at 1200°C is $\Delta G^o = 15.03\,kJ/mol$. (Roine 2002) In contrast, the Gibbs energy for the reaction

$$Mg(s) + 2Al(s) + 2O_2(g) = MgAl_2O_4(s) \tag{15.44}$$

at 1200°C is $\Delta G^o = -1658.6\,kJ/mol$, almost two orders of magnitude larger (Roine 2002). This large difference in energy is because there are no electrons transferred in Equation 15.43, while there are in Equation 15.44. In fact, an estimate for the Gibbs energy for many ternary compounds formed from their constituent oxides can, if necessary, be approximated from the ideal Gibbs energy of mixing (Kubaschewski 1972). For example, $\Delta G^{id}_{mixing} = RT(X_1 \ln X_1 + X_2 \ln X_2)$ and at 1200°C for a simple 1:1 compound such as $MgAl_2O_4$, where X_i are the mole fractions,*

$$\Delta G^{id}_{mixing}(1200\,°C) = (8.314)(1473)(0.5\ln 0.5 + 0.5\ln 0.5) = -8.489\,kJ/mole$$

not exactly the same as the tabulated values, but within a factor of 2, and not very large. This is particularly true of spinel compounds because there is not much difference in the oxygen ion packing between the reactants and products. For other compounds, such as a perovskite, $CaTiO_3$, the Gibbs energies of formation are somewhat more negative than that for an ideal solid solution (Kubaschewski 1972).

15.4.4.3 Ion Flux Charge Neutrality

This is important and has been presented several times now. But the literature is a bit confusing for ternary compounds with cations of different charge. Again, it is assumed that the electronic conductivities in MgO, Al_2O_3, and $MgAl_2O_4$ are sufficiently small, which they probably are at the temperatures of interest, something around 1200°C, so only charge balance between the moving ions need be considered. Anyway, this is what the model assumes and the charge flux balance is given by

$$\begin{aligned} J^*_{Mg} + J^*_{Al} &= 0 \\ 2eJ'_{Mg} + 3eJ'_{Al} &= 0 \text{ or} \\ 2J'_{Mg} &= -3J'_{Al} \end{aligned} \tag{15.45}$$

where the subscripts refer to the ions Mg^{2+} and Al^{3+}. The last of these three equations can be written in the following form:

$$J'_{MgAl_2O_4}(\text{right}) = J'_{MgO} = J'_{Mg} = -3\left(\frac{1}{2}\right)J'_{Al} = -3J'_{Al_2O_3} = -3J'_{MgAl_2O_4}(\text{left}). \tag{15.46}$$

In words, "the number of moles of spinel produced on the right side of the original interface is *three times* that produced on the left." This is an important result and it is depicted in Figure 15.7, in which "right" refers to the volume or number of moles of $MgAl_2O_4$ produced on the "right" side of the original interface and "left" the amount produced on the "left" side of the initial interface. Figure 15.7 shows four moles each of MgO and Al_2O_3 reacting to form four moles of $MgAl_2O_4$, one mole to the left of the interface and three to the right, and the initial interface is stationary consistent with no diffusion of oxygen. In a real reaction, probably only a few millimoles will be reacted because the particle dimensions must be small enough for a large fraction of the reaction to be completed in a reasonable amount of time. This is considered in more detail later. Inert marker experiments have not been very successful in identifying the original interface in this reaction because the markers break and give an inconsistent result. However, there is always some residual porosity at the original interface that can be used to mark its position and a 3:1 ratio in the amounts of spinel formed is what is found experimentally (Carter 1961a; Pettit et al. 1966) consistent with Figure 15.7.

* In $MgAl_2O_4$, the magnesium ions are on tetrahedral sites and the aluminum ions are on octahedral sites for a total of $3N_A$ cation sites. If these ions were randomly distributed over these $3N_A$ sites, the entropy of mixing would be multiplied by 3 as shown in Chapter 1.

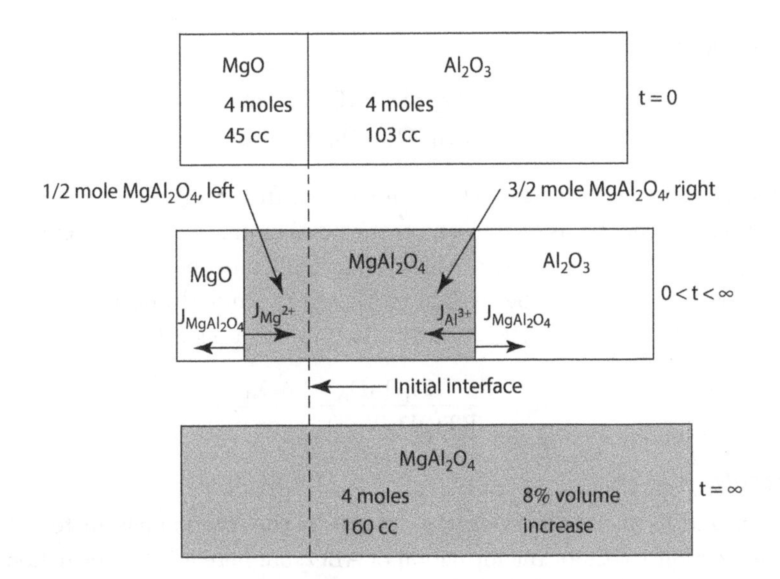

FIGURE 15.7 Microstructure of a diffusion couple of 4 moles of MgO reacting with 4 moles of Al_2O_3 to produce 4 moles of $MgAl_2O_4$. It is assumed that the oxygen ions are immobile. As a result, the cation charge neutrality requires that for each mole of spinel produced on the MgO side of the original interface, three moles of spinel are produced on the Al_2O_3 side. This is shown in the middle sketch in the figure. Overall, there is roughly an 8% increase in volume, but this is neglected in the analysis because it is a small effect.

15.4.4.4 Flux Density Equations

As usual, the number flux densities are given by

$$J'_M = \frac{\eta_M D_M}{k_B T}\left\{-\frac{1}{N_A}\frac{dG_M}{dx} + 2eE\right\}$$
$$J'_A = \frac{\eta_A D_A}{k_B T}\left\{-\frac{1}{N_A}\frac{dG_A}{dx} + 3eE\right\}. \tag{15.47}$$

15.4.4.5 Electric Field Term

Inserting Equations 15.47 into the electrical neutrality equation (15.45) after some algebra (see Equation A.45), the electric field term is

$$eE = \frac{1}{(4\eta_M D_M + 9\eta_A D_A)}\left\{2\frac{\eta_M D_M}{N_A}\frac{dG_M}{dx} + 3\frac{\eta_A D_A}{N_A}\frac{dG_A}{dx}\right\}. \tag{15.48}$$

15.4.4.6 Elimination of the Electric Field Term

Inserting the electric field term, Equation 15.48, into the expression for the Mg^{2+} flux density, J'_M, after some more algebra (Equation A.46), the flux density becomes

$$J'_M = -\frac{\eta_M D_M \eta_A D_A}{RT(4\eta_M D_M + 9\eta_A D_A)}\left\{9\frac{dG_M}{dx} - 6\frac{dG_A}{dx}\right\}. \tag{15.49}$$

15.4.4.7 Elimination of dG_A/dx

As shown in Equation A.55, in this case,

$$\frac{dG_A}{dx} = -\frac{1}{2}\frac{dG_M}{dx} \tag{15.50}$$

so Equation 15.49 becomes

$$J'_M = -\frac{12\eta_M D_M \eta_A D_A}{RT(4\eta_M D_M + 9\eta_A D_A)}\frac{dG_M}{dx}. \tag{15.51}$$

Now the diffusion coefficients D_M and D_A are for the diffusion of Mg^{2+} and Al^{3+} in $MgAl_2O_4$. Similarly, the number of Mg^{2+} ions per cm^3, η_M = the number of Mg^{2+} ions per cm^3 in $MgAl_2O_4$, so $\eta_M = \eta_{MgAl_2O_4} = \eta$ where η is the number of $MgAl_2O_4$ formula units per unit volume and can be calculated from the density and molecular weight of $MgAl_2O_4$. Similarly, $\eta_A = 2\eta_{MgAl_2O_4} = 2\eta$ because there are two Al^{3+} per formula unit. These substitutions into Equation 15.51 give

$$J'_M = -\frac{12\eta D_M D_A}{RT(2D_M + 9D_A)}\frac{dG_M}{dx}. \tag{15.52}$$

15.4.4.8 Result

To finish, dG_M/dx and J'_M need to be evaluated in terms of the reaction parameters. The procedure is the same as in Section 15.4.2 for the formation of ABO_2 but here it is the formation of $MgAl_2O_4$. The equilibrium constant, K_e, for this reactions is

$$K_e = \frac{a_{MgAl_2O_4}}{a_{MgO}a_{Al_2O_3}} = e^{-\frac{\Delta G^o_{MgAl_2O_4}}{RT}}. \tag{15.53}$$

The Gibbs energy of MgO at the MgO–$MgAl_2O_4$ interface is $G_{MgO} = G^o_{MgO}$ because $a_{MgO} = 1.0$ and $G_M = G_{MgO} = G^o_{MgO} + RT\ln a_{MgO}$. At the $MgAl_2O_4$–Al_2O_3 interface where the activities of both $MgAl_2O_4$ and Al_2O_3 are both 1.0, Equation 15.53 is then $G_{MgO} = G^o_{MgO} + \Delta G^o_{MgAl_2O_4}$. Again, if Y is the *thickness of the* $MgAl_2O_4$ *layer* formed, then

$$\frac{dG_M}{dx} = \frac{dG_{MgO}}{dx} = \frac{(G^o_{MgO} + \Delta G^o_{MgAl_2O_4}) - G^o_{MgO}}{Y} = \frac{\Delta G^o_{MgAl_2O_4}}{Y} \tag{15.54}$$

where this gradient is negative because $\Delta G^o_{MgAl_2O_4}$ is negative. Now the *total flux of* $MgAl_2O_4$ *is simply* $4/3J'_M$ because for every three moles that the flux of MgO produces to the right of the interface, there is one mole of spinel produced by the diffusion of Al_2O_3 to the left of the interface for a total of four moles. Therefore,

$$J_{MgAl_2O_4} = \frac{4}{3}J_{MgO} = \frac{4}{3}\frac{J'_M}{N_A} = -C_{MgO}\left(\frac{16D_{Al^{3+}}D_{Mg^{2+}}}{2D_{Mg^{2+}} + 9D_{Al^{3+}}}\right)\frac{1}{RT}\frac{\Delta G^o_{MgAl_2O_4}}{Y} \tag{15.55}$$

but because this is a planar reaction, the flux of spinel is also given by

$$J_{MgAl_2O_4} = \frac{\rho}{M}\frac{dY}{dt} = C_{MgAl_2O_4}\frac{dY}{dt} \tag{15.56}$$

and equating these last two equations, Equations 15.55 and 15.56, gives

$$Y\frac{dY}{dt} = -\left(\frac{16D_{Al^{3+}}D_{Mg^{2+}}}{2D_{Mg^{2+}} + 9D_{Al^{3+}}}\right)\frac{\Delta G^o_{MgAl_2O_4}}{RT}$$

because $C_{MgAl_2O_4} = C_{MgO}$, and can be integrated to give

$$Y^2 = kt$$

where the reaction rate constant, k, is given by

$$k = -2\left(\frac{16D_{Al^{3+}}D_{Mg^{2+}}}{2D_{Mg^{2+}} + 9D_{Al^{3+}}}\right)\frac{\Delta G^o_{MgAl_2O_4}}{RT}. \tag{15.57}$$

The value of k is, of course, positive because $\Delta G^{o}_{MgAl_2O_4}$ is negative and the reaction rate, as usual, depends on the diffusion coefficient of the slower diffusing species. Again, for a diffusion-controlled reaction, the thickness of the reacted layer depends on the square root of time. The overall reaction is schematically shown in Figure 15.8.

15.4.5 Reactions with Intermediate Products

Most phase diagrams with ternary compounds are not as simple as those for the two cases modeled above where there was only a single compound between the end members. This is generally not the case. Usually, there are several intermediate compounds—that may or may not have a compositional range of solid solubility—present between two end members A and B. Figure 15.9 schematically shows a typical phase diagram with four intermediate compounds: A_3B, AB, AB_2, and AB_3 none of which are shown to have any solid solubility for simplicity. Limited composition ranges are expected in the various compounds because point defects would need to form that requires considerable energy. Such phase diagrams are particularly common for perovskite titanates such as $BaTiO_3$ that have important technical applications. Usually, the desired compound is AB and is made by reacting pure A and pure B. As a result, assuming that the reactions are being limited by

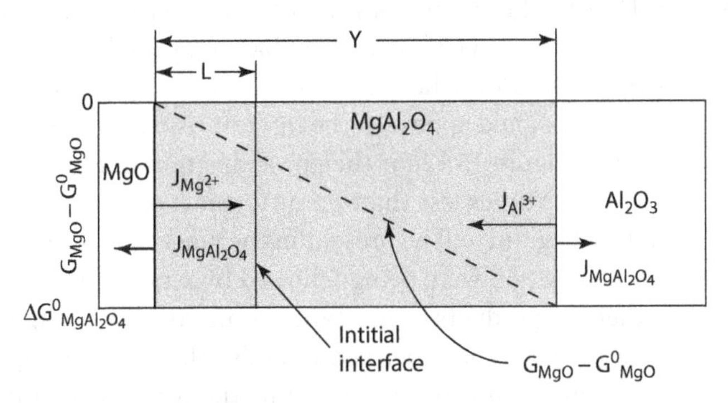

FIGURE 15.8 The reaction between MgO and Al_2O_3 to form $MgAl_2O_4$ showing the fluxes of Mg^{2+}, Al^{3+}, and $MgAl_2O_4$ in both directions. The dashed line shows the chemical potential for MgO, $G_{MgO} - G^{0}_{MgO}$ as a function of distance across the spinel phase beginning at zero at the MgO-$MgAl_2O_4$ interface and going to $\Delta G^{0}_{MgA1_2O_4}$ at the $MgAl_2O_4$–Al_2O_3 interface.

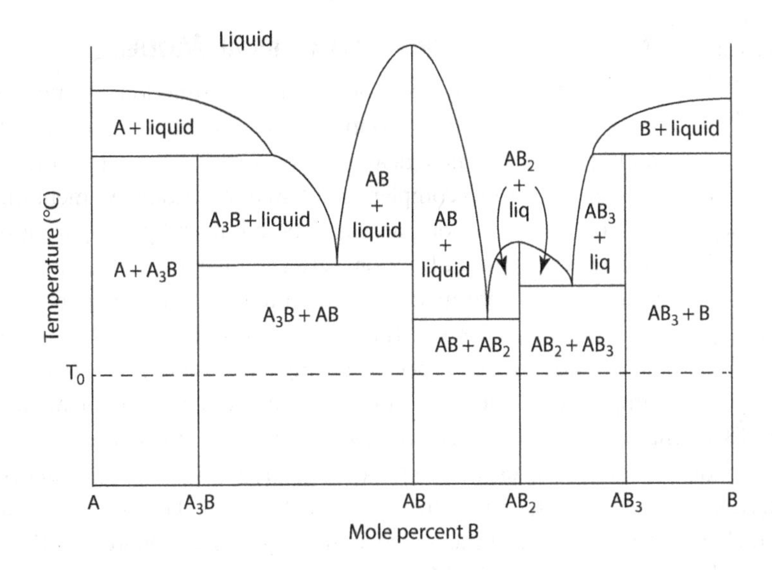

FIGURE 15.9 Hypothetical A–B phase diagram with several intermediate phases none of which exhibit any perceptible solid solubility. T_0 is a temperature at which A and B could be reacted to form any of the intermediate compounds by a solid-state reaction.

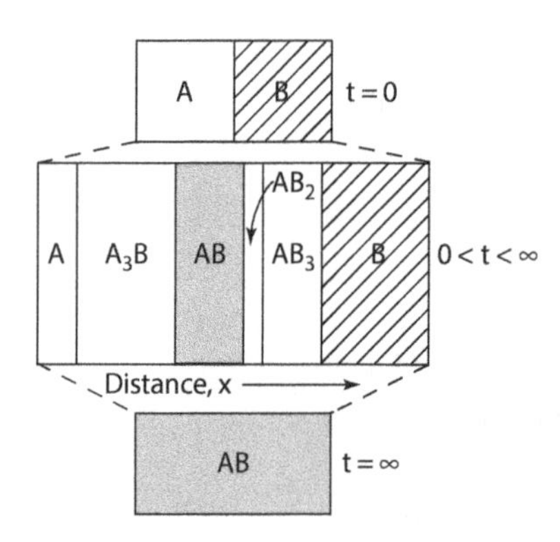

FIGURE 15.10 Hypothetical microstructures for an equimolar A–B diffusion couple with the phase diagram in Figure 15.9 at three different times. At time $0 < t < \infty$, all of the phases in the phase diagram will be present in the microstructure. The width—relative amount—of each phase depends on the effective interdiffusion coefficient across that phase: the larger the diffusion coefficient, the wider the phase field, for example, A_3B.

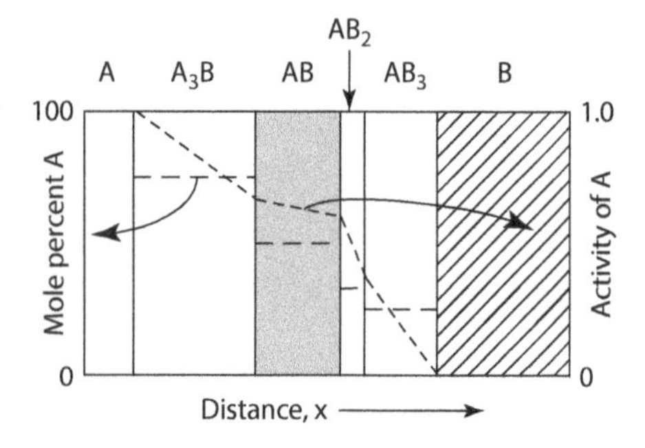

FIGURE 15.11 Essentially, the same hypothetical microstructure as in Figure 15.10 at some intermediate time except this figure shows how the concentration of A, m/o (heavy dashed lines), varies with distance as well as the thermodynamic activity of A (light dashed lines). It is important to note that at each phase interface (e.g. A_3B–AB), there is a *discontinuity* in the composition and the composition is *constant* across the single-phase region. Also note that *there are no two-phase regions* in the microstructure because equilibrium exists between a pair of phases *only* at their interface. Finally, the thermodynamic activity of A is the same in two phases at their mutual interface and the activity varies across each phase region going from $a_A = 1.0$ at pure A to $a_A = 0$ when pure B is reached; again, no solid solubility is present.

diffusion through the crystal lattices of the several compounds present, just as in the case of metals interdiffusing as discussed in Chapter 14, then *all of the phases in the phase diagram should be present* to some extent before the reaction goes to completion as depicted in Figure 15.10. Figure 15.10 makes the important point again that *there are no two-phase regions in the microstructure* of the interdiffusion zone. The two-phase equilibria occur *at each of the two phase interfaces*: A_3B–AB, for example. Furthermore, Figure 15.11 shows that at each phase interface there is a *discontinuity* in the composition shown in the figure while the thermodynamic activity *of an end-member compound* is *equal across each interface* and decreases through each phase going from left to right from pure A to pure B. A general assumption is that the fluxes of A and B through each phase are roughly equal as are the Gibbs energy (activity) differences (Figure 15.11); then $J' \approx D_{eff}/Y$ where D_{eff} is some effective diffusion coefficient across a given phase and Y is the thickness of a specific compound layer. As a result, those layers having higher diffusion coefficients will have greater thicknesses and vice versa. If the diffusion coefficient for a layer is very small, so will its thickness and may not be readily visible microscopically. Similarly, as mentioned in Chapter 14, a phase may not be present because it is hard to nucleate. Nevertheless, *thermodynamically*, all of the phases in the phase diagram should appear in the interdiffusion zone. This is shown schematically in Figure 15.12 for the phase diagram in Figure 15.9. As Figure 15.12 shows, for times less than infinity, some, or all of the intermediate phases in the diagram will be present in the reacting microstructure. If the extent of the reaction were being followed by x-ray diffraction peak intensities of either the products or reactants, some confusion might result by the presence some of the other compounds. This might suggest that "Maybe the phase diagram is incorrect," while the reality is that insufficient time has elapsed for the reaction to go to completion.

15.5 OTHER MODELS FOR THE EXTENT OF A REACTION

15.5.1 Different Geometrical Models

In Section 15.4.3, the *fraction reacted* as a function of time was modeled for a simple linear diffusion couple. The fraction reacted as a function of time is an important technological parameter that dictates the times and temperatures necessary to complete solid-state reactions. As mentioned above, the progress of a reaction can easily be followed by comparing the x-ray peak intensities in the diffraction patterns of the reactants and products. This is a very accurate method if the intensities are compared to the peaks of intentional mixtures of the products and reactants, if these mixtures have composition ratios expected during the reaction. As might be expected, the simple linear model is not very realistic for commercial calcination reactions. Figure 15.13 attempts to show the *conceptual* evolution of the simple linear diffusion model; to a cubic powder particle model of one of the reactants, MgO, in contact with powder particles of the other reactant, Al_2O_3; to the powder particle surrounded by the second reactant; to finally, a spherical geometry in which a spherical particle of the one reactant, MgO, is surrounded by the second, Al_2O_3 and a shell of $MgAl_2O_4$ growing around the MgO particle.

In Figure 15.14, the MgO spherical particle is isolated and its contribution to $MgAl_2O_4$ formation is shown as an inward growing shell into the original MgO particle. A few geometrical models applied to the formation of $MgAl_2O_4$ from MgO and Al_2O_3 are discussed to illustrate the different—essentially geometric—factors that require consideration in developing such models. The purpose is neither to present all such existing models nor to give the exact model, but rather to illustrate how these factors are considered in developing more precise models of solid-state reactions. The applicability and difficulties with various models have been examined by others (Frade and Cable 1992).

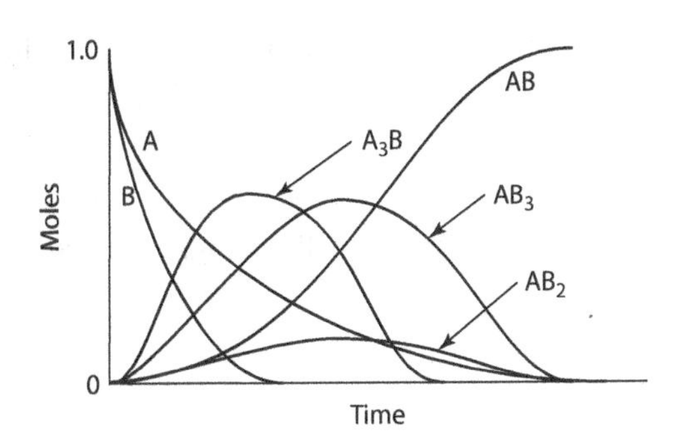

FIGURE 15.12 A schematic showing the number of moles of all the phases present as a function of time when an equimolar amount of A and B are reacted with the ultimate goal of forming AB for the phase diagram of Figure 15.9.

15.5.2 The Linear Model Again for $MgAl_2O_3$ Formation

Figure 15.8 gives the linear model for the reaction of MgO to form $MgAl_2O_4$. Consider only the consumption of MgO to the left of the original interface. In this case, Y is the total width of the $MgAl_2O_4$ reaction product, with 3/4 Y of it to the right of the original interface and only 1/4 Y to the left of the interface. Again, as in Section 15.4.3, the fraction reacted, f_R, is

$$f_R = \frac{V_0 - V(t)}{V_0} = 1 - \frac{V}{V_0} = 1 - \frac{(L - Y/4)}{L} \tag{15.58}$$

and so

$$f_R^2 = \frac{Y^2}{16L^2}.$$

Combining this with Equation 15.57

$$f_R^2 = -\left(\frac{\Delta G^o_{MgAl_2O_4}}{RT}\right)\left(\frac{2D_{Al^{3+}}D_{Mg^{2+}}}{2D_{Mg^{2+}} + 9D_{Al^{3+}}}\right)\frac{t}{L^2} = k't. \tag{15.59}$$

Note that the dimensionless parameter, $\frac{D_{eff}t}{L^2}$, appears as usual where

$$D_{eff} = \left(\frac{2D_{Al^{3+}}D_{Mg^{2+}}}{2D_{Mg^{2+}} + 9D_{Al^{3+}}}\right) \tag{15.60}$$

is the *effective* diffusion coefficient for the reaction.

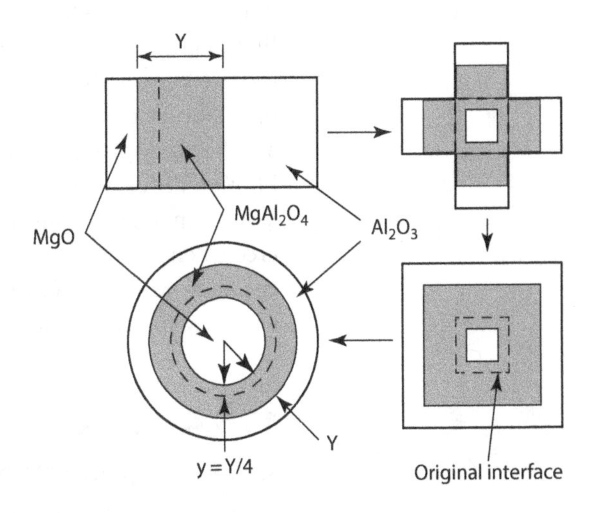

FIGURE 15.13 How the linear model of reaction to form $MgAl_2O_4$, upper left, conceptually progresses clockwise to the Jander model, lower left when the reaction is taking place in a mixture of powders of MgO and Al_2O_3. The white inner region is the unreacted MgO in the upper right and two bottom sketches.

15.5.3 The Jander Model

The Jander model (Jander 1927) is shown in Figure 15.14, in which a $MgAl_2O_4$ layer of thickness y (which is 1/4 that of the total $MgAl_2O_4$ thickness in Equation 15.55) has formed on a spherical MgO particle of original radius of a_0 leaving unreacted MgO at time t of radius a. Again, the fraction reacted, f_R, is then given by

$$f_R = \frac{V_0 - V(t)}{V_0}$$

so

$$f_R = 1 - \frac{V}{V_0} = 1 - \left(\frac{a}{a_0}\right)^3. \tag{15.61}$$

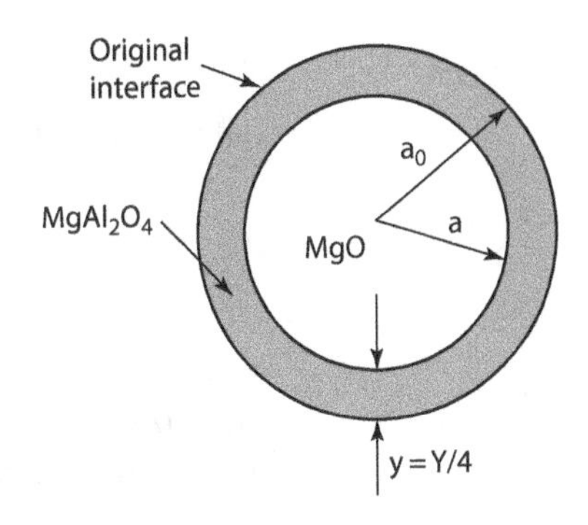

FIGURE 15.14 Sketch of the spherical Jander model only considering the reaction of the original MgO particle of radius a_0 with Al_2O_3 to form $MgAl_2O_4$ *within* the original MgO particle boundary. The total thickness of the $MgAl_2O_4$ is "Y," as shown in Figure 15.8 so only Y/4 of that occurs inside the original MgO–Al_2O_4 interface in the MgO.

Rearranging and taking the cube root,

$$(1-f_R)^{\frac{1}{3}} = \frac{a}{a_0} = \frac{a_0 - y}{a_0} = 1 - \frac{y}{a_0}.$$

Rearranging and squaring this,

$$\left\{1-(1-f_R)^{\frac{1}{3}}\right\}^2 = \frac{y^2}{a_0^2} = \frac{Y^2}{16a_0^2}$$

and substituting Y^2 from Equation 15.57 gives

$$F(f_R) = \left\{1-(1-f_R)^{\frac{1}{3}}\right\}^2 = -\frac{\Delta G^o_{MgAl_2O_4}}{RT}\left(\frac{2D_{Al^{3+}}D_{Mg^{2+}}}{2D_{Mg^{2+}}+9D_{Al^{3+}}}\right)\frac{t}{a_0^2} = k't. \tag{15.62}$$

Therefore, if the left-hand side of Equation 15.62, $\left\{1-(1-f_R)^{\frac{1}{3}}\right\}^2$, is plotted versus time, a straight line should result. As a result, if the Gibbs energy of formation of $MgAl_2O_4$ and the initial particle size are known, then D_{eff} can be calculated. Plotting the slopes of $F(f_R)$ versus time as a function of temperature can give an activation energy for D_{eff}. If the activation energies for diffusion of Mg^{2+} and Al^{3+} in $MgAl_2O_4$ are known, then the comparison of activation energies should suggest the rate-limiting diffusing ion. Note again that the dimensionless parameter $D_{eff}t/a_0^2$ appears in the model. Two main assumptions in the Jander model are as follows: (1) spherical coordinates are not considered when taking the thickness of the reacted layer, y, and (2) all of the particles are assumed to have the same size. The latter is certainly partially achievable by carefully sizing the reacting particles to a narrow size range: easily done in the laboratory but difficult in industrial practice.

15.5.4 The Ginstling and Brounshtein Model

15.5.4.1 Spherical Coordinates

The Ginstling and Brounshtein model (Ginstling and Brounshtein 1950) is somewhat of an improvement over the Jander model because it considers the spherical symmetry of the particles. The original model is based on concentration gradients but $MgAl_2O_3$ forms by nonideal diffusion. Therefore, the model needs a little adjustment. Whether or ideal or not, Fick's second law in spherical coordinates is given by

$$\frac{\partial C}{\partial t} = \frac{1}{r^2}\frac{\partial}{\partial r}\left(r^2 J_{MgO}\right) \tag{15.63}$$

and from Equation 15.52 because $J_{MgO} = \frac{J'_{MgO}}{N_A}$,

$$J_{MgO} = -\frac{12C_{MgO}D_{Mg^{2+}}D_{Al^{3+}}}{RT\left(2D_{Mg^{2+}}+9D_{Al^{3+}}\right)}\frac{dG_{MgO}}{dr} = -\alpha\frac{dG_{MgO}}{dr}. \tag{15.64}$$

where α is the constant containing all the factors in front of the gradient factor. As usual, the steady-state approximation is appropriate, so

$$\frac{\partial}{\partial r}\left(r^2 J_{MgO}\right) = 0$$

and integrating gives

$$J_{MgO} = \frac{A''}{r^2} = -\alpha\frac{dG_{MgO}}{dr}$$

where $A'' =$ constant. Integrating again between a and a_0

$$\alpha\{G_{a_0} - G_a\} = A''\left(\frac{1}{a_0} - \frac{1}{a}\right) = A''\left(\frac{a - a_0}{aa_0}\right)$$

where the G_i are the free energies of MgO at the two radii, a_0 and a. So the flux of MgO becomes

$$J_{MgO} = \frac{\alpha}{r^2}\frac{aa_0}{a - a_0}\{G_{a_0} - G_a\}$$

and at $r = a$ this is

$$J_{MgO} = \frac{\alpha}{a}\frac{a_0}{(a - a_0)}\{G_{a_0} - G_a\}. \tag{15.65}$$

Now $J_{MgO}/3 = -J_{MgAl_2O_4}$ inside the MgO particle where $J_{MgAl_2O_4}$ is the flux of $MgAl_2O_4$ *into the space originally occupied by MgO*, Equation 15.46, and three fourths of $MgAl_2O_4$ are being produced outside of the original interface of the MgO particle and not shown. Again, the flux of $MgAl_2O_4$ is given by

$$J_{MgAl_2O_4} = C_{MgAl_2O_4}\frac{1}{\text{Area}}\frac{dV}{dt} = C_{MgAl_2O_4}\frac{1}{4\pi a^2}\frac{d\left(\frac{4}{3}\pi a^3\right)}{dt} = C_{MgAl_2O_4}\frac{da}{dt}. \tag{15.66}$$

Combining Equations 15.65 and 15.66 gives

$$\left(a^2 - aa_0\right)da = -\frac{\alpha a_0}{3C_{MgAl_2O_4}}\{G_{a_0} - G_a\}dt \tag{15.67}$$

and integrating from a_0 to a and from zero to t,

$$\frac{1}{3}\left(a^3 - a_0^3\right) - \frac{1}{2}\left(a_0a^2 - a_0^3\right) = -\frac{\alpha a_0}{3C_{MgAl_2O_4}}\{G_{a_0} - G_a\}t.$$

Now multiply both sides by 3 and divide both by a_0^3,

$$\left(\frac{a}{a_0}\right)^3 - 1 - \frac{3}{2}\left\{\left(\frac{a}{a_0}\right)^2 - 1\right\} = -\frac{\alpha}{a_0^2C_{MgAl_2O_4}}\{G_{a_0} - G_a\}t.$$

$$\frac{1}{2} + \left(\frac{a}{a_0}\right)^3 - \frac{3}{2}\left(\frac{a}{a_0}\right)^2 = -\left(\frac{\Delta G^\circ_{MgAl_2O_4}}{RT}\right)\frac{3D_{Mg^{2+}}D_{Al^{3+}}}{\left(2D_{Mg^{2+}} + 9D_{Al^{3+}}\right)}\frac{1}{a_0^2}t = k''t \tag{15.68}$$

since $C_{MgO} = C_{MgAl_2O_4}$ and $\{G_{a_0} - G_a\} = \Delta G^\circ_{MgAl_2O_4}/4$ because the thickness of the $MgAl_2O_4$ layer in the original MgO particle is only 1/4 the total magnesium aluminate thickness over which the Gibbs energy difference exists.

15.5.4.2 Fraction Reacted

A little algebra needs to be applied to convert Equation 15.68 in an equation for the fraction reacted, f_R, as a function of time. Again, the fraction of MgO reacted is

$$f_R = \frac{V_0 - V(t)}{V_0} = 1 - \left(\frac{a}{a_0}\right)^3$$

so $1 - f_R = (a/a_0)^3$ or $a/a_0 = (1 - f_R)^{\frac{1}{3}}$ and $(a/a_0)^2 = (1 - f_R)^{\frac{2}{3}}$. Putting these values into Equation 15.68 gives

$$\frac{1}{2} + (1 - f_R) - \frac{3}{2}(1 - f_R)^{\frac{2}{3}} = k''t$$

and rearranging and multiplying by 2/3 gives

$$1-\frac{2}{3}f_R-(1-f_R)^{\frac{2}{3}}=\frac{2}{3}k''t$$

and finally,

$$F(f_R)=1-\frac{2}{3}f_R-(1-f_R)^{\frac{2}{3}}=kt \tag{15.69}$$

where F = function of the fraction reacted and

$$k=-\left(\frac{\Delta G^o_{MgAl_2O_4}}{RT}\right)\frac{2D_{Mg^{2+}}D_{Al^{3+}}}{\left(2D_{Mg^{2+}}+9D_{Al^{3+}}\right)}\frac{1}{a_0^2}. \tag{15.70}$$

Note again the dimensionless parameter $D_{eff}t/a_0^2$ where again

$$D_{eff}=\frac{2D_{Mg^{2+}}D_{Al^{3+}}}{\left(2D_{Mg^{2+}}+9D_{Al^{3+}}\right)}.$$

Note also that *k is the same* in all three models.

15.5.5 The Carter Model

The Carter modification of the Ginstling and Brounshtein model of the fraction reacted versus time takes into consideration that the volume of the products of a reaction are different than the volume of the reactants (Carter 1961b). The ratio of the volume of the products to the volume of reactants is designated by the variable z. For example, for the formation of $MgAl_2O_4$ on the MgO side of the original interface where, on reaction completion, one mole of $MgAl_2O_4$ replaces four original moles of MgO,

$$z=\frac{\overline{V}_{MgAl_2O_4}}{4\overline{V}_{MgO}}=\frac{40.07}{4\times 11.26}=0.89$$

where $\overline{V}_i$ = molar volume (cm³/mol). Let a_0 = initial radius of the reactant MgO, a = radius of reactant MgO at time t, and r_0 = the radius for total volume of product plus unreacted reactant, so that at any time t, the total volume = volume product + volume unreacted reactant (Figure 15.15):

$$r_0^3=z\left(a_0^3-a^3\right)+a^3$$
$$r_0=\left[za_0^3-za^3+a^3\right]^{\frac{1}{3}}. \tag{15.71}$$

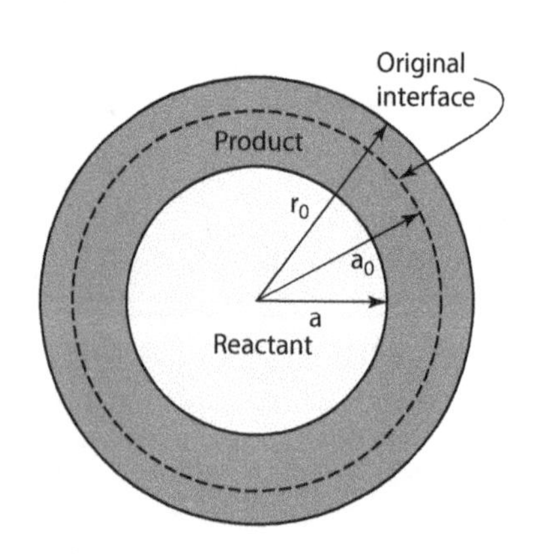

FIGURE 15.15 Similar to Figure 15.13 for the Jander model only in this case the dimensions are for the Carter model of a solid-state reaction by interdiffusion where the volume of the product phase can be different than the volume of the reactant phase—in this case, the outer radius of the product layer, r_0, is greater than the original radius a_0 because of the greater volume of the product than that of the reactant.

As in the case of the Ginstling and Brounshtein result (Ginstling and Brounshtein 1950), Equation 15.67 applies and in this case is

$$da=\frac{r_0}{(a-r_0)a}k'''dt.$$

The integration now is a little messier because of Equation 15.71. The integrations are carried out in Section A.3 with the final result, Equation A.36, the *Carter equation*:

$$F(f_R) = \left[1 + f_R(z-1)\right]^{\frac{2}{3}} + (z-1)(1-f_R)^{\frac{2}{3}} = z + (1-z)kt \tag{15.72}$$

with, again,

$$k = -\left(\frac{\Delta G^{o}_{MgAl_2O_4}}{RT}\right)\frac{2D_{Mg^{2+}}D_{Al^{3+}}}{\left(2D_{Mg^{2+}} + 9D_{Al^{3+}}\right)}\frac{1}{a_0^2}.$$

As a check, when $z \to 1$, that is, the products and reactants have the same volume, Equation 15.72 can be rewritten as

$$F(f_R) = \frac{1}{(1-z)}\left[1 + f_R(z-1)\right]^{\frac{2}{3}} + \frac{(z-1)}{(1-z)}(1-f_R)^{\frac{2}{3}} - \frac{z}{(1-z)} = kt. \tag{15.73}$$

Because $\left[1+f_R(z-1)\right]^{\frac{2}{3}} \cong \left[1+\frac{2}{3}(z-1)f_R\right]$ as $z \to 1$,* Equation 15.73 becomes

$$1 - \frac{2}{3}f_R - (1-f_r)^{\frac{2}{3}} = kt$$

the Ginstling–Brounshtein result, as it should. As noted earlier, for any of these models, a plot of $F(f_R)$ versus t should give a straight line, the slope of which is k. By repeating the reaction at different temperatures and plotting the values of k, hopefully an activation energy, Q, can be obtained, or even a value for the effective diffusion coefficient that can be compared to that of the two diffusing ions to determine which ion controls the rate of reaction.† Figure 15.16 illustrates this.

15.6 CHAPTER SUMMARY

In this chapter, models of solid-state reaction kinetics involving compounds are developed. These models are motivated by extremely important commercial processes best exhibited by the formation of Portland cement (the largest used engineering material by weight) from clay, sand, and calcium

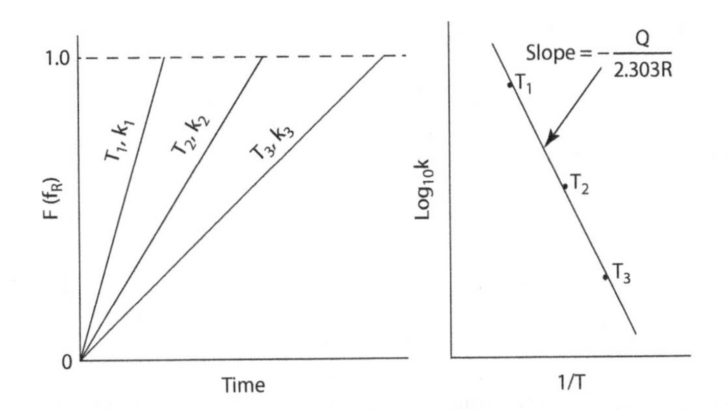

FIGURE 15.16 The left diagram shows how any of the functions of the fraction reacted based on the different models, $F(f_R)$, increase linearly with time up to F = 1.0. By doing the reactions at different temperatures, the reaction rate constants, k_i, are obtained as a function of temperature. By plotting these on a semi-log plot versus 1/T (right diagram), the activation energy for the reaction, Q, can be determined. Because the reaction rate constants all contain the effective diffusion coefficient, D_{eff}, it may be possible to determine the rate-determining diffusing ion from such data.

* The first two terms in the infinite series for $(1+x)^{\frac{2}{3}}$ when x is small.

† The only caveat to this is that, for a specific system, the likelihood of the diffusion coefficients having been measured is not great.

carbonate. First, the interdiffusion of MgO and NiO is modeled because they appear to form almost an ideal solution. The initial model presumes that oxygen ions—anions—are immobile as generally assumed. The combination of the chemical and electrical forces—zero charge flux—is used to model interdiffusion and an effective diffusion coefficient defined. The same reaction is modeled again but with mobile oxygen ions. This model shows *a Kirkendall effect and marker motion* under very restrictive conditions between the oxygen and cation diffusion coefficients. With mobile oxygen ions, MgO and NiO would interdiffuse essentially by the Darken model of interdiffusion for metals but now requiring *a Schottky defect flux* rather than a simple vacancy flux! However, the extent of expected marker motion is small since the oxygen ions (anions, in general) diffuse considerably more slowly than cations, primarily because of their larger sizes with concomitant difficulty in moving through a crystal lattice. Therefore, the assumption of no oxygen—anion—motion during interdiffusion in compounds is a realistic assumption during compound interdiffusion.

The reaction between two binary compounds forming a ternary compound is the second major topic covered. The reaction $MgO + Al_2O_3 \Rightarrow MgAl_2O_4$ is the example used for compound formation with cations of different charges diffusing. With the oxygen ions immobile, the cation motions are coupled by their charges again. The reaction rate constant obtained is proportional to the Gibbs energy of formation of the ternary compound with an effective interdiffusion coefficient, a function of the two cation diffusion coefficients. Some features of this model are generalized to systems in which more than one ternary compound can be formed from the two binary end-member compounds in the phase diagram. The importance of both the *discontinuity of the composition* at interphase boundaries and the *equality of the thermodynamic activities* at the phase boundaries is emphasized again. Finally, different geometric models that give the extent of a reaction with time are modeled: the simple linear model; the Jander model; the Ginstling and Brounshtein model; and finally, the Carter model that considers a volume change from the reactants to the products. Of course, all of these general models apply not only to oxides but also to other compounds.

APPENDIX

A.1 Chemical Potentials

A.1.1 Chemical Potential Relationships

The reaction to form a compound such as MgO can be written as

$$Mg(s) + 1/2\,O_2(g) \rightarrow MgO(s)\,;\,\Delta G^o \tag{A.1}$$

and the Gibbs energy for the reaction can be written as

$$\Delta G^o = \bar{G}_{MgO} = \bar{G}_{Mg} + 1/2\,\bar{G}_{O_2} \tag{A.2}$$

or in terms of the chemical potentials, μ_i,

$$\mu_{MgO} = \mu_{Mg} + 1/2\mu_{O_2}. \tag{A.3}$$

Normally for the formation of a compound from the elements in their standard states, the $\bar{G}_i$ are taken to be zero. However, the usage here is when a solid solution or compound is being formed and a given species is not likely to be in its standard state, equations such as (A.2) and (A.3) are useful to relate chemical potentials of the various species: Mg, Mg^{2+}, and MgO. For example, for the interdiffusion of MgO and NiO, where the gradients in chemical potential are of interest, Equation A.3 becomes*

$$\nabla\mu_{MgO} = \nabla\mu_{Mg} \tag{A.4}$$

* All of the models for interdiffusion are one-dimensional along the x-axis in general. However, to minimize the notation $\vec{\nabla} g_{Mg} = \nabla g_{Mg} = 1/N_A\, d\bar{G}/dx$ is used where g_i is the Gibbs energy per atom, ion, or molecule.

since $\nabla\mu_{O_2} = 0$ because the oxygen content in the atmosphere is assumed to be spatially uniform (oxygen or air) during the interdiffusion.

However, for interdiffusion, it is the *ion* chemical potential gradient that is used. This can be easily handled by writing

$$Mg \rightleftharpoons Mg^{2+} + 2e^-$$

so

$$\mu_{Mg} = \mu_{Mg^{2+}} + 2\mu_{e^-} \tag{A.5}$$

where μ_{e^-} is a *chemical potential* for electrons. So, differentiating Equation A.5 gives

$$\nabla\mu_{Mg} = \nabla\mu_{Mg^{2+}} + 2\nabla\mu_{e^-} \tag{A.6}$$

or

$$\nabla\mu_{Mg^{2+}} = \nabla\mu_{Mg} - 2\nabla\mu_{e^-} \tag{A.7}$$

A.1.2 Interdiffusion

A relation similar to Equation A.7 for NiO is

$$\nabla\mu_{Ni^{2+}} = \nabla\mu_{Ni} - 2\nabla\mu_{e^-}. \tag{A.8}$$

In Equation 15.4 in the text, the expression

$$\begin{aligned} J'_{Mg^{2+}} &\propto \left\{\frac{dG_{Ni^{2+}}}{dx} - \frac{dG_{Mg^{2+}}}{dx}\right\} = \nabla\mu_{Ni^{2+}} - \nabla\mu_{Mg^{2+}} \\ J'_{Mg^{2+}} &\propto \nabla\mu_{Ni} - \nabla\mu_{Mg} = \nabla\mu_{NiO} - \nabla\mu_{MgO} \end{aligned} \tag{A.9}$$

from Equations A.4, A.7, and A.8 leads to Equation 15.5 in the text.

A.1.3 Electron Chemical Potential Gradient

As was seen in Chapter 14, a Nernst–Planck-type equation can be written for electrons just as it can for ions, namely,

$$J'_e = \frac{\eta_e D_e}{k_B T}\left\{-\frac{1}{N_A}\frac{d\mu_e}{dx} - eE\right\}. \tag{A.10}$$

If the electronic conductivity is zero (the transference number for electrons, t_e, is zero), then $J'_e = 0$ and

$$\frac{d\mu_e}{dx} = -N_A eE = -\mathfrak{F}E. \tag{A.11}$$

A.2 Interdiffusion in MgO and NiO

A.2.1 All Three Ions Are Mobile

A.2.1.1 *Expression for Marker Velocity*

If it is assumed that the electrons are not mobile, then the electrical neutrality requirement is

$$\begin{aligned} &J^*_{Mg} + J^*_{Ni} + J^*_O = 0 \\ &2eJ'_{Mg^{2+}} + 2eJ'_{Ni^{2+}} - 2eJ'_{O^{2-}} = 0 \\ &J'_{Mg^{2+}} + J'_{Ni^{2+}} - J'_{O^{2-}} = 0 \end{aligned} \tag{A.12}$$

and with the change in notation given in the text,

$$\begin{aligned} J'_{Mg} &= \eta_{Mg} B_{Mg} \left(-\nabla g_{Mg} + 2eE \right) \\ J'_{Ni} &= \eta_{Ni} B_{Ni} \left(-\nabla g_{Ni} + 2eE \right) \\ J'_{O} &= \eta_{O} B_{O} \left(-\nabla g_{O} - 2eE \right). \end{aligned} \tag{A.13}$$

Now, allowing for marker movement, the fluxes relative to a fixed reference frame, J''_i, are

$$\begin{aligned} J''_{Mg} &= J'_{Mg} + \eta_{Mg} v \\ J''_{Ni} &= J'_{Ni} + \eta_{Ni} v \\ J''_{O} &= J'_{O} + \eta v \end{aligned} \tag{A.14}$$

where η_i = number of ions per cm^3 and $\eta = \eta_{Mg} + \eta_{Ni} = \eta_O$ and $1 = X_{Mg} + X_{Ni}$ where X_i = mole fraction and η = constant, independent of composition—constant volume. As shown in the text,

$$v = B_O 4eE \tag{A.15}$$

which implies that if there is a difference in the mobilities of the different ions giving rise to an electric field, E, as long as the oxygen has some mobility, $B_O > 0$, there *will be* marker movement and a Kirkendall effect during the interdiffusion of oxide compounds!

A.2.1.2 The Electric Field Term

As in other cases, the electric field term is obtained from the electrical neutrality equation, Equation A.12,

$$\begin{aligned} & J'_{Mg} + J'_{Ni} - J'_{O} = \\ & \eta_{Mg} B_{Mg} \left(-\nabla g_{Mg} + 2eE \right) + \eta_{Ni} B_{Ni} \left(-\nabla g_{Ni} + 2eE \right) + \eta_O B_O 2eE = 0 \end{aligned}$$

which gives on collecting like terms and rearranging,

$$\begin{aligned} & 2eE \left(\eta_{Mg} B_{Mg} + \eta_{Ni} B_{Ni} + \eta_O B_O \right) \\ & = \eta_{Mg} B_{Mg} \nabla g_{Mg} + \eta_{Ni} B_{Ni} \nabla g_{Ni} \end{aligned}$$

$$2eE = \frac{\left(\eta_{Mg} B_{Mg} \nabla g_{Mg} + \eta_{Ni} B_{Ni} \nabla g_{Ni} \right)}{\left(\eta_{Mg} B_{Mg} + \eta_{Ni} B_{Ni} + \eta_O B_O \right)}. \tag{A.16}$$

A.2.1.3 Expression for J'_{Mg}

Substituting this expression for 2eE in the equation for J'_{Mg},

$$\begin{aligned} J'_{Mg} &= \eta_{Mg} B_{Mg} \left(-\nabla g_{Mg} + 2eE \right) \\ &= \frac{\eta_{Mg} B_{Mg}}{\left(\eta_{Mg} B_{Mg} + \eta_{Ni} B_{Ni} + \eta_O B_O \right)} \left(-\nabla g_{Mg} \left(\eta_{Mg} B_{Mg} + \eta_{Ni} B_{Ni} + \eta_O B_O \right) + \eta_{Mg} B_{Mg} \nabla g_{Mg} + \eta_{Ni} B_{Ni} \nabla g_{Ni} \right) \\ J'_{Mg} &= \frac{\eta_{Mg} B_{Mg}}{\left(\eta_{Mg} B_{Mg} + \eta_{Ni} B_{Ni} + \eta_O B_O \right)} \left(-\nabla g_{Mg} \eta_{Ni} B_{Ni} - \nabla g_{Mg} \eta_O B_O + \eta_{Ni} B_{Ni} \nabla g_{Ni} \right). \end{aligned}$$

Now, from the Gibbs–Duhem relation $X_{MgO}dG_{MgO} + X_{NiO}dG_{NiO} = 0$ where X_i = mole fraction, but because

$$Mg^{2+} + O^{2-} = MgO$$

therefore

$$G_{Mg} + G_O = G_{MgO}$$

$$dG_{Mg} + dG_O = dG_{MgO}$$

$$dG_{Mg} = dG_{MgO}$$

since $dG_O = 0$ and the same for NiO so

$$X_{Mg}dG_{Mg} + X_{Ni}dG_{Ni} = 0 \tag{A.17}$$

and can then be written as $\eta_{Ni}\nabla g_{Ni} = -\eta_{Mg}\nabla g_{Mg}$, so

$$J'_{Mg} = \frac{\eta_{Mg}B_{Mg}}{\left(\eta_{Mg}B_{Mg} + \eta_{Ni}B_{Ni} + \eta_O B_O\right)}\left(-\nabla g_{Mg}\eta_{Ni}B_{Ni} - \nabla g_{Mg}\eta_O B_O - \eta_{Mg}B_{Ni}\nabla g_{Mg}\right)$$

$$J'_{Mg} = -\frac{\eta\eta_{Mg}B_{Mg}\left(B_{Ni} + B_O\right)}{\left(\eta_{Mg}B_{Mg} + \eta_{Ni}B_{Ni} + \eta_O B_O\right)}\nabla g_{Mg}. \tag{A.18}$$

A.2.1.4 *New Expression for v*

Substituting the expression for 2eE of Equation A.16 into the expression for v, Equation A.15,

$$v = B_O 2eE = B_O \frac{\left(\eta_{Mg}B_{Mg}\nabla g_{Mg} + \eta_{Ni}B_{Ni}\nabla g_{Ni}\right)}{\left(\eta_{Mg}B_{Mg} + \eta_{Ni}B_{Ni} + \eta_O B_O\right)}$$

and with the Gibbs–Duhem equation again, $\eta_{Ni}\nabla g_{Ni} = -\eta_{Mg}\nabla g_{Mg}$,

$$v = \frac{\eta_{Mg}B_O\left(B_{Mg} - B_{Ni}\right)}{\left(\eta_{Mg}B_{Mg} + \eta_{Ni}B_{Ni} + \eta_O B_O\right)}\nabla g_{Mg}. \tag{A.19}$$

A.2.1.5 *The Result*

From Equation A.14,

$$J''_{Mg} = J'_{Mg} + \eta_{Mg}v$$

so all that has to be done to get the final result is substitute Equations A.18 and A.19 into this expression and do a little more algebra.

$$J''_{Mg} = -\frac{\eta\eta_{Mg}B_{Mg}\left(B_{Ni} + B_O\right)}{\left(\eta_{Mg}B_{Mg} + \eta_{Ni}B_{Ni} + \eta_O B_O\right)}\nabla g_{Mg} + \frac{\eta_{Mg}\eta_{Mg}B_O\left(B_{Mg} - B_{Ni}\right)}{\left(\eta_{Mg}B_{Mg} + \eta_{Ni}B_{Ni} + \eta_O B_O\right)}\nabla g_{Mg}$$

$$J''_{Mg} = \frac{\eta_{Mg}\left(-\eta B_{Mg}\left(B_{Ni} + B_O\right) + \eta_{Mg}B_O\left(B_{Mg} - B_{Ni}\right)\right)}{\left(\eta_{Mg}B_{Mg} + \eta_{Ni}B_{Ni} + \eta_O B_O\right)}\nabla g_{Mg}$$

$$J''_{Mg} = -\eta\eta_{Mg}\left\{\frac{B_{Mg}B_{Ni} + B_O\left(X_{Mg}B_{Ni} + X_{Ni}B_{Mg}\right)}{\eta_{Mg}B_{Mg} + \eta_{Ni}B_{Ni} + \eta_O B_O}\right\}\nabla g_{Mg}$$

Writing the mobilities in terms of the diffusion coefficients and $\nabla g_{Mg} = (1/N_A)(dG_{Mg}/dx)$ this last equation takes the form

$$J''_{Mg} = -\frac{\eta_{Mg}}{RT}\left\{\frac{D_{Mg}D_{Ni} + D_O\left(X_{Mg}D_{Ni} + X_{Ni}D_{Mg}\right)}{X_{Mg}D_{Mg} + X_{Ni}D_{Ni} + D_O}\right\}\frac{dG_{Mg}}{dx}$$

Again, making the assumption that this is an ideal solution,

$$\frac{dG_{Mg}}{dx} = \frac{RT}{X_{Mg}}\frac{dX_{Mg}}{dx} = \frac{RT}{\eta_{Mg}}\frac{d\eta_{Mg}}{dx}$$

because $J_{Mg} = J''_{Mg}/N_A = -\tilde{D}(dC_{Mg}/dx)$ where J_{Mg} is the molar flux of either Mg or MgO, and the *interdiffusion coefficient* $\tilde{D}$ is given by

$$\tilde{D} = \frac{D_{Mg}D_{Ni} + D_O\left(X_{Mg}D_{Ni} + X_{Ni}D_{Mg}\right)}{X_{Mg}D_{Mg} + X_{Ni}D_{Ni} + D_O}. \tag{A.20}$$

For discussion and comparison purposes, these same changes made to the equation for the velocity, Equation A.19, give

$$v = \frac{D_O\left(D_{Mg} - D_{Ni}\right)}{\left(X_{Mg}D_{Mg} + X_{Ni}D_{Ni} + D_O\right)}\frac{dX_{Mg}}{dx}. \tag{A.21}$$

A.2.2 Interdiffusion with Mobile Electrons

A.2.2.1 The Flux Equations

The approach here is the same as used to obtain the Kirkendall effect with the assumption of mobile oxygen ions—here replaced by mobile electrons—with the same amount of algebra. Therefore, mobile electrons take the place of oxygen ions in Equations A.12 and A.13, so the relevant equations are

$$\begin{aligned} &J^*_{Mg} + J^*_{Ni} + J^*_e = 0 \\ &2eJ'_{Mg} + 2eJ'_{Ni} - eJ'_e = 0 \\ &2J'_{Mg} + 2J'_{Ni} - J'_e = 0 \end{aligned} \tag{A.22}$$

and

$$\begin{aligned} &J'_{Mg} = \eta_{Mg}B_{Mg}\left(-\nabla g_{Mg} + 2eE\right) \\ &J'_{Ni} = \eta_{Ni}B_{Ni}\left(-\nabla g_{Ni} + 2eE\right) \\ &J'_e = \eta_e B_e\left(-\nabla g_e - eE\right). \end{aligned} \tag{A.23}$$

Again it is necessary to solve for eE and substitute it back into the first equation in (A.23) to get the flux, J'_{Mg}. Therefore, substitution of Equations A.23 into the last of Equation A.22 gives

$$2J'_{Mg} + 2J'_{Ni} - J'_O = \eta_{Mg}B_{Mg}\left(-\nabla g_{Mg} + 2eE\right) + \eta_{Ni}B_{Ni}\left(-\nabla g_{Ni} + 2eE\right) - \eta_e B_e\left(-\nabla g_e - eE\right) = 0.$$

A.2.2.2 The Electric Field and Resulting Flux

Carrying out all of the algebra of solving for eE,

$$eE = \frac{\left(\eta_{Mg}B_{Mg}\nabla g_{Mg} + \eta_{Ni}B_{Ni}\nabla g_{Ni} - \eta_e B_e \nabla g_e\right)}{\left(2\eta_{Mg}B_{Mg} + 2\eta_{Ni}B_{Ni} + \eta_e B_e\right)} \tag{A.24}$$

and substituting this back into the first of Equations A.23 after some multiplication, canceling, and regrouping results in

$$J'_{Mg} = -\frac{2\eta_{Mg}B_{Mg}\eta_{Ni}B_{Ni}\left(\nabla g_{Mg} - \nabla g_{Ni}\right) - \eta_{Mg}B_{Mg}\eta_e B_e\left(\nabla g_{Mg} + 2\nabla g_e\right)}{\left(2\eta_{Mg}B_{Mg} + 2\eta_{Ni}B_{Ni} + \eta_e B_e\right)}. \quad (A.25)$$

Up until now, the η_i, B_i, and g_i have referred to the charged species. However, from Equation A.6

$$\nabla g_{Mg} = \nabla g_{Mg^{2+}} + 2\nabla g_{e^-}$$

so that the last factor in Equation A.25 is simply $\nabla g_{Mg} = \nabla g_{MgO}$ because the oxygen chemical potential is constant. Similarly, the last factor in the first term could be written as

$$\nabla g_{Mg} - \nabla g_{Ni} = \nabla g_{MgO} - \nabla g_{NiO}$$

because the electron chemical potential for these two terms just cancels. Therefore, Equation A.25 becomes

$$J'_{Mg} = -\frac{2\eta_{Mg}B_{Mg}\eta_{Ni}B_{Ni}\left(\nabla g_{Mg} - \nabla g_{Ni}\right) - \eta_{Mg}B_{Mg}\eta_e B_e \nabla g_{Mg}}{\left(2\eta_{Mg}B_{Mg} + 2\eta_{Ni}B_{Ni} + \eta_e B_e\right)} \quad (A.26)$$

where the subscripts now refer either to the metals or to the oxides.

A.2.2.3 The Results

Again, invoking the Gibbs–Duhem equation,

$$X_{Mg}\nabla g_{Mg} + X_{Ni}\nabla g_{Ni} = 0$$

solving for ∇g_{Ni}, replacing it in Equation A.26, and combining terms gives the desired result:

$$J'_{Mg} = -\frac{\eta_{Mg}B_{Mg}\left(2\eta B_{Ni} + \eta_e B_e\right)}{\left(2\eta_{Mg}B_{Mg} + 2\eta_{Ni}B_{Ni} + \eta_e B_e\right)}\nabla g_{Mg}. \quad (A.27)$$

This result could be expressed in many ways, such as Wagner's theory of oxidation in Chapter 14 in terms of conductivities and transference numbers; or the η_i and B_i could be expressed in terms of concentrations and diffusion coefficients as is the usual case. However, rather than go through all that additional algebra, Equation A.27 can be used to examine the results. For example, if the electronic conductivity is much higher than the ionic conductivity, then $\eta_e B_e \gg \eta_{Mg}B_{Mg} \cong \eta_{Ni}B_{Ni}$ and Equation A.27 reduces to

$$J_{MgO} = \frac{J'}{N_A} - \frac{\eta_{Mg}B_{Mg}}{N_A}\nabla g_{Mg} = -\frac{\eta_{Mg}D_{Mg}}{N_A k_B T}\left(\frac{RT}{N_A X_{Mg}}\nabla X_{Mg}\right) = -\frac{\eta_{Mg}D_{Mg}}{\left(\eta_{Mg}/\eta\right)}\frac{1}{N_A\eta}\frac{d\eta_{Mg}}{dx}$$

$$J_{MgO} = J_{Mg} = -D_{Mg}\frac{dC_{Mg}}{dx} \quad (A.28)$$

which is the same result as obtained in Section 15.3.2, Equation 15.23, for immobile oxygen ions letting the electric field go to zero, obtained in far fewer steps than Equation A.28.

Similarly, if the electronic conductivity is low, $\eta_e B_e \ll \eta_{Mg}B_{Mg} \cong \eta_{Ni}B_{Ni}$, then

$$J_{Mg} = -\tilde{D}\frac{dC_{Mg}}{dx}$$

where $\tilde{D}$, as obtained in Section 15.2.3 where an electric field couples ion motion, is

$$\tilde{D} = \frac{D_{Mg}D_{Ni}}{\left(X_{Mg}D_{Mg} + X_{Ni}D_{Ni}\right)}. \quad (A.29)$$

A.2.3 All Three Ions Plus Electrons are Mobile

A.2.3.1 The Model

The model now becomes

$$J'_{Mg} + J'_{Ni} - J'_O = 0 \quad (A.30)$$

with

$$\begin{aligned} J'_{Mg} &= -\eta_{Mg} B_{Mg} \nabla g_{Mg} \\ J'_{Ni} &= -\eta_{Ni} B_{Ni} \nabla g_{Ni} \\ J'_O &= -\eta_O B_O \nabla g_O \end{aligned} \quad (A.31)$$

and

$$\begin{aligned} J''_{Mg} &= J'_{Mg} + \eta_{Mg} v \\ J''_{Ni} &= J'_{Ni} + \eta_{Ni} v \\ J''_O &= J'_O + \eta v \end{aligned} \quad (A.32)$$

where v is the lattice velocity. But again, $J''_O = 0$ so $J'_O = -\eta v = -\eta B_O \nabla g_O$.

A.2.3.2 The Velocity

From Equations A.30 and A.32

$$\eta v = \eta_{Mg} B_{Mg} \nabla g_{Mg} + \eta_{Ni} B_{Ni} \nabla g_{Ni} \quad (A.33)$$

Because

$$\begin{aligned} &Mg + O \rightarrow MgO \\ &g_{Mg} + g_O = g_{MgO} \\ &\therefore \nabla g_{Mg} + \nabla g_O = \nabla g_{MgO} \end{aligned} \quad (A.34)$$

and the same for NiO. Invoking the Gibbs–Duhem equation again,

$$X_{Mg} \nabla g_{MgO} + X_{Ni} \nabla g_{NiO} = 0$$

and inserting the last of the equations in Equation A.34 for both MgO and NiO,

$$\begin{aligned} X_{Mg} \nabla g_{Mg} + X_{Mg} \nabla g_O &= -X_{Ni} \nabla g_{Ni} - X_{Ni} \nabla g_{Ni} \\ X_{Mg} \nabla g_{Mg} + (X_{Mg} + X_{Ni}) \nabla g_O &= -X_{Ni} \nabla g_{Ni} \\ \eta \{X_{Mg} \nabla g_{Mg} + (X_{Mg} + X_{Ni}) \nabla g_O\} &= -\eta_{Ni} \nabla g_{Ni}. \end{aligned} \quad (A.35)$$

Substitution of the last equation of Equation A.35 into Equation A.33 gives

$$-\eta v = -\eta_{Mg} B_{Mg} \nabla g_{Mg} + \eta B_{Ni} \{X_{Mg} \nabla g_{Mg} + \nabla g_O\}.$$

But $\nabla g_O = v/B_O$ so

$$-\eta v = -\eta_{Mg} B_{Mg} \nabla g_{Mg} + \eta B_{Ni} X_{Mg} \nabla g_{Mg} + \eta B_{Ni} \frac{v}{B_O} \quad (A.36)$$

and, after a little algebra, Equation A.36 becomes

$$v = \frac{X_{Mg}B_O\left(B_{Mg} - B_{Ni}\right)}{B_{Ni} + B_O}\nabla g_{Mg}. \tag{A.37}$$

A.2.3.3 The Flux

Substitution of Equation A.37 into the first of equation of Equation A.32,

$$J''_{Mg} = -\eta_{Mg}B_{Mg}\nabla g_{Mg} + \eta_{Mg}\frac{X_{Mg}B_O\left(B_{Mg} - B_{Ni}\right)}{B_{Ni} + B_O}\nabla g_{Mg} \tag{A.38}$$

and a little more algebra gives

$$J'' = \frac{-\eta_{Mg}B_{Mg}B_{Ni} + B_O\left[\eta_{Mg}X_{Mg}\left(B_{Mg} - B_{Ni}\right) - \eta_{Mg}B_{Mg}\right]}{B_O B_{Ni}}\nabla g_{Mg}. \tag{A.39}$$

Reducing the term in the brackets, [],

$$\eta_{Mg}X_{Mg}\left(B_{Mg} - B_{Ni}\right) - \eta_{Mg}B_{Mg}$$

$$\eta_{Mg}X_{Mg}B_{Mg} - \eta_{Mg}X_{Mg}B_{Ni} - \eta_{Mg}B_{Mg}$$

$$\eta_{Mg}B_{Mg}\left(X_{Mg} - 1\right) - \eta_{Mg}X_{Mg}B_{Ni}$$

$$-\eta_{Mg}\left(X_{Ni}B_{Mg} + X_{Mg}B_{Ni}\right)$$

and substituting the last of these equations into Equation A.39 gives the desired result:

$$J'' = -\eta_{Mg}\left\{\frac{B_{Mg}B_{Ni} + B_O\left(X_{Ni}B_{Mg} + X_{Mg}B_{Ni}\right)}{B_{Ni} + B_O}\right\}\nabla g_{Mg}. \tag{A.40}$$

A.2.3.4 The Result

Again, assuming an ideal solid solution,

$$\nabla g_{Mg} = \frac{k_B T}{X_{Mg}}\frac{dX_{Mg}}{dx}$$

Equation A.40 becomes

$$J_{MgO} = -\tilde{D}\frac{dC_{MgO}}{dx}$$

with

$$\tilde{D} = \frac{D_{Mg}D_{Ni} + D_O\left(X_{Ni}D_{Mg} + X_{Mg}D_{Ni}\right)}{D_{Ni} + D_O} \tag{A.41}$$

and the velocity, Equation A.37, becomes

$$v = \frac{D_O\left(D_{Mg} - D_{Ni}\right)}{D_{Ni} + D_O}\frac{dX_{Mg}}{dx}. \tag{A.42}$$

A.3 Formation of $MgAl_2O_4$ By Counter Diffusion

A.3.1 Charge Flux Balance

It is assumed that the electronic conductivities in MgO, Al_2O_3, and $MgAl_2O_4$ are necessarily small, which they probably are at the temperatures of interest, something above around 1200°C, so that charge balance between the moving ions is all that needs to be considered. In this case, the charge flux balance is given by

$$J^*_{Mg} + J^*_{Al} = 0$$

$$2eJ'_{Mg} + 3eJ'_{Al} = 0 \tag{A.43}$$

$$2J'_{Mg} = -3J'_{Al}$$

where the subscripts refer to the ions Mg^{2+} and Al^{3+}. As usual, the number flux densities are given by

$$J'_M = \frac{\eta_M D_M}{k_B T}\left\{-\frac{1}{N_A}\frac{dG_M}{dx} + 2eE\right\}$$

$$J'_A = \frac{\eta_A D_A}{k_B T}\left\{-\frac{1}{N_A}\frac{dG_A}{dx} + 3eE\right\} \tag{A.44}$$

So,*

$$2J'_M = 2\frac{\eta_M D_M}{k_B T}\left\{-\frac{1}{N_A}\frac{dG_M}{dx} + 2eE\right\}$$

$$-3J'_A = -3\frac{\eta_A D_A}{k_B T}\left\{-\frac{1}{N_A}\frac{dG_A}{dx} + 3eE\right\}$$

$$\left(4\eta_M D_M + 9\eta_A D_A\right)eE = 3\frac{\eta_A D_A}{N_A}\frac{dG_A}{dx} + 2\frac{\eta_M D_M}{N_A}\frac{dG_M}{dx}$$

resulting in

$$eE = \frac{1}{\left(4\eta_M D_M + 9\eta_A D_A\right)}\left\{2\frac{\eta_M D_M}{N_A}\frac{dG_M}{dx} + 3\frac{\eta_A D_A}{N_A}\frac{dG_A}{dx}\right\}. \tag{A.45}$$

Substituting this value eE into the expression for J'_M gives

$$J'_M = \frac{\eta_M D_M}{k_B T}\left\{-\frac{1}{N_A}\frac{dG_M}{dx} + \frac{1}{\left(4\eta_M D_M + 9\eta_A D_A\right)}\left\{4\frac{\eta_M D_M}{N_A}\frac{dG_M}{dx} + 6\frac{\eta_A D_A}{N_A}\frac{dG_A}{dx}\right\}\right\}$$

which can be simplified by getting a common dominator for the terms in the brackets,

$$J'_M = \frac{\eta_M D_M}{k_B T\left(4\eta_M D_M + 9\eta_A D_A\right)}\left\{-\frac{1}{N_A}\frac{dG_M}{dx}\left(4\eta_M D_M + 9\eta_A D_A\right) + 4\frac{\eta_M D_M}{N_A}\frac{dG_M}{dx} + 6\frac{\eta_A D_A}{N_A}\frac{dG_A}{dx}\right\}$$

and multiplying

$$J'_M = \frac{\eta_M D_M}{k_B T\left(4\eta_M D_M + 9\eta_A D_A\right)}\left\{-\frac{4\eta_M D_M}{N_A}\frac{dG_M}{dx} - \frac{9\eta_A D_A}{N_A}\frac{dG_M}{dx} + \frac{4\eta_M D_M}{N_A}\frac{dG_M}{dx} + \frac{6\eta_A D_A}{N_A}\frac{dG_A}{dx}\right\}$$

---

* To avoid significant extra letters that might make things more complex than they need be, Mg^{2+} is represented by M and Al^{3+} by A.

which simplifies to

$$J'_M = -\frac{\eta_{Mg^{2+}}D_{Mg^{2+}}\eta_{Al^{3+}}D_{Al^{3+}}}{RT\left(4\eta_{Mg^{2+}}D_{Mg^{2+}} + 9\eta_{Al^{3+}}D_{Al^{3+}}\right)}\left\{9\frac{dG_{Mg^{2+}}}{dx} - 6\frac{dG_{Al^{3+}}}{dx}\right\} \tag{A.46}$$

where the M and A have been replaced by their ionic designations Mg^{2+} and Al^{3+}.

A.3.2 Chemical Potentials

For the reaction $Mg \rightleftharpoons Mg^{2+} + 2e^-$ the chemical potentials can be written as

$$\frac{dG_{Mg^{2+}}}{dx} = \frac{dG_{Mg}}{dx} - 2\frac{dG_e}{dx}. \tag{A.47}$$

Similarly for $Al \rightleftharpoons Al^{3+} + 3e^-$

$$\frac{dG_{Al^{3+}}}{dx} = \frac{dG_{Al}}{dx} - 3\frac{dG_{e^-}}{dx} \tag{A.48}$$

Substitution of Equations A.47 and A.48 into Equation A.46 gives

$$J'_M = -\frac{\eta_{Mg^{2+}}D_{Mg^{2+}}\eta_{Al^{3+}}D_{Al^{3+}}}{RT\left(4\eta_{Mg^{2+}}D_{Mg^{2+}} + 9\eta_{Al^{3+}}D_{Al^{3+}}\right)}\left\{9\frac{dG_{Mg}}{dx} - 6\frac{dG_{Al}}{dx}\right\} \tag{A.49}$$

and the chemical potentials in the last term are now for the metals in $MgAl_2O_4$ because the chemical potential gradients for the electrons fortunately cancel. Again, the reaction to form magnesium aluminate spinel is

$$MgO(s) + Al_2O_3(s) \rightarrow MgAl_2O_4(s);\ \Delta G^o_{MgAl2O4} \tag{A.50}$$

or

$$G_{MgO} + G_{Al_2O_3} = \Delta G^o_{MgAl2O4} \tag{A.51}$$

and the latter is a constant throughout the thickness of the $MgAl_2O_4$. Differentiating Equation A.51 gives

$$\frac{dG_{MgO}}{dx} = -\frac{dG_{Al_2O_3}}{dx} \tag{A.52}$$

and because $Mg + 1/2O_2 \rightarrow MgO$,

$$\frac{dG_{MgO}}{dx} = \frac{dG_{Mg}}{dx} + \frac{dG_O}{dx} \tag{A.53}$$

where the last term is zero because the oxygen pressure is uniform over all three solids. Similarly,

$$2Al + 3/2O_2 \rightarrow Al_2O_3$$

so,

$$\frac{dG_{Al_2O_3}}{dx} = 2\frac{dG_{Al}}{dx} + \frac{3}{2}\frac{dG_O}{dx} \tag{A.54}$$

where the last term is again zero. Combining Equations A.52 through A.54 gives the desired result:

$$\frac{dG_A}{dx} = -\frac{1}{2}\frac{dG_M}{dx} \tag{A.55}$$

which when substituted into (A.49) gives the final result

$$J'_M = -\frac{12\eta_{Mg^{2+}}D_{Mg^{2+}}\eta_{Al^{3+}}D_{Al^{3+}}}{RT\left(4\eta_{Mg^{2+}}D_{Mg^{2+}} + 9\eta_{Al^{3+}}D_{Al^{3+}}\right)}\frac{dG_{Mg}}{dx}$$

A.4 The Carter Model

The Carter modification of the Ginstling and Brounshtein model of the fraction reacted versus time takes into consideration that the volume of the products of a reaction is different from the volume of the reactants (Carter 1961b). The ratio of the volume of the products to the volume of reactants is designated by the variable z. For example, for the formation of $MgAl_2O_4$ on the MgO side of the original interface where, on completion, one mole of $MgAl_2O_4$ replaces four original moles of MgO,

$$z = \frac{\bar{V}_{MgAl_2O_4}}{4\bar{V}_{MgO}} = \frac{40.07}{4\times 11.26} = 0.89$$

where $\bar{V}_i$ = molar volume (cm³/mol). Let a_0 = initial radius of the reactant MgO, a = radius of reactant MgO at time t, and r_0 = total volume of product plus unreacted reactant, so that at any time t, the total volume = volume product + volume unreacted reactant:

$$r_0^3 = z\left(a_0^3 - a^3\right) + a^3$$
$$r_0 = \left[za_0^3 - za^3 + a^3\right]^{\frac{1}{3}} \tag{A.56}$$

As in the case of the Ginstling and Brounshtein result (Ginstling and Brounshtein 1950),

$$da = \frac{r_0}{\left(a - r_0\right)a}k'''dt \tag{A.57}$$

where

$$k''' = -\left(\frac{\Delta G_{MgAl_2O_4}}{RT}\right)\left(\frac{2D_{Mg^{2+}}D_{Al^{3+}}}{4D_{Mg^{2+}} + 18D_{Al^{3+}}}\right)$$

or rearranging,

$$\begin{aligned} &\frac{\left(a - r_0\right)a}{r_0}da \\ &= \left(\frac{a^2}{r_0} - \frac{ar_0}{r_0}\right)da \\ &= \frac{a^2}{r_0}da - \frac{ar_0}{r_0}da \\ &= k'''dt \end{aligned} \tag{A.58}$$

Now, considering only the first integral,

$$\frac{a^2}{r_0}da = \frac{a^2}{\left[za_0^3 + a^3\left(1 - z\right)\right]^{\frac{1}{3}}}da \tag{A.59}$$

and let $y^3 = (1-z)a^3 + za_0^3$ so that

$$a^3 = \frac{y^3 - za_0^3}{1-z}$$

$$a = \left(\frac{y^3 - za_0^3}{1-z}\right)^{\frac{1}{3}}$$

$$a^2 = \left(\frac{y^3 - za_0^3}{1-z}\right)^{\frac{2}{3}}$$

$$da = \frac{1}{3}\left(\frac{y^3 - za_0^3}{1-z}\right)^{-\frac{2}{3}} \frac{3y^2}{1-z} dy$$

so

$$\frac{a^2}{y} da = \frac{\left(\frac{y^3 - za_0^3}{1-z}\right)^{\frac{2}{3}}}{y} \frac{1}{3}\left(\frac{y^3 - za_0^3}{1-z}\right)^{-\frac{2}{3}} \frac{3y^2}{1-z} dy = \frac{y}{1-z} dy$$

and integrating the last term we get

$$\left.\frac{1}{2}\frac{y^2}{(1-z)}\right]_{a_0}^{a} = \left.\frac{\left[(1-z)a^3 + za_0^3\right]^{\frac{2}{3}}}{2(1-z)}\right]_{a_0}^{a}$$

$$= \frac{\left[(1-z)a^3 + za_0^3\right]^{\frac{2}{3}}}{2(1-z)} - \frac{\left[(1-z)a_0^3 + za_0^3\right]^{\frac{2}{3}}}{2(1-z)} \tag{A.60}$$

$$= \frac{\left[(1-z)a^3 + za_0^3\right]^{\frac{2}{3}}}{2(1-z)} - \frac{a_0^2}{2(1-z)}$$

which is the first integral in Equation A.58. The second integral requires a lot less algebra

$$\int_{a_0}^{a} a da = \frac{1}{2}a^2 - \frac{1}{2}a_0^2, \tag{A.61}$$

and combining the last of Equations A.60 and A.61 gives

$$\frac{\left[(1-z)a^3 + za_0^3\right]^{\frac{2}{3}}}{2(1-z)} - \frac{a_0^2}{2(1-z)} - \frac{1}{2}a^2 + \frac{1}{2}a_0^2 = k''t \tag{A.62}$$

Multiplying both sides of the equation by $2(1-z)/a_0^2$, canceling, and rearranging, the final equation is finally obtained:

$$\left[(1-z)\left(\frac{a}{a_0}\right)^3 + z\right]^{\frac{2}{3}} - (1-z)\left(\frac{a}{a_0}\right)^2 = z + (1-z)kt. \tag{A.63}$$

where, again,

$$k = -\left(\frac{\Delta G^{o}_{MgAl_2O_4}}{RT}\right)\frac{2D_{Mg^{2+}}D_{Al^{3+}}}{\left(2D_{Mg^{2+}} + 9D_{Al^{3+}}\right)}\frac{1}{a_0^2}$$

Now, the fraction reacted, f_R, is given by $f_R = 1-(a/a_0)^3$ so $1-f_R = (a/a_0)^3$ and $\left(1-f_R\right)^{2/3} = (a/a_0)^2$, which with a little simplification gives the Carter equation

$$\left[1+f_R\left(z-1\right)\right]^{\frac{2}{3}} + \left(z-1\right)\left(1-f_R\right)^{\frac{2}{3}} = z + \left(1-z\right)kt. \quad \text{(A.64)}$$

To see what happens when $z \rightarrow 1$, that is, the products and reactants have the same volume, Equation A.64 can be rewritten as

$$\frac{1}{\left(1-z\right)}\left[1+f_R\left(z-1\right)\right]^{\frac{2}{3}} + \frac{\left(z-1\right)}{\left(1-z\right)}\left(1-f_R\right)^{\frac{2}{3}} - \frac{z}{\left(1-z\right)} = kt. \quad \text{(A.65)}$$

Because $\left[1+f_R\left(z-1\right)\right]^{\frac{2}{3}} \cong \left[1+\frac{2}{3}\left(z-1\right)f_R\right]$ as $z \rightarrow 1$, Equation A.65 becomes

$$1-\frac{2}{3}f_R - \left(1-f_r\right)^{\frac{2}{3}} = kt$$

the Ginstling–Brounshtein result as expected.

EXERCISES

15.1 Assume that two oxides AO and BO are interdiffusing with BO on the left. These two oxides diffuse and form an ideal solution so nonideality effects can be neglected and the oxygen ions are immobile. In fact, the overall concentration is a constant, $C_O = 6.25\times10^{-2}\ \text{mol/cm}^3$.

a. The diffusion coefficient of A as a function of mole fraction of BO, X_{BO}, is given by $D_A = 10^{-13}\left(10-9X_{BO}\right)\text{cm}^2/\text{s}$ and that of B by $D_B = 10^{-14}\left(10-9X_{BO}\right)\text{cm}^2/\text{s}$; that is, they are linear with X_B. In contrast, the diffusion coefficient of oxygen is $D_O = 10^{-17}\left(1+9X_{BO}\right)\text{cm}^2/\text{s}$. Calculate and plot D_A, D_B, D_O and the interdiffusion coefficient $\tilde{D}$ as a function of X_B for $0 < X_B < 1.0$ and do not consider the oxygen ion diffusion coefficient in the calculation of $\tilde{D}$.

b. Take the value of $\tilde{D}$ at $X_{BO} = 0.5$ and calculate and plot the concentration as a function of distance with the appropriate error function solution, $X_{BO} = 0.5\,\text{erfc}\left(x/\sqrt{4\tilde{D}t}\right)$ at $t = 10^4$ seconds between $-2\,\mu\text{m} < x < 2\,\mu\text{m}$ where $x = 0$ is the original interface. Use at least 20 data points.

c. Develop an expression for dX_{BO}/dx assuming that BO and AO form an ideal solid solution and calculate its value at $t = 10^4$ seconds at $X_{BO} = 0.5$.

d. Calculate the value of the electric field strength, E (V/m), at $x = 0$, $t = 10^4$ s assuming that AO and BO form an ideal solid solution on interdiffusing.

15.2 Similar to exercise 1 above, assume that these same two oxides AO and BO interdiffuse with BO on the left. These two oxides diffuse and form an ideal solution so nonideality effects can be neglected. Again, the overall concentration is a constant, $C_O = 6.25\times10^{-2}\ \text{mol/cm}^3$. But now the diffusion coefficients are slightly different.

a. The diffusion coefficient of A as a function of mole fraction of BO, X_{BO}, is given by $D_A = 10^{-13}(10-9X_{BO})cm^2/s$ and that of B by $D_B = 10^{-15}(10-9X_{BO})cm^2/s$; that is, they are linear with X_B. In contrast, the diffusion coefficient of oxygen is $D_O = 10^{-17}(1+9X_{BO})cm^2/s$. Calculate and plot $\log_{10}D_A$, D_B, D_O and the interdiffusion coefficient $\tilde{D}$ as a function of X_B for $0 < X_B < 1.0$ and do not consider the oxygen ion diffusion coefficient in the calculation of $\tilde{D}$.

b. Calculate $\tilde{D}$ again but now including D_O as a function of X_{BO} and plot it and the value of $\tilde{D}$ calculated in part a on the same plot. Only, this time just plot the values of $\tilde{D}$ and not their logs.

c. Calculate $\tilde{D}$ again with the same values of D_A and D_B but now with $D_O = 10^{-7}(1+9X_{BO})cm^2/s$. Assume that the oxygen diffusion coefficient has been increased by forming a solid solution in AO with Li_2O,

$$Li_2O \xrightarrow{AO} 2Li'_A + O_O^x + V_O^{\bullet\bullet} \text{ or } xLi_2O + (1-2x)AO = A_{1-2x}Li_{2x}O_{1-x}$$

because oxygen vacancies are created. Similarly for BO. So if AO were $A_{0.8}Li_{0.2}O_{0.9}$ and BO were $B_{0.8}Li_{0.2}O_{1.8}$, then (1) the oxygen diffusion coefficient would be high, and (2) it would increase as BO content indicated by the equation for oxygen diffusion given above because the oxygen vacancy concentration is higher in the lithium-doped BO. Neglect any concentration differences or diffusion of Li_2O.

d. At $t = 10^4$ seconds, calculate the Kirkendall velocity (μm/s) at the original interface, roughly at $X_{BO} = 0.5$ with the value of dX_{BO}/dx calculated in exercise 1c above.

e. If this velocity were constant for 10^4 seconds, calculate how far markers at the original interface have moved and indicate in which direction.

15.3 Aluminum oxide, Al_2O_3, and chromium oxide, Cr_2O_3, form almost an ideal solid solution. Develop the interdiffusion coefficient for a pure alumina-pure chromia diffusion couple. Assume that the oxygen ions are immobile and these two compounds do indeed form an ideal solid solution and the electronic conductivity is negligible.

15.4 Two oxides are reacting by interdiffusion to form a third by the following reaction:

$$AO(s) + BO_2(s) = ABO_3(s); \quad \Delta G^\circ(1200°C) = -25.5\,kJ/mol$$

implying that A^{2+} and B^{4+} are interdiffusing to form the compound because the oxygen ions are essentially immobile.

a. From electrical neutrality, determine the relationship between the number flux densities, $ions/cm^2\,s$, of A^{2+} and B^{4+}.

b. The densities and molecular weights are M(AO) = 35 g/mol, ρ(AO) = 3.5 g/cm^3; $M(BO_2)$ = 80 g/mol, ρ = 4.0 g/cm^3; and $M(ABO_3)$ = 115 g/mol, $\rho(ABO_3)$ = 3.83 g/cm^3. Calculate the increase in volume when this reaction goes to completion.

c. From a and b above, generate a diagram such as Figure 15.7 that shows the positions of the interfaces for $t = 0$, $0 < t < \infty$, and $t = \infty$.

d. Develop an expression for the electric field strength, E, produced by the interdiffusion of A and B thorough ABO_3.

e. Develop an expression for the reaction rate constant, k, in $L^2 = kt$ where L is the thickness of the reaction product layer at time t, in terms of the diffusion coefficients of A and B through the product layer, ABO_3.

f. If $D(A) = 10^{-10}\,cm^2/s$ and $D(B) = 10^{-12}\,cm^2/s$ at 1200°C, calculate the value of the L at $t = 10^4$ s in micrometers.

g. Calculate a *numerical value* of k for the Jander model if 100 μm spheres are reacting.

REFERENCES

Appel, M. and J. A. Pask. 1971. Interdiffusion and Moving Boundaries in NiO-CaO and NiO-MgO Single-Crystal Couples. *J. Am. Ceram. Soc.* 54(3): 152–158.

Bogue, R. H. 1955. *The Chemistry of Portland Cement*, 2nd ed. New York: Reinhold Publishing Company.

Carter, R. E. 1961a. Mechanism of Solid-State Reaction Between Magnesium Oxide and Aluminum Oxide and Between Magnesium Oxide and Ferric Oxide. *J. Am. Ceram. Soc.* 44(3): 116–120.

Carter, R. E. 1961b. Kinetic Model for Solid-State Reactions. *J. Chem. Phys.* 34(6): 2010–2015.

Crow, J. M. 2008. The Concrete Conundrum. *Chemistry World*. March 2008: 62–66. http://www.chemistry-world.org.

Frade, J. R. and M. Cable. 1992. Reexamination of the Basic Theoretical Model for the Kinetics of Solid-State Reactions. *J. Am. Ceram. Soc.* 75(7): 1949–1957.

Ginstling, A. M. and V. I. Brounshtein. 1950. Concerning the Diffusion Kinetics of Reactions in Spherical Particles. *J. Appl. Chem. USSR (English Transl.).* 23(12): 1327–1338.

Gopalan, S. and A. V. Virkar. 1995. Interdiffusion and Kirkendall Effect in Doped Barium Titanate-Strontium Titanate Diffusion Couples. *J. Am. Ceram. Soc.* 78(4): 993–998.

Haynes, W. M. Editor-in-Chief. 2013. *Handbook of Chemistry and Physics*, 94th ed. Boca Raton, FL: CRC Press.

Jander, W. 1927. Reactions in the Solid State at High Temperatures (in German). *Z. Anrg. Allg. Chem.* 163(1/2): 1–30.

Kingery, W. D., H. K. Bowen, and D. R. Uhlmann. 1976. *Introduction to Ceramics*, 2nd ed. New York: John Wiley and Sons.

Kubaschewski, O. 1972. The Thermodynamic Properties of Solid Oxides (A Review). *High Temperatures-High Pressures* 4: 1–12.

Levin, E. M., C. R. Robbins, and H. F. McMurdie. 1964. *Phase Diagrams for Ceramists*. Columbus, OH: The American Ceramic Society.

Nielsen, T. H. and M. H. Leipold. 1963. Thermal Expansion in Air of Ceramic Oxides to 2200°C. *J. Am. Ceram. Soc.* 46(8): 381–387.

Nielsen, T. H. and M. H. Leipold. 1965. Thermal Expansion of Nickel Oxide. *J. Am. Ceram. Soc.* 48(3): 164.

Norton, F. H. 1974. *Elements of Ceramics*, 2nd Ed. Reading, MA: Addison-Wesley Publishing Company.

Pettit, F. S., E. H. Randklev, and E. J. Felten. 1966. Formation of $NiAl_2O_4$ by Solid State Reaction. *J. Am. Ceram. Soc.* 49(4): 199–203.

Ramisetty, M., S. Sastri, U. Kashalikar, L. M. Goldman, and N. Nag. 2013. Transparent Polycrystalline Cubic Spinels Protect and Defend. *Am. Ceram. Soc. Bull.* 92(2): 20–25.

Review of REACH with regard to the Registration Requirements on Polymers. Draft Final Report. Part A: Polymers. December 2012. http://ec.europa.eu/enterprise/sectors/files/reach/review2012/registr-req-final-report-part-a_en.pdf.

Rhodes, W. H. and R. E. Carter. 1966. Cationic Self-Diffusion in Calcia-Stabilized Zirconia. *J. Am. Ceram. Soc.* 49(5): 244–249.

Roine, A. 2002. *Outokumpu HSC Chemistry for Windows*, Ver. 5.11, thermodynamic software program, Outokumpu Research Oy, Pori, Finland.

Schmalzried, H. 1981. *Solid State Reactions*. Weinheim Germany: Verlag Chemie.

USGS Minerals Information. 2014. Commodity Statistics and Information. http://minerals.usgs.gov/minerals/pubs/commodity/.

Wells, A. F. 1984. *Structural Inorganic Chemistry*, 5th ed. Oxford: Clarendon Press.

Worrell, E., et al. 2001. Carbon Dioxide Emissions from the Global Cement Industry. *Annu. Rev. Ener. Environ.* 26: 303–329.

Yurek, G. J. and H. Schmalzreid. 1975. Deviations from Local Thermodynamic Equilibrium During Interdiffusion of CoO-MgO and CoO-NiO. *Berichte der Bunsen-Gesellschaft*. 74(1): 255–262.

16

Spinodal Decomposition Revisited

16.1 INTRODUCTION

Spinodal decomposition kinetics are very strongly linked to the nonideal thermodynamic behavior of solid and liquid solutions. Spinodal decomposition occurs in metals, inorganic compounds, and particularly in polymers because polymer blends and solutions typically have limited liquid–liquid solubility. A phase diagram of a solid solution separating into two solid phases at low temperatures is shown in Figure 7.14, and some compositions will undergo phase separation by nucleation and growth, whereas others will phase separate by spinodal decomposition. Figure 16.1 shows the difference between the nucleation and growth mechanism and that of spinodal decomposition. In nucleation growth, nuclei of a very different composition than the average concentration with a clearly defined phase interface and interfacial energy form and grow by normal diffusion down a concentration gradient (as described in detail in Chapters 7 and 9). In contrast, spinodal decomposition occurs with small and continuous compositional fluctuations around the average composition, *lowering the Gibbs energy,* that increase in concentration until the final equilibrium concentrations are reached. This requires *diffusion up a concentration gradient,* so-called *up-hill diffusion* (Figure 16.1). An approach to explain this behavior is the major focus of this chapter.

The fundamentals of the thermodynamics and kinetics of spinodal decomposition were developed primarily in a series of papers by John Cahn (Cahn and Hilliard 1958; Cahn and

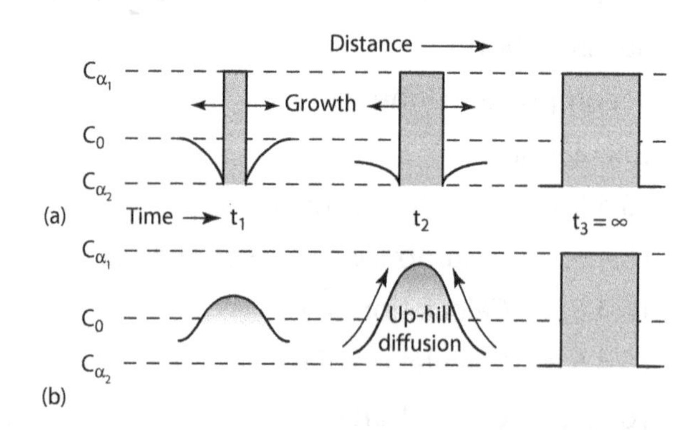

FIGURE 16.1 Schematic showing the differences between phase separation by (a) nucleation and growth and (b) spinodal decomposition. In nucleation and growth, small nuclei of the equilibrium composition with a compositionally sharp phase interface nucleate and grow by normal diffusion. In contrast, spinodal decomposition occurs by small fluctuations in composition growing by diffusion *up the concentration gradient* with a diffuse phase interface until phase separation is complete.

Hilliard 1959; Cahn 1961, 1966). An important consideration in spinodal decomposition is the energy in the concentration gradient associated with the development of the two-phase system (Hillert 1961). This approach is now part of what is called *the phase field* method to model microstructure changes with diffuse rather than sharp interfaces. This is a growing and important area of *computational materials* science (Moelans et al. 2008; Lesar 2013). Because of their generality and completeness, these original papers require significant effort to interpret and digest and, in some cases, require advanced mathematical techniques and thermodynamic considerations beyond those being used here. Several authors have distilled the essence of the theory to provide a more pedagogical approach to spinodal decomposition, often using quite different approaches and simplifications (Kingery et al. 1976; Schmalzried 1981; Ragone 1995; Balluffi et al. 2005; Porter et al. 2008; Jackson 2010). The approach here is to use some elements of these models, along with additions, that are hopefully more transparent and lead to a step-by-step explanation of spinodal decomposition thermodynamics and kinetics. The resulting model provides the ability to actually calculate the dimensions and kinetics of decomposition rather than just describe the principles used in applying the general theory. In doing so, a regular solution is used to make concepts more concrete and only one-dimensional decomposition is considered for the sake of simplicity.

16.2 REGULAR SOLUTION

16.2.1 Regular Solution of Uniform Composition

The thermodynamics of a *regular solution* were introduced in Section 14.4.4. Here, it is useful to make a connection between the thermodynamics of a regular solution and some very general aspects of the molecular or atomic composition and their relationship to the thermodynamic formalism of a regular solution. This is useful in the discussion of spinodal decomposition as a mechanism of phase separation kinetics. The Gibbs energy of mixing *per mole* of solution of a regular solution of components a and b is again given by* (Darken and Gurry 1953)

$$\bar{G}_{mix,reg} = RT\{X_a \ln X_a + X_b \ln X_b\} + \Gamma X_a X_b \tag{16.1}$$

where:

- Γ is a temperature-independent constant that may be positive or negative
- X_i is the mole fractions, and the first term is the Gibbs energy of mixing of an ideal solution, which is always negative.

The following microscopic development for Γ is taken largely from Denbigh (1964) and some of it is found in other texts (DeHoff 2006). The model is a *nearest-neighbor model* because only the interactions between neighboring atoms or molecules are considered. Longer range interactions are ignored. Let N_a be the number of a atoms and N_b the number of b atoms, and consider two adjacent sites in the solution. The probability of an a atom being on site number one, 1, in the solution (solid or liquid) is simply given by the mole (molecular) fraction

$$p_{a,1} = \frac{N_a}{N_a + N_b}.$$

Similarly, the probability that a b atom is on the neighboring site number two, 2, is

$$p_{b,2} = \frac{N_b}{N_a + N_b}.$$

---

* Lower case A and B are being used to not confuse Avogadro's number, N_A, with the number of a atoms, N_a.

The combined probability that site 1 is occupied by a and site 2 by b is just the product of the two probabilities

$$p_{ab} = \frac{N_a N_b}{(N_a + N_b)^2}.$$

Likewise, the probability of b on site 1 and a on site 2 is the same. So the probability of an a–b pair on neighboring sites is just the sum of these probabilities or

$$p_{a-b} = \frac{2N_a N_b}{(N_a + N_b)^2}. \tag{16.2}$$

If the number of nearest neighbors in the solution is z (maybe 12 or probably less), then the total number of pairs in the solution (a–a, b–b, and a–b) is simply $(1/2)z(N_a + N_b)$ because $N_a + N_b$ is the total number of atoms and the 1/2 comes about because each pair is counted only once. Therefore, the total number of a–b pairs, N_{ab}, is just the total number pairs times the probability of the pair being an a–b pair (Equation 16.2)

$$N_{ab} = \frac{1}{2}z(N_a + N_b) \times \frac{2N_a N_b}{(N_a + N_b)^2} = z\frac{N_a N_b}{N_a + N_b}. \tag{16.3}$$

Similarly, the numbers of a–a and b–b pairs are given by

$$\begin{aligned} N_{aa} &= \frac{1}{2}z(N_a + N_b) \times \frac{N_a^2}{(N_a + N_b)^2} = \frac{1}{2}z\frac{N_a^2}{N_a + N_b} \\ N_{bb} &= \frac{1}{2}z(N_a + N_b) \times \frac{N_b^2}{(N_a + N_b)^2} = \frac{1}{2}z\frac{N_b^2}{N_a + N_b}. \end{aligned} \tag{16.4}$$

Now the numbers of a–a pairs and b–b pairs in the pure a and b solutions before mixing were $(1/2)zN_a$ and $(1/2)zN_b$, respectively. The enthalpy increase, ΔH_{mix}, on mixing is given by (assuming no change in volume as usual) the total pair energy in the solution minus the pair energies in the pure components:

$$\Delta H_{mix} = z\frac{N_a N_b}{N_a + N_b}\varepsilon_{ab} + \frac{1}{2}z\frac{N_a^2}{N_a + N_b}\varepsilon_{aa} + \frac{1}{2}z\frac{N_b^2}{N_a + N_b}\varepsilon_{bb} - \frac{1}{2}zN_a\varepsilon_{aa} - \frac{1}{2}zN_b\varepsilon_{bb} \tag{16.5}$$

where:

ε_{ab} is the ab molecule pair energy
ε_{aa} is the aa molecule pair energy
ε_{bb} is the bb molecule pair energy.

Obtaining the common denominator for Equation 16.5, with some cancellation, gives

$$\Delta H_{mix} = \frac{zN_a N_b}{N_a + N_b}\left(\varepsilon_{ab} - \frac{1}{2}\varepsilon_{aa} - \frac{1}{2}\varepsilon_{bb}\right) \tag{16.6}$$

and if $N_a + N_b = N_A$, Avogadro's number (1 mol) and $E = N_A\left(\varepsilon_{ab} - \frac{1}{2}\varepsilon_{aa} - \frac{1}{2}\varepsilon_{bb}\right) = N_A\Delta\varepsilon$, then Equation 16.6 can be written as

$$\Delta\bar{H}_{mix} = zEX_aX_b = \Gamma X_aX_b \quad (16.7)$$

as in Equation 16.1, where $\Gamma = zE$. The important result *is that there is a relationship between the parameter for a regular solution,* Γ*, and the interaction energies between the a and b atoms.** It should be noted that, depending on the pairing energies, Γ can be positive or negative. If there is a strong tendency for a–b pairs to form, then $\varepsilon_{ab} < 0$, that is, it is negative, and if $|\varepsilon_{aa}|$ and $|\varepsilon_{bb}|$ are small, this leads to an overall *negative* $\Delta\bar{H}_{mix}$ and there is a tendency for compound formation. On the contrary, if the energies for a–a and b–b pairing are negative, $(\varepsilon_{aa} + \varepsilon_{bb}) \ll 0$, and ε_{ab} is small, then $\Delta\bar{H}_{mix}$ is positive and there is a tendency for phase separation that will begin when the $\Delta\bar{H}_{mix}$ term is just equal and of opposite sign to the entropy of ideal mixing term, which is always negative, in Equation 16.1. This forms the thermodynamic basis for solution phase separation and spinodal decomposition discussed in more detail below.

16.2.2 Regular Solution and a Composition Gradient

16.2.2.1 Becker Model

An important thermodynamic parameter in spinodal decomposition is the energy in a composition gradient that must be overcome before phase separation can occur. This is very much like the surface energy in nucleation and growth that must exceeded by the volume Gibbs energy of the nucleus as was discussed in Chapter 7. With the nearest-neighbor approach as applied in Section 16.2.1 for the Gibbs energy of a uniform composition regular solution, the energy in a concentration gradient can be approximated. What follow is basically a procedure developed by Becker (1938) and given in Kingery et al. (1976). Figure 16.2 schematically shows a concentration gradient between rows of a and b atoms both in the form of the individual atoms and a concentration profile that is chosen to be linear for simplicity. The technique used to calculate the Gibbs energy due to the concentration gradient is to calculate the energy of a plane, say the one labeled 0 in Figure 16.1, in a solution of this *uniform composition*. Then do the same for the neighboring plane labeled 1. Then the energy of neighboring planes 0 and 1 in the gradient is calculated. The difference between the energy of the two planes of different compositions minus those of the two planes in a solid of uniform compositions is an approximation to the energy in the gradient. A constant volume for all compositions and a simple cubic lattice are assumed for simplicity. The energy of a plane in a uniform solution of *atom fraction* x_0 is

$$E_0 = \frac{zm}{2}\left(x_0x_0\varepsilon_{aa} + (1-x_0)(1-x_0)\varepsilon_{bb} + 2x_0(1-x_0)\varepsilon_{ab}\right) \quad (16.8)$$

where:

- z is the number of nearest neighbors of an atom ($z = 6$ for simple cubic)
- m is the atom density per unit area, the ε_i are the bond energies between atoms,
- x_0 is the fraction of atoms of type a on the plane.

A term such as $1/2zmx_0x_0 = 1/2(mx_0)(zx_0)$ occurs because mx_0 = probability that an atom in a plane is of type a and the probability that one of its nearest neighbors is also type a is zx_0. Again, the 1/2 occurs to remove the double counting of this a–a pair. Equation 16.8 is very similar to Equation 16.5 for the regular solution. Likewise, for the plane at position 1 in Figure 16.1,

* The *Flory–Huggins interaction parameter,* χ, where $\chi = z\Delta\varepsilon/k_BT$ (so $\chi = \Gamma/RT$), is used to predict polymer–polymer and polymer–solvent interactions and phase separation (Elias 1997; Green 2005; Sperling 2006).

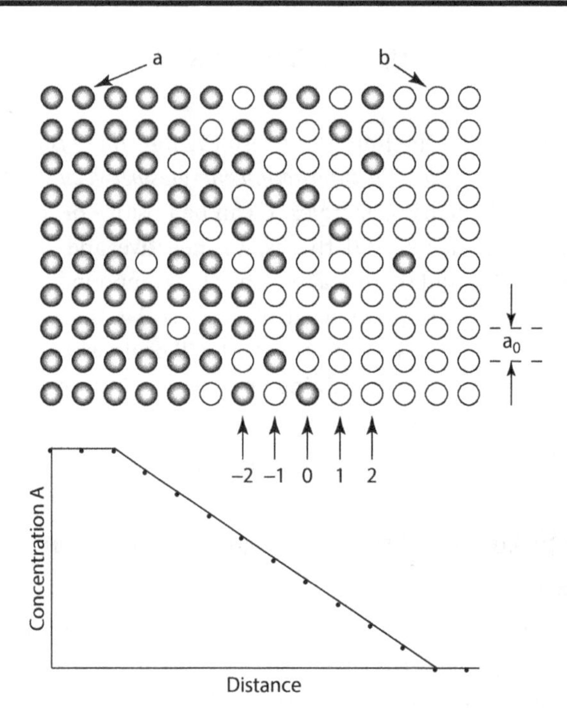

FIGURE 16.2 Schematic of a simple cubic crystal of interatomic spacing of a_0 consisting of atoms a and b in a concentration gradient shown in the lower part of the figure (Becker 1938; Kingery et al. 1976)

$$E_1 = \frac{zm}{2}\left(x_1 x_1 \varepsilon_{aa} + (1-x_1)(1-x_1)\varepsilon_{bb} + 2x_1(1-x_1)\varepsilon_{ab}\right) \tag{16.9}$$

and for the energy of adjacent planes 0 and 1 in Figure 16.1,

$$E_{0-1} = \frac{zm}{2}\left(x_0 x_1 \varepsilon_{aa} + (1-x_0)(1-x_1)\varepsilon_{bb} + \left(x_0(1-x_1) + x1(1-x_0)\right)\varepsilon_{ab}\right). \tag{16.10}$$

The *interfacial energy* is simply twice Equation 16.10 minus Equations 16.8 and 16.9, that is

$$E_{int} = 2E_{0-1} - (E_0 + E_1) \tag{16.11}$$

where the 2 simply comes from comparing two of the same planes in the gradient with two planes in the solutions of uniform compositions. The rest is just simple algebra! However, it leads to lengthy equations and is carried out in Section A.1 to conserve space here. The result is (Equation A.6)

$$E_{int} = zm\Delta\varepsilon(x_0 - x_1)^2 \tag{16.12}$$

where again $\Delta\varepsilon = \varepsilon_{ab} - \varepsilon_{aa}/2 - \varepsilon_{bb}/2$.

16.2.2.2 Interfacial Energy Compared to Surface Energy

Suppose that either x_0 or $x_1 = 1.0$ and the other is zero. This is essentially a sharp interface between two phases, so the interfacial energy is now a surface energy. Assume that $N_A z\Delta\varepsilon = 22\,\text{kJ/mol}$ (for reasons that will be obvious later). So,

$$z\Delta\varepsilon = \frac{22\times10^3\ \text{J/mol}}{6.022\times10^{23}\ \text{atoms/mol}} = 3.65\times10^{-20}\ \text{J/atom}.$$

Suppose that one of the phases were a metal simply to get some value of m and interfacial energy. (Caveat: most metals are not simple cubic but only approximate values are being sought not accurate values, so this is close enough. In fact, any element could be used.) For many metals, the molar volumes are on the order of $\bar{V} = 15\,\text{cm}^3/\text{mol}$ (Haynes 2013). So,

$$m \cong \eta^{\frac{2}{3}} = \left(\frac{N_A}{\bar{V}}\right)^{\frac{2}{3}}$$

$$= \left(\frac{6.022\times10^{23}}{15}\right)^{\frac{2}{3}}$$

$$m = 1.172\times10^{15}\,\text{atoms/cm}^2 \text{ or } 1.172\times10^{19}\,\text{atoms/m}^2$$

so

$$E_{int} = \left(3.65\times10^{-20}\right)\left(1.172\times10^{19}\right) = 0.428\,\text{J/m}^2$$

compared to the surface energy of metals of around 500 mJ/m^2 given in Table 6.1 in Chapter 6. Note that ε_{aa} is negative for phase separation making $\Delta\varepsilon$ positive. So, Equation 16.12 certainly seems to give reasonable values for the "interfacial" energy of a composition gradient.

16.2.2.3 Interfacial Gibbs Energy and Composition Gradient

Again, the energy per unit between planes of different composition in a solid or liquid is given by Equation 16.12. It is convenient to put this in terms of the concentration gradient in one dimension, y (rather than x, which is currently being used for fractional occupation of planes), $\partial C/\partial y$, C = concentration (mol/cm^3). Now, $x_1 \cong x_0 - a_0\left(dx/dy\right)$, where a_0 is the planar separation distance or the lattice parameter for a simple cubic crystal. So, Equation 16.12 becomes

$$E_{int} = zm\,\Delta\varepsilon\, a_0^2\left(\frac{dx}{dy}\right)^2.$$

Now, in the simple cubic lattice shown in Figure 16.1, $m \times a_0^2 = 1\,\text{atom}$ and $x = X$ = the mole fraction of B. The concentration, C (of B), is given by

$$C = \left(\frac{m}{a_0 N_A}\right)x \text{ or } x = \left(\frac{a_0 N_A}{m}\right)C$$

so Equation 16.12 now becomes

$$E_{int} = \frac{z\,\Delta\varepsilon\, a_0^4 N_A^2}{m}\left(\frac{dC}{dy}\right)^2. \qquad (16.13)$$

But $\bar{V} = \dfrac{a_0 N_A}{m}$ and also $\bar{V} = a_0^3 N_A$, so inserting these into Equation 16.13.

$$E_{int} = z\,\Delta\varepsilon\,\bar{V}^2\left(\frac{dC}{dy}\right)^2 \qquad (16.14)$$

where the Units(E_{int}) = J/m^2.

16.2.2.4 Comparison to Surface Energy

To ensure that the various substitutions in Section 16.2.2.3 are valid, the value of Equation 16.14 is calculated with values consistent with those used to calculate E_{int} in Section 16.2.2.2. The value of $z\Delta\varepsilon = 3.65\times10^{-20}$ J is the same, the molar volume is again $\bar{V} = 15\,\text{cm}^3/\text{mol}$ and the inverse of this $(1/\bar{V}) = (1/15) = 0.0667\,\text{mol/cm}^3$ is the concentration C in the pure metal. If the gradient is again taken over one interatomic distance,

$$a_0 = \left(\frac{\bar{V}}{N_A}\right)^{\frac{1}{3}} = \left(15/6.022\times10^{23}\right)^{1/3}$$

$$a_0 = 2.92\times10^{-8}\,\text{cm}.$$

Putting these values into Equation 16.14

$$E_{int} = z\Delta\varepsilon\bar{V}^2\left(\frac{dC}{dx}\right)^2 = \left(3.65\times10^{-20}\right)(15)^2\left(\frac{0.0667}{2.92\times10^{-8}}\right)^2$$

$$E_{int} = 0.429\times10^{-4}\,\text{J/cm}^2 = 0.429\,\text{J/m}^2$$

and shows that Equation 16.14 is correct, or at least consistent with Equation 16.12.

16.3 THERMODYNAMICS OF SPINODAL DECOMPOSITION

A nonideal solid or liquid solution can, under certain circumstances, decompose by *spinodal decomposition* rather than *nucleation and growth* as discussed in Chapter 7. However, the conditions for spinodal composition are rather restrictive and it does not occur as frequently as nucleation and growth, at least crystalline in solids.* Consider a regular solution, which is a nonideal solution that is mathematically manageable and easily understood (Equation 16.1)

$$\bar{G}_{mix,reg} = RT\{X_a\ln X_a + X_b\ln X_b\} + \Gamma X_a X_b$$

remembering that this is the Gibbs energy of *1 mol* of solution. This can be rewritten as

$$\bar{G}_{mix,reg} = RT\{X\ln X + (1-X)\ln(1-X)\} + \Gamma X(1-X). \tag{16.15}$$

where X can be the mole fraction of component b. With a value of $\Gamma = 21167.4$ J/mol, Equation 16.15 is plotted in Figure 16.3 for three different temperatures: 1273, 1073, and 973 K where a and b are now in the more typical form of A and B. The value of Γ was chosen so that the Gibbs energy at X = 0.5 would be at its minimum at 1273 K simply to provide some reasonable values for temperatures and free energies, and this is the reason why 22.5 kJ/mol was chosen in Section 16.2.2 earlier. A positive value of Γ indicates that the energies to form A-A and B-B bonds are stronger (more negative) than to form A–B bonds. As a result, there is a tendency for the solution to separate into two phases of different compositions, where $\left(dG_{mix,reg}/dX\right) = 0$, that is, at

$$\frac{d\bar{G}_{mix,reg}}{dX} = RT\left(\ln X + 1 - \ln(1-X) - 1\right) + \Gamma(1-X) - \Gamma X$$
$$0 = \ln\left(\frac{X}{1-X}\right) + \frac{\Gamma}{RT}(1-2X) \tag{16.16}$$

Clearly, at X = 0.5, this is always zero as can be seen in Figure 16.3. Solving (numerically and easily with a spreadsheet program) the second of Equation 16.16 for two other values of X give the

* In polymer solutions and inorganic glasses, it is much more common.

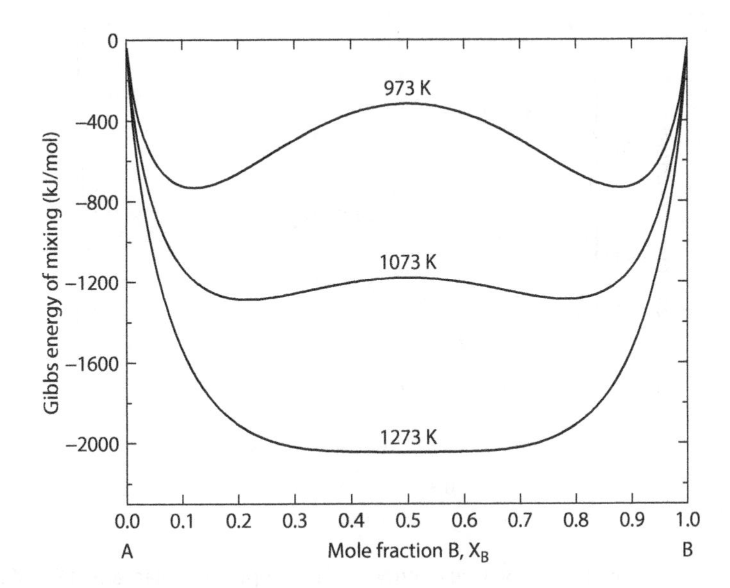

FIGURE 16.3 Gibbs energy Composition diagram for a regular solution system A–B, as a function of the mole fraction of B, that will phase separate at three different temperatures. The highest temperature, 1273 K, is the temperature below which phase separation will occur on cooling to lower temperatures.

minimum values at $X \cong 0.1$ and $X \cong 0.9$ in the 973-K curve in Figure 16.3. The other points of interest are where the second derivative of $\bar{G}_{mix,reg}$ is equal to zero. Differentiating the second equation in Equation 16.16 again

$$\frac{d^2\bar{G}}{dX^2} = \frac{1}{X} + \frac{1}{1-X} - 2\frac{\Gamma}{RT} = 0$$

and rearranging, a quadratic equation is obtained

$$X^2 - X + \frac{RT}{2\Gamma} = 0$$

whose solutions are, of course

$$X = \frac{1}{2} \pm \frac{1}{2}\sqrt{1 - \frac{2RT}{\Gamma}}. \tag{16.17}$$

Clearly, to have real solutions to Equation 16.17, $(2RT)/\Gamma \le 1$. If this were greater than 1, the solutions are imaginary, which just means that they have no physical meaning for the type of Gibbs energy curves in Figure 16.3; that is, curves with $(2RT)/\Gamma > 1$ have no second derivatives equal to zero. When $(2RT)/\Gamma = 1$, then there is only one second derivative at X = 0.5. This is why $\Gamma = 21167.4$ was chosen earlier so that at 1273 K and X = 0.5, the temperature at which phase separation just becomes possible on cooling $(2RT = (8.314)(1273)(2) = 21167.4 = \Gamma)$. For temperatures below 1273 K, there are two roots or X values where the second derivative is zero. For example, at 973 K, these values are $X_{\alpha 1} = 0.743$ and $X_{\alpha 2} = 0.257$, which seems about right looking at Figure 16.3. These two points are clearer in Figure 16.4 for the Gibbs energy at 973 K, the same curve as in Figure 16.3 only with the energy of mixing axis expanded. The value of X at which the second derivative becomes equal to zero is called a *spinodal point* as shown in the figure. Now any overall solution composition falling between the two minima $(X_{\alpha 1} = 0.1209$ and $X_{\alpha 2} = 0.8791)$ will reach equilibrium by separating into two phases with these two compositions. However, the important point is that some compositions will reach equilibrium by a nucleation and growth process and some by *spinodal decomposition.*

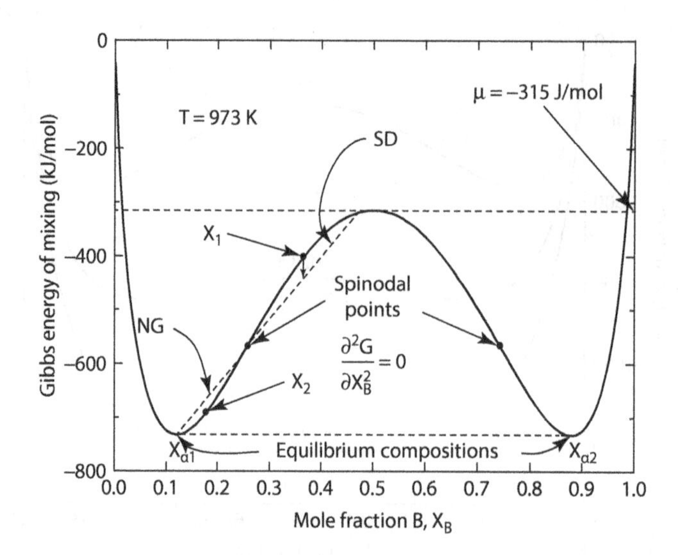

FIGURE 16.4 The same regular solution free energy versus composition diagram at 973 K as in Figure 16.3 with an expanded free energy range. The two equilibrium compositions, X_{α_1} and X_{α_2} are at the minima in the curve and give two points on the two-phase boundary. The spinodal points, where $\partial^2 G/\partial X_B^2 = 0$, are shown. Point X_1 is within the spinodal and will phase separate along the line SD by spinodal decomposition. Point X_2 is outside the spinodal and will decompose by nucleation and growth because it requires an increase in free energy to separate into two phases along the line NG. The horizontal dashed line shows the equal chemical potentials of A and B for X = 0.5, the curve maximum.

For Gibbs energies above the spinodal points, the second derivative of the Gibbs energy is negative, $\left(d^2\bar{G}/dx^2\right) < 0$. Therefore, a composition such as X_1 in Figure 16.4 can lower its energy by simply forming two liquids of slightly different composition, given by the dashed line labeled SD (spinodal decomposition), because the energy is lower on this dashed line than at X_1. On the contrary, composition X_2 is outside the spinodal points and if it tried to separate into two solutions of slightly different compositions, given by the dashed line labeled NG (nucleation and growth), the Gibbs energy increases and such a separation into two liquids of slightly different compositions is impossible and phase separation into liquids α_1 and α_2 must occur by nucleation and growth.

Figure 16.5 shows the phase diagram for separation of a single-phase, *regular*, α solution into two solutions of different compositions, α_1 and α_2, as a function of temperature with the value

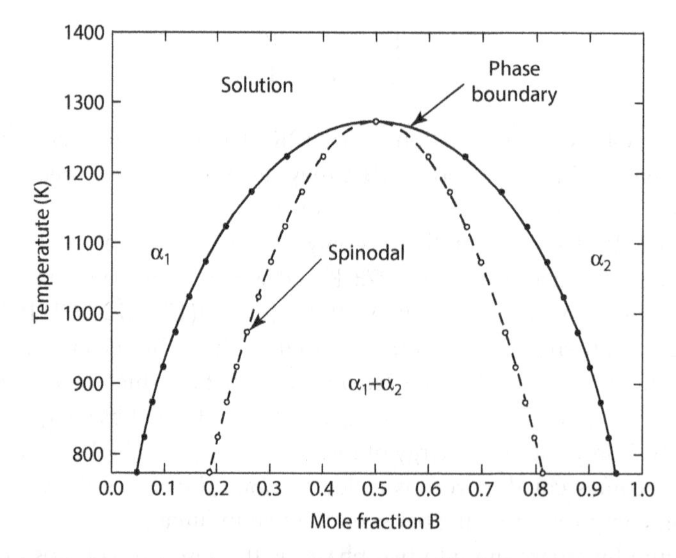

FIGURE 16.5 Calculated phase separation region and spinodal based on the same regular solution used to plot Figure 16.3; that is, $\Gamma = 2RT = (2)(8.314)(1273) = 21167.4$ J/mol. The calculated points are plotted because they must be determined numerically.

of Γ = 21167.4 used earlier. As pointed out in Figure 16.3, the phase decomposition begins at 1273 K on cooling. The phase boundary points (shown) were calculated numerically with Equation 16.16 and the spinodal points (shown), which define the *spinodal* within which spinodal decomposition will occur, were calculated numerically with Equation 16.17. Between the phase boundary and the spinodal, compositions phases separate by nucleation and growth. A regular solution gives a nice symmetrical curve for both the spinodal and the phase boundary and closely approximates phase separation in many solid systems. On the contrary, polymers can show more nonideality and phase separate in more complex ways as a function of composition and temperature. Figure 16.6 shows a polymer solution (polymer blend or polymer and solvent) as a function of composition and temperature that shows *two*-phase separation regions, the beginnings of which (temperature and composition) are called the *upper critical solution temperature*, UCST, and the *lower critical solution temperature*, LCST (Cowie 1991; Young and Lovell 1991). In both cases, spinodal decomposition as well as nucleation and growth of the two new phase compositions will occur as discussed above. The asymmetry of the two-phase regions and the existence two-phase separation regions emphasize the large deviation from ideality of these polymer solutions.

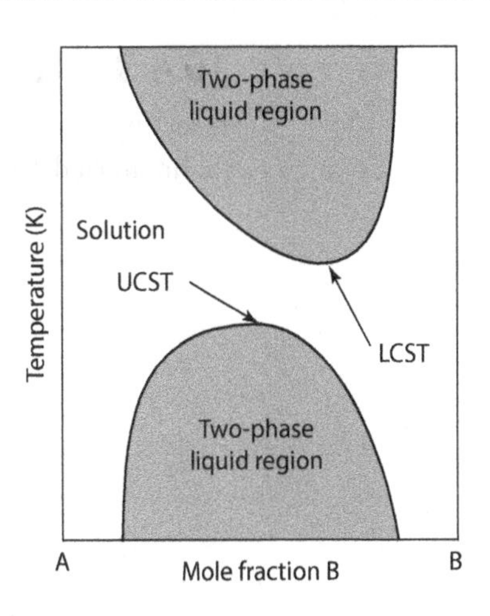

FIGURE 16.6 Liquid solution of two polymers, A and B, that show two-phase separation regions. One defined by the UCST (upper critical solution temperature) and one by the LCST (lower critical solution temperature). One of several more complex phase separation behaviors exhibited by polymer and polymer–solvent solutions (Cowie 1991; Young and Lovell 1991.)

16.4 KINETICS OF SPINODAL DECOMPOSITION

16.4.1 RATIONALE

The model to calculate the kinetics of phase separation via spinodal decomposition is limited to one dimension, yet gives results consistent with the more complete theory of Cahn and Hilliard (Cahn and Hilliard 1958; Cahn 1961). This more direct approach is, hopefully, more transparent than the development and application of the Cahn–Hilliard model that requires more advanced mathematical techniques. This more direct model draws fragments from a variety of literature sources none of which follow entirely the procedure presented here. Hopefully, this approach is reasonably clear and that the rationale for certain steps is obvious. This development of the kinetics of spinodal decomposition brings together many of the kinetic concepts covered in previous chapters.

16.4.2 CHEMICAL POTENTIAL OF THE SPINODAL SOLUTION

In a compositional gradient, there is a contribution to the total chemical potential, call it μ_{int}, where the subscript *int* refers to the interfacial contribution. The chemical potential for the uniform regular solution of composition X (or C), call it μ_{reg}, is given by Equation 16.15. However, μ_{int} must be subtracted from the Gibbs energy of the uniform composition since it represents energy no longer available to do work. Therefore, the chemical potential of the nonuniform solution, μ, is given by (Schmalzried 1981).

$$\mu = \mu_{reg} - \mu_{int} \tag{16.18}$$

where μ_{reg} is the chemical potential of the regular solution with respect to component 2 or B, Equation A.19 developed in Section A.3.3. Likewise, μ_{int} is the chemical potential contribution to the nonuniform composition given as Equations A.10 and A.11 of Section A.2. As a result

$$\mu = \left(\frac{\partial G_{reg,mix}}{\partial n_2}\right)_{T,p,n_1} - 2\kappa\frac{\partial^2 C}{\partial x^2}. \tag{16.19}$$

16.4.3 Flux Equation

Inserting this value for the chemical potential into the flux equation for component B (Schmalzried 1981)

$$J = \frac{CD}{kT}\left\{-\frac{1}{N_A}\frac{\partial\mu}{\partial x}\right\}$$

$$J = -\frac{CD}{RT}\left\{\frac{\partial\mu}{\partial x}\right\} = -\frac{CD}{RT}\left\{\frac{\partial}{\partial x}\left(\mu_{reg} - \mu_{int}\right)\right\}$$

$$J = -\frac{CD}{RT}\left\{\frac{\partial}{\partial x}\mu_{reg} - 2\kappa\frac{\partial^3 C}{\partial x^3}\right\}$$

$$J = -\frac{CD}{RT}\left\{\frac{\partial}{\partial C}\frac{\partial C}{\partial x}\mu_{reg} - 2\kappa\frac{\partial^3 C}{\partial x^3}\right\}$$

$$J = -\frac{CD}{RT}\left\{\bar{V}\frac{\partial}{\partial X}\frac{\partial C}{\partial x}\mu_{reg} - 2\kappa\frac{\partial^3 C}{\partial x^3}\right\}$$

results in

$$J = -\frac{XD}{RT}\left\{\left(\frac{\partial u_{reg}}{\partial X}\right)\frac{\partial C}{\partial x} - \frac{2\kappa}{\bar{V}}\frac{\partial^3 C}{\partial x^3}\right\}. \tag{16.20}$$

where C is the concentration of component 2 or B. It is convenient to take the derivative of uniform solution chemical potential, μ_{reg}, with respect to the mole fraction, X, because in Section A.3.5 the chemical potential is given in Equation A.25 in terms of X. That is

$$\mu_{reg} = RT\ln X + (1-X)^2\Gamma \tag{16.21}$$

where X is the mole fraction of component 2 and

$$\mu_{reg} = \left(\frac{\partial G_{mix,reg}}{\partial n_2}\right)_{T,p,n_1}.$$

Ignoring the second term in Equation 16.20 for the moment, the first term can be written as

$$J = -\frac{XD_{eff}}{RT}\frac{\partial C}{\partial x}$$

where $D_{eff} = D\left(\frac{\partial\mu_{reg}}{\partial X}\right)$. Now, taking the derivative of Equation 16.21 with respect to X,

$$\frac{\partial u_{reg}}{\partial X} = \frac{RT}{X} - 2(1-X)\Gamma \tag{16.22}$$

which means that for $\partial u_{reg}/\partial X < 0$ or negative requires that

$$\frac{RT}{X} - 2(1-X)\Gamma < 0$$

or

$$X^2 - X + \frac{RT}{2\Gamma} > 0$$

and the solution to this equation is again Equation 16.17 or

$$X = \frac{1}{2} \pm \frac{1}{2}\sqrt{1 - \frac{2RT}{\Gamma}} > 0. \qquad (16.23)$$

which is true for $0 < (2RT/\Gamma) < 1$ or anywhere inside of the spinodal. Therefore, inside the spinodal, D_{eff} is *negative* and material moves *up the concentration gradient*, as shown in Figure 16.1, so-called *up-hill diffusion*. As a result, a uniform concentration is unstable and small composition fluctuations will separate into solutions of different compositions to lower the overall Gibbs energy of the system.

16.4.4 Decomposition Kinetics

Put Equation 16.20 into the equivalent of equivalent of Fick's second law (mass conservation)

$$\frac{\partial C}{\partial t} = -\frac{\partial J}{\partial x}$$

which becomes (assuming XD/RT is essentially constant for small composition fluctuations):

$$\frac{\partial C}{\partial t} = -\frac{\partial J}{\partial x} = \frac{XD}{RT}\left\{\left(\frac{\partial \mu_{reg}}{\partial X}\right)\frac{\partial^2 C}{\partial x^2} - \frac{2\kappa}{\bar{V}}\frac{\partial^4 C}{\partial x^4}\right\} \qquad (16.24)$$

and can be written as*

$$\frac{\partial C}{\partial t} = A\frac{\partial^2 C}{\partial x^2} - B\frac{\partial^4 C}{\partial x^4} \qquad (16.25)$$

which is a fourth-order linear partial differential equation with

$$A = \frac{XD}{RT}\left(\frac{\partial \mu_{reg}}{\partial X}\right) \text{ and } B = 2\frac{\kappa XD}{\bar{V}RT}.$$

In the *linear solution approximation*, A and B are considered to be constants, so the solution of Equation 16.25 gives C(x,t) for the growth of the composition fluctuations for the decomposition of those compositions that lie within the spinodal composition range.

Ignoring the fourth-order term in Equation 16.25, this is just Fick's second law that was solved by separation of variables in Chapter 12, by assuming that $C(x,t) = X(x)T(t)$ and obtaining two simple ordinary differential equations. It is worth trying this procedure to see if it will solve Equation 16.25. Inserting $C(x,t) = X(x)T(t)$ into Equation 16.25 results in

$$X\frac{dT}{dt} = AT\frac{d^2C}{dx^2} - BT\frac{d^4C}{dx^4}$$

and partial derivatives are no longer needed. As before, dividing both sides by XT gives

$$\frac{1}{T}\frac{dT}{dt} = \frac{1}{X}\left(A\frac{d^2C}{dx^2} - B\frac{d^4C}{dx^4}\right) = -\alpha^2$$

because one side is only a function of t and the other only a function of x, which means that each of them is constant regardless of how x and t vary, so they both can be set equal to a constant $-\alpha^2$ as was done earlier. This results in two ordinary differential equations:

* This is confusing in the literature because different authors use similar notation for different groupings of the constants in this equation.

$$\frac{dT}{dt} = -\alpha^2 T$$

$$A\frac{d^2X}{dx^2} - B\frac{d^4X}{dx^4} + \alpha^2 X = 0 \tag{16.26}$$

and, as before, the solution to the first of these is

$$T = T_0 e^{-\alpha^2 t}.$$

The solution to the second of Equation 16.26 is less obvious. However, if the fourth-order term was not present, a solution would be $\cos(\beta x)$ (or $\sin\beta x$), $(\beta = 2\pi/\lambda)$, where λ = wavelength. So try one of these by substituting $\cos(\beta x)$ into the second of Equation 16.26, which results in

$$-A\beta^2\cos(\beta x) - B\beta^4\cos(\beta x) = -\alpha^2\cos(\beta x)$$

$$-A\beta^2 - B\beta^4 = -\alpha^2$$

so $\cos(\beta x)$ works, that is, it is a solution, and the total solution to Equation 16.25 is

$$C(x,t) = C_0 + D_m e^{\left(-A\beta^2 - B\beta^4\right)t}\cos(\beta x) \tag{16.27}$$

where $D_m\cos(\beta x)$ is some small periodic composition fluctuation around the original mean composition, C_0. Call the term in the exponent, qt. If $q < 0$, this exponential term decreases to zero with time so the composition fluctuations are unstable and disappear. If $q > 0$, the fluctuations will grow, which is the condition of interest for spinodal decomposition. Now

$$A = \frac{XD}{RT}\left(\frac{\partial\mu_{reg}}{\partial X}\right) \text{ and } B = 2\frac{\kappa XD}{\bar{V}RT} = 2\frac{\Gamma a_0^2 XD}{RT} \tag{16.28}$$

because $\partial\mu_{reg}/\partial X$ is negative inside the spinodal, Equation 16.22, the term A is negative, whereas term B, Equation A16.5, is positive because all of the factors are positive.

16.4.5 Calculations

16.4.5.1 Why Do Calculations?

Surprisingly, in spite of the many different developments of spinodal decomposition in the literature, relatively few of them include any quantitative calculations and conclusions about the process. Certainly, there are three major things of interest. First, how large are the $\cos(\beta x)$ compositional fluctuations in terms of their size or wavelength, $\lambda = 2\pi/\beta$; that is, what does the microstructure look like? Second, how long does it take to get there? Third, what is the effect of the interatomic attractive energy, Γ, on the first two? Some calculations are performed to try to answer these questions, and the answers are interesting.

16.4.5.2 Assumptions and Parameters

The temperature is taken to be T = 1263 K only slightly below the T = 1273 K, where spinodal decomposition begins to occur on cooling. As a result, the Gibbs energy composition plot for this temperature looks virtually the same as that for 1273 K as shown in Figure 16.3. At this temperature, the compositions of the two phases in equilibrium are calculated with Equation 16.16 to be $X_{\alpha_1} = 0.425$ and $X_{\alpha_2} = 0.575$. Keep $\Gamma = 21167.4\,J = R \times 1273$. Take a reasonable value for diffusion coefficient in a solid, $D \cong 2\times10^{-10}\,cm^2/s$, and $a_0 \cong 2\times10^{-8}\,cm$, and take the composition that is decomposing to be $X_0 = 0.5$. Finally, what should be the value of D_m? It must be nonzero, otherwise there is no solution. D_m is supposed to represent random compositional fluctuations around the average composition that will grow during spinodal decomposition. For the sake of concreteness, a value of $D_m \cong 10^{-6}X_0$ seems reasonable: the fluctuations in composition are on the order of a part per million (ppm) of the average composition.

16.4.5.3 Calculation of A

From Equations 16.28 and 16.22

$$A = \frac{XD}{RT}\left\{\frac{RT}{X} - 2(1-X)\Gamma\right\} \tag{16.29}$$

so

$$A = \frac{(0.5)(2\times10^{-10})}{(8.314)(1263)}\left\{\frac{(8.314)(1263)}{0.5} - 2(1-0.5)(21167.4)\right\}$$

or

$$A = -1.5898\times10^{-12}\ \mathrm{cm}^2/\mathrm{s}.$$

Notice that A is negative as promised if the composition is inside of the spinodal.

16.4.5.4 Calculation of B

Again from Equation 16.28

$$B = 2\frac{\Gamma a_0^2 XD}{RT} \tag{16.30}$$

so

$$B = 2\frac{(21167.4)(2\times10^{-8})^2(0.5)(10^{-10})}{(8.314)(1263)}$$

or

$$B = 8.04\times10^{-26}\ \mathrm{cm}^4/\mathrm{s}.$$

16.4.5.5 Calculation of β_{crit}

The critical wavenumber, β_{crit}, or critical wavelength, $\lambda_{crit} = (2\pi/\beta_{crit})$, is when β goes to zero. For values of $\beta > \beta_{crit}$, the exponent in Equation 16.27 is negative and any compositional fluctuations will decay away. This exponent, $-A\beta^2 - B\beta^4$, must be greater than zero for the fluctuations to grow and decomposition to take place. So, β_{crit} occurs when $-A\beta^2 - B\beta^4 = 0$, or when

$$\beta_{crit} = \left(\frac{A}{-B}\right)^{\frac{1}{2}} \tag{16.31}$$

so, in this case

$$\beta_{crit} = \left(\frac{-1.59\times10^{-12}}{-8.04\times10^{-26}}\right)^{\frac{1}{2}}$$

or

$$\beta_{crit} = 4.44\times10^{6}\mathrm{cm}^{-1}$$

or

$$\lambda_{crit} = \frac{2\pi}{\beta_{crit}} = \frac{2\pi}{4.44 \times 10^6}$$

$$\lambda_{crit} = 142 \times 10^{-8}\text{cm}.$$

This wavelength is only about 71 interatomic diameters. However, it is the *shortest* wavelength that will grow. For any compositional fluctuation with a shorter wavelength, the interfacial energy will overcome the volume energy created by the fluctuation, the exponent will be negative, and the fluctuation will decay rather than grow.

16.4.5.6 Calculation of the Term in the Exponent

The term in the exponent of Equation 16.27, q, is given by

$$q = -A\beta^2 - B\beta^4 \tag{16.32}$$

and has units of s^{-1}. It is useful at this juncture to calculate q as a function of β. This is shown in Figure 16.7. It is interesting to note the similarity of this curve of q versus β to the Gibbs energy versus radius for a nucleus during nucleation and growth, with the maximum being the minimum size nucleus that will grow as discussed in Chapter 7. Here too, there is competition between a volume energy term proportional to β^2 competing with an interfacial term proportional to β^4. In this case, the maximum q value, q_{max}, essentially gives the fastest growing wavelength, $\lambda_{max} = 2\pi/\beta_{max}$ of the decomposing solution. As such, it can be used to estimate the times for decomposition. The maximum in the q values, q_{max}, is determined by setting the derivative of Equation 16.32 equal to zero,

$$-2A\beta + 4B\beta^2 = 0$$

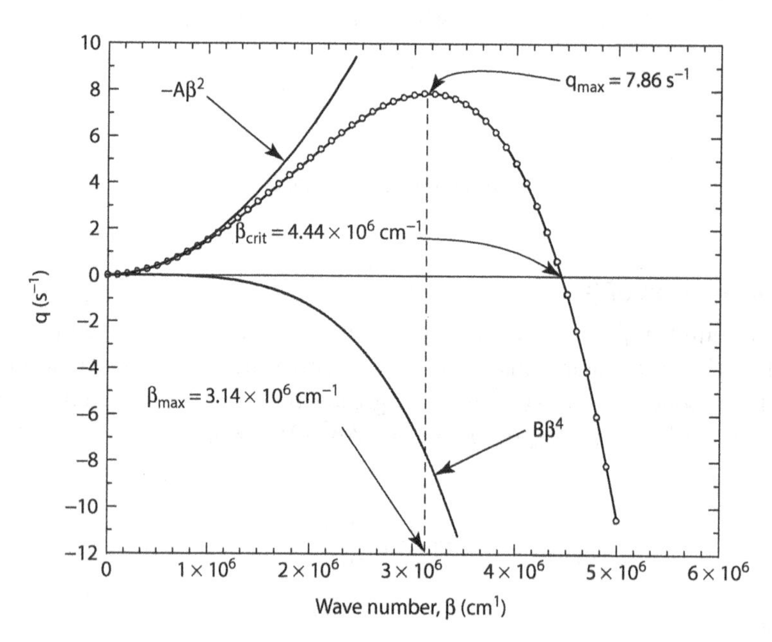

FIGURE 16.7 Plot of the calculated exponential growth parameter, q, versus the wavenumber, β, of the compositional fluctuations leading to spinodal decomposition at 1263 K for the regular solution of X = 0.5 that begins decomposition at 1273 K. The $-A\beta^2$ curve shows the effect of the increasing β on q, or the effect of shorter diffusion distances. In contrast, the $B\beta^4$ term shows the increase in interfacial free energy with shorter wavelengths. β_{crit} is the critical wavenumber at which these competing terms just balance and it gives the maximum β, the minimum wavelength, for which spinodal decomposition will occur. β_{max} is the wavenumber for the fastest growing compositional fluctuation and is used to determine the rate of phase change.

or

$$\beta_{max} = \left(\frac{2A}{-4B}\right)^{\frac{1}{2}} = \frac{\beta_{crit}}{\sqrt{2}} \tag{16.33}$$

from Equation 16.31. So, $\beta_{max} = 4.44\times10^6/\sqrt{2} = 3.14\times10^6\,\text{cm}^{-1}$ or $\lambda_{max} = 2\pi/3.14\times10^6 = 200\times10^{-8}\,\text{cm}$ or about 100 atoms. So, q_{max} in this case is

$$q_{max} = -A\beta_{max}^2 - B\beta_{max}^4$$

$$= 1.59\times10^{-12}\left(3.14\times10^6\right)^2 - 8.04\times10^{-26}\left(3.14\times10^6\right)^4$$

$$q_{max} = 7.86\ \text{s}^{-1}.$$

The values of q_{max}, β_{crit}, and β_{max} are all indicated in Figure 16.7.

16.4.5.7 Putting All Together

All the parameters have been either assumed or calculated that are necessary to put into Equation 16.27 to calculate the growth rate of most probable wavelength, λ_{max}

$$X(x,t) = X_0 + D_m e^{\left(-A\beta^2 - B\beta^4\right)t}\cos(\beta x) \tag{16.34}$$

which can be written in terms of the mole fraction of say component B because $C = X/\overline{V}$ and the molar volume has been assumed to be constant. Of course, the composition can vary with time only until the composition reaches the two equilibrium values, which are $X_1 = 0.425$ and $X_2 = 0.575$ at this temperature, 1263 K, for the chosen value of Γ. Figure 16.8 plots Equation 16.34 with the parameters calculated above as a function of time. At 1.51 s, the maximum reaches the equilibrium values of X. As can be seen, even at 1.0 s, there is very little variation in the concentration (left axis) with distance.

With only a single wavelength for the cosine function, of course, the rectangular equilibrium profile cannot be obtained as was observed in Chapter 12. More than a single term like Equation 16.34 is necessary and really not worth the effort because spinodal decomposition is really three-dimensional and not one-dimensional. However, the intent here has not been to get an exact result but to demonstrate how the thermodynamics and kinetics lead to spinodal decomposition.

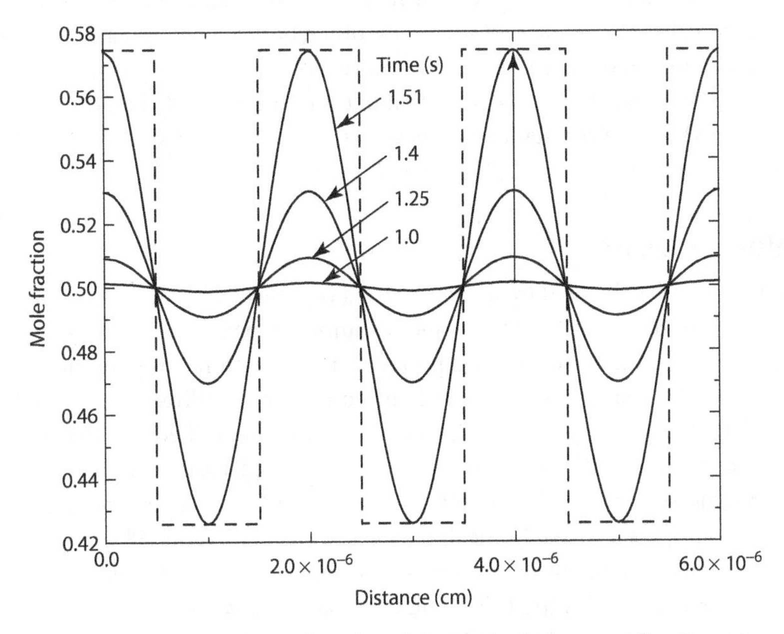

FIGURE 16.8 A plot of Equation 16.34 as a function of time with the q_{max} and β_{max} from Figure 16.7. The dashed rectangular function is the expected final composition variation with distance.

FIGURE 16.9 Schematic depiction of the three-dimensional co-continuous two-phase microstructure resulting from spinodal decomposition based on actual and calculated microstructures in the literature (Elmer 1992; Balluffi et al. 2005.)

Importantly, given all of the approximations made, the results give reasonably quantitative predictions.

In this particular one-dimensional numerical case, the wavelengths are small, a few tens or hundreds of atomic diameters, and this is generally the case in spinodal decomposition. Also, the times are very short, suggesting that it readily occurs. Because the dimensions of each phase are small, prolonged holding at an elevated temperature will cause coarsening driven by surface tension similar to Ostwald ripening discussed in Chapters 6 and 10. One consideration left out of this analysis is a strain energy term in the chemical potential in solids that needs to be included and will modify the results but not the general principles (Schmalzried 1981). This strain energy term is neglected here to simplify the overall analysis, and strain energy is certainly not a consideration for liquid–liquid separation. A schematic microstructure of the two phases after spinodal decomposition is depicted in Figure 16.9 (Elmer 1992; Varshneya 1994; Balluffi et al. 2005). Finally, one important result of spinodal decomposition that the one-dimensional analysis does not predict, and the microstructure does not illustrate, is that each phase of the decomposition is *three-dimensionally continuous.* This feature is utilized in some practical applications of spinodal decomposition.

16.5 A PRACTICAL APPLICATION: POROUS AND VYCOR® GLASS

16.5.1 Composition and Phase Separation

Pure SiO_2 glass has many desirable properties including a high softening point, excellent electrical insulation, and a very low thermal expansion coefficient of about 0.7×10^{-6} °C^{-1}, making it very resistant to rapid temperature changes. About 70 years ago, Corning Glass Company developed a glass with properties very similar to pure SiO_2, but one that could be melted and formed much more easily than pure SiO_2. This glass carries the trade name Vycor®, and it is made by a spinodal decomposition process.

Some compositions in the Na_2O–B_2O_3–SiO_2 system form glasses on cooling from the liquid. One of the most useful of these is a composition of about 67.7 m/o SiO_2, 28.4 m/o B_2O_3, and 9.1 m/o Na_2O. Compositions are melted around 1500 C and formed into shapes by conventional glass-forming process such as pressing, blowing, and drawing. These forming processes are much easier with this composition compared with pure SiO_2 because of the borate glass' much lower viscosity at this temperature. Upon cooling, the glass separates into two co-continuous phases by spinodal decomposition. One phase is about 96 m/o SiO_2 with about 4 m/o B_2O_3 and the other is a phase rich in boron and sodium oxide. The resultant two-phase glass is held at about 500 C to fully develop the two-phase microstructure (Elmer 1992; Varshneya 1994).

16.5.2 Porous Glass

After cooling, the glass is subjected to an acid etching treatment in HCl, H_2SO_4, or HNO_3 at about 90 C, which leaches or dissolves the boron and sodium-rich phase but does not attack the silica-rich phase. This leaves behind a continuous porous network with mean pore diameters between 4 and 6 nm. The etch rate is slow because the dissolved ions must diffuse out through the very fine porous network left after the soluble phase has been dissolved or leached. This leads to a linear leach rate on the order of 1.5 mm/h ($Dt \cong L^2$ and $D \cong 10^{-5}$ cm^2/s, so $t \cong (0.15)^2/10^{-5} = 2250\,s$). As a result, practical wall thicknesses are usually limited to about 10–20 mm. After leaching and washing in pure water, the porous glass is dried. In the dry state, the glass has about 28 v/o porosity with a typical pore diameter of about 5 nm and a surface area of about 200 m^2/g (Elmer 1992). The mean pore diameter can be increased by heating in HF solution that dissolves some of the SiO_2 and increases the pore diameter up to perhaps 30 nm. The resulting high-porosity glass, after heating to about 180 C, becomes capable of absorbing significant quantities of moisture because water vapor reacts with the SiO_2 surface to form silanol ($\equiv Si - OH$) surface groups. This chemically adsorbed water can be removed by subsequent slow heating to about 200 C. With such a small pore size, the mean free path and diffusion of water out of the pores is determined by Knudsen diffusion, Chapter 9, because the pore size is less than the mean free path of gas molecules at 1 atm, about 0.1 μm. As a result, this porous or "thirsty glass" is used as a desiccant in a number of applications. Furthermore, it is also used as a gas separation membrane because of the small pore size, Chapter 10 (Elmer 1992).*

16.5.3 High-Temperature Glass

The other option for the 4 m/o B_2O_3–96 m/o SiO_2 glass is to heat it at temperatures greater than 1200°C at which the high surface area glass densifies (sinters, the pores shrink) under the surface tension driving force of the small pore size: the pore diameter decreases because of the surface tension force of γ/r, where r is the pore radius. If done properly, all of the porosity can be removed and a clear glass having many of the properties very close to pure silica glass is obtained; that is, thermal expansion coefficient $\alpha \cong 0.75 \times 10^{-6}\ K^{-1}$, almost the same as pure silicon dioxide glass (Elmer 1992). Again, the main advantage is that the Vycor® glass is shaped at much lower temperatures prior to the removal of the acid-soluble, boron oxide-rich phase. One caveat, however, is that there is about 30 vol% shrinkage during sintering that must be taken into consideration when a part is being made so that, after consolidation or pore removal, the part is near the desired size and shape.

16.6 CHAPTER SUMMARY

This entire chapter is devoted to the single topic of spinodal decomposition. Although it occurs in metals, ceramics, and polymers both in the liquid and solid states, it is not a phenomenon that so dominates the processing and properties of materials that it justifies its own chapter. Rather, the main reason it is given some emphasis is to try to show all the essential thermodynamic and kinetic concepts in a coherent and understandable way, based on concepts from earlier chapters, that leads to results from which kinetic predictions about the process can be made: microstructures and times of reaction. The concept of why spinodal decomposition occurs can be readily shown with a *regular* solution model. The kinetics are more complicated. Many references begin with the three-dimensional Cahn–Hilliard model (Cahn and Hilliard 1958) and develop some of the kinetics from it. However, to do this properly requires thermodynamic and mathematical subtleties beyond the scope of this book. In contrast, a one-dimensional approach is chosen based on a regular solution model that easily gives both bulk and interfacial chemical potential terms. However, the details of the modeling requires many steps and, often, rather tedious algebra and calculus. Once the appropriate chemical potentials are obtained, then the flux equation and the equivalent of Fick's second law logically follow. A solution to the resulting partial differential equation gives reasonably satisfactory microstructural and kinetic parameters consistent with the more complex models. Finally, the technically important example of spinodal decomposition and production of nanoporous and Vycor® glasses is presented.

APPENDIX

A.1 Energy in a Concentration Gradient

From Section 16.2.2.1, the values for the energies of the planes in pure solutions of compositions 0 and 1 are

$$E_0 = \frac{zm}{2}\left(x_0x_0\varepsilon_{aa} + (1-x_0)(1-x_0)\varepsilon_{bb} + 2x_0(1-x_0)\varepsilon_{ab}\right) \tag{A.1}$$

$$E_1 = \frac{zm}{2}\left(x_1x_1\varepsilon_{aa} + (1-x_1)(1-x_1)\varepsilon_{bb} + 2x_1(1-x_1)\varepsilon_{ab}\right) \tag{A.2}$$

* It is rather interesting that this is another example of a material that has microstructural dimensions on the order of nanometers that has many interesting properties and had been used commercially for almost 60 years before the term "nanostructure" was even considered.

and the energy for a plane in a concentration gradient

$$E_{0-1} = \frac{zm}{2}\left(x_0 x_1 \varepsilon_{aa} + (1-x_0)(1-x_1)\varepsilon_{bb} + \left(x_0(1-x_1) + x_1(1-x_0)\right)\varepsilon_{ab}\right). \tag{A.3}$$

Then the energy per plane in the concentration gradient is

$$E_{int} = 2E_{0-1} - (E_0 + E_1) \tag{A.4}$$

where the 2 comes from the fact that there two of each of the 0 and 1 planes used to calculate their interaction in the uniform solutions. Equations A.1 through A.3 are substituted into Equation A.4 and the whole thing simplified. This is all algebra, but tedious. Making these substitutions, Equation A.4 becomes

$$\begin{aligned} E_{int} = {} & zm\left(x_0 x_1 \varepsilon_{aa} + (1-x_0)(1-x_1)\varepsilon_{bb} + \left(x_0(1-x_1) + x_1(1-x_0)\right)\varepsilon_{ab}\right) \\ & - \frac{zm}{2}\left(x_0 x_0 \varepsilon_{aa} + (1-x_0)(1-x_0)\varepsilon_{bb} + 2x_0(1-x_0)\varepsilon_{ab}\right) \\ & - \frac{zm}{2}\left(x_1 x_1 \varepsilon_{aa} + (1-x_1)(1-x_1)\varepsilon_{bb} + 2x_1(1-x_1)\varepsilon_{ab}\right) \end{aligned}$$

and rearranging,

$$E_{int} = \frac{zm}{2}\left\{\begin{array}{l} \left(-x_0^2 + 2x_0 x_1 - x_1^2\right)\varepsilon_{aa} \\ -\left(-[1-x_0]^2 + 2[1-x_0][1-x_1] - [1-x_1]^2\right)\varepsilon_{bb} \\ +\left\langle -2x_0[1-x_0] + 2\left[x_0(1-x_1)\right] + x_1(1-x_0) - 2x_1(1-x_1)\right\rangle \varepsilon_{ab} \end{array}\right\} \tag{A.5}$$

and with some more algebra and rearranging, the first term in Equation A.5 is

$$\begin{aligned} & \left(-x_0^2 + 2x_0 x_1 - x_1^2\right) \\ & = -(x_0 - x_1)^2 \end{aligned}$$

and doing likewise with the second term

$$\begin{aligned} & -\left(-[1-x_0]^2 + 2[1-x_0][1-x_1] - [1-x_1]^2\right) \\ & = -\left(1 - 2x_0 + x_0^2\right) + 2\left(1 - x_1 - x_0 + x_0 x_1\right) - \left(1 - 2x_1 + x_1^2\right) \\ & = -\cancel{1} + \cancel{2x_0} - x_0^2 + \cancel{2} - \cancel{2x_1} - \cancel{2x_0} + 2x_0 x_1 - \cancel{1} + \cancel{2x_1} - x_1^2 \\ & = -(x_0 - x_1)^2 \end{aligned}$$

and, yet again, the third term,

$$\begin{aligned} & \left\langle -2x_0[1-x_0] + 2\left[x_0(1-x_1)\right] + x_1(1-x_0) - 2x_1(1-x_1)\right\rangle \\ & = -2x_0 + 2x_0^2 + 2\left(x_0 - x_0 x_1 + x_1 - x_0 x_1\right) - 2x_1 + 2x_1^2 \\ & = -\cancel{2x_0} + 2x_0^2 + \cancel{2x_0} - 2x_0 x_1 + \cancel{2x_1} - 2x_0 x_1 - \cancel{2x_1} + 2x_1^2 \\ & = 2(x_0 - x_1)^2 \end{aligned}$$

giving the final result when substituting these last three simplifications into Equation A.5

$$E_{int} = \frac{zm}{2}(2\varepsilon_{ab} - \varepsilon_{aa} - \varepsilon_{bb})(x_0 - x_1)^2 = zm\Delta\varepsilon(x_0 - x_1)^2 \quad (A.6)$$

where $\Delta\varepsilon = \varepsilon_{ab} - \varepsilon_{aa}/2 - \varepsilon_{bb}/2$.

A.2 Contribution to the Chemical Potential by the Interfacial Energy Term

In this chapter, it is shown that the interfacial energy or compositional gradient term is given by

$$E_{int} = zm\Delta\varepsilon a_0^2\left(\frac{dx}{dy}\right)^2 \quad (A.7)$$

where:

z is the number of nearest neighbor atoms
m is the atoms in a plane (atoms/cm^2)
$\Delta\varepsilon$ is the overall interaction energy per atom (J/atom)
a_0 is the lattice parameter or separation between atoms (planes) (cm)
x is the fractional occupation of a plane of atoms
y is the distance (cm).

So, the Units(E_{int}) = J/cm^2

To determine the contribution of this term to the chemical potential of the system, its units must be in Joules. Therefore, if Equation A.7 is multiplied by a_0^2, then the units are in Joules. So call

$$G_{int} = a_0^2 E_{int} = zm\Delta\varepsilon a_0^4\left(\frac{dx}{dy}\right)^2 \quad (A.8)$$

where Units(G_{int}) = J. In other words, it is not a *molar* Gibbs energy (Darken and Gurry 1953). Now,

$$n = \frac{x}{N_A}\left(\frac{m}{a_0}\right)a_0^3 = \frac{1}{\text{atoms/mol}}\left(\text{atoms/cm}^3\right)\left(\text{cm}^3\right) = \text{moles}$$

or

$$x = \frac{nN_A}{ma_0^2}$$

and when substituted into Equation A.8 gives

$$G_{int} = z\Delta\varepsilon\frac{N_A^2}{m}\left(\frac{dn}{dy}\right)^2.$$

To get the contribution of this interfacial energy to the total chemical potential, take the derivative of G_{int} with respect to n:

$$\mu_{int} = \left(\frac{\partial G_{int}}{\partial n}\right)_{T,p,n_i\neq n} = z\Delta\varepsilon\frac{N_A^2}{m}\frac{\partial}{\partial n}\left(\frac{dn}{dy}\right)^2$$

$$\mu_{int} = z\Delta\varepsilon\frac{N_A^2}{m}\frac{\partial y}{\partial n}\frac{\partial}{\partial y}\left(\frac{dn}{dy}\right)^2 = z\Delta\varepsilon\frac{N_A^2}{m}\frac{\partial y}{\partial n}2\frac{dn}{dy}\frac{\partial^2 n}{\partial y^2}$$

giving the result because $(\partial y/\partial n)(dn/dy) = 1$

$$\mu_{int} = 2z\Delta\varepsilon \frac{N_A^2}{m}\frac{\partial^2 n}{\partial y^2}. \tag{A.9}$$

Now, the concentration, C (Units(C) = mol/cm³), is given by $C = n/a_0^3$ or $a_0^3 C = n$ and substituting this into Equation A.9 gives

$$\mu_{int} = 2z\Delta\varepsilon \frac{N_A^2}{m} a_0^3 \frac{\partial^2 C}{\partial y^2} = 2z\Delta\varepsilon a_0^2 \left(\frac{N_A a_0}{m}\right)(N_A)\frac{\partial^2 C}{\partial y^2}$$

and because $\bar{V} = N_A(a_0/m)$ and $\Gamma = z\Delta\varepsilon N_A$

$$\mu_{int} = 2\Gamma a_0^2 \bar{V}\frac{\partial^2 C}{\partial y^2}. \tag{A.10}$$

Again, $\Gamma = zN_A\left(\varepsilon_{ab} - \frac{1}{2}\varepsilon_{aa} - \frac{1}{2}\varepsilon_{bb}\right)$ as before. Equation A.10 can be written as

$$\mu_{int} = 2\Gamma a_0^2 \bar{V}\frac{\partial^2 C}{\partial y^2} = 2\kappa\frac{\partial^2 C}{\partial y^2} \tag{A.11}$$

where $\kappa = \Gamma a_0^2 \bar{V}$ to use notation that is consistent with the more rigorous three-dimensional theory (Cahn and Hilliard 1958). As a check, the units of μ_{int} are

$$\mu_{int} = 2\Gamma a_0^2 \bar{V}\frac{\partial^2 C}{\partial y^2} = \left(\frac{J}{mol}\right)(cm^2)\left(\frac{cm^3}{mol}\right)\left(\frac{mol}{cm^5}\right) = J/mol$$

as it should, so Equation A.11 should be correct.

A.3 Chemical Potential of a Molar Solution

A.3.1 Rationale

Although what follows really does not impact the results in Section 16.4, it is instructive and useful because, throughout Chapter 16, a regular molar (Gibbs energy per mole) solution is developed and used to discuss the *thermodynamics* of spinodal decomposition. Similarly, it is useful to use this same solution in discussing the *kinetics* of spinodal decomposition. However, the chemical potential is defined as *the effect that one mole of a given constituent has on the **total** Gibbs energy of the system*, G, not the Gibbs energy of a system containing a total of 1 mole of constituents, $\bar{G}$. So, it is useful to have the chemical potential for a 1 M solution. What follows is essentially the approach of Darken and Gurry (1953) and is typical of some of the sometimes arcane mathematical manipulations of thermodynamics.

A.3.2 Thermodynamics

If G is the total Gibbs energy of the system, then the chemical potential, μ_i, or partial molar Gibbs energy, $\bar{G}_i$, of constituent i is defined by

$$\mu_i = \bar{G}_i = \left(\frac{\partial G}{\partial n_i}\right)_{T,p,i\neq j} \tag{A.12}$$

so the change in the total Gibbs energy of a two-component system, such as the one under consideration, is

$$dG = \mu_1 dn_1 + \mu_2 dn_2. \tag{A.13}$$

Integrating with respect to the extensive variables dn_i, the total Gibbs energy is

$$G = \mu_1 n_1 + \mu_2 n_2. \tag{A.14}$$

Taking the total differential of G′ gives

$$dG = \mu_1 dn_1 + n_1 du_1 + \mu_2 dn_2 + n_2 d\mu_2$$

and subtracting Equation A.13 and dividing by $n = n_1 + n_2$ gives

$$\frac{n_1}{n_1 + n_2} du_1 + \frac{n_2}{n_1 + n_2} du_2 = X_1 d\mu_1 + X_2 d\mu_2 = 0. \tag{A.15}$$

This, of course, is the Gibbs–Duhem equation that has been invoked several times previously and developed in Chapter 1. Now divide Equation A.14 by $n = n_1 + n_2$ to get

$$\bar{G} = \mu_1 X_1 + \mu_2 X_2 \tag{A.16}$$

the Gibbs energy per mole of solution; that is, Units($\bar{G}$) = J/mol. The total differential of $\bar{G}$ is

$$d\bar{G} = \mu_1 dX_1 + X_1 d\mu_1 + \mu_2 dX_2 + X_2 d\mu_2$$

and subtracting the Gibbs–Duhem relation, Equation A.15 results in

$$d\bar{G} = \mu_1 dX_1 + \mu_2 dX_2. \tag{A.17}$$

Now $dX_1 = -dX_2$ and $X_1 + X_2 = 1$, multiplying Equation A.17 by X_1/dX_2 gives

$$X_1 \frac{d\bar{G}}{dX_2} = -X_1\mu_1 + X_1\mu_2. \tag{A.18}$$

Adding Equations A.16 and A.18 and rearranging, the result sought is

$$\mu_2 = \left(\frac{\partial G}{\partial n_2}\right)_{T,p,n_1} = \bar{G} + (1 - X_2)\frac{d\bar{G}}{dX_2} \tag{A.19}$$

the chemical potential of one of the constituents in terms of the parameters of molal solution. The result for μ_1 is of course the same simply replacing 2 by 1 in Equation A.19.

A.3.3 Ideal Solution

Most of the discussion involving solutions in this chapter has been implicit that X was essentially the same as X_2 in the discussion in Section A.3.2. With that in mind, the calculation of the chemical potential of component 2 in an ideal 1 M solution is calculated from Equation A.19. For an ideal solution, the Gibbs energy per mole of solution is given by

$$\bar{G} = RT\{X \ln X + (1 - X)\ln(1 - X)\}$$

so $d\bar{G}/dX$ is

$$\frac{d\bar{G}}{dX} = RT\{\ln X + 1 - \ln(1 - X) - 1\}$$

$$\frac{d\bar{G}}{dX} = RT\{\ln X - \ln(1 - X)\}$$

and substituting this into Equation A.19

$$\mu = RT\{X\ln X + (1-X)\ln(1-X)\} + RT\{(1-X)\ln X - (1-X)\ln(1-X)\}$$

and after canceling terms gives

$$\left(\frac{dG}{dn_2}\right)_{T,p,n_1} = \mu_2 = RT\ln X \tag{A.20}$$

a rather unsurprising result for an ideal solution.

A.3.4 Ideal Solution Another Way

Instead of the Gibbs energy per mole of an ideal solution, the G for an ideal solution can be used and differentiated with respect to n_2 to see if Equation A.20 is reproduced. In this case,

$$G = RT\left\{n_1 \ln\left(\frac{n_1}{n_1+n_2}\right) + n_2 \ln\left(\frac{n_2}{n_1+n_2}\right)\right\}$$

which can be written as

$$G = RT\{n_1 \ln(n_1) - n_1 \ln(n_1+n_2) + n_2 \ln(n_2) - n_2 \ln(n_1+n_2)\}. \tag{A.21}$$

Differentiating Equation A.21 with respect to n_2 with n_1 = constant gives the chemical potential of component 2.

$$\mu_2 = \left(\frac{\partial G}{\partial n_2}\right)_{T,p,n_1} = RT\left\{0 - \frac{n_1}{n_1+n_2} + \ln n_2 + 1 - \ln(n_1+n_2) - \frac{n_2}{n_1+n_2}\right\}$$

and canceling terms leads to

$$\frac{dG}{dn_2} = \mu_2 = RT\ln\left(\frac{n_2}{n_1+n_2}\right) = RT\ln X \tag{A.22}$$

as before.

A.3.5 Regular Solution

Equation 16.1 gives for a regular solution of two components, where $X = X_2$

$$\bar{G}_{mix,reg} = \bar{G} = RT\{X\ln X + (1-X)\ln(1-X)\} + \Gamma X(1-X) \tag{A.23}$$

and again, Equation A.19 is appropriate. In this case,

$$\frac{d\bar{G}}{dX} = RT\{\ln X - \ln(1-X)\} + (1-X)\Gamma - X\Gamma. \tag{A.24}$$

Substituting Equations A.23 and A.24 into Equation A.19 yields

$$\mu_2 = \left(\frac{\partial G_{mix,reg}}{\partial n_2}\right)_{T,p,n_1} = RT\{X\ln X + (1-X)\ln(1-X)\} + \Gamma X(1-X)$$

$$+ (1-X)\left[RT\{\ln X - \ln(1-X)\} + (1-X)\Gamma - X\Gamma\right]$$

$$\mu_2 = RT\{X\ln X + (1-X)\ln(1-X) + (1-X)\ln X - (1-X)\ln(1-X)\}$$
$$+\Gamma X(1-X) + (1-X)^2\Gamma - X(1-X)\Gamma$$

and simply canceling terms gives the desired result:

$$\mu_2 = RT\ln X + (1-X)^2\Gamma. \quad (A.25)$$

At X = 0.5, both chemical potentials are equal and at 973 K are given by

$$\mu_1 = \mu_2 = RT\ln X + (1-X)^2\Gamma$$
$$= (8.314)(973)\ln(0.5) + (1-0.5)^2 \times 21167.4$$
$$= -5607 + 5292$$
$$\mu = -315 \text{ J/mol}.$$

The partial molar free energies or the chemical potentials of the two constituents are given by the intersection on the two vertical axes at X = 0 and X = 1 of a line drawn tangent to the composition point X, where the chemical potentials are being determined. In this case, at X = 0.5, as seen in both Figures 16.3 and 16.4, the slope here is zero so that chemical potentials are given by the intersections of a *horizontal line* through the point X = 0.5 of the Gibbs energy curve. This is shown in Figure 16.4, and it can be seen that the intersection with the two vertical axes is at $\mu = -315$ J/mol as calculated.

A.3.6 Regular Solution Another Way

The procedure here will be the same as in Equation A.24 with moles rather than mole fraction. Now the enthalpy of mixing for a regular solution is given in Equation 16.6

$$\Delta H_{mix} = \frac{zN_aN_b}{N_a + N_b}\left(\varepsilon_{ab} - \tfrac{1}{2}\varepsilon_{aa} - \tfrac{1}{2}\varepsilon_{bb}\right) \quad (A.26)$$

where N_a and N_b are the number of *atoms*. Multiplying top and bottom of Equation A.26 by Avogadro's number squared, N_A^2, it becomes

$$\Delta H_{mix} = \Gamma\frac{n_1n_2}{n_1 + n_2} \quad (A.27)$$

in the current notation. The contribution to the chemical potential of the regular solution due to entropy of mixing of an ideal solution is given by Equation A.22, so only the contribution of the enthalpy of mixing to the chemical potential needs to be calculated, which is nothing more than the derivative of Equation A.27 by n_2. Taking this derivative

$$\partial\frac{\Delta H_{mix}}{\partial n_2} = \Gamma\left\{\frac{n_1}{n_1 + n_2} - \frac{n_1n_2}{(n_1 + n_2)^2}\right\}$$
$$= \frac{\Gamma}{(n_1 + n_2)^2}\left(n_1^2 + n_1n_2 - n_1n_2\right)$$
$$= \Gamma\left(\frac{n_1}{n_1 + n_2}\right)^2$$
$$\partial\frac{\Delta H_{mix}}{\partial n_2} = \Gamma(1-X)^2.$$

So, as before, the total chemical potential is the same as Equation A.25,

$$\mu_2 = RT\ln X + (1-X)^2\Gamma$$

which is again comforting and provides two ways of determining the chemical potential from either the total Gibbs energy of the system or the Gibbs energy per mole.

EXERCISES

16.1 a. Given that there is a regular solution between components A and B, plot the Gibbs energy-composition curve versus X_B at 973 K if $\Gamma = -21{,}167.4$ J/mol.
b. Calculate the chemical potentials of A and B at $X_B = 0.5$.
c. Calculate the chemical potentials of A and B at $X_B = 0.25$.

16.2 a. Given that there is a regular solution between components A and B, plot the Gibbs energy-composition curve (Figure 16.4) versus X_B at 973 K if $\Gamma = 20{,}000$ J/mol.
b. Calculate the values of X_B at the two equilibrium concentrations.
c. Calculate the values of X_B at the two spinodal points.
d. Calculate the chemical potentials of A and B at $X_B = 0.05$.
e. Calculate the values of the spinodal points and the phase boundary from T = 800 K to the maximum temperature at 10 K intervals.
f. Plot the temperatures of the phase boundary and spinodal compositions as a function of the mole fraction from 800 K to the composition maximum in the two-phase field, at 10 K intervals (Figure 16.5).

16.3 From the data given in Exercise 2 above and given that at 900 K the diffusion coefficient is $D \cong 10^{-8}$ cm²/s and the intermolecular spacing is 0.3 nm, at X = 0.5 and T = 900 K, calculate:
a. The value of A in Equation 16.29.
b. The value of B in Equation 16.30.
c. The value of β_{crit}.
d. The value of β_{max}.
e. The value of q_{max}.
f. The value of the wavelength at β_{max}.

16.4 a. Given that there is a regular solution between two polymers A and B, plot the Gibbs energy-composition curve (Figure 16.4) versus X_B at 300 K if $\Gamma = 5504$ J/mol.
b. Calculate the values of X_B at the two equilibrium concentrations.
c. Calculate the chemical potentials of A and B at $X_B = 0.1$.
d. Calculate the values of the spinodal and the phase boundary points from T = 200 K to the maximum temperature at 5 K intervals.
e. With the points calculated in d, plot the spinodal and phase boundary versus X_B from T = 200 K to the maximum temperature.

16.5 From the data given in Exercise 16.4 and given that at 300 K the diffusion coefficient is $D \cong 10^{-7}$ cm²/s and the intermolecular spacing is 0.5 nm, at X = 0.5 and T = 300 K, calculate:
a. The value of A in Equation 16.29.
b. The value of B in Equation 16.30.
c. The value of β_{crit}.
d. The value of β_{max}.
e. The value of q_{max}.
f. The value of the wavelength at β_{max}.

REFERENCES

Balluffi, R. W., S. M. Allen, and W. C. Carter. 2005. *Kinetics of Materials*. New York: Wiley-Interscience.

Becker, R. 1938. The Nucleation of Precipitation in Metallic Solid Solutions (in German). *Ann. Phys.* 32(5): 128–140.

Cahn, J. W. 1961. On Spinodal Decomposition. *Acta Metallurgica* 9(9): 795–801.

Cahn, J. W. 1966. The Later Stages of Spinodal Decomposition and the Beginnings of Particle Coarsening. *Acta Metallurgica* 14(12): 1685–1692.

Cahn, J. W. and J. E. Hilliard. 1958. Free Energy of a Nonuniform System: I. Interfacial Free Energy. *J. Chem. Phys.* 28(2): 258–267.

Cahn, J. W. and J. E. Hilliard. 1959. Free Energy of a Nonuniform System. III. Nucleation on a Two-Component Incompressible Fluid. *J. Chem. Phys.* 31(3): 688–699.

Cowie, J. M. G. 1991. *Polymers: Chemistry and Physics of Modern Materials*, 2nd ed. New York: Chapman & Hall.

Darken, L. S. and R. W. Gurry. 1953. *Physical Chemistry of Metals*. New York: McGraw-Hill.

DeHoff, R. 2006. *Thermodynamics in Materials Science*, 2nd ed. Boca Raton: CRC Publishing, Taylor & Francis Group.

Denbigh, K. 1964. *The Principles of Chemical Equilibrium*. Cambridge, UK: Cambridge University Press.

Elias, H.-G. 1997. *In Introduction to Polymer Science*. New York: VCH Publishers.

Elmer, T. H. 1992. Porous and Reconstructed Glasses. *Engineered Materials. Handbook, Vol. 4, Ceramics and Glasses*. Metals Park, OH: ASM International: 427–432.

Green, P. F. 2005. *Kinetics, Transport, and Structure in Hard and Soft Material*. Boca Raton, FL: Taylor & Francis Group.

Haynes, W. M. editor-in-chief. 2013. *Handbook of Chemistry and Physics*, 94th ed. Boca Raton, FL: CRC Press

Hillert, M. 1961. A Solid-Solution Model for Inhomogeneous Systems. *Acta Metallurgica* 9(6): 525–535.

Jackson, K. A. 2010. *Kinetic Processes*, 2nd ed. Weinheim, Germany: Wiley-VCH.

Kingery, W. D., H. K. Bowen, and D. R. Uhlmann. 1976. *Introduction to Ceramics*, 2nd ed. New York: John Wiley & Sons.

Lesar, R. 2013. *Introduction to Computational Materials Science*. Cambridge, UK: Cambridge University Press.

Moelans, N., B. Blanpain, and P. Wollants. 2008. An Introduction to Phase-Field Modeling of Microstructure Evolution. *Calphad* 12(2): 268–294.

Porter, D. A., K. E. Easterling, and M. Y. Sherif. 2009. *Phase Transformations in Metals and Alloys*, 3rd ed. Boca Raton, FL: Taylor & Francis Group.

Ragone, D. A. 1995. *Thermodynamics of Materials*, Volume II. New York: John Wiley & Sons.

Schmalzried, H. 1981. *Solid State Reactions*. Weinheim, Germany: Verlag Chemie.

Sperling, L. H. 2006. *Introduction to Physical Polymer Science*, 4th ed. Hoboken, NJ: Wiley.

Varshneya, A. K. 1994. *Fundamentals of Inorganic Glasses*. San Diego, CA: Academic Press.

Young, R. J. and P. A. Lovell. 1991. *Introduction to Polymers*, 2nd ed. New York: Chapman & Hall.

Appendix I: List of Symbols

LATIN/ENGLISH

A thermodynamic system (1.7.2)
A component (1.7.7)
A constant
A Helmholtz energy (1.7.5)
A area (1.7.3)
A pre-exponential (3.6.1)
A^* molecules of *A* in activated state (4.7.1)
A' constant (5.2.2)
A'' constant (13.3.1)
A_n Fourier series coefficient (Chapter 12, A.2.1)
A coefficient relating $\partial C/\partial t$ to $\partial^2 C/\partial x^2$ in spinodal decomposition (16.4.4)
a atom jump distance (7.8.5)
a factor in $ax^2 + bx + c = 0$ (Chapter 3, A.2.2)
a side of a cube (Chapter 6, A.1)
a sphere radius (5.2.6)
a thermodynamic activity (1.7.6)
a interatomic spacing in a crystal (9.3.4)
a polymer repeating unit length (13.10.4)
a_0 interplanar distance (16.2.2)
B component (1.7.7)
B constant
B thermodynamic system (1.7.2)
B coefficient relating $\partial C/\partial t$ to $\partial^4 C/\partial x^4$ in spinodal decomposition (16.4.4)
B absolute mobility (13.3.1)
B_n Fourier series coefficient (Chapter 12, A.4)
Bq Becquerel (2.7.3)
b factor in $ax^2 + bx + c = 0$ (Chapter 3, A.2.2)
c term in $ax^2 + bx + c = 0$ (Chapter 3, A.2.2)
c heat capacity per gram (8.5)
C concentration (8.3)
C constant

C	number of components (1.7.8)
C	thermodynamic system (1.7.2)
C'	constant (7.11.7)
Ci	curie (2.7.3)
C_P	heat capacity at constant pressure (1.7.3)
C_T	total concentration (14.4.3)
C_V	heat capacity at constant volume (1.7.3)
d	sphere diameter (6.9)
D	constant
D	diffusion coefficient (8.3)
$\tilde{D}$	interdiffusion coefficient (14.4.3)
D_i^*	tracer diffusion coefficient of species *i* (14.4.3)
D_0	diffusion coefficient pre-exponential (8.4.3)
D_{eff}	effective diffusion coefficient (13.9.3)
D_K	Knudson diffusion coefficient (9.6.5)
D_m	random composition fluctuation amplitude in spinodal decomposition (16.4.4)
da	differential length (Chapter 6, A.2)
d_{AB}	collision diameter between gas atoms *A* and *B* (9.6.1)
dl	differential displacement (1.7.3)
d_p	capillary length (6.5.1)
e	electron charge
E	energy (4.6)
E	energy of a plane in a solid solution (16.2.10)
E	Young's modulus (7.11.8)
E_A	activation energy (3.6.1)
eV	electron volt (2.7.2)
f	surface stress (6.2.2)
f	atom jump frequency (7.8.5)
f	fraction (12.6.20)
f	function (Chapter 12, A.1)
F	degrees of freedom (1.7.8)
F	force (1.7.3)
$f(\theta)$	function of the wetting angle, θ, relating the Gibbs energies between a heterogeneous and homogeneous nucleus (7.9.2)
f(x)	steady-state or equilibrium solution to Fick's second law (8.7.4)
f_{LV}	fractional liquid–vapor interface (6.15)
f_{LS}	fractional liquid–solid interface (6.15)
f_R	fraction reacted (2.6.2)
f_t	fraction transformed (7.8.6)
f_U	fraction unreacted (2.6.2)
f_{unrel}	unrelaxed free volume (7.11.7)
g	function (Chapter 12, A.1)
g	acceleration of gravity (4.5.1)
g	gas (2.1)
G	Gibbs energy (1.7.5)
$\bar{G}_A^o$	molar Gibbs energy of component *A* in its standard state (°) (1.7.6)
$G^* = \bar{G} - \bar{G}^o$	molar Gibbs energy less that in the standard state (1.7.8)
G	particle growth rate (7.8.5)
G	shear modulus (7.11.6)
g(x, t)	transient part of the solution to Fick's second law (8.7.4)
g_i	Gibbs energy per atom of species *i* (15.2.5)
g_V	Gibbs energy to form a vacancy (9.4.2)

h height (4.5.1)
h dimension (8.5)
h Planck's constant (9.3.4)
h mass transfer coefficient (10.9)
H enthalpy (1.6.2)
$\bar{H}$ molar enthalpy; bar represents molar or partial molar quantities $(\partial H/\partial n_1)_{T,P,n_2}$ (1.7.4)
H mean curvature of a surface (6.4)
I electrical current (1.6.2)
I gas atom impingement rate (5.2.4)
I function (Chapter 12, A.1)
$I^{\cdot}$ reaction initiator (3.6.3)
J Jacobian (Chapter 5, A.1)
J molar flux density (5.5.3)
J′ atomic/molecular flux density (5.9.2)
J″ molar flux relative to fixed coordinates (15.2.5)
k reaction rate constant (2.4.2)
k thermal conductivity (8.3)
k parabolic oxidation coefficient (14.2.2)
k′ interdiffusion rate constant (15.5.3)
k″ interdiffusion rate constant (15.5.4)
k‴ interdiffusion rate constant (15.5.5)
K degree kelvin (1.7.3)
K parabolic rate constant (10.6.4)
K partition coefficient (10.7.3)
k_0 reaction rate constant pre-exponential (2.4.2)
k_B Boltzmann's constant (1.6.2)
K_e equilibrium constant (1.7.6)
KE kinetic energy (5.2.1)
K_L Langmuir absorption coefficient (5.2.7)
K_S Schottky product (9.4.7)
K_{OW} octanol–water partition coefficient (10.7.3)
l liquid (4.3.2)
L liquid (1.7.8)
L length (5.2.1)
L dimension (8.7.1)
L defect length in a polymer molecule (9.5.2)
L contour length of a polymer (13.10.4)
m mass (4.5.1)
m integer (12.4.5)
m atoms per unit area (16.2.1)
M molecular weight (2.4.3)
M monomer molecule (3.6.3)
m^* mass of radioactive material (2.7.1)
m_A mass of atom of A (4.6)
M_A molecular weight of A (4.6)
MeV million electron volts (2.7.2)
n number of moles (1.7.7)
n number (Chapter 1, A.2.1)
n number of repeating units in a polymer molecule (9.5.2)
n atoms per unit area (9.3.1)
n integer (12.4.5)
N number (Chapter 1, A.2.1)

N	number of atoms (4.6)
n^*	activated atoms per unit area (9.3.2)
N_A	Avogadro's number (Chapter 1, A.3.1)
p	partial pressure (2.4.3)
p	probability (16.2.1)
P	pressure (1.7.3)
P	number of phases (1.7.8)
P	permeability coefficient (10.3.1)
P	partition coefficient (10.7.3)
P^o	standard state pressure = 1 bar (1.7.6)
$p(x, t)$	probability of finding an atom at position x at time t (11.7.2)
q	inverse of the relaxation time for spinodal decomposition (16.4.5)
$\dot{q}$	heat flux density (8.3)
Q	quantity of heat (1.7.4)
Q	activation energy (2.4.2)
Q	diffusion source, quantity of material (11.6.1)
q_P	heat transferred at constant pressure (1.7.3)
q_V	heat transferred at constant volume (1.7.3)
r	radial distance (Chapter 5, A.1)
r	radius of curvature (6.4)
R	electrical resistance (1.6.2)
R	gas constant (1.7.3)
r^*	critical nucleus size (7.8.1)
R_I	rate of initiator formation (3.6.3)
R_P	rate of polymerization (3.6.3)
R_n	length of polymer chain of n-repeating units (13.10.4)
R_{gx}	radius of gyration of a body about the x-axis (13.10.5)
s	solid (2.2)
S	entropy (1.6.2)
S	solid (1.7.8)
S	empty surface site (5.2.7)
S	specific surface area, m^2/g (6.9)
S	supersaturation (7.8.2)
S	solubility (10.3.1)
S_{surr}	entropy of the surroundings (1.7.4)
S_{sys}	entropy of a system (1.7.4)
S'	occupied surface site (5.2.7)
$\bar{S}$	specific surface area, m^2/mol (6.9)
soln	solution (2.2)
t	time (2.4.1)
T	temperature (1.7.3)
T	function of t only (12.3.1)
$t_{1/2}$	half-life (2.6.3)
t_e	electronic transference number (14.2.1)
t_i	ionic transference number (14.2.1)
T_g	glass transition temperature (7.11.1)
U	internal energy (1.6.2)
v	speed (5.2.1)
V	volume (1.7.3)
$\bar{v}$	mean speed (5.2.3)
V_L	volume liquid (1.7.8)
V_S	volume solid (1.7.8)

$V_{Cl}^{\bullet}$	chlorine vacancy in NaCl (9.4.6)
V_{Mg}''	magnesium vacancy in MgO (9.4.6)
v_{rms}	root mean square speed (5.2.1)
W	number of configurations (1.6.2)
W	work (Chapter 6, A.1)
x	variable (3.4.1)
x	concentration (3.4.1)
x	distance (4.5.1)
x	thickness (5.5.4)
X	mole fraction (1.7.7)
X	function of *x* only (12.3.1)
x′	distance (14.4.3)
x_0	fraction of atoms on plane at zero (16.2.1)
X_i	mole fraction of component *i* (1.7.7)
y	dimensionless variable (11.2.3)
Y	thickness of reacted layer (15.4.2)
z	function (11.2.3)
z	number of nearest neighbors atoms (16.2.1)
z	volume of solid reaction product/volume of reactants (15.5.5)
Z	collisions of atoms s^{-1} (4.7.1)
ze	charge on an ion (9.4.7)

GREEK

α	a phase (1.7.8)
α	reaction order (2.4.2)
α	accommodation or sticking coefficient (5.2.6)
α	thermal expansion coefficient (7.11.5)
α	integer in a compound $A_\alpha B_\beta$ (Chapter 13, A.1)
α	coefficient for a regular solution (14.4.4)
α_f	thermal expansion coefficient of the free volume (7.11.7)
β	a phase (1.7.8)
β	reaction order (2.4.2)
β	spring constant between vibrating atoms (9.3.4)
β	amplitude of sine function (12.2.3)
β	integer in a compound $A_\alpha B_\beta$ (Chapter 13, A.1)
β	spatial composition wavelength in spinodal decomposition (16.4.4)
β^-	high energy electron (2.7.2)
β^+	high energy positron (2.7.4)
β_{crit}	critical wavenumber for spinodal decomposition (16.4.5)
Γ	atom jumps per second (9.3.1)
Γ	regular solution constant (16.2.1)
$\Gamma_{B,A}$	Gibbs adsorption isotherm coefficient of segregation of *B on the surface of A* (6.14.1)
$\Gamma(x)$	gamma function of *x* (11.7.3)
γ	surface/interfacial energy (1.7.3)
γ	shear strain (7.11.6)
γ	viscous drag coefficient (13.3.1)
$\dot{\gamma}$	shear strain rate (7.11.6)
Δ	difference (1.7.3)
$\Delta H_m = \Delta H_f = \Delta H_{fusion}$	heat of melting or fusion (1.7.8, 6.8.1)
ΔH_{vap}	enthalpy of vaporization (7.8.4)

ΔG^* Gibbs energy of the activated state (4.7.1)
ΔG_V Gibbs energy per unit volume (7.8)
ΔG_{r^*} Gibbs energy of the critical nucleus (7.8.1)
ΔG_{hom} Gibbs energy to form a homogeneous nucleus (7.9.2)
ΔG_{het} Gibbs energy to form a heterogeneous nucleus (7.9.2)
$D\varepsilon$ energy difference between unlike and like pairs of atoms (16.2.1)
δ a phase (7.12)
δ grain boundary thickness (9.5.4)
δ boundary layer thickness (10.9)
δq path dependent heat (1.7.3)
δw path dependent work (1.7.3)
ε energy per atom/molecule (5.2.2)
ε strain (6.2.2)
ε bond energy between atoms (6.3.4)
$\dot{\varepsilon}$ tensile strain rate (7.11.6)
ε_0 permittivity of free space (10.5.2)
η number per unit volume (2.4.3)
η volume viscosity (7.11.7)
Θ nucleation rate (7.8.4)
θ angle (Chapter 5, A.1)
θ fraction of occupied surface sites (5.2.7)
θ_D Debye temperature (9.3.4)
θ_E Einstein temperature (9.3.4)
κ curvature of a surface (6.4)
κ factor relating interfacial chemical potential and concentration gradient (Chapter 16, A.3.2)
λ characteristic for linear ODE (Chapter 3, A.3.3)
λ mean free path (8.4.1)
λ wavelength (9.3.4)
λ diffusion distance (11.2.2)
λ_{crit} critical wavelength for spinodal decomposition (16.4.5)
μ viscosity (7.11.4)
μ_i chemical potential of component i (1.7.7)
μ_e electrical mobility of an electron (13.5)
ν Poisson's ration (7.11.8)
ξ volume strain (7.11.7)
ξ electrochemical potential (13.9.1)
ξ viscous drag on a single repeating unit of a polymer (13.10.3)
$\dot{\xi}$ volume strain rate (7.11.7)
π osmotic pressure (14.3.4)
ρ density (2.4.3)
σ electrical conductivity (6.11)
τ relaxation time (2.6.4)
τ shear stress (7.11.4)
τ time of nucleation (7.8.6)
ϕ electrical potential (13.9.1)
χ roughness factor of a surface (6.15)
χ Flory–Huggins interaction parameter (16.2.1)
Ω fractional solid angle (Chapter 5, A.4)
Ω number of broken surface bonds (6.3.4)
Ω_{AB} collision integral in the Chapman–Enskog equation (9.6.4)
ω angular frequency (7.11.6)

OTHER

Units()	the units of the quantity in parentheses (2.4.3)
[]	concentration, molar, molal (2.2)
$[]_0$	concentration at t = 0 (2.2)
$Cl^{\cdot}$	free radical (3.6.2)
w/o	weight percent
m/o	mole percent
v/o	volume percent
$\bar{\nabla}$	gradient operator (8.3)
$\tilde{D}$	diffusion coefficient tensor (8.3)
$\vec{J}$	vector (8.6)
∇^2	Laplacian operator (8.6)
[V]	fraction of crystal sites vacant (9.3.4)
$\langle\ \rangle$	the mean or average of a quantity (11.7.3)
$\mathfrak{F}$	the Faraday (13.9.1)
$^{238}_{92}U$	uranium isotope of atomic number 92 and atomic weight of 238 g/mol (2.7.2)
°C	degrees Celsius (1.2)
°F	degrees Fahrenheit (1.6.2)

Appendix II: Constants

INTERNATIONAL SYSTEM OF UNITS (SI)

Quantity	SI Units	SI Abbreviation
Length	meter	m
Volume	liter	L
Mass	kilogram	kg
Time	second	s
Temperature	kelvin	K
Electric current	ampere	A
Amount of substance	mole	mol

PHYSICAL CONSTANTS

Acceleration of gravity	$g = 9.80665\ \text{m·s}^{-2}$
Avogadro's number	$N_A = 6.0221367 \times 10^{23}\ \text{mol}^{-1}$
Boltzmann's constant	$k_B = 1.386503 \times 10^{-23}\ \text{J/K}$
Electron charge	$e = 1.602177 \times 10^{-19}\ \text{C}$
Electron mass	$m_e = 9.1093897 \times 10^{-31}\ \text{kg}$
Faraday	$\mathfrak{F} = \text{NA} \times \text{e} = 9.64853 \times 10^4\ \text{C mol}^{-1}$
Gas constant	$R = 8.314510\ \text{J·K}^{-1}\ \text{mol}^{-1}$
Planck's constant	$h = 6.62606901 \times 10^{-34}\ \text{J s}$
Speed of light	$c = 299792458\ \text{m s}^{-1}$ (exactly)
Vacuum permittivity	$\varepsilon_0 = 8.845 \times 10^{-12}\ \text{C}^2\text{·N}^{-1}\text{·m}^{-2}$

SOME SI DERIVED UNITS

Quantity	Name	Units
Electric charge	coulomb	A·s
Electric potential	volt	J·C^{-1}
Capacitance	farad	C·V^{-1}
Force	newton	$\text{kg·m}^2\text{·s}^{-2}$
Pressure	pascal	N·m^{-2}
Stress	pascal	N·m^{-2}
Energy	joule	N·m
Power	watt	J·s^{-1}

USEFUL CONVERSION FACTORS

1Å = 10^{-8} cm = 10^{-10} m = 0.01 nm
1 atm = 760 torr = 1.01325 × 10^5 Pa
1 bar = 10^5 Pa
1 cal = 4.184 J (defined)
1 eV = 1.602 × 10^{-19} J = 96485 J·mol^{-1}
1 R = 8.314 J·mol^{-1}· K^{-1} = 82.05 cm^3·atm·mol^{-1}·K^{-1}

Index

Note: Page numbers followed by f and t refer to figures and tables, respectively.

D

E

F

G

T